Nanotechnology in Plant Health

Editors

Mahendra Rai

Department of Biotechnology
SGB Amravati University
Amravati, Maharashtra, India

Graciela Avila-Quezada

Facultad de Ciencias Agrotecnologicas
Universidad Autonoma de Chihuahua
Chihuahua, Mexico

CRC Press
Taylor & Francis Group
Boca Raton London New York

CRC Press is an imprint of the
Taylor & Francis Group, an **informa** business

A SCIENCE PUBLISHERS BOOK

First edition published 2024
by CRC Press
2385 NW Executive Center Drive, Suite 320, Boca Raton FL 33431

and by CRC Press
4 Park Square, Milton Park, Abingdon, Oxon, OX14 4RN

© 2024 Mahendra Rai and Graciela Avila-Quezada

CRC Press is an imprint of Taylor & Francis Group, LLC

Library of Congress Cataloging-in-Publication Data (applied for)

ISBN: 978-1-032-45036-0 (hbk)
ISBN: 978-1-032-45038-4 (pbk)
ISBN: 978-1-003-37510-4 (ebk)

DOI: 10.1201/9781003375104

Typeset in Times New Roman
by Radiant Productions

Preface

According to an estimate by FAO, there is a need for augmentation in global food production by 60% up to 2050 to feed the world population. To achieve this target, controlling pathogens and pests is necessary, which are responsible for major loss in food crops. The excessive use of chemical fungicides/pesticides and fertilizers to improve plant health has produced toxicity to humans and the environment. Moreover, the development of resistance to pathogens and pests and emerging pathogens are the challenges that need to be tackled by alternative technologies such as nanotechnology.

Nanotechnology is a novel platform and plays a vital role in the promotion of plant health. On one hand, the nanomaterials can be applied as nutrients and nanofertilizers; on the other hand, they have proved to be potential tools in the diagnosis of plant diseases, delivery of fungicides/pesticides, and also for therapy of the diseases caused by pathogens and parasites. In this context, green nanotechnology is promising in plant health fitness that includes the synthesis of nanoparticles by microbes such as fungi, bacteria, algae, and plants since no pollution is generated during the synthesis of nanoparticles.

This book has been divided into three sections. Section I discusses the strategic role of nanotechnology in plant health fitness, particularly plant protection in agriculture and forestry. Section II incorporates the application of nanomaterials in plant health fitness. Section III includes the nanomaterials and their role in the management of plant diseases. It also discusses the toxicity problems generated by nanomaterials. The present book furnishes research-based up-to-date literature for postgraduate and research students. It would be highly useful for plant pathology, nanotechnology, and biotechnology researchers. In addition, it will be useful for plant protectors engaged in the development of sustainable alternative technologies for food production. The contributors are experts in their field and represent different parts of the world.

We earnestly acknowledge the authors for providing up-to-date chapters. Their sincere efforts will update the readers' knowledge of the role of nanotechnology in plant health fitness by enhanced understanding of nanonutrition and plant protection. Further, we would like to wholeheartedly thank everyone on the CRC/Taylor Francis team who continuously helped us.

Finally, Mahendra Rai wishes to thank the financial support by the Polish National Agency for Academic Exchange (NAWA) (Project No. PPN/ULM/2019/1/00117/U/00001) to visit the Department of Microbiology, Nicolaus Copernicus University, Toruń, Poland.

Mahendra Rai
India
Graciela Avila-Quezada
Mexico

Contents

Section I
General

Chapter 1

The Strategic Role of Nanotechnology in Plant Health Fitness
An Overview

Graciela Avila-Quezada,[1,][*] *Luis Guillermo Hernandez-Montiel*[2] *and*
Mahendra Rai[3,4,][*]

◇◇

Introduction

The world population is increasing enormously and is estimated to be 9.8 billion by 2050. As a result, there is a pressing need for additional food to feed the increasing population. Moreover, it is also necessary to redirect agricultural practices toward sustainable agriculture that is environment-friendly, socially just, and economically beneficial. Nanotechnology is emerging as an important tool that will have a positive impact on agriculture by improving food safety and reducing the adverse current problems of agriculture on the environment and human health. The positive impacts of nanomaterials are the production of disease-free food because of the timely diagnosis of pathogens by nanobiosensors and the effective absorption of nanonutrients, leading to enhanced production of crops (Shafi et al. 2020). Nanotechnology also helps maintain the soil microbiome (Farooq et al. 2022). The soil by nature is rich in microorganisms (González-Escobedo et al. 2022) that contribute to fertility for plants (Figure 1).

Due to leaching, precipitation, and volatilization, large volumes of chemical fertilizer formulations are required to achieve optimum performance. In addition, traditional fertilizers contain active ingredients with low water solubility, therefore their availability to the plant is low (Avila-Quezada et al. 2022b).

Since deficiencies of some minerals in the soil are a problem throughout the world, various attempts have been made to develop methods for biofortification of crops with essential nutrients, such as Fe, Zn, Mg, and others for being involved in the growth, metabolism, and yield of plants (Sida-Arreola et al. 2015; Sánchez et al. 2017; Babajani et al. 2019; Ciscomani-Larios et al. 2021).

Thus, in this chapter, we present relevant discussion about nano-sized materials as nutrients and plant protectors, which are more efficient compared to bulk fertilizers. To cite some examples, research by Yang et al. (2018) demonstrates that wheat seedlings exhibit an increase in

[1] Universidad Autonoma de Chihuahua, Escorza 900, col. Centro, Chihuahua 31000, Mexico.
[2] Centro de Investigaciones Biológicas del Noroeste (Cibnor), Km. 1 Carr. a San Juan de La Costa El Comitan. La Paz, Baja California Sur, 23205, Mexico.
[3] SGB Amravati University, Department of Biotechnology, Nanobiotechnology Lab., Amravati-444602, Maharashtra, India.
[4] Nicolaus Copernicus University, Department of Microbiology, Toruń, 87-100 Poland.
* Corresponding author: pmkrai@hotmail.com; gdavila@uach.mx

lateral root formation when exposed to zinc oxide nanoparticles (ZnONPs) at doses of 500 mg kg^{-1}. Such results can be applied to better support the seedlings. Also, cotton increases in root and shoot length and total biomass at doses of 25 mg L^{-1}, 50 mg L^{-1}, 75 mg L^{-1}, 100 mg L^{-1}, and 200 mg L^{-1} ZnONPs (Venkatachalam et al. 2017).

In this chapter, the readers can also find out how NPs application results without positive effects on the plant. NPs do not always have a favorable effect on plant growth. It depends on shape, size, and concentration. For instance, in *Arabidopsis thaliana* with ZnONPs at doses of 50 mg L^{-1}, 100 mg L^{-1}, and 200 mg L^{-1} caused chlorosis, reduction in leaf size, and primary root elongation; however, lateral root formation increased at 20 mg L^{-1} and 50 mg L^{-1} (Nair and Chung 2017). Also, reduction in growth and photosynthesis occurs in *Arabidopsis* with ZnONPs at doses of 200–300 mg L^{-1} (Wang et al. 2016).

Thus, the study in particular of precise doses is essential to achieve adequate nutrition in each crop, along with many other factors that affect the use of nutrients, all of which are addressed in this chapter. Either way, nanotechnology solves most agricultural problems. Nanoformulations, for instance, can solve several problems by the use of additives or NPs carriers that offer a more controlled release of nutrients according to the absorption time of the crop or by utilizing the NPs themselves as active ingredients (Younis et al. 2021).

Indeed, hundreds of commercial products based on nanomaterials with applications in various sectors are currently available on the market (Ravichandran 2010). These NMs present particular physicochemical and biological properties compared to bulk materials due to a larger surface area (Navya and Daima 2016). It is proven that various NMs stimulate plant health. For example, natural siliceous NMs, such as natural zeolites and diatomaceous earth, are agricultural inputs that protect plants and are biostimulants. When applied to the aerial part, they protect the plant from stress, while application to the soil stimulates root development. Its form of action is based on the large active siliceous surfaces that dry out insects and foliar pathogens and have a fungistatic effect in addition to stimulating photosynthesis (Constantinescu-Aruxandei et al. 2020). Moreover, NMs have great

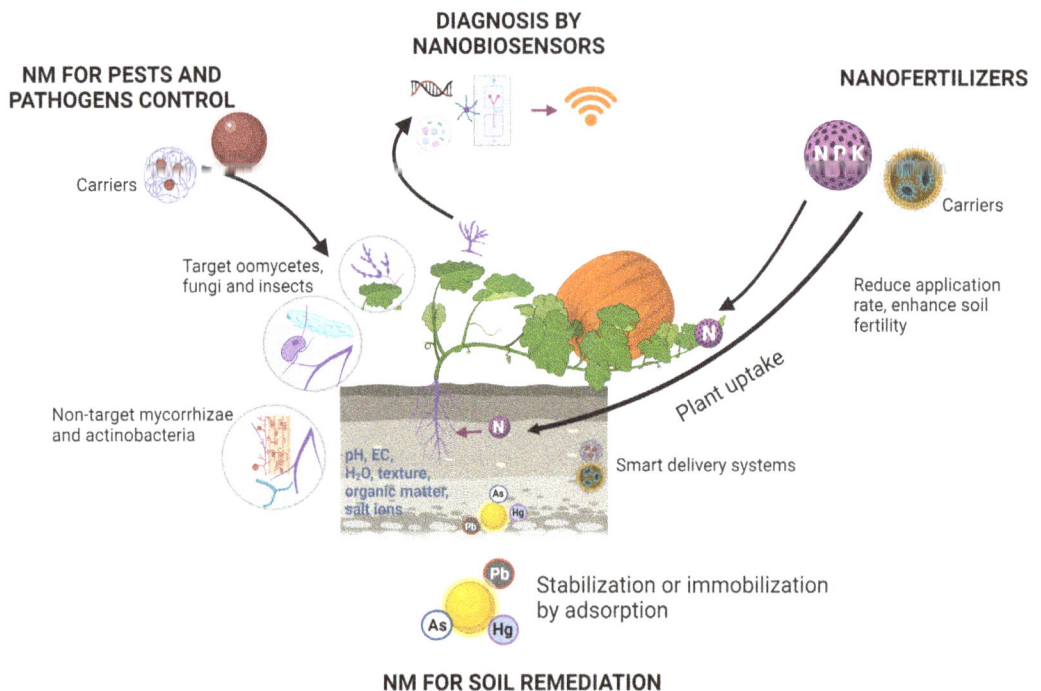

Figure 1. Nanotechnology contributes in many areas of agriculture to improve food production.

potential to protect plants from pathogens, such as the widely studied metallic nanoparticles that have antimicrobial properties (Avila-Quezada et al. 2022a).

Another advantage of using NPs is that they improve plant fitness even under metal stress conditions, as occurred in sunflower (*Helianthus annuus*) plants in soil with high chrome levels. FeONPs promoted the growth of sunflower plants under chrome toxicity and had an effect in reducing chrome uptake by cells (Mohammadi et al. 2020). Likewise, rice seedlings grown in cadmium-contaminated soil and subjected to foliar application of titanium dioxide (TiO$_2$) and silicon (Si) nanoparticles increased plant biomass and decreased cadmium concentration in plant tissues. Among other attributes, chlorophyll concentration and gas exchange of the rice leaf improved (Rizwan et al. 2019).

Since plants are exposed to various types of stress depending on the place on the planet where they are grown, the search for solutions is required to mitigate the global food crisis. Nanotechnology, specifically nanoparticles, is a useful tool to modulate sustainable agricultural production under biotic and abiotic stress conditions. This book chapter addresses these issues with results from recent applied research.

Nanomaterials in the Management of Plant Pathogens

World food production is at risk due to the presence of various pathogens, such as fungi, bacteria, and viruses, among others, that cause diseases in plants. These diseases result in a decrease and loss of the harvest, ultimately putting world food sovereignty at risk (Agathokleous and Calabrese 2021; Abumchukwu Okoli et al. 2022). The application of agrochemicals has been the traditional way to control phytopathogens; however, they develop resistance in pathogenic microbes, causing environmental problems as well as posing risks to human and animal health (Thind 2022; Sharma et al. 2022).

At present, nanomaterials are important alternatives for managing plant diseases. These act as more effective technological tools for the control of phytopathogens, maintaining the quality and production of crops compared to agrochemicals (Dash et al. 2022; Ashfaq et al. 2022). Various NPs have been evaluated for the control of several phytopathogens in crops of agricultural importance (Table 1).

Table 1. Nanoparticles that are used for controlling plant pathogens.

Nanoparticle	Phytopathogen	Disease	Medium	Control	Reference
Silver Nanoparticles (AgNPs)	*Colletotrichum gloeosporioides*	Anthracnose	*In vitro*	89% inhibition of mycelial growth	Aguilar-Méndez et al. 2011
AgNPs	*Xanthomonas oryzae* pv. *oryzae*	Bacterial leaf blight	Rice	Disease reduction by 49.2%	Namburi et al. 2021
Chitosan-NPs	*Botrytis cinerea*	Gray mold	Strawberry	On leaves, the NPs reduced the severity down to 3%	El-Naggar et al. 2022
AgNPs	*Fusarium oxysporum* f. sp. *radicis-lycopersici*	Crown and root rot	*In vitro*	90% inhibition of mycelium growth	Akpinar et al. 2021
Magnesium Oxide (MgO)-NPs	*Phytophthora nicotianae/ Thielaviopsis basicola*	Black shank and black root rot	Tobacco	Suppressed fungal invasion in 50 and 62%	Chen et al. 2020
Silver chloride (AgCl)-NPs	*Ralstonia solanacearum*	Bacterial wilt disease	Eggplant	Caused rupture of the cell wall and cytoplasmic membranes	Abd Alamer et al. 2021
Zinc oxide (ZnO)-NPs	*R. solanacearum*	Bacterial wilt disease	Tomato	57% protection against wilt disease	Konappa et al. 2022
ZnONPs	*Clavibacter michiganensis* (Cm)/*Pseudomonas syringae* (Ps)	Ring rot/leaf and fruit lesions	*In vitro*	Reduced the bacterial growth of Cm and Ps by 90% and 67% respectively	Vera-Reyes et al. 2019

The application of nanoparticles for the control of phytopathogens does not require specialized equipment; therefore, it is an alternative to the conventional use of agrochemicals, reducing crop production costs (Bapat et al. 2022).

Nanonutrients for Plant Growth and Fitness

NPs have been investigated in several studies due to the damage they cause in the cells of phytopathogenic fungi and bacteria (Bernardo-Mazariegos et al. 2019). In addition, they have been used to improve agricultural productivity through high nutrient efficiency in the form of nanofertilizers (NF). NFs play a key role in plant nutrition by providing nutrients to plants and increasing their bioavailability through slow and steady release over 30 days (El-Ramady et al. 2018). Moreover, their use reduces nutrient leaching in agroecosystems.

Nanotechnology has great potential to increase agricultural productivity with an ecological approach. Various NPs due to their unique physicochemical properties help the absorption of nutrients by plants, such as silica nanoparticles (Awad-Allah 2022). The synthesis of NMs is carried out by various processes, including biological processes that occur in nature. These are carried out by peptides, plants, or microorganisms such as fungi, bacteria, and algae (Durán et al. 2011). NPs can be managed/modulated to increase or decrease the acquisition of beneficial or toxic minerals (Le Van et al. 2016) and increase or control levels of plant hormones (Le Van et al. 2016). Also, depending on the concentration and type of application, the enzymes can be more or less active, as is the case of the increase in superoxide dismutase (SOD) activity with CuONPs applied to rice with 30 days of exposure in hydroponics (Da Costa and Sharma 2016). On the contrary, SOD activity decreased in corn exposed for three weeks to CuONPs in roots and leaves (Adhikari et al. 2016). Moreover, CuONPs reduced chloramphenicol acetyltransferase (CAT) activity in alfalfa determined after 15 days of root exposure (Hong et al. 2015).

In relation to plant growth, this is generally favored by the application of NPs, although it is not always the case. Indeed, Haghighi et al. (2014) reported that selenium nanoparticles (1–12 μM) in 10 days of exposure did not increase the growth of tomato plants, although it did promote tolerance under high and/or low-temperature stress. Besides, Le Van et al. (2016) reported that the biomass did not increase or decrease when CuONPs (10 mg L^{-1}, 200 mg L^{-1}, and 1,000 mg L^{-1}) were applied to cotton within 10 days of exposition.

Overcoming stress and achieving production is a challenge that is sought in agriculture, especially due to the large extensions of semi-arid areas in the world destined to produce food. Metallic NPs induce in plants resistance to drought and increase yield. Ahmed et al. (2021) demonstrated that AgNPs (10 mg L^{-1}) and CuNPs (3 mg L^{-1}) in greenhouse wheat plants under drought conditions stimulated nutrient uptake and water retention.

Also, metallic NPs may increase seed yield. Under field conditions, gold nanoparticles applied through foliar spray enhanced growth at concentrations of 10 ppm, 25 ppm, 50 ppm, and 100 ppm and increased seed yield at 10 ppm in *Brassica juncea* (Arora et al. 2012). Many other metal NPs increase yield as SeNPs. After five applications of 10 mg L^{-1} of SeNPs, the yield of tomato plants increased up to 21% in the experiment (Hernández-Hernández et al. 2019).

The doses and type of NPs, application method, and plant species must be studied and modulated to find the exact parameters to increase plant production. Even when the increase in the production of fruits or biomass is the same as that offered by synthetic fertilization, NFs are more efficient than bulk fertilizers due to the reduced environmental effects.

Nanomaterials Toxicity and Risks

The toxicity of metallic NPs has been studied by various researchers on numerous plants (Ahmad et al. 2021; Jia et al. 2022), although studies on their effects on human health are still lacking. The sizes of the nanoparticles and their effect on the human body when consumed cause great concern among the population and government officials. In addition, there is still a lack of regulatory laws for the use of NMs in agriculture that are specific for crops, edible parts of the plants, and sizes of the NPs applied as well as residuality time before harvesting among other aspects.

Although the assessment of health risks is still being evaluated and investigated, there are already some products on the market with NMs intended for fertilizing plants and controlling insect pests and phytopathogenic organisms by controlled release and additives, as well as active components. These NMs are nanoemulsions, capsules with the pyrethroid lambda-cyhalothrin, silicon oxide, titanium oxide, and zero-valent iron nanopowder, all with molecules smaller than 100 nm (Gogos et al. 2012).

A major concern that needs extensive investigation is the effect of NMs on the soil microbiome. As is known, this is abundant and diverse (González-Escobedo et al. 2023), therefore various processes such as nitrogen fixation, mineralization, and stimulation of exudates related to plant health are carried out. Besides high concentrations of NPs could be toxic for plants. Indeed, 2,000 mg L^{-1} of cerium dioxide (CeO$_2$) NPs in the reproductive stage of bean plants applied via foliar, provoked severe structural damage to pollen grains, causing pollen abortion and yield losses (Salehi et al. 2021). Results such as these lead us to suggest low doses of NPs to avoid toxicity in plants and accumulation of NPs in soil and underground water sources. Other examples are presented in this book.

Moreover, much has been published that NMs due to their distinctive characteristics, such as nano-size, large surface area, and high adsorption affinity, have great potential for phytoremediation of soils contaminated with heavy metals (Mohammadi et al. 2020). In our opinion, the benefits that the NMs will provide to clean up contaminated soils are greater than those to contaminate agricultural soils.

A subject little addressed in research is the adsorption, internalization, translocation, and accumulation of NMs in plants and their edible parts. For instance, in a 72-week experiment of soil and tomato plants, Das et al. (2018) found that earthworms exposed to 50 ppm AgNPs with PEG coating 7–14 nm due to high oxidative stress and reduced protein synthesis lost weight, although they did not have problems in reproduction. This dose of AgNPs reduced microbial growth. In the soil the AgNPs agglomerated, in addition, the availability of silver was reduced because Ag$_2$S or Ag$_3$PO$_4$ was formed; consequently, affecting the availability of P and S. The Ag was absorbed by the plant roots from the dose of 10 mg kg^{-1} soil of AgNPs causing normal plant growth, low production of tomato fruits, increased fruit Ag, and significantly affecting photosynthesis and CO$_2$ assimilation efficiency. Low concentrations of NPs can alter the physiology of plants. Moreover, the route of application also influences the physiological effects of the plant (Ahmad et al. 2021).

Conclusion

Since the use of NPs in agriculture to obtain healthy plants is promising, it should be noted that the ecosystem of each particular crop must be studied for the correct application in form and time of the NMs and that these cause the desired effect on the plant. The NMs can be used for the development of sophisticated tools, such as nanosensors for early detection of the disease as a carrier for the delivery of agrochemicals and also in therapeutical applications for the management of diseases. Moreover, the nanomaterials can be applied as nutrients for maintaining the plants' fitness. As synthetic agrochemicals, the NMs must be used with the correct indications, there must be regulation for its use, and proper application time before harvesting. Finally, the toxicity of NMs to the ecosystems and humans warrants thorough systematic studies to understand the host-nanomaterials interactions.

References

Abd Alamer, I.S., Tomah, A.A., Ahmed, T., Li, B. and Zhang, J. 2021. Biosynthesis of silver chloride nanoparticles by rhizospheric bacteria and their antibacterial activity against phytopathogenic bacterium *Ralstonia solanacearum*. Molecules 27(1): 224. https://doi.org/10.3390/molecules27010224.

Abumchukwu Okoli, F., Mbachu, N.A., Ogbonna, U.S., Obianom, A.O. and Ebele Mbachu, A. 2022. Mode of attack of microbiological control agents against plant pathogens for sustainable agriculture and food security. Asian J. Agricul. Horticul. Res. 1–16. https://doi.org/10.9734/ajahr%2F2022%2Fv9i130132.

Adhikari, T., Sarkar, D., Mashayekhi, H. and Xing, B.S. 2016. Growth and enzymatic activity of maize (*Zea mays* L.) plant: solution culture test for copper dioxide nano particles. J. Plant Nutr. 39(1): 99–115. https://doi.org/10.1080/01904167.2015.1044012.

Agathokleous, E. and Calabrese, E.J. 2021. Fungicide-induced hormesis in phytopathogenic fungi: A critical determinant of successful agriculture and environmental sustainability. J. Agric Food Chem. 69(16): 4561–4563. https://doi.org/10.1021/acs.jafc.1c01824.

Aguilar-Méndez, M.A., San Martín-Martínez, E., Ortega-Arroyo, L., Cobián-Portillo, G. and Sánchez-Espíndola, E. 2011. Synthesis and characterization of silver nanoparticles: effect on phytopathogen *Colletotrichum gloesporioides*. J. Nanopart. Res. 13: 2525–2532. https://doi.org/10.1007/s11051-010-0145-6.

Ahmad, A., Hashmi, S.S., Palma, J.M. and Corpas, F.J. 2021. Influence of metallic, metallic oxide, and organic nanoparticles on plant physiology. Chemosphere 133329. https://doi.org/10.1016/j.chemosphere.2021.133329.

Ahmed, F., Javed, B., Razzaq, A. and Mashwani, Z.U.R. 2021. Applications of copper and silver nanoparticles on wheat plants to induce drought tolerance and increase yield. IET Nanobiotech. 15(1): 68–78. https://doi.org/10.1049/nbt2.12002.

Akpinar, I., Unal, M. and Sar, T. 2021. Potential antifungal effects of silver nanoparticles (AgNPs) of different sizes against phytopathogenic *Fusarium oxysporum* f. sp. *radicis-lycopersici* (FORL) strains. SN Appl. Sci. 3(4): 506. https://doi.org/10.1007/s42452-021-04524-5.

Arora, S., Sharma, P., Kumar, S., Nayan, R., Khanna, P.K. and Zaidi, M.G.H. 2012. Gold-nanoparticle induced enhancement in growth and seed yield of *Brassica juncea*. Plant Growth Regul. 66: 303–310. https://doi.org/10.1007/s10725-011-9649-z.

Ashfaq, A., Khursheed, N., Fatima, S., Anjum, Z. and Younis, K. 2022. Application of nanotechnology in food packaging: Pros and Cons. J. Agricul. Food Res. 7: 100270. https://doi.org/10.1016/j.jafr.2022.100270.

Avila-Quezada, G.D., Golinska, P. and Rai, M. 2022a. Engineered nanomaterials in plant diseases: can we combat phytopathogens? Appl. Microbiol. Biotechnol. 106: 117–129. https://doi.org/10.1007/s00253-021-11725-w.

Avila-Quezada, G.D., Ingle, A.P., Golińska, P. and Rai, M. 2022b. Strategic applications of nano-fertilizers for sustainable agriculture: Benefits and bottlenecks. Nanotechnol. Rev. 11(1): 2123–2140. https://doi.org/10.1515/ntrev-2022-0126.

Awad-Allah, E.F. 2022. Effectiveness of silica nanoparticle application as plant nano-nutrition: a review. J. Plant Nutr. 1–14. https://doi.org/10.1080/01904167.2022.2160735.

Babajani, A., Iranbakhsh, A., Ardebili, Z.O. and Eslami, B. 2019. Differential growth, nutrition, physiology, and gene expression in *Melissa offcinalis* mediated by zinc oxide and elemental selenium nanoparticles. Environ. Sci. Poll Res. 26(24): 24430–24444. https://doi.org/10.1007/s11356-019-05676-z.

Bapat, M.S., Singh, H., Shukla, S.K., Singh, P.P., Vo, D.V.N., Yadav, A. and Kumar, D. 2022. Evaluating green silver nanoparticles as prospective biopesticides: An environmental standpoint. Chemosphere 286: 131761. https://doi.org/10.1016/j.chemosphere.2021.131761.

Bernardo-Mazariegos, E., Valdez-Salas, B., González-Mendoza, D., Abdelmoteleb, A., Camacho, O.T., Duran, C.C. and Gutiérrez-Miceli, F. 2019. Silver nanoparticles from *Justicia spicigera* and their antimicrobial potentialities in the biocontrol of foodborne bacteria and phytopathogenic fungi. Rev. Argentina Microbiol. 51(2): 103–109. https://doi.org/10.1016/j.ram.2018.05.002.

Chen, J., Wu, L., Lu, M., Lu, S., Li, Z. and Ding, W. 2020. Comparative study on the fungicidal activity of metallic MgO nanoparticles and macroscale MgO against soilborne fungal phytopathogens. Front Microbiol. 11: 365. https://doi.org/10.3389/fmicb.2020.00365.

Ciscomani-Larios, J.P., Sánchez-Chávez, E., Jacobo-Cuellar, J.L., Sáenz-Hidalgo, H.K., Orduño-Cruz, N., Cruz-Alvarez, O. and Ávila-Quezada, G.D. 2021. Biofortification efficiency with magnesium salts on the increase of bioactive compounds and antioxidant capacity in snap beans. Ciência Rural. 51. https://doi.org/10.1590/0103-8478cr20200442.

Constantinescu-Aruxandei, D., Lupu, C. and Oancea, F. 2020. Siliceous natural nanomaterials as biorationals-Plant protectants and plant health strengtheners. Agronomy 10(11): 1791. https://doi.org/10.3390/agronomy10111791.

Da Costa, M.V.J. and Sharma, P.K. 2016. Effect of copper oxide nanoparticles on growth, morphology, photosynthesis, and antioxidant response in *Oryza sativa*. Photosynthetica 54(1): 110–119. https://doi.org/10.1007/s11099-015-0167-5.

Das, P., Barua, S., Sarkar, S., Chatterjee, S.K., Mukherjee, S., Goswami, L., Das, S., Bhattacharya, S., Karak, N. and Bhattacharya, S.S. 2018. Mechanism of toxicity and transformation of silver nanoparticles: Inclusive assessment in earthworm-microbe-soil-plant system. Geoderma 314: 73–84. https://doi.org/10.1016/j.geoderma.2017.11.008.

Dash, K.K., Deka, P., Bangar, S.P., Chaudhary, V., Trif, M. and Rusu, A. 2022. Applications of inorganic nanoparticles in food packaging: A comprehensive review. Polymers 14(3): 521. https://doi.org/10.3390/polym14030521.

Durán, N., Marcato, P.D., Durán, M., Yadav, A., Gade, A. and Rai, M. 2011. Mechanistic aspects in the biogenic synthesis of extracellular metal nanoparticles by peptides, bacteria, fungi, and plants. Appl. Microbiol. Biotechnol. 90: 1609–1624. https://doi.org/10.1007/s00253-011-3249-8.

El-Naggar, N.E.A., Saber, W., Zweil, A.M. and Bashir, S.I. 2022. An innovative green synthesis approach of chitosan nanoparticles and their inhibitory activity against phytopathogenic *Botrytis cinerea* on strawberry leaves. Scientific Rep. 12(1): 3515. https://doi.org/10.1038/s41598-022-07073-y.

El-Ramady, H., Abdalla, N., Alshaal, T., El-Henawy, A., Elmahrouk, M., Bayoumi, Y. and Schnug, E. 2018. Plant nano-nutrition: perspectives and challenges. pp. 129–161. *In*: Gothandam, K., Ranjan, S., Dasgupta, N., Ramalingam, C. and Lichtfouse, E. (eds.). Nanotechnology, Food Security and Water Treatment, Environmental Chemistry for a Sustainable World. Springer, Cham. https://doi.org/10.1007/978-3-319-70166-0_4.

Farooq, B., Anjum, S., Farooq, M., Rather, G.A., Nazir, A., Nayak, B.K. and Nanda, A. 2022. Role of nanotechnology in soil microbiome and agricultural development. pp. 230–248. *In*: Parray, J.A., Shameem, N., Abd-Allah, E.F. and Mir, M.Y. (eds.). Core Microbiome: Improving Crop Quality and Productivity. John Wiley & Sons Ltd. https://doi.org/10.1002/9781119830795.ch14.

Gogos, A., Knauer, K. and Bucheli, T.D. 2012. Nanomaterials in plant protection and fertilization: Current state, foreseen applications, and research priorities. J. Agric Food Chem. 60(39): 9781–9792. https://doi.org/10.1021/jf302154y.

González-Escobedo, R., Muñoz-Castellanos, L.N., Muñoz-Ramirez, Z.Y., Guigón-López, C. and Avila-Quezada, G.D. 2023. Rhizosphere bacterial and fungal communities of healthy and wilted pepper (*Capsicum annuum* L.) in an organic farming system. Ciência Rural. 53(7): e20220072. https://doi.org/10.1590/0103-8478cr20220072.

Haghighi, M., Abolghasemi, R. and Teixeira da Silva, J.A. 2014. Low and high temperature stress affect the growth characteristics of tomato in hydroponic culture with Se and nano-Se amendment. Sci. Hortic. 178: 231–240. https://doi.org/10.1016/j.scienta.2014.09.006.

Hernández-Hernández, H., Quiterio-Gutiérrez, T., Cadenas-Pliego, G., Ortega-Ortiz, H., Hernández-Fuentes, A. D., Cabrera de la Fuente, M., Valdes-Reyna, J. and Juárez-Maldonado, A. 2019. Impact of selenium and copper nanoparticles on yield, antioxidant system, and fruit quality of tomato plants. Plants 8(10): 355. https://doi.org/10.3390/plants8100355.

Hong, J., Rico, C.M., Zhao, L., Adeleye, A.S., Keller, A.A., Peralta-Videa, J.R. and Gardea-Torresdey, J.L. 2015. Toxic effects of copper-based nanoparticles or compounds to lettuce (*Lactuca sativa*) and alfalfa (*Medicago sativa*). Environ. Sci. Proc. Impacts 17(1): 177–185. https://doi.org/10.1039/C4EM00551A.

Jia, Y., Klumpp, E., Bol, R. and Amelung, W. 2022. Uptake of metallic nanoparticles containing essential (Cu, Zn and Fe) and non-essential (Ag, Ce and Ti) elements by crops: A meta-analysis. Crit. Rev. Environ. Sci. Technol. 1–22. https://doi.org/10.1080/10643389.2022.2156225.

Konappa, N., Joshi, S.M., Dhamodaran, N., Krishnamurthy, S., Basavaraju, S., Chowdappa, S. and Jogaiah, S. 2022. Green synthesis of *Callicarpa tomentosa* routed zinc oxide nanoparticles and their bactericidal action against diverse phytopathogens. Biomass Conv. Bioref. 1–12. https://doi.org/10.1007/s13399-022-03438-5.

Le Van, N., Ma, C., Shang, J., Rui, Y., Liu, S. and Xing, B. 2016. Effects of CuO nanoparticles on insecticidal activity and phytotoxicity in conventional and transgenic cotton. Chemosphere 144: 661–670. https://doi.org/10.1016/j.chemosphere.2015.09.028.

Mohammadi, H., Amani-Ghadim, A.R., Matin, A.A. and Ghorbanpour, M. 2020. FeO nanoparticles improve physiological and antioxidative attributes of sunflower (*Helianthus annuus*) plants grown in soil spiked with hexavalent chromium. 3 Biotech 10: 19. https://doi.org/10.1007/s13205-019-2002-3.

Nair, P.M.G. and Chung, I.M. 2017. Regulation of morphological, molecular and nutrient status in *Arabidopsis thaliana* seedlings in response to ZnO nanoparticles and Zn ion exposure. Sci. Total Environ. 575: 187–198. https://doi.org/10.1016/j.scitotenv.2016.10.017.

Namburi, K.R., Kora, A.J., Chetukuri, A. and Kota, V.S.M.K. 2021. Biogenic silver nanoparticles as an antibacterial agent against bacterial leaf blight causing rice phytopathogen *Xanthomonas oryzae* pv. *oryzae*. Bioprocess Biosyst. Eng. 44(9): 1975–1988. https://doi.org/10.1007/s00449-021-02579-7.

Navya, P.N. and Daima, H.K. 2016. Rational engineering of physicochemical properties of nanomaterials for biomedical applications with nanotoxicological perspectives. Nano Converg. 3,1. https://doi.org/10.1186/s40580-016-0064-z.

Ravichandran, R. 2010. Nanotechnology applications in food and food processing: innovative green approaches, opportunities and uncertainties for global market. Int. J. Green Nanotechnol: Physics and Chem. 1(2): 72–96. https://doi.org/10.1080/19430871003684440.

Rizwan, M., Ali, S., ur Rehman, M.Z., Malik, S., Adrees, M., Qayyum, M.F., Alamri, S.A., Alyemeni, M.N. and Ahmad, P. 2019. Effect of foliar applications of silicon and titanium dioxide nanoparticles on growth, oxidative stress, and cadmium accumulation by rice (*Oryza sativa*). Acta Physiol. Plant. 41: 35. https://doi.org/10.1007/s11738-019-2828-7.

Salehi, H., Chehregani Rad, A., Raza, A. and Chen, J.T. 2021. Foliar application of CeO_2 nanoparticles alters generative components fitness and seed productivity in Bean crop (*Phaseolus vulgaris* L.). Nanomaterials 11(4): 862. https://doi.org/10.3390/nano11040862.

Sánchez, E., Sida-Arreola, J.P., Ávila-Quezada, G.D., Ojeda-Barrios, D.L., Flores-Córdova, M. A., Preciado-Rangel, P. and Márquez-Quiroz, C. 2017. Can biofortification of zinc improve the antioxidant capacity and nutritional quality of beans? Emir J. Food Agric. 237–241. https://doi.org/10.9755/ejfa.2016-04-367.

Shafi, A., Qadir, J., Sabir, S., Zain Khan, M. and Rahman, M.M. 2020. Nanoagriculture: A holistic approach for sustainable development of agriculture. pp. 1–16. *In*: Kharissova, O.V., Martínez, L.M.T. and Kharisov, B.I. (eds.). Handbook of Nanomaterials and Nanocomposites for Energy and Environmental Applications. Springer, Cham, https://doi.org/10.1007/978-3-030-11155-7_48-1.

Sharma, A., Abrahamian, P., Carvalho, R., Choudhary, M., Paret, M.L., Vallad, G.E. and Jones, J.B. 2022. Future of bacterial disease management in crop production. Ann. Rev. Phytopathol. 60: 259–282. https://doi.org/10.1146/annurev-phyto-021621-121806.

Sida-Arreola, J.P., Sánchez, E., Ávila-Quezada, G.D., Zamudio-Flores, P.B. and Acosta-Muñiz, C.H. 2015. Can improve iron biofortification antioxidant response, yield and nutritional quality in green bean? Agric Sci. 6(11): 1324. https://www.scirp.org/html/4-3001278_61350.htm.

Thind, T.S. 2022. New insights into fungicide resistance: A growing challenge in crop protection. Indian Phytopathol. 75: 927–939. https://doi.org/10.1007/s42360-022-00550-4.

Venkatachalam, P., Priyanka, N., Manikandan, K., Ganeshbabu, I., Indiraarulselvi, P., Geetha, N. and Sahi, S.V. 2017. Enhanced plant growth promoting role of phycomolecules coated zinc oxide nanoparticles with P supplementation in cotton (*Gossypium hirsutum* L.). Plant Physiol. Biochem. 110: 118–127. https://doi.org/10.1016/j.plaphy.2016.09.004.

Vera-Reyes, I., Esparza-Arredondo, I.J.E., Lira-Saldivar, R.H., Granados-Echegoyen, C.A., Alvarez-Roman, R., Vásquez-López, A. and Díaz-Barriga Castro, E. 2019. *In vitro* antimicrobial effect of metallic nanoparticles on phytopathogenic strains of crop plants. J. Phytopathol. 167(7-8): 461–469. https://doi.org/10.1111/jph.12818.

Wang, X., Yang, X., Chen, S., Li, Q., Wang, W., Hou, C., Gao, X., Wang, L. and Wang, S. 2016. Zinc oxide nanoparticles affect biomass accumulation and photosynthesis in *Arabidopsis*. Front Plant Sci. 6: 1243. https://doi.org/10.3389/fpls.2015.01243.

Yang, K.Y., Doxey, S., McLean, J.E., Britt, D., Watson, A., Al Qassy, D., Jacobson, A. and Anderson, A.J. 2018. Remodeling of root morphology by CuO and ZnO nanoparticles: effects on drought tolerance for plants colonized by a beneficial pseudomonad. Botany 96(3): 175–186. https://doi.org/10.1139/cjb-2017-0124.

Younis, S.A., Kim, K.H., Shaheen, S.M., Antoniadis, V., Tsang, Y.F., Rinklebe, J., Deep, A. and Brown, R.J. 2021. Advancements of nanotechnologies in crop promotion and soil fertility: Benefits, life cycle assessment, and legislation policies. Renew. Sust. Energ. Rev. 152: 111686. https://doi.org/10.1016/j.rser.2021.111686.

Chapter 2

Nanomaterials for Improving Health of Agricultural Crops and Forest Trees

Veeran Sethuraman,[1] *Pachaiyappan Saravana Kumar,*[2]
Dharmalingam Kirubakaran[3] and *Muthugounder Subramanian Shivakumar*[1,*]

Introduction

Nanotechnology is the most promising branch of science, which has found applications in multiple disciplines in the development of new technologies in the field of biology, chemistry, physics, medicines, electronics, biomedical energy, and environment, including agriculture in the 21st century, and is still in its infancy. The nanomaterials, ranging from 1–100 nm, are the foundation of nanotechnology, having unique physicochemical properties that enable their use in the development of various inorganic and organic substances (Tarafdar et al. 2013; He et al. 2019). In addition, it also plays a vital role in various fields viz., parasitology, pest science, drug delivery diagnostics, and others (Elechiguerra et al. 2005; Panneerselvam et al. 2016; Aziz et al. 2018; Teimouri et al. 2018). The idea that nanomaterials could be of interest in agricultural and forest development may help to achieve an environmentally friendly and cost-efficient solution in forest protection and crop production (Figure 1).

The word "nanotechnology" is concerned with microscopic particles, which are defined by their specialized physical and chemical processes. Recently, nanomaterials are used widely due to their distinctive characteristic features than more commonly used and conventional materials (Mu et al. 2010). Nanoparticles reduce the number of active ingredients used in paints, textiles, and cosmetics (Moloi et al. 2021). During the past decade, nanoparticles are widely used in agro-industries in improving the fertilizing process, nutrient optimization, and plant protection fields as nanofertilizers and nanopesticides (Klaine et al. 2008). With the increasing use of nanotechnology in agriculture, environmental remediation, and the biomedical field, it is imperative to understand the environmental fate of nanomaterials. As the effect of nanomaterial on the food chain and its chronic effect on the living system, water quality, and soil fertility need to be addressed. The regulation of the use and application of nanomaterials has to be regularly framed and widely implemented.

Nanoparticles are classified based on their size, morphology, and physicochemical properties. Most nanoparticles are classified as nanotubes and multi-walled carbon nanotubes, which contain hexagonal rings of carbon atoms linked by covalent bonds. Carbon nanotubes, metallic nanoparticles, semiconductors, and lipid-based nanoparticles are all different types of nanoparticles (Khan et al.

[1] Molecular Entomology Laboratory, Department of Biotechnology, Periyar University, Salem – 636 011, Tamil Nadu, India.
[2] Xavier Research Foundation, St. Xavier's College, Palayamkottai 627002, Tamil Nadu, India.
[3] Department of Botany, Periyar University, Salem – 636 011, Tamil Nadu, India.
* Corresponding author: skentomology@gmail.com

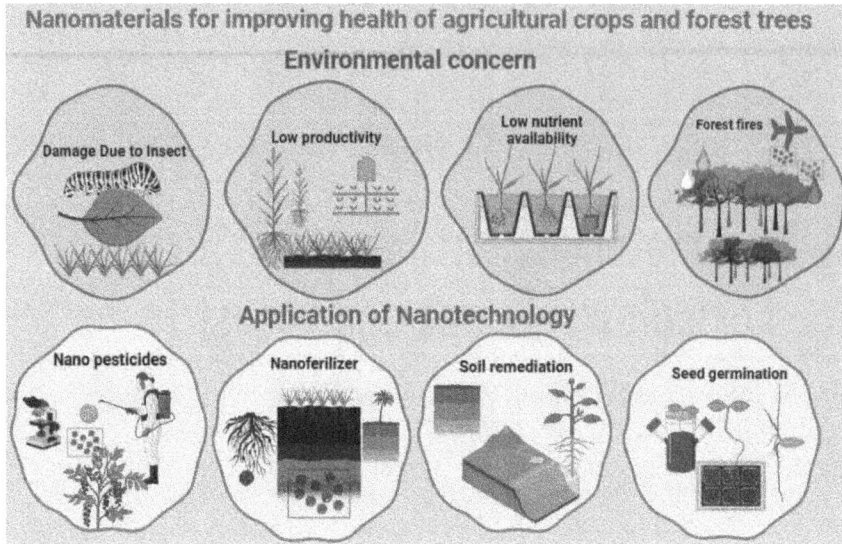

Figure 1. Overview of nanomaterials for improving health of agricultural crops and forest trees.

2022). Electrical conductivity and electron affinity properties of nanoparticles have been used in various stages of therapeutics, biomedical imaging biosensors, tissue engineering, and cancer therapy (Ishtiaq et al. 2020).

Fullerenes are an allotrope of carbon and its arrangement looks like a soccer ball-shaped molecule containing hexagonal rings of sixty carbon atoms linked by covalent bonds, which have electrical conductivity and electron affinity properties (Horlick et al. 1981; Ozkan et al. 2015). It is commonly used in the delivery of therapeutics, biomedical imaging, biosensors, tissue engineering, and cancer therapy (Harish et al. 2022). Ceramic nanoparticles, also called inorganic nanoparticles, are composed of carbides, carbonates, oxides, and phosphates. Ceramic nanoparticles play a major role in the drug delivery system and treatment of cancer and bacterial infections (Thomas et al. 2015; Kumar et al. 2014). It also has applications in photo-catalysis and imaging, which are used in environmental bioremediation and biomedical applications.

The lipid-based nanoparticles are spherical and with a diameter of around 10 100 nm. The bio-polymeric nanoparticles are organic-based in the form of nanospheres and nanocapsules. These nanoparticles play an important role in drug delivery and diagnostics (Kumari et al. 2010). The metallic nanoparticles have a wide range of applications in catalysis, electronics, biology, and physiology (Virkutyte et al. 2011). Metal nanoparticles are usually made up of metals such as gold, silver, titanium, zinc, and copper and their compounds like oxides, and sulfides. In general, biologically synthesized metal nanoparticle are mostly produced by the reduction of metal salts and has wide applications in the treatment of cancer, microbial pathogenic infections, gene delivery, and drug delivery (Mohanta et al. 2020).

Agriculture Crops and Forest Productivity

Damage Due to Insects

Globally, crop losses from insect pests and other factors have been 13.6% in post green revolution era and 10.8% at the beginning of this century. In India, crop losses have reduced from 23.3% in post green revolution era to 15.7% at present. In terms of financial value Indian agriculture currently suffers an annual loss of around 36 billion USD (Dhaliwal et al. 2015). Arthropod pests contribute to an estimated 18–20% of the annual crop loss with an estimated global value of > 470 billion USD (Sharma et al. 2017). Crop losses in different continents have been reported, and the estimated loss due to insect pests in Asia is 18.7%, nearly double the world average (Oerke et al. 1994). In Indian agriculture, losses due to insect pests are larger compared to those recorded at a global level (Sharma et al. 2017).

India occupies only 2.5% of the earth's geographical area and 1.85% of the world's forest area. Forests provide an extensive variety of vital ecosystem services but are increasingly affected by insects. Invasions by non-native insect pests also harmfully affect ecosystem services. Worldwide, forests are gradually affected by non-native insects and diseases, some of which cause large tree mortality (Fei et al. 2019). Forests and trees like other plants are attacked by insect pests and diseases, which cause a lot of damage, resulting in poor timber quality, poor tree growth, and, in some cases, complete destruction and reduction of forest cover (Sharma 2016). Insect pests of forest plants and agricultural crops and their management in the Indian arid zone were briefly reported by Sharma (2016). In addition, a study by Pal et al. (2016) has highlighted the role of insect pests of forest seeds and their management in natural and storage environments.

Invasive insect pests are one of the most serious and quickly spreading problems in forestry, human and animal health, agricultural biodiversity, and cause enormous economic losses. The prevalence of invading insect pests, such as lantana bug (*Orthezia insignis* Browne), San Jose scale [*Quadraspidiotus perniciousus* (Comstock)], woolly apple aphid [*Eriosoma lanigerum* (Hausmann)], potato tuber moth [*Phthorimaea operculella* (Zeller)], cottony cushion scale (*Icerya purchasi* Maskell), diamondback moth [*Plutella xylostella* (Linn.)], pine woolly aphid [*Pineus pini* (Macquart)], subabul psyllid (*Heteropsylla cubana* Crawford), serpentine leaf miner [*Liriomyza trifolii* (Burgess)], coffee berry borer [*Hypothenemus hampei* (Ferrari)], spiraling whitefly (*Aleurodicus disperses* Russell), silver leaf whitefly (*Bemisia argentifolii* Bellows), blue gum chalcid (*Leptocybe invasa* Fisher and La Salle), coconut eriophid mite (*Aceria gurreronis* Keifer), papaya mealy bug (*Paracoccus marginatus* Williams and Granara de Willink), cotton mealy bug (*Phenococcus solenopsis* Tinsley), erythrina gall wasp (*Quadrastichus erythrinae* Kim), South American tomato leaf miner (*Tuta absoluta* Meyrick), and fall armyworm [*Spodoperda frugiperda* (J.E. Smith)], has been reported (Gupta et al. 2019).

For the control of insect pests in crops and forest trees, the use of growth regulators, beneficial insects, and pheromone-lure has been studied as a possible alternative to chemical insecticides. Among the most promising biological control techniques, the use of predators and parasitoids has been used. However, the release of non-native insects may have an impact on the food web (Bale et al. 2008). The use of hormones and growth factors, which is essential for insect metamorphosis, has potential in insect control programs. These chemical compounds inhibit important physiological processes and have low toxicity to mammals (Casida et al. 2004). The use of novel biocontrol agents is instantly needed, opening increasing possibilities for the development of biopesticides. Biopesticides offer important social benefits when compared with conventional pesticides (Popp et al. 2013).

Forest Fires

Fires are a natural disturbance in many temperate forests as the temperate trees have thick bark, which makes them susceptible to low-intensity fires and hence less well-adapted (Nasi et al. 2002). Forest fires have huge implications for biological diversity and are a significant source of emitted carbon, contributing to global warming (Thompson et al. 2009). One of the most important environmental effects of burning is the increased probability of subsequent years as dead trees topple to the ground, the forest drying by sunlight, and building up the fuel load with an increase in fire-prone species (Nyamadzawo et al. 2013).

Tropical forests are also subject to fires started by humans for agricultural clearing. Deforestation due to fires, which are more common in disturbed forests, can vary in intensity and burn standing trees, at worst completely burning the forest (Purwar et al. 2020). In the natural forests of the northern hemisphere and sparsely stocked forest tundra, particularly on permafrost sites, surface fires, and large fires are becoming a near-annual occurrence in many regions globally (Zamolodchikov et al. 2020). The wildfires in Northwest forests will likely be a main challenge facing resource managers of public and private lands in future decades. In dry forests, forest thinning prescriptions may need to reduce forest density to forest resistance and resilience to fire, insects, and drought (Halofsky et al. 2020). Fewer options exist for reducing fire severity in wetter, high-elevation, and coastal forests of the Northwest, historically characterized by infrequent, stand-replacement fire regimes (Arno et al. 2000).

Forest fire in the world makes a global problem for humankind, hence the issue of controlling such natural disasters is tremendously urgent. The application and requirements of nanotechnology in fire protection have been elaborately described by Rabajczyk et al. (2021). Some of the chemical compositions used for fighting forest fires are patented (Rueda et al. 2014). Mosina et al. (2020) described nanocolloids based on aluminum hydroxide for the application of fire fighting. They used nanocolloids based on aluminum hydroxide, which was proposed as the extinguishing composition for two major purposes: fire prevention and fire suppression with alumina barrier formation.

Low Productivity Due to Low Nutrient Value

The usage of fertilizers, including organic, inorganic, and bio-fertilizers, is emphasized. The inorganic fertilizers typically with sizes more than 100 nm are easily lost due to volatilization and leaching, although organic matter utilization is vulnerable to its low mineral content and a prolonged period of nutrient release. Various efforts made to increase the efficiency of nutrient uptake of agricultural crops but have not been so successful earlier. Therefore, the time is risen to apply nanotechnology in solving some of these problems (Elemike et al. 2019). The low nutrient usage competence may be attributed to fertilizer overuse and the high nutrient loss resulting from inappropriate timing and methods of fertilizer application, especially in high-yielding fields (Fan et al. 2012). The major role of nanotechnology in the protection of plant nutrients and the improvement of crop production was elaborately reported by Elemike et al. (2019). Nutrient deficiency in food crops affects human health, particularly in rural areas, and nanotechnology may help in a more sustainable approach to overcome this challenge. Plants absorb nutrients from fertilizers, but most conventional fertilizers have low nutrient use and uptake efficiency. Therefore, nanofertilizers are engineered to be target oriented and not easily lost when compared to fertilizers. Applying insufficient amounts of nutrients contributes to low levels of crop production and crop establishment is the cause of soil degradation in many parts of the world (Gruhn et al. 2000).

Low Bioavailability of Nutrient for Plant

Nutrient uptake and subsequent translocation in the plant are critical for plant growth and development (Rana et al. 2021). Nanoparticles easily adsorb to the plant surfaces and uptakes by plants via nano to micrometer scale through natural openings of plants. Uptake rates depend on the surface properties and size of the nanoparticles. Several reports of evidence suggest that nanoparticles less than 5 nm in diameter are efficient in crossing the wall of the undamaged plant cell (Tarafdar et al. 2015).

The nutrient concentration of plant foliar materials is the result of nutrient uptake, re-translocation, and losses to foliar leaching. These processes are dependent both on the plant growth form and on the chemical and physical characteristics of the soil environment (Muller et al. 2003). Some of the alternative methods may be more costly than conventional chemical-intensive agricultural practices, but often these comparisons fail to account for the high environmental and social costs of pesticides (Gomiero et al. 2011).

Nanotechnology has also shown its ability in modifying the genetic structure of crop plants, thereby helping in the further improvement of crop plants. Generally metal-based nanomaterials and carbon-based nanomaterials are produced for their assimilation, translocation, and storage and specifically for their impacts on development and improvement in crop yield (Rani et al. 2020). In many crop plants, due to the positive morphological impacts of nanotechnology improved germination percentage, root and shoot length, and vegetative biomass of seedlings (Kole et al. 2013). It has also been reported that metal-based nanomaterials enhance numerous physiological parameters, such as increased photosynthetic rate and nitrogen assimilation in some crop plants (Nair et al. 2010). Nanobiotechnology provides tools and technology platforms to improve agricultural productivity by genetically modifying plants and transporting genes and drug molecules to particular locations on the cellular stage (Sastry et al. 2013).

However, in many situations now, soil degradation has progressed past the point of quick recovery, and the prevalence of invasive species means that the vegetation that recovers may resemble little to the original native community with its accompanying biodiversity advantage (Jensen et al. 2020). Non-crop habitats perform vital life-sustaining tasks that are necessary for a variety of natural enemies. As a result, non-crop habitats can increase the variety and richness of natural enemy species in the agricultural landscape. However, ecosystems other than crops can also attract pest species depending on their plant makeup (Bianchi et al. 2006). Complex landscapes comprising dense networks of non-crop habitats provide favorable conditions and requisites for natural enemy populations, often resulting in increased natural enemy activity in crop fields (Marshall 2004).

Nanotechnology for Crop Protection and Improving Crop Yield

Nanopesticides

Although there are many other ways in which nanotechnology is used in crop protection, slow-release encapsulating agrochemicals is the most promising approach. Novel nanomaterials that use metallic, polymeric, and inorganic nanoparticles have been developed. Utilizing nanoparticles for the creation of novel crop protection products has gained scientific and technological significance during the past two decades globally (Leon-Silva et al. 2016). This might be because of their unique qualities, such as their high surface area to volume ratio, cation exchange capacity, increased reactivity, uncommon structure, high ion adsorption ratio, and high stability and aggregation capacities (Nel et al. 2006; Carrillo et al. 2015; Sharon et al. 2010). These key characteristics help to develop a new product that differs from the molecules and bulk materials originally used (Li et al. 2001; Guo et al. 2004). Numerous industry sectors, including those in the disciplines of health, electronics, energy, genetics, aerospace, and agriculture, have generated nanoparticles, nanomaterials, and nanodevices. Additionally, the development of nanosensors enables the identification of pests and illnesses in crops (Dubey and Mailapalli 2016; Fraceto et al. 2016). Many different metallic nanoparticles, including Ag, Fe, Cu, and Zn, can be utilized in two ways: as pesticides or fungicides to combat some pathogenic microbes and diseases and as nanofertilizers to improve seed germination and plant growth (Le Van et al. 2016).

In comparison to the usage of conventional pesticides, numerous studies published recently demonstrate that nano-based pesticides have potential in the field of agricultural pest management, providing advantages to both the environment and human health (Parisi et al. 2015). Controlled releases for the delivery of agrochemicals, reduce phytotoxicity, volatilization, drift, leaching losses, and soil degradation, as well as greater safety during application, which are advantages of nano-based pesticides (Aouada et al. 2015). The use of nanomaterials in agriculture has several benefits, including increased pesticidal activity, increased crop production and quality, reduction in costs and consumption of agrochemicals for plant protection, lowered nutrient losses during fertilization, and reduced environmental contamination (Chhipa et al. 2017; Sweta et al. 2018).

Chemical pesticide nanoformulations have been designed efficiently to replace traditional pesticides (Agathokleous et al. 2021). Nanoparticles may act as plant growth promotors and enhancers for stress tolerance to an unconditional environment. Nanoformulation functions and efficacy not only depend on their physiochemical properties but includes application methods viz., foliar delivery, hydroponics, soil, etc. Nanoformulations amounts of the application make great effects on effectiveness (Zhao et al. 2020). A study reported that a nano-scaled formulation of a pesticide provided a better spatial distribution of the pesticide and uniformity of coverage on leaf surfaces during enhancing insecticidal efficiency (Liu et al. 2018). Therefore, alternatives of nanopesticide formulation with biomaterials have aroused great interest recently.

In this regard, nanochitin whisker enhances the toxicity and insecticidal activity of chemical pesticides. It has a strong enhanced effect on insecticidal effectiveness of chemical insecticides. It was further absorbed easily by plants, transported, and distributed from the mouth to other tissues of the insects while sucking plant fluid. Low acute oral and dermal toxicity to Sprague Dawley rats indicated that it is safe to apply in the agriculture and food industry (Li et al. 2021).

Applications of nanotechnology in agriculture can be generally categorized into a variety of groups, viz. nanopesticides/nanoherbicides, nanofertilizers for crop nutrition, seed science, water management, nanoscale carriers, biosensors to detect nutrients and contaminants, and crop-monitoring decisions. Nanoencapsulation plays an important role in the protection of the environment by reducing leaching and the evaporation of harmful substances as well as its specific potential for pesticide delivery. This may be due to such nanoencapsulated products would reduce the dosage of pesticides and becoming environmentally friendly for crop protection (Nuruzzaman et al. 2016).

A number of recent research publications have reported on the development of biopesticides' effectiveness and a reduction in losses using nanotechnologies (Bakry et al. 2016; De Oliveira et al. 2014; Giongo et al. 2016; Ramkumar et al. 2016; Sowndarya et al. 2017; Vivekanandhan et al. 2018; Rajkumar et al. 2018). Nanotechnology contributes to the development of less toxic biopesticides with favorable safety concerns, increased stability of active agents, enhanced activity on target pests, and further increased adoption by users (Prasad et al. 2014; Khot et al. 2012; Agrawal and Rathore 2014; Loganathan et al. 2021; Yashkamal et al. 2021). Nanobiotechnology has a potential role to develop formulations that might be utilized to increase the stability and effectiveness of natural products (Ghormade et al. 2011; Perlatti et al. 2013; Loganathan et al. 2021; Loganathan et al. 2022). These formulations can provide controlled release of the molecules at the target site of action, reduce potentially toxic effects on non-target organisms and protect against degradation of the active agent through microorganisms (Gogos et al. 2012; Durán and Marcato 2013).

Nanopesticides of biological origin viz., nanobiopesticides could be made-up using any metal, such as Ag, Cu, SiO_2, or ZnO with broad-spectrum pest management. Though research and spectra of various field studies are required to fully understand interactions between nanoparticles, microorganisms, soils, plants, and humans (Lade et al. 2017; Guru Bharathi et al. 2022; Loganathan et al. 2022; Vengateswari et al. 2022). The preparation of nanobiopesticides and their mode of action on insects are described in (Figure 2).

Nanoparticle syntheses by entomopathogenic microbes are considered novel methods in the field of agricultural pest control (Bharathi et al. 2022). The insecticidal and antifeedant activities of mycosynthesized selenium nanoparticle using *Trichoderma* sp. to control *Spodoptera litura* was reported, and mycosynthesized using fungal extract was reported as a reliable alternative biopesticide from a microbial resource for controlling agricultural pests (Arunthirumeni et al. 2022). Another important study described that mycosynthesis of bimetallic zinc oxide and titanium dioxide nanoparticles effectively controlled *Spodoptera frugiperda* (Kumaravel et al. 2021).

There are several biological applications for nanoemulsions, including in the pesticide, pharmaceutical, cosmetic, and food industries. Previous research shows that nanoemulsion has emerged promising method for the control of medical and agricultural insect pests (Wang et al. 2007). A botanically derived nanoformulation product has demonstrated potent larvicidal efficacy against larvae of *A. aegypti, C. quinquefasciatus, A. stephensi* (Suganya et al. 2014; Veerakumar et al. 2014; Angajala et al. 2014; Rawani et al. 2013). Water-soluble formulations that use nanoemulsion technology as a pesticide delivery system offer environmentally favorable properties and low toxicity to non-target organisms (Nuchuchua et al. 2009). The botanical-derived nanoemulsion is kinetically stable in field conditions and contains a sufficient amount of surfactants, with a small particle size range from 15–96 nm.

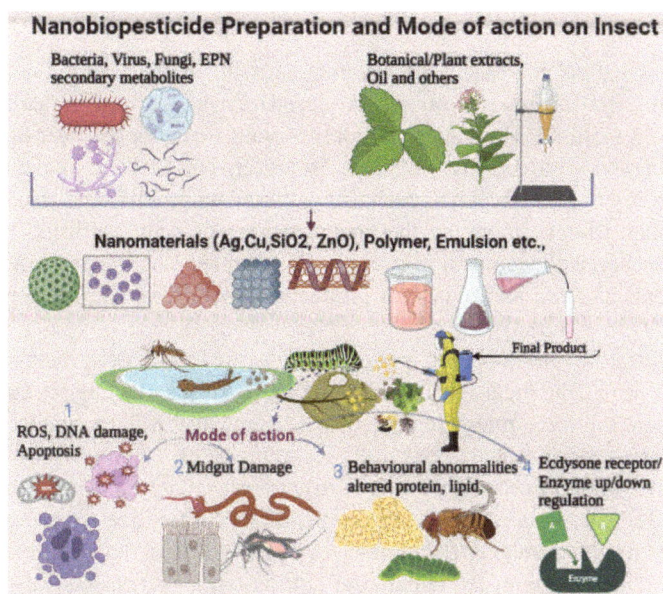

Figure 2. Preparation of nanobiopesticides and their mode of action on insects.

Microbial biocontrol agents are frequently used to manage plant diseases and pest control in a sustainable, long-lasting manner. The use of nanotechnology in the field of biopesticides has become a potentially appropriate solution nowadays. The nanobiopesticides basically in the form of biologically derived active compounds integrated as nanoparticles and integrated into a suitable polymer have applications in insect-pest management. The role of microbial biocontrol agents and microbes-based nanoformulations against plant diseases and for pest management with emphasis on bacteria-based nanoparticles, especially those derived from *Bacillus* species were elaborately reported by Kumar et al. (2021). Enterobacteriaceae, *Bacillaceae,* and *Pseudomonadaceae* families have the majority of entomopathogenic bacteria. Gram-positive, rod-shaped bacteria, heterotrophic that can create endospores make up the Bacillaceae family. *Bacillus thuringiensis*, *B. papillae*, *B. sphaericus*, *B. pumilus*, *Brevibacillus laterosporus*, and other members of this genus (Mampallil et al. 2017). Several bacterial species (*Corynebacterium* sp., *Bacillus* sp., *Pseudomonas* sp., and *Shewanella* sp.) have been utilized for the synthesis of NPs using various inorganic metals (viz., Ag, Al, Au, MnO, ZnO, and TiO_2) over the last decade. By minimizing losses during crop processing, shipping, and storage, the use of these important bacterial-mediated NPs has increased agricultural yield (Kumari et al. 2019; Kumar et al. 2021). Nanoscale *Bt* chitinases showed excellent nematicidal activity against *Caenorhabditis elegans*. Therefore, the nanoscale *B. thuringiensis* chitinases acted as a biopesticidal delivery system to extend farming applications (Qin et al. 2020).

Likewise *Photorhabdus luminescens*, a symbiotic bacteria from entomopathogenic nematode *Heterorhabditis* species bi-product syntheses into nanoparticles and found that its efficacy improved significantly in pest management (Kulkarni et al. 2017). Bacterial secretions of *P. luminescens* have been used as a biopesticide against a wide range of agricultural insects. Usually, it secretes an array of toxins and enzymes which act as poison to insects. The bacterial secretions along with the nanoparticles have a wide range of insecticidal actions against both sucking and chewing arthropod pests of agricultural crops. Biosynthesis of copper nanoparticles using symbiotic bacterium *Xenorhabdus* sp., isolated from entomopathogenic nematode, and its antimicrobial and insecticidal activity against *Spodoptera litura* were also reported recently (Bharathi et al. 2022).

Nanofertilizers

Bio-fertilizers are used to increase crop production with safeguarding the environment and non-target organisms. Bio-fertilizers are mostly produced industrially from particular microorganisms that either develop a mutually beneficial relationship with plants or are part of their rhizosphere and form an association with soil microbiomes. However, bio-fertilizers have a few drawbacks, and thus it is necessary to find novel nanotechnology-based nanofertilizers. Nanotechnology offers tremendous advantages for customizing the manufacturing of nanofertilizers. Nanofertilizers are coated with desired chemical composition have controlled release and targeted delivery of effective nanoscale ingredients and can improve plant productivity and minimize environmental pollutants (Fatima et al. 2020).

Without harming the environment, nanotechnology offers a new method to increase the production of fertilizers. Chemical fertilizers are being used in Indian agriculture systems due to rising demand. There are three major ingredients of chemical fertilizers available namely (NPK) potassium, nitrogen, and phosphate. Naturally, a number of complex processes cause these elements to become immobilized within soil obstructing its timely and adequate availability for their uptake by the plants. As a result of low usage, chemical fertilizers create significant human health and environmental problems (Bindraban et al. 2020). This has led to the acceptance of bio-fertilizers and nanofertilizers. By integrating cutting-edge instruments for precision farming, disease detection, nutrient absorption, and their treatment, nanotechnology has the potential to advance the agriculture sector (Prasad et al. 2017). Nanofertilizer coated with nanosilica forms a dual film on the microbial cell wall, which prevents infections and improves plant resistance to pathogenic infections (Rastogi et al. 2019).

Nanoscale additives, nanoscale coatings, and nanoscale fertilizers are the three primary categories of nanofertilizers (Mikkelsen et al. 2018). The release of nutrients immobilized or encapsulated into a particular nanocarrier namely biological, chemical, and physical are activated by three different factors. The biological factors are bacteria, fungi, and other beneficial microorganisms that degrade the coating based on a synthetic polymeric material, therefore allowing the release of nutrients and its fixation into the soil. The chemical-activated mechanisms such as solubilization, soil type, moisture, pH variation (Weeks and Hettiarachchi 2019; Ramzan et al. 2020), and ion exchange reactions were reported (Ribeiro and Carmo 2019). Diffusion-controlled release physical factors such as ultra-sound and magnetic field are also described (Mikkelsen et al. 2018; Ribeiro and Carmo 2019).

Nanosilica assists in the growth of seeds and roots for plants. It has been demonstrated that many nanofertilizers increase the soil's ability to retain water. Nanofertilizers are nanoparticles, mixed with bio-fertilizers such as *P. fluorescens, Bacillus subtilis* and *P. elgii* have shown a significant increase in the growth of plants under *in-vitro* conditions. These multiple positive characteristics, of nanoparticles bionanofertilizers could be an effective alternative in the agricultural sector for plant production (Karunakaran et al. 2016). A detailed description of a method for increasing nitrogen consumption and efficiency in grasslands using nanofertilizers was published (Mejias et al. 2021). In-depth reports on the functions of sustainable nanofertilizers for enhancing plant quality and quantity have been reported by Mohamed et al. (2021).

Copper, silica, zinc, and iron have all been synthesized at the nanoscale and used for plant growth development. The effect of nano Fe-chelated plant growth-promoting rhizobacteria (PGPR) on maize growth, physiological responses grain yield were investigated by Heidari et al. (2018), and they discovered that foliar application of *Azospirillum brasilens* improved maize plant growth and yield. Another notable recent study described that inoculating plants with PGPR, Arbuscular mycorrhizal with a low dose of Fe-NPs significantly increased heavy metal phytoremediation, improving the root zone and leaf space of young plants (Mokarram et al. 2019).

An understanding of the interaction between nanomaterials and plant mechanisms is necessary to develop new products. The nanofertilizers should be applied so that they do not lose their vital properties like time-controlled release, efficiency, solubility, improved targeted activity, stability, and furthermost importantly less toxicity due to their safe and simple mode of deliverance and disposal. However, their impact and effectiveness are mainly influenced by their mode of application. The delivery of these nanofertilizers to plants must be considered for field application. Consequently, the foliar application is one of the most efficient methods of correcting nutrient deficiencies and increasing crop yield and quality (Roemheld and El-Fouly 1999; Semida et al. 2021). Reducing environmental contamination and increase in nutrient use efficacy by decreasing the amount of fertilizer applied in the soil are the salient features of nanofertilizers (Abou- El-Nour 2002; Schwab et al. 2015).

Soil Remediation Using Nanotechnology

Advancements in nanotechnology may help accelerate the removal of toxic components from contaminated soils in a sustainable and environmentally friendly manner. Soil is mainly contaminated by toxic chemicals pollutants and anthropogenic sources at concentrations capable of making great risks to the environment and human health. Contaminated soils raised concern in the environmental sector due to the existence of a huge number of polluted spots mainly in industrialized and urban regions (Hu et al. 2006). The advancements in nanotechnology open a window worldwide to remediate or restore polluted soil in an effective way by mitigating the toxicities of various chemicals and metalloids (Kumari et al. 2021; Raffaet et al. 2021). In general, nanotechnology has been considered a potential tool for the remediation of pollutants in a range of environmental matrixes including soils (Rajput et al. 2021).

With regard to soils and groundwater remediation, there are many available technologies available, which include *ex-situ* technologies in which the contaminated soils or groundwater must be removed from the site and treated on-site or off-site and *in-situ* technologies, the contaminated soils or groundwater are treated directly within the subsurface (Sharma and Reddy 2004). An *in-situ* treatment technology is often preferred because of the major technical and economic advantages when compared to *ex-situ* technologies (Karn et al. 2011; Reddy 2013).

Nanotechnologies that utilize nanoparticles (NPs) for the remediation of contaminated soil have been rapidly developed in recent years but the majority of these studies are carried out on a bench or lab scale. A few studies have reported the use of NPs at the field scale (Karn et al. 2011; Mueller et al. 2012). Various efforts have been made to increase the efficiency of phytoremediation through the inclusion of chemical additives, the application of rhizobacteria, and genetic engineering. From this perspective, the integration of nanotechnology with bioremediation has introduced new dimensions for restoring the remediation methods. Therefore, the advanced remediation approaches combine nanotechnological and biological remediation methods in which the nanoscale process regulation supports the adsorption and weakening of pollutants were elaborately discussed in the recent study conducted by Rajput et al. (2022) that how nanotechnologies help in the remediation of polluted soil.

The use of NPs for removing pollutants from water bodies began in the 1990s; it is considered a new technology, and its development is still in progress (Thomé et al. 2015). Wang and Zang (1997) and Zang et al. (1998) reported the use of NPs for the decontamination of groundwater contaminated with organochlorines. Additionally, Thomé et al. (2015) synthesized bimetallic NPs (Pd/Fe, Pd, Pt/Fe/Zn, Ni/Fe) and tested in the bench-scale tests for remediation of numerous aromatic chlorinated and some organ chlorine pollutants in water.

Additionally, heavy metal nanobioremediation has been extensively covered in reviews by Rajput et al. (2022). Elements namely As, Cd, Cr, Hg, and Pb have no biological functions to participate in the biological system. Heavy metals comprise inorganic pollutants as they exhibit substantial toxic impacts on biota even at lower concentrations (Alissa et al. 2011; Rasmussen et al. 2011). The toxicity of heavy metals also rests on their bioavailability and absorption (Rasmussen et al. 2011). Acidic environments instigate the toxicity of heavy metals, particularly if the soil structure is poor and has low nutrients, e.g., in mining and industrial areas (Mukhopadhyay et al. 2010).

As a result, the application of biogenic nanoparticles for the removal of heavy metals and toxic materials has received a lot of interest. Biogenic nanoparticles, such as silver nanoparticles (AgNps) are synthesized using plant *Morganella psychrotolerans* (Capeness et al. 2019; Sharma et al. 2019). Although nanotechnology has many uses, there is still concern regarding its presence in environmental contexts, its future, and the ensuing toxicological effects. Moreover, the currently available research on nano-bioremediation is limited to laboratory experiment conditions and computational modeling. Therefore, there is a pressing need to develop new technologies to overcome this long-time problem with multidisciplinary approaches.

Nanotechnology for Forestry

Nanotechnology is crucial for the effective and sustainable production of new generations of materials derived from forests for society and the economy based on biomass. The forest products industry has the potential to improve almost into nanotechnology from the production of raw materials to new methods for producing engineered wood and wood-based materials to novel composite and paper product uses (Atalla et al. 2005). In order to assess loads, moisture levels, forces, temperature, pressure, chemical emissions, and attack by wood decay fungus; a variety of nanosensors are being included in intelligent wood and paper-based products (Moon et al. 2006). Recent years have seen the development of new products, economic growth, and social effects for nanomaterials in the field of nanobiotechnology forestry. The development of nanomaterials for use in the forest industry has a great deal of promise as a result of derived components (Vishwakarma et al. 2018). Most of them used for wood-based nanoproducts, such as laminates, composites, resins, and nanofibers are still in the development stages (Cai et al. 2003).

Protection of Trees From Insect Pests

Nanopesticides are formulation that contains components with a diameter of less than 100 nm and claims novel properties specific to this small size range, a few nanopesticides have already been made commercially available (Pavoni et al. 2019). Organic materials that can be used to make nanopesticides are several distinctions between intensively managed agricultural ecosystems and perennial forest ecosystems some of which may be advantageous to success (Shekhar et al. 2021). Invasive forest insect pests have historically been a top focus for traditional biological control as have woody plant pests more generally. The fastest and most expensive method of control is to use nanotechnology to control plant diseases and manage pests using fungicides, insecticides, and herbicides (Guyot et al. 2015). Whenever pesticides are applied, residual toxicity and other remedial procedures are also typically used. In this line, the use of pesticides has resulted in a number of harmful effects on human health, domestic animals, and pollinating insects. In addition, it also has adverse effects directly or indirectly on ecosystems by contaminating the soil and water (Ali et al. 2021).

Recent climate change has led to the dynamics of disturbances caused by insect pest invasions and local forest pathogens, such as facilitating the establishment and spread of introduced pest species (Sexton et al. 2007). Currently, the insecticidal value of plants and trees' active components from environmental sectors can be added to nanotechnology. More specifically, the insecticides were treated with active ingredients in the early stages (Rai et al. 2012).

The active ingredients in pesticides are contained in extremely small quantities in nanoformulations. Active ingredients of pesticides can also be coated with other materials of different sizes at the nanoscale in a process known as nanoencapsulation (Ghasemnezhad et al. 2019). The nanoformulations or the encapsulations of chemical pesticides assist in the controlled release of active ingredients in the root zones or inside plants without compromising effectiveness (Kumar et al. 2019). On the other hand, conventional formulations of solubility of pesticides can cause toxic effects in other organisms and also lead to increase resistance in target insect pests (Onyesolu et al. 2021). Therefore, one of the most efficient and cost-effective methods for controlling insect pests is to use active substances on the surface that is being treated. In particular, nanoencapsulation can be used to increase the insecticidal effects by protecting the active ingredient from adverse circumstances and promoting retention (Manjunatha et al. 2017). Bioactive products such as insecticides, fungicides, and nematicides are used in agriculture to increase crop efficiency and productivity and can be nanoencapsulated to develop formulations that effectively manage deleterious organisms (Kumar et al. 2017).

Application of Nanotechnology in Forestry

Since the beginning of civilization, wood products have been used to make furniture, timber, papers, and a number of other useful items. In general, wood and non-wood forest products are produced mostly in part using forests example rattan, firewood, dammar, charcoal, and bamboo (Jasmani et al. 2020). Since it has been the source of lignocelluloses materials for numerous product developments and the possibility that the forest played (Pramanik et al. 2016). Nanotechnology could be used to make wood-based goods that are stronger, more adaptable, and lighter. From this perspective, nanotechnology is crucial to the forestry sector, especially in terms of products made from wood or forests. Therefore, recent developments in nanotechnology applications in a few specific wood-based industries including wood composites, wood coating, wood durability, and paper (Sandberg et al. 2016). Nanocellulose is a new type of cellulose used in the manufacturing of wood-based products, in fields including energy and sensors. Therefore, there is a high chance for modern uses can expand or change traditional forest products, like pulp and paper, wood composites, wood coatings, and wood preservatives into new versions (Lin et al. 2014). Consequently, nanoparticles could be used to enhance the functionality of the current wood-based products.

Seed Germination

Since seeds are a gift from nature to mankind, nanotechnology can currently be employed to enhance seed growth. Seed production is a tedious process, especially in wind-pollinated crops. Seed nano-priming is an effective process that can change seed signaling pathways and metabolism affecting not only germination, and seedling establishment, and also the entire plant life cycle (Bovand et al. 2022). Seeds kept for a long period will eventually lose viability due to sudden biochemical damage occurring at the cellular level, which will cause natural seed aging and subsequently reduce crop productivity (Sano et al. 2013). Recent studies have shown that seed nano-priming may activate a wide range of genes during germination especially those responsible for plant stress resistance. All these factors together can result in safer seedlings for farmers and consumers with respect to the environment (Verma et al. 2022). In interacting with their biochemical and physiological processes treatments with heavy metals are known to inhibit seed germination, growth, and plant development (Solanki et al. 2011; Singh et al. 2021). Seed germination is a crucial stage in a plant's life cycle that supports seedling growth, survival, and population dynamics. However, a variety of factors including the environment, genetic makeup, moisture availability, and soil fertility have a major impact on seed germination. In addition to germination, metallic nanoparticles, such as CuO, Fe, FeO, TiO_2, ZnO, Zn, and hydroxy fullerenes have been shown to improve crop quality in a variety of crop species including mustard, onion, spinach, tomato, potato, and wheat (Eriksson et al. 2008).

Conclusion and Future Perspectives

Our modern society depends on a wide range of products produced by the forest products industry, which draws on a significant base of renewable resources (Hatti-Kaul et al. 2007). Companies that produce wood and paper products employ a million people globally and produce a billion dollars worth of goods each year. The advancement of nanotechnologies presents an opportunity to enhance completely new processes for the production of engineered wood and fiber-based products (Kamel et al. 2007). Additionally, they can encourage the creation of novel and improved wood-based products and materials that can be used as reasonable pricing substitutes for the non-renewable resources used to create metallic, plastic, and ceramic products (Schwarzkopf et al. 2016).

Nanotechnology research and development are necessary for the low-cost and environmentally friendly production of these next-generation forest-based resources, which will satisfy societal needs while enhancing forest health and advancing the biomass-based economy (Huttunen et al. 2013). In this context, nanotechnology has the potential to transform the forest product industry innumerable ways, including the production of raw materials, innovative applications for paper and composite materials, and new generations of effective machinery (Puurunen et al. 2007). Nanotechnology can be used in the processing of wood-based materials into paper products and a myriad of wood by improving water elimination, and reducing energy utilization in tagging fibers, flakes, drying, and particles to allow modified property enhancement while processing (Wegner et al. 2005).

The production of current products will be very efficient, opening the door for the development of various new products as well as significant improvements in functionality and surface properties (Wang et al. 2009). The variety of nanosensors built in to detect forces, chemical emissions, moisture levels, loads, pressure, temperature, and assault by wood-decaying fungus bacteria is one of the best innovations of nanotechnology (Shabani et al. 2017). Applying nano-dimensional building blocks will make it possible to synthesize substrates and functional materials with much-improved strength, resulting in the production of lighter, more energy-efficient goods from fewer resources (Leydecker 2008). The manufacturing of current goods will be very effective, facilitating the creation of many new products as well as considerable advancements in surface quality and functionality (Singh et al. 2009).

Nanotechnology has gained popularity in recent years as a result of its numerous uses in industries like energy and materials, pharmaceuticals, and medicine including agriculture and forestry. The development of nanotechnology like nanoparticles has led to the development of novel technologies for plant protection, plant growth, and fertilizers. In recent years the use of nanomaterials has become an alternative solution to control plant pests. In the future, the application of nanotechnology in agriculture and forestry is expected to play a major role in a country's economy. Nanotechnology has helped in improving agricultural and forestry productivity and yield quality through proper delivery mechanisms and decreased the usage of enormous quantities of agrochemicals. Utilization of different nanofertilizers has become a major role in enhancing crop production and also has led to a reduction in the cost of fertilizers and reduced pollution hazards. The usage of nanofertilizers will help feed a growing population.

The market for wood-based products has a large potential for using nanotechnology. Many nanocellulose products for energy storage and energy harvester utilization, as well as sensor, have been developed at a laboratory scale. The usage of nanomaterial other than nanocellulose has already benefited and has led improvement to in wood-based products area viz., paper and pulp, wood coating, wood composite, and wood preservative. In the future, the commercial application for nanomaterials in development fertilizers, pesticides, sensors, and forest applications are likely to increase and could provide further impetus in our resolve to slow climate change and for a sustainable future.

Acknowledgment

We thank our Department of Biotechnology, Periyar University, Salem, for providing the necessary facility. One of the authors V. Sethuraman would like to acknowledge the University of Grant Commission (UGC), Government of India, for providing Dr. D. S. Kothri Postdoctoral Fellowship (Ref No: BL/20-21/0106).

References

Abou El-Nour, E.A.A. 2002. Can supplemented potassium foliar feeding reduce? Pakistan J. Biol. Sci. 5: 259–262.

Agathokleous, E., Feng, Z., Iavicoli, I. and Calabrese, E.J. 2020. Nano-pesticides: A great challenge for biodiversity? The need for a broader perspective. Nano Today 30: 100808.

Agrawal, S. and Rathore, P. 2014. Nanotechnology pros and cons to agriculture: a review. Int. J. Curr. Microbiol. App. Sci. 3: 43–55.

Ali, S., Ullah, M.I., Sajjad, A., Shakeel, Q. and Hussain, A. 2021. Environmental and health effects of pesticide residues. pp. 311–336. *In*: Inamuddin, Mohd Imran Ahamed and Eric Lichtfouse (eds.). Sustainable Agriculture Reviews 48, Springer, Cham.

Alissa, E.M. and Ferns, G.A. 2011. Heavy metal poisoning and cardiovascular disease. J. Toxicol. doi: 10.1155/2011/870125.

Angajala, G., Ramya, R. and Subashini, R. 2014. *In-vitro* anti-inflammatory and mosquito larvicidal efficacy of nickel nanoparticles phytofabricated from aqueous leaf extracts of *Aegle marmelos* Correa. Acta Trop. 135: 19–26.

Aouada, F.A. and Moura, M.R.D. 2015. Nanotechnology applied in agriculture: controlled release of agrochemicals. pp. 103–118. *In*: Nanotechnologies in Food and Agriculture, Springer, Cham.

Arno, S.F. 2000. Fire in western forest ecosystems. Wild Land Fire in Ecosystems: Effects of Fire on Flora 2: 97–120.

Arunthirumeni, M., Veerammal, V. and Shivakumar, M.S. 2022. Biocontrol efficacy of mycosynthesized selenium nanoparticle using *Trichoderma* sp. on insect pest *Spodopteralitura*. J. Clust. Sci. 33: 1645–1653.

Aziz, Z.A., Ahmad, A., Setapar, S.H.M., Karakucuk, A., Azim, M.M., Lokhat, D. and Ashraf, G.M. 2018. Essential oils: extraction techniques, pharmaceutical and therapeutic potential—a review. Curr. Drug Metab. 19: 1100–1110.

Bakry, A.M., Abbas, S., Ali, B., Majeed, H., Abouelwafa, M.Y., Mousa, A. and Liang, L. 2016. Microencapsulation of oils: A comprehensive review of benefits, techniques, and applications. Compr. Rev. Food Sci. Food Saf. 15: 143–182.

Bale, J.S., Van Lenteren, J.C. and Bigler, F. 2008. Biological control and sustainable food production. Philosophical Transactions of the Royal Society B: Biol. Sci. 363: 761–776.

Bianchi, F.J., Booij, C.J.H, and Tscharntke, T. 2006. Sustainable pest regulation in agricultural landscapes: a review on landscape composition, biodiversity and natural pest control. Proc. Royal Soc. B, Biological Sciences 273: 1715–1727.

Bindraban, P.S., Dimkpa, C.O. and Pandey, R. 2020. Exploring phosphorus fertilizers and fertilization strategies for improved human and environmental health. Biol. Fertil. Soils 56: 299–317.

Blades, M.W. and Horlick, G. 1981. Interference from easily ionizable element matrices in inductively coupled plasma emission spectrometry a spatial study. Spectrochimica Acta Part B: At. Spectrosc. 36: 881–900.

Bovand, F., Chavoshi, S. and Ghorbanpour, M. 2022. Titanium dioxide and silicon dioxide nanoparticles differentially affect germination and biochemical traits in marigold (*Calendula officinalis* L.) and lemon balm (*Melissa officinalis* L.). Nanotechnol. Environ. Eng.

C Thomas, S., Kumar Mishra, P. and Talegaonkar, S. 2015. Ceramic nanoparticles: fabrication methods and applications in drug delivery. Curr. Pharm. Des. 21: 6165–6188.

Cai, Z., Rudie, A.W., Stark, N.M., Sabo, R.C. and Ralph, S.A. 2013. New products and product categories in the global forest sector. pp. 129–149. *In*: Hansen, E., Panwar, R. and Vlosky, R. (eds.). The Global Forest Sector: Changes, Practices, and Prospects. CRC Press: Boca Raton.

Capeness, M.J., Echavarri-Bravo, V. and Horsfall, L.E. 2019. Production of biogenic nanoparticles for the reduction of 4-Nitrophenol and oxidative laccase-like reactions. Front. Microbiol.

Carrillo Gonzalez, R., Martínez Gomez, M.A. and Gonzalez Chavez, M. 2014. Nanotecnología en la actividadagropecuaria y el ambiente. doi:10.13140/2.1.4534.5600.

Casida, J.E. and Quistad, G.B. 2004. Why insecticides are more toxic to insects than people: the unique toxicology of insects? Pestic. Sci. 29: 81–86.

Chhipa, H. 2017. Nanofertilizers and nanopesticides for agriculture. Environ. Chem. Lett. 15: 15–22.

de Oliveira, J.L., Campos, E.V.R., Bakshi, M., Abhilash, P.C. and Fraceto, L.F. 2014. Application of nanotechnology for the encapsulation of botanical insecticides for sustainable agriculture: prospects and promises. Biotechnol. Adv. 32: 1550–1561.

Dhaliwal, G.S., Jindal, V. and Mohindru, B. 2015. Crop losses due to insect pests: global and Indian scenario. Indian J. Entomol. 77: 165–168.

Dimkpa, C.O. and Bindraban, P.S. 2016. Fortification of micronutrients for efficient agronomic production: A review. Agronomy for Sustainable Development 36: 7.

Dubey, A. and Mailapalli, D.R. 2016. Nanofertilisers, nanopesticides, nanosensors of pest and nanotoxicity in agriculture. pp. 307–330. *In*: Sustain. Agric. Res. Springer, Cham.

Duran, N. and Marcato, P.D. 2013. Nanobiotechnology perspectives. Role of nanotechnology in the food industry: a review. Int. J. Food Sci. Technol. 48: 1127–1134.

Eichert, T., Kurtz, A., Steiner, U. and Goldbach, H.E. 2008. Size exclusion limits and lateral heterogeneity of the stomatal foliar uptake pathway for aqueous solutes and water-suspended nanoparticles. Physiol. Plant. 134: 151–160.

Elechiguerra, J.L., Burt, J.L., Morones, J.R., Camacho-Bragado, A., Gao, X., Lara, H.H. and Yacaman, M.J. 2005. Interaction of silver nanoparticles with HIV-1. Journal of Nanobiotechnology 3: 1–10.

Elemike, E.E., Uzoh, I.M., Onwudiwe, D.C. and Babalola, O.O. 2019. The role of nanotechnology in the fortification of plant nutrients and improvement of crop production. Appl. Sci. 9: 499.

Eriksson, O. and Ehrlen, J. 2008. Seedling recruitment and population ecology. pp. 239–254. *In*: Seedling Ecology and Evolution. Cambridge: Cambridge University Press.

Fan, M., Shen, J., Yuan, L., Jiang, R., Chen, X., Davies, W.J. and Zhang, F. 2012. Improving crop productivity and resource use efficiency to ensure food security and environmental quality in China. J. Exp. Bot. 63: 13–24.

Fantke, P. and Juraske, R. 2013. Variability of pesticide dissipation half-lives in plants. Environ. Sci. Technol. 47: 3548–3562.

Fatima, F., Hashim, A. and amp Anees, S. 2021. Efficacy of nanoparticles as nanofertilizer production: a review. Environ. Sci. Pollut. Res. 28: 1292–1303.

Fei, S., Morin, R.S., Oswalt, C.M. and Liebhold, A.M. 2019. Biomass losses resulting from insect and disease invasions in US forests. Proc. Natl. Acad. Sci. 116: 17371–17376.

Faceto, L.F., Grillo, R., de Medeiros, G.A., Scognamiglio, V., Rea, G. and Bartolucci, C. 2016. Nanotechnology in agriculture: which innovation potential does it have? Front. Environ. Sci. 20.

Ghasemnezhad, A., Ghorbanpour, M., Sohrabi, O. and Ashnavar, M. 2019. A general overview on application of nanoparticles in agriculture and plant science. Compr. Anal. Chem. 87: 85–110.

Gholami-Shabani, M., Gholami-Shabani, Z., Shams-Ghahfarokhi, M., Jamzivar, F. and Razzaghi-Abyaneh, M. 2017. Green nanotechnology: biomimetic synthesis of metal nanoparticles using plants and their application in agriculture and forestry. pp. 133–175. *In*: Nanotechnology Singapore: Springer.

Ghormade, V., Deshpande, M.V. and Paknikar, K.M. 2011. Perspectives for nano-biotechnology enabled protection and nutrition of plants. Biotechnol. Adv. 29: 792–803.

Giongo, A.M.M., Vendramim, J.D. and Forim, M.R. 2016. Evaluation of neem-based nanoformulations as alternative to control fall armyworm. Cienciae Agrotecnologia 40: 26–36.

Gogos, A., Knauer, K. and Bucheli, T.D. 2012. Nanomaterials in plant protection and fertilization: current state, foreseen applications, and research priorities. J. Agric. Food Chem. 60: 9781–9792.

Gomiero, T., Pimentel, D. and Paoletti, M.G. 2011. Environmental impact of different agricultural management practices: conventional vs. organic agriculture. Crit. Rev. Plant Sci. 30: 95–124.

Gruhn, P., Goletti, F. and Yudelman, M. 2000. Integrated nutrient management, soil fertility, and sustainable agriculture: current issues and future challenges. Intl. Food Policy Res. Inst.

Günal, H., Korucu, T., Birkas, M., Özgöz, E. and Halbac-Cotoara-Zamfir, R. 2015. Threats to sustainability of soil functions in Central and Southeast Europe. Sustainability 7: 2161–2188.

Guo, J. 2004. Synchrotron radiation, soft-X-ray spectroscopy and nanomaterials. Int. J. Nanotechnol. 1: 193–225.

Gupta, N., Verma, S.C., Sharma, P.L., Thakur, M., Sharma, P. and Devi, D. 2019. Status of invasive insect pests of India and their natural enemies. J. of Entomology and Zoology Studies 7(1): 482–489.

Guru Bharathi, B., Lalitha, K. and Shivakumar, M.S. 2022. Biosynthesis of copper nanoparticles using symbiotic bacterium Xenorhabdus sp, isolated from entomopathogenic nematode and its antimicrobial and insecticidal activity against *Spodoptera litura*. Inorg. Nano-Met. Chem. 1–13.

Guyot, V., Castagneyrol, B., Vialatte, A., Deconchat, M., Selvi, F., Bussotti, F. and Jactel, H. 2015. Tree diversity limits the impact of an invasive forest pest. PLoS One 10(9): e0136469.

Halofsky, J.E., Peterson, D.L. and Harvey, B.J. 2020. Changing wildfire, changing forests: the effects of climate change on fire regimes and vegetation in the Pacific Northwest, USA. Fire Ecol. 16: 1–26.

Harish, V., Tewari, D., Gaur, M., Yadav, A.B., Swaroop, S., Bechelany, M. and Barhoum, A. 2022. Review on nanoparticles and nanostructured materials: Bioimaging, biosensing, drug delivery, tissue engineering, antimicrobial, and agro-food applications. Nanomaterials 12: 457.

Hatti-Kaul, R., Törnvall, U., Gustafsson, L. and Börjesson, P. 2007. Industrial biotechnology for the production of bio-based chemicals—a cradle-to-grave perspective. Trends Biotechnol. 25: 119–124.

He, X., Deng, H. and Hwang, H.M. 2019. The current application of nanotechnology in food and agriculture. J. Food Drug Anal. 27: 1–21.

Ishtiaq, F., Bhatti, H.N., Khan, A., Iqbal, M. and Kausar, A. 2020. Polypyrole, polyaniline and sodium alginate biocomposites and adsorption-desorption efficiency for imidacloprid insecticide. Int. J. Biol. Macromol. 147: 217–232.

Jasmani, L., Rusli, R., Khadiran, T., Jalil, R. and Adnan, S. 2020. Application of nanotechnology in wood-based products industry: A review. Nanoscale Res. Lett. 15: 1–31.

Jensen, J.L., Schjønning, P., Watts, C.W., Christensen, B.T., Obour, P.B. and Munkholm, L.J. 2020 Soil degradation and recovery—Changes in organic matter fractions and structural stability. Geoderma 364: 114181.

Kamel, S. 2007. Nanotechnology and its applications in lignocellulosic composites, a mini review. Express Polym. Lett. 1: 546–575.

Karn, B., Kuiken, T. and Otto, M. 2011. Nanotechnology and *in situ* remediation: a review of the benefits and potential risks. Ciência and Saúde Coletiva 16: 165–178.

Khan, S. and Hossain, M.K. 2022. Classification and properties of nanoparticles. pp. 15–54. *In*: Nanoparticle-based Polymer Composites. Woodhead Publishing.

Khot, L.R., Sankaran, S., Maja, J.M., Ehsani, R. and Schuster, E.W. 2012. Applications of nanomaterials in agricultural production and crop protection: a review. J. Crop Prot. 35: 64–70.

Klaine, S.J., Alvarez, P.J., Batley, G.E., Fernandes, T.F., Handy, R.D., Lyon, D.Y. and Lead, J.R. 2008. Nanomaterials in the environment: behavior, fate, bioavailability, and effects. Environ. Toxicol. Chem. 27: 1825–1851.

Kole, C., Kole, P., Randunu, K.M., Choudhary, P., Podila, R., Ke, P.C. and Marcus, R.K. 2013. Nanobiotechnology can boost crop production and quality: first evidence from increased plant biomass, fruit yield and phytomedicine content in bitter melon (*Momordica charantia*). BMC Biotechnology 13: 1–10.

Kulkarni, R.A., Prabhuraj, A., Ashoka, J., Hanchinal, S.G. and Hiregoudar, S. 2017. Generation and evaluation of nanoparticles of supernatant of *Photorhabdus luminescens* (Thomas and Poinar) against mite and aphid pests of cotton for enhanced efficacy. Curr. Sci. 112: 2312–2316.

Kumar, K., Dasgupta, C.N. and Das, D. 2014. Cell growth kinetics of Chlorella sorokiniana and nutritional values of its biomass. Bioresour. Technol. 167: 358–366.

Kumar, M., Shamsi, T.N., Parveen, R. and Fatima, S. 2017. Application of nanotechnology in enhancement of crop productivity and integrated pest management. pp. 361–371. *In*: Nanotechnology. Singapore: Springer.

Kumar, S., Ahlawat, W., Bhanjana, G., Heydarifard, S., Nazhad, M.M. and Dilbaghi, N. 2014. Nanotechnology-based water treatment strategies. J. Nanosci. Nanotechnol. 14: 1838–1858.

Kumar, S., Nehra, M., Dilbaghi, N., Marrazza, G., Hassan, A.A. and Kim, K.H. 2019. Nano-based smart pesticide formulations: Emerging opportunities for agriculture. J. Control. Release 294: 131–153.

Kumaravel, J., Lalitha, K., Arunthirumeni, M. and Shivakumar, M.S. 2021. Mycosynthesis of bimetallic zinc oxide and titanium dioxide nanoparticles for control of *Spodoptera frugiperda*. Pestic. Biochem. Physiol. 178: 104910.

Kumari, A., Yadav, S.K. and Yadav, S.C. 2010. Biodegradable polymeric nanoparticles based drug delivery systems. Colloids Surf. B: Biointerfaces 75: 1–18.

Kumari, A., Kumari, P., Rajput, V.D., Sushkova, S.N. and Minkina, T. 2022. Metal(loid) nanosorbents in restoration of polluted soils: Geochemical, eco-toxicological, and remediation perspectives. Environ. Geochem. Health 44: 235–246.

Lade, B.D., Gogle, D.P. and Nandeshwar, S.B. 2017. Nanobiopesticide to constraint plant destructive pests. J. Nanomed. Res. 6: 1–9.

Le Van, N., Ma, C., Shang, J., Rui, Y., Liu, S. and Xing, B. 2016. Effects of CuO nanoparticles on insecticidal activity and phytotoxicity in conventional and transgenic cotton. Chemosphere 144: 661–670.

León-Silva, S., Fernández-Luqueño, F. and López-Valdez, F. 2016. Silver nanoparticles (AgNP) in the environment: a review of potential risks on human and environmental health. Water, Air, and Soil Pollution 227: 1–20.

Leydecker, S. 2008. Nano Materials. pp. 1–192. *In*: Architecture, Interior Architecture and Design. Basel: Birkhauser.

Li, L.S., Hu, J., Yang, W. and Alivisatos, A.P. 2001. Band gap variation of size-and shape-controlled colloidal CdSe quantum rods. Nano Letters 1: 349–351.

Li, Z., Gao, Y. and Wang, K. 2021. Vertical vibration of an end bearing pile interacting with the radially inhomogeneous saturated soil. Ocean Engineering 219: 108009.

Li, Z., Wang, H., An, S. and Yin, X. 2021. Nanochitin whisker enhances insecticidal activity of chemical pesticide for pest insect control and toxicity. J. Nanobiotechnology 19: 1–13.

Lin, N. and Dufresne, A. 2014. Nanocellulose in biomedicine: Current status and future prospect. Eur. Polym. J. 59: 302–325.

Liu, Y., Tong, Z. and Prud homme, R.K. 2008. Stabilized polymeric nanoparticles for controlled and efficient release of bifenthrin. Pest Manag. Sci. 64: 808–812.

Loganathan, S., Shivakumar, M.S., Karthi, S., Nathan, S.S. and Selvam, K. (2021). Metal oxide nanoparticle synthesis (ZnO-NPs) of Knoxia sumatrensis (Retz.) DC. Aqueous leaf extract and It's evaluation of their antioxidant, anti-proliferative and larvicidal activities. Toxicol. Rep. 8: 64–72.

Loganathan, S., Selvam, K., Padmavathi, G., Shivakumar, M.S., Senthil-Nathan, S., Sumathi, A.G. and Almutairi, S.M. 2022. Biological synthesis and characterization of Passiflora subpeltata Ortega aqueous leaf extract in silver nanoparticles and their evaluation of antibacterial, antioxidant, anti-cancer and larvicidal activities. J. King Saud. Univ Sci. 34(3): 101846.

Loganathan, S., Selvam, K., Shivakumar, M.S., Senthil-Nathan, S., Vasantha-Srinivasan, P., Gnana Prakash, D. and Krutmuang, P. 2022. Phytosynthesis of silver nanoparticle (AgNPs) using aqueous leaf extract of *Knoxia sumatrensis* (Retz.) DC. and their multi-potent biological activity: an eco-friendly approach. Molecules 27(22): 7854.

Mampallil, L.J., Faizal, M.H. and Anith, K.N. 2017. Bacterial bioagents for insect pest management. J. Entomol. Zool. Stud. 5: 2237–2244.

Manjunatha, S.B., Biradar, D.P. and Aladakatti, Y.R. 2016. Nanotechnology and its applications in agriculture: A review. Int. J. Farm Sci. 29: 1–13.

Marshall, E.J.P. 2004. Agricultural landscapes: field margin habitats and their interaction with crop production. J. Crop Improv. 12: 365–404.

Mejias, J.H., Salazar, F., Peerez Amaro, L., Hube, S., Rodriguez, M. and Alfaro, M. 2021. Nanofertilizers: a cutting-edge approach to increase nitrogen use efficiency in grasslands. Front. Environ. Sci. p.52.

Mikkelsen, R. 2018. Nanofertilizer and nanotechnology: a quick look. Better Crops with Plant Food 102: 18–19.

Mohanta, Y.K., Hashem, A., Abd_Allah, E.F., Jena, S.K. and Mohanta, T.K. 2020. Bacterial synthesized metal and metal salt nanoparticles in biomedical applications: an up and coming approach. Appl. Organomet. Chem. 34(9): e5810.

Moloi, M.S., Lehutso, R.F., Erasmus, M., Oberholster, P.J. and Thwala, M. 2021. Aquatic environment exposure and toxicity of engineered nanomaterials released from nano-enabled products: Current status and data needs. Nanomaterials 11: 2868.

Moon, R.J., Frihart, C.R. and Wegner, T. 2006. Nanotechnology applications in the forest products industry. For. Prod. J. 56: 4–10.

Mosina, K.S., Nazarova, E.A., Vinogradov, A., Vinogradov, V.V., Krivoshapkina, E.F. and Krivoshapkin, P.V. 2020. Alumina nanoparticles for firefighting and fire prevention. ACS Appl. Nano Mater. 3: 4386–4393.

Mu, L. and Sprando, R.L. 2010. Application of nanotechnology in cosmetics. Pharm. Res. 27: 1746–1749.

Mukhopadhyay, S. and Maiti, S.K. 2010. Phytoremediation of metal mine waste. Appl. Ecol. Environ. Res. 8: 207–222.

Muller, R. N. 2003. Nutrient Relations of the Herbaceous Layer in Deciduous Forest Ecosystems. The Herbaceous Layer in Forests of Eastern North America. Oxford University Press, New York, pp. 15–37.

Nair, R., Varghese, S.H., Nair, B.G., Maekawa, T., Yoshida, Y. and Kumar, D.S. 2010. Nanoparticulate material delivery to plants. Plant Science 179: 154–163.

Nasi, R., Dennis, R., Meijaard, E., Applegate, G. and Moore, P. 2002. Forest Fire and Biological Diversity. UNASYLVA-FAO, 36–40.

Nel, A., Xia, T., Madler, L. and Li, N. 2006. Toxic potential of materials at the nanolevel. Science 311: 622–627.

Nuchuchua, O., Sakulku, U., Uawongyart, N., Puttipipatkhachorn, S., Soottitantawat, A. and Ruktanonchai, U. 2009. *In vitro* characterization and mosquito (*Aedes aegypti*) repellent activity of essential-oils-loaded nanoemulsions. Aaps Pharmscitech 10. 1234–1242.

Nuruzzaman, M.D., Rahman, M.M., Liu, Y. and Naidu, R. 2016. Nanoencapsulation, nano-guard for pesticides: a new window for safe application. J. Agric. Food Chem. 64: 1447–1483.

Nyamadzawo, G., Gwenzi, W., Kanda, A., Kundhlande, A. and Masona, C. 2013. Understanding the causes, socio-economic and environmental impacts, and management of veld fires in tropical Zimbabwe. Fire Sci. Rev. 2: 1–13.

Oerke, E.C., Dehne, H.W., Schönbeck, F. and Weber, A. 2012. Crop Production and Crop Protection: Estimated Losses in Major Food and Cash Crops. Elsevier. 72–741

Okey-Onyesolu, C.F., Hassanisaadi, M., Bilal, M., Barani, M., Rahdar, A., Iqbal, J. and Kyzas, G.Z. 2021. Nanomaterials as nanofertilizers and nanopesticides: An overview. Chemistry Select 6: 8645–8663.

Oliveira, T., Thomas, M. and Espadanal, M. 2014. Assessing the determinants of cloud computing adoption: An analysis of the manufacturing and services sectors. Inf. Manag. J. 51: 497–510.

Ozkan, E. Ozkan, F.T. Allan, E. and Parkin, I.P. 2015. The use of zinc oxide nanoparticles to enhance the antibacterial properties of light-activated polydimethylsiloxane containing crystal violet. RSC Advances 5(12): 8806–8813.

Pal, A., Kundu, A. and Evans, M.R. 2016. Diffusion under time-dependent resetting. J. Phys. A Mathematical and Theoretical 49: 225001.

Pal, S. and Satpathi, S.K. 2016. Species richness and relative abundance of pest faunal complex infesting okra under mid-hill conditions of Eastern Himalayas. J. Agric. Sci. 3: 74–78.

Palop, J.J., Mucke, L. and Roberson, E.D. 2010. Quantifying biomarkers of cognitive dysfunction and neuronal network hyper excitability in mouse models of Alzheimer's disease: depletion of calcium-dependent proteins and inhibitory hippocampal remodeling. pp. 245–262. *In*: Alzheimer's Disease and Frontotemporal Dementia. Totowa: Humana Press.

Panneerselvam, C., Murugan, K., Roni, M., Aziz, A.T., Suresh, U., Rajaganesh, R. and Benelli, G. 2016. Fern-synthesized nanoparticles in the fight against malaria: LC/MS analysis of *Pteridium aquilinum* leaf extract and biosynthesis of silver nanoparticles with high mosquitocidal and antiplasmodial activity. Parasitol. Res. 115: 997–1013.

Parisi, C., Vigani, M. and Rodríguez-Cerezo, E. 2015. Agricultural nanotechnologies: what are the current possibilities? Nano Today 10: 124–127.

Pavoni, L., Benelli, G., Maggi, F. and Bonacucina, G. 2019. Green nanoemulsion interventions for biopesticide formulations. pp. 133–160. *In*: Nano-biopesticides Today and Future Perspectives. Academic Press.

Pérez-de-Luque, A. 2017. Interaction of nanomaterials with plants: what do we need for real applications in agriculture, Front. Environ. Sci. 5: 12.

Perlatti, B., de Souza Bergo, P.L., Fernandes, J.B. and Forim, M.R. 2013. Polymeric nanoparticle-based insecticides: a controlled release purpose for agrochemicals. *In*: Insecticides-Development of Safer and More Effective Technologies. Intech Open.

Popp, J., Pető, K. and Nagy, J. 2013. Pesticide productivity and food security. A review. Agron. Sustain Dev. 33: 243–255.

Pramanik, S. and Pramanik, G. 2016. Nanotechnology for sustainable agriculture in India. J. Sci. Food Agric. Springer. pp. 243–280.

Prasad, R., Kumar, V. and Prasad, K.S. 2014. Nanotechnology in sustainable agriculture: present concerns and future aspects. Afr. J. Biotechnol. 13: 705–713.

Puurunen, K. and Vasara, P. 2007. Opportunities for utilizing nanotechnology in reaching near-zero emissions in the paper industry. J. Clean. Prod. 15: 1287–1294.

Qin, J., Tong, Z., Zhan, Y., Buisson, C., Song, F., He, K. and Guo, S. 2020. A *Bacillus thuringiensis* chitin-binding protein is involved in insect peritrophic matrix adhesion and takes part in the infection process. Toxins 12: 252.

Rabajczyk, A., Zielecka, M., Popielarczyk, T. and Sowa, T. 2021. Nanotechnology in Fire Protection-Application and Requirements. Materials (Basel, Switzerland), 14: 7849.

Raffa, C.M., Chiampo, F. and Shanthakumar, S. 2021. Remediation of metal/metalloid-polluted soils: A short review. Appl. Sci. 11: 4134.

Rai, M. and Ingle, A. 2012. Role of nanotechnology in agriculture with special reference to management of insect pests. Appl. Microbiol. Biotechnol. 94: 287–293.

Raj, T.G. and Khan, N.A. 2016. Designer nanoparticle: Nanobiotechnology tool for cell biology. Nano Convergence 3: 22.

Rajkumar, R., Shivakumar, M.S., Senthil Nathan, S. and Selvam, K. (2018). Pharmacological and larvicidal potential of green synthesized silver nanoparticles using *Carmona retusa* (Vahl) Masam leaf extract. J. Clust. Sci. 29(6): 1243–1253.

Rajput, V.D., Minkina, T., Kumari, A., Shende, S.S., Ranjan, A., Faizan, M. and Kızılkaya, R. 2022. A review on nanobioremediation approaches for restoration of contaminated soil. Eurasian J. Soil Sci. 11: 43–60.

Rajput, V.D., Minkina, T., Upadhyay, S.K., Kumari, A., Ranjan, A., Mandzhieva, S. and Verma, K.K. 2022. Nanotechnology in the restoration of polluted soil. Nanomaterials 12: 769.

Ramkumar, G., Karthi, S., Suganya, R. and Shivakumar, M.S. 2016. Evaluation of silver nanoparticle toxicity of *Coleus aromaticus* leaf extracts and its Larvicidal toxicity against dengue and Filariasis vectors. Bionanoscience 6(4): 308–315.

Ramzan, F., Khan, M.U.G., Rehmat, A., Iqbal, S., Saba, T., Rehman, A. and Mehmood, Z. 2020. A deep learning approach for automated diagnosis and multi-class classification of Alzheimer's disease stages using resting-state fMRI and residual neural networks. J. Med. Syst. 44: 1–16.

Rana, R.A., Siddiqui, M., Skalicky, M., Brestic, M., Hossain, A., Kayesh, E. and Islam, T. 2021. Prospects of nanotechnology in improving the productivity and quality of horticultural crops. Horticulturae 7: 332.

Rani, A., Rani, K., Tokas, J., Singh, A., Kumar, R., Punia, H. and Kumar, S. 2020. Nanomaterials for agriculture input use efficiency. pp. 137–175. In Resources Use Efficiency in Agriculture. Singapore: Springer.

Rasmussen, L.D., Sorensen, S.J., Turner, R.R. and Barkay, T. 2000. Application of a mer-lux biosensor for estimating bioavailable mercury in soil. Soil Biol. Biochem. 32: 639–646.

Rastogi, A., Tripathi, D.K., Yadav, S., Chauhan, D.K., Živčák, M., Ghorbanpour, M. and Brestic, M. 2019. Application of silicon nanoparticles in agriculture. 3Biotech 9: 1–11.

Rawani, A., Ghosh, A. and Chandra, G. 2013. Mosquito larvicidal and antimicrobial activity of synthesized nano-crystalline silver particles using leaves and green berry extract of *Solanum nigrum* L. (Solanaceae: Solanales). Acta Tropica 128: 613–622.

Reddy, K.R. 2014. Evolution of geo environmental engineering. Environ. Geotech, 1: 136–141.

Ribeiro, A.M., de Sousa, A.M., Vicente, E.F.R. and do Carmo, C.H.S. 2019. Income smoothing e comparabilidade dos relatórios financeiros: evidências em empresas brasileiras de capital aberto. In Anais do XVII Congresso Internacional de Contabilidade e Auditoria, Porto, Portugal. pp. 1–26.

Rueda-Núñez, J.L. 2014. Chemical Composition for Fighting Forest Fires and Process for Obtaining Thereof. U.S. Patent 8,647,524 B2, 11 February 2014.

Sandberg, D. 2016. Additives in wood products-Today and future development. pp. 105–172. In Environmental Impacts of Traditional and Innovative Forest-based Bio Products. Singapore: Springer.

Sano, N., Rajjou, L., North, H.M., Debeaujon, I., Marion-Poll, A. and Seo, M. 2016. Staying alive: Molecular aspects of seed longevity. Plant Cell Physiol. 57: 660–674.

Sano, R. and Reed, J.C. 2013. ER stress-induced cell death mechanisms. Biochimica et Biophysica Acta 1833: 3460–3470.

Sastry, R.K. and Rao, N.H. 2013. Emerging technologies for enhancing indian agriculture-case of nano-biotechnology. Asian Biotechnol. Dev. Rev. 15: 1–19.

Schwarzkopf, M.J. and Burnard, M.D. 2016. Wood-plastic composites performance and environmental impacts. Environmental Impacts of Traditional and Innovative Forest-Based Bioproducts 19–43.

Semida, W.M., Abdelkhalik, A., Mohamed, G.F., Abd El-Mageed, T.A., Abd El-Mageed, S.A., Rady, M.M. and Ali, E.F. 2021. Foliar application of zinc oxide nanoparticles promotes drought stress tolerance in eggplant (*Solanum melongena* L.). Plants 10: 421.

Sexton, S.E., Lei, Z. and Zilberman, D. 2007. The economics of pesticides and pest control. Int. Rev. Environ. Resour. Econ. 1: 271–326.

Sharma, D. and Bisht, D. 2017. *M. tuberculosis* hypothetical proteins and proteins of unknown function: hope for exploring novel resistance mechanisms as well as future target of drug resistance. Front. Microbiol. 8: 465.

Sharma, D., Kanchi, S. and Bisetty, K. 2019. Biogenic synthesis of nanoparticles: a review. Arab. J. Chem. 12: 3576–3600.

Sharma, M. 2016. Insect pests of forestry plants and their management. Int. J. Adv. Res. 4: 2099–2116.

Sharma, S., Kooner, R. and Arora, R. 2017. Insect pests and crop losses. pp. 45–66. In Breeding Insect Resistant Crops for Sustainable Agriculture. Singapore: Springer.

Sharon, M., Choudhary, A.K. and Kumar, R. 2010. Nanotechnology in agricultural diseases and food safety. J. Phytol. 2: 78–82.

Shekhar, S., Sharma, S., Kumar, A., Taneja, A. and Sharma, B. 2021. The framework of nano-pesticides: a paradigm in biodiversity. Adv. Mater. 2: 6569–6588.

Singh, R.P., Handa, R. and Manchanda, G. 2021. Nanoparticles in sustainable agriculture: An emerging opportunity. J. Control. Release 329: 1234–1248.

Solanki, R. and Dhankhar, R. 2011. Biochemical changes and adaptive strategies of plants under heavy metal stress. Biologia 66: 195–204.

Sowndarya, P., Ramkumar, G., and Shivakumar, M.S. 2017. Green synthesis of selenium nanoparticles conjugated *Clausena dentata* plant leaf extract and their insecticidal potential against mosquito vectors. Artif Cells Nanomed Biotechnol. 45(8). 1490–1495.

Suganya, G., Karthi, S. and Shivakumar, M.S. 2014. Larvicidal potential of silver nanoparticles synthesized from *Leucas aspera* leaf extracts against dengue vector *Aedes aegypti*. Parasitol. Res. 113: 875–880.

Sweta, B., Lalit, M. and Srivastava, C.N. 2018. Nanopesticides: a recent novel ecofriendly approach in insect pest management. J. Entomol. Res. 42: 263–70.

Tarafdar, J.C., Sharma, S. and Raliya, R. 2013. Nanotechnology: Interdisciplinary science of applications. Afr. J. Biotechnol, 12: 219–226.

Tarafdar, J.C. 2015. Nanoparticle Production, Characterization and its Application to Horticultural Crops. Jodhpur, Rajasthan: ICAR, pp. 222–229.

Thomé, A., Reddy, K.R., Reginatto, C. and Cecchin, I. 2015. Review of nanotechnology for soil and groundwater remediation: Brazilian Perspectives. Water, Air, and Soil Pollution 226: 1–20.

Thompson, I., Mackey, B., McNulty, S. and Mosseler, A. 2009. Forest resilience, biodiversity, and climate change. pp. 1–67. In Secretariat of the Convention on Biological Diversity, Montreal. Technical Series no. 43.

Veerakumar, A., Challis, C., Gupta, P., Da, J., Upadhyay, A., Beck, S.G. and Berton, O. 2014. Antidepressant-like effects of cortical deep brain stimulation coincides with pro-neuroplastic adaptations of serotonin systems. Biological Psychiatry 76: 203–212.

Vengateswari, G., Arunthirumeni, M., Shivaswamy, M.S. and Shivakumar, M.S. 2022. Effect of host plants nutrients, antioxidants, and phytochemicals on growth, development, and fecundity of *Spodoptera litura* (Fabricius) (Lepidoptera: Noctuidae). International Int. J. Trop. Insect Sci. 42(4): 3161–3173.

Verma, K.K., Song, X.P., Joshi, A., Tian, D.D., Rajput, V.D., Singh, M. and Li, Y.R. 2022. Recent trends in nano-fertilizers for sustainable agriculture under climate change for global food security. Nanomaterials 12(1): 173.

Virkutyte, J. and Varma, R.S. 2011. Green synthesis of metal nanoparticles: biodegradable polymers and enzymes in stabilization and surface functionalization. Chem. Sci. 2: 837–846.

Vishwakarma, K., Upadhyay, N., Kumar, N., Tripathi, D.K., Chauhan, D.K., Sharma, S. and Sahi, S. 2018. Potential applications and avenues of nanotechnology in sustainable agriculture. pp. 473–500. *In*: Nanomaterials in Plants, Algae, and Microorganisms. Academic Press.

Vivekanandhan, P., Deepa, S., Kweka, E.J. and Shivakumar, M.S. 2018. Toxicity of Fusarium oxysporum-VKFO-01 derived silver nanoparticles as potential insecticide against three mosquito vector species (Diptera: Culicidae). J. Clust. Sci. 29(6): 1139–1149.

Wang, C.B. and Zhang, W.X. 1997. Synthesizing nanoscale iron particles for rapid and complete dechlorination of TCE and PCBs. Environ. Sci. Technol. 31: 2154–2156.

Weeks Jr, J.J. and Hettiarachchi, G.M. 2019. A review of the latest in phosphorus fertilizer technology: possibilities and pragmatism. J. Environ. Qual. 48: 1300–1313.

Wegner, T.H., Winandy, J.E. and Ritter, M.A. 2005. Nanotechnology opportunities in residential and non-residential construction. In 2nd International Symposium on Nanotechnology in Construction, 13–16 November 2005, Bilbao, Spain [CD-ROM]. Bagneux, France: RILEM, p9.

Yashkamal, K., Vivekanandhan, P., Muthusamy, R., Vengateswari, G. and Shivakumar, M.S. 2021. May nanoparticles offer chances to avoid the development of insecticide resistance in mosquitoes? pp. 549–563. In Applications of Nanobiotechnology for Neglected Tropical Diseases. Academic Press.

Zamolodchikov, D., Shvidenko, A., Bartalev, S., Kulikova, E., Held, A., Valentini, R. and Lindner, M. 2020. State of Russian Forests and Forestry. Russian Forests and Climate Change.

Zhao, L., Lu, L., Wang, A., Zhang, H., Huang, M., Wu, H. and Ji, R. 2020. Nano-biotechnology in agriculture: use of nanomaterials to promote plant growth and stress tolerance. J. Agric. Food Chem. 68: 1935–1947.

Section II
Application of Nanomaterials in Plant Health Fitness

Chapter 3

Nanofertilizers for Sustainable Agricultural Practices

Nadun H. Madanayake[1] and *Nadeesh M. Adassooriya*[2,*]

Introduction

Agriculture has been the backbone of the globe, playing a significant role in fulfilling the global food demand. Therefore, it is necessary to improve crop productivity to fulfill the daily requirement of the ever-increasing global population. Fertilizers play a significant role in enhancing plant crop productivity. Fertilizers simply supply the nutrients required for vegetative growth. These nutrients play a vital role in growth and development as well as in soil fertility, regardless of the crop type and environmental conditions. Most conventional fertilizers have a poor nutrient use efficiency, typically surpassing 30–35% for nitrogen, 18–20% for phosphorus, and 35–40% for potassium. The low nutrient use efficiency of conventional fertilizers not only has an economic impact but also incurs many environmental concerns. For instance, nitrogen fertilizer can reach water via leaching, drainage, or flow, leading to water pollution. Nitrification of nitrogen-containing fertilizers by soil microbes can leach nitrates into groundwater, thus reducing the quality of drinking water. Additionally, surface runoff of nitrogen and phosphorus can cause algal blooming, which can badly influence aquatic life. Overuse of conventional fertilizers can destroy the soil structure and conditions. For example, the application of fertilizers containing higher levels of sodium and potassium, as well as acid-forming nitrogen fertilizers, can alter the soil pH. This alteration makes arable farming lands infertile and, as a result, difficult to use in agriculture. Moreover, the application of ammonium fertilizer can volatilize to release ammonia, badly affecting ecosystems and vegetation. Furthermore, the synthesis of nitrogen fertilizers extensively utilizes fossil fuels, and greenhouse gas emission is a serious issue to deal with (Savci 2012; Khan and Mohammad 2014; Subramanian et al. 2015; Chhipa and Joshi 2016; Samavini et al. 2018). Therefore, the development of effective platforms to control the release of fertilizers is utterly needed. This development can enhance their nutrient use efficiency while simultaneously mitigating the environmental impacts.

Nanotechnology is a rapidly growing field of research and development that is revolutionizing the potential of science and technology, and its potential applications are virtually endless. Nanotechnology is the manipulation of matter on an atomic, molecular, and supramolecular scale. Mainly, nanotechnology deals with nanoparticles (NPs), which are the fundamental building blocks.

[1] Department of Botany, Faculty of Science, University of Peradeniya, 20400 Peradeniya, Sri Lanka.
[2] Department of Chemical and Process Engineering, Faculty of Engineering, University of Peradeniya, Peradeniya Sri Lanka.
* Corresponding author: nadeeshm@eng.pdn.ac.lk

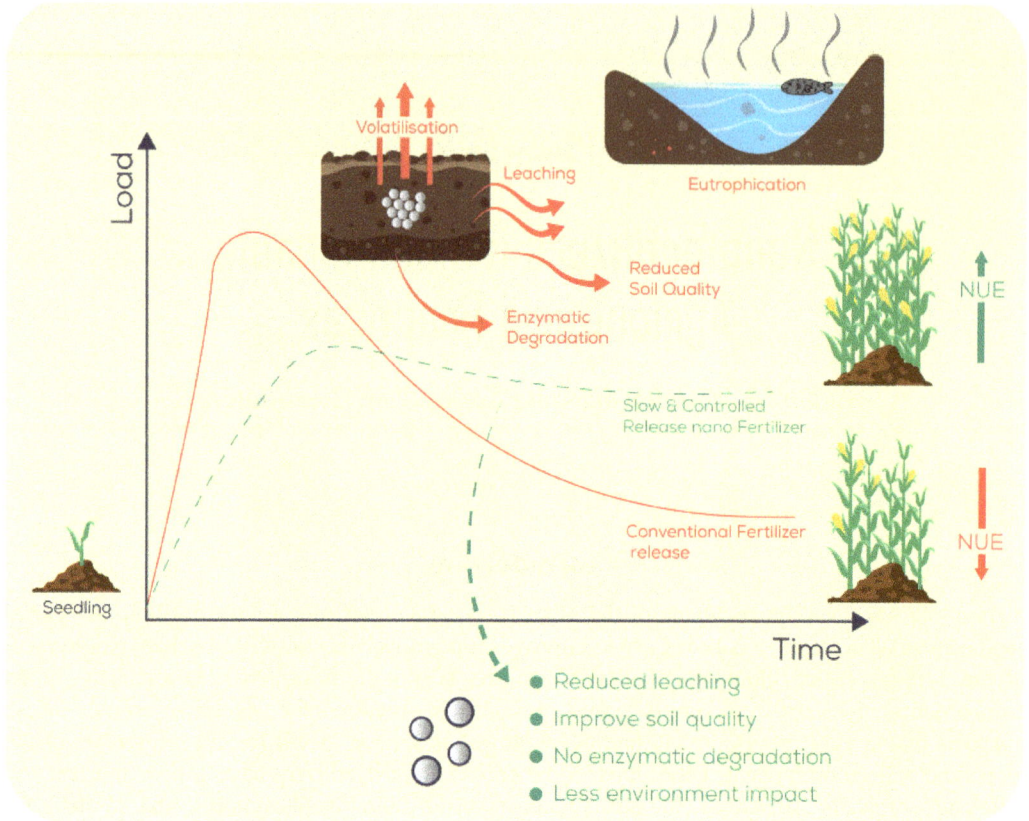

Figure 1. Effect of nanofertilizers on crop productivity.

Nanomaterials (NMs) or NPs are materials in the size range of 1–100 nm with either one of its dimensions in this limit. This is an interdisciplinary approach that augments diverse fields to create materials, devices, and systems with unique and novel properties and functions. This has been a rapidly advancing field of research and development, encompassing a broad range of applications in electronics, energy, medicine and pharmaceuticals, environmental remediation, food, and agriculture, etc. (Khin et al. 2012; Jariwala et al. 2013; Mei et al. 2013; Zhang et al. 2013; He et al. 2019; Madanayake et al. 2019).

Agriculture has been immensely blessed with nanotechnological applications. This has been used to enhance the efficiency of numerous agricultural applications. Nanofertilizers (NFs) are a type of fertilizer containing nano-sized particles of plant nutrients. These particles are designed to be more easily absorbed by plants, leading to increased crop yields and improved soil health by enhancing nutrient use efficiency. Also, the small size of these particles allows for more precise application and targeted delivery of nutrients to plants, thus reducing the risk of nutrient runoff. Moreover, this mitigates the deleterious effects of conventional fertilizers on the environment as they can be engineered to fabricate slow and controlled released fertilizers. Therefore, this would play a huge role in sustainable agricultural practices. Sustainable agriculture prefers the use of natural resources in a way that does not deplete them and utilizes them economically to enhance agricultural productivity. Figure 1 elaborates on the NFs and their key benefits in crop productivity and agriculture. Therefore, this chapter mainly highlights the NFs, their types, methods of fabrication, and release kinetics in the environment.

Nanofertilizer Types

NFs can be classified either as macro and micronutrient fertilizers or based on the method of fabrication. Macro and micronutrients are essential elements that require for effective vegetative growth. These components play a significant role in plant physiology as it directly involves in plant metabolism. Simply, macronutrients are required by greens in larger quantities, whereas micronutrients are essential in minute amounts (Römheld and Marschner 1991). Therefore, the fabrication of NFs using either macro or micronutrient as their principal component can be categorized as macro- or micronutrient NFs. NMs can be synthesized in different approaches, such as encapsulation, intercalation, and surface modification as carriers, and they can be utilized as a fertilizer (Kottegoda et al. 2011; Manikandan and Subramanian 2016; Azarian et al. 2018; Marchiol et al. 2019; Rios et al. 2021; Yoon et al. 2020; Sharma et al. 2022).

Macronutrient Nanofertilizers

Macronutrient NFs include nitrogen (N), phosphorus (P), potassium (K), magnesium (Mg), sulfur (S), and calcium (Ca) as principal components. As mentioned elsewhere in the text, conventional fertilizers have poor nutrient use efficiency, macronutrients are applied in large quantities and a greater proportion gets leached out. This could have detrimental effects on the environment and human life influencing the balance of the ecosystem. Hence, several studies have tried to develop different forms of NFs for the sustained release of plant nutrients.

Hydroxyapatite has been reported to utilize as a better phosphorus supplier by certain studies (Marchiol et al. 2019; Madanayake et al. 2021). Elsayed et al. (2022) utilized hydroxyapatite NPs as a phosphorus supplier and applied them to *Rosmarinus officinalis*. Hydroxyapatite NPs at 0.5 g/L and 1.0 g/L concentration had significantly increased the growth parameters and essential oil contents, such as α-pinene, eucalyptol, camphor, endo-borneol, and verbenone. Also, citric acid and humic acid functionalized hydroxyapatite NPs are reported as phosphorus NFs (Samavini et al. 2018; Yoon et al. 2020). Ramírez-Rodríguez et al. (2020) developed urea-doped calcium phosphate NFs and tested them with *Triticum durum* plants. Bioavailability studies showed that lower contents (40%) NFs do not show any difference with respect to their yield and quality of the test plant in comparison to controls. Ha et al. (2019) developed nitrogen, phosphorus, and potassium-loaded chitosan NFs as a sustained-release fertilizer platform. It was depicted that NPK NFs had improved the nutrient uptake (17.04% nitrogen, 16.31% phosphorus, and 67.50% potassium content in the leaves), chlorophyll content (30.68%), and rate of photosynthesis (71.7%) of coffee plants.

Micronutrient Nanofertilizers

Micronutrient NFs are specifically focused on delivering micronutrients required for plant growth. These are required in minute quantities for plant metabolisms. Micronutrients include iron (Fe), Manganese (Mn), zinc (Zn), copper (Cu), molybdenum (Mo), nickel (Ni), boron (B), and chlorine (Cl). Typically, micronutrient-based composites are prepared with these elements at very low dosages such as 5 ppm (Chhipa and Joshi 2016). Hanon Mohsen et al. (2022) performed field trials using variable dosages of iron-based NFs on four different varieties of wheat. It was reported that field application of iron NPs at 3 g/L dosages showed the maximum grain production with tested varieties. Iron humate NPs are currently in application to overcome the low iron use efficiency of synthetic iron chelates. Cieschi et al. (2019) evaluated the potential of iron-humic NFs to mitigate iron chlorosis on *Glycine max* in calcareous soils. Authors have confirmed that the novel formulation of iron NPs had improved plant growth with no leaf chlorosis. Furthermore, the slow and incessant release of iron from iron-humic NF validates their potential to deliver micronutrients for extended durations in iron-deficient soils. Also, this would confirm the bioavailability of iron for plants at different stages of the plant lifecycle. AL-Abody et al. (2021) performed a field experiment on the effect of iron NPs on *Triticum aestivum* L. in Iraq. It was found that foliar-applied iron NPs had improved the yield and certain growth parameters of wheat plants, such as height, leaf area, and spike length.

Cota-Ruiz et al. (2020) reported that nano-Cu can improve the physiology of *Medicago sativa*. Here, the plants treated with the nano-Cu had improved the abundance of beneficial soil microorganisms and increased the iron and zinc contents in the roots and leaves of the tests. Moreover, the application of nano-Cu significantly improved the leaf-superoxide dismutase levels. Leonardi et al. (2021) developed copper oxide-loaded chitosan/alginate NPs as NFs and evaluated the seedling and germination of *Fortunella margarita*. It was shown that the composite had a sluggish release of Cu from the nanocomposite, where 80% Cu was released in 22 days and enhanced seed germination.

Zn is another essential plant nutrient that plays a vital role in plant metabolism. Yusefi-Tanha et al. (2020) investigated the effect of the size, shape, and dosage of zinc oxide NPs on yield and plant vigor. Pot trials using *Rhizobium japonicum* inoculated soybean seeds showed a size, shape, and concentration-dependent impact on yield, lipid peroxidation, and antioxidant activity, thus proving its potential to apply as a novel NF in Zn-deficient soil. Bala et al. (2019) explored the effect of foliar-applied zinc oxide NPs on rice grown in Zn lacking soils for remediation and Zn fortification. Pot experiments reported that the foliar presentation of zinc oxide NPs at 5 g/L had noticeably amplified the growth and yield characteristics of rice. In addition, soil microbe population and dehydrogenase activity were at higher levels at 5 g/L zinc oxide NPs treatments. Despite being the nutrient that has received the most research, the effects of zinc leaching from the soil after crop harvest have received almost no attention. Sheoran et al. (2021) examined the influence of zinc oxide NPs on the growth parameters of wheat. At 80 ppm concentration, zinc oxide NPs showed the highest increment in height, seeds per spike, seed mass, yield, and biomass accumulation in comparison to control and zinc nitrate experiments. Also, the application of zinc oxide NPs resulted in better uptake of Zn, and a lesser amount remained in soil likened to the control and zinc nitrate treatments. Hence, this proves that the application of NPs as NFs exhibits lesser effects on the environment compared to metallic salts, which are highly water-soluble.

Samsoon et al. (2022) biologically synthesized manganese dioxide NPs as a potential NF on *Vigna unguiculata*. Both soil and foliar application of NPs showed a substantial increment in *V. unguiculata* growth, chlorophyll content, and non-enzymatic antioxidant potential. Also, ground application of manganese oxide NPs showed better results with respect to plant growth in contrast to the foliar pathway proving its potential to apply in Mn-deficient soils. Foliar application of green synthesized manganese zinc ferrite NPs on *Cucurbita pepo* L. gave a significant improvement in plant growth and yield (Shebl et al. 2019). Salama et al. (2022) showed that foliar application of 30 ppm manganese oxide NPs improved the growth conditions and the yield of common dry beans. The chemical quality of leaves and seeds varied in their response to manganese oxide NPs. Also, genomic studies have shown that foliar application of NPs had upregulated the gene encoding the protein synthesis was profoundly influenced by manganese dioxide NPs.

Non-Fertilizer Nanomaterials to Enhance Crop Productivity

In addition, plant essential nutrients certain NMs have been reported to show a significant impact on plant growth, hence, which act as phytostimulants. Titanium dioxide, silver, gold, ceria NPs, carbon-based NMs, silicon-based NPs, and certain biobased NMs, such as lignin NPs are reported to affect plant physiology, thus improving the vigor and productivity (Arora et al. 2021; Andersen et al. 2016; Mohammadi et al. 2016; Jasim et al. 2017; Dağhan et al. 2018; Pandey et al. 2018; Sadak 2019; Khan et al. 2020; Del Buono et al. 2021; Bueno et al. 2022). Although these materials have the potential to enhance plant growth conditions and response to biotic abiotic stresses, all these effects on the plant rely upon the dose of application and physicochemical properties of the NM.

Nanofertilizer Delivery Methods

The astounding properties of NMs enable the design of delivery platforms to enhance the efficiency of drug delivery. Tunable characteristics of NMs, such as size, shape, reactivity, quantum confinement, inherent electronic and magnetic properties as well easiness to functionalize enable them to engineer as nanocarriers. Several categories of nanocarriers have been reported in fertilizer delivery in agriculture. These carriers can be mainly grouped based on the type of NM incorporated. Metal-based nanocarriers, carbon-based nanocarrier, polymer-based nanocarriers and their hybrids, and other types are extensively studied to deliver fertilization effectively for vegetative growth (Talreja et al. 2020). Also, NFs are designed as encapsulated or intercalated platforms for slow and controlled release.

Nanofetilizer Delivery as Nanomaterial Itself

Metallic NPs, including copper, zinc, iron, magnesium, manganese, and nickel, have been reported as metal-based NMs for fertilization (Delfani et al. 2014; Shende et al. 2017; Dimkpa et al. 2018; López-Vargas et al. 2018; Abbasifar et al. 2020; Pisani et al. 2020; Yusefi-Tanha et al. 2020; Elbasuney et al. 2022; Raiesi-Ardali et al. 2022). These materials are applied either as foliar sprays or directly to the soil in field application. Many of the studies reported the application of NMs in their pristine forms or as surface-functionalized models to enhance bioavailability (Raiesi-Ardali et al. 2022).

Nanofertilizer Delivery as Coated or Encapsulated Materials

One of the promising approaches to control the release of fertilizers is the development of coated fertilizers. Therefore, controlled-release fertilizers can be created by encasing traditional fertilizers in a protective coating of substances, like sulfur or polymers, which leads to a slowdown in the release and dissolution rates of the nutrients (Li et al. 2019). Also, coating composite needs to have other properties which prevent its degradation. For instance, polymeric or coating materials utilized in the controlled release of urea might require urease inhibitory activity if it is applied directly to soil (Pereira et al. 2015). Polymeric NMs have been mainly utilized in the development of nanoencapsulated fertilizers for slow and controlled release. Chitosan, alginate, polyvinyl alcohol, starch-g-poly (L-lactide), and biodegradable polymers have been reported in the development of slow-release NFs (Yang et al. 2013; Kalia et al. 2017; Talreja et al. 2020; Zafar et al. 2021).

Hydrogels in agriculture have gained considerable attention as water management materials. Hydrogels are cross-linked hydrophilic polymers, which can absorb large amounts of water or other liquids. Therefore, these composite materials are also in an application for coated nanofertilizer development. However, due to their high price limitation, the development of hydrogel/clay composites has been tested for slow and controlled-release fertilizer fabrication (Rashidzadeh et al. 2015). Urea-coated polyacrylamide/methylcellulose/calcic montmorillonite nanocomposites are one such example of hydrogel/clay nanocomposite for slow-release fertilizers (Bortolin et al. 2013). Zafar et al. (2021) developed urea-encapsulated starch-polyvinyl alcohol nanocomposites for the slow release of urea. Field experiments using spinach have reported that the urea-encapsulated composite had significantly improved the yield, photosynthetic pigment content, and nitrogen uptake in comparison to control experiments.

Nanofertilizer Delivery as Intercalated Materials

NF fabrication via intercalation is another approach that has been extensively explored to develop slow and controlled-release fertilizers (Sirisena et al. 2013; Manikandan and Subramanian 2016; Golbashy et al. 2017; Kalia et al. 2017). Many studies have reported on utilizing silicates for intercalation purposes. For instance, montmorillonite is a type of phyllosilicate belonging to the smectite group. Typically, these materials are composed of tiny platelets of approximately 1 nm thick with an average lateral dimension ranging from 30 nm to several microns and beyond. The surface of these nanoplatelets has a net negative charge, allowing them to exchange cations and incorporate neutral molecules, like urea, between their layers by interacting with the structural water molecules that are present. Also, higher cation exchange capacity, swelling capacity, and large surface area of montmorillonite have much attention in the development of NFs (Golbashy et al. 2017). Studies on slow-release fertilizers based on zeolites are largely limited to nutrients in cationic forms such as NH_4^+ and K^+. However, the loading of anionic forms such as sulfates, nitrates, and phosphates is minimal on unmodified zeolites. To make use of SRFs, the material must have an affinity for anions so that they can be efficiently loaded. This can be achieved by surface modification using surfactants (Subramanian et al. 2015). Most of the studies have described the application of nano clay intercalated with urea for slow and controlled release. In addition, Sirisena et al. (2013) develop montmorillonite intercalated with potassium for rice cultivation. Furthermore, Manikandan and Subramanian (2016) developed urea-intercalated nanozeolite as a slow-release fertilizer for maize growth on different soil types. Pereira et al. (2012) synthesized a new nanocomposite containing urea, created from the intercalation of urea into a montmorillonite clay using an extrusion process at room temperature. Experiments measuring the release rate of urea into water showed that the nanocomposite possessed a slow-release behavior, even with low amounts of montmorillonite (20% in weight).

Nanofertilizer Delivery as Surface Functionalized Carriers

Hydroxyapatite-like NPs have been identified as a versatile platform to surface functionalize to deliver plant nutrients as nanocarriers. Kottegoda et al. (2011) developed urea hydroxyapatite NPs encapsulated on the soft wood of *Gliricidia sepium*. Here, the nitrogen release studies have proved that the engineered nanocomposite showed an initial spurt release followed by slow release of urea up to 60 days in comparison to commercial grade nitrogen fertilizers. Also, Pohshna and Mailapalli (2019) synthesized urea-doped hydroxyapatite NFs for sustained release of nitrogen. More recently, researchers have focused on utilizing biobased NMs, such as nanocellulose for slow and controlled release of fertilizers. For instance, Sherif and Hedia (2022) used rice straw-based nanocellulose to create a urea-loaded nanocellulose for testing its effect on *Triticum aestivum* seeds. Results indicated that the biobased NF had enhanced root length, surface area, and vigor index compared to control experiments. Wang et al. (2021) developed a pH-sensitive nanocellulose/sodium alginate/metal composite for a slow-release nitrogen. Carbon-based NMs have been recently reported to apply in engineering to develop slow-release micronutrient NFs. Kumar et al. (2018) fabricated poly vinyl alcohol-starch Cu–Zn carrying carbon nanofibers fertilizer for slow release of micronutrients. Also, Durgude et al. (2022) fabricated mesoporous nano silica and reduced graphene oxide to establish iron and zinc to enhance the micronutrient use efficiency. Thus, the foliar application of the nanocomposite developed significantly enhanced the grain yield of rice. Furthermore, Arikan et al. (2022) synthesized iron oxide-modified graphene oxide NPs to reduce the nanoplastic toxicity of *Triticum aestivum*. Figure 2 will pictorially represent the different forms of nanofertilizers based on the mode of fabrication.

Release Kinetics of Nanofertlizers

A study was conducted by Carmona et al. (2020) to assess the effect of size, shape, and crystallinity of CaP NPs in the dissolution behavior for its potential to pronounce as macronutrient NFs. Dissolution kinetics of the CaP and nitrate-doped amorphous CaP NPs showed that they are partially soluble in aqueous media decreasing the dissolution rates with time. The rate of dissolution is probably affected by carbonate ion content (obtained from chemical precursors during the synthesis), the structural disorder of the amorphous materials, and the pH of the medium. Also, the release of components

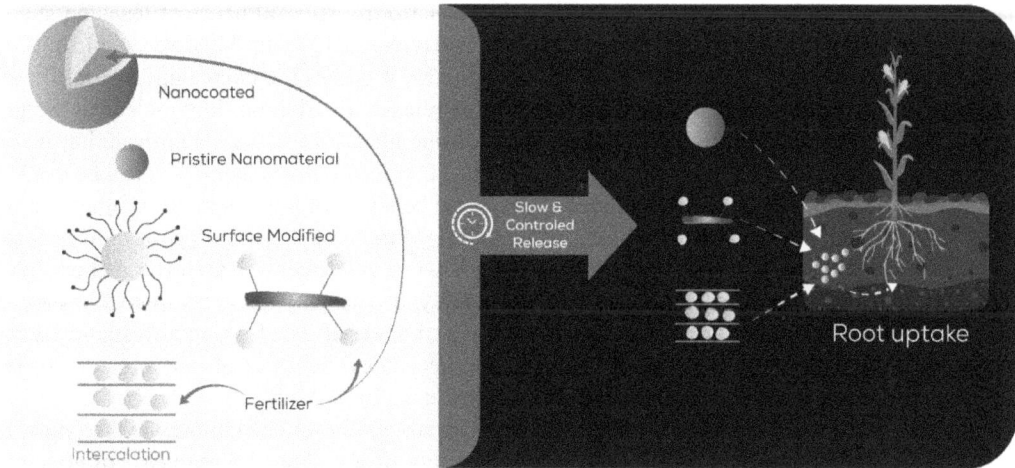

Figure 2. Different forms of nanofertilizers based on the mode of fabrication.

from the CaP NPs shows a pseudo-first-order kinetic model where Ca^{2+} release is proportional to $1 - e^{-kt}$ showing a clear difference between the release kinetics with fully amorphous and amorphous-crystalline NPs. Also, nitrate-doped CaP NPs had shown a faster dissolution rate compared to hybrid forms of CaP NPs proving that no significant effect from doping on the dissolution. Also, morphology has a prominent effect on release profile as CaP nanoplates dissolve more briskly than spherical-shaped NPs.

De Silva et al. (2022) studied the nitrogen release performance of urea-hydroxyapatite nanohybrids in water. It was shown that the nanohybrids have an excellent slow and controlled-release potential compared to urea granules, which are released to aqueous media in 30 minutes; in contrast, a delayed release was observed in nanohybrids with a progressive rise in urea release with nearly one-third of urea had a release in 50 minutes. Lately, these nanohybrids had followed a relatively slower release pattern emitting the remnant urea for more than 8-day time. This mode of release behavior was mainly evidenced due to moderately strong intermolecular bonding at the interphase. Release kinetics had shown that the nanohybrids, thus synthesized follow an Avrami-Erofe'ev kinetic model. This emphasizes that release of urea initiates with water diffusion into the nanocomposite followed by swelling to release urea. In addition, the urea transport mechanism described in the model shows that it follows a Super Case II transport mechanism. This involves nitrogen emission is a complex mechanism and slows down the rate with time. Hence, this can be related to a possible mechanism for urea release in soil applications. However, soil is a complex matrix, and several other factors would affect the release of urea. Factors such as soil pH, water retention rate, soil moisture content, and particle morphology on the solubility of urea. Equation 1 shows the Avrami relationship, which relates to the release kinetics of NFs.

$$X_{(t)} = X_{\infty} \left[1 - e^{1 - kt^n}\right] \tag{Eq. 1}$$

In this equation, X(t) is the volume fraction crystallinity at time t, X~ is the volume crystallinity after infinite time, K is the overall kinetic rate constant, and n is the Avrami exponent, which depends on the nucleation and growth mechanism of the crystal.

Furthermore, Carmona et al. (2021) synthesized a urea-doped amorphous CaP as an efficient nitrogen NF. Here the urea release pattern shows the reversing of the urea adsorption process. Initially, an explosive release of a greater urea fraction (89% in 15 minutes) was observed mainly attributed to weakly interacted urea molecules. This follows a steady emission of remaining urea for an extended period. Kottegoda et al. (2014) fabricated urea-intercalated nanocomposites using layered double hydroxides and montmorillonite. Analysis of release behavior depicts that nanohybrid types had released urea as a slow-release platform. After 40 hours of leaching study, both montmorillonite and layered double hydroxides had released 42% and 53% urea, while 82% of granular urea had leached out at the same duration. Hence, it proves the intercalation of fertilizers to develop NFs has a greater potential to control urea release. In addition, low molecular weight organics acids experimental system, pH of the medium, and ligand type have a significant impact on the dissolution of hydroxyapatite NPs. Hence, it implies that the hydroxyapatite NPs can be used as potential P NFs. Nido et al. (2019) evaluated the release behavior of potassium NF synthesized by integrating potassium in alginate-chitosan. Compared to Muriate of Potash, which is conventional potassium fertilizer the newly developed NFs had shown a controlled release. The authors have suggested that the controlled release of NF was mainly due to polymer relaxation following a Fickian diffusion model developing to Case II transport of material. The Fickian diffusion model is one of the frequently applied models to describe the mechanical behavior of polymer composites due to its simplicity and mathematical tractability (Roy et al. 2018).

Kottegoda et al. (2017) studied the release behavior of urea in water from urea-hydroxyapatite nanohybrids. It was reported that the nanohybrid showed a slow and persistent release of urea into the media and 86% of urea was released during the initial 3820s and remnants took 7 days to complete release. It was observed that the release rate of nanohybrids was 12 times slower compared to pristine urea. Same as in previous finding authors mainly speculated that moderate bonding of urea to hydroxyapatite mainly governs the release behavior of these NFs. Studies on release kinetics suggested that the nanohybrid show the Higuchi model, which follows a Fickian diffusion. Simply the diffusion of a given material can be described as the movement due to a concentration gradient. This has been described with Fick's laws which are defined as two laws. According to Fickian laws, it describes that the molar flux from diffusion is a function of the gradient in the concentration for a given material. Fick's second law is mainly attributed to the change in concentration of a substance with time and Equation 2 shows the Fickian relationship.

$$\frac{\partial \varphi}{\partial t} = D \frac{\partial^2 \varphi}{\partial x^2} \tag{Eq. 2}$$

In this equation, φ is the concentration of the material, x is the distance from the substratum interface, t is time, and D is the diffusion coefficient. Therefore, it can be speculated that once the NF is in soil, it will start to release urea to the immediate environment depending on the pH and moisture level in a slow and controlled approach.

Maduoanka et al. (2017) studied the release behavior and kinetics of urea-hydroxyapatite-montmorillonite nanocomposites synthesized using two different approaches. Authors have highlighted that the newly developed nanocomposite using solution phase synthesis showed a sluggish release of urea for an extended duration of up to 60 days. Composites synthesized using liquid-assisted grinding showed a similar pattern in a much slower manner for 140 days. However, the conventional form of urea stopped nitrogen release at the end of day 30. It is noteworthy to mention that the incorporation of nanoclays into this composite had a significant impact on the release pattern. Urea-hydroxyapatite NPs interact with the active sites on the montmorillonite layers and urea itself sometimes intercalate at the interlayer spaces of the clay. Hence, this is a control to their release as well as extra protection to urea molecules from photochemical, thermal, enzymatic, and other catalytic degradation in soil. Absorption of moisture from soil slowly releases urea from the composite via diffusion. If the released urea gets hydrolyzed to ammonia/ammonium ions, montmorillonite can swiftly recover them by adsorbing via physicochemical interaction. Also, mathematical modeling of the release behavior of urea from the nanocomposite followed the following relationship.

$$M_t / M_o = Kt^n \tag{Eq. 3}$$

In this equation, Mt is the amount of material released at time t, Mo is the total amount of material added, k is the rate constant, and n is the diffusion exponent related to the diffusion mechanism. According to the above relationship, it was speculated that all the nanoformulations had shown a non-Fickian diffusion mechanism. This mode of release mechanisms generally attributes when the diffusion and the rate of relaxation of the nanohybrid are compatible.

Beig et al. (2020) evaluated the release behavior and kinetics of biopolymer coated with urea for sustained release of nitrogen. According to the release kinetics studies, it was shown all the tested urea-coated biopolymers linked to a modified hyperbola formula, which follows a first-order leaching kinetics. Equation 4 here depicts the formula of the modified hyperbola formula.

$$Qt = \frac{at}{1+bt} \tag{Eq. 4}$$

Here, the terms Q and t denote the amount of urea leached at a given time t whereas a and b are release rate constants respectively (Al-Zahrani 2000; Beig et al. 2020).

A comparative assessment of urea release rates had shown that urea (10% starch/5% polyvinyl alcohol/5% molasses biopolymers) completely leached all the urea within 60 minutes. In addition, 10% starch, 5% PVA, and 5% paraffin wax biopolymers took 139 minutes to completely release urea from the matrix. The comparatively faster rate of release of urea from the 10% starch, 5% polyvinyl alcohol, and 5% molasses biopolymers was mainly attributed to the higher water solubility of molasses, whereas high hydrophobicity of paraffin wax had significantly affected the extended durational potential of urea release by 10% starch, 5% PVA, and 5% paraffin wax biopolymers. Therefore, it proves that the polarity of the coating materials too has a vital role in governing the release of fertilizers from a developed matrix.

Ekanayake and Godakumbura (2021) developed copper and zinc oxide NP-embedded alginate hydrogels for the sustained release of micronutrients. The tea bag experiment shows that zinc and copper ions were largely released during the first 8 hours, which was mainly due to the weak bonding of copper and zinc metals with the alginate matrix. Soil experiments show that the initial dosage of Cu is higher than that of Zn; however, the amount of Cu ions in the soil is low compared to Zn. This is mainly due to the higher affinity of Cu to organic matter and soil colloids due to its high sorption capacity, the decrease in copper mobilization in soils by heterotrophic bacteria, and the complexing of copper with other minerals in the soil.

Effects of Nanofertilizers on the Environment

Nanotechnology can be a double-edged sword. Although NFs have been proven to be effective in increasing crop yields, there are also some concerns about the potential negative effects of NFs on the environment. Materials implied in the NFs fabrication are utterly in the nanoscale and they are highly reactive moieties. It has been shown that long-term buildup in the soil over time, leads to potentially toxic levels of NMs. NFs specifically composed of metallic NPs which are micronutrients can positively or negatively influence soil microbial activity. Hence, NFs can have a significant impact on the physiology of beneficial soil microorganisms (Kalwani et al. 2022; Wijesooriya et al. 2023). However, this requires to be properly and extensively studied. In addition, NMs can influence the abundance and survival of soil invertebrates. For instance, earthworms are essential for the degradation of biogenic materials to keep the soil fertile and a negative impact on such a group can significantly influence the proper functioning of the ecosystem. The presence of metallic NPs has been linked to oxidative stress, changes in enzyme activity, and tissue apoptosis and, injury to the epidermal layer or intestinal tract of earthworms. This can disrupt their physiology negatively impacting their health. Furthermore, NFs can have phytotoxic effects on agricultural crops as it strictly depends upon the size, shape, and dosage of application (Madanayake et al. 2022). Overall, while NFs can be effective in increasing crop yields, they also come with several potential negative effects that need to be taken into consideration. It is important to take the necessary steps to minimize the negative impacts of NFs on the environment and to ensure that they are used properly, systematically, and responsibly.

Conclusion and Future Perspectives

Nanotechnology is seen as a promising interdisciplinary approach with many industrial applications, especially in agriculture. NMs can be used to enhance the nutrient use efficiency of fertilizers, making them ideal for slow or controlled-release fertilizers to deliver essential nutrients precisely and safely to plants. These fertilizers can be classified as macronutrients and micronutrients based on the quantity needed by plants. They can be delivered as pristine materials, nanocarriers, or by surface functionalization, intercalation, or encapsulation. The release behavior of these fertilizers can be controlled by the strength of the bond between the NM and the nutrient component. Additionally, surface functionalization of the pristine form can enhance biocompatibility and bioavailability. Different forms of NFs have varying release behaviors and kinetics, which are dependent on the physicochemical properties of NMs and environmental conditions, such as pH and moisture levels in the soil. Recently, there has been a focus on applying biobased NMs to deliver plant nutrients, making it a sustainable approach to enhance crop productivity. Despite the benefits of NFs, there is a lack of field trials related to the different forms of NFs and a need to assess their toxicity. It is also important to assess the economic feasibility of scaling up the production of NFs, otherwise, they may become a white elephant.

Acknowledgment

The authors would like to thank the National Research Council of Sri Lanka (Investigator Driven Grant-NRC IDG 22-114) for the financial support.

The authors greatly acknowledge Mr. L.L.D. Anuka Deveen, Department of Botany, Faculty of Applied Sciences, University of Sri Jayewardenepura, Sri Lanka for the assistance with preparing the figures.

References

Abbasifar, A., Shahrabadi, F. and Valizadeh Kaji, B. 2020. Effects of green synthesized zinc and copper nano-fertilizers on the morphological and biochemical attributes of basil plant. J. Plant Nutr. 43(8): 1104–1118.

AL-Abody, M.A.K., Abd Wahid, M.A. and Jamel, F.A. 2021. Effect of foliar application of nano-fertilizer of iron on growth and biological yield of varieties wheat (*Triticum aestivum* L.). Am. J. Life Sci. Res. 9(1): 8–17.

Al-Zahrani, S.M. 2000. Utilization of polyethylene and paraffin waxes as controlled delivery systems for different fertilizers. Ind. Eng. Chem. Res. 39(2): 367–371.

Andersen, C.P., King, G., Plocher, M., Storm, M., Pokhrel, L.R., Johnson, M.G. and Rygiewicz, P.T. 2016. Germination and early plant development of ten plant species exposed to titanium dioxide and cerium oxide nanoparticles. Environ. Toxicol. Chem. 35(9): 2223–2229.

Arikan, B., Alp, F.N., Ozfidan-Konakci, C., Balci, M., Elbasan, F., Yildiztugay, E. and Cavusoglu, H. 2022. Fe_2O_3-modified graphene oxide mitigates nanoplastic toxicity via regulating gas exchange, photosynthesis, and antioxidant system in *Triticum aestivum*. Chemosphere 307: 136048.

Arora, S., Sharma, P., Kumar, S., Nayan, R., Khanna, P.K. and Zaidi, M.G.H. 2012. Gold-nanoparticle-induced enhancement in growth and seed yield of *Brassica juncea*. Plant Growth Regul. 66: 303–310.

Azarian, M.H., Kamil Mahmood, W.A., Kwok, E., Bt Wan Fathilah, W.F. and Binti Ibrahim, N.F. 2018. Nanoencapsulation of intercalated montmorillonite-urea within PVA nanofibers: Hydrogel fertilizer nanocomposite. J. Appl. Polym. Sci. 135(10): 45957.

Bala, R., Kalia, A. and Dhaliwal, S.S. 2019. Evaluation of efficacy of ZnO nanoparticles as remedial zinc nanofertilizer for rice. J. Plant. Nutr. Soil Sci. 19: 379–389.

Beig, B., Niazi, M.B.K., Jahan, Z., Kakar, S.J., Shah, G.A., Shahid, M., Zia, M., Haq, M.U. and Rashid, M.I. 2020. Biodegradable polymer-coated granular urea slows down N release kinetics and improves spinach productivity. Polymers 12(11): 2623.

Bortolin, A., Aouada, F.A., Mattoso, L.H. and Ribeiro, C. 2013. Nanocomposite PAAm/methyl cellulose/ montmorillonite hydrogel: evidence of synergistic effects for the slow release of fertilizers. J. Agric. Food Chem. 61(31): 7431–7439.

Bueno, V., Wang, P., Harrisson, O., Bayen, S. and Ghoshal, S. 2022. Impacts of a porous hollow silica nanoparticle-encapsulated pesticide applied to soils on plant growth and soil microbial community. Environ. Sci. Nano 9(4): 1476–1488.

Carmona, F.J., Dal Sasso, G., Bertolotti, F., Ramírez-Rodríguez, G.B., Delgado-López, J.M., Pedersen, J.S., Masciocchi, N. and Guagliardi, A. 2020. The role of nanoparticle structure and morphology in the dissolution kinetics and nutrient release of nitrate-doped calcium phosphate nanofertilizers. Sci. Rep. 10(1): 12396.

Carmona, F.J., Dal Sasso, G., Ramírez-Rodríguez, G.B., Pii, Y., Delgado-López, J.M., Guagliardi, A. and Masciocchi, N. 2021. Urea-functionalized amorphous calcium phosphate nanofertilizers: optimizing the synthetic strategy towards environmental sustainability and manufacturing costs. Sci. Rep. 11(1): 1–14.

Chen, L., Xie, Z., Zhuang, X., Chen, X. and Jing, X. 2008. Controlled release of urea encapsulated by starch-g-poly (L-lactide). Carbohydr. Polym. 72(2): 342–348.

Chhipa, H. and Joshi, P. 2016. Nanofertilisers, nanopesticides and nanosensors in agriculture. J. Sci. Food Agric. pp. 247–282.

Cieschi, M.T., Polyakov, A.Y., Lebedev, V.A., Volkov, D.S., Pankratov, D.A., Veligzhanin, A.A., Perminova, I.V. and Lucena, J.J. 2019. Eco-friendly iron-humic nanofertilizers synthesis for the prevention of iron chlorosis in soybean (Glycine max) grown in calcareous soil. Front. Plant Sci. 10: 413.

Cota-Ruiz, K., Ye, Y., Valdes, C., Deng, C., Wang, Y., Hernández-Viezcas, J.A., Duarte-Gardea, M. and Gardea-Torresdey, J.L. 2020. Copper nanowires as nanofertilizers for alfalfa plants: Understanding nano-bio systems interactions from microbial genomics, plant molecular responses and spectroscopic studies. Sci. Total Environ. 742: 140572.

Dağhan, H.A.T.İ.C.E. 2018. Effects of TiO_2 nanoparticles on maize (Zea mays L.) growth, chlorophyll content and nutrient uptake. Appl. Ecol. Environ. Res. 16.

De Silva, M., Sandaruwan, C., Hernandez, F.C.R., Sahin, O., Ashokkumar, M., Ajayan, P.M., Karunaratne, V., Amaratunga, G.A. and Kottegoda, N. 2022. A greener mechanochemical approach to the synthesis of urea-hydroxyapatite nanohybrids for slow release of plant nutrients. Research Square [https://doi.org/10.21203/ rs.3.rs-1591563/v1].

Del Buono, D., Luzi, F. and Puglia, D. 2021. Lignin nanoparticles: A promising tool to improve maize physiological, biochemical, and chemical traits. J. Nanomater. 11(4): 846.

Delfani, M., Baradarn Firouzabadi, M., Farrokhi, N. and Makarian, H. 2014. Some physiological responses of black-eyed pea to iron and magnesium nanofertilizers. Commun Soil Sci. Plant Anal. 45(4): 530–540.

Dimkpa, C.O., Singh, U., Adisa, I.O., Bindraban, P.S., Elmer, W.H., Gardea-Torresdey, J.L. and White, J.C. 2018. Effects of manganese nanoparticle exposure on nutrient acquisition in wheat (*Triticum aestivum* L.). Agronomy 8(9): 158.

Durgude, S.A., Ram, S., Kumar, R., Singh, S.V., Singh, V., Durgude, A.G., Pramanick, B., Maitra, S., Gaber, A. and Hossain, A. 2022. Synthesis of Mesoporous Silica and graphene-based FeO and ZnO nanocomposites for nutritional biofortification and sustained the productivity of rice (Oryza sativa L.). J. Nanomater., 2022.

Ekanayake, S.A. and Godakumbura, P.I. 2021. Synthesis of a dual-functional nanofertilizer by embedding ZnO and CuO nanoparticles on an alginate-based hydrogel. ACS Omega 6(40): 26262–26272.

Elbasuney, S., El-Sayyad, G.S., Attia, M.S. and Abdelaziz, A.M. 2022. Ferric oxide colloid: towards green nano-fertilizer for tomato plant with enhanced vegetative growth and immune response against fusarium wilt disease. J. Inorg. Organomet. Polym. 32(11): 4270–4283.

Elsayed, A.A., Ahmed, E.G., Taha, Z.K., Farag, H.M., Hussein, M.S. and AbouAitah, K. 2022. Hydroxyapatite nanoparticles as novel nano-fertilizer for production of rosemary plants. Sci. Hortic. 295: 110851.

Gaiotti, F., Lucchetta, M., Rodegher, G., Lorenzoni, D., Longo, E., Boselli, E., Cesco, S., Belfiore, N., Lovat, L., Delgado-López, J.M. and Carmona, F.J. 2021. Urea-doped calcium phosphate nanoparticles as sustainable nitrogen nanofertilizers for viticulture: Implications on yield and quality of Pinot gris grapevines. Agronomy 11(6): 1026.

Golbashy, M., Sabahi, H., Allahdadi, I., Nazokdast, H. and Hosseini, M. 2017. Synthesis of highly intercalated urea-clay nanocomposite via domestic montmorillonite as eco-friendly slow-release fertilizer. Arch. Acker Pflanzenbau Bodenkd. 63(1): 84–95.

Ha, N.M.C., Nguyen, T.H., Wang, S.L. and Nguyen, A.D. 2019. Preparation of NPK nanofertilizer based on chitosan nanoparticles and its effect on biophysical characteristics and growth of coffee in greenhouse. Res. Chem. Intermed. 45: 51–63.

Hanon Mohsen, K., Alrubaiee, S.H. and ALfarjawi, T.M. 2022. Response of wheat varieties, *Triticum aestivum* L., to spraying by iron nano-fertilizer. Casp. J. Environ. Sci. 20(4): 775–783.

He, X., Deng, H. and Hwang, H.M. 2019. The current application of nanotechnology in food and agriculture. J. Food Drug Anal. 27(1): 1–21.

Hidayat, R., Fadillah, G., Chasanah, U., Wahyuningsih, S. and Ramelan, A.H., 2015. Effectiveness of urea nanofertilizer based aminopropyltrimethoxysilane (APTMS)-zeolite as slow release fertilizer system. Afr. J. Agric. Res. 10(14): 1785–1788.

Jariwala, D., Sangwan, V.K., Lauhon, L.J., Marks, T.J. and Hersam, M.C. 2013. Carbon nanomaterials for electronics, optoelectronics, photovoltaics, and sensing. Chem. Soc. Rev. 42(7): 2824–2860.

Jasim, B., Thomas, R., Mathew, J. and Radhakrishnan, E.K. 2017. Plant growth and diosgenin enhancement effect of silver nanoparticles in Fenugreek (*Trigonella foenum-graecum* L.). Saudi Pharm J. 25(3): 443–447.

Kalia, A., Luthra, K., Sharma, S.P., Singh Dheri, G., Sachdeva Taggar, M. and Gomes, C. 2017, September. Chitosan-urea nano-formulation: synthesis, characterization and impact on tuber yield of potato. pp. 97–106. In International Symposium on Horticulture: Priorities and Emerging Trends 1255.

Kalwani, M., Chakdar, H., Srivastava, A., Pabbi, S. and Shukla, P. 2022. Effects of nanofertilizers on soil and plant-associated microbial communities: Emerging trends and perspectives. Chemosphere 287: 132107.

Khan, M.N. and Mohammad, F. 2014. Eutrophication: challenges and solutions. Eutrophication: Causes, Consequences and Control 2: 1–15.

Khan, Z.S., Rizwan, M., Hafeez, M., Ali, S., Adrees, M., Qayyum, M.F., Khalid, S., ur Rehman, M.Z. and Sarwar, M.A. 2020. Effects of silicon nanoparticles on growth and physiology of wheat in cadmium contaminated soil under different soil moisture levels. Environ. Sci. Pollut. Res. 27: 4958–4968.

Khin, M.M., Nair, A.S., Babu, V.J., Murugan, R. and Ramakrishna, S. 2012. A review on nanomaterials for environmental remediation. Energy Environ. Sci. 5(8): 8075–8109.

Kottegoda, N., Munaweera, I., Madusanka, N. and Karunaratne, V. 2011. A green slow-release fertilizer composition based on urea-modified hydroxyapatite nanoparticles encapsulated wood. Curr. Sci. pp. 73–78.

Kottegoda, N., Sandaruwan, C., Perera, P., Madusanka, N. and Karunaratne, V. 2014. Modified layered nanohybrid structures for the slow release of urea. Nanosci. Nanotechnol. - Asia. 4(2): 94–102.

Kottegoda, N., Sandaruwan, C., Priyadarshana, G., Siriwardhana, A., Rathnayake, U.A., Berugoda Arachchige, D.M., Kumarasinghe, A.R., Dahanayake, D., Karunaratne, V. and Amaratunga, G.A. 2017. Urea-hydroxyapatite nanohybrids for slow release of nitrogen. ACS nano 11(2): 1214–1221.

Kumar, R., Ashfaq, M. and Verma, N. 2018. Synthesis of novel PVA–starch formulation-supported Cu–Zn nanoparticle carrying carbon nanofibers as a nanofertilizer: controlled release of micronutrients. J. Mater. Sci. 53(10): 7150–7164.

Leonardi, M., Caruso, G.M., Carroccio, S.C., Boninelli, S., Curcuruto, G., Zimbone, M., Allegra, M., Torrisi, B., Ferlito, F. and Miritello, M. 2021. Smart nanocomposites of chitosan/alginate nanoparticles loaded with copper oxide as alternative nanofertilizers. Environ. Sci. Nano 8(1): 174–187.

Li, T., Gao, B., Tong, Z., Yang, Y. and Li, Y. 2019. Chitosan and graphene oxide nanocomposites as coatings for controlled-release fertilizer. Wat. Air and Soil Poll. 230: 1–9.

López-Vargas, E.R., Ortega-Ortíz, H., Cadenas-Pliego, G., de Alba Romenus, K., Cabrera de la Fuente, M., Benavides-Mendoza, A. and Juárez-Maldonado, A. 2018. Foliar application of copper nanoparticles increases the fruit quality and the content of bioactive compounds in tomatoes. Appl. Sci. 8(7): 1020.

Madanayake, N.H., Rienzie, R. and Adassooriya, N.M. 2019. Nanoparticles in Nanotheranostics Applications. Nanotheranostics: Applications and Limitations pp. 19–40.

Madanayake, N.H., Adassooriya, N.M. and Salim, N. 2021. The effect of hydroxyapatite nanoparticles on *Raphanus sativus* with respect to seedling growth and two plant metabolites. Environ. Nanotechnol. Monit. Manag. 15: 100404.

Madanayake, N.H., Perera, N. and Adassooriya, N.M. 2022. Engineered nanomaterials: threats, releases, and concentrations in the environment. pp. 225–240. In Emerging Contaminants in the Environment. Elsevier.

Madusanka, N., Sandaruwan, C., Kottegoda, N., Sirisena, D., Munaweera, I., De Alwis, A., Karunaratne, V. and Amaratunga, G.A. 2017. Urea–hydroxyapatite-montmorillonite nanohybrid composites as slow-release nitrogen compositions. Applied Clay Science 150: 303–308.

Manikandan, A. and Subramanian, K.S. 2016. Evaluation of zeolite-based nitrogen nano-fertilizers on maize growth, yield and quality on inceptisols and alfisols. Int. J. Plant Soil Sci. 9(4): 1–9.

Marchiol, L., Filippi, A., Adamiano, A., Degli Esposti, L., Iafisco, M., Mattiello, A., Petrussa, E. and Braidot, E. 2019. Influence of hydroxyapatite nanoparticles on germination and plant metabolism of tomato (Solanum lycopersicum L.): Preliminary evidence. Agronomy 9(4): 161.

Mei, L., Zhang, Z., Zhao, L., Huang, L., Yang, X.L., Tang, J. and Feng, S.S. 2013. Pharmaceutical nanotechnology for oral delivery of anticancer drugs. Adv. Drug Deliv. Rev. 65(6): 880–890.

Mohammadi, H., Esmailpour, M. and Gheranpaye, A. 2016. Effects of TiO_2 nanoparticles and water-deficit stress on morpho-physiological characteristics of dragonhead (*Dracocephalum moldavica* L.) plants. Acta Agric. Slov. 107(2): 385–396.

Nido, P.J., Migo, V., Maguyon-Detras, M.C. and Alfafara, C. 2019. Process optimization potassium nanofertilizer production via ionotropic pre-gelation using alginate-chitosan carrier. In MATEC Web of Conferences (Vol. 268, p. 05001). EDP Sciences.

Pandey, K., Lahiani, M.H., Hicks, V.K., Hudson, M.K., Green, M.J. and Khodakovskaya, M. 2018. Effects of carbon-based nanomaterials on seed germination, biomass accumulation and salt stress response of bioenergy crops. PloS One 13(8): e0202274.

Pereira, E.I., Minussi, F.B., da Cruz, C.C., Bernardi, A.C. and Ribeiro, C. 2012. Urea–montmorillonite-extruded nanocomposites: a novel slow-release material. J. Agric. Food Chem. 60(21): 5267–5272.

Pereira, E.I., da Cruz, C.C., Solomon, A., Le, A., Cavigelli, M.A. and Ribeiro, C. 2015. Novel slow-release nanocomposite nitrogen fertilizers: the impact of polymers on nanocomposite properties and function. Ind. Eng. Chem. Res. 54(14): 3717–3725.

Pisani, C. and Rossi, L. 2020, August. Effects of foliar applications of zinc and nickel nano-fertilizers vs conventional fertilizers on plant physiology in pecan [Carya Illinoinensis (Wangenh.) K. Koch] Cvs.'Zinner'and 'Byrd'. In 2020 ASHS Annual Conference. ASHS.

Pohshna, C. and Mailapalli, D.R. 2019. Effect of Urea Doped Hydroxyapatite Nanofertilizer on Nutrients Movement and Rice Growth. pp. H51I–1607. *In*: AGU Fall Meeting Abstracts (Vol. 2019).

Raiesi-Ardali, T., Ma'mani, L., Chorom, M. and Moezzi, A. 2022. Improved iron use efficiency in tomato using organically coated iron oxide nanoparticles as efficient bioavailable Fe sources. Chem. Biol. Technol. Agric. 9(1): 59.

Ramírez-Rodríguez, G.B., Miguel-Rojas, C., Montanha, G.S., Carmona, F.J., Dal Sasso, G., Sillero, J.C., Skov Pedersen, J., Masciocchi, N., Guagliardi, A., Pérez-de-Luque, A. and Delgado-López, J.M. 2020. Reducing nitrogen dosage in Triticum durum plants with urea-doped nanofertilizers. Nanomaterials 10(6): 1043.

Rashidzadeh, A., Olad, A. and Reyhanitabar, A. 2015. Hydrogel/clinoptilolite nanocomposite-coated fertilizer: swelling, water-retention and slow-release fertilizer properties. Polym. Bull. 72: 2667–2684.

Rios, J.J., Lopez-Zaplana, A., Bárzana, G., Martinez-Alonso, A. and Carvajal, M. 2021. Foliar application of boron nanoencapsulated in almond trees allows B movement within tree and implements water uptake and transport involving aquaporins. Front. Plant Sci. 12: 752648.

Römheld, V. and Marschner, H. 1991. Function of micronutrients in plants. Micronutrients in Agriculture 4: 297–328.

Sadak, M.S. 2019. Impact of silver nanoparticles on plant growth, some biochemical aspects, and yield of fenugreek plant (Trigonella foenum-graecum). BNRC 43(1): 1–6.

Salama, D.M., Abd El-Aziz, M.E., Osman, S.A., Abd Elwahed, M.S. and Shaaban, E.A. 2022. Foliar spraying of MnO_2-NPs and its effect on vegetative growth, production, genomic stability, and chemical quality of the common dry bean. Rab J. Basic Appl. Sci. 29(1): 26–39.

Samavini, R., Sandaruwan, C., De Silva, M., Priyadarshana, G., Kottegoda, N. and Karunaratne, V. 2018. Effect of citric acid surface modification on solubility of hydroxyapatite nanoparticles. J. Agric. Food Chem. 66(13): 3330–3337.

Samsoon, S., Azam, M., Khan, A., Ashraf, M., Bhatti, H.N., Alshawwa, S.Z. and Iqbal, M. 2022. Green-synthesized MnO_2 nanofertilizer impact on growth, photosynthetic pigment, and non-enzymatic antioxidant of Vigna unguiculata cultivar. Biomass Convers. Biorefin. pp. 1–10.

Savci, S. 2012. An agricultural pollutant: chemical fertilizer. International Journal of Environmental Science and Development 3(1): 73.

Sharma, B., Shrivastava, M., Afonso, L.O., Soni, U. and Cahill, D.M. 2022. Zinc-and magnesium-doped hydroxyapatite nanoparticles modified with urea as smart nitrogen fertilizers. ACS Appl. Nano Mater. 5(5): 7288–7299.

Shebl, A., Hassan, A., Salama, D., Abd El-Aziz, M.E. and Abd Elwahed, M. 2019. Green synthesis of manganese zinc ferrite nanoparticles and their application as nanofertilizers for Cucurbita pepo L. Beilstein Archives 2019(1): 45.

Shende, S., Rathod, D., Gade, A. and Rai, M. 2017. Biogenic copper nanoparticles promote the growth of pigeon pea (*Cajanus cajan* L.). IET J. Nanobiotechnology 11(7): 773–781.

Sheoran, P., Grewal, S., Kumari, S. and Goel, S. 2021. Enhancement of growth and yield, leaching reduction in *Triticum aestivum* using biogenic synthesized zinc oxide nanofertilizer. Biocatal. Agric. Biotechnol, 32: 101938.

Sherif, F. and MR Hedia, R. 2022. Priming seeds with urea-loaded nanocellulose to enhance wheat (*Triticum aestivum*) germination. Alex. Sci. Exch. J. 43(1): 151–160.

Sirisena, D.N., Dissanayake, D.M.N., Somaweera, K.A.T.N., Karunaratne, V. and Kottegoda, N. 2013. Use of nano-K fertilizer as a source of potassium in rice cultivation. Annals of Sri Lanka Department of Agriculture 15: 257–262.

Subramanian, K.S., Manikandan, A., Thirunavukkarasu, M. and Rahale, C.S. 2015. Nano-fertilizers for balanced crop nutrition. Nanotechnologies in Food and Agriculture pp.69–80.

Talreja, N., Chauhan, D., Rodríguez, C.A., Mera, A.C. and Ashfaq, M. 2020. Nanocarriers: an emerging tool for micronutrient delivery in plants. Plant Micronutrients: Deficiency and Toxicity Management, pp. 373–387.

Wang, D., Xie, Y., Jaisi, D.P. and Jin, Y. 2016. Effects of low-molecular-weight organic acids on the dissolution of hydroxyapatite nanoparticles. Environ. Sci. Nano 3(4): 768–779.

Wang, Y., Shaghaleh, H., Hamoud, Y.A., Zhang, S., Li, P., Xu, X. and Liu, H. 2021. Synthesis of a pH-responsive nano-cellulose/sodium alginate/MOFs hydrogel and its application in the regulation of water and N-fertilizer. Int. J. Biol. Macromol. 187: 262–271.

Wijesooriya, S.N., Madanayake, N.H. and Adassooriya, N.M. 2023. Impact of nanomaterials on beneficial soil micro-organisms. pp. 367–385. In Nanotechnology in Agriculture and Agroecosystems. Elsevier.

Yang, Y., Tong, Z., Geng, Y., Li, Y. and Zhang, M. 2013. Biobased polymer composites derived from corn stover and feather meals as double-coating materials for controlled-release and water-retention urea fertilizers. J. Agric. Food Chem. 61(34): 8166–8174.

Yoon, H.Y., Lee, J.G., Esposti, L.D., Iafisco, M., Kim, P.J., Shin, S.G., Jeon, J.R. and Adamiano, A. 2020. Synergistic release of crop nutrients and stimulants from hydroxyapatite nanoparticles functionalized with humic substances: Toward a multifunctional nanofertilizer. ACS Omega 5(12): 6598–6610.

Yusefi-Tanha, E., Fallah, S., Rostamnejadi, A. and Pokhrel, L.R. 2020. Zinc oxide nanoparticles (ZnONPs) as a novel nanofertilizer: Influence on seed yield and antioxidant defense system in soil grown soybean (Glycine max cv. Kowsar). Sci. Total Environ. 738: 140240.

Zafar, N., Niazi, M.B.K., Sher, F., Khalid, U., Jahan, Z., Shah, G.A. and Zia, M. 2021. Starch and polyvinyl alcohol encapsulated biodegradable nanocomposites for environment friendly slow release of urea fertilizer. J. Adv. Chem. Eng. 7: 100123.

Zhang, Q., Uchaker, E., Candelaria, S.L. and Cao, G. 2013. Nanomaterials for energy conversion and storage. Chem. Soc. Rev. 42(7): 3127–3171.

Chapter 4

Agro-Nanotechnology
Zinc Oxide Nanoparticles as Nanofertilizers for Agriculture

Hernández-Adame, L.,[1,2,*] *Castellanos, T.,*[1] *Chiquito-Contreras, R.G.,*[3]
Loera-Muro A.,[2] *Ponce Pedraza, A.*[4] and *Hernández-Montiel L.G.*[1,*]

Introduction

Nanotechnology is a multidisciplinary science that allows controlling and transforming materials at the nanometric scale; thus, understanding nanomaterial as all matter with a dimension in the range of 1 to 100 nanometers. It is important to note that this miniaturization of materials induces a unique change in thermal, electrical, optical, mechanical, and magnetic properties relative to those observed for the same materials on a macroscopic scale (Hernández-Adame et al. 2018). This fact has motivated an exponential increase in the application of nanomaterials in recent years; specifically, metals, semiconductors, and biopolymer-based nanoparticles (NPs) have been utilized to manufacture new products or to substantially improve existing ones for various applications like industrial, commercial, and medical (Figure 1; El-Feky et al. 2014; Peters et al. 2016).

In the agricultural sector, the use of NPs is beginning to be a relevant practice for the manufacture of novel systems, among which the design of nanofertilizers and antimicrobial systems stands out. These nanostructured formulations are designed to improve the absorption of nutrients by plants (such as Zn, Fe, B, and Cu), increasing the quality and performance of products, reducing production costs, and being an alternative for reducing excessive use of agrochemicals (Liu et al. 2015; Raliya et al. 2017; Rajput et al. 2018).

Among the most widely used nanomaterials are zinc oxide nanoparticles (ZnONPs). These nanoparticles are the fourth most used worldwide in sectors that include the biomedical and technological fields (Sabir et al. 2014). The characteristics of ZnOPs are attributed to their optical properties (they absorb ultraviolet radiation), large surface area, low toxicity, and ability to transport active molecules on their surface (Laurenti et al. 2015).

In agriculture, ZnONPs are used as nanofertilizer and antimicrobial agents. However, it is currently possible to find many both positive and potentially negative results that may question

---

[1] Nanotechnology and Microbial Biocontrol Group, Centro de Investigaciones Biológicas del Noroeste, La Paz, Baja California Sur, 23090, México.

[2] CONACYT-Centro de Investigaciones Biológicas del Noroeste, La Paz, Baja California Sur, 23090, México.

[3] Facultad de Ciencias Agrícolas, Universidad Veracruzana, Xalapa, Veracruz, 91090, México.

[4] Department of Physics, University of Texas San Antonio, San Antonio, TX 78249, USA.

* Corresponding authors: ladame@cibnor.mx, lhernandez@cibnor.mx

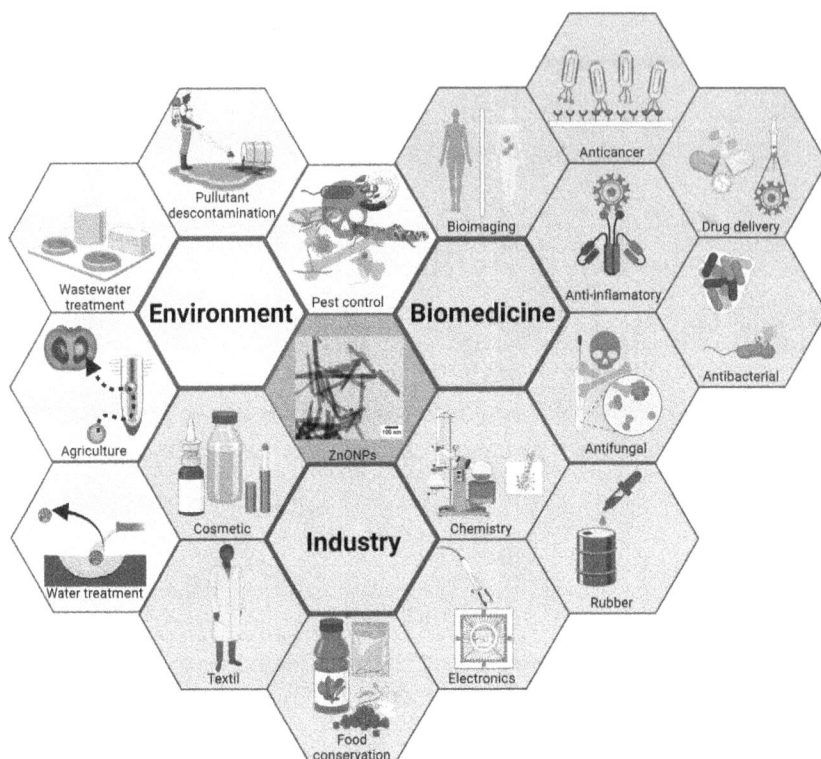

Figure 1. Main applications of ZnONPs in different sectors related to biomedicine, industry, agriculture, and environmental care.

its use and impact on soils and crops (Reddy et al. 2016; López-Moreno et al. 2018). Some works highlight promising results due to the increase in the quality of plants and fruits. However, it has been pointed out that the excessive use of ZnONPs, improper handling, or accidental release can cause serious damage to ecosystems. This fact has generated the need to evaluate their toxicity and negative effects where they are applied (Dimkpa et al. 2012). In this context, this work aims to establish a current overview of the use of ZnONPs and their effects on some crops of agricultural interest. Addressing in a general way, its main absorption mechanisms, concentration effect, and benefits of foliar application or directly on the soil. Finally, an analysis of the need to address possible negative impacts is put into perspective.

Zinc as an Element in Biological Systems

Zinc (Zn) is an essential micronutrient for the growth, development, and health of plants and humans. In biological systems, this element plays an important role in a wide variety of metabolisms, such as carbohydrates, lipids, nucleic acids, and proteins (Hafeez et al. 2013). In plants, it is an integral component of many enzymatic structures, such as oxidoreductases, hydrolases, lyases, isomerases, and ligases, among others (Auld 2001). Zn intervenes in the biosynthesis of chlorophyll, the precursor of various proteins, and regulates various mechanisms of recognition and response to abiotic stress in plants (Figure 2; Alloway 2009).

It is possible to find a large amount of Zn on the earth's crust. However, 90% of the Zn available in agricultural soils is in an insoluble form, which is why it is not suitable for plant absorption (Broadley et al. 2007). This fact, together with its poor bio-distribution, makes Zn^{2+} ions one of the main micronutrients that limit growth and development, both in crops and in all plant species (Longnecker and Robson 1993). This fact causes great economic losses to the agricultural sector,

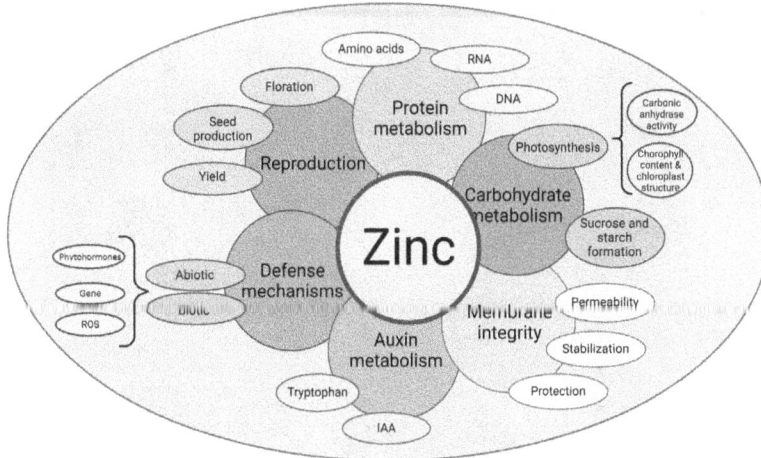

Figure 2. Schematic representation of main functions of Zn in plants.

in addition to generating a significant decrease in the nutritional quality of food for humans and livestock, affecting up to 73% of the world population (Nielsen 2012). These data have made Zn supplementation one of the priority areas of agricultural research, which aims to minimize health problems related to its deficiency in the soil, plants, animals, and humans (Cakmak et al. 2017).

Absorption and Transport of ZnONPs in Plants

Zn is one of the essential micronutrients, which is absorbed by plants in ionic form (Zn^{2+}) or in complex with chelates. In the form of ZnONP, its absorption mechanisms are still not very clear. However, it has been observed that it can enter the plant through the roots or the leaves (Figure 3). When ZnONPs are foliar applied, they can penetrate and adhere through the cuticle, stomata, hydathodes, epidermis, and/or leaf wounds. Subsequently, these are desorbed in the apoplast to finally be absorbed in the underlying cells (Shahid et al. 2020). Moreover, if its application is via the soil/substrate, the ZnONPs aggregate in the rhizosphere, and enter the root cell, either apoplastically or symplastically, reaching the vascular bundle to later be transported to the xylem. Zn molecules in the soil that are present in the soluble form are rapidly absorbed. However, the roots also exude organic acid, i.e., mucilage (hydrated polysaccharide) on the root surface, which aids in agglomeration and dissolution of the ZnONPs, so that they can be assimilated. Zn ions begin to move due to the concentration gradient through the ionic pores of the roots. After adsorption, ZnONPs increase the permeability of the cell wall by making "holes" in the wall and moving through the plasmodesmata, thus showing a symplastic movement and facilitating the transport of ZnONPs. Alternatively, it has also been observed that ZnONPs can enter via the apoplastic pathway through the epidermis and cortex. As well as entering the vascular tissue after passing through the protoplast of the endodermal cells (Lin and Xing 2008).

The translocation of ZnONPs from the root to the shoot occurs due to the decrease in the Zn concentration gradient. Most likely, there are two mechanisms of Zn transport from root to shoot: (1) radial movement of Zn and (2) axial transport from the xylem. The axial transport system carries ions from the lower region of the stem to the upper region through the xylem. The Zn content and velocity decrease as it is transported out of the root. Likewise, simultaneous radial movement transports Zn from the xylem to the cortical cells. The cortical tissue stores the transported Zn for the nutrition pathway and for the defense strategy, when Zn is not able to reach the leaves through the axial transport mode (Hussain et al. 2021). During the transport of Zn toward the petiole and leaflet, a Zn concentration gradient is created that decreases from the petiole to the tip of the leaf, where it occurs mainly as Zn phosphate, Zn-histidine complex, and Zn-malate.

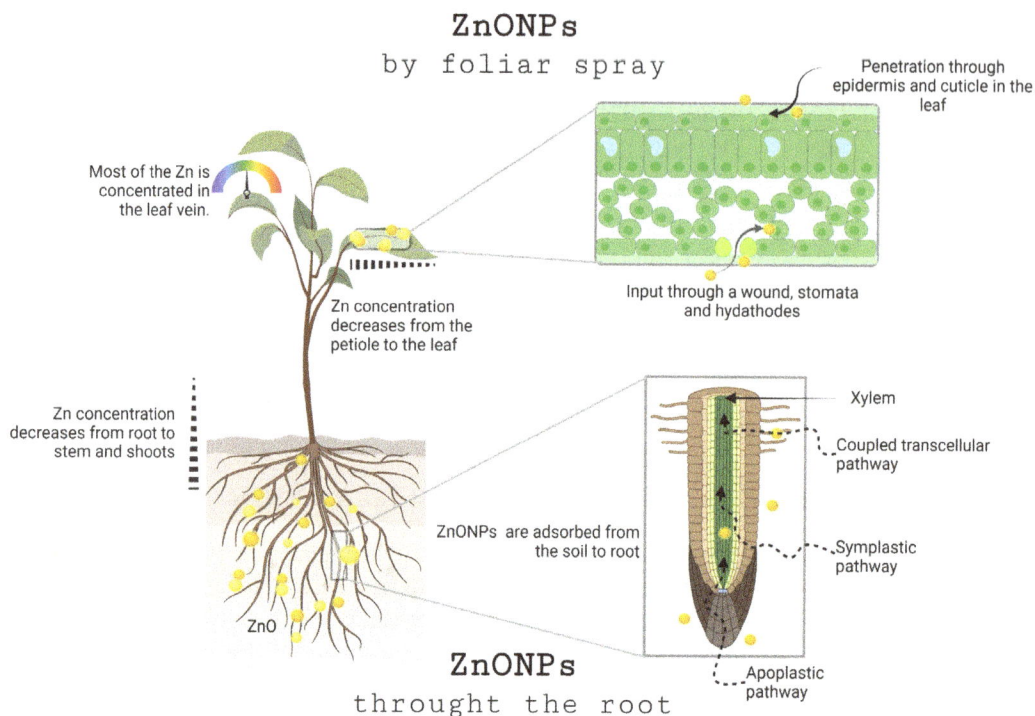

Figure 3. A diagrammatic representation of translocation ZnONPs inside the plants in two different applications.

Beneficial Effects of ZnONPs in Some Crops of Agricultural Interest

ZnONPs have been used as nanofertilizer in the cultivation of products of agricultural interest, showing important positive effects on different crops (Figure 4). For instance, in moringa (*Moringa peregrina* [Forssk.] Fiori), increased root growth was observed, along with elevated levels of total chlorophyll, carotenoids, proline, carbohydrates, crude protein levels, and antioxidants (Soliman et al. 2015). In tomato (*Solanum lycopersicum* L.), height, root length, and fruit weight increased (Raliya et al. 2015); increased chlorophyll, photosynthetic rate, protein content, catalase, peroxidase, lycopene, β-carotene, number of fruits, yield, and gas exchange by leaves (Alharby et al. 2016; Faizan et al. 2019); higher mRNA expression of the superoxide dismutase (SOD) and glutathione peroxidase (GPX) genes as well as a decrease in the harmful effects of salinity (Faizan et al. 2018).

In maize (*Zea mays* L.), several positive effects were observed, such as high root/shoot ratio and plant height (Liu et al. 2015), an increase in the weight of dry matter in branches (Adhikari et al. 2015) and length of roots (Zhang et al. 2015), in addition to acting as promoters by increasing the activity of ascorbate peroxidase (López-Moreno et al. 2017). With the use of alfalfa (*Medicago sativa* L.), there was also an increase in germination rate (Bandyopadhyay et al. 2015). In rice (*Oryza sativa* L.), ZnONPs stimulated the production of reactive oxygen species (ROS) and antioxidant enzymes (Chen et al. 2015). For its part, in wheat (*Triticum aestivum* L.), a significant increase in peroxidase activity was reported (Prakash and Chung 2016). In peas (*Pisum sativum* L.), it increased the concentrations of Chl-a and carotenoids (Mukherjee et al. 2014). In black mustard (*Brassica nigra* L. Koch.), an increase in shoot length was observed (Zafar et al. 2016). In tobacco (*Nicotiana tabacum* L.) and broad bean (*Vicia faba* L.) plants, intracellular ROS production, lipid peroxidation, and antioxidant enzyme activities were increased (Ghosh et al. 2016). In cucumber cultivation (*Cucumis sativus* L.), there was an increase in the dry biomass in the shoots (Moghaddasi et al. 2017). In lettuce (*Lactuca sativa* L.), the biomass produced and the net photosynthetic rate improved (Xu et al. 2018). It has also been observed in Lisianthus flowers (*Eustoma grandiflorum*

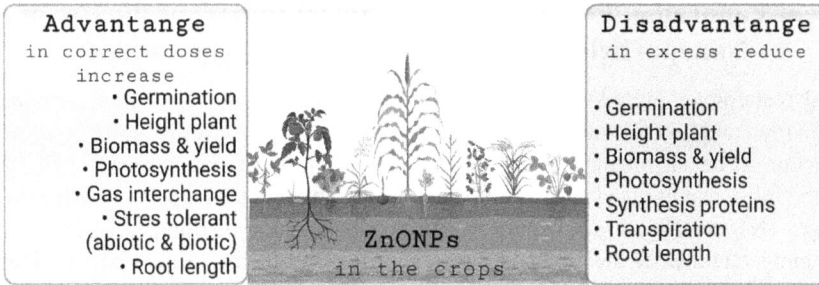

Figure 4. Diagram representing the positive and negative effects of ZnONPs in different plants and at different concentrations.

[Raf.] Shinn), a better quality of the floral stem and leaves, greater water absorption and increased rigidity, firmness, and color of the leaves (Soriano et al. 2018). In moha (*Setaria italica* L.), the content of oils and total nitrogen increased, the water stress index of the crops improved, and the nutritional parameters and enzymatic activities also enhanced (Kolenčík et al. 2019).

Negative Effects of the Use of ZnONPs in Agriculture

ZnONPs can be phytotoxic. These effects are usually associated with high concentrations and particle size (Kah 2015; Servin and White 2016). Regarding the concentration, some works have reported that concentrations greater than 50 mg/L of ZnONPs and sizes greater than 30 nm have negative effects on different crops. For example, in maize, they induced root elongation inhibition (Yang et al. 2015; Zhang et al. 2015), reduced germination, photosynthetic capacity, chlorophyll content, and yield (Zhao et al. 2015). In rice, inhibition in root elongation was observed (Yang et al. 2015), as well as a negative effect on shoot height and biomass (Chen et al. 2015). In alfalfa, ZnONPs reduced dry root biomass (Bandyopadhyay et al. 2015), as well as inhibition of root elongation in cucumber (Zhang et al. 2015), Chinese cabbage (*Brassica rapa* L. ssp. *pekinensis*) (Xiang et al. 2015), black mustard (Zafar et al. 2016), and carrot (*Daucus carota* L.), in addition to a decrease in its total biomass (Ebbs et al. 2016); the latter effect was also observed in tomato (Li et al. 2016) along with an inhibition of the growth rate in the stem (Alharby et al. 2016) and a decrease in the content of Chl-a and Chl-b (Wang et al. 2018).

In the green bean (*Macrotyloma uniflorum* [Lam.] Verdc.), ZnONPs delayed germination (Gokak et al. 2015). In wheat, they observed inhibition in root elongation (Watson et al. 2015) and shoot growth (Prakash and Chung 2016). With onion (*Allium cepa* L.), the mitotic index decreased (Taranath et al. 2015) and caused a loss in membrane integrity (Ghosh et al. 2016). Also, a reduction in the percentage of germination and root length was observed in lettuce (Liu et al. 2016). In the spinach (*Spinacia oleracea* L.), root length, shoots, total weight, as well as chlorophyll and carotenoid content were reduced (Singh et al. 2016). The decrease in growth and weakening of root rigidity was observed in soybean (*Glycine max* L.) (Hossain et al. 2016). Singh et al. (2018) observed a reduction in root and shoot length and seedling biomass in radish (*Raphanus sativus* L.) plants. In *in vitro* culture of *Arabidopsis thaliana* (L.) Heynh, growth, chlorophyll content, photosynthetic rate, and transpiration rate decreased (Wang et al. 2016). Regarding the issue of the shape and surface chemistry of the nanoparticle, it is important to mention that there is still not enough information to allow a precise trend to be developed. However, it is an issue that must be addressed because it is known that the interaction of NPs with biological systems strongly depends on the latter factors.

Current Panorama, Perspectives, and Social Impact of the Use of ZnO Nanoparticles and Nanotechnology in Agriculture

Agricultural technology includes the use of fertilizers and pesticides, which provide external nutrients to crops and protect them from pests and diseases. Likewise, its use has allowed global food production to be sufficient and support the growth of the human population (Erisman et al. 2008). Of the most widely used agrochemicals in the field, about 30–55% of fertilizers are made from nitrogen (N) and 18–20% phosphorus (P); however, much of this percentage is wasted by crops remaining retained in the soil and causing serious problems to ecosystems (Heffer 2013; Subramanian et al. 2015).

ZnO is a biocompatible compound approved by the FDA (Food and Drug Administration; 21CFR182.8991) for use in humans (Mishra et al. 2013; Colon et al. 2006; Padmavathy and Vijayaraghavan 2008). Since 2011, it was estimated that this compound would have a global production of over 1.2 million tons per year (Kumari et al. 2011). During 2015–2018, intense use of ZnO was registered in the agricultural sector, developing 33 agrochemicals that were recognized by the International Organization for Standardization (ISO), of which the majority were fungicides, herbicides, insecticides, acaricides, and nematicides (Jeanmart et al. 2021).

With the advent of nanotechnology in agriculture, the use of ZnO-based nanoparticles is emerging as an innovative solution to address farm problems while reducing environmental impacts (Chhipa 2017; Liu and Lal 2015). The background discussed in this document allows us to glimpse the use of ZnONPs and their effects on crops of agricultural interest. The literature review shows that the proper and ethical use of nanotechnology is allowing the development of sustainable agriculture to be supported. ZnONPs are a promising material due to their excellent physicochemical properties, low toxicity, high capacity to anchor active molecules, and fertilizer and antimicrobial activity (Rad et al. 2019). These findings mean there is a promising field of opportunity for the development of novel hybrid and multifunctional systems. For example, the use of biopolymers (chitosan, glucans, and gelatin) or beneficial microorganisms can allow ZnO to be used not only as nanofertilizer but also as antifungal, antibacterial compounds, or nutrient and drug carriers (Khan et al. 2013). However, it is also necessary to pay attention to the care and handling of these nanostructures to mitigate their negative effects. It has been seen in some crops that the concentration and particle size decrease the quality of the products, in addition to inducing a reduction in plant height, biomass, photosynthesis, yield, and transpiration, among other things (Jahan et al. 2022). Likewise, it is important to mention that in the NP-plant interaction, the effect of surface chemistry, shape, and size of the particle can also influence the positive or negative response of NPs in crops and agricultural fields (White and Gardea-Torresdey 2018). Currently, there is not enough information to accurately evaluate some of these last aspects that are necessary to make efficient and ethical use of nanomaterials. In the same way, we consider that the addition of specialists from different areas is necessary to be able to accurately evaluate the toxicological, environmental, and social impact that the use of ZnONPs in the agricultural sector has and will have in the future, as well as a regulation that guarantees the access and adequate use of agricultural users and for society (Behl et al. 2022).

Moreover, it is estimated that the population growth for 2050 will be 30%, about 9.7 billion people. Continuing with this growth trend, the inefficient use of agrochemicals will only increase the detrimental impacts on the environment. This fact makes it imperative to develop mechanisms that improve the efficiency of agrochemical delivery to support the growing demand for food supply in an environmentally sustainable efficient manner (Smith and Gilbertson 2018).

Conclusions

ZnONPs are one of the most widely used nanomaterials in nanotechnology. In agriculture, these nanomaterials have been used efficiently as nanofertilizers and in some cases as antimicrobial agents. In the following years, the development of hybrid and multifunctional materials that allow addressing various crop needs is expected. The beneficial effects on plants are attributed to the contribution of nutrients. However, the contradictory results arising from the concentration and possible environmental impact due to misuse highlight the need to incorporate research from other disciplines so as to gain a precise understanding of the toxicological, environmental, and social impact associated with the use of ZnONPs in agriculture, both presently and in the future.

References

Adhikari, T., Kundu, S., Biswas, A.K., Tarafdar, J.C. and Subba Rao, A. 2015. Characterization of zinc oxide nano-particles and their effect on growth of maize (*Zea mays* L.). J. Plant Nutr. 38(10): 1505–1515.

Alharby, H.F., Metwall, E.M., Fuller, M.P. and Aldhebiani, A.Y. 2016. The alteration of mRNA expression of SOD and GPX genes, and proteins in tomato (*Lycopersicon esculentum* Mill) under stress of NaCl and/or ZnO nanoparticles. Saudi J. Biol. Sci. 23(6): 773–781.

Alloway, B.J. 2009. Soil factors associated with zinc deficiency in crops and humans. Environ. Geochem. Health. 31(5): 537–548.

Auld, D.S. 2001. Zinc coordination sphere in biochemical zinc sites. *In*: Maret, W. (ed.). Zinc Biochemistry, Physiology, and Homeostasis. BioMetals 14: 271–313.

Bandyopadhyay, S., Plascencia-Villa, G., Mukherjee, A., Rico, C.M., José-Yacamán, M., Peralta-Videa, J.R. and Gardea-Torresdey, J.L. 2015. Comparative phytotoxicity of ZnO NPs, bulk ZnO, and ionic zinc onto the alfalfa plants symbiotically associated with *Sinorhizobium meliloti* in soil. Sci. Total Environ. 515: 60–69.

Behl, T., Kaur, I., Sehgal, Singh, S., Sharma, N., Bhatia, S., Al-Harrasi, A. and Bungau, S. 2022. The dichotomy of nanotechnology as the cutting edge of agriculture: Nano-farming as an asset versus nanotoxicity. Chemosphere 288(2): 132533.

Broadley, M.R., White P.J., Hammond, J.P., Zelko, I. and Lux, A. 2007. Zinc in plants. New Phytol. 173: 677–702.

Cakmak, I., McLaughlin, M.J. and White, P. 2017. Zinc for better crop production and human health. Plant Soil. 411: 1–4.

Chen, J., Liu, X., Wang, C., Yin, S.S., Li, X.L., Hu, W.J. and Peng, X.X. 2015. Nitric oxide ameliorates zinc oxide nanoparticles-induced phytotoxicity in rice seedlings. J. Hazard Mater. 297: 173–182.

Chhipa, H. 2017. Nanofertilizers and nanopesticides for agriculture. Environ. Chem. Lett. 15(1): 15–22.

Colon, G., Ward, B.C. and Webster, T.J. 2006. Increased osteoblast and decreased *Staphylococcus epidermidis* functions on nanophase ZnO and TiO₂. J. Biomed. Mater Res. 78A(3): 595–604.

Dimkpa, C.O., McLean, J.E., Britt, D.W. and Anderson, A.J. 2012. Bioactivity and biomodification of Ag, ZnO, and CuO nanoparticles with relevance to plant performance in agriculture. Ind. Biotechnol. 8(6): 344–357.

Ebbs, S.D., Bradfield, S.J., Kumar, P., White, J.C., Musante, C. and Ma, X. 2016. Accumulation of zinc, copper, or cerium in carrot (*Daucus carota*) exposed to metal oxide nanoparticles and metal ions. Environ. Sci. Nano 3(1): 114–126.

El Fekya, O.M., Hassan, E.A., Fadel, S.M. and Hassan, M.L. 2014. Use of ZnO nanoparticles for protecting oil paintings on paper support against dirt, fungal attack, and UV againg. J. Cult. Heritage 15: 165–172.

Erisman, J.W., Sutton, M.A., Galloway, J., Klimont, Z. and Winiwarter, W. 2008. How a century of ammonia synthesis changed the world. Nat. Geosci. 1(10): 636–639.

Faizan, M., Faraz, A., Yusuf, M., Khan, S.T. and Hayat, S. 2018. Zinc oxide nanoparticle-mediated changes in photosynthetic efficiency and antioxidant system of tomato plants. Photosynthetica 56(2): 678–686.

Faizan, M. and Hayat, S. 2019. Effect of foliar spray of ZnO-NPs on the physiological parameters and antioxidant systems of *Lycopersicon esculentum*. Pol. J. Nat. Sci. 34(1): 87–105.

Food and Drug Administration (FDA). Code of federal regulation sheet expided by the FDA. Title 21, volume 3, cite 21CFR182.8991. https://www.accessdata.fda.gov/scripts/cdrh/cfdocs/cfcfr/cfrsearch.cfm?fr=182.8991.

Ghosh, M., Jana, A., Sinha, S., Jothiramajayam, M., Nag, A., Chakraborty, A., Mukherjee, A. and Mukherjee, A. 2016. Effects of ZnO nanoparticles in plants: cytotoxicity, genotoxicity, deregulation of antioxidant defenses, and cell-cycle arrest. Mutat Res. Genet. Toxicol. Environ. Mutagen. 807: 25–32.

Gokak, I.B. and Taranath, T.C. 2015. Seed germination and growth responses of *Macrotyloma uniflorum* (Lam.) Verdc. exposed to Zinc and Zinc nanoparticles. Int. J. Environ. Sci. 5(4): 840–847.

Hafeez, B.M.K.Y., Khanif, Y.M. and Saleem, M. 2013. Role of zinc in plant nutrition—a review. Am. J. Exp. Agr. 3(2): 374–391.

Heffer, P. 2013. Assessment of Fertilizer Use by Crop at the Global Level 2010–2010/11. International Fertilizer Industry Association, Paris.

Hernández-Adame, L., Palestino, G., Meza, O., Hernandez-Adame, P.L., Vega-Carrillo, H.R. and Sarhid, I. 2018. Effect of Tb³⁺ concentration in the visible emission of terbium-doped gadolinium oxysulfide microspheres. Solid State Sci. 84: 8–14.

Hossain, Z., Mustafa, G., Sakata, K. and Komatsu, S. 2016. Insights into the proteomic response of soybean towards Al_2O_3, ZnO, and Ag nanoparticles stress. J. Hazar Mater. 304: 291–305.

Hussain, F., Hadi, F. and Rongliang, Q. 2021. Effects of zinc oxide nanoparticles on antioxidants, chlorophyll contents, and proline in *Persicaria hydropiper* L. and its potential for Pb phytoremediation. Environ. Sci. Pollut. Res. 28(26): 34697–34713.

Jahan, U., Kafeel, U., Naikoo, M.I. and Khan, F.A. 2022. Nanoparticles and their effects on growth, yield, and crop quality cultivated under polluted soil. pp. 333–352. *In*: Rajput, V.D., Verma, K.K., Sharma, N. and Minkina, T. (eds.). The Role of Nanoparticles in Plant Nutrition under Soil Pollution. Sustainable Plant Nutrition in a Changing World. Springer, Cham.

Jeanmart, S., Edmunds, A.J.F., Lamberth, C., Pouliot, M. and Morris, J.A. 2021. Synthetic approaches to the 2015–2018 new agrochemicals. Bioor. Med. Chem. 39: 116162.

Kah, M. 2015. Nanopesticides and nanofertilizers: emerging contaminants or opportunities for risk mitigation? Frontiers in Chemistry 3: 64.

Khan, A., Mehmood, S., Shafiq, M., Yasin, T., Akhter, Z. and Ahmad, S. 2013. Structural and antimicrobial properties of irradiated chitosan and its complexes with zinc. Radiat. Phys. Chem. 91: 138–142.

Kolenčík, M., Ernst, D., Komár, M., Urík, M., Šebesta, M., Dobročka, E. and Feng, H. 2019. Effect of foliar spray application of zinc oxide nanoparticles on quantitative, nutritional, and physiological parameters of foxtail millet (*Setaria italica* L.) under field conditions. Nanomaterials 9(11): 1559.

Kumari, M., Khan, S.S., Pakrashi, S., Mukherjee, A. and Chandrasekaran, N. 2011. Cytogenetic and genotoxic effects of zinc oxide nanoparticles on root cells of *Allium cepa*. J. Hazar Mater. 190(1): 613–621.

Laurenti, M., Canavese, G., Sacco, A., Fontana, M., Bejtka, K., Castellino, M., Pirri, C.F. and Cauda, V. 2015. Nanobranched ZnO structure: p-type doping induces piezoelectric voltage generation and ferroelectrice-photovoltaic effect. Adv. Mater. 27(28): 4218–4223.

Li, M., Ahammed, G.J., Li, C., Bao, X., Yu, J., Huang, C., Yin, H. and Zhou, J. 2016. Brassinosteroid ameliorates zinc oxide nanoparticles-induced oxidative stress by improving antioxidant potential and redox homeostasis in tomato seedling. Front Plant Sci. 7: 615.

Lin, D. and Xing, B. 2008. Root uptake and phytotoxicity of ZnO nanoparticles. Environ. Sci. Technol. 42(15): 5580–5585.

Lindsey, A.P.J., Murugan, S. and Renitta, R.E. 2020. Microbial disease management in agriculture: Current status and future prospects. Biocatal. Agric Biotechnol. 23: 101468.

Liu, R. and Lal, R. 2015. Potentials of engineered nanoparticles as fertilizers for increasing agronomic productions. Sci. Total Environ. 514: 131–139.

Liu, R., Zhang, H. and Lal, R. 2016. Effects of stabilized nanoparticles of copper, zinc, manganese, and iron oxides in low concentrations on lettuce (*Lactuca sativa*) seed germination: nanotoxicants or nanonutrients? Water Air Soil Pollut. 227(1): 42.

Liu, X., Wang, F., Shi, Z., Tong, R. and Shi, X. 2015. Bioavailability of Zn in ZnO nanoparticle-spiked soil and the implications to maize plants. J. Nanopart. Res. 17(4): 175.

Longnecker, N.E. and Robson, A.D. 1993. Distribution and transport of zinc in plants. *In*: Robson, A.D. (ed.). Zinc in Soils and Plants. Developments in Plant and Soil Sciences 55: 79–91.

López-Moreno, M.L., de la Rosa, G., Cruz-Jiménez, G., Castellano, L., Peralta-Videa, J.R. and Gardea-Torresdey, J.L. 2017. Effect of ZnO nanoparticles on corn seedlings at different temperatures; X-ray absorption spectroscopy and ICP/OES studies. Microchem J. 134: 54–61.

López-Moreno, M.L., Cedeño-Mattei, Y., Bailón-Ruiz, S.J., Vazquez-Nuñez, E., Hernandez-Viezcas, J.A., Perales-Pérez, O.J., de la Rosa, G., Peralta-Videa, J.R. and Gardea-Torresdey, J.L. 2018. Environmental behavior of coated NMs: Physicochemical aspects and plant interactions. J. Hazar Mater. 347: 196–217.

Mishra, M., Paliwal, J., Singh, S.K., Selvararajan, E., Subathradevi, C. and Mohanashrinivasan, V. 2013. Studies on the inhibitory activity of biologically synthesized and characterized ZnO nanoparticles using *Lactobacillus sporogens* against *Staphylococcus aureus*. J. Pure Appl. Microbiol. 7(2): 1–6.

Moghaddasi, S., Fotovat, A., Khoshgoftarmanesh, A.H., Karimzadeh, F., Khazaei, H.R. and Khorassani, R. 2017. Bioavailability of coated and uncoated ZnO nanoparticles to cucumber in soil with or without organic matter. Ecotoxicol. Environ. Saf. 144: 543–551.

Mukherjee, A., Peralta-Videa, J.R., Bandyopadhyay, S., Rico, C.M., Zhao, L. and Gardea-Torresdey, J.L. 2014. Physiological effects of nanoparticulate ZnO in green peas (*Pisum sativum* L.) cultivated in soil. Metallomics 6(1): 132–138.

Nielsen, F.H. 2012. History of zinc in agriculture. Advances in Nutrition 3(6): 783–789.

Padmavathy, N. and Vijayaraghavan, R. 2008. Enhanced bioactivity of ZnO nanoparticles—An antimicrobial study. Sci. Technol. Adv. Mater. 9(3): 035004.

Peters, R.J., Bouwmeester, H., Gottardo, S., Amenta, V., Arena, M., Brandhoff, P., Marvin, H.J.P., Mech, A., Botelho, M.F., Quiros, P.L., Rauscher, H., Schoonjans, R., Undas, A.K., Vettori, M.V, Weigel, S. and Aschberger, K. 2016. Nanomaterials for products and application in agriculture, feed and food. Trends Food Sci. Technol. 54: 155–164.

Prakash, M.G. and Chung, I.M. 2016. Determination of zinc oxide nanoparticles toxicity in root growth in wheat (*Triticum aestivum* L.) seedlings. Acta Biol. Hung. 67(3): 286–296.

Rad, S.S., Sani, A.M. and Mohseni, S. 2019. Biosynthesis, characterization and antimicrobial activities of zinc oxide nanoparticles from leaf extract of *Mentha pulegium* (L.). Microb Pathog. 131: 239–245.

Rajput, V.D., Minkina, T.M., Behal, A., Sushkova, S.N., Mandzhieva, S., Singh, R., Gorovtsov, A., Tsitsuashvili, V. S., Purvis, W.O, Ghazaryan, K.A. and Movsesyan, H.S. 2018. Effects of zinc-oxide nanoparticles on soil, plants, animals and soil organisms: A review. Environ. Nanotechnol. Monit. Manage. 9: 76–84.

Raliya, R., Nair, R., Chavalmane, S., Wang, W.N. and Biswas, P. 2015. Mechanistic evaluation of translocation and physiological impact of titanium dioxide and zinc oxide nanoparticles on the tomato (*Solanum lycopersicum* L.) plant. Metallomics 7(12): 1584–1594.

Raliya, R., Saharan, V., Dimkpa, C. and Biswas, P. 2017. Nanofertilizer for precision and sustainable agriculture: Current state and future perspectives. J. Agric Food Chem. 66(26): 6487–6503.

Reddy, P.V.L., Hernandez-Viezcas, J.A., Peralta-Videa, J.R. and Gardea-Torresdey, J.L. 2016. Lessons learned: Are engineered nanomaterials toxic to terrestrial plants? Sci. Total Environ. 568: 470–479.

Sabir, S., Arshad, M. and Chaudhari, S.K. 2014. Zinc oxide nanoparticles for revolutionizing agriculture: Synthesis and applications. Sci. World J. 2014: 925494.

Servin, A.D. and White, J.C. 2016. Nanotechnology in agriculture: Next steps for understanding engineered nanoparticle exposure and risk. NanoImpact 1: 9–12.

Shahid, M., Farooq, A.B.U., Rabbani, F., Khalid, S. and Dumat, C. 2020. Risk assessment and biophysiochemical responses of spinach to foliar application of lead oxide nanoparticles: A multivariate analysis. Chemosphere 245: 125605.

Singh, D. and Kumar, A. 2016. Impact of irrigation using water containing CuO and ZnO nanoparticles on *Spinach oleracea* grown in soil media. Bull Environ. Contam. Toxicol. 97(4): 548–553.

Singh, D. and Kumar, A. 2018. Investigating long-term effect of nanoparticles on growth of *Raphanus sativus* plants: A trans-generational study. Ecotoxicology 27(1): 23–31.

Smith, A.M. and Gilbertson, L.M. 2018. Rational ligand design to improve agrochemical delivery efficiency and advance agriculture sustainability. ACS Sustainable Chem. Eng. 6(11): 13599–13610.

Soliman, A.S., El-Feky, S.A. and Darwish, E. 2015. Alleviation of salt stress on *Moringa peregrina* using foliar application of nanofertilizers. J. Hortic. For. 7(2): 36–47.

Soriano Melgar, L.A., López-Guerrero, A.G., Cortéz-Mazatan, G., Mendoza-Mendoza, E. and Peralta-Rodríguez, R.D. 2018. Nanopartículas de óxido de zinc y óxido de zinc/grafeno empleadas en soluciones florero durante la vida poscosecha de lisianthus (*Eustoma grandiflorum*). Agro Productividad 11(8).

Subramanian, K.S., Manikandan, A., Thirunavukkarasu, M. and Rahale, C.S. 2015. Nano-fertilizers for balanced crop nutrition. pp. 69–80. *In*: Rai, M., Ribeiro, C., Mattoso, L. and Duran, N. (eds.). Nanotechnologies in Food and Agriculture. Springer, Cham.

Taranath, T.C., Patil, B.N., Santosh, T.U. and Sharath, B.S. 2015. Cytotoxicity of zinc nanoparticles fabricated by *Justicia adhatoda* L. on root tips of *Allium cepa* L.— a model approach. Environ. Sci. Pollut. Res. 22(11): 8611–8617.

Wang, X.P., Li, Q.Q., Pei, Z.M. and Wang, S.C. 2018. Effects of zinc oxide nanoparticles on the growth, photosynthetic traits, and antioxidative enzymes in tomato plants. Biol. Plant. 62(4): 801–808.

Wang, X., Yang, X., Chen, S., Li, Q., Wang, W., Hou, C., Gao, X., Wang, L. and Wang, S. 2016. Zinc oxide nanoparticles affect biomass accumulation and photosynthesis in *Arabidopsis*. Front Plant Sci. 6: 1243.

White, J.C. and Gardea-Torresdey, J. 2018. Achieving food security through the very small. Nat. Nanotechnol. 13(8): 627–629.

Watson, J.L., Fang, T., Dimkpa, C.O., Britt, D.W., McLean, J.E., Jacobson, A. and Anderson, A.J. 2015. The phytotoxicity of ZnO nanoparticles on wheat varies with soil properties. Biometals 28(1): 101–112.

Xiang, L., Zhao, H.M., Li, Y.W., Huang, X.P., Wu, X.L., Zhai, T., Cai, Q.Y. and Mo, C.H. 2015. Effects of the size and morphology of zinc oxide nanoparticles on the germination of Chinese cabbage seeds. Environ. Sci. Pollut. Res. 22(14): 10452–10462.

Xu, J., Luo, X., Wang, Y. and Feng, Y. 2018. Evaluation of zinc oxide nanoparticles on lettuce (*Lactuca sativa* L.) growth and soil bacterial community. Environ. Sci. Pollut. Res. 25(6): 6026–6035.

Yang, Z., Chen, J., Dou, R., Gao, X., Mao, C. and Wang, L. 2015. Assessment of the phytotoxicity of metal oxide nanoparticles on two crop plants, maize (*Zea mays* L.) and rice (*Oryza sativa* L.). Int. J. Environ. Res. Public Health 12(12): 15100–15109.

Zafar, H., Ali, A., Ali, J.S., Haq, I.U. and Zia, M. 2016. Effect of ZnO nanoparticles on Brassica nigra seedlings and stem explants: Growth dynamics and antioxidative response. Fron. Plant Sci. 7: 535.

Zhang, R., Zhang, H., Tu, C., Hu, X., Li, L., Luo, Y. and Christie, P. 2015. Phytotoxicity of ZnO nanoparticles and the released Zn (II) ion to corn (*Zea mays* L.) and cucumber (*Cucumis sativus* L.) during germination. Environ. Sci. Pollut. Res. 22(14): 11109–11117.

Zhao, L., Sun, Y., Hernandez-Viezcas, J.A., Hong, J., Majumdar, S., Niu, G., Duarte-Gardea, M., Peralta-Videa, J.R. and Gardea-Torresdey, J.L. 2015. Monitoring the environmental effects of CeO$_2$ and ZnO nanoparticles through the life cycle of corn (*Zea mays*) plants and *in situ* μ-XRF mapping of nutrients in kernels. Environ. Sci. Technol. 49(5): 2921–2928.

Chapter 5

Nanoencapsulation for Delivery of Agrochemicals to Boost Plant Health

Julia Helena da Silva Martins,[1,*] *Luiza Helena da Silva Martins,*[2]
Jonilson de Melo e Silva,[3] *Sandro Rodrigues de Almada,*[2]
Johnatt Allan Rocha de Oliveira[4] and *Mahendra Rai*[5]

Introduction

The global population is growing more and more each day with an estimated 9.7 billion people in 2050; there is a concern about the demand for food and with that, the agri-food industry must create ways to meet this demand. Among several technologies already studied, nanotechnology comes as an innovative means to bring a significant impact to this sector; factors such as safety, food sustainability, and nutritional security are the keywords for good development of the agro-food sector. Nanotechnology can be used in both the agricultural and food industries as a means of improving crop productivity and preserving food (Sampathkumar et al. 2020).

Even before it was firmly established in science, nanotechnology was already pervasive in the world around us. Examples include the smoke from a fire, the volcanic ash from an eruption, the casein molecules in milk, and other natural occurrences (Griffin et al. 2017).

Nanoparticles are tiny particles that combine with other particles to form an integrated system that performs functions inside experimental organisms, making the organism of interest functionally more helpful. Nanotechnology, as previously mentioned, is a young science that deals with alterations of matter with a specific size scale from 1 to 10 nm (Singh et al. 2021).

The use of sufficient fertilizer, herbicide, and pesticide treatments has become crucial for the success of fruit and vegetable production. However, misuse of these chemicals has emerged as one of the critical issues of high levels of soil pollution. The overuse of fertilizers and herbicides in agriculture leaves behind various harmful residues that are a significant source of soil and water pollution, spreading them over agricultural land and for irrigation by drainage channels. These residues left in ecosystems have harmful effects in several ways (Borišev et al. 2020).

The introduction of new pesticide formulations, advanced, sophisticated agricultural practices, and advancements in remediation methods depend on developing clean agriculture policies.

[1] Universidade Federal do Pará, Instituto de Física, Belém, Pará, Brazil.
[2] Universidade Federal Rural da Amazônia, UFRA, Instituto de Saúde e Produção Animal, Belém, Pará, Brazil.
[3] Universidade Federal do Pará, Programa de Pós-graduação em Ciência e Tecnologia de alimentos, Belém, Pará, Brazil.
[4] Universidade Federal do Pará. UFPA, Faculdade de Nutrição, Belém, Pará, Brazil.
[5] Department of Microbiology, Nicolaus Copernicus University, 87-100 Torun, Poland.
* Corresponding author: juliahelenamartins@gmail.com

Nanotechnology has evolved, and it is now receiving more and more attention in every field of human activity, including technology, engineering, agriculture, and environmental research (Srivastav et al. 2018).

Technologies that utilize nanoparticles in agricultural operations, such as the controlled release of pesticides, herbicides, and fertilizers, are referred to as nanoagriculture. Additionally, they engage in controlled environmental agriculture techniques (CEA), which include hydroponics, aeroponics, and aquaponics, as well as various types of agricultural packaging and more advanced delivery methods (Sunday et al. 2021).

Antimicrobials, insecticides, pesticides, and pesticide encapsulation have all been the subject of research to determine the amounts required for specific plants or crops to suppress microbial growth, illness, or stress. Since the issue stems from synthetic pesticides, this does not entirely resolve the issue. For agricultural nutrients and pesticides, there has been a demand for alternative, all-natural, economical, biodegradable, and non-toxic delivery systems during the past ten years (Mujtaba et al. 2020).

Due to its enormous success in pharmacy, nanoencapsulation has developed into an interdisciplinary discipline that touches on other industries, such as agriculture, to boost productivity and sustainability. During application, around 90% of pesticides leak, causing catastrophic and even deadly losses. Their nanoencapsulation guarantees the regulated release and precise delivery of pesticides, such as fertilizers, insecticides, fungicides, and herbicides. In a wide range of crops, these insecticides are essential for adequate nutrient intake, disease control, and growth stimulation (Wani et al. 2019). This chapter focuses on nanoencapsulation techniques for agrochemicals, which aims to increase plant health. It also discusses the classification of materials used and presents studies that have already been carried out on the subject.

Nanotechnology and Agrochemicals

Conventional crop protection might change due to the use of nanotechnology in or on plants. Nanotechnology is very useful for delivering biomolecules to specific locations and releasing insecticides in a regulated manner, using materials such as nucleotides, proteins, activators, and more. By 2027, public awareness and growth in the nanotechnology sector will be predicted to total $126.8 billion. This novel strategy has helped the agriculture industry category develop at a compound annual growth rate of 12.9% from 2020 to 2027 (Machado et al. 2022).

Depending on the pest in question, agrochemicals are compounds that have been technologically produced for use in the field and are categorized into several sorts, such as herbicides, insecticides, and fungicides (Guleria et al. 2022). Agrochemicals boost food output by stopping food spoilage brought on by pests and illnesses. Using conventional agrochemicals results in pesticide drift, decreased delivery to the target, and off-target buildup, which leads to a mutation in the genetic material that makes them resistant to these pesticides and poses environmental concerns (Guleria et al. 2022).

Conceptually, regulated medicine distribution to people and animals is comparable to the controlled delivery of agrochemicals and nutrients to plants. Contrary to the confined and regulated structure of the bloodstream, agricultural delivery occurs in an open field with unpredictable weather and geographic conditions and no defined transport pathway to the target. Nanocarriers can be delivered through the leaves where they are passively absorbed by the stomata and any wounds or through the soil and absorbed through the roots (Chariou et al. 2020).

Nanoencapsulation Concept

Nanoencapsulation controls medication or substances release while considering environmental factors and encapsulating items in an inert phase. Nanoencapsulation allows for the very small-scale encapsulation of substances (Gaur et al. 2022). Common names for encapsulated compounds include

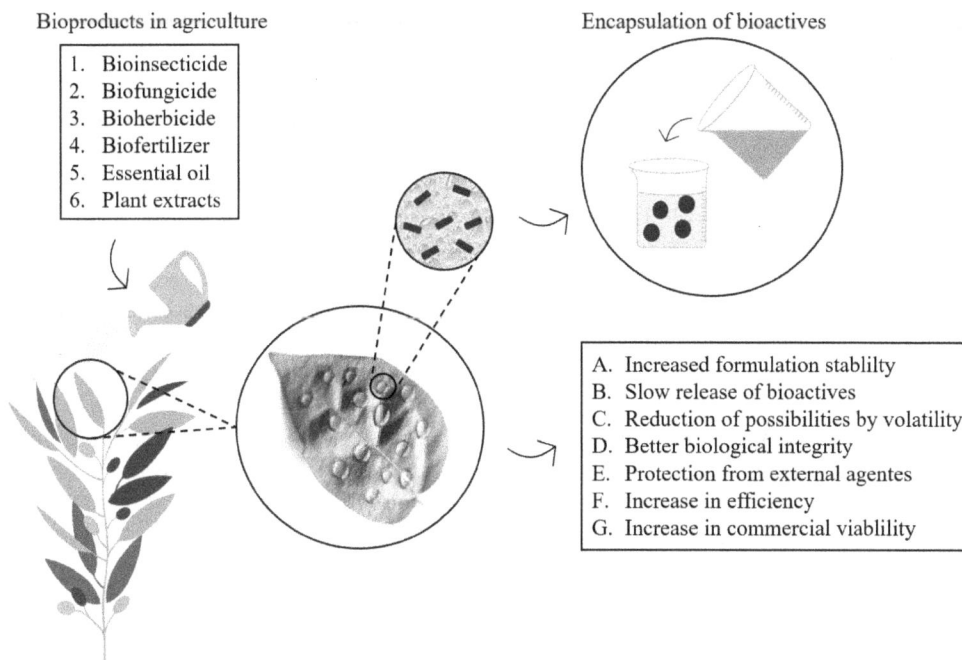

Figure 1. General scheme of nanoencapsulation system for plants adapted from Zabot et al. (2022) (Open access material).

inner phases, core components, fillers, and diluents, such as agrochemicals. External phases, such as nanocapsules, shells, coatings, or membranes, are called encapsulating materials (Nuruzzaman et al. 2016).

Commercial agrochemicals have attempted to be encapsulated in nanomaterials to enhance their physical properties and limit their widespread use. Agrochemicals can be loaded or confined inside nanoscale-diameter particles by the process of nanoencapsulation (Nuruzzaman et al. 2016).

Numerous materials may be used as encapsulating matrixes; however, the material used depends on its purpose and qualities. The majority of these carriers are either hydrophilic or lipophilic; the former category includes polysaccharides while the latter includes lipids, biopolymers, liposomes, micelles, etc. The goal of nanoencapsulation is to verify that the encapsulated material reaches the site of action without being impacted by the environment (Gaur et al. 2022).

The use of nanoencapsulation is justified for several reasons such as transforming liquids into freely flowing solids, separating incompatible materials, hiding the color, flavor, and aroma of substances, and enhancing their durability by avoiding oxidation or deactivation by environmental processes. The capsule is shielded, and the release is targeted (Gaur et al. 2022). A general nanoencapsulation scheme for the plant system application can be seen in Figure 1.

Nanocarrier and Nano-Delivery System Concepts

Due to their tiny size and capacity to alter their physical characteristics, such as charge and shape, to carry therapeutic chemicals to tissues, nanocarriers are valuable delivery agents. Drugs delivered by nanoparticles, dendrimers, and polymeric or lipid-based carriers like liposomes are examples of nanocarriers. Instead of active medication, nanocarriers function as carriers that control the pharmacokinetics of transport and distribution (Musante et al. 2017).

Nanocarriers significantly increase the chemical stability of pesticides and shield them from harmful environmental factors like radiation and high temperatures. Additionally, increasing the pesticides' wettability and dispersibility lowers the environmental danger posed by organic solvents.

High surface area, excellent stability, biodegradability, and pest control are all advantages of nano pesticide formulations (Mao et al. 2020).

In the field of agriculture, where they are loaded with pesticides, growth regulators, and fertilizers and ensure effective delivery of these loaded components into plants for crop protection and production purposes, nano-delivery systems are nano-sized materials (which are also known as nanocarriers) that can effectively deliver useful components to organisms (such as humans, animals, or plants) (Okeke et al. 2022). The nano-delivery system is superior to conventional formulas in a few ways, including potentiating and increasing the solubility of antibiotics, improving the efficiency of active ingredients, protecting the active ingredient against degradation through the use of encapsulation, controlling and sustained release of active substances that can be adjusted to a minimum effective dose, and decreasing the possibility of harm to non-target organisms/species. An excellent delivery system must have biocompatibility, large loading capacity, controlled distribution, stability, and a variety of capabilities (Okeke et al. 2022).

In order to accomplish the regulated release of nutrients or antibiotics based on the plant's requirement, it is crucial to adopt innovative stimulus-responsive nano-delivery systems. This helps to enhance the effective duration, decrease shedding, and reduce environmental toxicity (Okeke et al. 2022).

Nanomaterials Used as a Nanocarrier to Boost Plant Health

Nanopolymers

The covalent union of multiple smaller molecules known as monomers result in polymers, which are macromolecules. These substances are divided into synthetic polymers and natural polymers, depending on their nature. Polyesters, polyethers, poloxamers, and recombinant protein-based polymers are the different categories of synthetic polymers. Proteins and polysaccharides are the categories into which natural polymers may be separated (Pacios-Michelena et al. 2022). In this chapter, we will focus more on some natural-source polymers and combined materials.

Natural-source polymers are cheap, biodegradable, and create no byproducts of decomposition. They are also favorable to the environment and are an excellent encapsulating medium for active compounds because of these qualities. Researchers have recently been interested in amphiphilic block copolymers because of their capacity to combine polymers to generate a variety of nanoparticles. Typically, different kinds of monomers are polymerized to create block copolymers. The polymer should typically have a symmetrical structure, and one is more hydrophilic than the other. In aqueous solutions, block copolymers maintain their amphiphilic characteristics (Nuruzzaman et al. 2016).

It is a relatively new strategy to apply pesticides using polymeric nanoparticles. In order to enclose the active component in the polymer, polymeric nanocomposites (PNCs) typically consist of a polymer containing nanoparticles or nanofillers scattered in a polymeric matrix (Mujtaba et al. 2020).

According to Mujtaba et al. (2020), numerous biopolymers have been actively exploited as nanocarriers for the controlled release of pesticides, including cellulose, chitin, chitosan, collagen, alginate, and starch. Comparatively less cellulose has been used than starch or lignin derivatives. Polymer matrixes that are utilized as microencapsulation have to meet several requirements:

1. The biopolymer must have an appropriate glass transition state, molecular weight, and structure (long-term) for charge ejection.
2. Charged moieties and functional surface groups must not interact (such as pesticides).
3. It must be environmentally safe and biodegradable.
4. Polymers must be affordable and have enough storage and application stability.

The star polymer (SPc) has a particle size of 100.5 nm and comprises a hydrophilic shell that includes positively charged tertiary amines and a hydrophobic core. The hydrophilic coating helps make ecologically damaging active components more water soluble and stable upon their release (AIs) (Jiang et al. 2022).

To produce double-stranded RNA (dsRNA) and synthetic/botanical insecticides, star polymers (SPc) were created. To maximize plant absorption, decrease pesticide residues, and improve bioactivity against green peach aphids, we attempted to build a reasonably safe imidacrotizu nano-delivery method in this context (Jiang et al. 2022).

Imidaclostiz may self-assemble into nanoscale imidaclostiz/SPc complexes made up of roughly spherical particles through hydrophobic interaction with the hydrophobic core of SPc. Imidacrotiz droplet contact angles were absorbed by SPc, dramatically boosting plant absorption. Additionally, in the lab and field, SPc significantly increased the bioactivity and control efficacy of imidacrotiz. Interestingly, after seven days of treatment, imidacrotiz residues were still present despite the quicker breakdown of nanoscale imidacrotiz/SPc complexes with SPc. It has no adverse effects on the agronomic traits of tobacco plants. These researchers successfully developed a nano-delivery method for imidacrotiz that might boost pesticide effectiveness and decrease pesticide residues (Jiang et al. 2022).

Chitosan

Chitosan makes it possible to create intelligent pesticide administration systems for agriculture, decreasing the need to consume dangerous substances and minimizing adverse effects on humans and the environment. It is a flexible polysaccharide that harms people's health (Mujtaba et al. 2020).

Using organometallic systems, chitosan may create antifungal nanocapsules, such as charged tebuconazole. Through electrostatic interactions with chitosan and pectin, it is adsorbed and put together. This study demonstrates that the fungicidal nanocapsules display fungicidal activity and do not interfere with the development of Chinese cabbage, making their application safe. The release reacts to the action of pectinase in a slightly acidic media (Mujtaba et al. 2020).

Chitosan nanocarriers were loaded with the insecticides spinosad and permethrin by Sharma et al. (2019), cited by Machado (2022), and the nanocarriers were then evaluated against the model organism *Drosophila melanogaster* at various doses (10 g/mL, 50 g/mL, and 100 g/mL). Permethrin and spinosad had encapsulation efficiencies of over 99% and over 95%, respectively.

A focus of study in creating potent agrochemicals formulations is the controlled release and accurate targeting of agrochemicals. It is suggested that chitosan and its derivatives can act as intelligent pesticide medication delivery systems. It has been reported that biopesticides (like harpinPss) and synthetic pesticides (like imidacloprid, lambda-cyhalothrin, acetamiprid, and beauvericin) can be encapsulated in chitosan nanoparticles, demonstrating the suitability and effectiveness of chitosan for enhancing the delivery of both types of agrochemicals (Rani et al. 2020).

Chemically Modified Cellulose

Excellent hydrophilicity, film-forming, sticky, biocompatibility, and biodegradability are all features of carboxymethylcellulose (CMC). The keratin stability of denatured feathers and the accomplished pesticide load can both be enhanced by CMC. Additionally, it has been demonstrated that chromogenic groups, such as the carboxyl and hydroxyl groups of CMCS, have some anti-UV capabilities when generating UV light and the ability to absorb some UV radiation. Therefore, phosphorylated zein-based nano-pesticides can benefit from CMC to increase their performance in terms of stability, UV resistance, adhesion, etc. It is required to alter CMC by adding organic groups to the molecule since it is difficult for CMC to combine spontaneously with directly phosphorylated zein-based nano pesticides (Hao et al. 2020).

In situ encapsulation of hydrophobic fungicides via thiol-ene crosslinking in a miniemulsion was achieved by Machado et al. (2021) using chemically modified cellulose with under-10-enoic acid. The fungicides captan and pyraclostrobin encapsulation efficiency were between 80 and

100 percent. Additionally, cellulose nanocarriers were tested *in vitro* in antifungal assays against *Neonectria ditissima, Phaeoacremonium minimum,* and *Phaeomoniella chlamydospore* using the plain growth medium as a positive control and the drug Hygromycin as a positive control.

Lipid-Based Nanoparticles

According to Nadiminti et al. (2013), surfactants are frequently used in agrochemical spray formulations that are sprayed onto plants to aid in the active ingredient's delivery. Surfactants, however, have environmental off-target effects, including phytotoxicity. We suggest using nanostructured lipid crystalline particles (NLCP) as an alternative to pesticide delivery systems based on surfactants.

Regarding stabilized by solid and liquid lipid surfactants, lipid nanoparticles are a unique delivery technology that resembles the commonly used emulsions but differs in size and structure, water-insoluble, and nuclei are scattered. New and promising nanocarriers with unique features that are of significant interest for food applications include solid lipid nanoparticles (SLNs) and nanostructured lipid carriers (NLCs) (Kesharwani et al. 2014).

Lipid nanoformulations that circumvent natural chemicals' drawbacks include SLNs, solid lipid nanoemulsions, and NLCs. They can increase bioavailability, permeability, solubility, and stability (Piazzini et al. 2018).

The ability to achieve controlled or targeted release, improved bioavailability, versatility in encapsulating both lipophilic and hydrophilic drugs, high encapsulation efficiency, and long-term stability of encapsulated drugs are just a few of the favorable characteristics offered by nanostructured lipid transporters (NLCs), which can be used as effective delivery systems (Piazzini et al. 2018).

Tiny oil droplets encased in bigger oil droplets are known as emulsions. In a continuous aqueous phase, they scatter as droplets, creating a double emulsion. As a result, the encapsulation industry views nanoemulsions as a viable, straightforward nutraceutical and pharmaceutical delivery technology produced using natural and synthetic chemicals, including oils, surfactants, co-surfactants, and weighing ripening inhibitors, thickeners, and gelling agents. Nanoemulsions have been demonstrated to enhance the bioavailability of some encapsulated lipophilic compounds, and they offer potential benefits over macro- and microemulsions in several applications. Consequently, the bioactivity of the agrochemicals is enhanced. Additionally, nanoemulsions show exceptional stability against gravitational separation and particle aggregation (Wani et al. 2019).

Coacervation, commonly known as the phase separation process, is a very effective yet pricy method of encasing hydrophilic molecules. A homogenous solution of charged macromolecules is subjected to coacervation, which involves liquid-liquid phase separation to create a dense phase rich in polymers. Applying this method, biopolymer nanoparticle systems are developed using a variety of naturally occurring hydrophilic biopolymers, including sodium alginate, gelatin, and chitosan. Coacervation procedures produce nanocapsules with sizes ranging from 100 nm to 600 nm; however, they mostly rely on lyophilization and vacuum drying for drying. Understanding temperature, pH, mechanical stress, and other factors to build controlled release systems (Guha et al. 2015).

Following the treatment of leaves with NLCP, analysis of cuticular wax micromorphology revealed less wax solubilization in the monocot species. The results clearly show the advantages of using NLCP rather than surfactants for agrochemical delivery. To do this, they applied the commercial surfactants di (2-ethylhexyl) sulfosuccinate and alkyl dimethyl betaine to the adaxial surface of the leaves of four plant species: *Triticum aestivum* (wheat), *Zea mays* (maize), *Lupinus angustifolius* (lupin), and *Arabidopsis thaliana.* The concentrations of these surfactants were 0.1%, following treatment with NLCP, there was decreased phytotoxicity on the leaves of each species compared to when surfactants were applied (Nadiminti et al. 2013).

Liposomes

There are three different ways that liposomes can speed up drug delivery through the skin that include: adhesion to the skin's surface, subsequent drug transfer from vesicles to the skin, and the fusion with the stratum corneum's lipid matrix, which increases drug partitioning into the skin, and lipid exchange between the liposomal membrane and cell membrane, which speeds up drug diffusion across the membrane (Ophata et al. 2020).

Conventional liposomes do not, therefore, thoroughly enter the living skin and circulatory system, which is an issue. So, rather than for transdermal administration, liposomes have often been utilized as drug carriers for cutaneous distribution. Conventional liposomes also have drawbacks, such as ineffective hydrophilic drug encapsulation, a fragile membrane that causes loose behavior, and a brief half-life (Opatha et al. 2020).

It is possible to construct structures with the benefits of liposomes and emulsions to obtain the necessary protection and release of bioactive substances. Due to their advantageous characteristics, including simple manufacturing processes, low cost, and scalability, these nano-vehicles can be employed for industrial applications involving food and agroindustry. The pharmaceutical business and, more recently, the food industry uses these nanocarriers extensively, significantly impacting this industry's short-term commercialization (Kesharwani et al. 2014).

Dendrimers Nanocarriers

Dendron, which means "tree" in Greek, is where the name "dendrimer" (Figure 2) originates. This makes sense, given a typical structure that consists of a series of branching units. According to definitions, dendrimers are artificial macromolecules with high branching points, three-dimensional spheres, monodispersity, and size in the nanometer range. They are also referred to in the literature as "dendritic molecules," "arborous," and "cascade molecules," or "metric nanoarchitecture," which refers to nanoscopic scale and monodispersity (Kesharwani et al. 2014).

With sizes ranging from 5 kDa to 500 kDa, dendrimers are a distinct family of highly branched polymeric nanocarriers with ordered tree-like architectures and minimal polydispersity. They consist of a core that radiates a number of repeating branching units that end in chemical groups ready for functionalization (Chariou et al. 2020). A dendritic polymer called dendrimers may be constructed with clearly defined molecular architectures (Kesharwani et al. 2014).

There are new approaches to the use of dendrimer architecture in agriculture. Poly (etherhydroxylamine) PEHAM dendrimers have increased solubility in formulations of active agrochemicals. A range of plant activities has been enhanced by making active elements in plants or

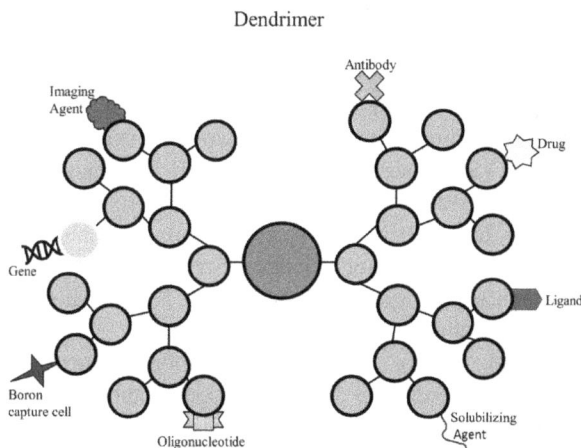

Figure 2. Scheme of a dendrimer, figure adapted from the work by Wani et al. (2019). With the permission of Elsevier.

seeds more water-resistant or by reducing the enzymatic breakdown of active elements. Researchers looked at the functional components of PETGE, PETriGE, and TMPTGE dendrimers. It was discovered that the PEHAM dendrimers improved the agriculturally active ingredient's efficacy (Hayes et al. 2010; Patel et al. 2022).

Peptides

Because of their protein-like chain structure, polypeptides and their derivatives have characteristics that make them both biodegradable and biocompatible. These encourage its usage in various applications, including gene therapy, regenerative medicine, and medication delivery. Due to its water solubility and simple manufacturing, polypeptide poly(aspartic acid) (PAsp) has received much attention among other polypeptides. As a sand fixing agent or an ecologically acceptable scaling inhibitor, PAsp is widely utilized due to its molecular chain's abundance of carboxylic and amide groups. By preventing the nitrification and ammoniation of urea in soil, PAsp may also be used as a controlled-release chemical in agriculture to improve plant nutrient absorption and lessen nitrogen loss (Lu et al. 2016).

L-aspartic acid-based crosslinked peptide nanoparticles have been used to distribute fertilizer slowly into the soil. The fertilizer's continuous release in the soil was accomplished and after 30 days, the nitrogen and phosphorus released were 79.8% and 64.4%, respectively. Additionally, the crosslinked polypeptide nanoparticles served as superabsorbents, enhancing the soil's ability to store and retain water. The authors reported a water-holding capacity of 81.8% for 200 g of soil with 1.5 g of superabsorbent, and the water-retention capacity remained at 22.6% after 23 days (Lu et al. 2016; Machado et al. 2022).

Lignin-based Nanocarrier

In order to create lignin-based nanocarriers for the prolonged release of a fungicide with better leaf adherence, Liang et al. (2022) described a unique method without chemical crosslinking. Nowadays, with the swift advancement of nanotechnology in several industries, it presents a promising strategy for enhancing the growth of sustainable agriculture. The design of nanocarriers exhibits improved performance in terms of water-dispersible permeability, duration, and effectiveness when compared to traditional pesticide formulations. As a result, many kinds of inorganic materials, including zinc oxide, metal-organic frameworks, and calcium carbonate, as well as synthetic polymers, chitosan from biobased sources, and cellulose have been employed to create pesticide nanocarriers.

Liang et al. (2022) prepared pesticide nanocarriers with improved water dispersion utilizing the raw material lignin sulfonates (LS), which are negatively charged. However, because of LS's potent hydrophilicity, nanoparticle production in water is not encouraged. Thus, to create benzoic acid-esterified lignin sulfonates (BLS) with variable hydrophobicity, LS was used to esterify benzoic acid, a generally safe and widespread agricultural and food preservative. By examining whether BLS can dissolve in DCM at different levels of substitution, it may be determined whether BLS is amphiphilic. The fungicide was loaded using two different techniques. The results of the scanning electron microscope (SEM) are shown in Figure 3. They carefully studied retention and foliar distribution in cucumber leaves and peanut leaves, selected the ideal nanoparticles with the best formulation stability, and tested their release performance, photostability, and biological activity *in vitro* or *in vivo* against the model pathogen strawberry anthracnose (*Colletotrichum gloeosporioides*), and biological activity *in vitro* or *in vivo*. This revealed the possibility of using BLS nanocarriers for pesticide delivery.

According to Liang et al. (2022), commercial fungicide microemulsions could not match the length, photostability, effectiveness, or foliar retention of the fungicide-loaded nanoparticle formulation created by benzoyloxylation of LS utilizing a solvent exchange or solvent evaporation approach. Although this formulation offered advantages over conventional pesticide formulations and was a "green" method because it was water-based, free of organic solvents, and included fewer

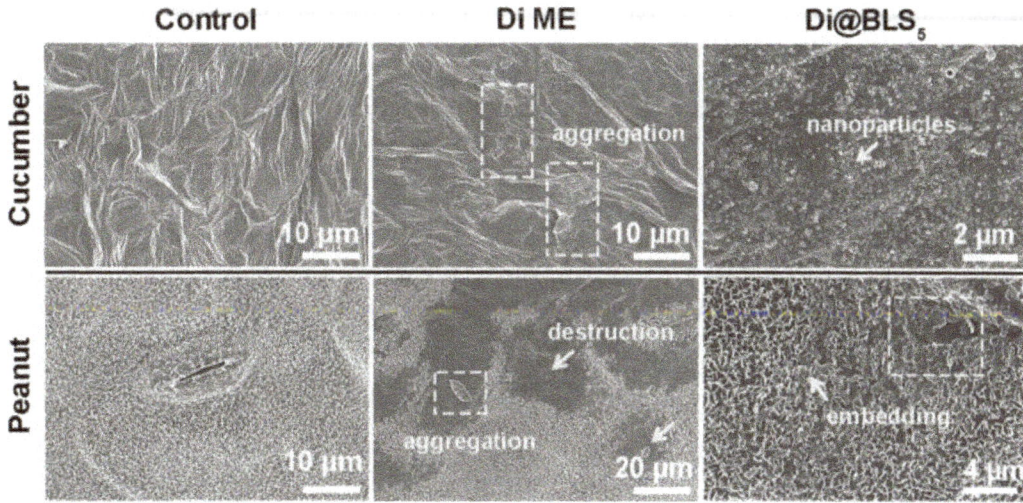

Figure 3. SEM of leaf surfaces treated with the same concentrations of Di ME and Di@BLS5 on cucumber and peanut plants. With permission from the International Journal of Biological Macromolecules from the work of Liang et al. (2022).

surfactants, the danger of contamination posed by this nanoformulation technology must also be considered.

Possible Toxicity of Nanomaterials of Natural Origin

Despite the fact anticipated that the exposure will be significantly reduced if the nanoparticle stays attached to the packaging materials or coats the surfaces of the food preparation equipment or packaging materials. A serious safety risk is the migration of nanoparticles within the meal, which causes their appearance. The main factor influencing attention to safety issues is the free forms of nanoparticles. Nanocapsules, which are frequently made of lipids and peptide monomers, are a type of nano-delivery mechanism.

Although there is a need for further studies on these materials and their full consolidation in use in agriculture and food production, efforts by several researchers have been made to discover the potential of these nanomaterials (Okeke et al. 2022). As a result of larger plant residues, these formulations may enhance human exposure (Figure 4). The continuous use of nanotechnologies

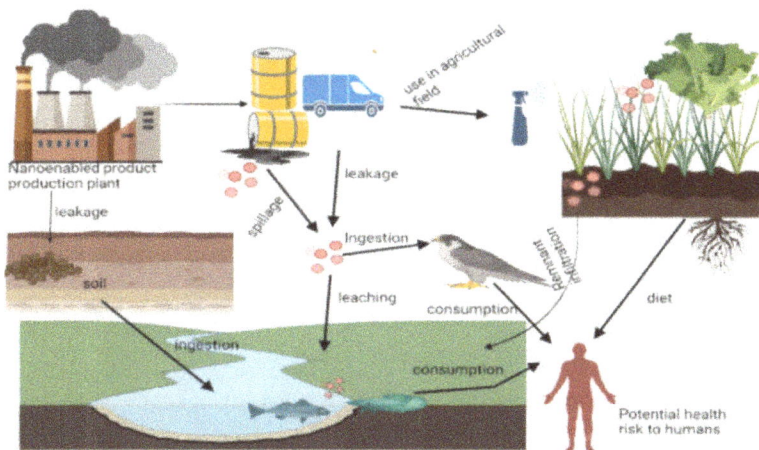

Figure 4. Routes via which products with nanotechnology come into touch with people. With the permission of Environmental Pollution (Okeke et al. 2022).

at many stages of the food production process is the most unexpected discovery. It is clear that a range of ways is being used, from processing procedures to the addition of inert or encapsulating nanoparticles to food items, including nanoformulation agricultural chemicals. The type/nature of the application may be used as a first estimate for the evaluation of consumers' varied risk assessments by determining their possible exposure (Okeke et al. 2022).

The majority of the risk analysis for nanomaterials used in agrochemicals is still being carried out in laboratories and not enough risk analysis has been done for actual application situations (An et al. 2022). Different application contexts might produce different outcomes. Zhang et al. (2021) suggested combining nano-informatics techniques with existing crop productivity and nutrient cycling models to enhance targeting, uptake, delivery, nutrient capture, and long-term effects on soil microbial communities in order to create nanoscale agrochemicals with the best safety and functional properties (An et al. 2022).

Conclusions

We have seen that many materials of natural origin or coupled with inorganic materials can be promising for the delivery of agrochemicals to plants, which can be good for increasing crop production, as well as avoiding contamination of groundwater by undesirable substances, focusing only on plant disease. Further studies on the possible toxic effects of these substances in humans should not be discarded before they are consolidated in the industry; more studies must be developed to give us greater security in the use of nanomaterials.

References

Ahmad, B., Zaid, A., Jaleel, H., Khan, M.M.A. and Ghorbanpour, M. 2019. Nanotechnology for phytoremediation of heavy metals: mechanisms of nanomaterial-mediated alleviation of toxic metals. pp. 315–327. In Advances in Phytonanotechnology. Academic Press.

An, C., Sun, C., Li, N., Huang, B., Jiang, J., Shen, Y. and Wang, Y. 2022. Nanomaterials and nanotechnology for the delivery of agrochemicals: strategies towards sustainable agriculture. J. Nanobiotechnology 20(1): 1–19.

Borišev, I., Borišev, M., Jović, D., Župunski, M., Arsenov, D., Pajević, S. and Djordjevic, A. 2020. Nanotechnology and remediation of agrochemicals. pp. 487–533. In Agrochemicals Detection, Treatment and Remediation. Butterworth-Heinemann.

Gaur, N., Chaudhary, R. and Diwan, B. 2022. Nanoencapsulation for chemical intermediate, biocides, and bio-based binder. pp. 267–279. In Smart Nanomaterials for Bioencapsulation. Elsevier.

Griffin, S., Masood, M.I., Nasim, M.J., Sarfraz, M., Ebokaiwe, A.P., Schäfer, K.H. and Jacob, C. 2017. Natural nanoparticles: a particular matter inspired by nature. Antioxidants 7(1): 3.

Guha, T., Gopal, G., Kundu, R. and Mukherjee, A. 2020. Nanocomposites for delivering agrochemicals: A comprehensive review. J. Agric. Food Chem. 2020 Mar 4; 68(12): 3691–702.

Guleria, G., Thakur, S., Shandilya, M., Sharma, S., Thakur, S. and Kalia, S. 2022. Nanotechnology for sustainable agro-food systems: The need and role of nanoparticles in protecting plants and improving crop productivity. Plant Physiol. Biochem.

Hao, L., Lin, G., Lian, J., Chen, L., Zhou, H., Chen, H., Xu, H. and Zhou, X. 2020. Carboxymethyl cellulose capsulated zein as pesticide nano-delivery system for improving adhesion and anti-UV properties. Carbohydr. Polym. 2020 Mar 1; 231: 115725.

Hayes, R.T., Owen, J.D., Chauhan, A.S. and Pulgam, V.R. 2010. PEHAM Dendrimer for Use in Agriculture. WO/2011/053605.

Jiang, Q., Peng, M., Yin, M., Shen, J. and Yan, S. 2022. Nanocarrier-loaded imidaclothiz promotes plant uptake and decreases pesticide residue. Int. J. Mol. Sci. 23(12): 6651.

Kesharwani, P., Jain, K. and Jain, N.K. 2014. Dendrimer as nanocarrier for drug delivery. Prog. Polym. Sci. 39(2): 268–307.

Liang, W., Zhang, J., Wurm, F.R., Wang, R., Cheng, J., Xie, Z. and Zhao, J. 2022. Lignin-based non-crosslinked nanocarriers: A promising delivery system of pesticide for development of sustainable agriculture. Int. J. Biol. Macromol. 220: 472–481.

Lu, S., Feng, C., Gao, C., Wang, X., Xu, X., Bai, X. and Liu, M. 2016. Multifunctional environmental smart fertilizer based on L-aspartic acid for sustained nutrient release. J. Agric. Food Chem. 64(24): 4965–4974.

Machado, T.O., Beckers, S.J., Fischer, J., Sayer, C., de Araújo, P.H., Landfester, K. and Wurm, F.R. 2021. Cellulose nanocarriers via miniemulsion allow Pathogen-Specific agrochemical delivery. J. Colloid Interface Sci. 601: 678–688.

Machado, T.O., Grabow, J., Sayer, C., de Araújo, P.H., Ehrenhard, M.L. and Wurm, F.R. 2022. Biopolymer-based nanocarriers for sustained release of agrochemicals: A review on materials and social science perspectives for a sustainable future of agri- and horticulture. Adv. Colloid Interface Sci. 102645.

Mujtaba, M., Khawar, K.M., Camara, M.C., Carvalho, L.B., Fraceto, L.F., Morsi, R.E. and Wang, D. 2020. Chitosan-based delivery systems for plants: A brief overview of recent advances and future directions. Int. J. Biol. Macromol. 154: 683–697.

Musante, C.J., Ramanujan, S., Schmidt, B.J., Ghobrial, O.G., Lu, J. and Heatherington, A.C. 2017. Quantitative systems pharmacology: a case for disease models. Clin. Pharmacol. Ther. 2017 Jan; 101(1): 24–7.

Nadiminti, P.P., Dong, Y.D., Sayer, C., Hay, P., Rookes, J.E., Boyd, B.J. and Cahill, D.M. 2013. Nanostructured liquid crystalline particles as an alternative delivery vehicle for plant agrochemicals. ACS Appl. Mater. Interfaces 5(5): 1818–1826.

Nuruzzaman, M.D., Rahman, M.M., Liu, Y., Naidu, R. 2016. Nanoencapsulation, nano-guard for pesticides: a new window for safe application. J. Agric. Food Chem. 2016 Feb 24; 64(7): 1447–83.

Okeke, E.S., Ezeorba, T.P.C., Mao, G., Chen, Y., Feng, W. and Wu, X. 2022. Nano-enabled agrochemicals/materials: Potential human health impact, risk assessment, management strategies and future prospects. Environmental Pollution 295: 118722.

Opatha, S.A.T., Titapiwatanakun, V. and Chutoprapat, R. 2020. Transfersomes: A promising nanoencapsulation technique for transdermal drug delivery. Pharmaceutics 12(9): 855.

Pacios-Michelena, S., García-García, J.D., Ramos-González, R., Chávez-González, M., Laredo-Alcalá, E.I., Govea-Salas, M. and Ilyina, A. 2022. Nanocarrier-based formulations: Concepts and applications. Bio-Based Nanoemulsions for Agri-Food Applications 413–439.

Piazzini, V., Lemmi, B., D'Ambrosio, M., Cinci, L., Luceri, C., Bilia, A.R. and Bergonzi, M.C. 2018. Nanostructured lipid carriers as promising delivery systems for plant extracts: The case of silymarin. Appl. Sci. 8(7): 1163.

Rani, T.S., Nadendla, S.R., Bardhan, K., Madhuprakash, J. and Podile, A.R. 2020. Chitosan conjugates, microspheres, and nanoparticles with potential agrochemical activity. pp. 437–464. In Agrochemicals Detection, Treatment and Remediation. Butterworth-Heinemann.

Sampathkumar, K., Tan, K.X. and Loo, S.C.J. 2020. Developing nano-delivery systems for agriculture and food applications with nature-derived polymers. Iscience 23(5): 101055.

Sharma, A., Sood, K., Kaur, J. and Khatri, M. 2019. Agrochemical loaded biocompatible chitosan nanoparticles for insect pest management. ISBAB 18: 101079.

Singh, R.P., Handa, R. and Manchanda, G. 2021. Nanoparticles in sustainable agriculture: An emerging opportunity. Journal of Controlled Release 329: 1234–1248.

Srivastav, A., Yadav, K.K., Yadav, S., Gupta, N., Singh, J.K., Katiyar, R. and Kumar, V. 2018. Nano-phytoremediation of pollutants from contaminated soil environment: current scenario and future prospects. pp. 383–401. In Phytoremediation. Springer, Cham.

Sunday, O.E., Chidike, E.T.P., Mao, G., Chen, Y., Feng, W. and Wu, X. 2021. Nano-enabled agrochemicals/materials: Potential human health impact, risk assessment, management strategies and future prospects. Environ. Pollut. 118722.

Wani, T.A., Masoodi, F.A., Baba, W.N., Ahmad, M., Rahmanian, N. and Jafari, S.M. 2019. Nanoencapsulation of agrochemicals, fertilizers, and pesticides for improved plant production. pp. 279–298. In Advances in Phytonanotechnology. Academic Press.

Zabot, G.L., Schaefer Rodrigues, F., Polano Ody, L., Vinícius Tres, M., Herrera, E., Palacin, H. and Olivera-Montenegro, L. 2022. Encapsulation of bioactive compounds for food and agricultural applications. Polymers 14(19): 4194.

Zhang, P., Guo, Z., Ullah, S., Melagraki, G., Afantitis, A. and Lynch, I. 2021. Nanotechnology and artificial intelligence to enable sustainable and precision agriculture. Nat. Plants 7(7): 864–876.

Chapter 6

Nanonutrients and Their Delivery in Crop Plants

Pratik Jagtap,[1] *Rajesh Raut,*[1,*] *Anup Sonawane,*[4] *Mahendra Rai*[2,3] and
Aniket Gade[2,3,4]

Introduction

Agriculture is a critical sector in many developing countries, contributing significantly to population dependency and economic values. However, agriculture is facing numerous unusual threats worldwide, including global warming, greenhouse effects, climate change, rapid urbanization eutrophication, and pollution across the biosphere. The situation is likely to intensify due to the population explosion, leading to deteriorating circumstances in the agricultural sector (Kaini 2020). This scenario is critical, especially for developing nations that rely on agriculture as the backbone of their national economy (Prasad et al. 2017).

To address these challenges, significant emphasis should be placed on adopting emerging technologies for efficient crop production systems, including nutrient management, good plant protection practices, enhancing photo-capturing mechanisms, precision agriculture methods, digital sensing, and more (Mittal et al. 2020). The agriculture industry faces several challenges, including increasing demand for food, climate change, and sustainability. However, technological advancements and a growing focus on sustainable agriculture practices offer opportunities to create a more efficient, productive, and sustainable industry (Bongiovanni and Lowenberg-Deboer 2004a; Mittal et al. 2020). Agriculture is confronted with numerous global challenges that impede its productivity, sustainability, and profitability. Changes in weather patterns and increasing temperatures are leading to droughts, floods, and unpredictable weather events, thus adversely impacting crop yields, soil health, and water availability (van Esse 2022). The overuse of chemical fertilizers, pesticides, and intensive farming practices have resulted in soil degradation diminishing soil fertility, therefore jeopardizing the long-term sustainability of agricultural production (Godfray et al. 2010). Water scarcity is a significant issue, particularly in major agricultural regions worldwide due to overuse and depletion of groundwater reserves, pollution, and climate change. Plant diseases and pests can also cause considerable crop damage, resulting in yield losses and decreased overall agricultural productivity. The loss of biodiversity is a growing concern that has negative effects on

[1] Department of Botany, The Institute of Sciences, Dr. Homi Bhabha State University, Mumbai-400032, Maharashtra, India.
[2] Nanobiotechnology Laboratory, Department of Biotechnology, Sant Gadge Baba Amravati University, Amravati, Maharashtra, India – 444602, Maharashtra, India.
[3] Department of Microbiology, Nicolaus Copernicus University, 87–100 Torun, Poland.
[4] Department of Biological Science and Biotechnology, DBT-ICT Centre for Energy Biosciences, Institute of Chemical Technology, Matunga, Mumbai – 400019.
* Corresponding author: rajesh.w.raut@gmail.com

soil health, crop production, and the ecosystem services provided by agriculture. These challenges pose substantial risks to agricultural productivity, food security, and the sustainability of the industry (Cappelli et al. 2022). Addressing these challenges requires a comprehensive approach, including sustainable land-use practices, innovative farming technologies, and policies that support small-scale farmers and natural resource conservation. Novel technologies such as precision agriculture, digital sensing, and remote sensing are being increasingly used to enhance crop management and achieve sustainable agriculture (Mittal et al. 2020).

In addition, the adoption of precision agriculture technologies, such as remote sensing, GPS, and variable-rate applications, can help farmers improve their efficiency, reduce waste, and maximize yields (Sishodia et al. 2020). Other emerging technologies, such as artificial intelligence and blockchain, can also provide significant benefits to the agriculture industry, such as improving supply chain management and ensuring food safety and traceability (Jha et al. 2019).

Furthermore, there is a growing interest in alternative and sustainable farming practices, such as organic farming, agroforestry, and regenerative agriculture. These methods aim to improve soil health, biodiversity, and ecosystem services, while reducing the use of chemicals and enhancing climate resilience (Elemike et al. 2019). Such practices can also improve the livelihoods of small-scale farmers and support rural development. In conclusion, the agriculture industry is facing several challenges that require urgent attention and action. A multifaceted approach is needed, including the adoption of innovative technologies, sustainable farming practices, and supportive policies to ensure the productivity, sustainability, and profitability of the industry while addressing global challenges such as climate change, food security, and environmental degradation.

The aim of this chapter is to provide an overview of nanonutrients and their delivery in crop plants. It will explore the concept of nanonutrients, their properties, and how they can be utilized to enhance crop growth and yield. The chapter will also discuss the various methods of nanonutrient delivery in crop plants, including foliar spraying, seed priming, and soil application. Additionally, the potential benefits and challenges of using nanonutrients in agriculture will be examined along with their impact on soil health, plant growth, and the environment. Overall, this chapter aims to provide a comprehensive understanding of nanonutrients and their potential as a tool for sustainable agriculture.

Challenges Related to Traditional Fertilizers

The use of traditional chemical fertilizers in agriculture can have significant benefits for plant growth and yield. However, their overuse and potential negative impacts on soil, water, biodiversity, and human health should be carefully considered. Sustainable agriculture practices that incorporate organic and natural fertilizers can be a viable alternative to chemical fertilizers, promoting long-term soil health and environmental sustainability (Kakar et al. 2020).

There are several potential disadvantages associated with the overuse of chemical fertilizers, including:

1. *Soil Degradation*: Overuse of chemical fertilizers can lead to soil degradation, reducing soil quality and decreasing the long-term productivity of farmland. This can be caused by an imbalance in soil nutrients, leading to soil acidity, salinity, or compaction (Dar et al. 2021).

2. *Environmental Pollution*: Excess nutrients from chemical fertilizers can leach into groundwater, causing water pollution and eutrophication. Additionally, the production and transportation of chemical fertilizers can result in greenhouse gas emissions and contribute to climate change (Dar et al. 2021).

3. *Negative Impact on Biodiversity*: The overuse of chemical fertilizers can lead to a reduction in soil biodiversity, including beneficial microbes and fungi that support plant growth and health (Ghormade et al. 2011).

4. *Health Risks*: The use of chemical fertilizers can pose health risks to farmers and other agricultural workers who handle them regularly, as well as to consumers who consume crops that have been treated with them (Carvalho 2006).

On the other hand, recent studies have shown that the use of nanonutrients can help address some of the challenges associated with traditional fertilizers. Nanonutrients are particles that range in size from 1 to 100 nanometers and can be applied to crops in very small quantities (Tarafdar et al. 2015). These nanonutrients can enhance plant growth, increase nutrient uptake, and improve crop yields while minimizing the negative environmental impacts associated with traditional fertilizers (Ali et al. 2021).

While traditional chemical fertilizers have played a critical role in agriculture, the potential negative impacts associated with their overuse cannot be ignored. Sustainable agriculture practices that incorporate organic and natural fertilizers, as well as emerging technologies like nanonutrients, can help promote long-term soil health and environmental sustainability and ensure food security.

To address the challenges associated with traditional fertilizers, researchers are exploring new approaches to fertilization, such as the use of nanonutrients. Nanonutrients are tiny particles that can deliver nutrients directly to plant cells, promoting plant growth and development while reducing the need for traditional fertilizers.

One advantage of nanonutrients is their ability to provide targeted delivery of nutrients, which reduces the amount of fertilizer needed and minimizes potential negative impacts on soil and water quality (Kurczynska et al. 2021). In addition, nanonutrients have the potential to improve the efficiency of fertilizer use, as they can be designed to release nutrients slowly over time, reducing the frequency of application (Hajihashemi and Kazemi 2022a).

Recent investigations have shown promising results for the use of nanonutrients in crop plants. For example, the application of zinc oxide nanoparticles has been shown to improve the growth and yield of wheat plants under drought-stress conditions (Abbas et al. 2023). Similarly, the use of copper oxide nanoparticles has been shown to increase the yield of tomato plants by enhancing their nutrient uptake efficiency (Bakshi and Kumar 2021). While nanonutrients show great promise for sustainable agriculture, more research is needed to fully understand their potential benefits and risks. Factors such as nanoparticle size, shape, and surface properties can all affect their behavior in soil and their interactions with plant cells (Kurczynska et al. 2021). Careful consideration of these factors will be essential in developing safe and effective nanonutrient-based fertilizers for widespread use in agriculture.

Next-Generation Crop Nutrients

Nanoparticles have been gaining attention as potential nanonutrients to improve plant growth and development. These nanoparticles have unique physical and chemical properties that allow for targeted delivery and controlled release of nutrients to crops. They can be used as carriers for micronutrients and macronutrients, improving their bioavailability and uptake by plants, thus improving the efficiency of nutrient use in agriculture and reducing the amount of fertilizer needed (Wanyika et al. 2012).

Studies have shown that zinc oxide nanoparticles (ZnO NPs) can increase the concentration of zinc in plants, which is an essential micronutrient for plant growth (Hussain et al. 2018). ZnO NPs can be delivered to plants via foliar application or as a component of fertilizer. A recent study published in the Journal of Plant Growth Regulation demonstrated that foliar application of ZnO NPs improved the growth and yield of tomato plants by enhancing the uptake of zinc and other nutrients (Ahmed et al. 2023). Iron oxide nanoparticles (FeO NPs) have also been shown to enhance the solubility and availability of iron in soil. A study published in the Journal of Environmental Science and Pollution Research showed that the use of FeO NPs as a fertilizer improved the growth and yield of capsicum, as well as the iron content in the grains (Yuan et al. 2018).

In addition to micronutrients, nanoparticles can also be used to deliver macronutrients such as nitrogen, phosphorus, and potassium. Mesoporous silica nanoparticles (MSNPs) have been used as carriers for nitrogen fertilizers, resulting in improved growth and yield of wheat plants (Wanyika et al. 2012).

Nanoparticles can also be used to remediate contaminated soil by delivering substances that can bind to heavy metals or other pollutants. A study published in the Journal of Hazardous Materials showed that the use of chitosan nanoparticles as a soil amendment reduced the uptake of lead and cadmium by plants, thereby improving their growth and health (Ingle et al. 2022). While the use of nanoparticles as nanonutrients has shown promising results, their safety and efficacy are still being evaluated (Jiang et al. 2021). The effects of nanoparticles on human health, soil health, and the environment must be carefully studied and understood before their widespread use in agriculture. Additionally, the use of nanoparticles in agriculture must be carefully regulated to ensure their safe and sustainable use.

In conclusion, nanoparticles have the potential to revolutionize agriculture by improving the efficiency of nutrient use, reducing fertilizer usage, and enhancing crop yield and quality. However, further research is needed to fully understand their potential benefits and risks. By carefully regulating and monitoring their use, nanoparticles can be used in a safe and sustainable manner to contribute to sustainable agricultural development.

Nanotechnology in Agriculture

Nanotechnology has emerged as a promising field in agriculture for the development of more efficient and sustainable farming practices. One of the main applications of nanotechnology in agriculture is the use of nanoscale fertilizers, which have the potential to minimize nutrient losses through leaching and enhance nutrient use efficiency. Additionally, these fertilizers can be formulated to release nutrients in a controlled manner, which can prolong nutrient availability in the agro-environment and improve Nutrient Utilizing Efficiency (NUE) while reducing application frequency and labor cost (Reddy and Chhabra 2022).

Precision farming, which involves the use of technology to monitor and manage crop production, is also being revolutionized by nanotechnology. By using controlled farming practices that include targeted nutrient delivery, precision farming can help improve crop yields and enhance sustainable agricultural practices (Chen and Yada 2011). Agro-nanotechnology has also emerged as a leading crop management system for controlling production and enhancing food productivity, safety, and quality while reducing resource use and augmenting nanoscale nutrient uptake from the supporting substratum (Khot et al. 2012).

In addition to improving nutrient delivery and management, nanotechnology can also help reduce the spread of toxic agrochemicals and minimize nutrient fatalities, leading to increased yields and well-organized pest and nutrient management. The development of smart nutrient delivery systems in the form of nanofertilizers, nanopesticides, nanoherbicides, and nanosensors has opened up a new mode of applications for sustainability in crop production (Ghormade et al. 2011; Kashyap et al. 2015). Nanofertilizers can serve as nanonutrients themselves or as carrier matrixes for controlled release, making them a smart and efficient delivery system that reduces environmental pollution. For example, nanonutrients can be encapsulated inside a thin protective polymer film or emulsion of nanoscale dimensions, which can help deliver essential nutrients to crops (Ghormade et al. 2011).

Nanotechnology has promisingly provided nanostructured materials, which served as nanofertilizers or carrier matrices of the nanonutrients for controlled release as a smart and efficient delivery system at the cost of a reduction in environmental pollution. A nanofertilizer is considered a nanonutrient itself or a matrix loaded with nanonutrients that delivers the essential nutrients to the crops. For example, a thin protective polymer film or emulsion of nanoscale dimensions encapsulated the nanonutrients inside or coated in it (Ghormade et al. 2011). The development of smart nutrient delivery systems in the form of nanofertilizer, nanopesticide, nanoherbicide, and nanosensor has opened up a new mode of applications for sustainability in the crop production of the agricultural practice (Kah and Hofmann 2014; Kah 2015).

Why Nanonutrients are Essential in Agriculture?

Nanonutrients are a new class of nutrient delivery systems that utilize nanotechnology to enhance the efficiency of nutrient uptake by plants. Nanoparticles are used as carriers for nutrients, allowing for targeted and controlled release of nutrients to plants. Nanonutrients have the potential to revolutionize agriculture by increasing crop yields, reducing fertilizer usage and environmental pollution, and improving plant resistance to abiotic and biotic stresses (Ingle et al. 2022).

The use of nanonutrients in agriculture is still in the early stages of research and development, but the promising results obtained from studies suggest that nanonutrients have great potential in improving plant growth and productivity. However, there are also concerns about the potential risks associated with the use of nanoparticles in agriculture, such as their potential toxicity and environmental impact. Therefore, it is important to carefully evaluate the benefits and risks of nanonutrients before widespread adoption in agriculture. Overall, nanonutrients have the potential to play a significant role in sustainable agriculture and further research and development in this area are needed to fully realize their potential (Ali et al. 2021).

Nanocarriers and Their Role in Agriculture

Nanonutrients, which utilize nanotechnology to enhance nutrient uptake by plants, have the potential to revolutionize agriculture. They are a new class of nutrient delivery systems that use nanoparticles as carriers for nutrients, allowing for targeted and controlled release of nutrients to plants. By improving plant growth and productivity, nanonutrients could increase crop yields, reduce fertilizer usage, and improve plant resistance to abiotic and biotic stresses (El-Saadony et al. 2021). While nanonutrients are still in the early stages of research and development, promising results from studies suggest that they could be beneficial for sustainable agriculture. However, there are also concerns about the potential risks associated with the use of nanoparticles in agriculture, such as their toxicity and environmental impact. Therefore, it is important to carefully evaluate the benefits and risks of nanonutrients before widespread adoption in agriculture (Nemc et al. 2021).

The development of nanocarriers for nanonutrient delivery systems in agriculture has the potential to improve the efficiency of nutrient delivery, reduce environmental impact, and increase crop yields. This could have a significant impact on food security and sustainability in the future. The development of nanocarriers for Nanonutrient delivery systems in agriculture has the potential to revolutionize the way we deliver and utilize nutrients in plants. Several Advantages of the Nanonutrient delivery are discussed in the Table 1 which potentially focused on the sustainable agricultural practices.

Nanocarriers, which are tiny particles designed to carry and deliver substances, such as nutrients and pesticides to plants, are used in agriculture to improve the delivery and efficacy of agricultural inputs and reduce waste and pollution associated with traditional practices. The development of nanocarriers for nanonutrient delivery systems in agriculture has the potential to further improve nutrient delivery efficiency, reduce environmental impact, and increase crop yields. This could have a significant impact on food security and sustainability in the future (Zulfiqar et al. 2019). Moreover, recent studies have shown that the use of nanonutrients and nanocarriers in agriculture can also improve plant resistance to biotic stresses such as diseases and pests (Ali et al. 2021; Dutta et al. 2022a). The targeted delivery of pesticides through nanocarriers can reduce the amount of chemicals required for effective pest control, minimizing environmental damage, and increasing the safety of agricultural workers.

Table 1. Advantages of nanocarriers as nanofertilizers (Cui et al. 2010).

Sr No.	Quality Factors	Advantages
1.	Improved Nutrient Uptake	Through engineering, nanocarriers can be customized to deliver nutrients directly to plant cells, leading to an increase in the efficiency of nutrient uptake. As a result, plants can grow faster and produce higher yields, while also reducing the amount of fertilizer waste.
2.	Reduced Environmental Impact	By design, nanocarriers can be developed to release nutrients gradually, which reduces the likelihood of nutrient leaching into groundwater or runoff into adjacent water sources. This feature can aid in minimizing the environmental impact of fertilizer use in agriculture.
3.	Targeted Delivery	Through intentional design, nanocarriers can be tailored to target specific tissues or organs within a plant, resulting in more precise nutrient delivery. This feature can prove to be especially valuable for delivering nutrients to roots, where they are most essential.
4.	Protection from Degradation	Nanocarrier-delivered nutrients are shielded from degradation by soil enzymes or other factors, enhancing their bioavailability and effectiveness.
5.	Reduced Fertilizer Use	Nanocarriers can aid in delivering nutrients with better efficiency, which can help decrease the quantity of fertilizer required to obtain equivalent crop yields. This can lead to cost savings for farmers and lessen environmental harm.

However, there are still concerns regarding the safety of nanonutrients and nanocarriers in agriculture. Recent research has highlighted the potential for nanoparticles to accumulate in the environment and have negative impacts on soil health and biodiversity (Bongiovanni and Lowenberg-Deboer 2004b). Additionally, there is limited knowledge of the potential long-term effects of nanonutrient use on plant growth and productivity.

Difference Between Nanofertilizers and Conventional Fertilizers

The use of nanofertilizers in agriculture has the potential to improve crop yields, reduce the amount of fertilizer needed, and minimize environmental pollution. However, more research is needed to fully understand the long-term effects of nanofertilizers on soil health, plant growth, and environmental impact before widespread adoption in agriculture. Nanofertilizers could be a substantial revolution for agricultural practices, which serves the larger surface area and minute size could consent for improved interaction with crop systems and efficient uptake ability of nutrients for crop fertilization (Bongiovanni and Lowenberg-Deboer 2004b; Ali et al. 2021).

Nanofertilizers in agriculture are a relatively new concept that holds great potential for revolutionizing traditional farming practices. Nanofertilizers, which utilize nanotechnology to improve nutrient delivery and uptake by crops, have been shown to have a much larger surface area than traditional fertilizers, allowing for more efficient interaction with crop systems. The minute size of nanofertilizers also enables more precise and targeted delivery of nutrients to crops, which can improve the efficiency of nutrient uptake by plants (Ali et al. 2021).

Recent studies have highlighted the benefits of using nanofertilizers in crop production. For example, research has shown that the use of zinc oxide nanoparticles as fertilizer can increase the yield and nutritional quality of wheat crops while reducing the amount of fertilizer needed (Abbas et al. 2023). Similarly, the use of iron oxide nanoparticles as a fertilizer has been shown to increase the yield and quality of crops, as well as improve the plant's resistance to environmental stresses (Rui et al. 2016).

Integrated nanotechnology in nutrient fertilizer production strengthens the uptake efficiency, cost efficiency, smart governance, and possible environmental benefits. A chief attributional property of nanoformulation, i.e., discussed in the Table 2 significantly emphasized, the surface area can lead to greater reactivity and faster dissolution capacity might exacerbate the inefficiency problems of conventional fertilizers that could cause a nuisance in the environment. This chapter emphasizes some

Table 2. Comparison of nanofertilizers formulations over the conventional chemical fertilizers application on the crops (Hussain et al. 2019).

Sr No.	Properties	Nanofertilizers Formulations	Conventional Chemical Fertilizers
1.	Bioavailability	Nanosizing nutrients can improve their solubility, dispersion, fixation, and bioavailability. This is due to the increased surface area of nanoparticles, which enhances their interactions with crop systems and improves plant uptake efficiency (Fernandes et al. 2021).	The bioavailability to crop plants can be hindered by large particle size and low solubility.
2.	Efficiency of Uptake	The use of nanostructured formulations can improve plant uptake efficiency by optimizing the particle size and uptake ratio.	Efficiency in nutrient uptake is created by structured composites that are bulky.
3.	Controlled Release Mechanism	The semipermeable matrix enabled the regulated discharge of water-soluble nutrients by encapsulating nanonutrients.	If used excessively, chemically active components can uncontrollably leach into the water and potentially lead to contamination.
4.	Extent of the Nutrient Release	Nanostructured matrixes or emulsions with controlled release systems can maintain sustained release for an extended period after application.	The extent to which unused chemical fertilizers are available for uptake depends on their efficiency, and any excess may leach out from the soil into groundwater.
5.	Loss Rate of Nutrients	By employing a controlled release system through nanostructured formulations, the extent of loss of nanofertilizers is minimized, and the occurrence of leaching and leaking can be prevented.	The most significant loss of applied fertilizers occurs through leaching and rain runoff carried by water currents.

renovated investigations regarding the effect of nanoparticles, nanomaterials, and nanocomposites on the cultivation of methods emerging with the newer techniques and implementation of efficient astonishing crop management. It also focused on the discussion of efficient delivery systems that are used in agricultural practices.

Recent research has shown that integrated nanotechnology in nutrient fertilizer production can enhance uptake efficiency, cost efficiency, and smart governance and offer potential environmental benefits (Rai et al. 2015; Bairwa et al. 2023). However, the increased surface area of nanoformulation, which can lead to greater reactivity and faster dissolution capacity, may exacerbate the inefficiency problems of conventional fertilizers and potentially cause environmental harm (Kumar et al. 2021). To address these concerns, several studies have investigated the effects of nanoparticles, nanomaterials, and nanocomposites on crop cultivation, employing newer techniques and efficient crop management strategies (Gogos et al. 2012; Jiang et al. 2021). Efficient delivery systems have also been explored in agricultural practices. This chapter provides an overview of the potential of nanoscience and emerging nanotechnology to revolutionize crop management and improve the quality of crop production.

Why Noncarriers Are Momentously Influencing the Agriculture?

The concerning attributes of nanoparticles may alter impending agricultural practices. The inclusion of nanocarriers has had a significant influence in the field, promoting the use of more sustainable practices and helping to maintain high yields to feed populations. Nanocarriers can be used to deliver pesticides and fertilizers to crops, which can increase their efficiency and reduce their environmental impact. For example, nanoparticles can be used to encapsulate and deliver herbicides or insecticides directly to the pests, while minimizing the impact on the surrounding environment. In addition, it can deliver nutrients and growth-promoting substances to seeds. For instance, nanoparticles can be used to deliver micronutrients, such as iron, zinc, and copper to plants, which can improve their growth and health. Nanocarriers can be functionalized with antibodies that bind to specific plant pathogens, allowing for early detection and prevention of plant diseases (Shakiba et al. 2020).

Nanoparticles have the potential to significantly impact agriculture practices. Nanocarriers can be used to deliver pesticides and fertilizers to crops, increasing their efficiency, and minimizing their environmental impact. Additionally, nanocarriers can deliver nutrients and growth-promoting substances to seeds, such as micronutrients like iron, zinc, and copper, which can enhance plant growth and health. Nanocarriers can also improve food safety by detecting and removing contaminants from food. For instance, nanoparticles can detect pathogens or toxins in food or remove harmful substances from food during processing (Mittal et al. 2020).

Nanocarriers can be used to improve food safety by detecting and removing contaminants from food. For example, nanoparticles can be used to detect pathogens or toxins in food or to remove harmful substances from food during processing. There are concerns about the potential environmental and health risks of nanoparticles, and regulatory frameworks will need to be developed to ensure their safe use in agricultural applications. Recent studies have shown that the use of nanocarriers in agriculture can improve the uptake of nutrients by plants and increase their resistance to disease and pests (Mittal et al. 2020). Additionally, functionalized nanocarriers have been shown to enhance the detection of plant pathogens, improving disease management (Dutta et al. 2022a).

Selection of Ideal Type of Nanocarriers for Agricultural Applications

The specific ideal type of nanocarriers would depend on the specific application and the properties required for the nanocarriers to effectively deliver the desired payload to the target site. In general, ideal nanocarriers for agricultural applications should have the following characteristics (Vega-Vásquez et al. 2020):

1. *Biocompatibility and non-toxicity*: The nanocarriers should be biocompatible and non-toxic to plants and the environment. It should not have any negative impact on soil or water quality and should be safe for human consumption if it is used to deliver nutrients to crops.
2. *Targeted delivery*: The nanocarriers should be able to deliver the payload (e.g., pesticides, nutrients, enzymes) to the targeted site, such as the roots of the plants or the contaminated soil. This can be achieved through the functionalization of the nanoparticles with targeting ligands or by designing the nanoparticles to be responsive to certain environmental cues.
3. *Controlled release*: The nanocarriers should be able to control the release of the payload over a desired period. This can be achieved by designing the nanoparticles to degrade or release the payload in response to certain triggers, such as pH or temperature.
4. *Stability*: The nanocarriers should be stable under environmental conditions and should not degrade or lose their properties over time. This is important to ensure that the payload is delivered in the desired amount and at the desired time.
5. *Cost-effectiveness*: The nanocarriers should be cost-effective and scalable to produce for large-scale agricultural applications.

Some types of nanocarriers, like liposomes, polymeric nanoparticles, and dendrimers, have demonstrated potential in agricultural applications. However, the selection of the appropriate nanocarriers for a particular application requires consideration of its unique characteristics, and more research is required to develop ideal nanocarriers for widespread use in agricultural applications. Recent studies have shown promising results with the use of magnetic nanoparticles and clay nanoparticles as nanocarriers for delivering pesticides and nutrients to plants, respectively. Additionally, advances in nanotechnology have led to the development of smart nanocarriers that respond to external stimuli for controlled and targeted delivery of payloads in agricultural applications.

Types of Nanocarriers for Agriculture

Nanocarriers, which can be made from a range of materials such as lipids, polymers, dendrimers, and carbon nanotubes, have shown potential in agriculture by improving crop yields, reducing waste and pollution, and promoting sustainable farming practices. However, more research is needed to fully understand their safety, efficacy, and long-term environmental impact before their widespread adoption in agriculture (Dutta et al. 2022b).

Nanocarriers are an important tool in agriculture, providing a way to deliver nutrients and other substances to plants more effectively. The ideal nanocarriers for agricultural applications should have characteristics such as biocompatibility, targeted delivery, controlled release, stability, and cost-effectiveness.

The use of nanocarriers in agriculture can revolutionize the delivery of nutrients and other substances to plants. Some common types of nanocarriers used in agriculture included are shown in Figure 1 (Dutta et al. 2022b):

1. *Liposomes*: Small, spherical nanoparticles made of a lipid bilayer that can encapsulate a variety of substances. They have been used to deliver pesticides and fertilizers to plants.

2. *Polymeric Nanoparticles*: Nanoparticles made of polymers such as polyethylene glycol (PEG) or poly (lactic-co-glycolic acid) (PLGA). They can deliver nutrients, growth regulators, and pesticides to plants.

3. *Dendrimers*: Highly branched spherical nanoparticles that can encapsulate a variety of substances. They have been used to deliver nutrients, growth regulators, and pesticides to plants.

4. *Nanoemulsions*: Oil-in-water or water-in-oil emulsions with droplet sizes typically in the nanometer range. They can deliver pesticides and nutrients to plants, as well as improve the solubility of poorly soluble compounds.

5. *Nanotubes*: Cylindrical nanoparticles with a high aspect ratio. They can deliver nutrients, growth regulators, and pesticides to plants.

6. *Nanocapsules*: Small and spherical nanoparticles that can encapsulate a variety of substances. They have been used to deliver nutrients and pesticides to plants.

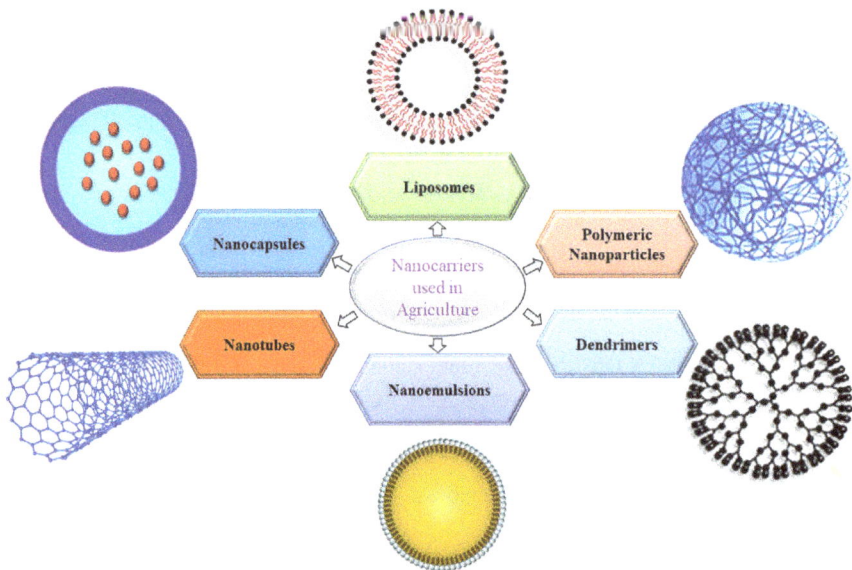

Figure 1. Types of nanocarriers commonly used in the agricultural practices.

Nanonutrients Delivery System in Agriculture

The nanonutrient delivery system is a rapidly growing field of nanotechnology-based approaches that involve the use of nanoparticles to deliver nutrients to plants in a controlled and targeted manner. This system has the potential to improve the efficiency and effectiveness of nutrient delivery to crops, leading to increased yields and improved crop quality. The nanoparticles used in nanonutrient delivery systems can be made from various materials, such as metals, metal oxides, and polymers. These nanoparticles can be functionalized with specific ligands that allow them to target specific parts of the plant, such as the roots or leaves. They can also be engineered to release nutrients in a controlled manner, based on environmental cues such as pH or temperature. Nanonutrient delivery systems have several potential advantages over traditional methods of nutrient delivery, such as foliar spraying or soil application. For example, recent studies have shown that nanonutrients can be more easily absorbed by plant roots, which can increase their bioavailability and effectiveness (Vega-Vásquez et al. 2020). Moreover, nanonutrients can reduce the number of nutrients required, as they are targeted to the specific part of the plant that needs them, reducing waste and potential environmental impacts (Karny et al. 2018). In addition, nanonutrient delivery systems can improve the uptake of essential micronutrients that are often deficient in the soil, such as iron (Yuan et al. 2018) and zinc (Hussain et al. 2018). This can lead to improved plant growth and crop yields. Therefore, nanonutrient delivery systems hold great promise for improving agricultural practices and promoting sustainable farming.

Methods of Application of Nanofertilizers in Crop Plants

Nanofertilizers can be applied to crops using various methods, including (Nongbet et al. 2022):

1. *Soil application*: Nanofertilizers can be mixed with soil or applied to the soil surface as a solution. This method allows for the gradual release of nutrients into the soil, providing a sustained supply of nutrients to plants.
2. *Seed treatment*: Nanofertilizers can be applied to seeds prior to planting, allowing for targeted delivery of nutrients to the developing seedlings.
3. *Foliar spray*: Nanofertilizers can be sprayed directly onto the leaves of plants. This method provides a quick supply of nutrients to the plants, but the effects may not last as long as with soil application.
4. *Hydroponics*: Nanofertilizers can be added to hydroponic nutrient solutions, providing a controlled and precise supply of nutrients to plants grown in soilless systems.

The choice of application method may depend on factors such as the type of nanofertilizers, the crop type, and the growth stage of the plant. It is important to follow recommended application dosage and timing to avoid over-application or under-application of nutrients, which can have negative impacts on plant growth and development (Goswami et al. 2019).

The specific mechanisms and pathways of nanoparticle uptake and transport in plants may vary depending on various factors, such as the type of nanoparticle, plant species, and environmental conditions. In general application of nanonutrients loaded on nanocarriers can be applied during crop plant management is illustrated as in Figure 2.

Nanoparticles can enter plant cells through various pathways, including via the root system or by foliar uptake. Once the nanoparticles encounter the plant surface, they can adhere to the outer surface or penetrate through the cell wall and plasma membrane. For root uptake, nanoparticles can be transported across the root cells and into the xylem or phloem vessels and then move upward into the plant tissue system. Nanoparticles can also be transported between cells through plasmodesmata, which are small channels that connect adjacent plant cells. Once inside the plant cells, nanoparticles can interact with various intracellular components, including organelles and proteins, which may result in changes in plant physiology, metabolism, and gene expression.

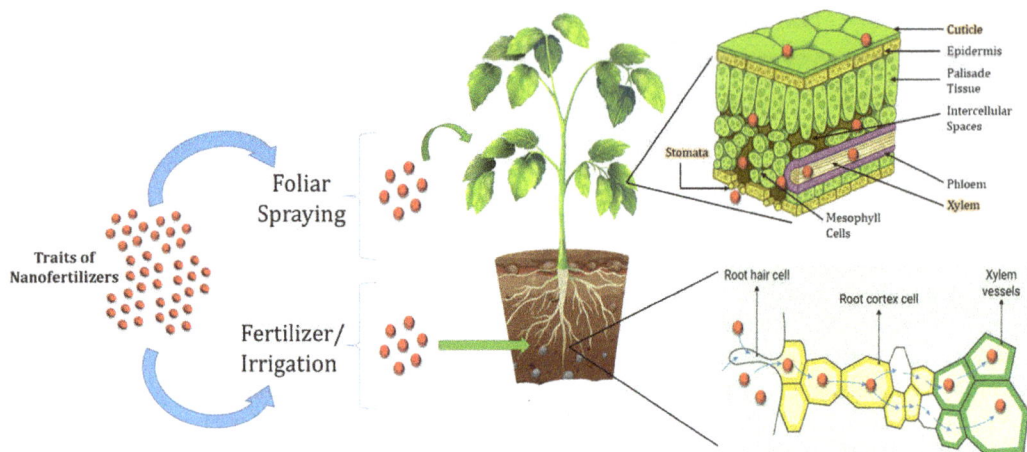

Figure 2. Basic modes of application of nanocarriers loaded with nanonutrients.

Recent research has shown that nanoparticles can enter plant cells through various pathways such as root uptake or foliar uptake (Ali et al. 2021). Upon contact with the plant surface, nanoparticles can either adhere to the outer surface or penetrate through the cell wall and plasma membrane.

Mechanism of Nanoparticle Uptake by Plants

The mechanism by which plants take up nanoparticles is still not fully understood, and there are several proposed mechanisms that may contribute to the uptake and translocation of nanoparticles by plants. One proposed mechanism is passive uptake, Figure 3 shows how probably nanoparticles enter plants through pores in the cell wall or cuticle (Pérez-de-Luque 2017). This mechanism is more likely to occur with smaller nanoparticles, which can more easily enter plant tissues.

Facilitated uptake is another proposed mechanism, in which nanoparticles are actively transported across the plant membrane through membrane transporters that are involved in the uptake of nutrients or other compounds. Demonstration cellular intake of nanoparticles are shown in Figure 4 which is a third proposed mechanism is endocytosis in which nanoparticles are engulfed by the plant cell membrane and transported through the endomembrane system (Khan et al. 2020).

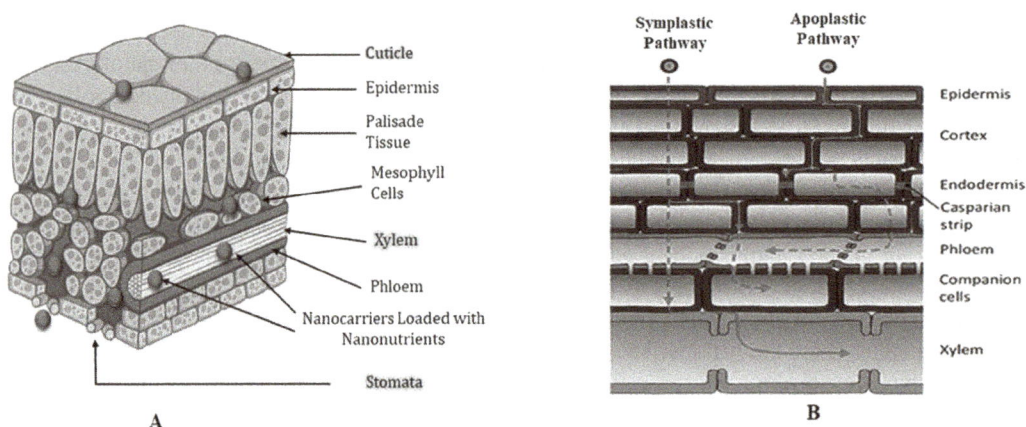

Figure 3. Nanoparticles enter plants through pores in the cell wall or cuticle. (A) Movement of nanocarrier nutrients passing through the cuticle or stomata toward the vascular bundles. (B) Schematic pathways of the nanocarrier nutrient's transport across cells.

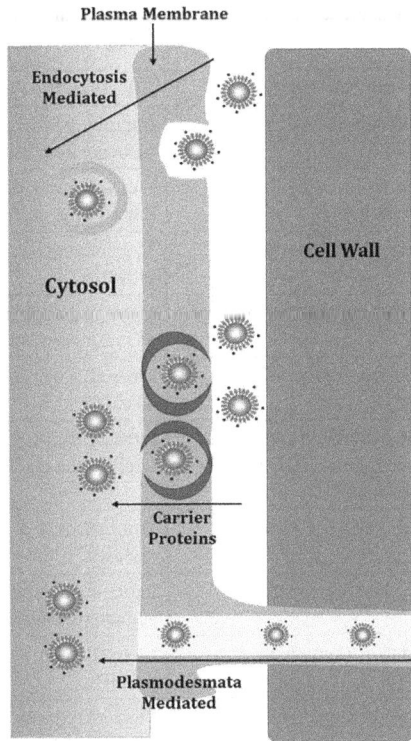

Figure 4. Transport of nanocarrier materials across the cell membrane for efficient delivery of nanonutrient to the biometabolic functions (Pérez-de-Luque 2017).

Once inside the plant, nanoparticles may be translocated through the xylem and/or phloem and distributed to different parts of the plant. The translocation of nanoparticles may be influenced by factors such as particle size, shape, surface charge, and chemical composition, as well as plant-specific factors, such as the structure and composition of plant tissues and the plant's physiological state (Lin et al. 2009).

Case Studies of Developing Nanocarriers Containing Nanonutrient/Organic Molecules

Studies have shown that nanocarriers containing nanonutrients and organic molecules have the potential to enhance plant growth and yield. The use of nano-chitosan-based fertilizers improved the uptake and utilization of nitrogen in maize plants, leading to increased plant growth, grain yield, and nitrogen content in the grain (Kurczynska et al. 2021). Another study published in the journal *Nanotechnology* developed carbon nanotube-based fertilizers as a carrier for phosphate to improve the phosphate nutrition of rice plants, resulting in increased plant growth, grain yield, and phosphate content in the grain (Lin et al. 2009). These case studies demonstrate the potential of nanocarriers as a delivery system for nanonutrients in agriculture. However, the safety and efficacy of these technologies are still being evaluated, and more research is needed to fully understand their benefits and risks. Table 3 provides a summary of some of the developed nanonutrient-containing nanocarriers for agricultural applications (Schiavi et al. 2021; Beig et al. 2022).

Table 3. Case studies of the effects of nanocarriers loaded with nanonutrients/fungicides/herbicides/plant growth regulators as compared to traditional fertilizer delivery system.

Nanocarriers	Crop Plant	Payload Nutrients	Effects on the Plants	Reference
Liposome	Rice	Micronutrients (Fe and Zn)	Improving plant growth and reducing iron and zinc deficiencies.	Dai et al. (2018)
	Tomato	Plant Growth Regulator Gibberellin	Increase fruit size and yield and improve the quality of the fruit.	Li et al. (2017)
	Wheat	Herbicide Glyphosate	More effective at controlling weed growth.	Kim et al. (2019)
	Soyabean	Plant Hormone Abscisic Acid	Improve drought tolerance and increase plant growth and yield.	Zhao et al. (2018)
	Maize	Fungicide Azoxystrobin	More effective at controlling fungal diseases.	Liu et al. (2019)
Polymeric nanoparticles	Tomato	Gibberellic Acid and Indole-3-Acetic Acid (Plant Growth Regulators)	Increase fruit size and yield, and improve the quality of the fruit.	Li et al. (2017)
	Cucumber	Fungicide Azoxystrobin	More effective at controlling fungal diseases.	Li et al. (2017)
	Lettuce	Iron Oxide Nanoparticles	Increase iron uptake by the plants and improve plant growth.	Zhang et al. (2018)
	Soyabean	Herbicide Imazethapyr	More effective at controlling weed growth.	Liu et al. (2018)
	Rice	Silicon Nanoparticles	More effective at improving plant growth and increasing rice yield.	Li et al. (2020)
Dendrimers	Wheat	Fungicide Thiabendazole	More effective at controlling fungal diseases.	Sharma et al. (2014)
	Maize	Deliver RNA Interference (RNAi) Molecules	Significantly reduce virus accumulation and disease severity in the plants.	Shepherd et al. (2018)
	Tomato	Silver Nanoparticles	Increase plant height, leaf area, and fruit yield.	Mukherjee et al. (2018)
	Soyabean	Zinc Oxide Nanoparticles	Enhance photosynthesis and nutrient uptake in plants.	Li et al. (2019)
	Rice	Herbicide Butachlor	More effective at controlling weed growth.	Lu et al. (2019)
Nanoemulsions	Grapevine	Mixture of Plant Essential Oils	More effective at controlling powdery mildew.	Panebianco et al. (2020)
	Tomato	Copper Nanoparticles	Increase plant height, leaf area, and fruit yield, and reduce bacterial population in the plants.	Huang et al. (2020)
	Strawberry	Plant Growth Regulator Gibberellin	Enhance plant height, flower number, and fruit yield.	Llorens et al. (2019)
	Maize	Herbicide Glyphosate	More effective at controlling weed growth.	Wang et al. (2019)
	Rice	Insecticide Imidacloprid	More effective at controlling rice plant hoppers.	Jiang et al. (2017)

Table 3 contd. ...

...Table 3 contd.

Nanocarriers	Crop Plant	Payload Nutrients	Effects on the Plants	Reference
Nanotubes	Wheat	Multi-Walled Carbon Nanotubes (MWCNT) With Chitosan,	Improved the germination and growth of wheat plants, as well as increased their tolerance to drought stress.	Tripathi et al. (2017)
	Arabidopsis	Single-Walled Carbon Nanotubes (SWCNT)	Enhanced plant growth and increased the activities of antioxidant enzymes, which help protect plants from oxidative stress.	Gupta et al. (2016)
	Tomato	Titanium Dioxide Nanotubes (TiO_2 Nts)	Improved the photosynthetic rate and chlorophyll content of the plants, leading to increased plant growth and yield.	Abdallah et al. (2019)
	Rice	Functionalized Carbon Nanotubes	Improved the root and shoot growth of rice plants, as well as increased grain yield.	Ahmed et al. (2017)
Nanocapsules	Lettuce	Polyelectrolyte Nanocapsules (Pncs)	Enhanced the root and shoot growth of lettuce plants, as well as increased the uptake of essential nutrients, such as nitrogen, phosphorus, and potassium.	Daud et al. (2019)
	Soyabean	Chitosan-Coated Liposomes (Csls)	Improved the growth and nitrogen metabolism of soybean plants, leading to increased plant biomass and nitrogen content.	Lu et al. (2018)
	Tomato	Biodegradable Polymeric Nanocapsules	Improved the photosynthetic rate and water use efficiency of the plants, leading to increased plant growth and yield.	Khatami et al. (2019)
	Wheat	Chitosan-Coated Nanocapsules	Improved the germination and growth of wheat plants, as well as increased their tolerance to salt stress.	Hussain et al. (2016)

Current Status of Nanofertilisers in Crop Production

The use of nanofertilizers in crop production is gaining popularity, and these fertilizers are becoming increasingly available in the market. However, major fertilizer companies have not yet designed specific agricultural nanofertilizers. In the past decade, several studies have investigated the uptake of nanofertilizers from the soil, their bioavailability, and potential toxicity, particularly for metal/metal oxide nanoparticles such as TiO_2, ZnO, Al_2O_3, FeO, and CeO_2 NPs, to improve agronomic productivity (Rajput et al. 2021). Some of the leading producers of nanofertilizers are listed below in the Table 4.

Recent advancements in agriculture have led to a growing interest in the development of sustainable and efficient nutrient delivery systems. Nano-based nutrient delivery systems have emerged as a promising solution to enhance agricultural productivity while minimizing environmental impact. The production of such systems could potentially mitigate food supply-related issues in society (Toksha et al. 2021). In fact, several studies have highlighted the potential of nanofertilizers to increase crop yields and improve nutrient uptake efficiency in plants (Dimkpa and Bindraban 2018; Chand Mali et al. 2020). As the world population continues to rise, there is an increasing demand for food production. Therefore, the development of sustainable agricultural practices that ensure food security and minimize environmental degradation is of utmost importance. The use of

Table 4. Some commercial product of nanofertilizers (Usman et al. 2020).

Sr. No.	Name of Commercial Product	Content of the Nanomaterial Serving Nanonutrient	Commercial Firm of the Source
Indian Sources			
1	Nano Green	Extracts of corn, grain, soybeans, potatoes, coconut, and palm	Nano Green Sciences, Inc., India
2	Nano Max NPK Fertilizer	Multiple organic acids chelated with major nutrients, amino acids, organic carbon, organic micronutrients/ trace elements, vitamins, and probiotics	JU Agri Sciences Pvt. Ltd, Janakpuri, New Delhi, India
3	TAG NANO (NPK, PhoS, Zinc, Cal, etc.) fertilizers	Proteino-lacto-gluconate chelated with micronutrients, vitamins, probiotics, seaweed extracts, humic acid	Tropical Agrosystem India (P) Ltd, India
Other Global Sources			
1	Nano-Gro™	Plant growth regulator and immunity enhancer	Agro Nanotechnology Corp., FL, United States
2	Nano-Ag Answer®	Microorganisms, sea kelp, and mineral electrolyte	Urth Agriculture, CA, United States
3	Biozar Nanofertilizer	Combination of organic materials, micronutrients, and macromolecules	Fanavar Nano-Pazhoohesh Markazi Company, Iran
4	Master Nano-chitosan Organic Fertilizer	Water-soluble liquid chitosan, organic acid, salicylic acids, and phenolic compounds	Pannaraj Intertrade, Thailand

nanotechnology in agriculture could offer several benefits, including improved nutrient delivery, pest and disease management, and increased crop yields (Kah et al. 2013). However, it is important to ensure the safety and efficacy of these technologies through rigorous testing and evaluation.

Future Prospectives

Nanostructured delivery systems/technologies hold great promise for the future of agriculture. Nanocarriers have the potential to deliver nutrients and other beneficial substances precisely to specific parts of plants, resulting in optimized plant growth and yield with minimal environmental impact. They can also be tailored to deliver substances that aid in soil remediation by directly targeting contaminated areas. Controlled release fertilizers that respond to specific environmental cues, such as temperature or pH, could be developed to reduce nutrient loss and increase plant nutrient uptake efficiency. Nanocarriers can also deliver substances that enhance plant resistance to pests and diseases, such as RNA interference molecules. These innovative technologies can efficiently manage crops, improve food safety, and reduce the need for synthetic supplements, thereby enhancing the nutritional quality of food.

Conclusion

Nanonutrients have the potential to contribute to sustainable agricultural development by improving nutrient uptake efficiency, reducing fertilizer usage, and increasing crop yields. The use of nanofertilizers has shown promise in improving the targeted delivery of nutrients to crops, increasing nutrient utilization efficiency, and enhancing crop quality. However, it is important to note that the safety and efficacy of nanofertilizers are still being evaluated and further research is necessary to fully understand their potential benefits and risks. The interaction of nanoparticles with living organisms may have unexpected effects, and their impact on human health, soil health, and the environment must be carefully studied before widespread use. Therefore, while the potential benefits of nanonutrients are significant, caution must be taken in their use and continued investment in research is crucial to understand their potential risks and benefits. Careful regulation, testing, and monitoring of nanonutrients use will be crucial to ensure that it is used in a safe and sustainable manner. Overall, nano-delivery system technology has the potential to revolutionize agriculture and address many of the challenges faced by the agri-sector.

Acknowledgement

PG and AG would like to acknowledge this research is part of project No. UMO-2022/45/P/ NZ9/01571 co-funded by the National Science Centre and the European Union Framework Programme for Research and Innovation Horizon 2020 under the Marie Skłodowska-Curie grant agreement No. 945339.

References

Abbas, S.F., Bukhari, M.A., Raza, M.A.S., Abbasi, G.H., Ahmad, Z., Alqahtani, M.D., Almutairi, K.F., Abd_Allah, E.F. and Iqbal, M.A. 2023. Enhancing drought tolerance in wheat cultivars through nano-zno priming by improving leaf pigments and antioxidant activity. Sustain 15: 1–15. https://doi.org/10.3390/su15075835.

Ahmed, R., Uddin, M.K., Quddus, M.A., Samad, M.Y.A., Hossain, M.A.M. and Haque, A.N.A. 2023. Impact of foliar application of zinc and zinc oxide nanoparticles on growth, yield, nutrient uptake and quality of tomato. Horticulturae 9: 1–23. https://doi.org/10.3390/horticulturae9020162.

Ali, S., Mehmood, A. and Khan, N. 2021. Uptake, translocation, and consequences of nanomaterials on plant growth and stress adaptation. J. Nanomater. 2021: 17. https://doi.org/10.1155/2021/6677616.

An, C., Sun, C., Li, N., Huang, B., Jiang, J., Shen, Y., Wang, C., Zhao, X., Cui, B., Wang, C., Li, X., Zhan, S., Gao, F., Zeng, Z., Cui, H. and Wang, Y. 2022. Nanomaterials and nanotechnology for the delivery of agrochemicals: strategies towards sustainable agriculture. J. Nanobiotechnology 20: 1–19. https://doi.org/10.1186/s12951-021-01214-7.

Bairwa, P., Kumar, N., Devra, V. and Abd-Elsalam, K.A. 2023. Nano-biofertilizers synthesis and applications in agroecosystems. Agrochemicals 2: 118–134. https://doi.org/10.3390/agrochemicals2010009.

Bakshi, M. and Kumar, A. 2021. Copper-based nanoparticles in the soil-plant environment: Assessing their applications, interactions, fate and toxicity. Chemosphere 281: 130940.

Beig, B., Niazi, M.B.K., Sher, F., Jahan, Z., Malik, U.S., Khan, M.D., Américo-Pinheiro, J.H.P. and Vo, D.V.N. 2022. Nanotechnology-based controlled release of sustainable fertilizers. A review. Environ. Chem. Lett. 20: 2709–2726. https://doi.org/10.1007/s10311-022-01409-w.

Bongiovanni, R. and Lowenberg-Deboer, J. 2004a. Precision agriculture and sustainability. Precis. Agric. 5: 359–387. https://doi.org/10.1023/B:PRAG.0000040806.39604.aa.

Bongiovanni, R. and Lowenberg-Deboer, J. 2004b. Precision agriculture and sustainability. Precis. Agric. 5: 359–387

Cappelli, S.L., Domeignoz-Horta, L.A., Loaiza, V. and Laine, A.L. 2022. Plant biodiversity promotes sustainable agriculture directly and via belowground effects. Trends Plant Sci. 27: 674–687. https://doi.org/10.1016/j.tplants.2022.02.003.

Carvalho, F.P. 2006. Agriculture, pesticides, food security and food safety. Environ. Sci. Policy 9: 685–692. https://doi.org/10.1016/j.envsci.2006.08.002.

Chand Mali, S., Raj, S. and Trivedi, R. 2020. Nanotechnology a novel approach to enhance crop productivity. Biochem. Biophys Reports 24: 100821. https://doi.org/10.1016/j.bbrep.2020.100821.

Chen, H. and Yada, R. 2011. Nanotechnologies in agriculture: New tools for sustainable development. Trends Food Sci. Technol. 22: 585–594. https://doi.org/10.1016/j.tifs.2011.09.004.

Dar, G.H., Bhat, R.A., Mehmood, M.A. and Hakeem, K.R. 2021. Microbiota and biofertilizers, Vol 2: Ecofriendly tools for reclamation of degraded soil environs.

Dimkpa, C.O. and Bindraban, P.S. 2018. Nanofertilizers: New products for the industry? J. Agric Food Chem. 66: 6462–6473. https://doi.org/10.1021/acs.jafc.7b02150.

Dutta, P., Kumari, A., Mahanta, M., Biswas, K.K., Dudkiewicz, A., Thakuria, D., Abdelrhim, A.S., Singh, S.B., Muthukrishnan, G., Sabarinathan, K.G., Mandal, M.K. and Mazumdar, N. 2022a. Advances in nanotechnology as a potential alternative for plant viral disease management. Front Microbiol. 13: 1–12. https://doi.org/10.3389/fmicb.2022.935193.

Dutta, S., Pal, S., Panwar, P., Sharma, R.K. and Bhutia, P.L. 2022b. Biopolymeric nanocarriers for nutrient delivery and crop biofortification. ACS Omega 7: 25909–25920. https://doi.org/10.1021/acsomega.2c02494.

El-Saadony, M.T., ALmoshadak, A.S., Shafi, M.E., Albaqami, N.M., Saad, A.M., El-Tahan, A.M., Desoky, E.S.M., Elnahal, A.S.M., Almakas, A., Abd El-Mageed, T.A., Taha, A.E., Elrys, A.S. and Helmy, A.M. 2021. Vital roles of sustainable nano-fertilizers in improving plant quality and quantity—an updated review. Saudi J. Biol. Sci. 28: 7349–7359.

Elemike, E.E., Uzoh, I.M., Onwudiwe, D.C. and Babalola, O.O. 2019. The role of nanotechnology in the fortification of plant nutrients and improvement of crop production. Appl. Sci. 9: 1–32. https://doi.org/10.3390/app9030499.

Ghormade, V., Deshpande, M.V. and Paknikar, K.M. 2011. Perspectives for nano-biotechnology enabled protection and nutrition of plants. Biotechnol. Adv. 29: 792–803. https://doi.org/10.1016/j.biotechadv.2011.06.007.

Godfray, H.C.J., Beddington, J.R., Crute, I.R., Haddad, L., Lawrence, D., Muir, J.F., Pretty, J., Robinson, S., Thomas, S.M. and Toulmin, C. 2010. Food security: the challenge of feeding 9 billion people. Science (80-) 327: 812–818. https://doi.org/10.1109/CIS.2016.52.

Gogos, A., Knauer, K. and Bucheli, T.D. 2012. Nanomaterials in plant protection and fertilization: Current state, foreseen applications, and research priorities. J. Agric Food Chem. 60: 9781–9792. https://doi.org/10.1021/jf302154y.

Goswami, P., Yadav, S. and Mathur, J. 2019. Positive and negative effects of nanoparticles on plants and their applications in agriculture. Plant Sci. Today 6: 232–242. https://doi.org/10.14719/pst.2019.6.2.502.

Guha, T., Gopal, G., Kundu, R. and Mukherjee, A. 2020. Nanocomposites for delivering agrochemicals: a comprehensive review. J. Agric Food Chem. 68: 3691–3702. https://doi.org/10.1021/acs.jafc.9b06982.

Hajihashemi, S. and Kazemi, S. 2022a. The potential of foliar application of nano-chitosan-encapsulated nano-silicon donor in amelioration the adverse effect of salinity in the wheat plant. BMC Plant Biol. 22: 1–15. https://doi.org/10.1186/s12870-022-03531-x.

Hajihashemi, S. and Kazemi, S. 2022b. The potential of foliar application of nano-chitosan-encapsulated nano-silicon donor in amelioration the adverse effect of salinity in the wheat plant. BMC Plant Biol. 22: 1–15. https://doi.org/10.1186/s12870-022-03531-x.

Hussain, A., Ali, S., Rizwan, M., Zia ur Rehman, M., Javed, M.R., Imran, M., Chatha, S.A.S. and Nazir, R. 2018. Zinc oxide nanoparticles alter the wheat physiological response and reduce the cadmium uptake by plants. Environ Pollut. 242: 1518–1526. https://doi.org/10.1016/j.envpol.2018.08.036.

Ingle, P.U., Shende, S.S., Shingote, P.R., Mishra, S.S., Sarda, V., Wasule, D.L., Rajput, V.D., Minkina, T., Rai, M., Sushkova, S., Mandzhieva, S. and Gade, A. 2022. Chitosan nanoparticles (ChNPs): A versatile growth promoter in modern agricultural production. Heliyon 8: e11893. https://doi.org/10.1016/j.heliyon.2022.e11893.

Jha, K., Doshi, A., Patel, P. and Shah, M. 2019. A comprehensive review on automation in agriculture using artificial intelligence. Artif. Intell. Agric. 2: 1–12. https://doi.org/10.1016/j.aiia.2019.05.004.

Jiang, M., Song, Y., Kanwar, M.K., Ahammed, G.J., Shao, S. and Zhou, J. 2021. Phytonanotechnology applications in modern agriculture. J. Nanobiotechnology 19: 1–20. https://doi.org/10.1186/s12951-021-01176-w.

Kah, M., Beulke, S., Tiede, K. and Hofmann, T. 2013. Nanopesticides: State of knowledge, environmental fate, and exposure modeling. Crit. Rev. Environ. Sci. Technol. 43: 1823–1867. https://doi.org/10.1080/10643389.2012.671750.

Kah, M. and Hofmann, T. 2014. Nanopesticide research: Current trends and future priorities. Environ. Int. 63: 224–235. https://doi.org/10.1016/j.envint.2013.11.015.

Kah, M. 2015. Nanopesticides and nanofertilizers: Emerging contaminants or opportunities for risk mitigation? Front Chem. 3: 1–6. https://doi.org/10.3389/fchem.2015.00064.

Kaini, M. 2020. Role of agriculture in ensuring food security Int. J. Humanit. Appl. Soc. Sci. 1–5. https://doi.org/10.33642/ijhass.v5n1p1.

Kakar, K., Xuan, T.D., Noori, Z., Aryan, S. and Gulab, G. 2020. Effects of organic and inorganic fertilizer application on growth, yield, and grain quality of rice. Agric 10: 1–11. https://doi.org/10.3390/agriculture10110544.

Karny, A., Zinger, A., Kajal, A., Shainsky-Roitman, J. and Schroeder, A. 2018. Therapeutic nanoparticles penetrate leaves and deliver nutrients to agricultural crops. Sci. Rep. 8: 1–10. https://doi.org/10.1038/s41598-018-25197-y.

Kashyap, P.L., Xiang, X. and Heiden, P. 2015. Chitosan nanoparticle based delivery systems for sustainable agriculture. Int. J. Biol. Macromol. 77: 36–51. https://doi.org/10.1016/j.ijbiomac.2015.02.039.

Khan, M.R., Adam, V., Rizvi, T.F., Zhang, B., Ahamad, F., Zhu, Y., Yang, M., Mao, C., Republic, C. and Life, S. 2020. Nanoparticle-plant interactions: a two-way traffic. Small 15: 1–37. https://doi.org/10.1002/smll.201901794. Nanoparticle-plant.

Khot, L.R., Sankaran, S., Maja, J.M., Ehsani, R. and Schuster, E.W. 2012. Applications of nanomaterials in agricultural production and crop protection: A review. Crop Prot. 35: 64–70. https://doi.org/10.1016/j.cropro.2012.01.007.

Kumar, Y., Tiwari, K.N., Singh, T. and Raliya, R. 2021. Nanofertilizers and their role in sustainable agriculture. Ann. Plant Soil Res. 23: 238–255. https://doi.org/10.47815/apsr.2021.10067.

Kurczynska, E., Sala, K., Godel-Jedrychowska, K. and Milewska-Hendel, A. 2021. Nanoparticles—plant interaction: what we know, where we are ? Appl. Sci. 11: 1–12.

Lin, S., Reppert, J., Hu, Q., Hudson, J.S., Reid, M.L., Ratnikova, T.A., Rao, A.M., Luo, H. and Ke, P.C. 2009. Uptake, translocation, and transmission of carbon nanomaterials in rice plants. Small 5: 1128–1132. https://doi.org/10.1002/smll.200801556.

Mittal, D., Kaur, G., Singh, P., Yadav, K. and Ali, S.A. 2020. Nanoparticle-based sustainable agriculture and food science: recent advances and future outlook. Front Nanotechnol. 2: 579954. https://doi.org/10.3389/fnano.2020.579954.

Muhammad Aslam, M., Waseem, M., Jakada, B.H., Okal, E.J., Lei, Z., Saqib, H.S.A., Yuan, W., Xu, W. and Zhang, Q. 2022. Mechanisms of abscisic acid-mediated drought stress responses in plants. Int. J. Mol. Sci. 23: 1–21. https://doi.org/10.3390/ijms23031084.

Mustafa, I.F. and Hussein, M.Z. 2020. Synthesis and technology of nanoemulsion-based pesticide. Nanomaterials 10: 1–26.

Neme, K., Nafady, A., Uddin, S. and Tola, Y.B. 2021. Application of nanotechnology in agriculture, postharvest loss reduction and food processing: food security implication and challenges. Heliyon 7: e08539. https://doi.org/10.1016/j.heliyon.2021.e08539.

Nongbet, A., Mishra, A.K., Mohanta, Y.K., Mahanta, S., Ray, M.K., Khan, M., Baek, K.H. and Chakrabartty, I. 2022. Nanofertilizers: A smart and sustainable attribute to modern agriculture. Plants 11: 1–20. https://doi.org/10.3390/plants11192587.

Pereira, A. do, E.S., Oliveira, H.C. and Fraceto, L.F. 2019. Polymeric nanoparticles as an alternative for application of gibberellic acid in sustainable agriculture: a field study. Sci. Rep. 9: 1–11. https://doi.org/10.1038/s41598-019-43494-y.

Pérez-de-Luque, A. 2017. Interaction of nanomaterials with plants: What do we need for real applications in agriculture? Front Environ. Sci. 5: 1–7. https://doi.org/10.3389/fenvs.2017.00012.

Prasad, R., Bhattacharyya, A. and Nguyen, Q.D. 2017. Nanotechnology in sustainable agriculture: Recent developments, challenges, and perspectives. Front Microbiol. 8: 1–13. https://doi.org/10.3389/fmicb.2017.01014.

Rai, M., Ribeiro, C., Mattoso, L. and Duran, N. 2015. Nanotechnologies in Food and Agriculture. Cham: Springer. doi: 10.1007/978-3-319-14024-7.

Rajput, V.D., Singh, A., Minkina, T., Rawat, S., Mandzhieva, S., Sushkova, S., Shuvaeva, V., Nazarenko, O., Rajput, P., Komariah, Verma, K.K., Singh, A.K., Rao, M. and Upadhyay, S.K. 2021. Nano-enabled products: Challenges and opportunities for sustainable agriculture. Plants 10: 1–12. https://doi.org/10.3390/plants10122727.

Reddy, S.S. and Chhabra, V. 2022. Nanotechnology: Its scope in agriculture. *In*: Journal of Physics: Conference Series. p. 012112.

Rui, M., Ma, C., Hao, Y., Guo, J., Rui, Y., Tang, X., Zhao, Q., Fan, X., Zhang, Z., Hou, T. and Zhu, S. 2016. Iron oxide nanoparticles as a potential iron fertilizer for peanut (Arachis hypogaea). Front Plant Sci. 7: 1–10. https://doi.org/10.3389/fpls.2016.00815.

Safdar, M., Kim, W., Park, S., Gwon, Y., Kim, Y.O. and Kim, J. 2022. Engineering plants with carbon nanotubes: a sustainable agriculture approach. J. Nanobiotechnology 20: 1–30. https://doi.org/10.1186/s12951-022-01483-w.

Sampathkumar, K., Tan, K.X. and Loo, S.C.J. 2020. Developing nano-delivery systems for agriculture and food applications with nature-derived polymers. iScience 23: 101055.

Schiavi, D., Balbi, R., Giovagnoli, S., Camaioni, E., Botticella, E., Sestili, F. and Balestra, G.M. 2021. A green nanostructured pesticide to control tomato bacterial speck disease. Nanomaterials 11: 1–20. https://doi.org/10.3390/nano11071852.

Shakiba, S., Astete, C.E., Paudel, S., Sabliov, C.M., Rodrigues, D.F. and Louie, S.M. 2020. Emerging investigator series: Polymeric nanocarriers for agricultural applications: Synthesis, characterization, and environmental and biological interactions. Environ. Sci. Nano 7: 37–67. https://doi.org/10.1039/c9en01127g.

Sishodia, R.P., Ray, R.L. and Singh, S.K. 2020. Applications of remote sensing in precision agriculture: A review. Remote Sens. 12: 1–31. https://doi.org/10.3390/rs12193136.

Tarafdar, J., Rathore, I. and Thomas, E. 2015. Enhancing nutrient use efficiency through nano technological interventions. Indian J. Fertil. 46.

Toksha, B., Sonawale, V.A.M., Vanarase, A., Bornare, D., Tonde, S., Hazra, C., Kundu, D., Satdive, A., Tayde, S. and Chatterjee, A. 2021. Nanofertilizers: A review on synthesis and impact of their use on crop yield and environment. Environ. Technol. Innov. 24: 101986.

van Esse, G.W. 2022. The quest for optimal plant architecture. Science 376(6589): 133–134. https://doi.org/10.1126/science.abo7429.

Vega-Vásquez, P., Mosier, N.S. and Irudayaraj, J. 2020. Nanoscale drug delivery systems: from medicine to agriculture. Front Bioeng. Biotechnol. 8: 1–16. https://doi.org/10.3389/fbioe.2020.00079.

Wanyika, H., Gatebe, E., Kioni, P., Tang, Z. and Gao, Y. 2012. Mesoporous silica nanoparticles carrier for urea: Potential applications in agrochemical delivery systems. J. Nanosci. Nanotechnol. 12: 2221–2228. https://doi.org/10.1166/jnn.2012.5801.

Yuan, J., Chen, Y., Li, H., Lu, J., Zhao, H., Liu, M., Nechitaylo, G.S. and Glushchenko, N.N. 2018. New insights into the cellular responses to iron nanoparticles in Capsicum annuum. Sci. Rep. 8: 1–9. https://doi.org/10.1038/s41598-017-18055-w.

Zulfiqar, F., Navarro, M., Ashraf, M., Akram, N.A. and Munné-Bosch, S. 2019. Nanofertilizer use for sustainable agriculture: Advantages and limitations. Plant Sci. 289: 110270. https://doi.org/10.1016/j.plantsci.2019.110270.

Chapter 7

Nano-Agri Products for Plant Health:
From Lab to Farms

Sunita Tanpure,[1,*] *Sarika Bhalerao,*[2] *Haridas Tanpure*[3] and *Vaishnavi Tattapure*[2]

Introduction

Nanotechnology has rapidly emerged as a promising field of science that has been applied in various disciplines, including physics, chemistry, pharmaceutical science, material science, medicine, and agriculture. Its success in other fields has led to increased interest in its potential applications in agriculture. The European Commission recognizes nanotechnology as one of the "key enabling technologies" that can contribute to sustainable competitiveness and growth in various industrial fields. Nanotechnology is expected to revolutionize agriculture, allowing for the development of products that were previously impossible through conventional methods. By improving management and conservation, nanotechnology has the potential to transform agricultural production. Using the smallest possible particles, it can solve agricultural problems that traditional methods have been unable to address. Researchers are currently exploring the potential of "smart seeds" that are coated with nanopolymer to germinate under favorable conditions. In controlled environment agriculture and precision farming, nanobiosensors and satellite systems help diagnose crop input requirements based on their needs, thus enabling delivery of the required quantities at the right time and place.

Nanotechnology is a promising solution for sustainable agriculture, especially in developing countries. Products such as nanoherbicides are being developed to tackle problems in weed management by targeting perennial weeds and reducing weed seed banks. Nano-structured formulations have targeted delivery or slow/controlled-release mechanisms and can respond more precisely to environmental triggers and biological demands. This reduces soil toxicity, minimizes negative effects associated with overdosage, and reduces the frequency of application.

Nanoparticles possess diverse mechanisms, including redox reactions, adsorption, ion exchange, surface complexation, and electrostatic interaction, which enable them to effectively adsorb and degrade pollutants. These properties make them highly valuable in environmental remediation applications. They are effective in degrading common industrial contaminants, like chlorinated organic compounds, petroleum nano aromatics, nitrates, heavy metals (arsenic, lead, copper, zinc, nickel, and cadmium), insecticides, and dyes (Abdullahi et al. 2021). Components like natural short-ordered aluminosilicate, the surface of titanium oxide, and humic acids can be

---
[1] Nano-Biotech Company, MIDC, Amravati, Maharashtra, India.
[2] Department of Plant Biotechnology, Vilasrao Deshmukh College of Agricultural Biotechnology, Latur, Vasantrao Naik Marathwada Krishi Vidyapith, Parbhani, Maharashtra, India.
[3] Shivaji College of Horticulture Amravati, Maharashtra, India.
* Corresponding author: sunita.bansod@gmail.com, htanpure00@gmail com

coupled with Ni through a multi-walled carbon nanotube to create effective nanobioremediation for sustainable agriculture.

Nanotechnology also plays a cost-effective role in the agricultural sector by protecting plants, soil, monitoring plant growth, and detecting diseases. However, scientists and researchers must work on reducing potential risks associated with the technology. Proper regulatory authorities and regulations must be put in place to increase public acceptance of the technology.

Nano-Agriculture Products: From Lab to Farm

Nanotechnology holds significant promise for enhancing crop production and protection in agriculture, primarily through the development of nanofertilizers, nanopesticides, nanobiosensors, and nano-enabled remediation strategies for contaminated soils. However, it is crucial to assess the fate, mobility, and toxicity of nanomaterials in soil matrixes to ensure their safe and sustainable use. The agricultural sector faces numerous challenges, including climate change, urbanization, natural resource depletion, and environmental pollution, which are expected to exacerbate as the global population grows from 7 billion to 9 billion by 2050. Therefore, innovative nanotechnology-based solutions can play a pivotal role in mitigating the adverse effects of these challenges and ensuring global food security. However, it is essential to adopt a responsible and cautious approach toward the development and implementation of nanotechnology in agriculture to ensure ecological and societal sustainability.

To overcome these challenges, new technologies are needed. Nanotechnology can address these problems by enhancing crop growth and quality through the use of nanoparticles, nanotubes, and nanozeolites. Farmers face challenges in the cultivation process, such as using fertilizers and water, but nanotechnology can facilitate farming procedures and improve the quality of the final products. Consumers prefer nanoparticle-mediated agricultural products because of their better size, shape, and scent.

Types of Nanoparticles Used in Agro-Nano-Biotech Lab

Nanoparticles (NPs) can be classified based on various characteristics, including their size, morphology, and physical, chemical, and biological properties. Based on their shape, NPs can be categorized into different types, such as quantum dots, nanotubes, nanofibers, nanorods, nanosheets, aerogel, and nanoballs, as identified by Wang et al. in 2016. NPs can also be classified as magnetic or non-magnetic. Furthermore, NPs can be categorized as organic, inorganic, and carbon-based nanomaterials, which exhibit superior properties compared to their bulk counterparts. Examples of these nanomaterials include lipid-based NPs, polymeric NPs, semiconductor-based NPs, metal NPs, ceramic NPs, and carbon-based NPs. Understanding the different types and properties of NPs is crucial for developing innovative applications in various fields, including medicine, electronics, and environmental remediation.

Classification of Nanoparticles

Organic Nanoparticles

Organic nanoparticles or polymers, such as ferritin, micelles, dendrimers, and liposomes, are promising vehicles for drug delivery due to their non-toxic and biodegradable properties. Some of these nanoparticles possess hollow centers that are sensitive to electromagnetic radiation, including heat and light, making them a suitable choice for delivering drugs to targeted sites. The efficacy of organic nanoparticles in drug delivery depends on several factors, such as their drug-carrying capacity, stability, and delivery systems. These nanoparticles are predominantly used in the biomedical sciences, particularly in targeted drug delivery systems, where they can be administered to specific body parts. Tiwari et al. (2008) and Jia et al. (2013) have documented the benefits of organic nanoparticles in drug delivery, highlighting their potential for revolutionizing therapeutic interventions in healthcare.

Inorganic Nanoparticles

Inorganic nanoparticles, which do not contain carbon, are commonly referred to as metal and metal oxide-based nanoparticles. These particles are typically synthesized at the nanoscale using constructive or destructive methods, from various metals such as aluminum, cadmium, cobalt, copper, gold, iron, lead, silver, and zinc. Additionally, metal oxide-based nanoparticles, like aluminum oxide, cerium oxide, iron oxide, magnetite, silicon dioxide, titanium oxide, and zinc oxide are commonly produced, and their properties differ from those of metal-based nanoparticles. For example, iron nanoparticles are highly reactive and rapidly oxidize in the presence of oxygen at room temperature, which is not observed in iron oxide nanoparticles. The unique properties of metal oxide nanoparticles make them an attractive option for various scientific and technological applications. Several studies conducted by Tai et al. (2007) and Niasari et al. (2008) have documented the beneficial properties of these nanoparticles, highlighting their potential for advanced applications in fields, such as nanoelectronics, nanomedicine, and environmental remediation.

Carbon-Based Nanoparticles

Nanoparticles composed entirely of carbon are referred to as carbon-based nanoparticles. Fullerenes, graphene, carbon nanotubes, carbon nanofibers, carbon black, and occasionally activated carbon at the nanoscale are examples of this type of nanoparticle (Bhaviripudi et al. 2007; Kumar and Kumbhat 2016).

a. *Fullerenes*

   Fullerene (C60) is a carbon-based molecule that has a spherical shape and is formed by the bonding of carbon atoms through sp2 hybridization. It is composed of approximately 28 to 1,500 carbon atoms, and its single-layer structures can have diameters as small as 8.2 nanometers, while the size of multilayered fullerenes can be as large as 436 nanometers.

b. *Graphene*

   Graphene is a carbon-based allotrope that has a unique hexagonal honeycomb lattice structure composed of carbon atoms arranged on a two-dimensional, flat surface. Typically, a single graphene sheet has a thickness of approximately 1 nm.

c. *Carbon Nanotubes*

   Carbon nanotubes are formed by wrapping a honeycomb arrangement of carbon atoms into a cylindrical shape with a hollow interior. They can be single-layered with diameters as small as 0.7 nm or multilayered with diameters up to 100 nm. Carbon nanotubes have variable lengths ranging from a few micrometers to several millimeters and can have either a hollow or half-fullerene molecule-capped end.

d. *Carbon Nanofiber*

   Carbon nanofibers are produced by using graphene nanofoil, which is also used to create carbon nanotubes. However, instead of forming cylindrical tubes, carbon nanofibers are coiled into a cone or cup shape.

e. *Carbon Black*

   This refers to a spherical carbon-based substance that lacks a well-defined crystalline structure. Its diameter usually falls within the range of 20 nm to 70 nm. Due to their tendency to agglomerate, the particles come into close contact with each other and form larger clusters, which can reach approximately 500 nm in size.

Industrial Scale Production of Nanomaterials

Nowadays, nanomaterials are present in a wide range of products and technologies with many being intentionally manufactured on an industrial scale, while others can be produced as by-products during the manufacturing of other materials (Kengar et al. 2019). For specific applications, it is crucial to obtain precisely defined particle sizes, also known as monodispersity. Various types of nanoparticles (NPs), coatings, dispersions, or composites can be synthesized through specific techniques that require well-defined production and reaction conditions. Particle size, chemical composition, crystallinity, and shape can be regulated by controlling parameters such as temperature, pH, concentration, chemical composition, surface changes, and process control. Two main approaches are used to produce NPs: top-down and bottom-up methods. The "top-down" approach involves mechanically crushing the source material through a milling process, while the "bottom-up" approach utilizes chemical processes to create structures (see Figure 1). The selection of the appropriate method depends on the desired properties and chemical composition of the NPs.

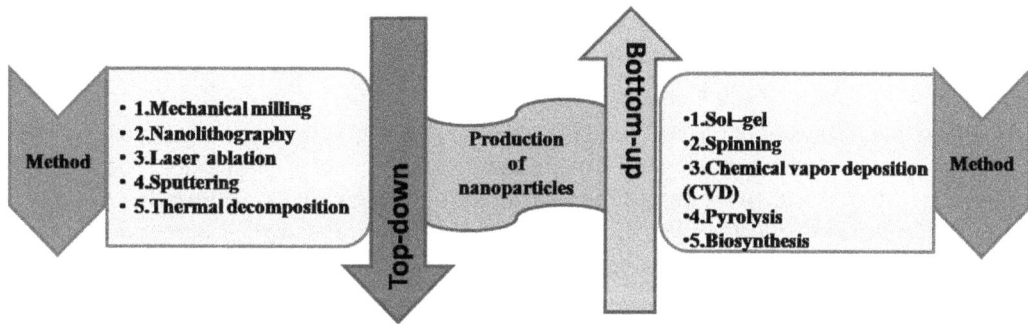

Figure 1. Production of nanomaterials.

Top-Down Approach

Richard Feynman proposed a method for reducing the size of objects called the "bottom-up" approach, which involves breaking down larger solids into smaller nanoparticles using techniques such as machining, templating, and lithography. However, this approach is slow, difficult to scale up, and not suitable for mass production. Alternatively, the top-down approach uses various methods, such as milling, chemical procedures, and lithography techniques, such as photo-lithography, electron beam lithography, and X-ray lithography to create nanoparticles.

Bottom-Up Approach

The bottom-up approach, first pioneered by Jean-Marie Lehn, involves the gradual addition of atoms and molecules. This technique allows for the condensation of atoms and molecules, while they are in either the gas phase or solution phase, ultimately leading to their enlargement into nanoparticles. The cost of producing nanoparticles is significantly lower when using this approach. Examples of bottom-up approaches include sol-gel processing, chemical vapor deposition, plasma spraying synthesis, and molecular condensation. Particle size and shape are crucial factors in the synthesis of nanoparticles, as their properties are heavily influenced by these variables. Often, nanoparticles must undergo several processes to eliminate interactions between particles and to manage the particle spacing within the matrix. Physical methods are typically used for creating nanoparticles in the solid or gaseous phases, while chemical methods are frequently employed in the solution phase. Finally, biological methods are used to synthesize nanoparticles from biological sources.

Methods Used in the Production of Nanoparticles

Nanoparticles (NPs) are synthesized using various methods, such as physical, chemical, physicochemical (aerosol), or biological processes. In contemporary times, NPs have become ubiquitous and are utilized in diverse products and technologies. Although they may also occur incidentally during the production of other materials, most nanosized products manufactured on an industrial scale are NPs, as highlighted in studies by Hett (2004), Allabashi et al. (2009), and Kengar et al. (2019).

Physical Methods

a) *Grinding*: After purging contaminants and subsequently grinding the particles in a ball mill or pot mill, the particles with natural deposits can be readily manufactured in nano form by physical means. For instance, rock phosphate can be used to physically create phosphate nanoparticles that are 28 nm and 70 nm in size, respectively, after the impurities have been removed.

b) *Thermal Evaporation*: The primary components are heated to the vapor point or higher. The evaporated base material deposits on the colder area of the substrate as a result of evaporation.

c) *Sputtering*: It is the energetic bombardment of a solid's surface layers by ions or other neutral particles that causes the loss of atomized material.

d) *Pulsed Laser Deposition Technique*: There are three phases to it. In the beginning, the laser beam is concentrated on the target's surface, heating it to the point of evaporation. The materials that are released in the second stage go in that direction. The thin film's quality is then evaluated. The completion of this thin layer is subsequently accomplished through numerous minor steps.

Chemical Methods

A very effective technique is chemical synthesis, which allows the creation of complex compounds from basic ingredients.

a) *Sol-gel Techniques*: In the sol-gel process, the initial stage involves preparing a uniform solution of one or more alkoxides, which act as organic precursors for materials such as silica, alumina, zirconia, etc. A catalyst is used to start the reaction and manage the pH. There are three phases to it.

   1. *Hydrolysis*: (OH) groups are used in place of the (OR) groups.

   2. *Condensation*: It involves the release of water or alcohol, as well as the production of dimer and trimer.

   3. *Growth and Agglomeration*: As the number of links rises, they begin to organize into a network, and as they dry, a gel is produced. Higher temperatures result in the production of larger particles.

b) *Co-Precipitation*: The simultaneous occurrence of nucleation, growth, coarsening, and agglomeration processes is known as co-precipitation. It causes the creation of many tiny particles. As a result of their high supersaturation during formation, the products are insoluble.

c) *Microwave Synthesis*: The source is heated using microwave radiation, which increases yield, promotes the uniform reaction, and opens up other paths. Acicular magnetite and goethite are a few examples.

d) *Microencapsulation*: The process of microencapsulation allows for the incorporation of nanoparticles that can serve as active ingredients into a matrix or coating that has a diameter of just a few microns. When and when it is necessary, this strategy is helpful for the regulated release of NPs.

e) *Hydrothermal Methods*: Solvents are heated to temperatures over their boiling points in a closed container to increase pressure. If water serves as the solvent, the chemical processes carried out under such circumstances are referred to as hydrothermal processing.

f) *Polyvinylpyrrolidone (PVP) Method*: A large number of colloidal particles are synthesized in the solution phase. An appropriate reductant for the aqueous production of palladium, platinum, silver, and gold nanoparticles is the hydroxyl group of PVP.

g) *Sonochemistry*: High-frequency sound waves are used in sonochemistry to drive chemical reactions. The ultrasound causes highly reactive radical species to develop in the form of small bubbles that burst quickly, initiating a chemical reaction.

Aerosol Methods (Physico-Chemical)

A mixture of solid or liquid particles suspended in air or another gaseous environment is referred to as an aerosol. Aerosols can be created artificially or naturally and can vary in size and composition. Gleiter and his colleagues' initial synthesis methods involved evaporating metal into a low-density gas, where the vapors then generated nanoparticles through a process known as homogeneous nucleation.

Aerosol is a metastable suspension of particles in a gas, therefore it requires meticulous synthesis in vast numbers under controlled circumstances to assure the expression of nanoparticle features. Aerosol technologies are favored over physical and chemical ones due to their capacity to meet the demands of high product purity, enhanced composition, and reduced environmental dangers. There are five distinct aerosol generation methods for nanoparticles: furnace methods (> 100 nm), flame methods include (with appropriate precaution) TiO_2, electro spray methods (> 1 g), CVD methods (> 100 nm), and PVD methods (> 100 nm). Although electrospray is the most accurate approach for producing nanoparticles, it is also the slowest.

Biological Methods

As an alternative to physical, chemical, and aerosol approaches for making nanoparticles, biological methods are a secure and environmentally responsible way. There are several ways to create nanoparticles biologically. Selected microbial proteins are employed to convert salts into their corresponding nanoforms (Rai et al. 2009; Bansod et al. 2015).

a) *Use of Bacteria*: Tri-calcium phosphate is converted into phosphorus nanoparticles by *Bacillus megaterium*.

b) *Use of Fungi*: In the biological manufacturing of nanoparticles, fungi have been a crucial component, producing a greater variety of nanoparticles than bacteria. For instance, *Aspergillus oryzae* can be used to produce nanoparticles of Zn, P, Ag, Au, Fe, and Ti. When compared to bacteria, fungi have higher productivity and yield due to their enormous protein secretions.

c) *Use of Plants*: Silver nanoparticles are produced by *Azadirachta indica*. A trifoliate-leaved wild herb called *Desmodium triflorum* that grows in meadows, wooded areas, and agricultural fields has been used to make silver nanoparticles. According to another study, the antibacterial efficacy of these nanoparticles against microbes is effective.

d) *Use of Biomolecules (Proteins)*: Mg, Zn, and Fe nanoparticles can be created using proteins (32 and 33 kDa weight group) from the corresponding oxide salts.

e) *Microwave-Assisted Biosynthesis*: In this method, stable poly-shaped gold nanoparticles are created using guava leaf anti-malignant plant with microwave assistance. This technique has greatly increased productivity while reducing the amount of work needed to synthesize using microorganisms.

Characterization of Nanoparticles (NPs) in Industrial Laboratory

Advanced microscopic techniques such as scanning electron microscopy (SEM), transmission electron microscopy (TEM), and atomic force microscopy (AFM) are typically used to characterize the size, shape, and surface charge of nanoparticles. The size distribution, average particle size, and surface charge can have an impact on the *in vivo* distribution and physical stability of nanoparticles. The shape of polymeric nanoparticles, which can be determined using electron microscopy techniques, may also contribute to their toxicity. Additionally, the surface charge of nanoparticles can affect their physical stability, dispersibility in polymer dispersions, and *in vivo* performance. Industrial laboratories use various techniques to characterize nanoparticles, as described by Bansod et al. (2019).

UV-Visible Spectrophotometric Analysis

To evaluate the properties of produced nanoparticles, an initial assessment is conducted using a UV-Visible spectrophotometer (Shimadzu UV-1700, Japan). The cell filtrate is analyzed within the range of 200–800 nm with a resolution of 1 nm. The characteristics of the absorbance peaks, such as peak width and peak shift, are utilized to determine the size and stability of the nanoparticles.

Fourier Transform Infrared Spectroscopy (FTIR) Analysis

The samples are dehydrated using potassium bromide (KBr), pulverized using a mortar and pestle, and then subjected to FTIR spectroscopy using a Perkin-Elmer FTIR-1600 instrument from the United States. The analysis is carried out over a range of 400–4,000 cm^{-1} with a resolution of 4 cm^{-1} to detect the presence of biomolecules responsible for reducing gold ions and stabilizing the nanoparticles present in the solution.

Nanoparticles Tracking and Analysis (NTA) by Nano-Sight (LM 20)

The sample is diluted and injected into the sample chamber of the NanoSight (LM-20, UK) to determine the size of the nanoparticles. A small amount of 0.5 µl is used for the injection.

Zeta Potential Measurement

In order to determine the zeta potential of the nanoparticles produced, the Malvern Zeta Sizer 90 (ZS 90, USA) is utilized. The sample is first mixed with nuclease-free water and sonicated for 15 minutes at 20 Hz to break down larger particles and prepare the solution. The sample is then filtered through a 0.2 µm filter before being used for zeta potential measurement. This procedure is described by using the techniques of the Malvern Zeta Sizer 90 in the scientific literature.

TEM Analysis

Moreover, conventional carbon-coated copper grids (400 meshes, Plano Gmbh, Germany) are used with TEM (Philips, CM 12) to observe the characteristics of NPs. To achieve a distinct topology representation, 5 µl of the sample is placed onto a copper grid, and three images are captured.

Industrial Large-Scale Production of Agri Nano-Products

Agriculture is currently facing a multitude of challenges, including nutrient deficiencies, stagnant crop yields, decreased soil organic matter, limited water availability, nitrogen depletion in the soil, reduced land area due to urbanization and land degradation, and labor shortages (Godfray et al. 2016). To address these issues, nanoscience is being applied in agricultural research to develop new techniques and materials for crop production and management. This concept is a crucial aspect of precision agriculture, which involves farmers using fertilizers and other inputs in an efficient manner.

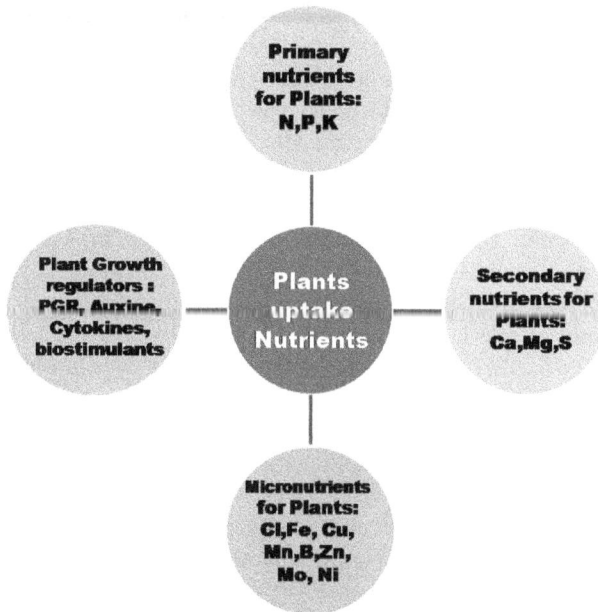

Figure 2. Discovered nano-agri-products depend on the uptake of nutrients by plants.

The effectiveness of fertilizer products is dependent on the plant's ability to absorb nutrients, as depicted in Figure 2.

Nanoparticles-Based Soil Conditioner

Soil fertility refers to the capacity of soil to support plant growth and productivity, which encompasses the supply of nutrients to plants and various chemical, physical, and biological properties (Panpatte and Jhala 2019). Clays, organic matter, iron oxides, and other minerals are naturally occurring colloids and macromolecules in soils that play a critical role in biogeochemical processes. Generally, soil contains two larger categories of nanoparticles: mineral NPs and nano-minerals. Mineral NPs are minerals that can also exist in larger sizes, such as the majority of known minerals, while nano-minerals are minerals that occur only in this size range, such as certain clays and Fe and Mn (oxyhydr) oxides. The origin of natural nanoparticles found in volcanic dust, most natural fluids, soils, and sediments are controlled by different soil, geological, and biological processes. Three types of naturally occurring nanoparticles are present in the soil system: nanosheets, nanorods, and nanoparticles. Nanosheets frequently appear as thin coatings on the surfaces of the primary soil matrix minerals. Another possible occurrence is the presence of thin platelet-like nanosheets, which facilitate the rapid and extensive breakdown of pre-existing minerals and the subsequent precipitation of neophases. The formation of nanorods resulted from intense feldspar and mica dissolving, followed by the subsequent precipitation of goethite, sodalite, and other secondary phases. Additionally, biotic processes in soils may contribute to the formation of NPs (Tarafdar and Adhikari 2015).

Each component of the soil structure, including the colloidal fraction, clay fraction, silt fraction, sand fraction, and gravel, has unique properties and functions within the soil matrix. The ultrafine fraction, which is the smallest fraction in soils, is of particular interest because it controls the physical and chemical properties of soil, such as tortuosity, cation exchange capacity, anion exchange capacity, soil water holding capacity, particle aggregation, and more. Although the largest particles have the highest percentage mass across the range of colloid particle sizes, the smallest particles have the highest number and percentage of total surface area. The ultra-fine fraction significantly affects the physicochemical properties of soil due to its extensive surface area, a high percentage of surface atoms with unbalanced charge, and a high number of surface functional groups per unit of mass. While various types of nanoparticles have been developed for use in several fields of study, more research and development are needed before they can be utilized as soil conditioners. Dr. Sunita Nano-Biotech Company, located in Amravati, Maharashtra, India, uses plant-derived silver nanoparticles coated bentonite granules with NPK, micronutrient, and soil-beneficial microbes to develop soil conditioner products, such as Morya Magic, Biozyme, and Money Maker. The company has also developed innovative nanoformulations from soil to fruits (see Figure 3).

Recent research has established a positive correlation between silver nanoparticles and crop productivity in agriculture. Studies have demonstrated that optimal concentrations of silver nanoparticles play a crucial role in enhancing seed germination and plant growth, increasing the effectiveness of water and fertilizer, and boosting the concentration of chlorophyll and photosynthetic quantum efficiency. However, excessive amounts of 25 nm silver nanoparticles have been observed to cause harm to *Oryza sativa* root cells by damaging vacuoles and rupturing cell walls. Although silver nanoparticles up to 30 g/mL could not penetrate *Oryza sativa* root cells, higher concentrations caused adverse effects, which were attributed to the destruction of the cell structure. Previous research suggests a strong correlation between the size of silver nanoparticles and their toxicity to plants, with smaller nanoparticles having more significant harmful effects on plants than larger ones (Lu et al. 2002; Tiede et al. 2008; Barrena et al. 2009; Wijnhoven et al. 2009; Mazumdar and Ahmed 2011; Sharma et al. 2012; Yin et al. 2012; Vannini et al. 2013; Hatami and Ghorbanpour 2013; Mirzajani et al. 2013; Jiang et al. 2014; Shelar and Chavan 2015; Cvjetko et al. 2017).

Nanotechnology-Based Micronutrient Fertilizers: Plant/Animal/Human Nutrition

Nano-based water-soluble fertilizers are an effective means of promoting rapid plant growth and development when compared to non-nanoparticle-containing fertilizers. Such fertilizers contain essential micronutrients, including boron, copper, iron, manganese, molybdenum, zinc, and others, which facilitate plant growth and development by improving nutrient uptake, penetration, and utilization. These micronutrients are essential for a well-balanced crop diet, and their presence is as critical as primary and secondary macronutrients. The lack of any micronutrients in the soil can inhibit plant growth, even when all other nutrients are present in adequate amounts. Multiple studies have highlighted the effectiveness of nano-based water-soluble fertilizers in boosting crop yield (Subramanian et al. 2015; Munir et al. 2018; Salama et al. 2019).

Mode of Action: The essential micronutrients play various roles in the metabolism of plants. These elements are important for participating in different metabolic activities of the plant, such as regulating the permeability of cell membranes, maintaining the osmotic pressure of cell sap, participating in electron transport systems, buffering action, maintaining electrical neutrality, and more.

Benefits: This substance promotes the structural and functional integrity of plant cell membranes and improves crop quality by providing necessary micronutrients. It also activates enzymes and catalyzes reactions involved in various plant growth processes, making it essential for crop growth and food production. Moreover, it plays a direct role in photosynthesis and several metabolic reactions. The response to this nutrient is fast, allowing for the timely correction of deficiencies during the growing season. Additionally, it helps reduce disease incidence and enhances yield.

Nanotechnology-Based Water-Soluble Fertilizers

NPK 19:19:19 is a type of water-soluble fertilizer that comes in the form of crystalline powder. It is composed of essential nutrients, such as nitrogen, phosphorus, and potassium that are easily accessible to plants. This makes it a convenient method for meeting the crop's demand for these crucial elements. The fertilizer can be applied to the crop through fertigation or by spraying it onto the leaves.

Benefits: The use of this product will lead to a rapid and significant improvement in crop yield due to the immediate absorption of nutrients by the plants. The crop's production capacity will also be enhanced, and the efficiency of nutrient uptake will be improved by reducing losses due to leaching and volatilization. The specially formulated N, P, and K nutrients will promote excellent growth at all stages of the crop. Compatibility with pesticides and fungicides is ensured, and labor costs can be minimized compared to conventional fertilizer application methods (Corradini et al. 2010) (SNBT Product).

Nanotechnology-Based Plant Growth Regulators

The controlled-release nanoformulation of plant hormones derived from naturally occurring plant growth-promoting bacteria is a nanoformulated plant growth regulator. It functions as a plant stimulant with natural hormones that can enhance the survival of seedlings or stem cuttings when transplanted into soil. It does this by promoting the rapid development of shoots and roots. Unlike other plant hormones and root promoters, this nanoformulation is enclosed in nanosized granules to prevent degradation in the soil and to ensure controlled release.

Nano-Root Booster

Dr. SNBT AgroHumic is a newly developed PGR formulation utilizing nanotechnology. It is composed of 12% humic acid, 1% fulvic acid, and 1% nanoparticles, making it a nutritional organic chemical activator. This product is suitable for both foliar and soil application to promote the growth of plants' roots, fruits, and flowers. It is best applied during the early stages of growth, critical stages such as plantlet establishment, pre-flowering, fruit growth, and during periods of climatic stress. Significant improvements can be achieved with the use of AgroHumic containing 12% humic acid as a seed/root dip treatment.

Benefits: This product enhances the absorption of foliar fertilizers and herbicides, accelerates seed germination and root development due to the presence of nanoparticles, boosts photosynthesis and respiration in plants, promotes the growth of beneficial microorganisms in the soil, facilitates rapid nutrient distribution throughout the plant, improves root absorption efficiency, increases disease resistance and shelf life, and enhances the yield and quality of fruits and seeds.

Dosage: For foliar application, mix 5–10 ml AgroHumic in 1 liter of water. For seed treatments, mix 5–10 ml AgroHumic in 1 kg of the seeds.

Recommended Crops: Applicable for all types of crops (agricultural crops, horticultural crops, lawns, flowering plants, ornamental plants, etc.) (Al-Hasany et al. 2019; Noaema et al. 2020).

Nano-Flower Booster

Nano-flower is a special product that utilizes nanotechnology to promote abundant flowering. It is quickly absorbed by plants and produces a fast response. It facilitates the plant's transition to the reproductive stage at the appropriate time and helps to maintain a male-to-female flower ratio of 1:8. It is suitable for use on all crops where flowering is a crucial part of the plant's life cycle.

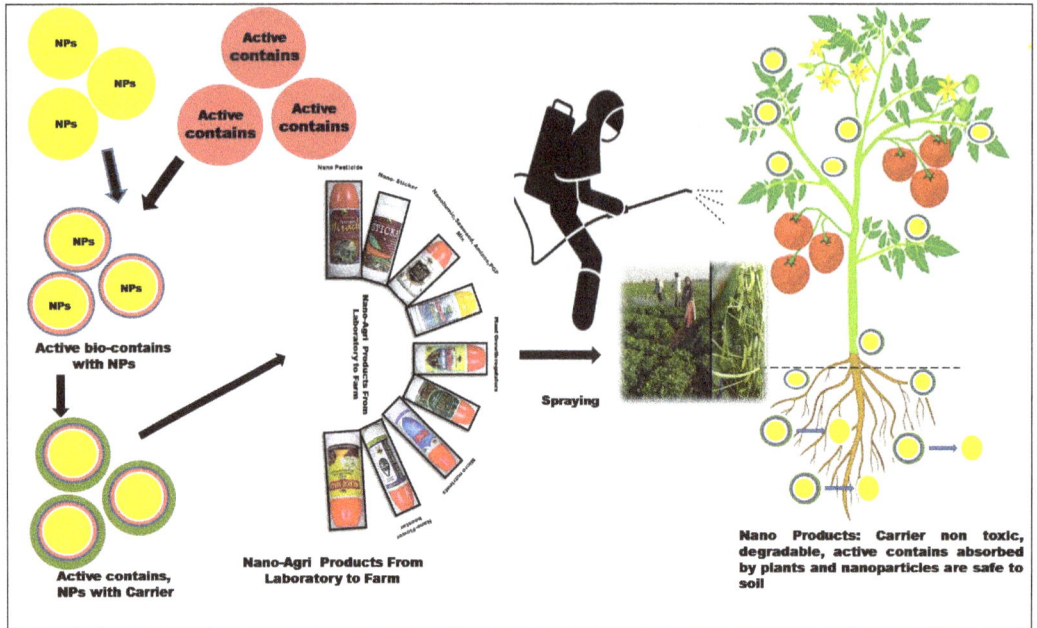

Figure 3. Novel nanoformulations from soil to fruits developed by Dr. Sunita Nano-Biotech Company.

Nano-flower can be applied together with pesticides, fungicides, and nutrients for optimal results (Rathinasamy 2012) (see Figure 3).

Dr. SNBT nitro-nano-bloom 20% is a scientifically developed product that aims to promote healthy flowering in crops. It comprises 20% nitrobenzene, nanoparticles, a sticking agent, and natural proteins that are essential for plant growth. As a growth regulator, it stimulates the plant to produce abundant flowers, making it useful as a plant energizer, flowering stimulant, and yield booster. Nitro-nano-bloom 20% is widely used in commercial farming to enhance plant growth and increase yield and quality by promoting flowering and fruiting in crops. It also helps improve the plant canopy by inducing more vigorous growth.

Benefits: Increase photosynthesis rate, triggers the growth of flowers, thus significantly enhancing the flowering rate of the plant and stimulating numerous processes like cell division, cell elongation, and formation of buds, roots, flowers, and fruits. Nanoparticles play an important role in the rapid action of formulation and protect plants from phytopathogens; moreover, they are compatible with pesticides and fungicides.

Dosage: For foliar application, mix 1ml nitro-nano-bloom 35% in 1 liter of water.

Recommended Crops: Applicable for all types of crops (crops, horticultural crops, flowering plants, ornamental plants, etc.).

Nanopesticides for Plant Protection

Nano-based pesticides are effective in protecting crops from pests, insects, and phytopathogens, which ultimately results in an increased yield of crops (Fernandes et al. 2014; Chhipa 2017). Some nano-based pesticides can be made by converting an oil-in-water emulsion system into organic nanoparticles, while others can be directly processed into nanoparticles or loaded onto nanocarriers for delivery (Park et al. 2006; Elek et al. 2010; Rai and Ingle 2012; Urkude 2019). There are already some nano-based insecticides available in the market.

Dr. SNBT nanoplex-miracle is a new research-oriented product that has been specifically designed to protect crops from insects, pests, and phytopathogens, while also promoting healthy flower growth. It contains nitrobenzene, antimicrobial oils and plant extracts, amino acids, plant nutrients, and green synthesized nanoparticles, as well as sticking agents that are essential for plants. It acts as a growth regulator that induces profuse flowering and increases yield. Additionally, it works as a plant energizer, flowering stimulant, and yield booster. Dr. SNBT nanoplex-miracle can be widely used on a commercial level to produce more vigorous growth, promote flowering and fruiting, and enhance the plant canopy, which ultimately leads to increased yield and better quality of crops.

Benefits: Preventive measures against insects, pests, and phytopathogens, increase the photosynthesis rate, and triggers the growth of flowers, thus significantly enhancing the flowering rate of the plant. Stimulates numerous processes like cell division, cell elongation, and the formation of buds, roots, flowers, and fruits. Nanoparticles play an important role in the rapid action of formulation and protect plants from phytopathogens. Compatible with pesticides and fungicides.

Dosage: For foliar application, mix 1 ml in 1 liter of water.

Recommended Crops: Applicable for all types of crops (agricultural crops, horticultural crops, flowering plants, ornamental plants, etc.).

Planto-Nano-Biozyme (PNB)

Being client-oriented, Dr. SNBT company is a manufacturer and exporter of high-quality planto-nano-biozyme (PNB). It is processed using high-grade raw acids, such as humic acid, a combination of enzymes, proteins, vitamins, carbohydrates, folic acid, and other ingredients. This is the safest component, which is widely used in the agricultural sector and is highly considered for the crop. Available at reasonable prices, it helps in growing fruits and flowers. Among the clients, the best in quality PNB is widely acclaimed for organic farming (Al-Hasany et al. 2019).

Benefits: Benefits of using the product include faster germination, healthier seedling growth, more extensive root development, improved soil microbe activity, greater nutrient absorption, increased branching and foliage, reduced loss of fruits and flowers, better grain and fruit development, bigger and heavier produce, higher yield, and improved quality.

Dosage: 2 ml per liter of water for foliar application.

Nano-Biostimulant

A plant biostimulant refers to a substance or microorganism that is applied in small quantities to optimize crop nutrition, enhance stress resistance, and improve the quality of crops. The term "biostimulant" was first introduced by Kauffman et al. (2007) (Anjum and Pradhan 2018) and has since been used increasingly to describe a wide range of substances, including bacteria, fungi, algae, embryophytes, animal-derived ingredients, and humate-containing ingredients (Zeitl et al. 2005; Singh and Rattanpal 2014; Manjunatha et al. 2016). Biostimulants can be found on the market and offer natural organic substances that help plants reach their maximum health potential and regenerate healthy soil. While not a fertilizer, biostimulants act as daily vitamins and supplements for plants, providing essential nutrients during times of stress. Our product range includes Sunami Biostimulant, a versatile biostimulant that has been shown to influence several metabolic processes, such as respiration, photosynthesis, nucleic acid synthesis, and ion uptake. Sunami Biostimulant has been proven to decrease problems, like chlorosis by enabling the plant, especially when under stress.

Mode of Action: Plants can experience a range of stresses, including insect infestations, heat, drought, and disease. These stresses can lead to nutrient depletion in the soil and impair the plant's ability to absorb necessary materials. As a result, the plant may not be able to produce enough vitamins, amino acids, or hormones. Biostimulants can help by supplementing these elements in the soil, supporting the plant's recovery and growth.

Benefits: The use of this product results in improved crop vigor, increased photosynthesis, and enhanced hormonal activities. As a result, there is a reduction in the dropping of flowers and fruits, an increase in flowering, and an improvement in fruit quality and yield. Applying this product once a month acts as a tonic for the crops and protects against insect and pest attacks as well as diseases. It also helps to minimize stress and disease levels. In addition, it promotes the development of a healthier root system, which enables plants to efficiently absorb applied products. As a result, the overall need for nitrate-rich fertilizers, pesticides, and fungicides is reduced.

Dosage: For foliar application, mix 1–2 ml Sunami Biostimulant in 1 liter of water.

Recommended Crops: Applicable for all types of crops (agricultural crops, horticultural crops, lawns, flowering plants, ornamental plants, etc.).

Nanoscale Carrier: Efficient Delivery of Fertilizers, Pesticides, and Herbicides

Nano-Based Sticker

The sticker is a product that serves as a non-toxic, spreader, sticker, and penetrator. It is made up of non-ionic surfactants and silicone compounds. Surfactants are adjuvants that help to improve the dispersing, emulsifying, absorbing, spreading, sticking, and/or pest-penetrating properties of the spray mixture. They are used to prevent the loss of components due to rain, dew, or irrigation, and are particularly useful when spraying plants with waxy or hairy leaves.

Surfactants work by adsorbing strongly onto the surface of particles and providing a charged or steric barrier to prevent the reaggregation of particles. The cuticle, a water-repellent wax, on a plant surface is the major barrier to the spreading, retention, and penetration of pesticides. Surfactants are mainly used to overcome this barrier by forming bridges between water and wax on a leaf surface or altering the permeability of the leaf cuticle (Shete 2012).

The sticker can be applied with insecticides, fungicides, herbicides, as well as foliar fertilizers. It enables the uniform and quick spreading of agrochemicals within a few seconds of spraying on the leaves, thereby improving the bio-efficacy of the agrochemicals. It is ten times more effective than carbon-based spreaders and stickers and can reduce the spray volume. It is highly compatible and can be used with a broad range of agrochemicals, with a higher and faster absorption rate.

Mode of Action: Reducing the surface tension of water that touches the leaves allows for even wetting of the leaves, leading to better absorption of fertilizers, insecticides, pesticides, and micronutrients by the crop.

Benefits: Extremely active on the surface of leaves and fruits of crops. Effectively improves absorption of crop-protection products. Easy to use and cost-effective, non-toxic, and residue-free product.

Nano-Based Crop Booster

Benefits: These are growth miracles for plants, which results in greener, thicker, richer crops. Enhances cell division and cell differentiation. Increase chlorophyll content and rate of photosynthesis. Accelerates natural growth and development. Facilitates nutrient assimilation, and translocation. Increases root respiration and root formation. Helps to keep diseases and stress down to a minimum. Improves efficiency of plant metabolism. Easy to apply and required in a very trace quantity.

Dosage: For foliar application, mix 2 ml Praharcrop Booster in 1 liter of water.

Recommendation: Applicable for all types of crops (agricultural crops, horticultural crops, lawns, flowering plants, ornamental plants, etc.).

Nano-Neem Power

The Prahar Crop Booster is a novel and effective method to achieve healthy crop growth. It is an indispensable power booster that facilitates growth and development in crops. Crop plants undergo various stages of development, starting from a single cell to a flowering plant. During each phase of growth, crops require specific nutrients, which the Prahar Crop Booster provides. This product has been carefully formulated using high-tech processes that involve the use of seaweed extracts, amino acids, vitamins, and other biological inputs. These inputs act via different pathways to enhance crop vigor, and quality yield and help crops withstand environmental stresses. The Prahar Crop Booster is readily utilized by plants due to its natural formula and can significantly accelerate crop growth.

Benefits: Preventive measures against insects, pests, and phytopathogens, and increase photosynthesis rate. It triggers the growth of flowers, thus significantly enhancing the flowering rate of the plant. Stimulates numerous processes like cell division, cell elongation, and the formation of buds, roots, flowers, and fruits. Nanoparticles play an important role in the rapid action of formulation and protect plants from phytopathogens. Compatible with pesticides and fungicides.

Dosage: Mix 1 ml nitro-nano-bloom 1 ml in 1 liter of water.

Recommendation: Applicable for all types of crops (agricultural crops, horticultural crops, flowering plants, ornamental plants, etc.).

Nanobiofertilizers

Nanofertilizers can be combined with bio-organic fertilizers to create nanobiofertilizers that offer numerous advantages, including enhancing plants' tolerance against abiotic stress. These fertilizers are considered a promising approach to the future of sustainable agriculture. In recent years, bio- and nanofertilizers have become increasingly popular as safer alternatives to chemical fertilizers, ensuring the safety of agriculture. Biofertilizers are composed of live formulations of beneficial microorganisms, including plant growth-promoting rhizobacteria, such as rhizobium, blue-green algae (BGA), mycorrhizae, bacterium azotobacter, azospirillum, and phosphate-solubilizing bacteria such as *Pseudomonas* sp. and *Bacillus* sp. These microorganisms enhance the nutrient supply to crops by increasing biological nitrogen fixation and solubilizing complex organic matter, making it more available to plants. Biofertilizers also improve soil moisture retention, soil nutrient availability, soil microbial status, and soil aeration. However, biofertilizers also have some significant drawbacks, including poor shelf life, on-field stability, and susceptibility to desiccation. In contrast, nanoparticle-based formulations of biofertilizers have shown superior performance under fluctuating environmental conditions, making them an exciting and innovative approach to combat major issues of crop production, food security, sustainability, and eco-safety. In nanobiofertilizer formulation, biofertilizer is coated with nanoscale polymers to protect the beneficial microorganisms and allow controlled release of the plant growth-promoting bacteria. This slow and steady release of nutrients to crop plants can improve nutrient utilization efficiency, enrich microbial population beneficial for soil, improve soil fertility, improve crop quality, and enhance the disease resistance of crops. Nanobiofertilizers are eco-sustainable, renewable, and cost-effective, benefiting farmers by reducing economic expenses and improving crop yields. An example of a nanobiofertilizer product is Morya Magic SNBT.

Nanobiopesticides

In agriculture, insect pests pose a significant biotic challenge, and conventional chemical pesticides are extensively used to manage them. However, the usage of chemical pesticides causes various issues, such as pest resistance, environmental pollution, elimination of natural enemies, loss of biodiversity, and human health hazards. Therefore, nanotechnology can be a promising alternative for pest control strategies in insect pest management. In India, commercialized biopesticides, including *T. harzianum, T. asperellum, Pseudomonas fluorescens,* and *Bacillus thuringiensis*, are regulated by the Central Insecticide Board (CIB) in Faridabad, India. Currently, 970 microbial-based biopesticides have been registered with the CIBRC under sections 9(3B) and 9(3) of the Insecticides Act of 1968. Although the effectiveness of consortia-based items has been demonstrated by their multifunctionality, insufficient attention is given to developing quality products for registration with CIBRC.

One of the most effective methods to combat fungal diseases is the creation of NPs. AgNPs are commonly used for disinfection due to their antibacterial properties (Baker et al. 2005). The antifungal activity of both CuNPs and AgNPs has been demonstrated against two fungal diseases, namely *Alternaria alternata* and *Botrytis cinerea*. ZnO NPs and MgO NPs exhibit antifungal properties against *Alternaria alternata, Rhizopus stolonifer, Fusarium oxysporum,* and *Mucor plumbeus* (Wani and Shah 2012). In a study, two common pesticides, zineb, and mancozeb, were enclosed in MWCNT-g-PCA hybrid material, and mancozeb was found to be a more effective fungicide against *A. alternata* (Sarlak et al. 2014).

Nanoherbicides

The use of herbicides in Indian agriculture is limited, and weeds can have a significant impact on crop yields in fragile agroecosystems. Indian agriculture relies on rainfall, and commercially available herbicides are designed to control or kill weeds that grow above the ground. However, these herbicides do not prevent the growth of rhizomes or tubers, which are belowground plant parts that can produce new weeds in the following season. Research has shown that weed infestations and seedlings can reduce crop productivity more than weed-free soil (Verma et al. 2019). Nanotechnology may offer a solution to enhance the effectiveness of herbicides and increase crop productivity. Specifically, encapsulated nanoherbicides that are protected in the natural environment and activated by rainfall can mimic the rainfed system. Target-specific herbicide chemicals enclosed in nanoparticles can be designed to bind to specific receptors in the roots of the target weeds. This binding can impede the hydrolysis of food reserves in the root system, leading to starvation and death of the weed plant (Chinnamuthu and Kokiladevi 2007). Currently, adjuvants are available for herbicide application, and the use of a "nanotechnology-derived surfactant," such as one made from soybean micelles, has been shown to render glyphosate-resistant crops susceptible to the herbicide.

Nanosensor: To Detect Nutrients and Contaminants

The utilization of nanosensors in plant science has opened up multiple avenues for nutrient evaluation, disease detection, and the detection of biological substances, such as proteins and hormones. This integration has the potential to significantly contribute to achieving the 2030 Sustainable Development Goals. Despite this, the commercial use of nanosensors is hindered by several factors, including the limited understanding of the health impacts of nanomaterials and the high cost of required raw materials.

Researchers have developed a nanosensor that reacts with auxin, which is a hormone that promotes root development and seedling establishment. This breakthrough in auxin research has allowed scientists to better understand how plant roots adapt to their environment. The nanosensor generates an electrical signal through interaction with auxin, which can be used to measure its concentration at different points along the root.

The development of biosensors is increasingly reliant on nanotechnology, and they are becoming more sensitive and effective in detecting various components in soil. Portable devices have been introduced that can analyze multiple components in soil, including nutrients and hazardous elements. This enables procedures to be planned to ensure that the soil composition is suitable for the corresponding crops and that hazardous chemicals like metals are correctly managed.

A microbial biosensor is an analytical tool that uses a biologically integrated transducer to produce a quantifiable signal showing the analyte concentration. This approach is most suited for environmental and extracellular chemical analysis, as well as metabolic sensory modulation. To overcome issues such as low sensitivity, poor selectivity, and impractical portability, microbial sensors are combined with numerous other micro/nanodevices (Qureshi et al. 2009; Sagadevan and Periasamy 2014; Stanisavljevic et al. 2015; Johnson et al. 2021; Voke et al. 2021; Zhang et al. 2021; Ang et al. 2022; Kulabhusan et al. 2022; Safdar et al. 2022).

Nanophotocatalysts: ZnO, TiO$_2$, CaCO$_3$, and MgO (Oxidizing Agent)

The use of nanomaterials, specifically nanofertilizers, is becoming increasingly popular in agricultural practices. Nanofertilizers are engineered to provide one or more types of nutrients to plants during their growth and development, which in turn can enhance their productivity. Nanofertilizers come in two forms: nanoparticles that serve as nutrient transporters and simply aid in the transport and release of nutrients, and those that act as direct sources of nutrients. By providing plants with essential nutrients, nanofertilizers have demonstrated the potential in improving crop growth and yield. Several studies have explored the benefits of nanofertilizers, including their ability to increase nutrient uptake efficiency, enhance plant resistance to stress, and promote soil health (Lal 2008; Liu and Lal 2015; Chhipa 2017; Qureshi et al. 2018; Xu 2022).

ZnO nanoparticles stand out for their remarkable capabilities and a broad variety of applications, but all metallic nanoparticles have an impact on how plants grow and develop. Zinc is a structural component of numerous enzymes and proteins and a regulatory cofactor that is crucial for photosynthesis, phytohormone production, and antioxidant processes in plants. Zinc must be applied and made available in the proper quantity because both deficits and excesses can be damaging to plants. ZnO nanoparticles have been found to possess outstanding properties, making them viable particles for preserving the required zinc concentration in plants (Pavani et al. 2014; Sabir et al. 2014; Milani et al. 2015; Hassan et al. 2020).

A well-known nanoparticle called titanium dioxide has been used in both human consumption and crop cultivation. The morphologic, biochemical, and physiological properties of the crop are significantly impacted by titanium dioxide nanoparticles in several noteworthy ways (Misra et al. 2016). They found that wheat seedlings treated with titanium dioxide nanoparticles had improved growth and production traits, including yield (Jaberzadeh et al. 2013). Furthermore, observed that canola plants treated with titanium dioxide nanoparticles exhibited superior radicle and plumule development as well as higher germination rates (Mahmoodzadeh et al. 2013).

Calcium carbonate (CaCO$_3$) is a ubiquitous component of the geosphere and is an essential ingredient in numerous engineering and technological applications. The industrial use of calcium carbonate in the food and beverage industries is already established (Hua et al. 2015). When applied to citrus tankan leaves, calcium carbonate has been demonstrated to effectively repel pests such as California red scales and oriental fruit flies. Similarly, the application of calcium carbonate combined with hydroxyl apatite nanoparticles to soybean plants under irrigation has been shown to result in the greatest yield, outperforming other treatments (El-Hady and Hussein 2021). Furthermore, the application of calcium carbonate nanoparticles, with a size of 20–80 nm, to groundnut seedlings has been found to significantly enhance seedling development and dry biomass, surpassing the control.

Magnesium oxide (MgO) nanoparticles have generated considerable interest due to their simple stoichiometry, strong ionic character, crystal structure, and surface structural defects. MgO nanoparticle dispersion has been shown to have a positive effect on peanut seed germination, growth, and photosynthetic pigments (Jhansi et al. 2017). At a concentration of 4 mg/L, the application of magnesium oxide nanoparticles has been found to result in quick germination and higher rates of seedling emergence and vigor index in mung bean seedlings, compared to other treatments (Ashok et al. 2016). Additionally, green gram exhibited the highest rates of germination, seedling emergence, and vigor index upon the application of magnesium oxide nanoparticles (Anand et al. 2020).

Nanoparticles Mediated Gene Delivery: For Treatment/Detection

In recent years, modern analytical techniques, such as fluorescence microscopy, spectroscopic measurement, wearable sensors, and smartphone-based microscopy, have been combined with nanoscale materials to detect plant diseases with the high specificity of engineered molecular recognition. Nanoscale materials, particularly nanoparticles, exhibit the potential for developing specific detection systems in plants and can be used for the passive delivery of DNA, RNA, and proteins across species.

Bansod et al. (2013) have demonstrated a highly selective nano-PCR strategy using biologically synthesized AuNPs and AgNPs, which significantly improves the yields and specificity of PCR with fewer cycles. The bioconjugate AuNP-based PCR assay successfully detected and distinguished Candida (ITS) sequences from those of other fungal infections, making it a useful tool for simultaneous detection with enhanced sensitivity and specificity of PCR. This method has applications in both medical and agricultural molecular biology.

Furthermore, the use of nanoplex as an efficient gene and protein carrier in plants has been found to have higher transformation efficiency compared to other methods. Amino-gold nanoplex can penetrate the plant cell wall, offering a novel and inexpensive gene transformation method that is rapid and safe (Bansod et al. 2019). This technique can be applied to transfer genes of interest or important industry proteins into tobacco plants or other crops for higher yield.

In conclusion, the combination of modern analytical techniques and nanoscale materials offers promising opportunities for developing specific detection systems and gene transformation methods in plants, with potential applications in both medical and agricultural fields.

Conclusion

The application of nanotechnology has had a transformative effect on the agriculture industry. Nanoparticles, which act as efficient catalysts in numerous chemical reactions utilized in industry formulations due to their small size, have proven to be advantageous in agriculture. Moreover, their ability to easily penetrate soil and plants has introduced new prospects for agricultural applications. With the aid of nanotechnology-based products and technologies, such as GPS and remote sensing devices, farmers can identify and overcome factors impeding optimal crop growth efficiency, including weather conditions, soil composition, plant development, fertilizers, chemicals, and water supply. This facilitates cost minimization and increased yields in production. Furthermore, the cost of precision farming is significantly reduced through the use of nanotechnology, with nanodevices and sensors monitoring environmental changes and alerting farmers to any crop growth issues.

The utilization of nanoparticle-based products is critical for enhancing the agricultural sector, augmenting food production and productivity, and ensuring global food and nutritional security. Agricultural manufacturers must acknowledge the potential of nanotechnology in improving their products, meeting farmers' expectations, and enhancing product sales. It is necessary to assess the effectiveness of novel nanoformulations through long-term field experimentation and multi-location trials to fully realize the potential of nanotechnology in agriculture. As a groundbreaking technology of the 21st and 22nd centuries, nanotechnology has far-reaching implications for the future of agriculture and food production.

References

Al-Hasany, Ali R.K., Mohammed, A.R., Aljaberi, Sundus, K.J. and Alhilfi. 2019. Effect of spraying with seaweed extraction growth and yield of two varieties of Wheat (*Triticum aestivum* L.). Basrah J. Agric. Sci. 32 (Special issue): 124–134.

Allabashi, R., Stach, W., Escosura-Muniz, A. de. la., Liste-Calleja, L. and Merkoci, A. 2009. ICPMS: A powerful technique for quantitative determination of gold nanoparticles without previous dissolving. J. Nanopart. Res. 11: 2003–2011.

Anand, K.V., Anugraga, A.R., Kannan, M., Singaravelu, G. and Govindaraju, K. 2020. Bio-engineered magnesium oxide nanoparticles as nano-priming agent for enhancing seed germination and seedling vigour of green gram (*Vigna radiata* L.). Mater. Lett. 271: 127792.

Ang, M.C.-Y. and Lew, T.T.S. 2022. Non-destructive technologies for plant health diagnosis. Front. Plant Sci. 13: 884454. doi: 10.3389/fpls.2022.884454.

Anjum, M. and Pradhan, S. 2018. Application of nanotechnology in precision farming: A review. IJCS 6(5): 755–760.

Ashok, C., Rao, K.V., Chakra, C.S. and Rao, K.G. 2016. Mgo nanoparticles prepared by microwave-irradiation technique and its seed germination application. Nano Trends A J. Nanotechnol. Appl. 18: 10–17.

Baker, C., Pradhan, A., Pakstis, L., Pochan, D.J. and Shah, S.I. 2005. Synthesis and antibacterial properties of silver nanoparticles. J. Nanosci. Nanotechnol. 5: 244–249. doi: 10.1166/jnn.2005.034.

Bansod, S. and Rai, M. 2008. Antifungal activity of essential oils from Indian medicinal plants against human pathogenic *Aspergillus fumigatus* and *A. niger.* World J. Medical Sci. 3(2): 81–88, ISSN 1817-3055.

Bansod, S., Bonde, S., Tiwari, V., Bawaskar, M., Deshmukh, S., Gaikwad, S., Gade, A. and Rai, M. 2013. Bioconjugation of gold and silver nanoparticles synthesized by *Fusarium oxysporum* and their use in rapid identification of candida species by using bioconjugate-nano-polymerase chain reaction. J. Biomed. Nanotechnol, (12): 1962–71.

Bansod, S., Bawaskar, M., Shende, S., Gade, A. and Rai, M. 2019. Novel nanoplex-mediated plant transformation approach. IET Nanobiotechnol. 13(6): 609–616.

Bansod, S.D., Bawaskar, M.S., Gade, A.K. and Rai, M.K. 2015. Development of shampoo, soap and ointment formulated by green synthesized silver nanoparticles functionalised with antimicrobial plants oils in veterinary dermatology: treatment and prevention strategies. IET Nanobiotechnol. 9(4): 165–171.

Bansod, S.D., Bawaskar, M.S., Shende, S., Gade, A.K., Rai, M.K. and Sunita, D. 2019. Novel nanoplex-mediated plant transformation approach. IET Nanobiotechnol. 13(6): 609–616.

Barrena, R., Casals, E., Colón, J., Font, X., Sánchez, A. and Puntes, V. 2009. Evaluation of the ecotoxicity of model nanoparticles. Chemosphere 75: 850–857.

Bhaviripudi, S., Mile, E., Iii, S.A.S., Zare, A.T., Dresselhaus, M.S. and Belcher, A.M. 2007. CVD synthesis of single-walled carbon nanotubes from gold nanoparticle catalysts. J. Am. Chem. Soc. 129(6): 1516–1517.

Calvo, P., Nelson, L. and Kloepper, J.W. 2014. Agricultural uses of plant biostimulants. Plant Soil 383: 3–41.

Chhipa, H. 2017. Nanofertilizers and nanopesticides for agriculture. Environmental Chemistry Letters 15(1): 15–22.

Chinnamuthu, C.R. and Kokiladevi, E. 2007. Weed management through nano herbicides. *In*: Chinnamuthu, C.R., Chandrasekaran, B. and Ramasamy, C. (eds.). Application of Nanotechnology in Agriculture. Tamil Nadu Agricultural University, Coimbatore, India.

Corradini, E., de Moura, M.R. and Mattoso, L.H.C. 2010. A preliminary study of the incorporation of NPK fertilizer into chitosan nanoparticles. eXPRESS Polymer Letters 4.8: 509–515.

Cvjetko, P., Milošić, A., Domijan, A.M., Vrček, I.V., Tolić, S., Štefanić, P.P., Letofsky-Papst, I., Tkalec, M. and Balen, B. 2017. Toxicity of silver ions and differently coated silver nanoparticles in Allium cepa roots. Ecotoxicol. Environ. Saf. 137: 18–28.

Dadmal, S.M., Pawar, N.P., Kale, K.B. and Shiva Sankar S.K. 2002. Efficacy of plant products and some insecticides against citrus psylla, Diaphorina citri. Insect Environment 8: 94.

Dhir, B. 2017. Biofertilizers and biopesticides: ecofriendly biological agents. pp. 167–188. In Advances in Environmental Biotechnology, Springer.

Ealias, A.M. and Saravanakumar, M.P. 2017. A review on the classification, characterisation, synthesis of nanoparticles and their application. IOP Conf. Ser. Mater. Sci. Eng. 263: 032019.

Elek, N., Hoffman, R., Raviv, U., Resh, R., Ishaaya, I. and Magdassi, S. 2010. Novaluron nanoparticles: Formation and potential use in controlling agricultural insect pests. Colloids and Surfaces A: Physicochemical and Engineering Aspects 372(1-3): 66–72.

El-Ghamry, A., Mosa, A.A., Alshaal, T. and ElRamady, H. 2018. Nanofertilizers vs. biofertilizers: new insights. Environment, Biodiversity, and Soil Security 2(2018): 51–72.

El-Hady, A. and Hussein, H. 2021. Effect of foliar nano fertilizers and irrigation intervals on soybean productivity and quality. J. Plant Prod. 12: 1007–1014.

Fernandes, C.P., de Almeida, F.B., Silveira, A.N., Gonzalez, M.S., Mello, C.B., Feder, D. and Falcão, D.Q. 2014. Development of an insecticidal nanoemulsion with *Manilkara subsericea* (Sapotaceae) extract. J. Nanobiotechnology 12(1): 1–9.

Godfray, H.C., Mason-D'Croz, D. and Robinson, S. 2016. Food system consequences of a fungal disease epidemic in a major crop. Philos. Trans. R. Soc Lond. B. Biol. Sci. 371: 20150467. Doi: 10.1098/rstb.2015.0467.

Golbashy, M., Sabahi, H., Allahdadi, I., Nazokdast, H. and Hosseini, M. 2017. Synthesis of highly intercalated urea-clay nanocomposite via domestic montmorillonite as eco-friendly slow-release fertilizer. Archives of Agronomy and Soil Science 63(1): 84–95.

Gouda, S., Kerry, R.G., Das, G., Paramithiotis, S., Shin, H.S. and Patra, J.K. 2018. Revitalization of plant growth promoting rhizobacteria for sustainable development in agriculture. Microbiological Research 206: 131–140.

Halpern, M., Bar-Tal, A., Ofek, M., Minz, D., Muller, T. and Yermiyahu, U. 2015. The use of biostimulants for enhancing nutrient uptake. Advances in Agronomy 129: 141–174.

Hassan, M., Aamer, M., Umer Chattha, M., Haiying, T., Shahzad, B., Barbanti, L., Nawaz, M., Rasheed, A., Afzal, A. and Liu, Y. 2020. The critical role of zinc in plants facing the drought stress. Agriculture 10: 396.

Hatami, M. and Ghorbanpour, M. 2013. Effect of nanosilver on physiological performance of pelargonium plants exposed to dark storage. J. Hortic. Res. 21: 15–20.

Hett, A. 2004. Nanotechnology. Small Matter, Many Unknown. SwissRe, Zurich.

http://www.lbl.gov/Science-Articles/Archive/sabl/2005/May/Tiniest-Motor.pdf surface-tension-driven nanoelectromechanical relaxation oscillator, Dr. Alex Zeitl and colleagues.

http://www.news.cornell.edu/releases/Nov99/molecules.ws.html Cornell News, Chemical bonding by assembling molecules one at a time.

Hua, K.H., Wang, H.C., Chung, R.S. and Hsu, J.C. 2015. Calcium carbonate nanoparticles can enhance plant nutrition and insect pest tolerance. J. Pestic. Sci. 40: 208–213.

Jaberzadeh, A., Moaveni, P., Moghadam, H.R.T. and Zahedi, H. 2013. Influence of bulk and nanoparticles titanium foliar application on some agronomic traits, seed gluten and starch contents of wheat subjected to water deficit stress. Not. Bot. Horti Agrobot. Cluj-napoca 41: 201–207.

Jhansi, K., Jayarambabu, N., Reddy, K.P., Reddy, N.M., Suvarna, R.P., Rao, K.V., Kumar, V.R. and Rajendar, V. 2017. Biosynthesis of MgO nanoparticles using mushroom extract: Effect on peanut (*Arachis hypogaea* L.) seed germination. 3 Biotech 7: 263.

Jia, F., Liu, X., Li, L., Mallapragada, S., Narasimhan, B. and Wang, Q. 2013. Multifunctional nanoparticles for targeted delivery of immune activating and cancer therapeutic agents. J. Controlled Release 172: 1020–1034.

Jiang, H., Qiu, X., Li, G., Li, W. and Yin, L. 2014. Silver nanoparticles induced accumulation of reactive oxygen species and alteration of antioxidant systems in the aquatic plant Spirodela polyrhiza. Environ. Toxicol. Chem, 33: 1398–1405.

Johnson, M.S., Sajeev, S. and Nair, R.S. 2021. Role of nanosensors in agriculture. pp. 58–63. In Proceedings of the 2021 International Conference on Computational Intelligence and Knowledge Economy (ICCIKE), Dubai, United Arab Emirates, 17–18 March 2021; IEEE: Piscataway, NJ, USA.

Kauffman, G.L., Khelvel, D.F. and Watsclike, T.L. 2007. Effects of a biostimulant on the heat tolerance associated with photosynthetic capacity, membrane thermostability, and polyphenol production of perennial ryegrass. Crop Sci. 47: 261–267.

Kengar, M.D., Jadhav, A.A., Kumbhar, S.B. and Jadhav, R.P. 2019. A review on nanoparticles and its application. Asian J. Pharm. Technol. 9(2): 115–124.

Khan, M.R. and Rizvi, T.F. 2017. Application of nanofertilizer and nanopesticides for improvements in crop production and protection. pp. 405–427. In Nanoscience and Plant–Soil Systems. Springer.

Kulabhusan, P.K., Tripathi, A. and Kant, K. 2022. Gold nanoparticles and plant pathogens: an overview and prospective for biosensing in forestry. Sensors 22: 1259.

Kumar, N. and Kumbhat, S. 2016. Carbon-based nanomaterials. Essentials in Nanoscience and Nanotechnology. John Wiley & Sons, Inc, Hoboken, NJ, pp. 189–236.

Lal, R. 2008. Soils and sustainable agriculture. A review. Agron. Sustain. Dev. 28: 57–64.

Lavicolia, L., Lesoa, V., Donald, H.B. and Shvedovab, A.A. 2017. Nanotechnology in agriculture: opportunities, toxicological implications, and occupational risks. Toxicol. Appl. Pharmacol. 15(329): 96–111.

Liu, C., Zhou, H. and Zhou, J. 2021. The applications of nanotechnology in crop production. Molecules 26: 7070.

Liu, R. and Lal, R. 2015. Potentials of engineered nanoparticles as fertilizers for increasing agronomic productions. Sci. Total Environ. 514: 131–139.

Lu, L., Wang, H., Zhou, Y., Xi, S., Zhang, H., Hu, J. and Zhao, B. 2002. Seed-mediated growth of large, monodisperse core-shell gold-silver nanoparticles with Ag-like optical properties. Chem. Commun. 144–145.

Mahmoodzadeh, H., Nabavi, M. and Kashefi, H. 2013. Effect of nanoscale titanium dioxide particles on the germination and growth of canola (*Brassica napus*). Ornam. Plants 3: 25–32.

Manjunatha, S.B., Biradar, D.P. and Aladakatti, Y.R. 2016. Nanotechnology and its applications in agriculture: A review. J. Farm Sci. 29: 1–13.

Mardalipour, M., Zahedi, H. and Sharghi, Y. 2014. In Evaluation of nano biofertilizer efficiency on agronomic traits of spring wheat at different sowing date. Biological Forum, 2014; Research Trend, p 349.

Mazumdar, H. and Ahmed, G.U. 2011. phytotoxicity effect of silver nanoparticles on *Oryza sativa*. Int. J. Chem. Tech. Res. 3: 1494–1500.

Milani, N., Hettiarachchi, G.M., Kirby, J.K., Beak, D.G., Stacey, S.P. and McLaughlin, M.J. 2015. Fate of zincoxide nanoparticles coated onto macronutrient fertilizers in an alkaline calcareous soil. PLoS ONE 10: e0126275.

Mirzajani, F., Askari, H., Hamzelou, S., Farzaneh, M. and Ghassempour, A. 2013. Effect of silver nanoparticles on *Oryza sativa* L. and its rhizosphere bacteria. Ecotoxicol. Environ. Saf. 88: 48–54.

Misra, P., Shukla, P.K., Pramanik, K., Gautam, S. and Kole, C. 2016. Nanotechnology for crop improvement. pp. 219–256. In Plant Nanotechnology; Springer: Berlin/Heidelberg, Germany.

Mukhopadhyay, R. and De, N. 2014. Nano clay polymer composite: synthesis, characterization, properties and application in rainfed agriculture. Global J. Bio Biotechnol. 3(2): 133–138.

Munir, T., Rizwan, M., Kashif, M. et al. 2018. Effect of zinc oxide nanoparticles on the growth and Zn uptake in wheat (*Triticum aestivum* L.) by seed priming method. Digest Journal of Nanomaterials & Biostructures (DJNB) (13): 1.

Noaema, Ali, H., Maitham, H., AlKafaji and Ali R. Alhasany. 2020. Impact of foliar application of seaweed extract and Nano humic acid on growth and yield of wheat varieties. Int. J. Agricult. Stat. Sci. 16: 169–1174.

Ouda, S.M. 2014. Antifungal activity of silver and copper nanoparticles on two plant pathogens, *Alternaria alternata* and *Botrytis cinerea*. Res. J. Microbiol. 9: 34–42. doi: 10.3923/jm.2014.34.42.

Panpatte, D.G. and Jhala, Y.K. 2019. Soil Fertility Management for Sustainable Development; Springer Nature Singapore Pte Ltd.: Singapore.

Park, H.J., Kim Sung-Ho, Kim Hwa-Jung and Seong Ho Choi. 2006. A new composition of nanosized silica-silver for control of various plant diseases. Plant Pathology 22.3: 295–302.

Pavani, K., Divya, V., Veena, I., Aditya, M. and Devakinandan, G. 2014. Influence of bioengineered zinc nanoparticles and zinc metal on *Cicer arietinum* seedlings growth. Asian J. Agric. Biol. 2: 216–223.

Qureshi, A., Kang, W.P., Davidson, J.L. and Gurbuz, Y. 2009. Review on carbon-derived, solid-state, micro, and nano sensors for electrochemical sensing applications. Diam. Relat. Mater 18: 1401–1420.

Qureshi, A., Singh, D.K. and Dwivedi, S. 2018. Nano-fertilizers: A novel way for enhancing nutrient use efficiency and crop productivity. Int. J. Curr. Microbiol. App. Sci. 7: 3325–3335.

Rai, M., Yadav, A. and Gade, A. 2009. Silver nanoparticles as a new generation of antimicrobials. Biotechnol. Adv. 27: 76–83.

Rai, M. and Ingle, A. 2012. Role of nanotechnology in agriculture with special reference to management of insect pests. Applied Microbiology and Biotechnology 94(2): 287–293.

Rathinasamy, T.C. 2012. [IN]DEVI CROP SCIENCE PRIVATE LTD [IN] Flowering stimulant composition using nitrobenzene. Patent: EP2476312A1.

Sabir, S., Arshad, M. and Chaudhari, S.K. 2014. Zinc oxide nanoparticles for revolutionizing agriculture: Synthesis and applications. Sci. World J. 1–8.

Safdar, M., Kim, W., Park, S., Gwon, Y., Kim, Y.O. and Kim, J. 2022. Engineering plants with carbon nanotubes: a sustainable agriculture approach. J. Nanobiotechnol. 20: 275.

Sagadevan, S. and Periasamy, M. 2014. Recent trends in nanobiosensors and their applications—a review. Rev. Adv. Mater. Sci. 36: 62–69.

Salama, D.M., Osman, S.A., Abd El-Aziz, M.E., Abd Elwahed, M.S.A. and Shaaban, E.A. 2019. Effect of zinc oxide nanoparticles on the growth, genomic DNA, production and the quality of common dry bean (*Phaseolus vulgaris*). Biocatalysis and Agricultural Biotechnology (18): 101083.

Salavati-Niasari, M., Davar, F. and Mir, N. 2008. Synthesis and characterization of metallic copper nanoparticles via thermal decomposition. Polyhedron 27(17): 3514–3518.

Sarlak, N., Taherifar, A. and Salehi, F. 2014. Synthesis of nanopesticides by encapsulating pesticide nanoparticles using functionalized carbon nanotubes and application of new nanocomposite for plant disease treatment. J. Agric. Food Chem. 62: 4833–4838.

Sharma, P., Bhatt, D., Zaidi, M.G.H., Saradhi, P.P., Khanna, P.K. and Arora, S. 2012. Silver nanoparticle-mediated enhancement in growth and antioxidant status of *Brassica juncea*. Appl. Biochem. Biotechnol. 167: 2225–2233.

Shelar, G.B. and Chavan, A.M. 2015. Myco-synthesis of silver nanoparticles from *Trichoderma harzianum* and its impact on germination status of oil seed. Biolife 3: 109–113.

Shete, M.V. 1970. A Super Spreading Silicon Base Non-Ionic Surfactant And Process For Making The Same Visilon 8083 Grade. 156/MUM/2012 PATENT ACT 1970.

Singh, G. and Rattanpal, H.S. 2014. Use of nanotechnology in horticulture: A review. Int. J. Agric. Sc. & Vet. Med. 2(1): 34–42.

Singh, R. and Zala, R.G. 2022. Nanotechnology in agriculture: e-magazine for agricultural articles, Agri Articles 2(03): 2582–9882.

Stanisavljevic, M., Krizkova, S., Vaculovicova, M., Kizek, R. and Adam, V. 2015. Quantum dots-fluorescence resonance energy transfer-based nanosensors and their application. Biosens. Bioelectron. 74: 562–574.

Subramanian, K.S., Manikandan, A., Thirunavukkarasu, M. and Rahale, C.S. 2015. Nano-fertilizers for balanced crop nutrition. pp. 69–80. *In*: Rai, M., Ribeiro, C., Mattoso, L. and Duran, N. (eds.). Nanotechnologies in Food and Agriculture. Springer International Publishing, Cham.

Tai, C.Y., Tai, C., Chang, M. and Liu, H. 2007. Synthesis of magnesium hydroxide and oxide nanoparticles using a spinning disk reactor. Ind. Eng. Chem. Res. 46(17): 5536–5541.

Tarafdar, J.C. and Adhikari, T. December 2015. Nanotechnology in Soil Science. Chapter.

Teng, Q., Zhang, D., Niu, X. and Jiang, C. 2018. In Influences of application of slow-release Nano-fertilizer on green pepper growth, soil nutrients and enzyme activity, IOP Conference Series: Earth and Environmental Science, IOP Publishing, 2018: p 012014.

The Science of Plant Biostimulants—A bibliographic analysis. Ad hoc Study Report to the European Commission DG ENTR. Retrieved from: http://ec.europa.eu/enterprise/sectors/chemicals/files/fertilizers/final_report_bio_2012_en.

Tiede, K., Boxall, A.B.A., Tear, S.P., Lewis, J., David, H. and Hassellöv, M. 2008. Detection and characterization of engineered nanoparticles in food and the environment. Food Addit. Contam. 25: 795–821.

Tiwari, D.K., Behari, J. and Sen, P. 2008. Application of nanoparticles in wastewater treatment. World Appl. Sci. J. 3(3): 417–433.

Urkude, R. 2019. Application of nanotechnology in insect pest management. International Research Journal of Science & Engineering 7(6): 151–156.

Vannini, C., Domingo, G., Onelli, E., Prinsi, B., Marsoni, M., Espen, L. and Bracale, M. 2013. Morphological and proteomic responses of Eruca sativa exposed to silver nanoparticles or silver nitrate. PLoS ONE (8): e68752.

Vejan, P., Abdullah, R., Khadiran, T., Ismail, S. and Nasrulhaq Boyce, A. 2016. Role of plant growth promoting rhizobacteria in agricultural sustainability—a review. Molecules 21(5): 573.

Verma, P., Bodh, S. and Thakur, S. 2019. Nanotechnology in agriculture: A review. International Journal of Chemical Studies 7(4): 488–491.

Voke, E., Pinals, R.L., Goh, N.S. and Landry, M.P. 2021. In planta nanosensors: understanding biocorona formation for functional design. ACS Sens. 6: 2802–2814.

Wang, H., Zhao, R., Li, Y., Liu, H. and Li, F. 2016. Aspect ratios of gold nanoshell capsules mediated melanoma ablation by synergistic photothermal therapy and chemotherapy. Nanomed. Nanotechnol. Biol. Med. 12: 439–448.

Wang, W.N., Tarafdar, J.C. and Biswas, P. 2013. Nanoparticle synthesis and delivery by an aerosol route for watermelon plant foliar uptake. Journal of Nanoparticle Research (5): 1417.

Wani, A.H. and Shah, M.A. 2012. A unique and profound effect of MgO and ZnO nanoparticles on some plant pathogenic fungi. J. Appl. Pharma. Sci. 2: 4.

Wijnhoven, S.W.P., Peijnenburg, W.J.G.M., Herberts, C.A., Hagens, W.I., Oomen, A.G., Heugens, E.H.W., Roszek, B., Bisschops, J., Gosens, I. and Van De Meent, D. 2009. Nano-silver—a review of available data and knowledge gaps in human and environmental risk assessment. Nanotoxicology 3: 109–138.

Xu, Z.P. 2022. Material nanotechnology is sustaining modern agriculture. ACS Agric. Sci. Technol. 2: 232–239.

Yin, L., Colman, B.P., McGill, B.M., Wright, J.P. and Bernhardt, E.S. 2012. Effects of silver nanoparticle exposure on germination and early growth of eleven wetland plants. PLoS ONE 7: e47674.

Zhang, Q., Ying, Y. and Ping, J. 2021. Recent advances in plant nanoscience. Adv. Sci. 2022(9): 2103414.

Section III
Nanotechnology for Management of Plant Diseases

Chapter 8

Biological Agents for Synthesis of Nanoparticles and Their Applications Against Plant Pathogens

Mayeen Uddin Khandaker[1],* and *Md. Habib Ullah*[2]

◇◇◇

Introduction

On December 29, 1959, during the American Physical Society meeting held at the California Institute of Technology, Nobel laureate Richard Feynman gave his lecture titled "There's Plenty of Room at the Bottom: An Invitation to Enter a New Field of Physics" (Goddard et al. 2003). The concept of nanotechnology introduced by Feynman was first published in Caltech's Engineering and Science magazine (Feynman 1960). Feynman revealed in his inspirational presentation that minuscule new structures can be formed from a small number of atoms, and that these new structures can have entirely different properties from atoms. Although Feynman could not use the term "nanotechnology," several researchers worked to validate his idea (Poole and Owens 2003). At the International Conference on Production Engineering in Tokyo, Japan, in 1974, Norio Taniguchi, a Japanese scientist, gave the first definition of "nanotechnology" after more than a decade (Taniguchi 1974). According to Taniguchi, the main components of nanotechnology are the processing of materials at the level of a single atom or molecule, resulting in their separation, consolidation, and deformation. The prefix "nano" is taken from the Greek word *nânos*, which means small. In the case of units of measurement, it signifies "one billionth" or 10^{-9} (examples: 1 nanometer (nm) = 1×10^{-9} meter, 1 nanosecond (ns) = 1×10^{-9} second). Many definitions of nanotechnology refer to dimensions. However, as defined by various international organizations, a material is generally called nanomaterial when its size is between 1 nm to 100 nm (Baig et al. 2021; Barhoum et al. 2022). Generally, nanoscience and nanotechnology are the study and application of nanoscale materials, and the nanomaterials can be used in various fields of science, such as physics, chemistry, biology, materials science, engineering, agriculture, and food sciences (Bayda 2020; Mansoori 2005). Nanoscale or nanostructured materials, however, have attracted much attention from researchers since the early 1980s due to the pioneering works of Ekimov et al. (1980), Henglein (1982), Brus (1983), Ijima (1991), and so on. Though Feynman is called the father of nanotechnology, it is good to admit that Michael Faraday first studied the preparation and size-dependent optical properties of colloidal suspensions of gold and other metals (Faraday 1857).

[1] Centre for Applied Physics and Radiation Technologies, School of Engineering and Technology, Sunway University, 47500, Bandar Sunway, Selangor, Malaysia.
[2] Department of Physics, Faculty of Science and Technology, American International University-Bangladesh (AIUB), 408/1, Kuratoli, Khilkhet, Dhaka 1229, Bangladesh.
* Corresponding author: mu_khandaker@yahoo.com; mayeenk@sunway.edu.my

The nanometer-sized objects are those, which have a nanometer size of at least one dimension (Krug and Wick 2011). Based on the literature survey, the classification of nanomaterials is shown in Figure 1. If all three dimensions of a particle of material are confined to the nanoscale regime (1–100 nm), the particle is known as a nanoparticle, nanocrystal, or quantum dot. The nanoparticles/nanocrystals/quantum dots are often called 0-D (zero-dimensional) nanomaterials. Examples include solid nanospheres, hollow nanospheres, nano-cubes, etc. The terms "quantum dot" or "quantum size," however, are usually used for small nanoparticles (less than 10 nm) (Ullah and Ha 2005; Park et al. 2007). If two dimensions of an object are in the nano-regime (typically, 1–10 nm), while another dimension is significantly longer (typically, 100 nm to a few micrometers) compared to the lateral dimensions, the object falls under the group of one-dimensional (1-D) nanomaterials. Examples include nanorods, nanotubes, nanofibers, etc. When only a dimension (such as thickness) of an object is in the nano-regime (typically, 1–10 nm) and the other two dimensions are beyond the nanoscale zone (typically, 100 nm to a few micrometers) then the object belongs to the two-dimensional (2-D) nanomaterial. Examples include nanosheets, nanolayers, etc. Another type of nanomaterial is nanoclusters or nano-dendrites. Nanoclusters are the building blocks of small nanoparticles or nanocrystals (Liu et al. 2021; Ullah et al. 2006). There are three types of nanoclusters are commonly observed: 1-D nanoclusters (chain-like structure), 2-D nanoclusters (sheet-like structures), and 3-D nanoclusters (dendritic-like sphere) (Ullah et al. 2021; Wang et al. 2015). The most striking feature of nanostructure or nanocluster materials is that their properties differ dramatically from those of bulk materials. The unique properties directing to electronic, electrical, optical, magnetic, and catalytic depend on the size, shape, and composition of nanoparticles (Baig et al. 2021; Ullah et al. 2021; Ullah et al. 2006; Ding et al. 2022). One of the vital reasons for changing their properties is the high surface area to volume ratio. Consequently, the number of surface atoms is a large fraction of the total atoms in nanoparticles. For example, if the size of a nanoparticle is 1.13 nm, the total number of atoms in the particle is 94, while the number of atoms on the surface of the particle is 48 (51.1%). This can be easily understood if we compare it with a larger nanoparticle of size 56.5 nm. The total number of atoms in this particle is 8.06×10^6, while the number of surface atoms is 1.2×10^5. Only 1.5% of atoms are on the surface (Poole and Owens 2003).

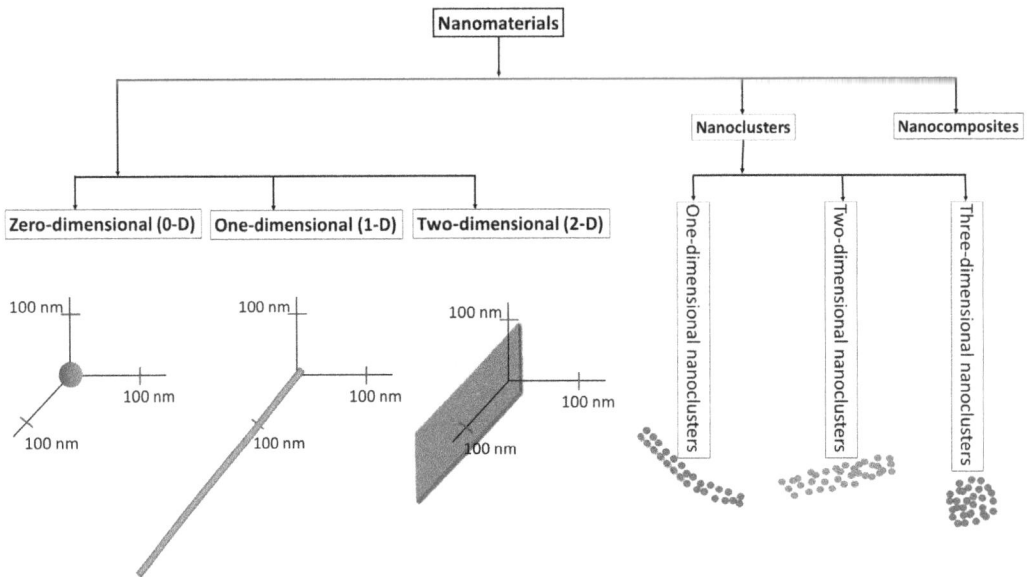

Figure 1. Classification of nanomaterials.

Therefore, the smaller the particle, the higher the fraction of atoms exposed on the surface. The concentration of surface atoms, their electron transfer rate and activation energy are dependent on shape (Narayanan and El-Sayed 2004). Furthermore, the intrinsic properties of the interior of nanoparticles are modulated by quantum size effects (Alivisatos 1996). Therefore, innovative designs of nanomaterials can be achieved by controlling the size and shape, based on requirements. Surface atoms of noble metal nanoparticles, such as Ag, Au, Cu, Pt, and Pd, play an important role in catalytic reactions as active sites (Ardakani 2021; Astruc 2020; Jiang 2018; Ullah et al. 2006; Aiken and Finke 1999). Additionally, the surfaces of the metal and semiconductor nanocrystals are ready to bond with different functional groups (such as amino, $-NH_2$; sulfhydryl, $-SH$; carboxyl, $-COOH$; hydroxyl, $-OH$) of organic molecules or biomolecules (Poole and Owens 2003; Goddard et al. 2003; Ullah et al. 2007; Roy et al. 2021).

Two approaches are widely used for the synthesis of nanomaterials. These are top-down and bottom-up approaches (Iravani 2011; Baig et al. 2021; Bayda et al. 2020). The bulk material is reduced to nanosized particles through the top-down approach. With bottom-up techniques, nanostructured materials are created atom-by-atom, molecule-by-molecule, cluster-by-cluster, or by self-assembly, as opposed to top-down techniques. Various techniques, including mechanical milling, electrospinning, lithography, the arc discharge method, laser ablation, sputtering, electro-explosion, etc., are included in the top-down approach. The bottom-up approach can be subdivided into two parts: chemical methods (such as sol-gel process, solvothermal and hydrothermal processes, spinning, supercritical fluid synthesis, laser pyrolysis, chemical vapor deposition, aerosol-based process, etc.) and green methods (biological synthesis via bacteria, fungi, algae, yeasts, actinomycetes, viruses, and plants) (Khanna et al. 2019; Khan et al. 2019).

Techniques under the top-down approach are energy-consuming and not suitable for the synthesis of nanoparticles using biological agents. Under this approach, most of the techniques do not require ligands or capping agents that are essential for biological synthesis. Moreover, the equipment needed for the top-down methods is expensive, and hazardous chemicals are often required to fabricate nanomaterials (Baig et al. 2021; Barhoum et al. 2022). Stabilizing or capping agents, such as polyvinylpyrrolidone, polypyrrole, polyaniline, sodium dodecyl sulfate, thioglycerol, cetyl trimethyl ammonium bromide, sodium hexametaphosphate, mercaptoethanol, ortho-, meta-para-phenylenediamine, etc., are needed for the synthesis of metal nanoparticles in most chemical processes under the bottom-up approach. In addition, for the synthesis of metal nanoparticles, various types of reducing agents (such as sodium citrate, ascorbate, and sodium borohydride) are used to reduce metal ions. By adjusting the parameters, such as the concentration of precursors, the concentration of stabilizing agents, the concentration of reducing agents, pHs, solvents, and temperatures, the size and shape of metal nanoparticles can be regulated (Khanna et al. 2019; Ullah et al. 2021; Yadav et al. 2019; Ullah et al. 2006; Khan et al. 2019). Although chemical synthesis techniques have several benefits for regulating the size, shape, and surface morphology of nanoparticles, they frequently include toxic chemicals that may release dangerous byproducts or precursor compounds into the environment (Khan et al. 2022; Villaseñor-Basulto et al. 2019; Roy et al. 2021). Therefore, it is important to avoid using hazardous organic solvents and toxic chemicals while synthesizing nanomaterials in order to build a greener planet.

The aforementioned drawbacks can be removed through an environmentally benign bottom-up biological production of nanoparticles. The synthesis of nanoparticles using biological agents, such as microorganisms and plant extracts, has some benefits, including availability, cost-effectiveness, non-toxic chemical precursors, ecologically acceptable products, and byproducts (Riaz et al. 2022; Roy et al. 2021; Ali et al. 2020). An emerging field of nanotechnology is the use of nanomaterials against plant diseases. An organism (such as bacteria, fungi, protists, nematodes, and viruses) that causes a disease in a plant is known as a plant pathogen. For the purpose of protecting plants, as-produced nanoparticles can be utilized directly or as pesticide carriers (Ali et al. 2020). This chapter discusses the synthesis of noble metal nanoparticles using a range of biological agents such as bacteria, fungi, and plants. The use of biosynthesized nanoparticles to control plant diseases is addressed.

Methods for Synthesis of Nanoparticles by Biological Agents

According to numerous research, various microbes and plants are used in the biosynthetic processes that produce nanoparticles (Pandit et al. 2022; Riaz et al. 2022; Khan et al. 2022; Roy et al. 2021; Srivastava et al. 2021; Ali et al. 2020; Chaudhary et al. 2020; Khanna et al. 2019; Singh et al. 2015; Rafique et al. 2017). Biological methods for the synthesis of metal nanoparticles employing microorganisms and plants are of great importance to biomaterial scientists because most of the bio-procedures can be performed at ambient conditions without using various types of expensive synthesized chemicals. In general, three elements must be chosen for the synthesis of nanoparticles using biological methods: the solvent to be used, the biological reducing agent(s) that convert metal ions into zero-valent metal atoms, and the biomolecules that serve as the capping agent(s) or stabilizing agents for nanoparticles (Patra and Baek 2014). The production of nanoparticles using biological agents can be accomplished in two key ways: (i) employing microbes to produce the particles and (ii) using plant extracts. In a typical synthesis, biological extracts are combined with metal salts, followed by the biological reduction of the metal ions into their constituent atoms, nucleation of the reduced metal atoms, and, in the final step, the agglomeration of smaller neighboring particles into bigger nanoparticles. This process is continued until thermodynamically stable nanoparticles are formed. The nature of biological entities, their concentrations, and the type of organic reducing agents all influence the thermodynamic stability of nanoparticles in terms of size and shape. Furthermore, the growth of nanoparticles in colloidal media is greatly affected by pH and temperature. Figure 2 depicts a conceptual biosynthesis procedure for the production of metallic nanoparticles (Patra and Baek 2014). Among the biological techniques of synthesizing nanoparticles, the use of microbes (such as bacteria and fungi) is one of the most popular approaches. In both extracellular and intracellular processes, the synthesis of noble metal nanoparticles can be done safely and affordably using microbes (Omar and Bendahou 2020; Ali et al. 2020). Not all microbes have the ability to produce metal nanoparticles. Bacteria that inhabit metal-rich ecosystems are typically extremely resistant to those metals due to metal uptake and chelation by intracellular and extracellular proteins (Bahrulolum et al. 2021). However, for extracellular synthesis, the culture filtrate is extracted by centrifugation and mixed with an aqueous metallic salt solution (Srivastava et al. 2021). The metal ions are trapped on the surface of the cells during this process, and the reduction of the metal ions is accomplished by microbial enzymes and proteins, bacterial or fungal cell wall components, or organic compounds contained in the filtrate (Lahiri et al. 2021; Bahrulolum et al. 2021; Ali et al. 2020; Khandel and Shahi 2018; Li et al. 2011). Research has demonstrated that several functional groups of microbial proteins, such as $-NH_2$, $-OH$, $-SH$, and $-COOH$, work as the primary reducing and/or capping agents during the preparation of metal nanoparticles (Lahiri et al. 2021).

For intracellular synthesis, after cultivating microorganisms under ideal growth conditions, the biomass is typically thoroughly washed with sterile water. Next, the biomass is incubated with a metal ion solution, while ions are transported into the microbial cell to form nanoparticles in the presence of enzymes (Ali et al. 2020; Li et al. 2011). The mixed solution's color change, however, makes it possible to see the reduction process and the formation of nanoparticles. For instance, the color range of pale yellow to wine red indicates the synthesis of gold nanoparticles, which have an average size of 20 nm (Herizchi et al. 2016). The ultrasonication, centrifugation, and washing processes can be used to gather the produced nanoparticles. However, in addition to using whole microorganisms, different approaches were used for the intracellular creation of metallic nanoparticles. For instance, the cell-free (CF) approach is a frequently employed method for intracellular production. In the CF method, the medium supernatant is obtained by centrifuging a liquid culture medium, and the medium supernatant is then incubated with an aqueous metal salt

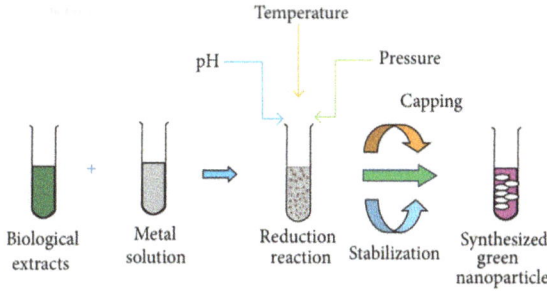

Figure 2. Scheme for the biosynthesis of metal nanoparticles (Patra and Baek 2014).

solution to produce metal nanoparticles (Bahrulolum et al. 2021; Vetchinkina et al. 2018). There are numerous techniques for extracting intracellular metabolites. An ideal technique for collecting intracellular metabolites must be reproducible, able to release intracellular metabolites of various classes in an equal amount, and capable of preventing chemical and biological decomposition (Canelas et al. 2009; Pinu et al. 2017).

An alternative, effective, affordable, and ecologically secure approach to synthesizing nanoparticles with desired characteristics is by utilizing plant extracts. The plant extracts can synthesize stable metallic nanoparticles and reduce metallic ions more quickly than microorganisms (Ali et al. 2020; Baker et al. 2013; Rai and Yadav 2013; Makarov et al. 2014). Plant extracts rich in flavonoids, polyphenols, terpenoids, tannins, alkaloids, polysaccharides, proteins, amino acids, and other compounds can reduce metal ions and stabilize nanoparticles (Shah et al. 2015; Ali et al. 2020). However, plant systems can be used to produce metal nanoparticles both intracellularly and extracellularly (Saim et al. 2021; Rai and Yadav 2013). For the synthesis of metal nanoparticles, extracts from various plant components, including leaf, stem, flower, fruit, root, and seed, can be employed. A sample of plant extract is mixed with a metal salt solution in a typical synthesis (Figure 3). The color of the reaction mixture changes as metal ions are reduced biochemically, nucleation occurs, and ultimately thermodynamically stable nanoparticles are formed (Dikshit et al. 2021). The rate of reaction and the progress of nanoparticle formation can be monitored by UV-visible spectroscopy. Plant-based synthesis of metal nanoparticles will be discussed later in another section.

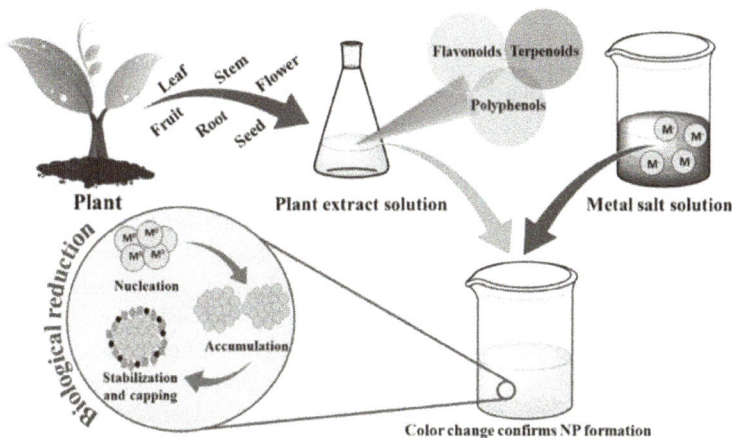

Figure 3. Schematic representation for plant-mediated biosynthesis of metal nanoparticles (Dikshit et al. 2021).

Bacteria-Based Synthesis of Noble Metal Nanoparticles and Their Applications Against Pathogens

Due to their ease of maintenance, high yield, and cheap cost of purification compared to other microorganisms, bacteria are regarded as a possible bioagent for the production of noble metal nanoparticles (such as silver, gold, platinum, palladium, etc.). Nature is full of bacteria, the majority of which are found in soil, water, plants, and animals. They can endure a range of soil conditions, including pH, salinity, temperature, and nutrient levels. *Pseudomonas stutzeri* and *Pseudomonas aeruginosa* are two bacteria species that can thrive and persist even in environments with high metal ion concentrations (Ali et al. 2020; Iravani 2014). When growing on elemental sulfur as an energy source, some of them, including *Thiobacillus ferrooxidans*, *Thiobacillus thiooxidans* (presently known as *Acidithiobacillus thiooxidans*), and *Sulfolobus acidocaldarius*, were able to convert ferric ions to the ferrous form (Brock and Gustafson 1976). They have the ability to oxidize metallic materials and generate energy for themselves (Ali et al. 2020). The surface area to volume ratio of bacteria is high, and they are good at removing metals from solutions by sorbing them. *Bacillus cereus*, *Bacillus subtilis*, *Escherichia coli*, and *Pseudomonas aeruginosa* were the four bacteria that Mullen et al. tested to see if they could remove Ag^+, Cd^{2+}, Cu^{2+}, and La^{3+} from solution using batch equilibration techniques (Mullen et al. 1989). They found that bacteria are particularly efficient at removing Ag^+ from the solution. A total of 89% of the entire Ag^+ was removed from the 1 mM solution, while 12%, 29%, and 27% of the total Cd^{2+}, Cu^{2+}, and La^{3+} were sorbed from the identical solutions (1 mM each), respectively. Thus, bacterial cells have the capacity to bind enormous amounts of silver ions.

The endophytic bacterium *Bacillus siamensis*, which was isolated from the medicinal plant *Coriandrum sativum*, was used to produce Ag nanoparticles, according to Ibrahim et al. (2019). The biosynthesized Ag nanoparticles were spherical in shape and ranged in size from 25 to 50 nm. They investigated the antibacterial effects of synthesized Ag nanoparticles at different concentrations (5 µg/mL, 10 µg/mL, 15 µg/mL, and 20 µg/mL) and found a potent antibacterial effect against the pathogens of rice bacterial leaf blight and bacterial brown stripe, while an inhibition zone of 17.3 mm and 16.0 mm was observed for *Xanthomonas oryzae* pv. *Oryzae* (Xoo) and *Acidovorax oryzae* (Ao), respectively (Figure 4). Additionally, the growth of rice seedlings was markedly accelerated by these Ag nanoparticles. The silver nanoparticle-treated rice seedlings had roots that were 7.4 cm long, shoots that were 11.6 cm long, fresh weights were 0.24 g, and dry weights were 0.04 g. The root, shoot, fresh, and dry weights of rice seedlings that were propelled by water, on the other hand, were around 5.1 cm, 6.5 cm, 0.12 g, and 0.03 g, respectively (Ibrahim et al. 2019). Ag nanoparticles were synthesized extracellularly using the bacterial strain *Bacillus cereus* which was isolated from wastewater-contaminated soil (Ahmed et al. 2020). Silver nitrate solution (5 mM) is poured into the supernatant of the isolated bacterial strain to synthesize silver nanoparticles. The reddish-brown color of the supernatant was evidence of Ag nanoparticles that were further confirmed by the UV-Vis spectrum (~ 419 nm). Spherical-shaped particles ranging in size from 18 nm to 39 nm were stabilized by protein functional groups in bacteria. Synthesized Ag nanoparticles showed substantial antibacterial activity against the rice pathogen *Xanthomonas oryzae* pv. *Oryzae*. Four different concentrations of Ag nanoparticles (5 µg/mL, 10 µg/mL, 15 µg/mL, and 20 µg/mL) were examined against the growth of pathogenic *Xanthomonas oryzae* pv. *Oryzae*. At 5 µg/mL, 10 µg/mL, 15 µg/mL, and 20 µg/mL, the Ag nanoparticles' efficacy in inhibiting growth was 38.85%, 45.14%, 57.71%, and 91.42%, respectively. The application of Ag nanoparticles was an effective and environmentally safe strategy to control bacterial leaf blight disease in rice plants. Additionally, bacterial synthesized Ag nanoparticles showed no toxicity to healthy rice plants.

Ibrahim et al. synthesized Ag nanoparticles using the endophytic bacterium *Pseudomonas poae* and successfully applied these nanoparticles against the pathogenic infection of wheat (Ibrahim et al. 2020). *Pseudomonas poae* strain was isolated from fresh leaves of a garlic plant (*Allium sativum*) and culture was carried out in nutrient broth. To synthesize Ag nanoparticles, 10 ml of

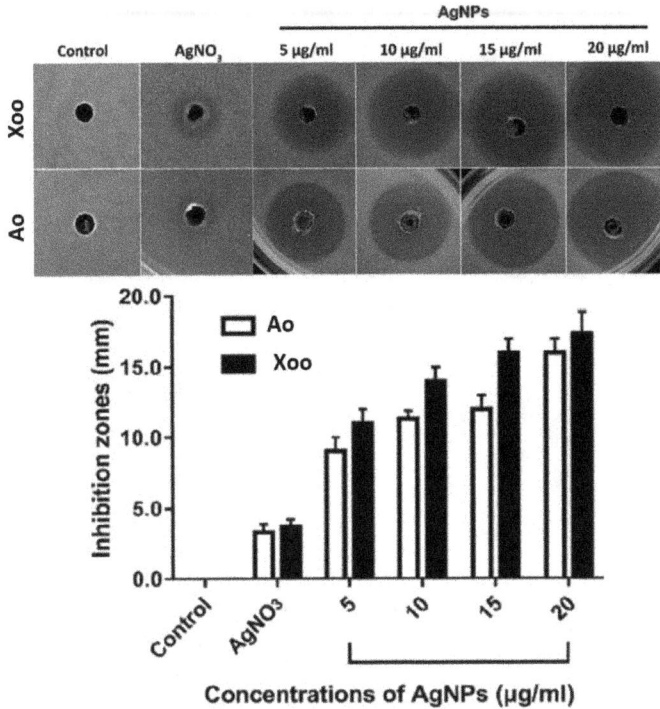

Figure 4. Biosynthesized Ag nanoparticles exhibit antibacterial activity against *Xanthomonas oryzae* pv. *oryzae* (Xoo) and *Acidovorax oryzae* (Ao) strains (Ibrahim et al. 2019).

culture filtrate colloidal was mixed with 90 ml of 3 mM aqueous $AgNO_3$ solution in a flask, and the flask was placed in a shaker at 200 rpm for four days at 30°C in darkness. The size of synthesized silver nanoparticles was between 19.8 nm and 44.9 nm, and they were spherical. The antimicrobial activity of Ag nanoparticles was tested against the wheat pathogenic fungal *Fusarium graminearum*. The mycelial growth of pathogenic fungal was potentially suppressed by the application of Ag nanoparticles both in potato dextrose agar (PDA) medium and potato dextrose broth (PDB) medium. The maximum percentage of inhibition against the mycelial growth of *F. graminearum* was recorded at 85.78 in PDB and 80.56 in PDA when the concentration of Ag nanoparticles was 20 µg/ml. The microscopic studies (Figure 5) revealed that the structure of mycelia was distorted and collapsed by the application of *Pseudomonas poae*-mediated Ag nanoparticles. As a result, DNA and proteins were released from *F. graminearum* cells. This group (Hossain et al. 2019) previously reported the synthesis of Ag nanoparticles by the bacterium *Pseudomonas rhodesiae*. The isolated bacterial strain from garlic plants was properly cultured in nutrient broth, and the CF culture supernatant was obtained through centrifugation. The *P. rhodesiae*-mediated Ag nanoparticles were spherical in shape with a size range, of 20–100 nm. In addition to the inhibition studies against the growth of phytopathogenic bacterial strain *Dickeya dadantii* with the treatment of biosynthesized Ag nanoparticles, transmission electron microscope (TEM) studies were conducted to confirm the collapsing cells of *Dickeya dadantii* (Figure 6). It is noted that *Dickeya dadantii* is a phytopathogen that is responsible for the root rot disease of sweet potato planting areas in China.

In a separate study, Gomaa et al. reported using the bacterial strain *Pseudomonas fluorescens* to biosynthesize Ag and Cu nanoparticles (Gomaa et al. 2019). With a centrifuge set at 10,000 rpm for 10 minutes, the bacterial supernatant was extracted. The bacterial supernatant was added to solutions (1 mM, pH 8.5) containing equal volumes of $AgNO_3$ and $CuSO_4$ in flasks to synthesize Ag and Cu nanoparticles. The mixtures were incubated in a dark place at room temperature for 48 hours. The brown and dark green colors of the mixtures represented the formation of Ag and Cu nanoparticles that were confirmed in UV-Vis spectra. However, the products were separated

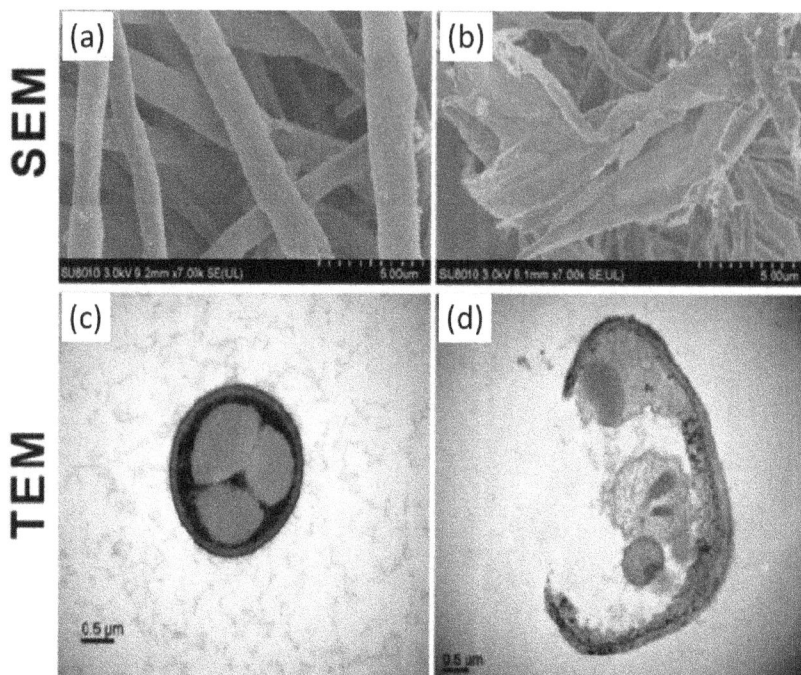

Figure 5. SEM images of the *Fusarium graminearum* strain in the absence (a) and presence (b) of biosynthesized Ag nanoparticles are shown at the top. TEM images of the *Fusarium graminearum* strain in the absence (a) and presence (b) of biosynthesized Ag nanoparticles are shown at the bottom (Ibrahim et al. 2020).

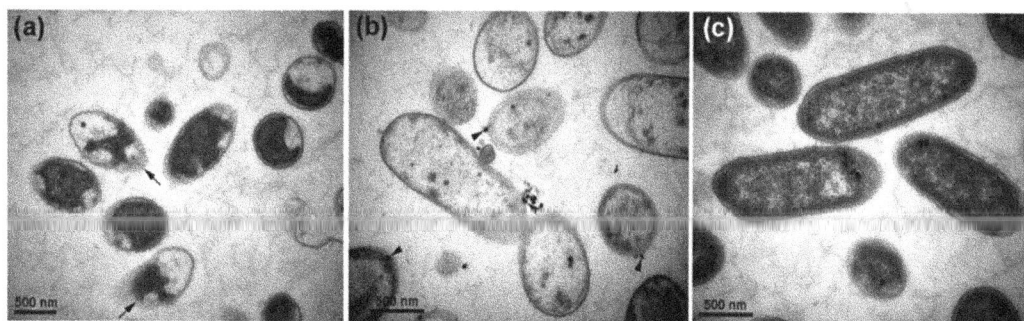

Figure 6. (a, b) TEM images of *Dickeya dadantii* cells treated with *Pseudomonas rhodesiae*-mediated Ag nanoparticles; (a) cell collapse after 2 hours of treatment, (b) cell collapse after 6 hours of treatment [the damage of cell envelops was marked by arrows], and (c) *Dickeya dadantii* cells not treated with Ag nanoparticles (Hossain et al. 2019).

through centrifugation and lyophilized for future investigations. The size ranges of copper and silver nanoparticles, as per TEM analysis, were 15.6–34.2 nm and 10.2–35.2 nm, respectively. The nanoparticles of Ag and Cu were tested against various species of fungal pathogens, *Fusarium*, *Aspergillus*, *Trichoderma*, and *Penicillium*. The overall inhibitory effects of *P. fluorescens*-mediated Ag and Cu nanoparticles were in the range of 70.26–92.90% and 57.79 to 72.78%, respectively.

Silver nanoparticles were synthesized by Monowar et al. (2018) using an extracellular extract of an endophytic bacterium, *Pantoea ananatis*. The spherical-shaped Ag nanoparticles with a size range of 8.06 nm to 91.32 nm were tested against various pathogenic microbes, such as *Staphylococcus aureus*, *Bacillus cereus*, *Escherichia coli*, and *Pseudomonas aeruginosa*. Significant antibacterial activity was observed for the Ag nanoparticles against the Gram-positive bacteria *S. aureus* and *B. cereus* with zones of inhibitions of 11.30 ± 0.07 mm (minimum inhibitory concentration,

MIC: 2.75 μg/mL) and 9.16 ± 0.05 mm (MIC: 2.25 μg/mL), respectively. Additionally, silver nanoparticles exhibited a pronounced antibacterial activity against multidrug-resistant strains, such as *S. pneumonia*, and *E. faecium*. It is noted that most of the plant-pathogenic bacteria are either Gram-positive or Gram-negative, classified in the phyla *Actinobacteria* and *Proteobacteria*, respectively (Vidaver and Lambrecht 2004). *S. aureus* and *Pseudomonas aeruginosa* are usually found in animals and humans, and *B. cereus* is a foodborne pathogen; however, they can simultaneously infect animals, humans and plants (Prithiviraj et al. 2005; Rahme et al. 1997).

In a different investigation, *Lactobacillus plantarum*, a probiotic, was used to produce Ag nanoparticles (Yusof et al. 2020). From locally fermented foods, *Lactobacillus plantarum* was isolated. In addition to producing Ag nanoparticles, it was found that the cell biomass of *L. plantarum* was capable of tolerating Ag^+ at a concentration of 2 mM. The nanoparticles, which had an average size of 14.0 ± 4.7 nm, exhibited concentration-dependent antibacterial activity against both Gram-positive and Gram-negative bacteria. According to Ibrahim et al. (2021), *Bacillus cereus*, which was isolated from contaminated soil, was used in the synthesis of Ag nanoparticles. The main elements that affected the synthesis of nanoparticles were the quantity of $AgNO_3$, inoculum size, temperature, duration, and pH during the experiment. Response Surface Methodology Design-Expert-12 software was used to calculate the individual and combined effects of these elements using the central composite design (CCD) approach. Traditional methods of optimization were eschewed because they took too long and were prone to mistakes. To make Ag nanoparticles, they employed the following computer-supported optimum values: $AgNO_3$ (1 mM) 10 ml, *Bacillus cereus* inoculum size 8.7 ml, temperature 48.5°C, time 69 hours, and pH 9. The produced silver nanoparticles have a huge surface area of 358.78 m²/g and a size range of 5–7 nm (Figure 7). Ag nanoparticles were tested for their ability to combat microbes. Gram-positive bacterial strains *Staphylococcus epidermidis* and *Staphylococcus aureus* as well as Gram-negative bacterial strains *Escherichia coli*, *Salmonella enterica*, and *Porteus mirabilis* both showed a significant antimicrobial response. The interactions between silver nanoparticles and the proteins in bacterial cell membranes and their DNA were what gave silver nanoparticles their antibacterial properties. The interactions could have damaged the bacterial cells' membranes, which would have ultimately resulted in cell death and slowed bacterial development (Ibrahim et al. 2021).

The effectiveness of gold nanoparticle adhesion to both Gram-positive (*Bacillus subtilis*, *Staphylococcus carnosus*) and Gram-negative (*Neisseria subflava*, *Stenotrophomonas maltophilia*) bacterial strains was studied by Pajerski et al. (2019). They observed that, even when the experimental parameters like temperature, time, and pH were held the same, the bacteria had distinct affinities for the attachment of gold nanoparticles to their surfaces. The findings revealed that Gram-positive bacteria were attracted to considerably more Au nanoparticles than Gram-negative bacteria. The

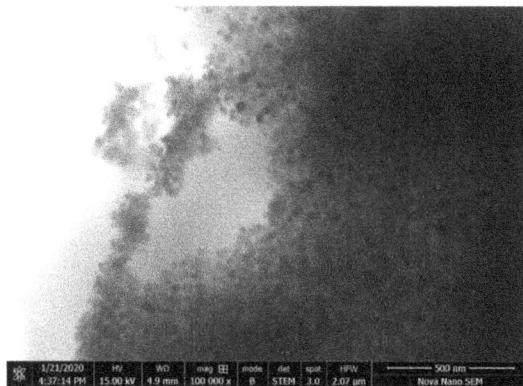

Figure 7. SEM image shows silver nanoparticles produced by *Bacillus cereus* (Ibrahim et al. 2021).

S. maltophilia   **N. subflava**   **S. carnosus**   **B. subtilis**

20 nm

50 nm

500 nm   500 nm   500 nm   500 nm

Gram-negative   **Gram-positive**

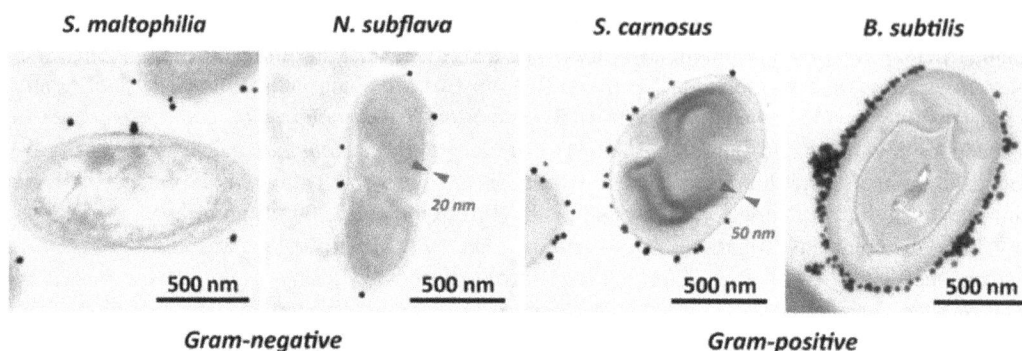

Figure 8. TEM images show single resin-embedded bacterial cells decorated with gold nanoparticles. The thickness of cell walls is indicated by arrow heads (Pajerski et al. 2019).

bacterial cell surface's molecular characteristics are crucial to the adhesion process. Compared to Gram-negative bacteria, which have cell walls that are only 20 nm thick, Gram-positive bacteria have thicker cell walls (30–50 nm). Gram-positive bacteria's multilayered surface section contains peptidoglycan, teichoic acid, and lipoteichoic acid, whereas Gram-negative bacteria's thin wall surface region contains peptidoglycan, periplasm, and lipopolysaccharides. Nevertheless, Pajerski et al. (2019) observed that the attachment efficiency to Au nanoparticles (30 nm) decreases in the following order: *B. subtilis* > *S. carnosus* > *N. subflava* > *S. maltophilia* (Figure 8).

Ahmad et al. used an alkalotolerant actinomycete *Rhodococcus* sp. to synthesize gold nanoparticles with dimensions ranging from 5 nm to 15 nm (Ahmad et al. 2003). The bacteria were found in fig trees (*Ficus carica*). Actinomycetes are a diverse group of Gram-positive bacteria that are similar to fungi in that they can form mycelium and dry spores like most fungi. This is because they have evolved to live on solid surfaces (Reponen et al. 1998). However, due to the presence of enzymes in the cell wall and on the cytoplasmic membrane, the reduction of gold ions took place on the cytoplasmic membrane as well as on the surface of the mycelia (Ahmad et al. 2003). The Gram-positive bacterium *Bacillus subtilis*, which was isolated from a gold mine, was used by Srinath et al. to synthesize Au nanoparticles (Srinath et al. 2018). A change in hue from yellow to pink served as a marker for the reduction of gold ions and the formation of Au nanoparticles with sizes ranging from 20 nm to 25 nm. Using UV-visible spectroscopy, a peak at about 545 nm proved the existence of the final product. Correa-Llantén et al. reported the synthesis of gold nanoparticles by the thermophilic bacterium *Geobacillus* sp. (Correa-Llantén et al. 2013). The gold nanoparticles had a quasi-hexagonal shape and ranged in size from 5 nm to 50 nm, with the bulk of them being between 10 nm and 20 nm. The gold nanoparticles were produced intracellularly.

Li et al. used *Deinococcus radiodurans*, an extreme bacterium known for its great tolerance to stressors such as radiation and oxidants, to synthesize Au nanoparticles (Li et al. 2016). To synthesize Au nanoparticles, many synthesis parameters, including the starting concentration of gold salt, the length of the bacterial growth period, the solution pH, and the temperature, were optimized. The size of the produced Au nanoparticles was determined by dynamic light scattering (DLS), and they were found to be dispersed throughout the cytosol, extracellular space, and cell membrane. With an average size of 43.75 nm, the particles were of different shapes. However, the Au nanoparticles exhibited antibacterial activity by damaging effects on the cytoplasmic membrane of both Gram-negative (*Escherichia coli*) and Gram-positive (*Staphylococcus aureus*) pathogenic bacteria.

Konishi et al. synthesized Pt nanoparticles using the bacterium *Shewanella algae*. Microbial reduction tests were conducted in an anaerobic glovebox. A suspension of *Shewanella algae* cells was added to an aqueous H_2PtCl_6 solution, and the gas phase was N_2-CO_2 (80:20, v/v). The suspension cells were able to convert aqueous platinum ions into elemental platinum in 60 minutes at room temperature and neutral pH, while lactate produced by the cells acted as an electron donor.

Table 1. Biosynthesis of metal nanoparticles using bacterial strains and their applications against plant pathogens.

Bacteria	Source of isolation	Types of Metal NPs	Size of NPs (nm)	Shape	Application against plant Pathogens	References
Bacillus siamensis	Medicinal plant *Coriandrum sativum*	Ag	25–50 nm	Spherical	*Xanthomonas oryzae* pv. *Oryzae* and *Acidovorax oryzae*	Ibrahim et al. 2019
Bacillus cereus	Wastewater-contaminated soil	Ag	18–39 nm	Spherical	*Xanthomonas oryzae* pv. *oryzae*	Ahmed et al. 2020
Pseudomonas poae	Garlic plants *Allium sativum*	Ag	20–45 nm	Spherical	*Fusarium graminearum*	Ibrahim et al. 2020
Pseudomonas rhodesiae	Garlic plants	Ag	20–100 nm	Spherical	*Dickeya dadantii*	Hossain et al. 2019
Calothrix elenkinii (cyanobacterium)	-	Ag	-	-	*Alternaria alternata*	Mahawar et al. 2020
Pseudomonas sp. and *Achromobacter* sp.	Soil (rhizosphere of chickpea plants)	Ag	20–50 nm	-	*Fusarium oxysporum* f. sp. *ciceri*	Kaur et al. 2018
Pseudomonas fluorescens	-	Ag and Cu	10.2–35.2 nm for Ag and 15.6–34.2 nm for Cu	Mostly spherical	*Fusarium* spp., *Aspergillus* spp., *Trichoderma* spp. and *Penicillium* spp.	Gomaa et al. 2019
Stenotrophomonas sp.	Soil (agricultural farm)	Ag	5–30 nm	Almost spherical	*Sclerotium rolfsii*, *Alternar a alternata*, *Bipolaris sorokiniana*, and *Curvularia lunata*	Mishra et al. 2017
Bacillus licheniformis, *Bacillus pumilus*, and *B. persicus*	Soil	Ag	77–92 nm	-	Bean Yellow Mosaic Virus (BYMV)	Elbeshehy et al. 2015
Stenotrophomonas sp.	Soil (agricultural farm)	Ag	12.7 nm	Spherical	*Xanthomonas oryzae* pv. *oryzae*	Mishra et al. 2020

Platinum nanoparticles were around 5 nm in size. The particles were deposited in the periplasmic region, which is located between the cell's outer and inner membranes (Konishi et al. 2007).

Pt nanoparticles were synthesized by reducing ionic platinum with hydrogen gas from a precursor solution of potassium tetrachloroplatinate (K_2PtCl_4) (Aritonang et al. 2014). As a substrate, a bacterial cellulose matrix (coconut water fermented by the bacterium *Acetobacter xylinum*) was used (Radiman and Yuliani 2008). The formation of Pt nanoparticles on the cellulose membrane was seen through a color change, as the cellulose membrane's color changed from transparent white to black. Based on X-ray diffraction patterns, the calculated size of Pt nanoparticles was between 6.3 and 9.3 nm (Aritonang et al. 2014).

The extremely acidophilic Fe(III)-reducing bacteria *Acidocella aromatica* and *Acidiphilium crytpum* were used in a recent study by Matsumoto et al. to create Pt nanoparticles from a highly acidic Pt(IV) solution (as H_2PtCl_6 $6H_2O$) utilizing a one-step microbiological reaction. The different membrane proteins and enzymes that are responsible for the formation of nanocrystals were evident in the size and distribution of the Pt nanoparticles between the two species. The finest particles were formed by *Acidocella aromatica* cells, with mean and median sizes of 16.1 nm and 8.5 nm, respectively (Matsumoto et al. 2021). Platinum nanoparticles demonstrated good antibacterial activity against bacterial strains *Escherichia coli* and *Klebsiella pneumoniae*, although their antibacterial effects were eventually overcome by time kinetics (Vukoja et al. 2022). Table 1 lists the bacterial strains used in the biosynthesis of metal nanoparticles and their applications against plant pathogens.

Fungi-Based Synthesis of Noble Metal Nanoparticles and Their Applications Against Pathogens

Fungi-based synthesis processes are advantageous over bacteria in many respects, such as the easy isolation process of fungi through dilution plating, the comparatively large surface area of fungi due to the presence of mycelia, easy downstream processing, and fungi can accumulate heavy metals into their body as well as their tolerance ability to heavy metals (Roy et al. 2021; Ali et al. 2020; Khan et al. 2017; Pantidos and Horsfall 2014). Moreover, metabolite production by fungi is abundant compared to other microorganisms (Narayanan and Sakthivel 2010). Due to all these features, fungi-mediated synthesis of metal nanoparticles has gained great popularity (Chauhan et al. 2022; Dikshit et al. 2021; Ghosh et al. 2021; Zhang et al. 2020; Rafique et al. 2017; Guilger-Casagrande and Lima 2019; Singh et al. 2018; Shah et al. 2015)

Wang et al. synthesized Ag nanoparticles using the fungus *Aspergillus sydowii* through an extracellular process (Wang et al. 2021). The isolated fungus from the soil was cultured in a Sabouraud agar medium. Time-dependent reaction kinetics were monitored by UV-Vis spectrometer when silver nitrate solution was mixed into cell filtrate of *Aspergillus sydowii*. The absorbance peak at 420 nm confirmed the formation of Ag nanoparticles while the shape and size were observed by TEM. Most of the Ag nanoparticles were spherical in shape and the average size was 12 ± 2 nm. Three factors (temperature, pH, and concentration of $AgNO_3$) were identified that played an important role in the production of Ag nanoparticles. The optimal conditions to produce Ag nanoparticles were 50°C, pH 8.0 and 1.5 mM $AgNO_3$. However, various pathogenic fungi were selected to test the antifungal activity of the fungal-stabilized Ag nanoparticles. The results are presented in Table 2. The results shown in the table indicate that Ag nanoparticles play an active role in antifungal activity. *Trichoderma longibrachiatum* fungus was used as a reducing and stabilizing agent to synthesize Ag nanoparticles extracellularly (Elamawi et al. 2018). The synthesis process was carried out in the dark without shaking at optimal conditions, which were at 28°C with 10 g of fungal biomass and 1 mM $AgNO_3$ solution. Both DLS and TEM were used to determine the average size of the particles. However, based on TEM analysis, the particles were spherical and ranged in size from 5 nm to 25 nm. The inhibition efficacy of fungal-stabilized Ag nanoparticles was tested against different plant-pathogenic fungi (Table 3). The data presented in Table 3 showed that Ag nanoparticles were

Table 2. Effectiveness of Ag NPs compared with the standard antifungal drugs (reproduced from Wang et al. 2021).

Tested fungal strains	MIC (µg/mL)		
	Itraconazole	Fluconazole	Ag Nanoparticles
Candida albicans	0.03	0.25	0.25
Candida glabrata	0.25	8	0.125
Candida parapsilosis	0.25	8	0.25
Candida tropicalis	0.25	0.25	0.125
Fusarium solani	> 16	> 64	1
Fusarium moniliforme	> 16	> 64	2
Fusarium oxysporum	> 16	> 64	4
Aspergillus favus	0.125	> 64	1
Aspergillus fumigatus	0.03	> 64	2
Aspergillus terreus	0.25	> 64	2
Sporothrix schenckii	0.125	> 64	0.25
Cryptococcus neoformans	0.0625	2	0.25

Table 3. Inhibition efficacy of *T. longibrachiatum*-stabilized Ag NPs against different fungal pathogens (reproduced concise data from Elamawi et al. 2018).

Fungal Pathogens	Inhibition Efficacy of Ag Nanoparticles (%)
Alternaria alternata	93.0
Fusarium verticillioides	96.4
Fusarium moniliforme	93.6
Aspergillus flavus	86.7
Aspergillus heteromorphus	83.6
Penicillium glabrum	75.7
Penicillium brevicompactum	92.9
Hirschmanniella oryzae	95.1
Pyricularia grisea	98.9

effective against fungal growth. More than 90% inhibition efficacy against six pathogenic fungi out of nine. *Trichoderma asperellum* stabilized Ag nanoparticles were applied against the growth of several plant pathogens, *Sclerotinia sclerotiorum*, *Sclerotium rolfsii*, *Rhizoctonia solani*, and *Fusarium oxysporum* (Kaman and Dutta 2017). They found that biosynthesized Ag nanoparticles were effective (more than 65%) against the mycelial growth of the pathogens.

Mishra et al. reported biosynthesis of Au nanoparticles using the fungus *Aspergillus terreus* as a reducing and stabilizing agent (Mishra et al. 2022). The Au nanoparticles were produced under different synthesis conditions, such as changing the temperature, changing the pH, changing the amount of gold salt in the solution, and changing the amount of fungal extract. However, the optimal conditions of synthesis were obtained at 40°C for 24 hours, pH 8, with an aqueous $HAuCl_4$ solution concentration of 1 mM, and an extracted fungal powder of 100 ppm. The biosynthesis reaction was conducted at 150 rpm in an orbital shaker. *A. terreus*-stabilized Au nanoparticles were nearly spherical and their size ranged from 9 nm to 14 nm based on TEM analysis. The authors studied the antimicrobial activity of Au nanoparticles against foodborne pathogens and plant pathogens. Two types of phytopathogenic (plant pathogens) fungi, *Fusarium oxysporum*, and *Rhizoctonia solani* were selected for the test in a PDA medium. Various amounts of colloidal solution of Au nanoparticles were added to the PDA medium so that the concentrations of Au solution reach 25 µg/ml, 50 µg/ml, 100 µg/ml, and 200 µg/ml PDA. The highest percentages of inhibition against the radial mycelial growth of *Fusarium oxysporum* and *Rhizoctonia solani* were 65.46 and 52.5 at the concentration of 200 µg/ml PDA. Priyadarshini et al. followed a quick synthesis process to produce Au nanoparticles using a similar fungus *A. terreus* at pH 10 (Priyadarshini et al. 2014). The nanoparticles they produced ranged in size from 10 nm to 19 nm, which were tested against the growth of Gram-negative bacteria, *Escherichia coli*, and found to have antibacterial activity. Molnár et al. conducted research on the synthesis of Au nanoparticles using a huge number of thermophilic fungi (Molnár et al. 2018). Various synthetic strategies were used to produce Au nanoparticles using different extracts (intracellular extract, extracellular extract, and autolysate) of fungi (Figure 9). The size of the nanoparticles ranged from 6 nm to 40 nm depending on the variety of fungi and experimental factors. It was remarked by conclusion that fungi-based chemical compounds that have a molecular weight greater than 3 kDa (such as proteins or other biomolecules) could successfully stabilize Au nanoparticles, but they were unable to reduce gold ions to zero-valent gold particles. However, biomolecules of fungi, those that have a molecular weight less than 3 kDa (such as amino acids, peptides, primary metabolites, mono- and oligosaccharides, etc.) could successfully produce Au nanoparticles from Au ions. Hemashekhar et al. (2019) reported the biosynthesis of Au nanoparticles using the endophytic fungus *Alternaria* spp. The fungus was isolated from the root of the *Rauvolfia tetraphylla* plant. The fungal biomass was collected through centrifugation at 10,000 rpm for 15 minutes after proper culture with incubation. The extracted biomass (5 ml) was mixed with gold chloride solution (5 mM, 50 ml) at room temperature for 30 minutes to obtain Au nanoparticles identified by color (ruby red) and UV-Vis spectrum (surface plasmin peak at 530 nm). However, Au nanoparticles were collected through centrifugation, and various shapes and sizes were observed on the TEM micrograph (average size 28 nm). The nanoparticles were tested against Gram-negative (*Staphylococcus aureus*) and Gram-positive (*Escherichia coli, Pseudomonas*) bacteria, and found antibacterial activities.

Pt nanoparticles were synthesized by the fungus *Fusarium oxysporum*. To produce Pt nanoparticles, hexachloroplatinic acid solution was added into the filtrate fungal mycelium

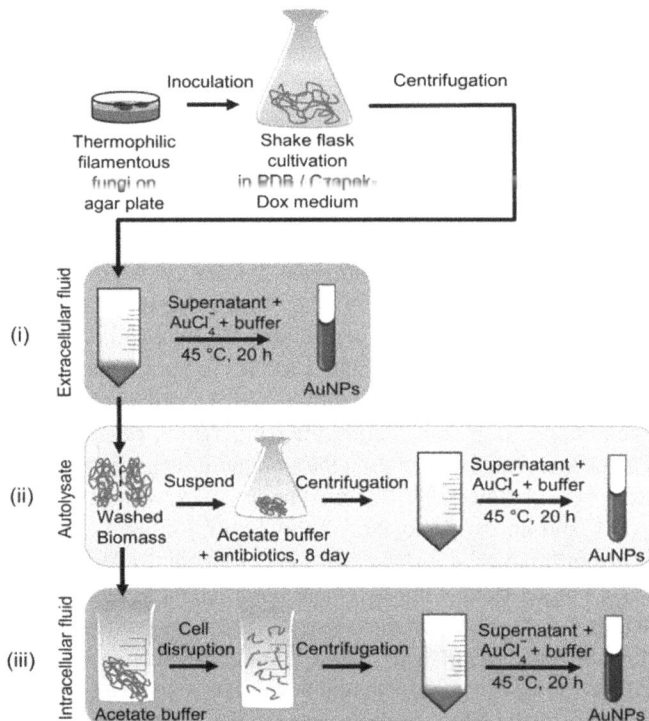

Figure 9. Synthesis processes of Au nanoparticles using different extracts of fungi (Molnár et al. 2018).

Table 4. Fungi-mediated synthesis of metal nanoparticles and their applications against plant pathogens.

Name of Fungus	Source of Isolation	Types of Metal NPs	Size of NPs (nm)	Shape	Application Against Plant Pathogens	References
Fusarium oxysporum	Wilted tomato plants	Ag	16–27 nm	Spherical	*Pectobacterium carotovorum*	Ghazy et al. 2021
Penicillium duclauxii	Corn grains	Ag	3–32	Spherical	*Bipolaris sorghicola*	Almaary et al. 2020
Aspergillus niger	Infected Fruits (grapes or soft dates)	Ag	10–100 nm	Undetected	*Aspergillus flavus, Fusarium oxysporum* and *Penicillium digitatum*	Al-Zubaidi et al. 2019
Setosphaeria rostrata	Leaves of *Solanum nigrum*	Ag	2–50 nm	Spherical, cylindrical, etc.	*Rhizoctonia solani, Fusarium udum, Aspergillus niger,* and *Fusarium graminearum*	Akther and Hemalatha 2019
Trichoderma harzianum	Soil	Ag	20–30 nm	Spherical	*Sclerotinia sclerotiorum*	Guilger et al. 2017
Penicillium citrinum	Grape	Ag	3–13 nm	Mostly spherical	*Aspergillus flavus* var. *columnaris*	Yassin et al. 2017
Trichoderma hazarium	Tomato	Ag	12.7 ± 0.8nm	Spherical	*Alternaria alternata, Helminthosporium* sp., *Betrytis* sp., and *Phytophthora arenaria*	EL-Moslamy et al. 2017
Arthroderma fulvum	Soil	Ag	15.5 ± 2.5 nm	Spherical	*Candida* spp., *Aspergillus* spp., and *Fusarium* spp.	Xue et al. 2016
Trichoderma sp.	Roots of oak trees	Ag	5–35 nm	Spherical	*Enterococcus perryi*	Qu et al. 2021
Cephalosporium sp. and *Trichoderma* sp.	Soil (rhizosphere of chickpea plants)	Ag	20–50 nm	-	*Fusarium oxysporum* f. sp.*ciceri*	Kaur et al. 2018
Penicillium verrucosum	Vegetable-cultivated greenhouse soil	Ag	10–12 nm	Irregular	*Fusarium chlamydosporum* and *Aspergillus flavus*	Yassin et al. 2021
Aspergillus terreus	Roots of *Datura metel*	Au	9–14 nm	Mostly spherical	*Fusarium oxysporum* and *Rhizoctonia solani*	Mishra et al. 2022
Fusarium oxysporum f. sp. *cubense*	Infected banana plants	Au	15–30 nm	Undetected	*Pseudomonas* sp.	Thakker et al. 2013
Phoma sp.	Vascular tissue of peach trees	Au	10–100 nm	Spherical	*Rhizoctonia solani* and *Xanthomonas oryzae* pv. *oryzae*	Nejad et al. 2022

while the overall solution concentration reached to 1 mM, and the pH of the solution was kept between 6 and 7 with the addition of sodium hydroxide (Gupta and Chundawat 2019). The reaction temperature was approximately 100°C on a shaker. The antimicrobial activity of fungal-stabilized Pt nanoparticles (average size 25 nm) was tested against the growth of specific bacteria and fungi. *Fusarium oxysporum* stabilized Pt nanoparticles were also synthesized at room temperature on a rotary shaker at 200 rpm (Syed and Ahmad 2012). Fungal mycelium was mixed into 1 mM aqueous solution of hexachloroplatinic acid, and the reaction continued for 96 hours to obtain fungal-stabilized Pt nanoparticles. The Pt nanoparticles produced by this method were spherical and their size range was between 5–30 nm. Previously, Ag (Ahmad et al. 2003) and Au (Mukherjee et al. 2002) nanoparticles were synthesized extracellularly by using the fungus *Fusarium oxysporum,* where the overall concentration both for the Ag^+ and Au^{3+} was 1mM. The synthesis processes for producing Ag and Au nanoparticles proceeded in the dark at ambient temperature for 72 hours. Based on TEM analysis, the size of the Ag nanoparticles was in the range, of 5–50 nm, whereas the size range for Au nanoparticles was 8–40 nm. However, a similar biosynthesis approach may be applied to the synthesis of other noble metal nanoparticles, such as Pd, Ru, etc. Fungi-mediated synthesis of metal nanoparticles and their applications against plant pathogens are provided in Table 4.

Plant-Based Synthesis of Noble Metal Nanoparticles and Their Applications Against Pathogens

Plant-mediated biosynthesis of metal nanoparticles is more advantageous over microorganism-mediated biosynthesis because plants are readily available, they are easy to process, and plant extracts can be obtained through one-step or two-step processes. Before collecting the biomass of microorganisms, two important steps are followed: one is the isolation of microorganisms and the other is the culture of microorganisms in a proper medium. However, these are not required in the case of plants. Thus, plant-based synthesis of metal nanoparticles is cost-effective and environmentally friendly. In the introduction part of this chapter, we discussed the different parts of plants and the biomolecules that contain in plant extracts. However, the biosynthesis of metal nanoparticles using various plant extracts has become widespread in recent years (Kumar et al. 2023; Adeyemi et al. 2022; Parmar and Sanyal 2022; Al-Radadi 2022; Sharma et al. 2022; Shreyash et al. 2021; Bao et al. 2021; Dikshit et al. 2021; Selvakesavan and Franklin 2021; Kushwah and Verma 2021). To synthesize plant-mediated metal nanoparticles, it is important to know the effects of biomolecules present in plant extracts because they play effective roles in converting metal ions to zero-valent metal, growth kinetics of metallic embryos, and the stabilization of metal nanoparticles. Plant extract is a complex composition composed of various biochemicals as mentioned earlier. Among these bio-compounds, phenolic acids and flavonoids with variable phenolic structures are usually found in various parts of plants, such as leaves, fruits, grains, roots, stems, bark, flower, etc. (Panche et al. 2016; Khoddami et al. 2013). Hydroxyl-rich phenolics are generally considered as reducing agents for metal ions. For instance, Pradeep et al. explored the insights of a medicinal plant (*Hypericum perforatum* Linn known as St. John's wort) extract-mediated synthesis of Ag nanoparticles (Pradeep et al. 2022). They found that among the phytochemicals of the *Hypericum perforatum* extract, polar parts of phenolic acids and flavonoids, such as quercetin, kaempferol-3-glucoside, and quercetin-3-glucoside acted as reducing agents for Ag^+ ions.

Moreover, they explored that comparatively low polar parts, available in phloroglucinols (hyperforin) and xanthones (mangiferin), were unable to reduce Ag^+ ions, but they played a vital role in stabilizing the Ag nanoparticles. It was noted that naphthodianthrones, such as hypericin, pseudohypericin, and protohypericin acted as reducing as well as capping agents. Different leaf extracts contain phytochemicals that may have different types and amounts that have effects on the formation, growth, and stabilization of metal nanoparticles. Alex et al. studied the effects of leaf extracts of neem (*Azadirachta indica*), aloe vera (*Aloe vera*), Indian mint (*Coleus amboinicus*), and guava (*Psidium guajava*) on the formation of Ag nanoparticles (Alex et al. 2020). They found that

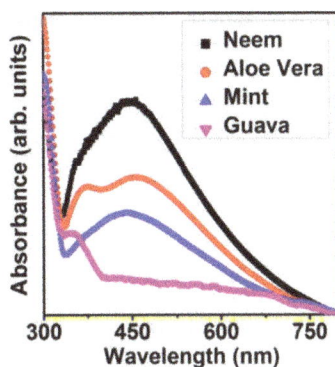

Figure 10. UV-Visible absorption spectra of Ag nanoparticles synthesized with the extracts of neem, aloe vera, mint, and guava (Alex et al. 2020).

the leaf extract of neem contain 44.65% phytol, which was the major element in the extract that vitally contributed to the reduction of Ag^+ ions. The major elements of phytochemicals present in leaf extracts of aloe vera, mint, and guava are alpha-iridescence (10.50%), 2-p-Cymenol (55.28%) and caryophyllene (34.89%), respectively. The UV-Vis absorption spectra of Ag nanoparticles synthesized with leaf extracts of neem, aloe vera, mint, and guava are presented in Figure 10. Hemlata et al. used a leaf extract of *Cucumis prophetarum* to synthesize Ag nanoparticles, where flavonoids were absent in the leaf extract, but tannins, alkaloids, triterpenoids, phenol, and saponins were present (Hemlata et al. 2020). However, plant extract-mediated synthesis of Ag nanoparticles and their applications in plant disease control are being researched. A huge number of plants and their parts, such as roots, stems, leaves, fruits, and flowers, are still unexplored. The phytochemicals present in various parts of plants, the interactions between these phytochemicals and metal ions, and the antimicrobial activities of phytochemical stabilized metal nanoparticles have aroused great scientific interest.

The medicinal plant *Mentha pulegium* was studied widely among the *Mentha* species (El Hassani 2020). The leaf extract of *Mentha pulegium* was used to synthesize Ag nanoparticles at various conditions, such as pH, concentrations, and incubation time, while the biochemicals of the leaf extract acted as reducing and stabilizing agents (Wang and Wei 2022; Gholami and Azarbani 2021; Kelkawi et al. 2017). The biosynthesized Ag nanoparticles (stabilized by the leaf extract of *Mentha pulegium*) showed potential antibacterial activities. In another study, Ag nanoparticles synthesized using the extract of neem leaf were applied against various plant-pathogenic fungi such as *Fusarium oxysporum*, *Fusarium graminearum*, and *Aspergillus niger* (Bawskar et al. 2021). Mankad et al. also synthesized Ag nanoparticles using neem leaf extract. These nanoparticles were tested against the phytopathogenic bacteria *Xanthomonas oryzae* pv. *Oryzae* showed strong antibacterial efficacy compared to the standard antibacterial antibiotic streptocycline (Mankad et al. 2020). The extracts of *Piper nigrum* leaf and stem were used to synthesize Ag nanoparticles (Paulkumar et al. 2014). The synthesized Ag nanoparticles using leaf and stem were in the size range of 7–50 nm and 9–30 nm, respectively. Both the biosynthesized Ag nanoparticles showed potential antibacterial activities against the phytopathogenic bacteria *Erwinia cacticida* and *Citrobacter freundii*. Javed et al. used leaf extract of *Mentha longifolia* to synthesize Ag nanoparticles and applied these Ag nanoparticles against a variety of phytopathogenic bacteria, such as *Xanthomonas vesicatoria*, *Pectobacterium carotovorum*, *Ralstonia solanacearum*, and *Xanthomonas oryzae* (Javed et al. 2020). Ag nanoparticles were produced when the reaction mixture of $AgNO_3$ and the leaf extract of *Polyalthia longifolia* was kept in dark for 24 hours at pH 7 and 60°C (Dashor et al. 2022). The Ag nanoparticles were spherical in shape and their sizes ranged from 10–40 nm, which were tested against the pathogenic fungal *Alternaria alternata* and showed significant inhibition of the fungal growth. *Melia azedarach* leaf extract was used as a reducing and stabilizing agent for the synthesis of Ag nanoparticles (Jebril et al. 2020). The biosynthesized Ag nanoparticles were

Figure 11. Untreated eggplants (top), eggplants treated with Ag nanoparticles (bottom) (Jebril et al. 2020).

spherical in shape and their size range was 18–30 nm. The nanoparticles of Ag were tested (*in vitro* and *in vivo*) against the phytopathogenic fungal *Verticillium dahlia* available in eggplants. It was observed that the eggplants treated with the Ag nanoparticles were in good health compared to the untreated plants (Figure 11). Ag nanoparticles (3.7–29.3 nm) were synthesized using rice (*Oryza sativa*) leaf extract and applied *in vitro* against the fungus *Rhizoctonia solani* that causes sheath blight disease in rice (Kora et al. 2020). The maximum percentage of inhibition against the mycelial growth of *R. solani* was 96.7% at the concentration of 10 μg/ml of Ag nanoparticles. Namburi et al. synthesized Ag nanoparticles using leaf extract of rice and applied these nanoparticles successfully against phytopathogenic bacterial strain (*Xanthomonas oryzae* pv. *oryzae*) that causes leaf blight disease in rice (Namburi et al. 2021).

Khan et al. used aqueous fruit extract (strawberry waste after fruit juice extraction) to synthesize Ag nanoparticles (Khan et al. 2021). The pure aqueous strawberry extract was obtained followed by cleaning of waste, drying in sunlight, grinding, refluxing in water for 30 minutes, and cooling to room temperature. To synthesize Ag nanoparticles, 10 ml of purified strawberry extract was mixed with 1 mM aqueous solution of $AgNO_3$ (90 ml), and the mixture was stirred for 30 minutes while the colorless solution turns to wine red. The changing color indicated the formation of silver particles from the ionic nature of silver. The wine-red solution was tested spectroscopically. The UV-Vis spectrum with an absorption band at 415 nm confirmed the formation of Ag nanoparticles. Based on TEM images, the particle size was determined in the range of 55–70 nm, and the particles were rectangular in shape. The strawberry extract stabilized Ag nanoparticles were tested *in vitro* against the plant-pathogenic fungus *Fusarium oxysporum*, bacterium *Ralstonia solanacearum*, and plant parasitic root-knot nematode *Meloidogyne incognita*, and the particles showed antifungal, antibacterial, and antinematode activities. Various parts of strawberry plants, such as strawberry fruit (Umoren et al. 2017; Kulkarni 2015), strawberry seed (Ali et al. 2022), strawberry fruit pomace (Alam 2022), etc. were used to synthesize Ag nanoparticles. Fruit extract of *Phyllanthus emblica*

(Amla) was used to synthesize Ag nanoparticles (Masum et al. 2019). The biosynthesized Ag nanoparticles ranged in size from 19.8 nm to 92.8 nm and the nanoparticles showed antimicrobial activity against the pathogen *Acidovrax oryzae*, which is one of the pathogenic bacteria responsible for the brown stripe disease of rice. Tian et al. synthesized Ag nanoparticles using the leaf extracts of *Taraxacum mongolicum* and *Solanum melongena*, and the fruit extract of *Arctium lappa* (Tian et al. 2022). The synthesized Ag nanoparticles were spherical in shape and their size ranged from 20 nm to 40 nm. The antimicrobial activity of all three classes of nanoparticles was tested *in vitro* against the leaf blight pathogen (*Xanthomonas oryzae* pv. *Oryzae*) of rice. The maximum number of bacterial pathogens decreased by the application of Ag nanoparticles (synthesized from the fruit extract of *Arctium lappa*) was 70.10%. Saha et al. synthesized Ag nanoparticles using the peel extract of the Satkara (*Citrus macroptera*) fruit. The synthesized Ag nanoparticles were spherical in shape and their average diameter was measured at 11 nm (Saha et al. 2021). Riaz et al. reported the synthesis of Ag nanoparticles using the extracts of *Cucumis sativus* (cucumber) and *Aloe vera* (Riaz et al. 2022). Pod extract of *Pisum sativum* L. (pea) was used to synthesize Ag nanoparticles that showed antibacterial activity (Alarjani et al. 2022). Acharya et al. used onion extract to synthesize Ag and Au nanoparticles, and these nanoparticles were applied to onion seeds to enhance germination (Acharya et al. 2019). Bark extracts of *Zanthozylum armatum* and *Aesculus indica* were used to produce Ag nanoparticles that showed potential antimicrobial activities (Riaz et al. 2022). The fruit of *Capsicum annuum* (chili pepper) was used to synthesize Au nanoparticles (Qais et al. 2021). The Au nanoparticles were obtained when the mixture of aqueous salt of $HAuCl_4$ (1 mM) was mixed with aqueous extract of *Capsicum annuum* followed by stirring for 2 hours using a magnetic stirrer. The formation of Au nanoparticles was confirmed by the UV-visible absorption band at 521 nm. The Au nanoparticles ranged in size from 7–36 nm that were tested on bacterial strains *Pseudomonas aeruginosa* and *Serratia marcescens*. The various virulent factors such as colonization and adherence, and different activities such as exoprotease, elastase, pyoverdine, pyocyanin, swimming, and rhamnolipids production were reduced when the Au nanoparticles were applied to the bacterial strains. It can be noted by findings with various reports that bacteria *Pseudomonas aeruginosa* and *Serratia marcescens* are pathogenic to plants (Wang et al. 2015; Walker et al. 2004; Fu et al. 2021), as well as they are the cause of various diseases in humans and animals. Au nanoparticles were synthesized using onion peel extract containing quercetin which served as a reducing agent for Au ions and a capping agent for Au nanoparticles (Phukan et al. 2021). Extracts of different fruits, such as *Musa acuminate* (banana), *Malus domestica* (apple), *Prunus persica* (peach) and *Actini diadeliciosa* (kiwi), were used to synthesize Au nanoparticles at various pH and reaction conditions (Yang et al. 2017). The size of Au nanoparticles varied from 1.5 nm to 7.5 nm by adjusting the pH of the reactant solution after the reduction of Au ions by various fruit extracts. Au nanoparticles were synthesized using extracts of various parts of different plants such as *Azadirachta indica* (flower), *Ocimum tenuiflorum* (flower and leaf), *Mentha spicata* (leaf), and *Citrus sinensis* (peel) (Rao et al. 2017). The size range of Au nanoparticles synthesized by different extracts was 20–30 nm. The antibacterial activity of the synthesized Au nanoparticles was studied against *Staphylococcus aureus*, *Pseudomonas aeruginosa*, and *Klebsiella pneumoniae*. The efficacy of inhibition of Au nanoparticles against the bacterial strains was compared with the standard drug and found that maximum efficacy was 99% when the concentration of Au nanoparticles was in the range of 512–600 μg/ml. It can be mentioned that *Klebsiella pneumoniae* inhibits the respiration of the roots of the plant leading to the death of the plant as well as inhibits human respiration (Ajayasree and Borkar 2018).

Platinum nanoparticles were synthesized using neem leaf extract and these nanoparticles were spherical and had a size range of 5–50 nm (Thirumurugan et al. 2016). Pt nanoparticles were synthesized using the plant extract of *Taraxacum laevigatum* (Tahir et al. 2017). The biosynthesized Pt nanoparticles were spherical in shape and their size was in the range of 2–7 nm. *T. laevigatum* extract-mediated Pt nanoparticles were tested against the bacterial strains *Pseudomonas aeruginosa* and *Bacillus subtilis*. The zones of inhibition of Pt nanoparticles against *Pseudomonas aeruginosa*

Table 5. Metal nanoparticles synthesized from plant extracts and their applications against plant pathogens.

Plant Name	The Part of the Plant Used	Types of Metal NPs	Size of NPs (nm)	Shape	Application Against Pathogens	References
Fragaria × ananassa (Strawberry plant)	Strawberry fruits (waste after juice extraction)	Ag	55–70 nm	Rectangular	*Fusarium oxysporum* (fungus), *Ralstonia solanacearum* (bacterium), and *Meloidogyne incognita* (root-knot nematode)	Khan et al. 2021
Azadirachta indica (Neem)	Leaf	Ag	15–35 nm	Spherical	*Fusarium oxysporum, Fusarium graminearum,* and *Aspergillus niger*	Bawskar et al. 2021
Piper nigrum	Leaf and Stem	Ag	7–50 nm (leaf) and 9–30 nm (stem)	Spherical, hexagonal, cluster	*Erwinia cacticida* and *Citrobacter freundii*	Paulkumar et al. 2014
Phyllanthus emblica	Fruit	Ag	19.8–92.8 nm	Spherical	*Acidovrax oryzae*	Masum et al. 2019
Taraxacum mongolicum, Solanum melongena, and *Arctium lappa*	Leaf and fruit	Ag	20–40 nm	Spherical	*Xanthomonas oryzae* pv. *oryzae*	Tian et al. 2022
Mentha longifolia	Leaf	Ag	~ 20–100 nm	Spherical, cubical and triangular	*Xanthomonas vesicatoria, Pectobacterium carotovorum, Ralstonia solanacearum,* and *Xanthomonas oryzae*	Javed et al. 2020
Taraxacum mongolicum, Solanum melongena and *Arctium lappa*	Leaf and fruit	Ag	20–40 nm	Spherical	*Xanthomonas oryzae* pv. *oryzae*	Tian et al. 2022
Polyalthia longifolia	Leaf	Ag	10–40 nm	Spherical	*Alternaria alternata*	Dashor et al. 2022
Melia azedarach	Leaf	Ag	18–30 nm	Spherical	*Verticillium dahlia*	Jebril et al. 2020
Oryza sativa	Leaf	Ag	3.7–29.3 nm	Spherical	*Rhizoctonia solani*	Kora et al. 2020
Oryza sativa	Leaf	Ag	5–33 nm	Spherical	*Xanthomonas oryzae* pv. *oryzae*	Namburi et al. 2021
Pongamia glabra (Karanja)	Leaf	Ag	6–80 nm	Spherical	*Rhizopus nigricans*	Sahayaraj et al. 2020
Azadirachta indica (Neem)	Leaf	Ag	67.94–133.2 nm	–	*Xanthomonas oryzae* pv. *oryzae*	Mankad et al. 2020
Piper betle	Leaf	Ag	6–14 nm	Spherical	*Alternaria brassicae* and *Fusarium solani*	Khan et al. 2020
Capsicum annuum	Fruit	Au	7–36 nm	Spherical, anisotropic	*Pseudomonas aeruginosa* and *Serratia marcescens*	Qais et al. 2021
Capsicum annuum	Fruit	Au	7–36 nm	Spherical, anisotropic	*Pseudomonas aeruginosa* and *Serratia marcescens*	Qais et al. 2021
Azadirachta indica, Ocimum tenuiflorum, Mentha spicata, and *Citrus sinensis*	Flower, leaf, and peel	Au	20–30 nm	Spherical, anisotropic	*Pseudomonas aeruginosa* and *Klebsiella pneumoniae*	Rao et al. 2017
Taraxacum laevigatum	Plant	Pt	2–7 nm	Spherical	*Pseudomonas aeruginosa*	Tahir et al. 2017

and *Bacillus subtilis* were (15 ± 0.5) mm and (18 ± 0.8) mm, respectively. However, these zones of inhibition were lower than the zones of inhibition measured by the standard drug. As mentioned earlier, *Pseudomonas aeruginosa* is a plant-pathogenic Gram-negative bacterium that is also pathogenic to animals. However, *Bacillus subtilis* (Gram-positive bacterium) promotes plant growth by suppressing diseases caused by pathogens (Hashem et al. 2019). An alkaloid fraction of *Peganum harmala* seed was used to synthesize Pt, Pd, and Pt-Pd nanoparticles (Fahmy et al. 2021). Based on TEM analysis, the sizes of Pt, Pd and Pt-Pd nanoparticles were in the range of 18.4–22.2 nm, 16.8–28.2 nm, and 28.1–38.9 nm, respectively. Table 5 shows a list of metal nanoparticles synthesized from various plant extracts and their applications against plant pathogens.

Conclusions

It has been observed in recent years that green synthesis of nanoparticles is growing rapidly because green synthesis has some advantages such as availability, cost-effectiveness, non-toxic precursors/chemicals, ecologically friendly products, and byproducts. Metal nanoparticles are produced using green synthesis methods using plant extracts and microorganisms (such as bacteria and fungi) that are widely available and processable. The use of microbes or microorganisms is one of the prominent approaches to biological methods of nanoparticle synthesis, especially, for the synthesis of noble metal nanoparticles extracellularly and intracellularly. Studies have shown that various functional groups (such as $-NH_2$, $-OH$, $-SH$, and $-COOH$) of microbial proteins act as major reducing or capping agents during the synthesis of metal nanoparticles. Plant-mediated biosynthesis of metal nanoparticles is more advantageous than microbial-mediated biosynthesis because plants are readily available, they are easy to process, and plant extracts can be obtained through one-step or two-step processes. However, plant extracts are a complex composition composed of various biochemicals substances such as polysaccharides, proteins, amino acids, organic acids, polyphenols, flavonoids, terpenoids, alkaloids, tannins, and alcoholic substances that can reduce and stabilize the metal nanoparticles. A few drawbacks of biosynthesized nanoparticles compared to chemically synthesized nanoparticles have been noticed. (i) Since different functional groups of biochemicals can, in a given synthesis, act as reducing and stabilizing agents, their size range can be wide, and their shape can be irregular. (ii) Physical and chemical properties of plants (e.g., plant growth, the chemical composition of plants) are affected by various factors such as climatic conditions, sunlight intensity and exposure time. The factors usually vary due to seasons like summer, monsoon, autumn and winter. However, since the amount and nature of plant chemical constituents are not the same throughout the year, the synthesis conditions (such as concentrations of metal ions and plant extracts, solution pH, temperature, etc.) need to be optimized in different seasons.

The application of nanomaterials against plant pathogens is an emerging area of nanotechnology. The biosynthesized metal nanoparticles were tested against various plant pathogens and showed strong antimicrobial activity. But in most cases, nanoparticles were tested against pathogens in the laboratory (*in vitro*), so nanoparticles should be applied against pathogens in the plant field (*in vivo*) to know the real situation. To evaluate the efficacy of nanoparticles against various crop diseases caused by pathogens, the production of nanoparticles should be in a bulk state. A holistic evaluation of biosynthesized metal nanoparticles in terms of environmental safety, cost-effectiveness, efficacy of crop disease management, and reproducibility is required compared to commercial insecticides and biocontrol agents.

References

Acharya, P., Jayaprakasha, G.K., Crosby, K., Jifon, J. and Patil, B.S. 2019. Green-synthesized nanoparticles enhanced seedling growth, yield, and quality of onion (*Allium cepa* L.). ACS Sustainable Chem. Eng. 7(17): 14580–14590.

Adeyemi, J.O., Oriola, A.O., Onwudiwe, D.C. and Oyedeji, A.O. 2022. Plant extracts mediated metal-based nanoparticles: synthesis and biological applications. Biomolecules 12: 627.

Ahmad, A., Mukherjee, P., Senapati, S., Mandal, D., Khan, M.I., Kumar, R. and Sastry, M. 2003. Extracellular biosynthesis of silver nanoparticles using the fungus *Fusarium oxysporum*. Colloids Surf. B. 28: 313–318.

Ahmad, A., Senapati, S., Khan, M.I., Kumar, R., Ramani, R., Srinivas, V. and Murali, S. 2003. Intracellular synthesis of gold nanoparticles by a novel alkalotolerant actinomycete, *Rhodococcus* species. Nanotechnology 14: 824–828.

Ahmed, T., Shahid, M., Noman, M., Niazi, M.B.K., Mahmood, F., Manzoor, I., Zhang, Y., Li, B., Yang, Y., Yan, C. and Chen, J. 2020. Silver nanoparticles synthesized by using *Bacillus cereus* SZT1 ameliorated the damage of bacterial leaf blight pathogen in rice. Pathogens 9: 160.

Aiken III, J.D. and Finke, R.G. 1999. A review of modern transition-metal nanoclusters: their synthesis, characterization, and applications in catalysis. J. Mol. Catal. A Chem. 145: 1–44.

Ajayasree, T.S. and Borkar, S.G. 2018. Pathogenic potentiality of the bacterium of *Klebsiella pneumoniae* strain Borkar on different plants. J. Appl. Biotechnol. Bioeng. 5(4): 233–235.

Akther, T. and Hemalatha, S. 2019. Mycosilver nanoparticles: synthesis, characterization and its efficacy against plant pathogenic fungi. BioNanoSci. 9: 296–301.

Alam, M. 2022. Analyses of biosynthesized silver nanoparticles produced from strawberry fruit pomace extracts in terms of biocompatibility, cytotoxicity, antioxidant ability, photodegradation, and *in-silico* studies. J. King Saud. Univ. Sci. 34: 102327.

Alarjani, K.M., Huessien, D., Rasheed, R.A. and Kalaiyarasi, M. 2022. Green synthesis of silver nanoparticles by *Pisum sativum* L. (pea) pod against multidrug-resistant foodborne pathogens. J. King Saud. Univ. Sci. 34: 101897.

Alex, K.V., Pavai, P.T., Rugmini, R., Prasad, M.S., Kamakshi, K. and Sekhar, K.C. 2020. Green synthesized ag nanoparticles for bio-sensing and photocatalytic applications. ACS Omega 5: 13123–13129.

Ali, F., Younas, U., Nazir, A., Hassan, F., Iqbal, M., Hamza, B., Mukhtar, S., Khalid, A. and Ishfaq, A. 2022. Biosynthesis and characterization of silver nanoparticles using strawberry seed extract and evaluation of their antibacterial and antioxidant activities. J. Saudi Chem. Soc. 26: 101558.

Ali, M.A., Ahmed, T., Wu, W., Hossain, A., Hafeez, R., Masum, M.M.I., Wang, Y., An, Q., Sun, G. and Li, B. 2020. Advancements in plant and microbe-based synthesis of metallic nanoparticles and their antimicrobial activity against plant pathogens. Nanomaterials 10: 1146.

Alivisatos, A.P. 1996. Perspectives on the physical chemistry of semiconductor nanocrystals. J. Phys. Chem. 100: 13226–13239.

Almaary, K.S., Sayed, S.R.M., Abd-Elkader, O.H., Dawoud, T.M., El Orabi, N.F. and Elgorban, A.M. 2020. Complete green synthesis of silver-nanoparticles applying seed-borne *Penicillium duclauxii*. Saudi J. Biol. Sci. 27: 1333–1339.

Al-Radadi, N.S. 2022. Laboratory-scale medicinal plants mediated green synthesis of biocompatible nanomaterials and their versatile biomedical applications. Saudi J. Biol. Sci. 29: 3848–3870.

Al-Zubaidi, S., Al-Ayafi, A. and Abdelkader, H. 2019. Biosynthesis, characterization and antifungal activity of silver nanoparticles by *Aspergillus niger* isolate. J. Nanotechnol. Res. 1(1): 23–36.

Ardukuni, L.O., Ourendar, A., Thangavelu, L. and Mandal, T. 2021. Silver nanoparticles (Ag NPs) as catalyst in chemical reactions. Synthetic Communications 51(10): 1516–1536.

Aritonang, H.F., Onggo, D., Ciptati, C. and Radiman, C.L. 2014. Synthesis of platinum nanoparticles from K2PtCl4 solution using bacterial cellulose matrix. J. Nanomater. Volume 2014, Article ID 285954.

Astruc, D. 2020. Introduction: Nanoparticles in catalysis. Chem. Rev. 120: 461–463.

Bahrulolum, H., Nooraei, S., Javanshir, N., Tarrahimofrad, H., Mirbagheri, V.S., Easton, A.J. and Ahmadian, G. 2021. Green synthesis of metal nanoparticles using microorganisms and their application in the agrifood sector. J. Nanobiotechnol. 19: 86.

Baig, N., Kammakakam, I. and Falatha, W. 2021. Nanomaterials: a review of synthesis methods, properties, recent progress, and challenges. Mater. Adv. 2: 1821–1871.

Baker, S., Rakshith, D., Kavitha, K.S., Santosh, P., Kavitha, H.U., Rao, Y. and Satish, S. 2013. Plants: Emerging as nanofactories towards facile route in synthesis of nanoparticles. BioImpacts 3(3): 111–117.

Bao, Y., He, J., Song, K., Guo, J., Zhou, X. and Liu, S. 2021. Plant-extract-mediated synthesis of metal nanoparticles. J. Chem. 2021: Article ID 6562687.

Barhoum, A., García-Betancourt, M.L., Jeevanandam, J., Hussien, E.A., Mekkawy, S.A., Mostafa, M., Omran, M.M., Abdalla, M.S. and Bechelany, M. 2022. Review on natural, incidental, bioinspired, and engineered nanomaterials: history, definitions, classifications, synthesis, properties, market, toxicities, risks, and regulations. Nanomaterials 12(2): 177.

Bawskar, M., Bansod, S., Rathod, D., Santos, C.A.D., Ingle, P., Rai, M. and Gade, A. 2021. Silver nanoparticles as nanofungicide and plant growth promoter: evidences from morphological and chlorophyll 'a' fluorescence analysis. Adv. Mater. Lett. 12(10): 2115702.

Bayda, S., Adeel, M., Tuccinardi, T., Cordani, M. and Rizzolio, F. 2020. The history of nanoscience and nanotechnology: from chemical–physical applications to nanomedicine. Molecules 25: 112.

Brock, T.D. and Gustafson, J. 1976. Ferric iron reduction by sulfur- and iron-oxidizing bacteria. Appl. Environ. Microbiol. 32: 567–571.

Brus, L.E. 1983. A simple model for the ionization potential, electron affinity, and aqueous redox potentials of small semiconductor crystallites. J. Chem. Phys. 79(11): 5566–5571.

Canelas, A.B, Pierick, A.T., Ras, C., Seifar, R.M., Dam, J.C.V., Gulik, W.M.V. and Heijnen, J.J. 2009. Quantitative evaluation of intracellular metabolite extraction techniques for yeast metabolomics. Anal. Chem. 81: 7379–7389.

Chaudhary, R., Nawaz, K., Khan, A.K., Hano, C., Abbasi, B.H. and Anjum, S. 2020. An overview of the algae-mediated biosynthesis of nanoparticles and their biomedical applications. Biomolecules 10: 1498.

Chauhan, A., Anand, J., Parkash, V. and Rai, N. Published online: 19 Jan 2022. Biogenic synthesis: a sustainable approach for nanoparticles synthesis mediated by fungi. https://doi.org/10.1080/24701556.2021.2025078.

Correa-Llantén, D.N., Muñoz-Ibacache, S.A., Castro, M.E., Muñoz, P.A. and Blamey, J.M. 2013. Gold nanoparticles synthesized by Geobacillus sp. strain ID17 a thermophilic bacterium isolated from Deception Island, Antarctica. Microb. Cell Factories 12: 75.

Dashor, A., Rathore, K., Raj, S. and Sharma, K. 2022. Synthesis of silver nanoparticles employing *Polyalthia longifolia* leaf extract and their *in vitro* antifungal activity against phytopathogen. BB Reports 31: 101320.

Dikshit, P.K., Kumar, J., Das, A.K., Sadhu, S., Sharma, S., Singh, S., Gupta, P.K. and Kim, B.S. 2021. Green synthesis of metallic nanoparticles: applications and limitations. Catalysts 11: 902.

Ding, R., Espinosa, I.M.P., Loevlie, D., Azadehranjbar, S., Baker, A.J., Mpourmpakis, G., Martini, A. and Jacobs, T.D.B. 2022. Size-dependent shape distributions of platinum nanoparticles. Nanoscale Adv. 4: 3978–3986.

Efros, Al.L. and Efros, A.L. 1982. Interband absorption of light in a semiconductor sphere. Sov. Phys. Semicond. 16(7): 772–775.

Ekimov, A.I., Onushcenko, A.A. and Tzekhomskii, V.A. 1980. Exciton light absorption by CuCl microcrystals in glass matrix. Sov. Glass Phys. Chem. 6: 511–512.

El Hassani, F.Z. 2020. Characterization, activities, and ethnobotanical uses of Mentha species in Morocco. Heliyon. 6: e05480.

Elamawi, R.M., Al-Harbi, R.E. and Hendi, A.A. 2018. Biosynthesis and characterization of silver nanoparticles using *Trichoderma longibrachiatum* and their effect on phytopathogenic fungi. Egypt. J. Biol. Pest Control. 28: 28.

Elbeshehy, E.K.F., Elazzazy, A.M. and Aggelis, G. 2015. Silver nanoparticles synthesis mediated by new isolates of *Bacillus* spp., nanoparticle characterization and their activity against Bean Yellow Mosaic Virus and human pathogens. Front Microbiol. 6: 453.

EL-Moslamy, S.H., Elkady, M.F., Rezk, A.H. and Abdel-Fattah, Y.R. 2017. Applying Taguchi design and largescale strategy for mycosynthesis of nano-silver from endophytic *Trichoderma harzianum* SYA.F4 and its application against phytopathogens. Sci. Rep. 7: 45297.

Fahmy, S.A., Fawzy, I.M., Saleh, B.M., Issa, M.Y., Bakowsky, U. and Azzazy, H.M.E. 2021. Green synthesis of platinum and palladium nanoparticles using Peganum harmala L. seed alkaloids: biological and computational studies. Nanomaterials 11: 965.

Faraday, M. 1857. On the relations of gold and other metals to light. Proc. R. Soc. Lond. 8: 356–361.

Feynman, R.P. 1960. There's plenty of room at the bottom: An Invitation to Enter a New Field of Physics. Eng. Sci. 23(5): 22–36.

Fu, M., Zhang, X., Chen, B., Li, M., Zhang, G. and Cui, L. 2021. Characteristics of isolates of *Pseudomonas aeruginosa* and *Serratia marcescens* associated with post-harvest Fuzi (Aconitum carmichaelii) rot and their novel loop-mediated isothermal amplification detection methods. Front. Microbiol. 12: 705329.

Ghazy, N.A., El-Hafez, O.A.A., El-Bakery, A.M. and El-Geddawy, D.I.H. 2021. Impact of silver nanoparticles and two biological treatments to control soft rot disease in sugar beet (*Beta vulgaris* L.). Egypt. J. Biol. Pest Control. 31: 3.

Gholami, M., Azarbani, F. and Hadi, F. 2021. Silver nanoparticles synthesised by using Iranian Mentha pulegium leaf extract as a non-cytotoxic antibacterial agent. Mater Technol. 37(9): 934–942.

Ghosh, S., Ahmad, R. Zeyaullah, M. and Khare, S.K. 2021. Microbial nano-factories: synthesis and biomedical applications. Front. Chem. 9: 626834.

Goddard, W.A., Brenner, D.W., Lyshevski, S.E. and Iafrate, G.J. 2003. Handbook of Nanoscience, Engineering, and Technology. CRC Press, Washington, D.C.

Gomaa, E.Z., Housseiny, M.M. and Omran, A.A.A.-K. 2019. Fungicidal efficiency of silver and copper nanoparticles produced by *Pseudomonas fluorescens* ATCC 17397 against four *Aspergillus* species: A molecular study. J. Clust. Sci. 30: 181–196.

Guilger, M., Pasquoto-Stigliani, T., Bilesky-Jose, N., Grillo, R., Abhilash, P.C., Fraceto, L.F. and Lima, R.D. 2017. Biogenic silver nanoparticles based on Trichoderma harzianum: synthesis, characterization, toxicity evaluation and biological activity. Sci. Rep. 7: 44421.

Guilger-Casagrande, M. and Lima, R.D. 2019. Synthesis of silver nanoparticles mediated by fungi: a review. Front. Bioeng. Biotechnol. 7: 287.

Gupta, K. and Chundawat, T.S. 2019. Bio-inspired synthesis of platinum nanoparticles from fungus *Fusarium oxysporum*: its characteristics, potential antimicrobial, antioxidant and photocatalytic activities. Mater. Res. Express. 6: 1050d6.

Hashem, A., Tabassum, B. and Abd_Allah, E.F. 2019. *Bacillus subtilis*: A plant-growth promoting rhizobacterium that also impacts biotic stress Saudi J. Biol. Sci. 26: 1291–1297.

Hemashekhar, B., Chandrappa, C.P., Govindappa, M. and Chandrashekar, N. 2019. Endophytic fungus *Alternaria* spp. isolated from *Rauvolfia tetraphylla* root arbitrate synthesis of gold nanoparticles and evaluation of their antibacterial, antioxidant and antimitotic activities. Adv. Nat. Sci.: Nanosci. Nanotechnol. 10: 035010.

Hemlata, Meena, P.R., Singh, A.P. and Tejavath, K.K. 2020. Biosynthesis of silver nanoparticles using Cucumis prophetarum aqueous leaf extract and their antibacterial and antiproliferative activity against cancer cell lines. ACS Omega 5: 5520−5528.

Henglein, A. and Ber. Bunsenges. 1982. Photo-degradation and fluorescence of colloidal-cadmium sulfide in aqueous solution. Phys. Chem. 86(4): 301–305,

Herizchi, R., Abbasi, E., Milani, M. and Akbarzadeh, A. 2016. Current methods for synthesis of gold nanoparticles. Artif Cells Nanomed. Biotechnol. 44(2): 596–602.

Hossain, A., Hong, X., Ibrahim, E., Li, B., Sun, G., Meng, Y., Wang, Y. and An, Q. 2019. Green synthesis of silver nanoparticles with culture supernatant of a bacterium *Pseudomonas rhodesiae* and their antibacterial activity against soft rot pathogen *Dickeya dadantii*. Molecules 24: 2303.

Ibrahim, E., Fouad, H., Zhang, M., Zhang, Y., Qiu, W., Yan, C., Li, B., Mo, J. and Chen, J. 2019. Biosynthesis of silver nanoparticles using endophytic bacteria and their role in inhibition of rice pathogenic bacteria and plant growth promotion. RSC Adv. 9: 29293–29299.

Ibrahim, E., Zhang, M., Zhang, Y., Hossain, A., Qiu, W., Chen, Y., Wang, Y., Wu, W., Sun, G. and Li, B. 2020. Green-synthesization of silver nanoparticles using endophytic bacteria isolated from garlic and its antifungal activity against wheat *Fusarium* head blight pathogen *Fusarium graminearum*. Nanomaterials 10: 219.

Ibrahim, S., Ahmad, Z., Manzoor, M.Z., Mujahid, M., Faheem, Z. and Adnan, A. 2021. Optimization for biogenic microbial synthesis of silver nanoparticles through response surface methodology, characterization, their antimicrobial, antioxidant, and catalytic potential. Sci. Rep. 11: 770.

Iijima, S. 1991. Helical microtubules of graphitic carbon. Nature 354: 56–58.

Iravani, S. 2011. Green synthesis of metal nanoparticles using plants. Green Chem. 13: 2638–2650.

Iravani, S. 2014. Bacteria in nanoparticle synthesis: current status and future prospects. Int. Sch. Res. Notices 2014: Article ID 359316.

Javed, B., Nadhman, A. and Mashwani, Z.-R. 2020. Optimization, characterization and antimicrobial activity of silver nanoparticles against plant bacterial pathogens phyto-synthesized by *Mentha longifolia*. Mater. Res. Express. 7: 085406.

Jebril, S., Jenana, R.K.B. and Dridi, C. 2020. Green synthesis of silver nanoparticles using Melia azedarach leaf extract and their antifungal activities: *In vitro* and *in vivo*. Mater. Chem. Phys. 248: 122898.

Jiang, S.-F., Ling, L.-L., Xu, Z., Liu, W.-J. and Jiang, H. 2018. Enhancing the catalytic activity and stability of noble metal nanoparticles by the strong interaction of magnetic biochar support. Ind. Eng. Chem. Res. 57(39): 13055–13064.

Kaman, P.K. and Dutta, P. 2017. *In vitro* evaluation of biosynthesized silver nanoparticles (Ag NPs) against soil-borne plant pathogens. IJNA 11(3): 261–264.

Kaur, P., Thakur, R., Duhan, J.S. and Chaudhary, A. 2018. Management of wilt disease of chickpea in vivo by silver nanoparticles biosynthesized by rhizospheric microflora of chickpea (Cicer arietinum). J. Chem. Technol. Biotechnol. 93(11): 3233–3243.

Kelkawi, A.H.A., Kajani, A.A. and Bordbar, A.-K. 2017. Green synthesis of silver nanoparticles using Mentha pulegium and investigation of their antibacterial, antifungal and anticancer activity. IET Nanobiotechnol. 11(4): 370–376.

Khan, I., Saeed, K. and Khan, I. 2019. Nanoparticles: Properties, applications and toxicities. Arab. J. Chem. 12: 908–931.

Khan, M., Khan, A.U., Bogdanchikova, N. and Garibo, D. 2021. Antibacterial and antifungal studies of biosynthesized silver nanoparticles against plant parasitic nematode *Meloidogyne incognita*, plant pathogens *Ralstonia solanacearum* and *Fusarium oxysporum*. Molecules 26: 2462.

Khan, N., Ali, S., Latif, S. and Mehmood, A. 2022. Biological synthesis of nanoparticles and their applications in sustainable agriculture production. Nat. Sci. 14(6): 226–234.

Khan, N.T., Khan, M.J., Jameel, J., Jameel, N. and Rheman, S.U.A. 2017. An overview: biological organisms that serves as nanofactories for metallic nanoparticles synthesis and fungi being the most appropriate. Bioceram Dev Appl. 7(1): 1–4.

Khan. S., Singh, S., Gaikwad, S., Nawani, N., Junnarkar, M. and Pawar, S.V. 2020. Optimization of process parameters for the synthesis of silver nanoparticles from Piper betle leaf aqueous extract, and evaluation of their antiphytofungal activity. Environ. Sci. Pollut. Res. 27: 27221–27233.

Khandel, P. and Shahi, S.K. 2018. Mycogenic nanoparticles and their bio-prospective applications: current status and future challenges. J. Nanostructure Chem. 8: 369–391.

Khanna, P., Kaur, A. and Goyal, D. 2019. Algae-based metallic nanoparticles: Synthesis, characterization and applications. J. Microbiol Methods 163: 105656.

Khoddami, A., Wilkes, M.A. and Roberts, T.H. 2013. Techniques for analysis of plant phenolic compounds. Molecules 18: 2328–2375.

Konishi, Y., Ohno, K., Saitoh, N., Nomura, T., Nagamine, S., Hishida, H., Takahashi, Y. and Uruga, T. 2007. Bioreductive deposition of platinum nanoparticles on the bacterium Shewanella algae. J. Biotechnol. 128(3): 648–653.

Kora, A.J, Mounika, J. and Jagadeeshwar, R. 2020. Rice leaf extract synthesized silver nanoparticles: An *in vitro* fungicidal evaluation against *Rhizoctonia solani*, the causative agent of sheath blight disease in rice. Fungal Biol. 124(7): 671–681.

Krug, H. F. and Wick, P. 2011. Nanotoxicology: An interdisciplinary challenge. Angew. Chem. Int. Ed. 50: 1260–1278.

Kulkarni, S.R. 2015. Biosynthesis and size control of Ag nanoparticles using strawberry fruit extract. J. Chem. and Cheml. Sci. 5(7): 377–383.

Kumar, M., Dibyajyoti, P. and Duarah, H.P. 2023. Chapter 10 - Applications of plant-derived metal nanoparticles in pharmaceuticals. Book: Advances in Extraction and Applications of Bioactive Phytochemicals. Elsevier B.V. https://doi.org/10.1016/B978-0-443-18535-9.00001-6.

Kushwah, K.S. and Verma, D.K. 2021. Book chapter 12: Biological synthesis of metallic nanoparticles from different plant species. Book: 21st Century Nanostructured Materials - Physics, Chemistry, Classification, and Emerging Applications in Industry, Biomedicine, and Agriculture. Edited by Phuong V. Pham. IntechOpen Limited. DOI: 10.5772/intechopen.101355.

Lahiri, D., Nag, M., Sheikh, H.I., Sarkar, T., Edinur, H.A., Pati, S. and Ray, R.R. 2021. Microbiologically-synthesized nanoparticles and their role in silencing the biofilm signaling cascade. Front. Microbiol. 12: 636588.

Li, J., Li, Q., Ma, X., Tian, B., Li, T., Yu, J., Dai, S., Weng, Y. and Hua, Y. 2016. Biosynthesis of gold nanoparticles by the extreme bacterium Deinococcus radiodurans and an evaluation of their antibacterial properties. Int. J. Nanomedicine 11: 5931–5944.

Li, X., Xu, H., Chen, Z.-S. and Chen, G. 2011. Biosynthesis of nanoparticles by microorganisms and their applications. J. Nanomater. 2011: Article ID 270974.

Liu, X., Yue, X., Yan, N. and Jiang, W. 2021. Self-assembled 3D free-standing superlattices of gold nanoparticles driven by interfacial instability of emulsion droplets. Mater. Chem. Front. 5: 7306–7314.

Mahawar, H., Prasanna, R., Gogoi, R., Singh, S.B., Chawla, G. and Kuma, A. 2020. Synergistic effects of silver nanoparticles augmented *Calothrix elenkinii* for enhanced biocontrol efficacy against *Alternaria* blight challenged tomato plants. 3 Biotech. 10: 102.

Makarov, V.V., Love, A.J., Sinitsyna, O.V., Makarova, S.S., Yaminsky, I.V., Taliansky, M.E. and Kalinina, N.O. 2014. "Green" nanotechnologies: synthesis of metal nanoparticles using plants. Acta naturae 6(1): 35–44.

Mankad, M., Patil, G., Patel, D., Patel, P. and Patel, A. 2020. Comparative studies of sunlight mediated green synthesis of silver nanoparaticles from Azadirachta indica leaf extract and its antibacterial effect on *Xanthomonas oryzae* pv. *Oryzae*. Arab. J. Chem. 13(1): 2865–2872.

Mansoori, G.A. and Soelaiman, T.a.f. 2005. Nanotechnology—An introduction for the standards community. J. ASTM Int. 2(6): 1–22.

Masum, M.M.I., Siddiqa, M.M., Ali, K.A., Zhang, Y., Abdallah, Y., Ibrahim, E., Qiu, W., Yan, C. and Li, B. 2019. Biogenic synthesis of silver nanoparticles using *Phyllanthus emblica* fruit extract and its inhibitory action against the pathogen *Acidovorax oryzae* strain RS-2 of rice bacterial brown stripe. Front. Microbiol. 10: 820.

Matsumoto, T., Phann, I. and Okibe, N. 2021. Biogenic platinum nanoparticles' production by extremely acidophilic Fe(III)-reducing bacteria. Minerals 11: 1175.

Mishra, R.C., Kalra, R., Dilawari, R., Goel, M. and Barrow, C.J. 2022. Bio-synthesis of *Aspergillus terreus* mediated gold nanoparticle: antimicrobial, antioxidant, antifungal and *in vitro* cytotoxicity studies. Materials 15: 3877.

Mishra, S., Singh, B.R., Naqvi, A.H. and Singh, H.B. 2017. Potential of biosynthesized silver nanoparticles using Stenotrophomonas sp. BHU-S7 (MTCC 5978) for management of soil-borne and foliar phytopathogens. Sci. Rep. 7: 45154.

Mishra, S., Yang, X., Ray, S., Fraceto, L.F. and Singh, H.B. 2020. Antibacterial and biofilm inhibition activity of biofabricated silver nanoparticles against *Xanthomonas oryzae* pv. *oryzae* causing blight disease of rice instigates disease suppression. World J. Microbiol. Biotechnol. 36: 55.

Molnár, Z., Bódai, V., Szakacs, G., Erdélyi, B., Fogarassy, Z., Sáfrán, G., Varga, T., Kónya, Z., Tóth-Szeles, E., Szűcs, R. and Lagzi, I. 2018. Green synthesis of gold nanoparticles by thermophilic filamentous fungi. Sci. Rep. 8: 3943.

Monowar, T., Rahman, M.S., Bhore, S.J., Raju, G. and Sathasivam, K.V. 2018. Silver nanoparticles synthesized by using the endophytic bacterium *Pantoea ananatis* are promising antimicrobial agents against multidrug resistant bacteria. Molecules 23: 3220.

Mukherjee, P., Senapati, S., Mandal, D., Ahmad, A., Khan, M.I., Kumar, R. and Sastry, M. 2002. Extracellular synthesis of gold nanoparticles by the fungus *Fusarium oxysporum*. ChemBioChem. 5: 461–463.

Mullen, D., Wolf, D.C., Ferris, F.G., Beveridge, T.J., Flemming, C.A. and Bailey, G.W. 1989. Bacterial sorption of heavy metals. Appl. Environ. Microbiol. 55(12): 3143–3149.

Namburi, K.R., Kora, A.J., Chetukuri, A.V. and Kota, S.M.K. 2021. Biogenic silver nanoparticles as an antibacterial agent against bacterial leaf blight causing rice phytopathogen *Xanthomonas oryzae* pv. *Oryzae*. Bioprocess Biosyst. Eng. 44(9): 1975–1988.

Narayanan, K.B. and Sakthivel, N. 2010. Biological synthesis of metal nanoparticles by microbes. Adv. Colloid. Interface. Sci. 156(1-2): 1–13.

Narayanan, R. and El-Sayed, M.A. 2004. Shape-dependent catalytic activity of platinum nanoparticles in colloidal solution. Nano Lett. 4(7): 1343–1348.

Nejad, M.S., Najafabadi, N.S., Aghighi, S., Pakina, E. and Zargar, M. 2022. Evaluation of *Phoma* sp. biomass as an endophytic fungus for synthesis of extracellular gold nanoparticles with antibacterial and antifungal properties. Molecules 27: 1181.

Omar, M. and Bendahou, M. 2020. Book chapter: Biological Synthesis of Nanoparticles Using Endophytic Microorganisms: Current Development. Book: Nanotechnology and the Environment. Publisher: IntechOpen.

Pajerski, W., Ochonska, D., Brzychczy-Wloch, M., Indyka, P., Jarosz, M., Golda-Cepa, M., Sojka, Z. and Kotarba, A. 2019. Attachment efficiency of gold nanoparticles by Gram-positive and Gram-negative bacterial strains governed by surface charges. J. Nanopart. Res. 21: 186.

Panche, A.N., Diwan, A.D. and Chandra, S.R. 2016. Flavonoids: an overview. J. Nutr. Sci. 5(47): 1–15.

Pandit, C., Roy, A., Ghotekar, S., Khusro, A., Islam, M.N., Emran, T.B., Lam, S.E., Khandaker, M.U. and Bradley, D.A. 2022. Biological agents for synthesis of nanoparticles and their applications. J. King Saud Univ. Sci. 34: 101869.

Pantidos, N. and Horsfall, L.E. 2014. Biological synthesis of metallic nanoparticles by bacteria, fungi and plants. J. Nanomed. Nanotechnol. 5(5): 1000233.

Park, J. W., Ullah, M.H., Park, S.S. and Hu, C.-G. 2007. Organic electroluminescent devices using quantum-size silver nanoparticles. J. Mater. Sci.: Mater. Electron. 18(1): 393–397.

Parmar, M. and Sanyal, M. 2022. Extensive study on plant mediated green synthesis of metal nanoparticles and their application for degradation of cationic and anionic dyes. Environ. Nanotechnol. Monit. Manag. 17: 100624.

Patra, J.K. and Baek, K.-H. 2014. Green Nanobiotechnology: Factors Affecting Synthesis and Characterization Techniques. 2014. Article ID 417305.

Paulkumar, K., Gnanajobitha, G., Vanaja, M., Rajeshkumar, S., Malarkodi, C., Pandian, K. and Annadurai, G. 2014. Piper nigrum Leaf and Stem Assisted Green Synthesis of Silver Nanoparticles and Evaluation of Its Antibacterial Activity Against Agricultural Plant Pathogens. 2014: Article ID 829894.

Phukan, K., Devi, R. and Chowdhury, D. 2021. Green synthesis of gold nano-bioconjugates from onion peel extract and evaluation of their antioxidant, anti-inflammatory, and cytotoxic studies. ACS Omega 6: 17811–17823.

Pinu, F.R., Villas-Boas, S.G. and Aggio, R. 2017. Analysis of intracellular metabolites from microorganisms: quenching and extraction protocols. Metabolites 7: 53.

Poole, C.P. and Owens, F.J. 2003. Introduction to Nanotechnology. John Wiley and Sons, Inc., Hoboken, New Jersey. USA.

Pradeep, M., Kruszka, D., Kachlicki, P., Mondal, D. and Franklin, G. 2022. Uncovering the phytochemical basis and the mechanism of plant extract-mediated eco-friendly synthesis of silver nanoparticles using ultra-performance liquid chromatography coupled with a hotodiode array and high-resolution mass spectrometry. ACS Sustainable Chem. Eng. 10: 562–571.

Prithiviraj, B., Bais, H.P., Jha, A.K. and Vivanco, J.M. 2005. *Staphylococcus aureus* pathogenicity on *Arabidopsis thaliana* is mediated either by a direct effect of salicylic acid on the pathogen or by SA-dependent, NPR1-independent host responses. The Plant Journal 42: 417–432.

Priyadarshini, E., Pradhan, N., Sukla, L.B. and Panda, P.K. 2014. Controlled synthesis of gold nanoparticles using *Aspergillus terreus* IF0 and its antibacterial potential against gram negative pathogenic bacteria. J. Nanotechnol. 2014: Article ID 653198.

Qais, F.A., Ahmad, I., Altaf, M. and Alotaibi, H. 2021. Biofabrication of gold nanoparticles using Capsicum annuum extract and its antiquorum sensing and antibiofilm activity against bacterial pathogens. ACS Omega 6: 16670−16682.

Qu, M., Yao, W., Cui, X. Xia, R., Qin, L. and Liu, X. 2021. Biosynthesis of silver nanoparticles (AgNPs) employing *Trichoderma* strains to control empty-gut disease of oak silkworm (*Antheraea pernyi*). Mater. Today Commun. 28: 102619.

Radiman, C. and Yuliani, G. 2008. Coconut water as a potential resource for cellulose acetate membrane preparation. Polym. Int. 57: 502–508.

Rafique, M., Sadaf, I., Rafique, M.S. and Tahir, M.B. 2017. A review on green synthesis of silver nanoparticles and their applications. Artif Cells Nanomed. Biotechnol. 45(7): 1272–1291.

Rahme, L.G., Tan, M.-W., Le, L., Wong, S.M., Tompkins, R.G., Stephen, B.C. and Ausubel, F.M. 1997. Use of model plant hosts to identify *Pseudomonas aeruginosa* virulence factors. Proc. Natl. Acad. Sci. USA. 94: 13245–13250.

Rai, M. and Yadav, A. 2013. Plants as potential synthesiser of precious metal nanoparticles: progress and prospects. IET Nanobiotechnol. 7(3): 117–124.

Rao, Y., Inwati, G.K. and Singh, M. 2017. Green synthesis of capped gold nanoparticles and their effect on Gram-positive and Gram-negative bacteria. Future Sci. OA. 3(4): FSO239.

Reponen, T.A., Gazenko, S.V., Grinshun, S.A., Willeke, K., Cole, E.C. 1998. Characteristics of airborne actinomycete spores. Appl. Environ. Microbiol. 64(10): 3807–3812.

Riaz, M., Altaf, M., Ahmad, P., Khandaker, M.U., Osman, H., Eed, E.M. and Shakir, Y. 2022. Biogenic synthesis of Ag nanoparticles of 18.27 nm by zanthozylum armatum and determination of biological potentials. Molecules 27: 1166.

Riaz, M., Sharafat, U., Zahid, N., Ismail, M., Park, J., Ahmad, B., Rashid, N., Fahim, M., Imran, M. and Tabassum, A. 2022. Synthesis of biogenic silver nanocatalyst and their antibacterial and organic pollutants reduction ability. ACS Omega 7: 14723−14734.

Riaz, M., Suleman, A., Ahmad, P., Khandaker, M.U., Alqahtani, A., Bradley, D.A. and Khan, M.Q. 2022. Biogenic synthesis of AgNPs using aqueous bark extract of Aesculus indica for antioxidant and antimicrobial applications. Crystals 12: 252.

Roy, A., Elzaki, A., Tirth, V., Kajoak, S., Osman, H., Algahtani, A., Islam, S., Faizo, N.L., Khandaker, M.U., Islam, M.N., Emran, T.B. and Bilal, M. 2021. Biological synthesis of nanocatalysts and their applications. Catalysts 11(12): 1494.

Saha, P., Mahiuddin, M., Islam, A.B.M.N. and Ochiai, B. 2021. Biogenic synthesis and catalytic efficacy of silver nanoparticles based on peel extracts of Citrus macroptera fruit. ACS Omega 6: 18260−18268.

Sahayaraj, K., Balasubramanyam, G. and Chaval, M. 2020. Green synthesis of silver nanoparticles using dry leaf aqueous extract of Pongamia glabra Vent (Fab.), Characterization and phytofungicidal activity. Environ. Nanotechnol. Monit. Manag. 14: 100349.

Saim, A.K., Kumah, F.N. and Oppong, M.N. 2021. Extracellular and intracellular synthesis of gold and silver nanoparticles by living plants: a review. Nanotechnol. Environ. Eng. 6: 1.

Selvakesavan, R.K. and Franklin, G. 2021. Prospective application of nanoparticles green synthesized using medicinal plant extracts as novel nanomedicines. Nanotechnol. Sci. Appl. 14: 179–195.

Shah, M. Fawcett, D., Sharma, S., Tripathy, S.K. and Poinern, G.E.J. 2015. Green synthesis of metallic nanoparticles via biological entities. Materials 8: 7278–7308.

Sharma, N.K., Vishwakarma, J., Rai, S., Alomar, T.S., AlMasoud, N. and Bhattarai, A. 2022. Green route synthesis and characterization techniques of silver nanoparticles and their biological adeptness. ACS Omega 7(31): 27004–27020.

Shreyash, N., Bajpai, S., Khan, M.A., Vijay, Y., Tiwary, S.K. and Sonker, M. 2021. Green synthesis of nanoparticles and their biomedical applications: a review. ACS Appl. Nano Mater. 4(11): 11428–11457.

Singh, J., Dutta, T., Kim, K-H., Rawat, M., Samddar, P. and Kumar, P. 2018. 'Green' synthesis of metals and their oxide nanoparticles: applications for environmental remediation. J. Nanobiotechnol. 16: 84.

Singh, R., Shedbalkar, U.U., Wadhwani, S.A. and Chopade, B.A. 2015. Bacteriagenic silver nanoparticles: synthesis, mechanism, and applications. Appl. Microbiol. Biotechnol. 99: 4579–4593.

Srinath, B.S., Namratha, K. and Byrappa, K. 2018. Eco-friendly synthesis of gold nanoparticles by *Bacillus subtilis* and their environmental applications. Adv. Sci. Lett. 24(8): 5942–5946.

Srivastava, S., Usmani, Z., Atanasov, A.G., Singh, V.K., Singh, N.P., Abdel-Azeem, A.M., Prasad, R., Gupta, G., Sharma, M. and Bhargava, A. 2021. Biological nanofactories: using living forms for metal nanoparticle synthesis. Mini. Rev. Med. Chem. 21(2): 245–265.

Syed, A. and Ahmad, A. 2012. Extracellular biosynthesis of platinum nanoparticles using the fungus *Fusarium oxysporum*. Colloids Surf. B. 97: 27–31.

Tahir, K., Nazir, S., Ahmad, A., Li, B., Khan, A.U., Khan, Z.U.H., Khan, F.U., Khan, Q.U., Khan, A. and Rahman, A.U. 2017. Facile and green synthesis of phytochemicals capped platinum nanoparticles and *in vitro* their superior antibacterial activity. J. Photochem. Photobiol. B, Biol. 166: 246–251.

Taniguchi, N. 1974. "On the Basic Concept of 'Nano-Technology'". 26–29 August, 1974. Proceedings of the International Conference on Production Engineering, Part II. Tokyo, Japan.

Thakker, J.N., Dalwadi, P. and Dhandhukia, P.C. 2013. Biosynthesis of gold nanoparticles using *Fusarium oxysporum* f. sp. *cubense* JT1, a plant pathogenic fungus. 2013: Article ID 515091.

Thirumurugan, A., Aswitha, P., Kiruthika, C., Nagarajan, S. and Christy, A.N. 2016. Green synthesis of platinum nanoparticles using *Azadirachta indica* – An eco-friendly approach. Mater. Lett. 170: 175–178.

Tian, Y., Luo, J., Wang, H., Zaki, H.E.M., Yu, S., Wang, X., Ahmed, T., Shahid, M.S., Yan, C., Chen, J. and Li, B. 2022. Bioinspired green synthesis of silver nanoparticles using three plant extracts and their antibacterial activity against rice bacterial leaf blight pathogen *Xanthomonas oryzae* pv. *Oryzae*. Plants 11: 2892.

Ullah, M.H. and Ha, C.-S. 2005. Size-controlled synthesis and optical properties of doped nanoparticles prepared by soft solution processing. J. Nanosci. Nanotechnol. 5(9): 1376–1394.

Ullah, M.H., Kim I. and Ha, C.-S. 2006. *In-Situ* preparation of binary-phase silver nanoparticles at a high Ag+ concentration. J. Nanosci. Nanotechnol. 6: 1–6.

Ullah, M.H., Kim, I. and Ha, C.-S. 2006. Preparation and optical properties of colloidal silver nanoparticles at a high Ag+ concentration. Mater. Lett. 60: 1496–1501.

Ullah, M.H., Chung, W.-S., Kim, I. and Ha, C.-S. 2006. pH-selective synthesis of monodisperse nanoparticles and 3D dendritic nanoclusters of CTAB-stabilized platinum for electrocatalytic O_2 reduction. Small. 2(7): 870–873.

Ullah, M.H., Kim, I. and Ha, C.-S. 2007. pH selective synthesis of ZnS nanocrystals and their growth and photoluminescence. Mater. Lett. 61: 4267–4271.

Ullah, M.H. Chon, B., Joo, T., Sona, M., Kim, I. and Ha, C.-S. 2007. Synthesis of ligand-selective ZnS nanocrystals exhibiting ligand-tunable fluorescence. J. Colloid Interface Sci. 316: 939–946.

Ullah, M.H., Moon, H. and Ha, C.-S. 2021. Effect of pHs on the structure evolution of platinum nanoclusters and their surface plasmon resonance properties. J. Nanosci. Nanotechnol. 21: 4700–4704.

Umoren, S.A., Nzila, A.M., Sankaran, S., Solomon, M.M. and Umoren, P.S. 2017. Green synthesis, characterization and antibacterial activities of silver nanoparticles from strawberry fruit extract. Pol. J. Chem. Tech. 19(4): 128–136.

Vetchinkina, E., Loshchinina, E., Kupryashina, M., Burov, A., Pylaev, T. and Nikitina, V. 2018. Green synthesis of nanoparticles with extracellular and intracellular extracts of basidiomycetes. PeerJ 6: e5237.

Vidaver, A.K. and Lambrecht, P.A. 2004. Bacteria as plant pathogens. Department of Plant Pathology, University of Nebraska, Lincoln, NE. DOI. 10.1094/PHI-I-2004-0809-01.

Villaseñor-Basulto, D.L., Pedavoah, M.-M. and Bandala, E.R. 2019. Handbook of Ecomaterials (Chapter 5). Springer Nature Switzerland AG.

Vukoja, D., Vlainić, J., Bilić, V.L., Martinaga, L., Rezić, I., Gorski, D.B. and Kosalec, I. 2022. Innovative insights into *in vitro* activity of colloidal platinum nanoparticles against ESBL-producing strains of *Escherichia coli* and *Klebsiella pneumoniae*. Pharmaceutics 14: 1714.

Walker, T.S., Bais, H.P., De´ziel, E., Schweizer, H.P., Rahme, L.G., Fall, R. and Vivanco, J.M. 2004. *Pseudomonas aeruginosa*-plant root interactions. Pathogenicity, biofilm formation, and root exudation. Plant Physiol. 134: 320–331.

Wang, D., Xue, B., Wang, L., Zhang, Y., Liu, L. and Zhou, Y. 2021. Fungus-mediated green synthesis of nano-silver using *Aspergillus sydowii* and its antifungal/antiproliferative activities. Sci. Rep. 11: 10356.

Wang, F., Wang, Y., Liu, Y.-H., Morrison, P.J., Loomis, R.A. and Buhro, W.E. 2015. Two-dimensional semiconductor nanocrystals: properties, templated formation, and magic-size nanocluster intermediates. Acc. Chem. Res. 48: 13−21.

Wang, X.-Q, Bi, T., Li, X.-D., Zhang, L.-Q. and Lu, S.-E. 2015. First report of corn whorl rot caused by Serratia marcescens in China. J. Phytopathol. 163(11-12): 1059–1063.

Wang, Y. and Wei, S. 2022. Green fabrication of bioactive silver nanoparticles using Mentha pulegium extract under alkaline: an enhanced anticancer activity. ACS Omega. 7: 1494−1504.

Xue, B., He, D., Gao, S., Wang, D., Yokoyama, K. and Wang, L. 2016. Biosynthesis of silver nanoparticles by the fungus *Arthroderma fulvum* and its antifungal activity against genera of *Candida*, *Aspergillus* and *Fusarium*. Int. J. Nanomedicine. 11: 1899–1906.

Yadav, S., Shrivas, K. and Bajpai, P.K. 2019. Role of precursors in controlling the size, shape and morphology in the synthesis of copper sulfide nanoparticles and their application for fluorescence detection. J. Alloys Compd. 772: 579–592.

Yang, B., Chou, J., Dong, X., Qu, C., Yu, Q., Lee, K.J. and Harvey, N. 2017. Size-controlled green synthesis of highly stable and uniform small to ultrasmall gold nanoparticles by controlling reaction steps and pH. J. Phys. Chem. C. 121(16): 8961–8967.

Yassin, M.A., El-Samawaty, A.E.M.A., Dawoud, T.M., Abd-Elkader, O.H., Al Maary, K.S., Hatamleh, A.A. and Elgorban, A.M. 2017. Characterization and anti-*Aspergillus flavus* impact of nanoparticles synthesized by Penicillium citrinum. Saudi J. Biol. Sci. 24: 1243–1248.

Yassin, M.A., Elgorban, A.M., El-Samawaty, A.E.M.A. and Almunqedhi, B.M.A. 2021. Biosynthesis of silver nanoparticles using *Penicillium verrucosum* and analysis of their antifungal activity. Saudi J. Biol. Sci. 28: 2123–2127.

Yusof, H.M., Rahman, N.A.A., Mohamad, R. and Zaidan, U.H. 2020. Microbial mediated synthesis of silver nanoparticles by *Lactobacillus plantarum* TA4 and its antibacterial and antioxidant activity. Appl. Sci. 10: 6973.

Zhang, D., Ma, X.-L., Gu, Y., Huang, H. and Zhang, G.-W. 2020. Green synthesis of metallic nanoparticles and their potential applications to treat cancer. Front. Chem. 8: 799.

Chapter 9

Silver Nanoparticles in Sustainable Management of Plant Pathogens

Yalavarthi Nagaraju,[1,]* *Praveen S. Patted*[2] and *Duppala Manoj Kumar*[3]

Introduction

Nanotechnology is an emerging scientific field considered to have the potential to generate new and innovative materials (Otunola et al. 2018). Nanomaterials find their way into solving the global environmental perturbation challenges and found solutions to the trapping solar energy, conversion and catalysis, delivery of drugs, gene transfer, mitigation of insect pests damage, production of bioelectronic devices, and as an antibacterial agent against many bacterial pathogens, including multidrug-resistant bacteria (Cohen-Karni et al. 2012; Hayles et al. 2017). Research in the interdisciplinary field of nanotechnology focuses on the controlled monodispersed production of nanoparticles with various chemical compositions and sizes.

Metal nanoparticles are nanosized and exhibit unique physical, chemical, and biological properties compared to their macroscale counterparts (Aadil et al. 2016; Aabed and Mohammed 2021). Due to their unusual surface-to-volume ratio and ability to significantly alter physical, chemical, and biological characteristics, nanosized metallic particles have been used for various applications. Metals such as silver or argentium (Ag), copper or cyprium (Cu), and lead or plumbum (Pb) are toxic to several bacterial species, and Pb and tin (Sn) are considered harmful to humans. Silver (Ag) is prominently utilized in medicine to control infectious diseases and antibiotic-resistant microorganisms (Sharma et al. 2009). Nanotechnology provides the ability to open new avenues to prevent and treat diseases by tailoring materials on an atomic scale. Most services are because the matter at the nanoscale scale differs from matter in the bulk state in terms of its properties.

For this reason, several research organizations worldwide are experimenting with novel techniques for the nanoscale production of various materials for broad applications. One objective is to regulate the size, shape, and ordering of atomic clusters or nanoparticles in different dimensional arrays. Other synthesis techniques have been used to meet the need for nanoparticles (NPs). Generally, traditional physical and chemical procedures are highly costly and dangerous, limiting medical applications due to subsequent toxicity (Mittal et al. 2014). The production of NPs using green chemistry has great potential (Gurunathan et al. 2015). For instance, one of the current techniques crystallizes nanoparticles in microemulsions while utilizing many surfactants as stabilizing agents

---
[1] ICAR-National Bureau of Agriculturally Important Microorganisms, Maunath Bhanjan, Uttar Pradesh – 275103.
[2] ICAR-Indian Institute of Seed Science, Maunath Bhanjan, Uttar Pradesh – 275103.
[3] Acharya N.G. Ranga Agricultural University, Agricultural College, Bapatla, Andhra Pradesh – 522101.
* Corresponding author: nagarajulvrth62@gmail.com; yalavarthi.nagaraju@icar.gov.in

and a range of chemical precursors. Metal, dielectric, semiconductor, and magnetic nanoparticles have all been successfully synthesized using various preparation techniques. However, excessive synthetic reactants and solvents are unhealthy for the environment and ecology. Therefore, it is highly desired to develop alternative, "green" nanomaterial production techniques that use eco-friendly methods. The green synthesis process produced silver nanoparticles suitable for biological systems (Baklanov 2012). Interestingly, biologically produced NPs exhibit excellent yield, solubility, and stability (Fouad et al. 2016). Among the several synthetic techniques for NPs, biological processes appear to be the most straightforward, quick, safe, trustworthy, and environmentally friendly ways to generate well-defined sizes and shapes under the ideal circumstances for translational research (Abdallah et al. 2019).

The principal components producing AgNPs include plant extracts, bacteria, and fungi (Behravan et al. 2019). This adds an incentive to improve cost-effective technologies, which helps reduce environmental damage (Mie et al. 2014). Biological substances such as honey, milk, coconut water, and egg white can be used to bio-synthesize silver nanoparticles under *in vitro* conditions. Utilizing biological resources in procedures and industrial processes, which must be done in tandem with environmental sustainability utilizing natural resources, is a problem for developing nations with a wealth of biodiversity (Zuin et al. 2020). The use of modern green process engineering (MGPE) has led to the adoption of several strategies in this area, such as the creation of nanoparticles (NPs) from biological systems (Mondal et al. 2020).

The research established the presence of silver ions and AgNPs in the solution named "Holy water," known from the beginning of the first millennium as a protection tool against infection by microorganisms (Henglein 1989). Notably, using silver nanoparticles as bactericides, antimycotics, and anticancer drugs has shown encouraging results (Das et al. 2013; Zhang et al. 2012). However, avoiding the situation rather than curing it is always better; preventing bacterial attachment to material surfaces is the best way to abate these problems. Typically, Ag^+ solutions are used to grow the nanocrystals for silver particles. From a salt called silver nitrate, silver ions are produced ($AgNO_3$). Initially, a reducing agent is used to decrease the ions to atoms. The atoms that have been gathered then form tiny clusters that eventually expand into particles. The size and form of the nanoparticles can be regulated based on the atoms' accessibility, which in turn depends on the ratio of the concentration of silver salt to the reducing agent. This procedure requires a reducing agent and a silver salt to develop the nanoparticle (Durán and Marcato 2012).

Because a particle's physicochemical characteristics may significantly influence those particles' biological characteristics, accurate particle characterization is required after synthesis. The manufactured nanoparticles must be characterized before addressing the safety concern and utilizing the full potential of any nanomaterial for human welfare, nanomedicines, the healthcare sector, etc. Before determining if a substance is poisonous or biocompatible, it is necessary to consider its distinctive properties, such as size, shape, distribution, surface area, form, solubility, aggregation, surface chemistry, coating/capping, dissolution rate, particle reactivity in solution, the efficiency of ion release, and cell type. Cytotoxicity is also greatly influenced by the reducing agents used to create AgNPs. The physicochemical characteristics of nanoparticles increase the bioavailability of therapeutic agents following systemic and local administration.

On the other hand, they can impact cellular uptake, biological distribution, penetration through physical barriers, and the subsequent therapeutic effects. Consequently, creating AgNPs with well-planned structures homogeneous in size, shape, and functionality is crucial for various biological applications (Jo et al. 2015). In recent years the disease incidence and morbidity of plants increased due to the increased resistance to the commonly used fungicides and immuno-compromization of plants due to harsh environmental conditions (Stephenson 2001). The administration of large agrochemicals to control pests and diseases envisages alternative management methods. Silver nanoparticles received immense importance during the current decade due to their antibacterial and antifungal activity with promising results (Cătălin Balaure et al. 2017).

Additionally, one of the least studied aspects of nanoparticles is how they affect plant growth and development. AgNPs (now the most frequently manufactured nanomaterials) have been linked to increased crop output and growth in agriculture. According to several publications, AgNPs at the proper concentrations are crucial for improved seed germination, more significant plant development, higher photosynthetic quantum efficiency and chlorophyll content, and improved water and fertilizer usage. The unique biological and physical qualities and the capacity to increase plant metabolism and antioxidant enzyme activities to combat diverse diseases were also reported for the nanoparticles. Numerous analytical methods, such as UV-Vis spectroscopy, X-ray diffractometry, Fourier transform infrared spectroscopy (FTIR), X-ray photoelectron spectroscopy (XPS), dynamic light scattering (DLS), Transmission electron microscopy (TEM), atomic force microscopy (AFM), and scanning electron microscopy (SEM) have been employed to assess the produced nanomaterial's structure, agglomeration, distribution, surface properties, etc. As NPs are required for usage in sectors that are directly linked to people, there is an urgent need to create sustainable processes and methods for using nanoparticles (Shams et al. 2013).

Microbial Biosynthesis of Silver Nanoparticles

Nanotechnology is a crucial area of contemporary science that deals with the synthesis and manipulation of nanomaterials or structures of matter with a size between 1 nm and 100 nm. The tiny nanoparticles offer benefits like a more surface volume ratio and antimicrobial activity with low animal toxicity (Yang et al. 2008). The noble metal nanoparticles displayed distinctive and significantly altered physical and chemical characteristics compared to their macro-sized counterparts. Noble metals, such as ruthenium, rhodium, palladium, iridium, gold, copper, and silver offer tremendous antimicrobial activity. They are chemically resistant to oxidation and reduction even at higher temperatures in the presence of air, and their acid resistance has high densities and optical reflectiveness. Among the noble metals, the corrosion resistance is high for iridium and less for silver.

Noble metals are commonly used in jewelry, coins, medals, and medical treatments. Since the dawn of human civilization, silver has been recognized for its antibacterial properties; nevertheless, due to its numerous practical uses, interest in silver has increased over time. Silver, either alone or in combination, has therapeutic value against several pathogenic microorganisms. The silver treatment inhibits the bacterial growth of *Staphylococcus aureus, Alcaligenes viscolactis* (Divya et al. 2019), *Curvularia lunata, Candida albicans, Rhizoctonia solani, Trichophyton rubrum, Collototrichum* sp., *Trichosporon rubrum, Aspergillus niger, Fusarium* sp., *Alternaria alternata, Sclerotinia sclerotium, Macrophomina phaseolina,* and *Botrytis cinerea* (Krishnaraj et al. 2012). The notion of biological synthesis is based on the fact that many species have evolved to withstand environments with elevated metal concentrations. Such microorganisms can transform hazardous and poisonous chemicals and substances into less damaging components or entirely harmless compounds (Sharma and Bisen 1992).

Bio-fabricating silver nanoparticles from microorganisms is essential as they are environmentally benign, practical, and cost-effective. Microorganisms grow on cheap raw materials from primary or secondary byproducts of industry or agricultural waste. This waste is a source of food for the growth of microbes like algae, higher plants, fungi, bacteria, and lichens as a part of green technology. The conditions are congenial for the optimum development and production of silver nanoparticles in the fermenter. There are two main divisions in the biosynthesis of "natural" biogenic metallic NP.

a) *Bioreduction*: Using biological resources, a bioreduction procedure that uses different metals may produce rather powerful models of metallic ions. Specific enzymes have oxidized and reduced the metal ion (Deplanche et al. 2010).

b) *Biosorption*: The metallic ions from a soil or water sample have been linked to the organism. After bacteria and fungi produced the peptides or cell wall chains, the metal ions bound to them, and these peptide chains then formed sturdy NP structures (Yong et al. 2002).

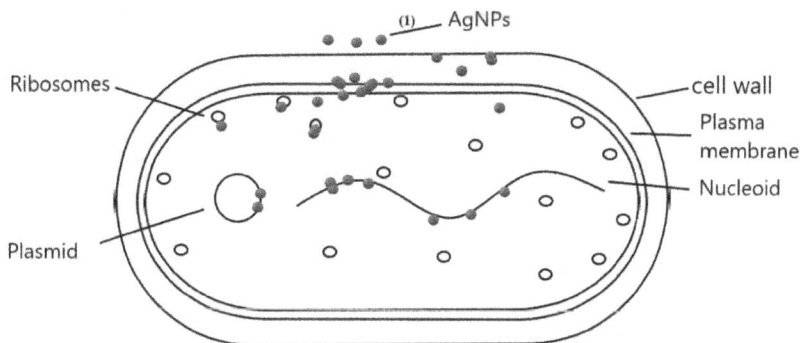

Figure 1. Antifungal action of AgNPs.

A protein with amino acids with -SH bonds are thought to be the source of the biological reduction of different metallic NP, and cysteine likely underwent the process of dehydrogenation after reacting with silver nitrate to create AgNPs. The free amino acid will cap AgNPs (Mukherjee et al. 2008). In some, pili are conducive to the extracellular biosynthesis of silver nanoparticles, *Geobacter sulfurreducens* pilin is used to synthesize uranium nanoparticles (UNPs). Without pilin proteins, they were formed in the periplasmic space (Reguera 2012). The transfer of electrons is critical for the formation of nanoparticles; electron shuttlers like NADH, coenzyme Q, and oxygen/ superoxide acts as carriers and transporters (Figure 1).

1. AgNPs attachment to the cell membrane alters its permeability and structure, hindering transportation and ATP production.
2. The AgNPs destabilize DNA, proteins, and ribosomes inside the cell by binding with phosphorous and sulfhydryl groups.
3. Oxidation of proteins and lipids results in cellular toxicity and ROS generation.
4. Altering the phosphotyrosine signal transduction pathways ultimately causes cell death.

Photoinduced electron transfer is commonly employed to create AgNPs. The creation of Ag^0 clusters and the production of AgNPs by photo-induction are enabled via electron transfer through amino acids that possess binding side chains, such as amides, aliphatic amines, alkyl groups, or alcohols, which led to the discovery that efficient AgNPs eventually needed quick exergonic aggregation of Ag^0 into Ag^n groups (Reguera 2012).

The silver nanoparticles were biosynthesized using rhizospheric bacteria *Bacillus subtilis*, *Bacillus licheniformis* (Kalishwaralal et al. 2008), *Escherichia coli* (Shahverdi et al. 2007), *Enterobacter cloacae, Klebsiella pneumoniae* (Mokhtari et al. 2009), *Pseudomonas stutzeri, Aeromonas* sp. *Corynebacterium* sp. (Mokhtari et al. 2009), *Lactobacillus* sp. (Korbekandi et al. 2012), and endophytic bacteria such as *Bacillus siamensis* (Ibrahim et al. 2019). The site of bio-fabrication may be inside or outside the cell based on the production of the metabolites and redox reactions (Tanzil et al. 2016). Enzymes like nitrate reductase and sulfite reductase internally bio-fabricate the nanoparticles. Previous reporters identified a nitrate reductase enzyme from the *Streptomyces* sp. LK3 converts the Ag^+ to Ag^0 (Karthik et al. 2013).

More stable silver nanoparticle production was observed in the non-pathogenic bacterium *Bacillus licheniformis*. Combining *Bacillus subtilis* supernatant and microwave irradiation produced 5–50 nm nanoparticles with low aggregation (Saifuddin et al. 2009). A silver-tolerant *Pseudomonas stutzeri* was used for the biosynthesis of silver nanoparticles, wherein the larger particles were synthesized under higher concentrations, and smaller sizes were observed under lower concentrations (Klaus-Joerger et al. 2001). The silver is precipitated inside the cells and detoxification. Enzymes like chitinase are involved in stabilizing AgNPs through particle coating and were identified as crucial agents (Costa Silva et al. 2017). Nanoparticles that fall in the size range of 3–40 nm are authenticated by characterizing them through TEM, UV-visible (UV-Vis) absorption spectroscopy, X-ray diffraction, absorption spectroscopy, differential scanning calorimetry, and EPR spectroscopy.

Scientists are interested in using plants to synthesize nanoparticles since it offers a simple, one-step biosynthetic process. This synthesis technique is a better choice since procedures employing plant materials are toxicant-free because plants readily supply natural capping agents. When silver in an aqueous solution is exposed to plant extract from various organs, silver is reduced, producing bimetallic silver nanoparticles. Such publications on the use of *Plumeria rubra* in the manufacture of silver nanoparticles are available (plant latex), *Syzygium aromaticum* (bud extract), with *Murraya koenigii* (leaf extract). Eugenol or carbazoles, natural reducing agents in the extracts, make it feasible to do this synthesis. In the polysaccharide method, cellulosic extracts from the water hyacinth produce desirable NPs at variable sizes of interest through pH control (Mochochoko et al. 2013). Using the triazole sugars, a size of 5–8 nm can be achieved; conversely, irradiation was used without sugars bio-fabricating NPs (Flores-Rojas et al. 2020).

In recent years, attempts have been made to produce nanoparticles biologically using fungus because of their low toxicity, strong bioaccumulation, relative affordability, easy synthesis technique, straightforward downstream processing, and ease of managing biomass. Fungi are well known for extracellular biosynthesis as they produce a wide range of enzymes outside (Alghuthaymi et al. 2015; Rai et al. 2021). The metabolites actively involve metal reduction and oxidation (Zhang et al. 2016). The fungi can be used to produce on a large scale since they are highly responsive to bioreactor conditions and produce enzymes externally (Balakumaran et al. 2016). The extracellular production of silver nanoparticles employs various fungi, including *Aspergillus niger*. The yeast strain MKY3 is silver-tolerant and capable of producing silver nanoparticles with a size range of 2 nm to 5 nm by inoculating the culture with aqueous silver nitrate. Mourato et al. (2011) have also looked into the biosynthesis of gold and silver nanoparticles using an extremophile strain of yeast obtained from acid mine drainage. *Rhodosporidium diobovatum*, another aquatic yeast, has been investigated to manufacture stable lead sulfide nanoparticles intracellularly.

Physiochemical Methods of Synthesis of Silver Nanoparticles

The silver nanoparticles can be synthesized using microemulsion techniques, chemical reductions, UV-initiated photoreduction, photoinduced reduction, synthetic electrochemical method, irradiation methods, microwave-assisted synthesis, polymers and polysaccharides, and Tollens methods (Korbekandi et al. 2013). Different approach yields vary because of the method's inherent capacity. The silver nanoparticles are synthesized using two primary processes—physical and chemical. The mechanical grinding of bulk metals using the top-down method stabilizes the resultant nanosized materials by adding colloidal protective agents. The top-down strategy involves mechanically grinding bulk metals and stabilizing the resultant nanosized materials by adding colloidal protective agents, whereas the bottom-up approaches use electrochemical techniques, metal reduction, and sono-decomposition. In the electrochemical production technique, $AgNO_3$ in an aqueous solution is reduced by an electric current in the presence of polyethylene glycol. Sono-decomposition, in which ultrasonic vibrations are used to create cavitations, is another method for producing silver nanoparticles. Because sonication produces hydrogen radicals, these procedures for making silver nanoparticles need sonochemical reduction of an aqueous $AgNO_3$ solution in an argon-hydrogen environment. Silver ions and an anionic surfactant aerosol are electrostatically combined to create highly stable liquid foam, which is then used to manufacture nanoparticles using the foam as a template. Following draining out, $NaBH_4$ is added to decrease the foam. Given that they have a diameter of 5–40 nm and are stable in solution, these silver nanoparticles may benefit from the stabilizing effects of the aerosol.

Microwaves are also used to synthesize silver nanoparticles, which includes reducing silver nanoparticles using a range of microwave radiation frequencies. With the same temperature range and exposure time, this process produces a relatively quicker response and a larger concentration of silver nanoparticles. Additionally, it can be deduced from the studies that an increase in the concentration of silver nitrate results in an extension of reaction time, an increase in temperature, and a reduction in particle size (between 15 nm and 25 nm). In contrast, an increase in the concentration of poly (N-vinylpyrrolidone) has the opposite effect.

In the microemulsion method, two-phase aqueous organic immiscible reducing agents synthesize the silver nanoparticles at the interface and are later coated with stabilizer molecules for stabilization (Krutyakov et al. 2008). The UV irradiated method is simple and reliable, and the process offers a competitive synthesis of nanoparticles in the presence of collagen, citrate, polyvinylpyrrolidone, and poly acrylic acid (Huang and Yang 2008). The silver nanoparticles were synthesized using the UV photo inactivation method in the presence of Triton X-100, which was used as a reducing agent. Photo-induced reduction synthesis of silver nanoparticles achieved 8 nm-sized nanoparticles using polyelectrolyte capsules as microreactors. It was also noted that the shape of the sphere is modulated to triangular (Shchukin et al. 2003). A well-defined shape and size distribution can be achieved through the laser irradiation method, where the laser is used the photo-sensitize the material to produce 20 nm-sized NPs (Eustis et al. 2005). Higher degrees of crystallization and yield can be achieved through microwave-assisted with carboxy methyl cellulose sodium as a reducing and stabilizing agent (Chen et al. 2020).

The most popular method for creating silver nanoparticles is chemical reduction using reducing agents that are both organic and inorganic. Metal precursors, reducing agents, and stabilizing/capping agents are the typical primary components used in this process. For instance, sodium citrate, ascorbate, sodium borohydride ($NaBH_4$), elemental hydrogen, the polyol process, the Tollens reagent, N, N-dimethylformamide (DMF), and poly (ethylene glycol)-block copolymers are used to reduce silver ions (Ag^+) in aqueous or non-aqueous solutions. In addition, poisonous and dangerous compounds, including citrate, borohydride, this glycerol, and 2-mercaptoethanol, are used to manufacture AgNPs. These reducing substances cause the reduction of Ag^+ to metallic silver (Ag^0), which then aggregates into oligomeric clusters. Eventually, these clusters cause the emergence of metallic colloidal silver particles (Merga et al. 2007). When preparing metal nanoparticles, it is crucial to apply stabilizing agents so to keep dispersive NPs from clumping together and safeguard NPs that may adhere to or absorb onto nanoparticle surfaces (Oliveira et al. 2005). Surfactants having functionalities for interacting with particle surfaces, such as thiols, amines, acids, and alcohols, can prevent sedimentation, agglomeration, and loss of surface characteristics while stabilizing particle development.

The initial $AgNO_3$ concentration, reducing agent, $AgNO_3$ molar ratios, and stabilizer concentrations all impact various characteristics, including the particle size and aggregation state of silver nanoparticles when formed using the chemical reduction process. The reduction of silver salts essentially happens in two stages: (1) initial nucleation and (2) subsequent growth. Generally, there are two "top-down" and "bottom-up" techniques for producing silver nanoparticles (Deepak et al. 2011). The mechanical grinding of bulk metals using the "top-down" technique is followed by stabilization using colloidal stabilizers (Mallick et al. 2004). Sono-decomposition, electrochemical processes, and chemical reduction are "bottom-up" techniques. High yield, as opposed to physical methods' poor yield, is the main benefit of chemical procedures. The plans mentioned above are rather pricey.

It was discovered that the generated particles' surfaces were covered in chemical silt, detracting from their intended purity. AgNPs with a precise size must also be prepared with great difficulty; this requires an additional step to prevent particle aggregation. Too many harmful and dangerous byproducts are removed throughout the synthesis process. Cryochemical synthesis, laser ablation, lithography, electrochemical reduction, laser irradiation, sono-decomposition, thermal decomposition, and chemical reduction are a few techniques used in chemical procedures (Gurunathan et al. 2015). The simplicity, cheap cost, and high yield of chemically synthesized nanoparticles are advantages; nevertheless, employing chemical reducing agents is detrimental to living things.

Physiochemical Properties of Silver Nanoparticles

An essential first step in understanding and introducing controlled nanoparticle manufacturing and its uses is nanoparticle characterization. AFM, DLS, powder X-ray diffractometry (XRD, SEM, and TEM), FTIR, XPS, and UV-Vis spectroscopy are just a few of the numerous techniques used to characterize nanoparticles. These analytical tools help with criterion resolutions, such as pore size, surface area, crystallinity, fractal dimensions, particle size, and form. These methods might also be used to analyze the size and morphology of nanocomposite materials using TEM, SEM, and AFM and to identify other requirements, including orientation, intercalation, and dispersion of nanoparticles and nanotubes. In order to determine the particle height and volume in 3D pictures, the AFM method is used, and DLS may be used to determine the particle size distribution. X-ray diffraction is used to analyze crystallinity, and UV-Vis spectroscopy is used to analyze surface Plasmon resonance to validate sample formation.

The biological effects of nanoparticles, such as cellular absorption, activation, and intercellular distribution, depend on their size, shape, surface charge, functionalization, and core structure (Cha et al. 2008). Nano-silver forms cubes, platelets, rods, rings, and bipyramids (Chung et al. 2008). However, there is still little information available about the impact of particle shape on the kinetics of their dissolution and the toxicity of such particles in biological mediums. Extraordinary toxicity results from the higher silver ions concentration due to the release of silver ions more quickly. Depending on the silver nanoparticle size and silver content, they discovered an antibacterial effect at a silver concentration between 1 ppm and 30 ppm (Sotiriou and Pratsinis 2011). According to (Hong et al. 2015), compared to spheres and wires, silver nano-cubes with a size of 55 nm showed the strongest antibacterial efficacy against *E. coli*.

Nanoparticles as Carriers

In order to create successful agricultural formulations, nanoparticles are frequently utilized as carriers to entrap, encapsulate, absorb, or attach active chemicals. The list below includes the typical nanoparticles employed as delivery systems for RNAi-inducing compounds, insecticides, fungicides, and herbicides.

Silica nanoparticles are excellent delivery vehicles because they are simple to produce and have regulated size, shape, and structure (Mody et al. 2014). For instance, porous hollow silica nanoparticles (PHSNs) or mesoporous silica nanoparticles are frequently manufactured spherically with pore-like pores (MSNs). Pesticides are commonly loaded into the inner core of PHSN and MSN to preserve the active molecules and enable a prolonged release. The PHSNs' shell structure shields the active chemicals inside the nanoparticles from UV radiation deterioration. Silica nanoparticles appear to be the obvious choice for creating agricultural goods for pest control because silicon has already been utilized to increase plant tolerance against numerous abiotic and biotic stressors (Barik et al. 2008).

Chitosan nanoparticles due to their hydrophobic characteristics, chitosan nanoparticles have limited solubility in aqueous solutions (Li et al. 2011). To increase its solubility, chitosan is frequently combined with an organic and inorganic copolymer (Kashyap et al. 2015). Because chitosan possesses reactive amine and hydroxyl groups, it may be modified, and graft reactions, ionic interactions, and other processes can improve its characteristics. Chitosan sticks firmly to the epidermis of stems and leaves, extending contact duration and promoting the absorption of bioactive compounds (Malerba and Cerana 2016).

Solid lipid nanoparticles (SLNs) are composed of solid lipids at ordinary temperatures, similar to emulsions. One benefit of SLNs is that they may entrap lipophilic active molecules in a matrix without organic solvents (Ekambaram et al. 2012). SLNs can provide controlled release of various lipophilic components because of the limited mobility of the active molecule in the solid matrix (Borel et al. 2014). Surfactants are included to stabilize the SLN after it has been dissolved in water.

Their major drawback is their low loading efficiency, and the action may leak while being kept outside the framework (Ekambaram et al. 2012).

Layered double hydroxides (LDHs) clays are hexagonal-shaped sheets of clay that include layers of active molecules trapped in the interlayer gap (Xu et al. 2006). Under acidic circumstances, such as the addition of moisture and carbon dioxide from the environment, LDH nanoparticles degrade (Mitter et al. 2017).

It has been demonstrated that positively charged delaminated LDH lactate nanoparticles let biologically active substances get through the barrier of the plant cell wall (Bao et al. 2016). Common carriers for managing plant diseases include silica, chitosan, SLN, and LDH nanoparticles. The sections organized by insecticide, fungicide, herbicide, or RNAi detail the specific research employing these nanoparticles and less often used nanoparticle kinds.

Silver Nanoparticles for the Management of Plant Diseases

The attachment of pathogens is the primary step of initiating pathogenesis, and attachment prevention is considered the viable option for abating the disease occurrence. Research on antimicrobial compounds produced several materials, such as plastics, ceramics, cloths, and stainless steel materials coated with silver (Ag), lead (Pb), and copper (Cu). The newest method for managing plant diseases involves using nanoparticles, which might be quite successful. The most fundamental way of protecting plants from pathogen assault is the direct application of nanoparticles in the soil, treatment of seeds before sowing/storage, or foliar spray that suppresses the pathogens similarly to chemical pesticides. When nanoparticles are applied directly to the soil, their impact on non-target species will be significant, particularly the microbes that fix or solubilize minerals. It is necessary to analyze the implications of nanosized particles on bacteria and their use in synthesizing pesticides to fully leverage the benefit of nanotechnology in plant disease protection and control. Due to their small size and strong reactivity, nanoparticles may be handy diagnostic tools for plant infections and diseases and for analyzing pesticide residues.

The higher antifungal activity of silver nanoparticles stabilized by polymers and surfactants was due to their increased aggregate stability. Due to the breakdown of their cell walls by surfactant activity, yeasts also develop an increased sensitivity to silver nanoparticles. The maximum antifungal activity of the silver NPs was achieved using PVP 360, Tween 80, Brij, and SDS stabilizers. Their findings agree with a previous study examining how stabilizing silver nanoparticles affected their antibacterial activity (Kvitek et al. 2008). Since the physiochemical properties of nanosized particles differ significantly from those of their microform, it is crucial to investigate how nanoparticles affect microbes to fully utilize this technology for plant protection, particularly against phytopathogens. Due to their extremely small size-even smaller than a virus particle and significant reactivity, nanoparticles may impact the pathogenicity of microbes. The many components that form the basis of the nanoparticles' antimicrobial action include concentration, physiology, metabolism, intracellular selective permeability of membranes, and the kind of microbial cell.

Surfactants act on the organisms' cell walls, which increases the organisms' sensitivity to the nanoparticles. Silver nanoparticles were utilized to control Xoo (*Xanthomonas oryzae* pv. *oryzae*) and *Acidovorax oryzae,* which are responsible for the cause of bacterial leaf blight and bacterial brown stripe since they have high inhibitory activity on both gram-positive and gram-negative bacteria. Biofilm formation is found to be the major cause of the disease progression, targeted molecules are very few, and the results were unsatisfactory. A study by (Ibrahim et al. 2019) found inhibition of the Xoo and *A. oryzae* at 20 µg mL^{-1} concentration of AgNPs. Inhibition of biofilm formation and motility were the chief mechanisms identified in their study, and cell wall damage was also reported in the cells (Ibrahim et al. 2019). They result in the loss of DNA, proteins, and cell contents and the death of the cells. Several reports on inhibiting plant pathogenic bacteria such as *Dickey* spp., *Ralstonia solanacearum* (Chen et al. 2016), *Pseudomonas syringae* pv. *tomato* (Marpu

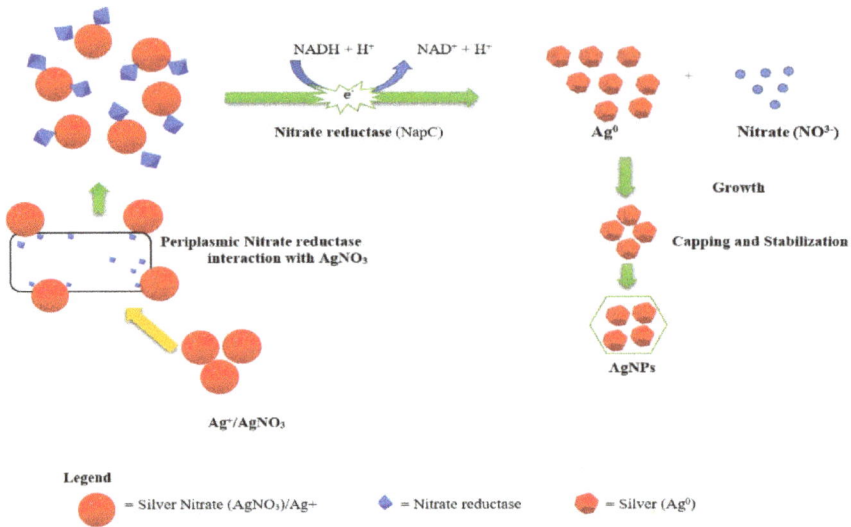

Figure 2. Flow diagram of the proposed mechanism of *Pseudomonas fluorescence*-mediated synthesis of AgNPs.

et al. 2017), *Pectobacterium* spp. (Dzimitrowicz et al. 2018), and *Clavibacter michiganensis* (Rivas-Cáceres et al. 2018).

Several microbes and other organisms, such as plants and invertebrates, are tolerant of heavy metals and capable of accumulating metal ions in their body. Many organisms, including fungi, bacteria, yeast, and algae, are said to sequester heavy metals extracellularly. Extracellular polymeric substances produced by microbes were reportedly involved in extracellular heavy metal accumulation. The metal ions are transported across the cell membrane in the metabolism-dependent category and accumulate intracellularly, while in the non-metabolism-dependent category of metal. Ion uptake is accomplished by physicochemical interaction between the cell membrane and the metal ions, such as physical adsorption or chemical sorption (Figure 2).

The accumulated metal ions affect microbes in two ways, i.e., direct and indirect ways. The dissolved ions indirectly affect bacteria by incorporating them into bacterial protein, making the protein non-functional and/or malfunctioning. On the other hand, metal dissolution forms active radicals, which affect bacteria directly, resulting in the rupture of the cell wall and death. Silver nanoparticles can stop the activity of bacteria through a variety of techniques.

Silver is also used in a variety of forms for the treatment of wounds and the management of infections. Silver sulphadiazine is applied as a cream or aqueous solution to treat some burn wounds. However, the effectiveness of these silver compounds has been questioned in the last few years. Diseases brought on by several plant pathogens, including bacteria, fungi, viruses, and nematodes, are major limiting factors for the production of food material. Several methods are used to manage plant diseases, but none offer ideal defense because they harm the environment and people. Therefore, there is much potential for using nanotechnology to manage plant diseases.

Additionally, it has been discovered that the nanoparticles inhibit the growth of fungi. The fungicidal activity of the silver nanoparticles was evaluated against various yeasts and molds, including *Candida albicans*, *C. krusei*, *C. tropicalis*, *C. glabrata*, and *Aspergillus brasiliensis* (Xia et al. 2016). The hybrid materials validated the powerful antifungal activities against these microorganisms. It has been proven that *Botrytis cinerea* and *Penicillium expansum*, two fungal diseases prevalent in the post-harvest phases, are susceptible to zinc oxide nanoparticles (ZnO) antifungal properties. By suppressing the growth of conidiophores and conidia in *P. expansum*, the treatment with nanoparticles also caused deformation in the hyphae of *B. cinerea*, which further contributed to the destruction of fungal hyphae. A palladium-based photo-dynamic therapy agent was approved in 2019 for the treatment of prostate cancer by the European Medicines Agency (EMA).

Antifungal Activity

Combined with nystatin and chlorhexidine digluconate, silver nanoparticles could enhance antifungal activity against *Candida albicans*, and *Candida glabrata*. The release of silver ions from the crystalline core of silver nanoparticles contributes to the antimicrobial activity of these nanomaterials, making them useful for many applications in antimicrobial treatment. It is believed that AgNP attaches and anchors to the surface of the fungus and produces an increase in reactive oxygen species (ROS). This interaction causes structural changes and damage, markedly disturbing vital cell functions, such as permeability and the membrane potential, forming pores causing ion leakage and other materials, depressing the activity of respiratory chain enzymes, and finally leading to cell death (Zhang et al. 2018). Also, it was shown that the accumulation of extracellular AgNP suggests a dynamic release of silver ions (Ag^+) by adjacent AgNP that actively penetrate the cell and lead to the intracellular biosynthesis of AgNP. The interaction of AgNP with phosphorus- or sulfur-containing compounds as DNA and thiol groups of proteins can cause further damage to yeasts by inhibiting DNA replication and protein inactivation. Furthermore, the gradual release of Ag^+ by AgNP could have special relevance, as they may act as a reservoir, increasing the duration of the antimicrobial effects (Sorensen and Baun 2015).

Control of Seed-Borne Diseases by Using Silver Nanoparticles

Chemical pesticide usage caused environmental pollution and the emergence of insect resistance. It has become increasingly difficult for the agricultural sector to safeguard plants, prompting researchers to find effective and non-toxic alternatives. The most effective and environmentally acceptable nanopesticides are made of metals (Chhipa and Joshi 2016; Werdin González et al. 2014). The development of biobased nanopesticides offered hope for replacing chemical-based pesticides, which have formulations that release slowly and active components that degrade slowly (Kashyap et al. 2015). In particular, because of their outstanding antibacterial capabilities, fungal-based AgNPs are the ideal choice (Borase et al. 2014). Unlike untreated Tomato seeds, seeds treated with biogenic AgNPs showed greater germination and seedling development. *Clavibacter michiganensis* subsp. *michiganensis* was successfully eradicated by it. Increases in plant height (cm), tomato yield per plant (g), fresh biomass per plant (g), number of shoots per plant (g), and dry biomass per plant were seen in plants treated with AgNPs (Noshad et al. 2019).

Nano-Priming for Sustainable Growth and Disease Management in Plants

Rapid and uniform seed germination and seedling emergence are crucial elements in commercial agriculture that determine the viability of stand development (Rajjou et al. 2012; Chen and Arora 2013). Imbibition, the process by which a mature, dry seed absorbs water, starts germination. Elongation of the embryonic axis, typically the radicle, through the seed envelope results in the protrusion of the root and eventually the shoot (Rajjou et al. 2012). According to Dragicevic et al. (2013), seeds stored in ambient conditions will have undergone some degree of deterioration, and seeds kept in long-term storage will eventually lose their viability due to spontaneous biochemical damage occurring at the cellular level (Butler et al. 2009), which will cause natural seed aging and subsequently reduce crop productivity.

Priming substances such as polyethylene glycol, inorganic salts, fertilizers, and plain water are frequently utilized (Butler et al. 2009; Horii et al. 2007; Hussain et al. 2015). For each crop species, priming agent optimization is necessary since different priming solutions have varied features and levels of efficacy (Horii et al. 2007). To improve seed germination of numerous agricultural plants, new priming agents are thus increasingly needed. Numerous metal-based NPs have been developed recently, including AgNPs (Mohamed et al. 2017), AuNPs (Mahakham et al. 2016), CuNPs (Taran et al. n.d.), FeNPs (Panyuta et al. 2016), FeS2NPs (Srivastava et al. 2014), and TiO$_2$NPs (Abdel Latef et al. 2017). To encourage seed germination, seedling development, and stress tolerance in various agricultural plants, carbon-based NPs have been used as seed pre-treatment agents, such as fullerene (Kole et al. 2013) and carbon nanotubes (Ratnikova et al. 2015). One of the finest methods for treating seeds with AgNPs is seed priming. It is a method in which seeds are partly hydrated in natural or synthetic substances in particular settings to the point where germination-related metabolic processes start but radicle emergence does not (Ibrahim et al. 2016). Priming seeds benefits seed quality, seedling emergence, agricultural yields, and increased resistance to environmental challenges (Chen and Arora 2013; Ibrahim et al. 2016). According to Dragicevic et al. (2013), seed priming can even boost the germination of seeds in unfavorable environments (Ibrahim et al. 2016). Only a few researchers have employed the seed priming technique in these investigations, requiring seeds to be dried again to their original moisture content before planting. As a result, seed nano-priming differs from pre-sowing seed treatment without drying seeds in terms of its process.

By boosting α-amylase activity, nano-priming might produce more soluble sugar, which would assist the development of seedlings. Additionally, nano-priming promoted the up-regulation of aquaporin genes in germination seeds. Compared to unprimed control and other priming treatments, greater ROS generation was seen in the developing seeds of the nano-priming treatment, indicating that ROS and aquaporins are significant in promoting seed germination. Various mechanisms have been proposed to explain how nano-priming-induced seed germination occurs. These include developing nanopores for improved water uptake, restarting ROS/antioxidant systems in seeds, producing hydroxyl radicals to loosen cell walls, and using nanocatalysts to accelerate starch hydrolysis.

AgNPs significantly impacted lentil seed germination and seedling development. Despite an increase in the mean germination time for lentil plants, the germination percentage for lentils improved at 10 g mL^{-1} AgNP concentrations. Applying 10 mL^{-1} of nano-silver can be feasible to raise and strengthen lentil seed germination under drought stress. Nano-silver improved the germination parameters and caused higher drought tolerance. The findings of this study can be used to assess the biocompatibility of AgNPs and to identify potential agricultural uses for nanoparticles in crop improvement. The effects of nano-silver on lentils' performance in the field must be evaluated for quality and quantity parameters as emphasized by Hojjat (2016).

AgNPs are used in agriculture to protect plants against fungi since their antibacterial impact is one of their best-known and most well-documented characteristics. Further evidence for their applicability as anti-phytopathogenic agents comes from study findings showing that they can promote the growth and development of agricultural plants. In the presence of AgNPs, maize (*Zea mays* L.), watermelon (*Citrullus lanatus* Thunb.), and zucchini (*Cucurbita pepo* L.) showed increased seed germination characteristics and plant development; however, maize root elongation was reduced under these circumstances.

Effect of silver nanoparticles (AgNPs) on bean seed germination, field emergence, and physiological characteristics of seedlings at normal and cold temperatures. As a quick pre-sowing treatment, AgNP solutions (0.25 mg dm^3, 1.25 mg dm^3, and 2.5 mg dm^3) were administered to seeds with the microbial preparation Nitragina, which contains *Rhizobium leguminosarum* bv. *phaseoli*. In both lab and field settings, low concentrations of AgNPs (0.25 mg dm^3 and 1.25 mg dm^3) had an immediate positive effect that led to quick and uniform germination. They also positively affected seedling development later on, as evidenced by an increase in the average seedling height, fresh and dry weight, and net photosynthesis. Particularly, favorable effects were noted in suboptimal temperature conditions, suggesting that AgNPs activate plant tolerance mechanisms to environmental stress. The highest concentration tested of AgNPs was not particularly effective for the plants but had a strong antimicrobial effect, which was beneficial in the period of seed germination, but at the later stage of plant development, was unfavorable, probably due to disruption of symbiosis between the bean seedlings and rhizobia (Prażak et al. 2020).

Diverse observations on the uptake, translocation, accumulation, biotransformation, and toxicity of nanoparticles on various plant species are available, and these results are frequently contradictory. One of the nanomaterials whose impacts are being studied is silver nanoparticles (AgNPs) (Kaegi et al. 2010). AgNPs' effect on higher plants appears to rely on various factors, including the kind and age of the plants, the size and concentration of the nanoparticles, the temperature of the experiment, the length of exposure, and the exposure technique. For example, 10 mg/L AgNPs decreased shoot length in *Linum usitatissimum* and *Hordeum vulgare* and impeded seed germination in Hordeum vulgare (El-Temsah and Joner 2012). Additionally, soluble and reducing sugar levels, lipase activity, and seed germination in *Brassica nigra* germinating seeds and seedlings were all decreased by an AgNPs dose ranging from 0.2 mg/L to 1.6 mg/L (Amooaghaie et al. 2015). Even at greater concentrations of AgNPs, castor bean, *Ricinus communis* L. (Yasur and Rani 2013), and *Vicia faba* (Abdel-Azeem and Elsayed 2013), however, did not significantly respond to AgNPs in terms of seed germination, root, or shoot length. Similarly, *Cucumis sativus* and *Lactuca sativa* seed germination were unaffected by 100 mg/L AgNPs. Following other research, AgNPs positively influence plant development in *Brassica juncea*, *Panicum virgatum*, *Phytolacca americana*, *Phaseolus vulgaris*, and *Zea mays*. Treatment with AgNPs had a beneficial impact on seed germination in Pennisetum glaucum and *Boswellia ovalifoliolata*.

Formulations Based on Silver Nanoparticles

Fungal bioinoculants are applied as an active component in the root region at the beginning of crop growth to help defend the plant against encroaching diseases. This ingredient's gradual release is preferred because it prevents all spores from invading the host at once and reduces the amount of pesticide required to control the disease, both of which automatically shield people and the environment from the adverse effects of the accidental release of chemical pesticides into the atmosphere. A nanotechnological method can further enhance the controlled release of the bioinoculants' active components by employing nanoparticles as a carrier for these compounds. These formulations may significantly reduce the intake of pesticides and related environmental risks. Because the amount of genuinely effective chemicals is 10 to 15 times lower than that of conventional formulations, nanopesticides will minimize the application rate. The features of nano-emulsion, such as enhanced kinetic stability, small size, reduced viscosity, and increased optical transparency, make it one of the dependable pesticide delivery technologies. When used as pesticide delivery vehicles, micro- or nano-emulsions can increase the solubility and bioavailability of the active components. Small particles of the active substances or other tiny designed structures with beneficial pesticidal characteristics are present in nanopesticides. Nanopesticides can speed up the dispersion and wettability of agricultural formulations, which will slow down unwanted pesticide movement and discharge from organic solvents. The properties of nanopesticides, such as their permeability, crystallinity, stiffness, thermal stability, solubility, and biodegradability, are well shown by nanomaterials. The enormous particular surface area that nanopesticide formulations provide also contributes to the target's enhanced affinity. Some of the nanopesticides delivery methods used and might be successful in plant protection programs include nano-emulsions, nano-encapsulated, nanocontainers, and nanocages.

Conclusions

Nanoparticles (NPs) have broader applications in the different fields of medicine, agriculture, and industry. The development of antibiotic resistance and lack of alternatives for disease management leads to the food crisis; hence, holistic approaches are required for alternative and effective control of diseases. Nanoparticles are eco-friendly and achieve maximum control in the necessary time without resulting in resistance. In addition to several advantages, research has shown that such NPs adversely affect various ecosystem parameters. Studies showed that nanoparticles might be used to monitor or manage chemicals that cause plant diseases. The plant pathogen-controlling NPs may have spread from the agricultural field to the soil, water, and atmosphere. Leaching, surface run-off, and airborne transportation of the dispersion are all possible. As a result, regular employment of nanomaterials in agriculture is necessary for a safe and sustainable environment. The use of NPs in managing plant diseases is currently very limited; establishing pilot projects with necessary objectives is the need of the hour.

References

Aabed, K. and Mohammed, A.E. 2021. Phytoproduct, Arabic gum and *Opophytum forsskalii* seeds for bio-fabrication of silver nanoparticles: antimicrobial and cytotoxic capabilities. Nanomaterials 11(10): 2573.

Aadil, K.R., Barapatre, A., Meena, A.S. and Jha, H. 2016. Hydrogen peroxide sensing and cytotoxicity activity of Acacia lignin stabilized silver nanoparticles. Int. J. Biol. Macromol. 82: 39–47.

Abdallah, Y., Ogunyemi, S.O., Abdelazez, A., Zhang, M., Hong, X., Ibrahim, E. and Chen, J. 2019. The green synthesis of MgO nano-flowers using *Rosmarinus officinalis* L. (Rosemary) and the antibacterial activities against *Xanthomonas oryzae* pv. oryzae. BioMed Res. Inter, 2019.

Abdel-Azeem, E.A. and Elsayed, B.A. 2013. Phytotoxicity of silver nanoparticles on *Vicia faba* seedlings. NY Sci. J. 6(12): 148–155.

Alghuthaymi, M.A., Almoammar, H., Rai, M., Said-Galiev, E. and Abd-Elsalam, K.A. 2015. Myconanoparticles: synthesis and their role in phytopathogens management. Biotechnol. Biotechnol. Equip, 29(2): 221–236.

Baklanov, M.R. 2012. Nanoporous dielectric materials for advanced micro- and nanoelectronics. NATO Science for Peace and Security Series B: Physics and Biophysics, 3–18. https://doi.org/10.1007/978-94-007-4119-5_1.

Balakumaran, M.D., Ramachandran, R., Balashanmugam, P., Mukeshkumar, D.J. and Kalaichelvan, P.T. 2016. Mycosynthesis of silver and gold nanoparticles: Optimization, characterization and antimicrobial activity against human pathogens. Microbiol. Res. 182: 8–20. https://doi.org/10.1016/J.MICRES.2015.09.009.

Bao, W., Wang, J., Wang, Q., O'Hare, D. and Wan, Y. 2016. Layered double hydroxide nanotransporter for molecule delivery to intact plant cells. Sci. Rep. 6(1): 1–9. https://doi.org/10.1038/srep26738.

Barik, T.K., Sahu, B. and Swain, V. 2008. Nanosilica—from medicine to pest control. Parasitol. Res. 103(2): 253–258. https://doi.org/10.1007/S00436-008-0975-7.

Behravan, M., Hossein Panahi, A., Naghizadeh, A., Ziaee, M., Mahdavi, R. and Mirzapour, A. 2019. Facile green synthesis of silver nanoparticles using Berberis vulgaris leaf and root aqueous extract and its antibacterial activity. Int. J. Biol. Macromol. 124: 148–154. https://doi.org/10.1016/J.IJBIOMAC.2018.11.101.

Butler, L.H., Hay, F.R., Ellis, R.H., Smith, R.D. and Murray, T.B. 2009. Priming and re-drying improve the survival of mature seeds of Digitalis purpurea during storage. Ann. Bot. 103(8): 1261–1270. https://doi.org/10.1093/AOB/MCP059.

Cătălin Balaure, P., Gudovan, D. and Gudovan, I. 2017. Nanopesticides: a new paradigm in crop protection. New Pesticides and Soil Sensors, 129–192. https://doi.org/10.1016/B978-0-12-804299-1.00005-9.

Cha, K., Hong, H.W., Choi, Y.G., Lee, M.J., Park, J.H., Chae, H.K., Ryu, G. and Myung, H. 2008. Comparison of acute responses of mice livers to short-term exposure to nanosized or micro-sized silver particles. Biotechnol. Lett. 30(11): 1893–1899. https://doi.org/10.1007/S10529-008-9786-2.

Chen, J., Li, S., Luo, J., Wang, R. and Ding, W. 2016. Enhancement of the Antibacterial activity of silver nanoparticles against phytopathogenic bacterium *Ralstonia solanacearum* by stabilization. J. Nanomater, https://doi.org/10.1155/2016/7135852.

Chen, K. and Arora, R. 2013. Priming memory invokes seed stress-tolerance. Environ. Exp. Bot. 94: 33–45. https://doi.org/10.1016/J.ENVEXPBOT.2012.03.005.

Chen, T.L., Kim, H., Pan, S.Y., Tseng, P.C., Lin, Y.P. and Chiang, P.C. 2020. Implementation of green chemistry principles in circular economy system towards sustainable development goals: Challenges and perspectives. Science of The Total Environment 716: 136998. https://doi.org/10.1016/J.SCITOTENV.2020.136998.

Chhipa, H. and Joshi, P. 2016. Nanofertilisers, nanopesticides and nanosensors in agriculture. *In*: Ranjan, S., Dasgupta, N. and Lichtfouse, E. (eds.). Nanoscience in Food and Agriculture 1. Sustainable Agriculture Reviews, vol 20. Springer, Cham. https://doi.org/10.1007/978-3-319-39303-2_9.

Chung, Y.C., Chen, I.H. and Chen, C.J. 2008. The surface modification of silver nanoparticles by phosphoryl disulfides for improved biocompatibility and intracellular uptake. Biomaterials 29(12): 1807–1816. https://doi.org/10.1016/J.BIOMATERIALS.2007.12.032.

Cohen-Karni, T., Langer, R. and Kohane, D.S. 2012. The smartest materials: The future of nanoelectronics in medicine. ACS Nano 6(8): 6541–6545. https://doi.org/10.1021/NN302915S.

Costa Silva, L.P., Pinto Oliveira, J., Keijok, W.J., da Silva, A.R., Aguiar, A.R., Guimarães, M.C.C., Ferraz, C.M., Araújo, J.V., Tobias, F.L. and Braga, F.R. 2017. Extracellular biosynthesis of silver nanoparticles using the cell-free filtrate of nematophagous fungus *Duddingtonia flagrans*. Int. J. Nanomedicine 12: 6373. https://doi.org/10.2147/IJN.S137703.

Das, S., Das, J., Samadder, A., Bhattacharyya, S.S., Das, D. and Khuda-Bukhsh, A.R. 2013. Biosynthesized silver nanoparticles by ethanolic extracts of *Phytolacca decandra*, *Gelsemium sempervirens*, *Hydrastis canadensis* and *Thuja occidentalis* induce differential cytotoxicity through G2/M arrest in A375 cells. Colloids and Surfaces B: Biointerfaces 101: 325–336. https://doi.org/10.1016/J.COLSURFB.2012.07.008.

Deepak, V., Umamaheshwaran, P.S., Guhan, K., Nanthini, R.A., Krithiga, B., Jaithoon, N.M.H. and Gurunathan, S. 2011. Synthesis of gold and silver nanoparticles using purified URAK. Colloids and Surfaces B: Biointerfaces 86(2): 353–358. https://doi.org/10.1016/J.COLSURFB.2011.04.019.

Deplanche, K., Caldelari, I., Mikheenko, I.P., Sargent, F. and Macaskie, L.E. 2010. Involvement of hydrogenases in the formation of highly catalytic Pd(0) nanoparticles by bioreduction of Pd(II) using *Escherichia coli* mutant strains. Microbiology 156(9): 2630–2640. https://doi.org/10.1099/MIC.0.036681-0/CITE/REFWORKS.

Divya, M., Kiran, G.S., Hassan, S. and Selvin, J. 2019. Biogenic synthesis and effect of silver nanoparticles (AgNPs) to combat catheter-related urinary tract infections. Biocatal. Agric. Biotechnol. 18: 101037. https://doi.org/10.1016/J.BCAB.2019.101037.

Dragicevic, V., Spasic, M., Simic, M., Dumanovic, Z. and Nikolic, B. 2013. Stimulative influence of germination and growth of maize seedlings originating from aged seeds by 2,4-D potencies. Homeopathy 102(3): 179–186. https://doi.org/10.1016/J.HOMP.2013.05.005/ID/JR000398-33.

Durán, N. and Marcato, P.D. 2012. Biotechnological routes to metallic nanoparticles production: Mechanistic aspects, antimicrobial activity, toxicity and industrial applications. Nano-Antimicrobials: Progress and Prospects 337–374. https://doi.org/10.1007/978-3-642-24428-5_12/COVER.

Dzimitrowicz, A., Motyka, A., Jamroz, P., Lojkowska, E., Babinska, W., Terefinko, D., Pohl, P. and Sledz, W. 2018. Application of silver nanostructures synthesized by cold atmospheric pressure plasma for inactivation of bacterial phytopathogens from the genera *Dickeya* and *Pectobacterium*. Materials 11(3): 331. https://doi.org/10.3390/MA11030331.

El-Temsah, Y.S. and Joner, E.J. 2012. Impact of Fe and Ag nanoparticles on seed germination and differences in bioavailability during exposure in aqueous suspension and soil. Environ. Toxicol. 27(1): 42–49. https://doi.org/10.1002/TOX.20610.

Eustis, S., Krylova, G., Eremenko, A., Smirnova, N., Schill, A.W. and El-Sayed, M. 2005. Growth and fragmentation of silver nanoparticles in their synthesis with as laser and CW light by photo-sensitization with benzophenone. Photochem. Photobiol. Sci. 4(1): 154–159. https://doi.org/10.1039/B411488D.

Flores-Rojas, G.G., López-Saucedo, F. and Bucio, E. 2020. Gamma-irradiation applied in the synthesis of metallic and organic nanoparticles: A short review. Radiat. Phys. Chem. 169: 107962. https://doi.org/10.1016/J.RADPHYSCHEM.2018.08.011.

Fouad, H., Hongjie, L., Yanmei, D., Baoting, Y., El-Shakh, A., Abbas, G. and Jianchu, M. 2016. Synthesis and characterization of silver nanoparticles using *Bacillus amyloliquefaciens* and *Bacillus subtilis* to control filarial vector *Culex pipiens* pallens and its antimicrobial activity. Http://Dx.Doi.Org/10.1080/21691401.2016.1241793, 45(7): 1369–1378. https://doi.org/10.1080/21691401.2016.1241793.

Gurunathan, S. and Park, J.H.I. n.d. Comparative assessment of the apoptotic potential of silver nanoparticles synthesized by *Bacillus tequilensis* and *Calocybe indica* in MDA-MB-231. *Ncbi.Nlm.Nih.Gov*. https://www.ncbi.nlm.nih.gov/pmc/articles/PMC4494182/.

Gurunathan, S., Park, J.H., Han, J.W. and Kim, J. 2022. Comparative assessment of the apoptotic potential of silver nanoparticles synthesized by *Bacillus tequilensis* and *Calocybe indica* in MDA-MB-231 human breast cancer cells: Targeting p53 for anticancer therapy [Corrigendum]. Inter J. Nanomed. 17(1): 5207–5208. https://doi.org/10.2147/IJN.S395879.

Hayles, J., Johnson, L., Worthley, C. and Losic, D. 2017. Nanopesticides: a review of current research and perspectives. New Pesticides and Soil Sensors 193–225. https://doi.org/10.1016/B978-0-12-804299-1.00006-0.

Henglein, A. 1989. Small-Particle Research: physicochemical properties of extremely small colloidal metal and semiconductor particles. Chem. Rev. 89(8): 1861–1873. https://doi.org/10.1021/CR00098A010/ASSET/CR00098A010.FP.PNG_V03.

Hojjat, S.S. 2016. The Effect of silver nanoparticle on lentil Seed Germination under drought stress. International Journal of Farming and Allied Sciences www.ijfas.com.

Hong, X., Wen, J., Xiong, X. and Hu, Y. 2015. Shape effect on the antibacterial activity of silver nanoparticles synthesized via a microwave-assisted method. Envi. Sci. Pollu. Res. 23(5): 4489–4497. https://doi.org/10.1007/S11356-015-5668-Z.

Horii, A., Wang, X., Gelain, F. and Zhang, S. 2007. Biological designer self-assembling peptide nanofiber scaffolds significantly enhance osteoblast proliferation, differentiation and 3-D migration. PLOS ONE 2(2): e190. https://doi.org/10.1371/JOURNAL.PONE.0000190.

Huang, H. and Yang, Y. 2008. Preparation of silver nanoparticles in inorganic clay suspensions. Comp. Sci. Tech. 68(14): 2948–2953. https://doi.org/10.1016/J.COMPSCITECH.2007.10.003.

Ibrahim, E., Fouad, H., Zhang, M., Zhang, Y., Qiu, W., Yan, C., Li, B., Mo, J. and Chen, J. 2019. Biosynthesis of silver nanoparticles using endophytic bacteria and their role in inhibition of rice pathogenic bacteria and plant growth promotion. RSC Advances 9(50): 29293–29299. https://doi.org/10.1039/C9RA04246F.

Ibrahim, I., Ali, I.O., Salama, T.M., Bahgat, A.A. and Mohamed, M.M. 2016. Synthesis of magnetically recyclable spinel ferrite (MFe_2O_4, M = Zn, Co, Mn) nanocrystals engineered by sol gel-hydrothermal technology: High catalytic performances for nitroarenes reduction. Applied Catalysis B: Environmental 181: 389–402. https://doi.org/10.1016/J.APCATB.2015.08.005.

Jo, D.H., Kim, J.H., Lee, T.G. and Kim, J.H. 2015. Size, surface charge, and shape determine therapeutic effects of nanoparticles on brain and retinal diseases. Nanomed. Nanotech. Bio. Med. 11(7): 1603–1611. https://doi.org/10.1016/J.NANO.2015.04.015.

Kaegi, R., Sinnet, B., Zuleeg, S., Hagendorfer, H., Mueller, E., Vonbank, R., Boller, M. and Burkhardt, M. 2010. Release of silver nanoparticles from outdoor facades. Environ. Poll. 158(9): 2900–2905. https://doi.org/10.1016/J.ENVPOL.2010.06.009.

Kalishwaralal, K., Deepak, V., Ramkumarpandian, S., Nellaiah, H. and Sangiliyandi, G. 2008. Extracellular biosynthesis of silver nanoparticles by the culture supernatant of *Bacillus licheniformis*. Mat. Lett. 62(29): 4411–4413. https://doi.org/10.1016/J.MATLET.2008.06.051.

Karthik, L., Kumar, G., Kirthi, A.V., Rahuman, A.A. and Bhaskara Rao, K.v. 2013. *Streptomyces* sp. LK3 mediated synthesis of silver nanoparticles and its biomedical application. Biopro. Biosys. Eng. 37(2): 261–267. https://doi.org/10.1007/S00449-013-0994-3.

Kashyap, P.L., Xiang, X. and Heiden, P. 2015. Chitosan nanoparticle based delivery systems for sustainable agriculture. Inter. J. Bio. Macromol. 77: 36–51. https://doi.org/10.1016/J.IJBIOMAC.2015.02.039.

Klaus-Joerger, T., Joerger, R., Olsson, E. and Granqvist, C.G. 2001. Bacteria as workers in the living factory: metal-accumulating bacteria and their potential for materials science. Tren. Biotech 19(1): 15–20. https://doi.org/10.1016/S0167-7799(00)01514-6.

Korbekandi, H., Iravani, S. and Abbasi, S. 2012. Optimization of biological synthesis of silver nanoparticles using *Lactobacillus casei* subsp. *casei*. J. Chemical Tech. Biotech 87(7): 932–937. https://doi.org/10.1002/JCTB.3702.

Korbekandi, H., Ashari, Z., Iravani, S. and Abbasi, S. 2013. Optimization of biological synthesis of silver nanoparticles using *Fusarium oxysporum*. Iranian Journal of Pharm. Res. 12(3): 289. /pmc/articles/PMC3813263/.

Krutyakov, Yu.A., Olenin, A.Yu., Kudrinskii, A.A., Dzhurik, P.S. and Lisichkin, G.v. 2008. Aggregative stability and polydispersity of silver nanoparticles prepared using two-phase aqueous organic systems. Nanotechnologies in Russia 3(5): 303–310. https://doi.org/10.1134/S1995078008050054.

Li, M., Huang, Q. and Wu, Y. 2011. A novel chitosan-poly(lactide) copolymer and its submicron particles as imidacloprid carriers. Pest Manag. Sci. 67(7): 831–836. https://doi.org/10.1002/PS.2120.

Mahakham, W., Theerakulpisut, P., Maensiri, S., Phumying, S. and Sarmah, A.K. 2016. Environmentally benign synthesis of phytochemicals-capped gold nanoparticles as nanopriming agent for promoting maize seed germination. Sci. Envir. 573: 1089–1102. https://doi.org/10.1016/J.SCITOTENV.2016.08.120.

Malerba, M. and Cerana, R. 2016. Chitosan effects on plant systems. Inter. J. Mol Sci. 17(7): 996. https://doi.org/10.3390/IJMS17070996.

Mallick, K., Witcomb, M.J. and Scurrell, M.S. 2004. Polymer stabilized silver nanoparticles: A photochemical synthesis route. J. Mat. Sci. 39(14): 4459–4463. https://doi.org/10.1023/B:JMSC.0000034138.80116.50.

Marpu, S., Kolailat, S.S., Korir, D., Kamras, B.L., Chaturvedi, R., Joseph, A., Smith, C.M., Palma, M.C., Shah, J. and Omary, M.A. 2017. Photochemical formation of chitosan-stabilized near-infrared-absorbing silver Nanoworms: A "Green" synthetic strategy and activity on Gram-negative pathogenic bacteria. J. Coll. Inter. Sci. 507: 437452. https://doi.org/10.1016/J.JCIS.2017.08.009.

Merga, G., Wilson, R., Lynn, G., Milosavljevic, B.H. and Meisel, D. 2007. Redox catalysis on "naked" silver nanoparticles. J. Physical Chem. 111(33): 12220–12226. https://doi.org/10.1021/JP074257W/SUPPL_FILE/JP074257WSI20070601_110535.PDF.

Mie, R., Samsudin, M.W., Din, L.B., Ahmad, A., Ibrahim, N. and Adnan, S.N.A. 2014. Synthesis of silver nanoparticles with antibacterial activity using the lichen *Parmotrema praesorediosum*. Inter. J. Nanomed. 9(1): 121. https://doi.org/10.2147/IJN.S52306.

Mittal, A., Bhaumik, J., Kumar, S. and U. B.-J. of colloid, and 2014, undefined. 2014. Biosynthesis of silver nanoparticles: elucidation of prospective mechanism and therapeutic potential. Elsevier, 415: 39–47. https://doi.org/10.1016/j.jcis.2013.10.018.

Mitter, N., Worrall, E.A., Robinson, K.E., Li, P., Jain, R.G., Taochy, C., Fletcher, S.J., Carroll, B.J., Lu, G.Q. and Xu, Z.P. 2017. Clay nanosheets for topical delivery of RNAi for sustained protection against plant viruses. Nature Plants 3(2): 1–10. https://doi.org/10.1038/nplants.2016.207.

Mochochoko, T., Oluwafemi, O.S., Jumbam, D.N. and Songca, S.P. 2013. Green synthesis of silver nanoparticles using cellulose extracted from an aquatic weed; water hyacinth. Carbohydrate Polymers 98(1): 290–294. https://doi.org/10.1016/J.CARBPOL.2013.05.038.

Mody, V.v., Cox, A., Shah, S., Singh, A., Bevins, W. and Parihar, H. 2014. Magnetic nanoparticle drug delivery systems for targeting tumor. Applied Nanosci. (Switzerland) 4(4): 385–392. https://doi.org/10.1007/S13204-013-0216-Y/FIGURES/7.

Mohamed, S.A., Al-Harbi, M.H., Almulaiky, Y.Q., Ibrahim, I.H. and El-Shishtawy, R.M. 2017. Immobilization of horseradish peroxidase on Fe_3O_4 magnetic nanoparticles. Electronic J. Biotech 27: 84–90. https://doi.org/10.1016/J.EJBT.2017.03.010.

Mokhtari, N., Daneshpajouh, S., Seyedbagheri, S., Atashdehghan, R., Abdi, K., Sarkar, S., Minaian, S., Shahverdi, H.R. and Shahverdi, A.R. 2009. Biological synthesis of very small silver nanoparticles by culture supernatant of *Klebsiella pneumonia*: The effects of visible-light irradiation and the liquid mixing process. Materials Res. Bull. 44(6): 1415–1421. https://doi.org/10.1016/J.MATERRESBULL.2008.11.021.

Mondal, P., Anweshan, A. and Purkait, M.K. 2020. Green synthesis and environmental application of iron-based nanomaterials and nanocomposite: A review. Chemosphere 259: 127509. https://doi.org/10.1016/J.CHEMOSPHERE.2020.127509.

Mourato, A., Gadanho, M., Lino, A.R. and Tenreiro, R. 2011. Biosynthesis of crystalline silver and gold nanoparticles by extremophilic yeasts. Bioinorganic Chem. Appli, 2011. https://doi.org/10.1155/2011/546074.

Mukherjee, P., Roy, M., Mandal, B.P., Dey, G.K., Mukherjee, P.K., Ghatak, J., Tyagi, A.K. and Kale, S.P. 2008. Green synthesis of highly stabilized nanocrystalline silver particles by a non-pathogenic and agriculturally important fungus *T. asperellum*. Nanotechnology 19(7): 075103. https://doi.org/10.1088/0957-4484/19/7/075103.

Noshad, A., Hetherington, C. and Iqbal, M. 2019. Impact of AgNPs on seed germination and seedling growth: A focus study on its antibacterial potential against *Clavibacter michiganensis* subsp. *michiganensis* infection in *Solanum lycopersicum*. J. Nanomat, 2019. https://doi.org/10.1155/2019/6316094.

Oliveira, M.M., Ugarte, D., Zanchet, D. and Zarbin, A.J.G. 2005. Influence of synthetic parameters on the size, structure, and stability of dodecanethiol-stabilized silver nanoparticles. J. Coll. Interface Sci. 292(2): 429–435. https://doi.org/10.1016/J.JCIS.2005.05.068.

Otunola, G. 2018. *In vitro* antibacterial, antioxidant and toxicity profile of silver nanoparticles green-synthesized and characterized from aqueous extract of a spice blend formulation. Taylor and Francis 32(3): 724–733. https://doi.org/10.1080/13102818.2018.1448301.

Panyuta, O., Belava, V., Fomaidi, S., Kalinichenko, O., Volkogon, M. and Taran, N. 2016. The effect of pre-sowing seed treatment with metal nanoparticles on the formation of the defensive reaction of wheat seedlings infected with the eyespot causal agent. Nanoscale Res. Lett. 11(1): 1–5. https://doi.org/10.1186/S11671-016-1305-0/FIGURES/3.

Prażak, R., Święciło, A., Krzepiłko, A., Michałek, S. and Arczewska, M. 2020. Impact of Ag nanoparticles on seed germination and seedling growth of green beans in normal and chill temperatures. Agriculture 10(8): 312. https://doi.org/10.3390/AGRICULTURE10080312.

Rai, M., Bonde, S., Golinska, P., Trzcińska-Wencel, J., Gade, A., Abd-Elsalam, K.A., Shende, S., Gaikwad, S. and Ingle, A.P. 2021. *Fusarium* as a novel fungus for the synthesis of nanoparticles: mechanism and applications. J. Fungi 7(2): 139. https://doi.org/10.3390/jof7020139.

Rajjou, ic, Duval, M., Gallardo, K., Catusse, J., Bally, J., Job, C. and Job, D. 2012. Seed germination and vigor. Ann. Rev. Plant Biol. 63: 507–540. https://doi.org/10.1146/annurev-arplant-042811-105550.

Ratnikova, T.A., Podila, R., Rao, A.M. and Taylor, A.G. 2015. Tomato seed coat permeability to selected carbon nanomaterials and enhancement of germination and seedling growth. Sci. World J. https://doi.org/10.1155/2015/419215.

Reguera, G. 2012. Electron transfer at the cell–uranium interface in *Geobacter* spp. Bioch. Soc. Tran 40(6): 1227–1232. https://doi.org/10.1042/BST20120162.

Rivas-Cáceres, R.R., Luis Stephano-Hornedo, J., Lugo, J., Vaca, R., del Aguila, P., Yañez-Ocampo, G., Mora-Herrera, M.E., Camacho Díaz, L.M., Cipriano-Salazar, M. and Alaba, P.A. 2018. Bactericidal effect of silver nanoparticles against propagation of *Clavibacter michiganensis* infection in *Lycopersicon esculentum* Mill. Microbial Path 115: 358–362. https://doi.org/10.1016/J.MICPATH.2017.12.075.

Shahverdi, A.R., Minaeian, S., Shahverdi, H.R., Jamalifar, H. and Nohi, A.A. 2007. Rapid synthesis of silver nanoparticles using culture supernatants of Enterobacteria: A novel biological approach. Process Bioch. 42(5): 919–923. https://doi.org/10.1016/J.PROCBIO.2007.02.005.

Shams, S. and Pourseyedi, S. 2013. Green synthesis of Ag nanoparticles in the present of *Lens culinaris* seed exudates. Cabdirect.Org. https://www.cabdirect.org/cabdirect/abstract/20133341210.

Sharma, S.K. and Bisen, P.S. 1992. Hg^{2+} and Cd^{2+} induced inhibition of light induced proton efflux in the cyanobacterium *Anabaena flos-aquae*. Biometals 5(3): 163–167. https://doi.org/10.1007/BF01061323.

Shchukin, D.G., Radtchenko, I.L. and Sukhorukov, G.B. 2003. Photoinduced reduction of silver inside microscale polyelectrolyte capsules. Physical Chem. 4(10): 1101–1103. https://doi.org/10.1002/CPHC.200300740.

Sorensen, S.N. and Baun, A. 2015. Controlling silver nanoparticle exposure in algal toxicity testing-A matter of timing. Nanotoxicology 9(2): 201–209. https://doi.org/10.3109/17435390.2014.913728/SUPPL_FILE/INAN_A_913728_SM5042.DOCX.

Sotiriou, G.A. and Pratsinis, S.E. 2011. Engineering nanosilver as an antibacterial, biosensor and bioimaging material. Curr. Opi. Chem. Eng. 1(1): 3–10. https://doi.org/10.1016/J.COCHE.2011.07.001.

Tanzil, A.H., Sultana, S.T., Saunders, S.R., Shi, L., Marsili, E. and Beyenal, H. 2016. Biological synthesis of nanoparticles in biofilms. Enzyme Micro. Tech 95: 4–12. https://doi.org/10.1016/J.ENZMICTEC.2016.07.015.

Taran, M., Monazah, A. and Alavi, M. n.d. Using petrochemical wastewater for synthesis of cruxrhodopsin as an energy capturing nanoparticle by *Haloarcula* sp. IRU1. Progress in Biological Sci. 6(2). https://doi.org/10.22059/PBS.2016.590017.

Werdin González, J.O., Stefanazzi, N., Murray, A.P., Ferrero, A.A. and Fernández Band, B. 2014. Novel nanoinsecticides based on essential oils to control the German cockroach. J. Pest Science 88(2): 393–404. https://doi.org/10.1007/S10340-014-0607-1.

Xia, Z.K., Ma, Q.H., Li, S.Y., Zhang, D.Q., Cong, L., Tian, Y.L. and Yang, R.Y. 2016. The antifungal effect of silver nanoparticles on *Trichosporon asahii*. J. Micro, Immunology and Infection 49(2): 182–188. https://doi.org/10.1016/J.JMII.2014.04.013.

Xu, Z.P., Stevenson, G.S., Lu, C.Q., Lu, G.Q., Bartlett, P.F. and Gray, P.P. 2006. Stable suspension of layered double hydroxide nanoparticles in aqueous solution. Journal of the American Chemical Society 128(1): 36–37. https://doi.org/10.1021/JA056652A/SUPPL_FILE/JA056652ASI20051119_092741.PDF.

Yang, W., Peters, J.I. and Williams, R.O. 2008. Inhaled nanoparticles—A current review. International J. Pharm. 356(1–2): 239–247. https://doi.org/10.1016/J.IJPHARM.2008.02.011.

Yong, P., Rowson, N.A., Farr, J.P.G., Harris, I.R. and Macaskie, L.E. 2002. Bioaccumulation of palladium by *Desulfovibrio desulfuricans*. J. Chem. Tech. Biotech 77(5): 593–601. https://doi.org/10.1002/JCTB.606.

Zhang, H., Wu, M. and Sen, A. 2012. Silver nanoparticle antimicrobials and related materials. Nano-Antimicrobials: Progress and Prospects, 3–45. https://doi.org/10.1007/978-3-642-24428-5_1/COVER.

Zhang, L., Wu, L., Si, Y. and Shu, K. 2018. Size-dependent cytotoxicity of silver nanoparticles to *Azotobacter vinelandii*: Growth inhibition, cell injury, oxidative stress and internalization. PLOS ONE 13(12): e0209020. https://doi.org/10.1371/JOURNAL.PONE.0209020.

Zhang, X.F., Liu, Z.G., Shen, W. and Gurunathan, S. 2016. Silver nanoparticles: synthesis, characterization, properties, applications, and therapeutic approaches. Inter. J. Molecular Sci. 17(9): 1534. https://doi.org/10.3390/IJMS17091534.

Zuin, V.G., Stahl, A.M., Zanotti, K. and Segatto, M.L. 2020. Green and sustainable chemistry in Latin America: Which type of research is going on? And for what? Current Opinion in Green Sus. Chem. 25: 100379. https://doi.org/10.1016/J.COGSC.2020.100379.

Chapter 10

Plant Pathogens and Detection Approaches With Gold Nanoparticles

Anugrah Tripathi,[1] *Garima Mishra*[2] and *Krishna Kant*[3],*

Introduction

Forests supply a variety of ecosystem services, including a healthy watershed, a habitat for wildlife, and support for many economic sectors through timber harvesting and tourism. These complex ecosystems provide socio-ecological and economic services. Forests contribute to a major portion of the global carbon dioxide sink (Pan et al. 2011). However, climate change (i.e., changes in global climatic factors) has increased forest vulnerabilities, including diseases caused by pathogens, such as bacteria, fungi, viruses, and nematodes (Boyd et al. 2013). The disease is an outcome of interaction among a host, pathogen, and environment. Pathogens are infectious and transmissible organisms that can spread from one host to infect another. They attack plants and trees to obtain nutrients and energy to complete the life cycle and cause disease in the host tree. These pathogenic organisms are of several types viz. bacteria, viruses, nematodes, and fungi. According to Whetzel (1929), diseases in plants are the plants' collective responses to the pathogenic agents that are operating upon them. Forest infections are among the most important and evident dangers to the health of forests, and contemporary problems like climate change and global trade amplify these effects (Allen et al. 2010; Wingfield et al. 2016). It is crucial for society to comprehend and counteract the risks to these resources.

In modern times, revolutionary uses of nanomaterials have been defined in fields of forestry to protect trees from various diseases in the forest and balance the ecological system (Strange et al. 2005). Along with several types of nanomaterials metal-based nanoparticles, AuNPs have been used for diverse purposes involving nanodevices and pathogen detection tools. The remarkable properties of AuNPs, make them amenable to a variety of recognition platforms (as presented in Figure 1). The application of metal nanomaterial in forest plant plants comes with several advantages and challenges. Since the forest has its ecosystem to maintain itself (Mother Nature), nanomaterial can only be allowed as a supporting supplement to maintain the forest plants. It is first important to understand plants and their diseases, as well as the main challenge of detection of disease. There are numerous illnesses in the forest for which no preventive measures have been developed or for which

---

[1] Monitoring and Evolution Division, Directorate of Research, Indian Council of Forestry Research and Education, Dehradun 248006, India.

[2] Division of Genetics and Tree Improvement, Forest Research Institute, Dehradun 248006, India.

[3] Departamento de Química Física, Campus Universitario, CINBIO Universidade de Vigo, 36310 Vigo, Spain.

* Corresponding author: krishna.kant@uvigo.es

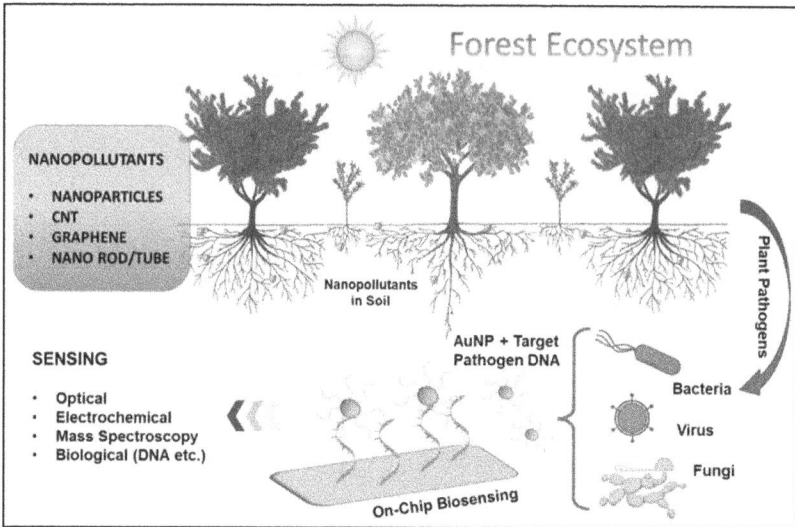

Figure 1. Graphic demonstration of nanomaterial uptake and their use for plant pathogen detection (Kulbhushan et al. 2022).

a large-scale management effort is neither logistically nor economically possible. However, diseases that are treatable have been managed using the following strategies:

- *Appropriate Site Choice*: The right planting location will encourage healthy growth and natural resistance, reducing damage from most native diseases in a stand.
- *Disease Management Techniques*: Certain illnesses can be managed using techniques including thinning, trimming, and cutting. For instance, harvesting practices are largely responsible for controlling dwarf mistletoe.
- *Measures for Disease Control*: Direct command by taking precautions like fungicide treatment, post-felling treatment of trees, and removal of infected stumps can help avoid infection and eradicate recurrent disease centers. Each of these therapies contributes to managing root illness.
- *Quarantine*: To stop the introduction or control the spread of a pest, it is legal to prohibit the transportation of trees from sick areas to disease-free areas.
- *Breeding for Resistance*: Choosing trees that exhibit disease resistance and using these trees in breeding programs raises the possibility that some pest issues may be genetically resolved.

Forest Plants and Pathogens

Forest tree diseases are linked to three major groups of fungi and one group of fungus-like organisms. Various types of diseases are indicative of the organism group, such as bracket fungi and mushrooms are associated with wood and root deterioration. There are some recognized exceptions and overlaps.

Water Molds (Oomycetes)

Formerly classed as fungi, the oomycetes are microscopic soil and water creatures that are now included in a class called the "Stramenopiles." Their cell wall composition and the characteristics of the motile, swimming spores set them apart from real fungi. They resemble diatoms and brown algae more closely than actual mushrooms. In forestry, several species are linked to root infections, particularly those that affect seedlings. Fruit bodies are too small for the unassisted eye to see clearly. The asexual spores, known as "zoospores," are aquatic and motile and are created inside sporangia. For their spread, water is necessary. Oospores, the sexual spores, have thick walls and are non-motile. Typically, laboratory culturing methods and the use of a microscope are needed for the diagnosis of water mold root rots and damping-off produced by oomycetes.

Molds and Sac Fungi

These include ascomycetes and mitosporic fungus, sometimes known as "fungi imperfecti" or "deuteromycetes." The sexual spores of the ascomycetes are produced in sacs called asci. The asci typically possess eight ascospores and are generated in fruit bodies. These fruit bodies range from microscopic to huge disks up to 15 cm in diameter in size (such as flask-shaped perithecia or cup-shaped apothecia). Ascospores, which are frequently discharged by an ascus explosion, are carried by the wind. Numerous ascomycetes also produce an anamorph, which is an asexual (conidial, or mitosporic) stage of spore. The anamorph develops a mold-like growth and development before the teleomorph or sexual stage. Only the anamorph frequently appears in nature. In the condition of the unclear identity of the teleomorph, man-made collection viz. *mitosporic* fungi may be considered (former names include deuteromycetes and fungi imperfecti). Mitosporic fungi produce conidia on conidiophores, specialized hyphae that may form in fruit bodies with specific functions, such as pycnidia. Leaf blights, cankers, needle casts, and vascular wilts are major diseases caused by these fungi, which obstruct and harm conductive tissues.

Conks With Mushrooms

Holobasidiomycetes (hymenomycetes) are the common mushroom, and bracket fungi are two examples of hymenomycete fruit bodies. All of the phylum's members, the hymenomycetes, microscopic, typically club-shaped, one-celled basidia are what distinguish Basidiomycota from other fungi. Millions of basidiospores are produced by basidia, which are found on the surfaces of fruit bodies, fungal spores, and mushroom gills. Examples of several of these structures are shown. Many of these fungi frequently cause root rot and wood degradation. It is usual to see mushrooms that fruit under forest trees. Numerous varieties of mushrooms are beneficial to trees rather than pathogenic. Their mycelium joins with tree roots in a symbiotic relationship that aids the roots in absorbing water and minerals for the tree. These roots of the fungus are known as mycorrhizae.

Rusts: Basidiomycota's Urediniomycetes

The Urediniomycetes, often known as rusts, make up a distinct class of obligate parasites. On two unrelated host plants, many rust fungi finish their life cycle. One of their five spore stages, the thick-walled teliospore, germinates to develop basidium. They can produce up to five different spore states. Depending on which spore state is currently being created, the same disease may seem different from one spore state to the next. Rust fungus often has four-celled basidia, which sets them apart from hymenomycetes under the microscope. The term "rust" refers to the distinctive symptom of these infections in which diseased parts of the host are frequently covered in obvious masses of loose spores that vary from rusty orange to yellow. Some rusts also cause cankers that resemble blisters on stems, while others affect the leaves and needles of trees.

Effect of Tree Pathogens on the Environment and Forest Ecosystem

In recent years, the emergence of new diseases on many economically important tree species posed severe negative impacts on our forest ecosystems. Disease and pest outbreaks in plants and trees are among the worst stressors in forests. Forests are exposed to multiple pathogens, pest attacks invasion of exotics, as well as nutrient deficiencies. Various plant pathogens, such as fungi, bacteria, viruses, oomycetes, phytoplasma, and plant parasitic nematodes cause disease (foliar, stem, culm, emerging buds, and fruits) in various plants that are mostly caused by fungi (Figure 2). Forest diseases result in severe landscape-level mortality of economically important trees and associated plant species, which can pose secondary effects on other landscape-level factors, such

Figure 2. Different diseases in forest tree and bamboo species. Foliar diseases (a–d), fruit disease (e), stem diseases (f–h), culm and bud rot disease in bamboo (i–j).

as soil fertility, wind velocity, land-use patterns, endemism, and invasion of species. For instance, large-scale mortality of tree beans (*Parkia timoriana*) has been recently observed in Northeast India. The frequently associated pathogen with sudden death or decline of tree beans was identified as *Botrydiplodia theobromae* (Singh et al. 2018). Earlier, large-scale mortality of *Cedrus deodara* due to root rot caused by *Phytophthora cinnamomic* was reported by Singh and Lakhanpal (2000) in Chail, Himachal Pradesh. Similarly, common ash (*Fraxinus excelsior*) in Poland has been severely impacted by a leaf and twig dieback caused by *Chalara fraxinea* (currently *Hmenoscyphus fraxineus*) (Kowalski 2006). A large part of the area in Europe in which *F. excelsior* is native is currently affected by *H. fraxineus* (Kowalski et al. 2017).

In Hawaii, an emerging tree fungus has killed hundreds of thousands of Hawaii's iconic and native ohia lehua (*Metrosideros polymorpha*) trees. The fungus identified as *Ceratocystis fimbrata* poses a serious threat to Hawaii's flagship native tree species whose loss would be catastrophic for the diversity, structure, and function of Hawaii's remaining native forests and the services they provide (Keith et al. 2015). In the declining ash stands in Poland and Denmark, Orlikowski et al. (2011) reported the presence of four *Phytophthora species*, viz., *P. cactorun*, *P. plurivora*, *Phytophthora* sp., and *P. gonapodyides*. The reasons for its current devastating outbreak, however, remain unclear. It is important to mention here that several factors including anthropogenic disturbances, such as inappropriate management practices and extreme adverse environmental fluxes, disease, and pest outbreaks also resulted in complex disease situations. As a result of the interaction of all these adverse factors, a disease known as "oak decline" is dominating the Mediterranean basin (Moricca et al. 2016).

Detection of Plant Pathogens

Timely detection and management of pathogens are of paramount importance for the sustainable productivity of forests. This also addresses the most critical issues and finding remedial measures along with management of emerging plant and tress diseases throughout the world. Traditional pathology as well as new biotechnological and molecular diagnostic approaches to disease and pathogens are being considered (Tripathi and Sharma 2021). Diseases caused by fungal pathogens pose a significant challenge to the survival and efficacy of both natural and man-made forests, and sustainable and environment-friendly approaches have received a lot of attention in this respect. However, we must first understand them, including what they are, where they originate from, and how they evolved, all of which must be correctly identified to develop a successful management strategy (Idnurm and Howlett 2001; Chakraborty et al. 2014). Until recently, the only means to identify new species was through specimen-based taxonomic research, which is time-consuming and may not always offer species-level resolution (Pecnikar et al. 2014; Truong et al. 2017). Therefore, DNA-based identification of fungal pathogens gained importance. For example, DNB is a serious disease of pines (*Pinus* spp.), with a worldwide distribution. It is caused by the ascomycete fungi *Dothistroma septosporum* (teleomorph: *Mycospharella pini*) and *Dothistroma pini*. The DNB was found on *Pinus peuce* in Austria, *Pinus pallasiana* in Ukraine, and the European part of southwestern Russia, as well as on *Pinus radiata* and *P. wallichiana* in Bhutan. Morphological examination of selected specimens from different hosts and countries showed that *D. septosporum* and *D. pini* overlap in the length of their conidia, while the width is slightly wider in *D. pini* than in *D. septosporum*. The differences in conidial width are so small; however, the identification of the two *Dosthistroma* species solely based on morphology is virtually impossible. Based on DNA sequence comparisons of the internal transcribed spacer (ITS) and b-tubulin gene regions, isolates from Austria and Bhutan were identified as *D. septosporum*, while isolates from Ukraine and southwestern Russia were identified as *D. pini*.

There are several laboratory-based techniques viz. Polymerase chain reaction (PCR), fluorescence *in-situ* hybridization (FISH), enzyme-linked immunosorbent assay (ELISA), immunofluorescence (IF), flow cytometry (FCM), and gas chromatography-mass spectrometry (GC-MS), which show direct detection of the plant pathogens. Indirect methods of disease detection include thermography, hyperspectral, and fluorescence imaging. Biosensors based on bio-recognition factors viz. enzyme, antibody, DNA/RNA, and bacteriophage are some of the new tools for identification of the plant disease. There are several techniques with advantages and disadvantages for disease detection. However, the use of nanomaterial further improves the detection limits and helps to improve the strategies. Nanomaterial is expanding in disease detection and plant safety (Gogos et al. 2012). Nanoparticles are employed with DNA/RNA or protein-based biomarkers at the molecular level to detect diseases to benefit the sensitivity of the detection for plant diseases. Nano-based microfluidic devices are being investigated in the field of plant propagation and genetic transformation (Torney et al. 2007). Nanoparticles and their application in sensing various plant diseases are summarized in Table 1.

Metal nanoparticles, like gold nanoparticles, iron oxide nanoparticles, zinc oxide nanoparticles, silver nanoparticles, etc., are usually non-toxic and biocompatible with high stability in the sensing experiments. Gold and silver are considered antibacterial materials in their low concentrations. However, these nanoparticles have a high affinity toward the capturing biomolecules and are easy to perform the surface chemistry to apply in the biosensors. Nanoparticles bind to thiol or the amine moiety to further bind with cellular proteins and lead to the capturing of target pathogen proteins. Recently researchers developed a sensitive diagnostic methodology for plant pathogen recognition employing a synergistic blend of recombinase polymerase amplification (RPA) and AuNP-enhanced biosensor. This sensing approach was applied for the detection of the plant pathogen, *Leptosphaeria maculans*. The achieved limit of detection was approximately one copy of the pathogen per sample, which is significantly smaller than the real-time fluorescence RPA assay as presented in Figures 3 A, B, and C (Lei et al. 2021). In another approach, the target DNA was captured in a sandwich assay using AuNP-probe. DNA and a capture probe were used in a lateral flow immune assay or, in other words, a paper-based microfluidic system. The measurable assessment is achieved by the optical intensity of the color band of a biosensor. The system is able to detect as little as 0.1 pg μL^{-1} genomic DNA. Additionally, the selectivity of the biosensor was established by identifying multiplexed three *Phytophthora* species and two pathogenic fungi.

It is claimed by the researcher that the developed method has the capability for application in the early forecast of potato late blight disease (Zhan et al. 2018). Researchers developed a cost-efficient smartphone-based volatile organic compound (VOC) sensing system for sensitive detection of late blight triggered by *Phytophthora infestans*. The sensing device examines leaf volatile molecules in the air. The portable hand-held device incorporates a disposable colorimetric sensor array consisting

Table 1. Different sensing approaches and nanomaterial used for plant pathogen detection.

Sl. No.	Technique	LOD (CFU/mL)/ Accuracy (%)	Advantages	Disadvantages	References
1	Fluorescence *in-situ* Hybridization (FISH)	10^3	High single-cell sensitivity due to the high affinity and specificity of DNA probes	False positive results with autofluorescence materials	Lopez et al. 2003
2	Enzyme-Linked Immunosorbent Assay	10^5–10^6	Detection by color changes and cost-effectiveness	Possesses low sensitivity level of bacterial detection	Lopez et al. 2003
3	Polymerase Chain Reaction (PCR)	10^3–10^4	1. On-site and rapid diagnosis of plant pathogens diseases based on their nucleic acids by real-time PCR 2. High specificity and sensitivity due to DNA amplification fidelity	Effectiveness depends upon DNA extraction, inhibitors, polymerase activity, and PCR reagents	Lopez et al. 2003
4	Flow Cytometry (FCM)	10^4	Simultaneous measurement of several parameters rapid identification of cells	Costly and overwhelming information is gathered	Lopez et al. 2003
5	Immunofluorescence	10^3	Targeted distribution and high sensitivity	False negative results due to photobleaching	Lopez et al. 2003
6	Multispectral imaging	90%	Identifying defects in images	Visual detection only	Sankaran et al. 2010
7	Near-infrared Based technique	97.2%	Spectroscopy-based imaging technique	High scattering level and false readings	Sankaran et al. 2010
8	Support Vector Machine	97%	1. Plants inoculated with pathogens 2. Comparison between healthy and diseased leaves	It does not perform well when we have a large data set because the required training time is higher	Rumf et al. 2010
9	Hyperspectral imaging (VNIR and SWIR)	90%	1. Collect hypercubes of capsicum plant leaves in the VNIR and SWIR range 2. Disease detection, identification, and quantification on different scales from the tissue to the canopy level	Cost and complexity. Fast computers, sensitive detectors, and large data storage capacities are needed	Moghadam et al. 2017
10	Spectral Information Divergence	95.2%	Classification or discrimination of pigments based on the obtained spectral reflectance information	Only A point-analysis tool	Qin et al. 2009

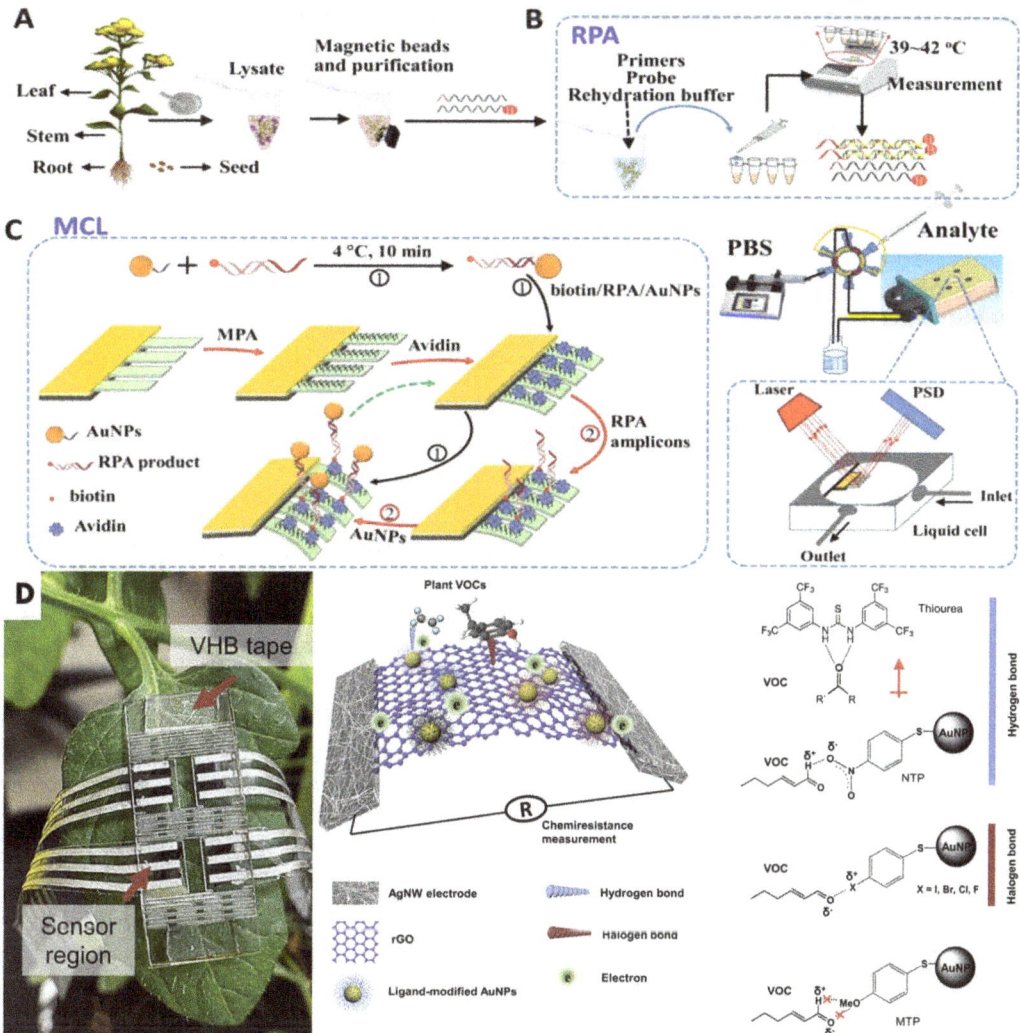

Figure 3. Schematic image of the combined RPA and microcantilever (MCL) assay strategy. (A) Presents DNA and purification by a compact magnetic bead-based DNA extraction kit. (B) Defines the RPA process. (C) Demonstrates the MCL analysis process and MCL assay approach. Two different processes were used to detect RPA amplicons. Cantilever deflection and resonance frequency shift were quantified as a function of amplicons and AuNP adsorption (Lei et al. 2021). (D) Volatile organic compound sampling and recognition of tomato leaf with flexible sensor array as a wearable sensor for VOC (Li et al. 2021).

of plasmonic nano-colorants and chemo-responsive organic dyes to identify key plant volatiles with a detection accuracy of ≥ 95% in the diagnosis of *P. infestans* in the laboratory, as well as in real-world condition collected tomato leaves in blind pilot tests (Li et al. 2019). In another study, researchers developed a wearable sensor for the tomato leaf to detect VOC chemical compounds in real-time monitoring. This early detection approach is for late blight disease caused by *Phytophthora infestans*. The was integrated with graphene-based sensing materials along with AuNPs. This sensor is consisting of flexible silver nanowire electrodes, which minimize strain interference. The sensor is attached to the leaf of tomato plant as presented in Figure 3D, to detect VOC at low concentrations, the sensor performed the sensing with > 97% classification accuracy (Li et al. 2021).

Conclusions and Future Directions

Nanomaterials are capable enough to be used as an effective material for the improvement in pathogen identification and to apply for forestry. However, it has some difficulties related to the nanotoxicological effects and parameters that are not yet quite clear. The properties of nanomaterials such as morphology, size, and surface functionalities define the level of toxicity for plants. With the integration of new tools and techniques, it might be the key to getting an early detection approach for plant disease. It is essential for material science and forest biologists to work in close collaboration to develop a deeper understanding of the fundamental mechanisms of this complex forest bio-nano ecosystem. Regardless of application in forest or agriculture, it is possible to apply similar approaches for the identification of the disease for big trees as well as for agricultural plants. A broad sense of the structural properties of the nanoparticles might offer a valuable route to start using suitable nanoparticles in developing plant biosensors. The reliability and reproducibility of the system, with efficiency to monitor the interaction of pests with host plants. The present research and progress in the agricultural field are encouraging, as the opportunities presented by using nanoparticles for creating efficient biosensors. Likewise, disease diagnostic methods for a large-scale plant disease to estimate disease demands the consideration of cost-effective testing kits and their portability along with on-site assessment. Advanced biosensing approaches are required to control the spread of plant diseases across the area in a time-effective manner. The authors consider that multidisciplinary and collective research will offer a specific solution to apply nanomaterial for plant pathogen detection as a reliable and robust tool.

References

Allen, C.D., Macalady, A.K., Chenchouni, H., Bachelet, D., McDowell, N., Vennetier, M., Kitzberger, T., Rigling, A., Breshears, D.D., Hogg, E.T. and Gonzalez, P. 2010. A global overview of drought and heat-induced tree mortality reveals emerging climate change risks for forests. For. Ecol. Manag. 259(4): 660–684.

Boyd, I.L. Freer-Smith, P.H. Gilligan, C.A. and Godfray, H.C.J. 2013. The consequence of tree pests and diseases for ecosystem services. Science 342: 1235773.

Chakraborty, C., Doss, C.G.P., Patra, B.C. and Bandyopadhyay, S. 2014. DNA barcoding to map the microbial communities: current advances and future directions. Appl. Microbiol. Biotechnol. 98(8): 3425–3436.

Gogos, A., Knauer, K. and Bucheli, T.D. 2012. Nanomaterials in plant protection and fertilization: Current state, foreseen applications, and research priorities. J. Agric. Food Chem. 60: 9781–9792.

Idnurm, A. and Howlett, B.J. 2001. Pathogenicity genes of phytopathogenic fungi. Mol. Plant Pathol. 2(4): 241–255.

Keith, L.M., Hughes, R.F., Sugiyama, L.S., Heller, W.P., Bushe, B.C. and Friday, J.B. 2015. First report of Ceratocystis wilts on Ohia. Plant Dis. 99(9): 1276.

Kowalski, T. 2006. *Chalarafraxinea* sp. nov. associated with dieback of ash (*Fraxinus excelsior*) in Poland. For. Pathol. 36: 264–270.

Kowalski, T., Bilanski, P. and Kraj, W. 2017. Pathogenicity of fungi associated with ash dieback towards *Fraxinus excelsior*. Plant Pathol. 66: 1228–1238.

Kulabhusan, P.K., Tripathi, A. and Krishna Kant. 2022. Gold nanoparticles and plant pathogens: an overview and prospective for biosensing in forestry. Sensors 22(3): 1259.

Lei, R., Wu, P., Li, L., Huang, Q., Wang, J., Zhang, D., Li, M., Chen, N. and Wang, X. 2021. Ultrasensitive isothermal detection of a plant pathogen by using a gold nanoparticle-enhanced microcantilever sensor. Sens. Actuators B Chem. 338: 129874.

Li, Z., Paul, R., Tis, T.B., Saville, A.C., Hansel, J.C., Yu, T., Ristaino, J.B. and Wei, Q. 2019. Non-invasive plant disease diagnostics enabled by smartphone-based fingerprinting of leaf volatiles. Nat. Plants 5(8): 856–866.

Li, Z., Liu, Y., Hossain, O., Paul, R., Yao, S., Wu, S., Ristaino, J.B., Zhu, Y. and Wei, Q. 2021. Real-time monitoring of plant stresses via chemiresistive profiling of leaf volatiles by a wearable sensor. Matter. 4(7): 2553–2570.

Lopez, M.M., Bertolini, E., Olmos, A., Caruso, P., Corris, M.T., Llop, P., Renyalver, R. and Cambra, M. 2003. Innovative tools for detection of plant pathogenic viruses and bacteria. Int. Microbiol. 6: 233–243.

Moghadam, P., Ward, D., Goan, E., Jayawardena, S., Sikka, P. and Hernandez, E. 2017. International Conference on Digital Image Computing: Techniques and Applications (DICTA). Plant Disease Detection using Hyperspectral Imaging, Sydney, NSW pp. 1–8.

Moricca, S., Linaldeddu, B.T., Ginetti, B., Scanu, B., Franceschini, A. and Ragazzi, A. 2016. Endemic and emerging pathogens threatening cork Oak trees: Management options for conserving a unique forest ecosystem. Plant Dis. 100(11): 2184–2193.

Orlikowski, L.B., Ptaszek, M., Rodziewicz, A., Nechwatal, J., Thinggaard, K. and Jung, T. 2011. *Phytophthora* root and collar rot of mature *Fraxinus excelsior* in forest stands in Poland and Denmark. For. Pathol. 41: 510–519.

Pan, Y., Birdsey, R.A., Fang, J. and Houghton, R.A. 2011. A large and persistent carbon sink in the World's forests. Science 333(6045): 988–93.

Pecnikar, Z.F. and Buzan, E.V. 2014. 20 years since the introduction of DNA barcoding: from theory to application. J. Appl. Genet. 55(1): 43–52.

Qin, J., Burks, T.F., Ritenour, M.A. and Bonn, W.G. 2009. Detection of citrus canker using hyperspectral reflectance imaging with spectral information divergence. J. Food Eng. 93(2): 183–191.

Rumf, T., Mahlein, A.K., Steiner, U., Oerte, E.C., Dehne, H.W. and Plumer, L. 2010. Early detection and classification of plant disease with support vector machine based on hyperspectral reflectance. Comput. Electron. Agric. 74(2010): 91–99.

Sankaran, S., Mishra, A., Ehsani, R. and Davis, C. 2010. A review of advanced techniques for detecting plant diseases. Comput. Electron. Agric. 72(1): 1–13.

Singh, A.R., Dutta, S.K., Singh, S.B. and Boopathi, T. 2018. Occurrence, severity and association of fungal pathogen, *Botrydiplodia theobromae* with sudden death or decline of tree bean (*Parkia timoriana* (DC.) Merr) in North Eastern India. Curr. Sci. 115(6): 1133–1142.

Singh, L. and Lakhanpal, T.N. 2000. *Cedrus deodara* root rot disease-threat to the Himalayan forestry and environment. Indian Phytopathol. 53: 50–56.

Strange, R.N. and Scott, P.R. 2005. Plant disease: a threat to global food security. Annu. Rev. Phytopathol. 43(1): 83–116.

Torney, F., Trewyn, B.G., Lin, V.S.Y. and Wang, K. 2007. Mesoporous silica nanoparticles deliver DNA and chemicals into plants. Nat. Nanotechnol. 2: 295–300.

Truong, C., Mujie, A.B., Healy, R., Kuhar, F., Furci, G., Torres, D., Niskanen, T., Sandoval-Leiva, P.A., Fernández, N., Escobar, J.M. and Moretto, A. 2017. How to know the fungi: combining field inventories and DNA-barcoding to document fungal diversity. New Phytol. 214(3): 913–919.

Tripathi, R.M. and Sharma, P. 2021. Gold nanoparticles-based point-of-care colorimetric diagnostic for plant diseases. Biosensors in Agriculture: Recent Trends and Future Perspectives 191–204.

Whetzel, H.H. 1929. The terminology of phytopathology. Proc. Int. Congr. Plant Sci. 2: 1204–1215.

Wingfield, M.J., Garnas, J.R., Hajek, A., Hurley, B.P., de Beer, Z.W. and Taerum, S.J. 2016. Novel and co-evolved associations between insects and microorganisms as drivers of forest pestilence. Biol. Invasions 18(4): 1045–1056.

Zhan, F., Wang, T., Iradukunda, L. and Zhan. J. 2018. A gold nanoparticle-based lateral flow biosensor for sensitive visual detection of the potato late blight pathogen, *Phytophthora infestans.* Anal. Chim. Acta. 1036: 153–161.

Chapter 11

Facile Biosynthesis of Gold Nanoparticles Using Microbes and Their Applications in the Management of Economically Important Plant Pathogens

Tesleem Taye, Bello[1,2] and *Oluwatoyin Adenike Fabiyi*[3,*]

Introduction

Gold is one of the most important metals that ever existed. It occurs in a number of forms and serves several purposes in different facets of human lives. Gold and its compounds have been used in medicine, automobile production, biotechnology, electrical, pharmaceutical, and agricultural fields. Nanotechnology is one emerging area of research that is currently getting attention (Fabiyi et al. 2020a). It involves the combination of science, technology, and engineering so as to achieve nanoscale materials mostly between 1–100 nm in range. The nanoscale materials (nanoparticles) obtained therefrom possess enhanced properties and have been applied in various fronts, which include but are not limited to health care, biomedical, engineering, pharmaceutical, optics, mechanics, space industry, drug delivery, microbiology, agriculture, and food industry (Mohanpuria et al. 2008; Ikram 2015; Abdel-Aziz et al. 2018, 2018; Fabiyi et al. 2018; Fabiyi and Olatunji 2018; Prasad et al. 2018; Marooufpour et al. 2019; Fabiyi et al. 2020b). Gold and silver nanoparticles are among the most important and extensively studied metal nanoparticles due to their unique electrical and photothermal properties, surface functionalities, stability to oxidation, and enhanced flexibility for modification (Islam et al. 2015; Shahzad et al. 2017; Fabiyi 2021; Fabiyi et al. 2021).

Due to its wide range of applications, gold nanoparticles (AuNPs) like other metal nanoparticles have been synthesized in different ways using both chemical and biological means. Chemical synthesis of metal nanoparticles has been the most commonly utilized approach, which is usually costly, consumes high amounts of energy, and involves the use of toxic reagents as stabilizing agents during synthesis; this has given rise to a series of environmental concerns. Furthermore, harmful effects of chemically synthesized nanoparticles have been reported from biomedical applications as a result of an accumulation of chemical residues (Shanker et al. 2004; Noruzi et al. 2011). Thus, there is a need to develop easier, ecologically sound, and cost-effective alternatives to chemical methods. Recently, the use of naturally occurring reducing agents, such as plant tissues, microbes (fungi, bacteria, and actinomyces); enzymes has provided a cost-effective and ecologically friendly

---

[1] Department of Plant Soil and Microbial Sciences, Michigan State University, USA.
[2] Department of Agricultural Science Education, Federal College of Education, PMB 2096 Abeokuta, Ogun State, Nigeria.
[3] Department of Crop Protection, Faculty of Agriculture, University of Ilorin, Kwara State, Nigeria.
* Corresponding author: fabiyitoyinike@hotmail.com

alternative in this regard (Abdel-Aziz et al. 2018; Marooufpour et al. 2019). The biogenic synthesis of metal nanoparticles using microbes is achieved through a green chemistry approach, which operates at low ambient temperatures and pressures and favors simpler downstream processing at a reduced cost (Abdel-Aziz et al. 2018; Prasad et al. 2018). Microbial synthesis of metal nanoparticles is achieved in two ways: intracellularly and extracellularly (Jain et al. 2011). The former method utilizes suitable detergents and special ultrasound treatment to produce synthesized nanoparticles (Kalimuthu et al. 2008), while the latter method is cheaper requiring simpler downstream reactions, which explains the reason why more researchers now concentrate more on the use of extracellular methods in achieving a microbial synthesis of nanoparticles (Prasad et al. 2018). Agricultural crops are devasted by a variety of pests, which constitute a major constraint to crop production worldwide (Bello et al. 2020). The continuous use of chemical pesticides, like fungicides, bactericides, etc., to manage agricultural pathogens has resulted in serious environmental concerns due to the hazardous effects of these chemicals on human health and the environment. This chapter shall focus on the biological synthesis of gold nanoparticles using several microbial agents and the applications of these in economically important crop disease management.

Biosynthesis of Gold Nanoparticles

Biological agents like plant tissues, fungi, bacteria, and enzymes have been largely employed in the biosynthesis of AuNPs. Available information has shown that the biological synthesis of AuNPs when compared with conventional methods has the advantages of being facile, eco-friendly, and cost-effective. A good example of this is the use of fungi which apart from being easy to culture both in the laboratories and industries, requires a short period to produce high yields with high ion concentration tolerance to mix and incubate with gold salts to produce AuNPs (Tetgure et al. 2015). Furthermore, the biosynthesis of AuNPs using microbial agents provides a single-step synthesis approach at ambient temperature and pressure conditions (Figure 1). As against the large volumes of reports in respect of microbial-mediated synthesis of AuNPs, scanty reports are available on the various mechanisms involved in the process. Most available studies were only able to explain

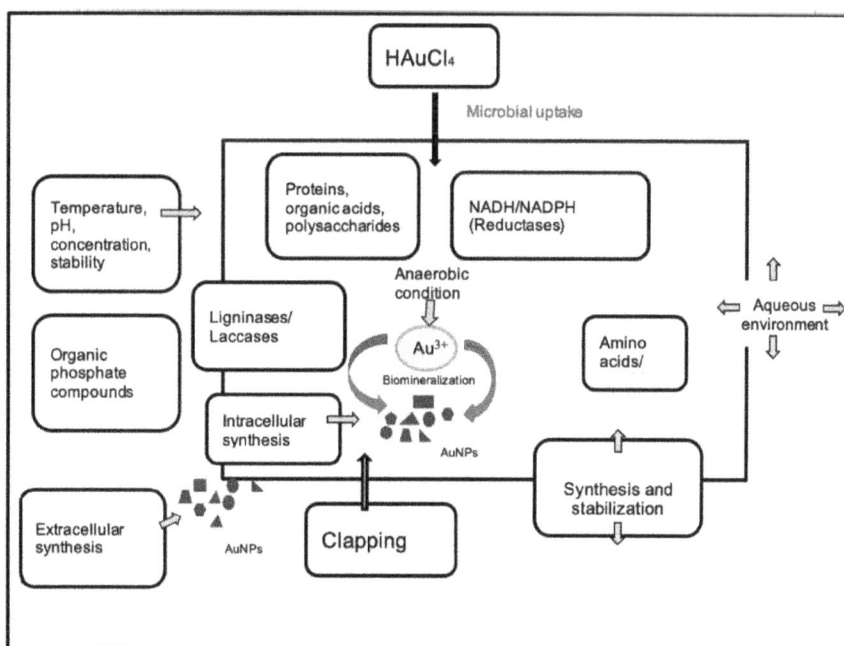

Figure 1. A schematic summary to represent the mechanisms involved in the microbial synthesis of gold nanoparticles.

the mechanisms involved in the microbial synthesis of silver nanoparticles by exposing the protein-dependent-hydrogenase and nitrate reductase-mediated pathways (Das et al. 2012; Shedbalkar et al. 2014).

Mechanisms of Biosynthesized AuNPs

Although information regarding the mechanisms involved in the synthesis of AuNPs using different microbial agents is scanty, it is largely suggested that the proteins and enzymes (naphthoquinones, anthraquinones, flavonoids, and reductases) present in individual microbes act as reducing agents on metal ions, which results into reduction and synthesis of gold metal to nanoparticles (Tamuly et al. 2013; Paul et al. 2015; Singh et al. 2016b). Several physicochemical parameters such as temperature, pH, and substrate concentrations play significant roles in the intracellular synthesis of AuNPs (Gericke and Pinches 2006). Furthermore, alteration in these parameters resulted in the manipulation of basic morphological features of AuNPs. It was suggested that the monodispersity of synthesized AuNPs may be achieved by optimizing these conditions. In addition to these, some amino acid and protein residues, such as tyrosine, and cysteine are known to perform important roles in the stabilization of biosynthesized AuNPs, while the free amino or cysteine groups in protein act as surface binding agents in the stabilization process (Gole et al. 2001; Sastry et al. 2003). However, different proteins have been reported to contribute substantially to the stabilization and capping processes of most AuNPs. The success of biosynthesis of AuNPs is usually revealed by the UV-visible spectrum of the reaction mixture, which is announced by a change in color from pale yellow to deep red or dark purple of the reaction mixture as dictated by their surface plasmon resistance properties (He et al. 2007).

Biosynthesis of AuNPs Using Fungi and Yeasts

Abundant worldwide, fungi are members of the group of eukaryotic organisms which include molds, some yeasts, and mushrooms. They perform important roles such as organic matter decomposition, nutrient recycling, and exchange within the environment since they exist mostly as symbionts of animals, plants, and other fungi as well as parasites. Several species of fungi have been employed in the microbial synthesis of metal nanoparticles due to their ability to secrete large amounts of proteins required for this purpose (Manjunath et al. 2017).

The process involved in the biosynthesis of AuNPs using fungi involves culturing to obtain a mycelia-free filtrate, which would be mixed and incubated with gold salts so as to produce AuNPs. Several species of penicillium and aspergillus have been utilized to achieve biosynthesis of AuNPs. Common examples were reported from fungi-mediated biosynthesis of AuNPs by *Penicillium citrinum, P. waksmanii* and *P. aurantiogriseum* (Honary et al. 2013). Others include the use of *Penicillium citrinum* an endophytic fungus to synthesize AuNPs, which produced irregularly shaped nanosized particles of gold between 60–80 nm (Manjunath et al. 2016). *Penicillium aculeatum* was found to produce spherical nanoparticles of gold with a size of approximately 60 nm (Barabadi et al. 2017). However, lower-sized AuNPs (8.7–15.6 nm) were obtained by using high concentrations of *Aspergillus sydowii* (Vala 2015).

Also, a mitosporic species of the *Aspergillus* strain produced AuNPs with a size of 4–29 nm after 7 days of incubation (Shen et al. 2017). In a study by Das et al. (2009), the fungus: *Rhizopus oryzae* was utilized in synthesizing AuNPs that had a stability of up to 6 months. In addition to these, different kinds of common mushrooms such as *Ganoderma lucidum, Lentinus edodes*, and *Pleurotus ostreatus* have been used to synthesize AuNPs (Vetchinkina et al. 2017). Furthermore, the chaga mushroom (*Inonotus obliquus*) has been used to achieve varying (triangular, rod, hexagonal, and spherical) shapes of AuNPs within a very relatively short incubation time (Lee et al. 2015). Although the presence of certain proteins and polyphenols present in most fungi are largely responsible for their bioreduction and stabilizing potentials, many authors believed that certain electrostatic forces play a vital role in determining the stability of most nanoparticles produced

Table 1. Biosynthesis of gold nanoparticles by fungi and yeasts.

SN	Fungal Strain	Shapes of AuNPs Produced	AuNPs Sizes (nm)	References
1	*Alternaria alternata*	Spherical and triangular	2–30	Sarka et al. (2012)
2	*Aspergillus foetidus*	Spherical	30–50	Roy and Das (2016)
	Aspergillus niger	Polydisperse	10–20	Xie et al. (2007)
3	*Aureobasidium pullulans*	Spherical	35–23	Barabadi et al. (2014)
4	*Colletotrichum* sp.	Decahedral and icosahedral	20–40	Shankar et al. (2003)
5	*Botrytis cinerea*	Spherical, pyramidal, hexagonal triangular and decahedral	1–100	Castro et al. (2014)
6	*Fusarium semitectum*	Spherical	10–35	Sawle et al. (2008)
7	*Neurospora crassa*	Spherical	20–50	Castro-longoria et al. (2011)
8	*Fusarium solani*	Spherical	3–100	Castro et al. (2014)
	Fusarium oxysporum	Aggregates	70–128	Narayanan and Sakthivel (2013)
9	*Penicillium crustosum*	Spherical	100	Barabadi et al. (2014)
10	*Penicillium rugulosum*	Spherical, triangular, and hexagonal	30–70 and 20–80	Mishra et al. (2012)
11	*Penicillium chrysogenum*	Spherical, triangular, and rod	5–100	Sheikhloo and Salouti (2011)
12	*Phanerochaete chrysosporium*	Spherical	10–100	Sanghi et al. (2011)
13	*Rhizopus oryzae*	Spherical	5–65	Das et al. (2012)
14	*Sclerotium rolfsii*	Spherical	25	Narayanan and Sakthivel (2011)
15	*Trichoderma viride hypocrea lixii*	Spherical	20–30	Narayanan and Sakthivel (2010)
16	*Trichoderma konningi*	Spherical	10–14	Maliszewska (2013)
17	*Penicillium citrinum*	Irregular	60–80	(Manjunath et al. (2016)
18	*Penicillium aculeatum*	Spherical	60	Barabadi et al. (2017)
19	*Aspergillus sydowii*		8.7–15.6	Vala (2015)
20	*Inonotus obliquus*	Triangular, rod, hexagonal, and spherical		Lee et al. (2015)
	Yeasts	**Shapes of AuNPs Produced**	**AuNPs Sizes (nm)**	**References**
21	*Saccharomyces cerevisiae*	Spherical	20–100	Lim et al. (2011)
22	*Candida guilliermondii*	Spherical	50–70	Mishra et al. (2011)
23	*Candida albicans*	Spherical	20–40	Chauhan et al. (2011)
24	*Yarrowia lipolytica*	Triangular and hexagonal	15	Agnihotri et al. (2009)
25	*Yarrowia lipolytica*	Spherical	9–27	Pimprikar et al. (2009)
26	*Magnusiomyces ingens*	Spherical, triangular, and hexagonal	80.1	Zhang et al. (2016)

from microbial biosynthesis. The different shapes produced on biosynthesized nanoparticles were found to be largely due to the employed capping agents as determined by their individual binding properties (Graily-Moradi et al. 2020).

Yeasts are single-cell eukaryotic organisms that are grouped into the kingdom of fungi. The biosynthesis of AuNPs by *Saccharomyces cerevisiae* (Lim et al. 2011) and *Candida guilliermondii* (Mishra et al. 2011) has been reported. Another example is the production of 15 nm triangular and hexagonal AuNPs by the intracellular activity of *Yarrowia lipolytica* (Agnihotri et al. 2009). A more recent report is the biosynthesis of AuNPs by the non-conventional yeast *Magnusiomyces ingens*, which produced spherical, triangular, and hexagonal nanoparticles with an average size of 80.1 nm (Zhang et al. 2016). A summary of fungal-mediated biosynthesis of AuNPs is presented in Table 1.

Biosynthesis of AuNPs Using Bacteria

Bacteria are grouped as Prokaryotes, which are unicellular organisms with a simple structure that lacks a nucleus. Different types of bacteria have been employed in the biosynthesis of several metal nanoparticles (Table 2). Biosynthesis of AuNPs using bacteria is achieved through two pathways: intracellular and extracellular based on the site of the nanoparticle's synthesis. The intercellular method involves the transfer of Au ions into the bacterial cell to form nanoparticles in the presence of enzymes while in the extracellular method, nanoparticles are formed by trapping the Au ions on the bacterial cell membrane by an enzyme (Kasthuri et al. 2009; Hulkoti and Taranath 2014). However, intracellular methods have produced more small-sized nanoparticles when compared with those produced using extracellular methods (Narayanan and Sakthivel 2010). Beveridge and Murray (1980) provided the first report on the bacterial synthesis of AuNPs using *Bacillus subtilis* to release octahedral-shaped nanoparticles of 5–25 nm in size.

Since then, several common pathogenic bacteria strains have been employed by many authors to synthesize AuNPs of varying shapes and sizes. *Pseudomonas fluorescens* was used to synthesize AuNPs within a size range of 50–70 nm in size (Rajasree and Suman 2012). Furthermore, bacterial-mediated synthesis of AuNPs using a strain of *Delftia* sp. under a controlled laboratory

Table 2. Biosynthesis of gold nanoparticles by bacteria.

SN	Bacteria Strain	Shapes of AuNPs Produced	Sizes (nm)	References
1	*Bacillus subtilis*	Octahedral	5–25	Beveridge and Murray (1980)
2	*Bacillus stearothermophilus*	Triangular	5–30	Luo et al. (2014)
3	*Delftia* sp.	Spherical	11.3	Kumar et al. (2014)
4	*Escherichia coli*	Spherical	5–20	Gopal et al. (2013)
5	*Escherichia coli* DH5a	Spherical	20	Suganya et al. (2015)
6	*Geobacillus* sp.	Hexagonal	10–20	Correa Llanten et al. (2013)
7	*Geobacillus stearothermophilus*	Spherical	12–14	Gomathy and Sabarinathan (2015)
8	*Halomonas salina*	Spherical	30–100	
9	*Klebsiella pneumoniae*	Spherical	5–65	Malarkodi et al. (2014)
10	*Magnetospirillum gryphiswaldense* MSR-1	Spherical	10–40	Cai et al. (2011)
11	*Pseudomonas aeruginosa*		15–30	Husseiny et al. (2007)
12	*Pseudomonas fluorescens*	Spherical	50–70	Rajasree and Suman (2012)
13	*Pseudomonas fluorescens* 417	Spherical	5–50	Husseiny et al. (2007)
14	*Rhodopseudomonas capsulata*	Spherical	10–20	Syed et al. (2016)
15	*Salmonella enterica*		5–50	Mortazavi et al. (2017)
16	*Shewanella algae*	Spherical	9.6	Mishra et al. (2011)
17	*Shewanella oneidensis*	Spherical	2–50	Suresh et al. (2011)
18	*Sporosarcina koreensis* DC4	Spherical	30–50	Singh et al. (2016a)
19	*Staphylococcus epidermidis*	Spherical	20–25	Ogi et al. (2010)
20	*Stenotrophomonas maltophilia*	Spherical	40	Srinath and Rai (2015)

condition (pH 8, temperature of 45°C, and reaction time of 7 hours) has been used to achieve spherical-shaped small-sized AuNPs of about 11.3 nm (Kumar et al. 2014). Another prominent example of bacterial-mediated biosynthesis of AuNPs was reported from the use of *Pseudomonas aeruginosa*, which produced AuNPs in the range of 15–30 nm in size (Husseiny et al. 2007). *Halomonas salina*, the highly salt-tolerant proteobacteria, was used to successfully synthesize spherical-shaped AuNPs of size range: 30–100 nm at pH 9 and a regulated temperature of 30°C (Shah et al. 2012). In a recent study, the human pathogenic bacterium, *Salmonella enterica* was used to synthesize with sizes ranging between 5–50 nm (Mortazavi et al. 2017). Also, the photosynthetic bacterium: *Rhodopseudomonas capsulata* was used to successfully synthesize AuNPs of different shapes and within a range of 10–20 um in size (He et al. 2007). A detailed compilation of bacterial-mediated biosynthesis of AuNPs is presented in Table 2. These reports have demonstrated that successful bacterial-mediated synthesis of AuNPs is largely dependent on pH and temperature (Marooufpour et al. 2019). However, unlike fungi, bacteria culturing is tedious and requires extra precautions. Also, the reaction time required for the reduction process is long and ranges from hours to days, and this has been a major constraint that is limiting the bacterial-mediated synthesis of nanoparticles on a commercial scale.

Biosynthesis of AuNPs Using Actinomycetes

Actinomycetes are a unique group of gram-positive fungus-like facultative anaerobic bacteria. They are known to produce several bioactive substances like antibiotics, proteins, enzymes, and vitamins. Members of the group have been successfully employed as capping and stabilizing agents in the microbial synthesis of gold as well as some other metal nanoparticles (Golinska et al. 2014; Prasad et al. 2018; Aswani et al. 2019).

It has been established that the free amine groups contained in the protein of actinomycetes play a prominent role in the biosynthesis process (Gole et al. 2001; Golinska et al. 2014). An earlier example of actinomycetes in the biogenic synthesis of AuNPs includes the use of *Rhodococcus* sp. to achieve spherical AuNPs within the size range of 5–15 nm. Also, *Thermomonospora* sp., an alkali thermophilic actinomycete was used to achieve extracellular synthesis of AuNPs of 8 nm in size (Ahmed et al. 2003; Fayaz et al. 2011). Waghmare et al. (2014) effectively utilized *Streptomyces hygroscopicus* to synthesize spherical AuNPs of 10–20 nm in size while Oza et al. (2012) utilized *Nocardia farcinica* to synthesized spherical-shaped AuNPs with 15–20 nm in size. Furthermore, *Streptomyces fulvissimus* was also used to achieve extracellular biosynthesis of AuNPs within the size range of 20–50 nm (Soltani Nejad et al. 2015). In addition to these, an actinomycetes-derived compound with free radical scavenging activities has proved its efficacy in the biosynthesis of AuNPs: methyl esters derived from the actinomycetes *Gordonia amicalis* HS-11 produced spherical AuNPs within the range of 5–25 nm (Sowani et al. 2016). A list of common examples of actinomycetes-mediated biosynthesis of AuNPs is described in Table 3.

Biosynthesis of AuNPs Using Algae

Algae are a group of photosynthetic eukaryotic microorganisms. They possess a cell wall that contains proteins, enzymes, and polysaccharides with characteristics of reducing agents, which have made them ideal for use in the biosynthesis of metal nanoparticles (Table 4). Common examples of biosynthesis of AuNPs include the use of *Plectonema boryanum* to extracellularly and intracellularly produce cubic-shaped AuNPs of sizes ranging between 9–25 nm (Lengke et al. 2006). Also, planar-shaped AuNPs of sizes between 8–12 nm were obtained from the extracellular synthesis of gold using *Sargassum wightii* (Singaravelu et al. 2007). Furthermore, the algae: *Spirulina platensis* was employed in the intracellular synthesis of AuNPs to produce nanoparticles of 5 nm in size (Suganya et al. 2015). A more recent report by Abdel-Raouf et al. (2017) detailed a successful biosynthesis of spherical, 3.9–77 nm in size AuNPs using *Galaxaura elongata*.

Table 3. Biosynthesis of gold nanoparticles by actinomycetes.

SN	Actinomycete Strain	Shapes of AuNPs Produced	AuNPs Sizes (nm)	References
1	*Gordonia amicalis*	Spherical	5–25	Sowani et al. (2016)
2	*Nocardia farcinica*	Spherical	10–20	Oza et al. (2012)
3	*Streptomyces fulvissimus*	Spherical	20–50	Soltani Nejad et al. (2015)
4	*Streptomyces hygroscopicus*	Spherical	20	Husseiny et al. (2007)
5	*Streptomyces hygroscopicus*	Spherical	20	Khan et al. (2016)
6	*Streptomyces hygroscopicus*	Spherical	10–20	Waghmare et al. (2014)
7	*Streptomyces* sp. VITDDK3	Hexagonal, cubical, brick, and irregular	90	Arumugan and Berchmans (2011)
8	*Streptomyces viridogens* (HM10)	Spherical and rod	18–20	Balagurunthan et al. (2011)
9	*Thermonospora* sp.	Spherical	8	Ahmed et al. (2003)
10	*Streptomyces griseoruber*	Spherical, triangular and hexagonal	5–15	Ranjitha and Rai (2017)

Table 4. Biosynthesis of gold nanoparticles by algae.

SN	Algal Strain	Shapes of AuNPs Produced	AuNPs Sizes (nm)	References
1	*Brown, Eecklonia cava*	Triangular and spherical	30	Kathiraven et al. (2015)
2	*Brown, Sargassum muticum*	Spherical	5.4	Namvar et al. (2014)
3	*Chlorella vulgaris*	Spherical	2–10	Annamalai and Nallamuthu (2015)
4	*Fucus vesiculosus*	Spherical and rodlike	–	Mata et al. (2009)
5	*Padina gymnospora*	Spherical	53–35	Singh et al. (2013)
6	*Tetraselmis kochinensis*	Triangular and spherical	5–35	Venkatesan et al. (2014)
7	*Galaxaura elongata*	Spherical	3.9–77	Abdel-Raouf et al. (2017)
8	*Plectonema boryanum*	Cubic	9–25	Lengke et al. (2006)
9	*Sargassum wightii*	Planar	8–12	Singaravelu et al. (2007)
10	*Spirulina platensis*	Spherical	5	Suganya et al. (2015)
11	*Turbinaria conoides*	Cubic	–	Vijayaraghavan et al. (2011)

Applications of Biosynthesized Gold Nanoparticles in the Management of Economically Important Plant Pathogens

On a global scale, plant diseases stand as a major limiting factor in crop production. This menace is further aggravated by an upsurge in the resistance of plant pathogens to conventional pesticides, which is largely due to excessive continuous use over time with their devastating consequences on human health and the environment. In order to ensure food sufficiency and eradicate poverty through increased crop production, especially in the developing world, efforts must be put in place to explore all available alternatives to chemical pesticides used in agricultural disease management. Several organic compounds have been studied for their potential as suitable alternatives to chemical pesticides in plant disease management. Nanoparticles are made up of small molecules with large multiple-surface compatibility, which ensures low toxicity in biomolecular interactions. AuNPs achieve their antibacterial and antifungal effects by attaching to the surface of the microbe and utilizing their surface charge and electrostatic reaction to interrupt cell membrane permeability causing DNA destruction and protein denaturation, which results in the eventual death of the host pathogen (Vanaraj et al. 2017). This attribute has positioned nanoparticles as having the potentials for use as an ecologically sound approach to pest and disease management. Gold nanoparticles have displayed antibacterial and antifungal properties that are useful in agricultural pest and disease management (Jayaseelan et al. 2013). Biosynthesized AuNPs have been reported to display antibacterial activities against some strains of *Bacillus* sp., *Pseudomonas* spp., and *Klebsiella* sp. (Reddy et al. 2013; Muthuvel et al. 2014).

AuNPs as Biosensors for Plant Disease Detection

The application of nanoparticle-based biosensors has played a significant role in plant disease detection worldwide (Kulabhusan et al. 2022). The utilization of several nano-based materials, such as carbon, metal, polymers, composites, allotropes, etc., have proved very useful in the accurate detection and reproducibility of many plant diseases. Biosensors in different forms, such as electrochemical biosensors, immunosensors, enzyme biosensors, and DNA sensors play a significant role in the rapid, timely, and accurate detection of several quarantine pathogens of crops, which has helped to prevent the introduction of alien pathogens into new environments (Kulabhusan et al. 2022). AuNPs have been specifically used to label antibodies that are specific to a target pathogen and developed as a diagnostic device. This approach has been applied in managing several plant diseases within different agricultural cropping systems in many parts of the world.

Detection of Fungal Disease in Crops

AuNPs have been modified in the biosensing process to detect several fungal pathogens causing damage to several agricultural crops. An example is in the rapid detection of the black sigatoka disease of banana caused by the fungus *Pseudocercospora fijiensis*, which was detected using a highly specific surface plasmon resonance (SPR) immuno-sensor (Luna-Moreno et al. 2019). A recent study reported the utilization of AuNPs enhanced dynamic microcantilever in an isothermal polymerase amplification reaction, which was highly effective for the detection of oil seed rape fungus (*Leptosphaeria maculans*) causing the blackleg disease of brassica crops (Lei et al. 2021). Furthermore, rapid detection of the fungus *Aspergillus niger* causing black mold disease in several vegetable crops like apricots, grapes, onions, and peanuts was achieved by developing a peptide-modified probe with AuNPs serving as the label (Lee et al. 2021).

Detection of Bacterial Pathogens in Crops

AuNPs employed as nano-based biosensors have also yielded significant success in the detection of several bacterial pathogens of crops, such as *Pseudomonas syringe* pathovars, which are responsible for causing large-scale bacterial diseases in agricultural crops worldwide. This pathogen was detected by using AuNPs labeled DNA probes (Vaseghi et al. 2013). Also, onsight detection of the fruit blotch bacteria (*Acidovorax avenae* subsp. *citrulli*) was achieved using colloidal AuNPs labeled single-stranded DNA probes (Zhao et al. 2011). The direct detection of *Ralstonia solanacearum*, a gram-negative, non-spore-forming, aerobic bacterium that is responsible for causing bacterial wilt diseases in crops like pepper, tomato, eggplant, and potato was easily achieved using the AuNP-probe-based assay (Khaledian et al. 2017). A multiplex point-of-care system was developed by Lau et al. (2016) for the detection of two highly destructive bacterial pathogens (*Botrytis cinerea* and *Pseudomonas syringae*) and a fungal pathogen (*Fusarium oxysporum*) on tomato and *Arabidopsis thaliana*.

Detention of Viral Diseases of Crops

Compared to the other types of common nanoparticles, AuNPs have recently received a lot of attention in the detection of several viral diseases of crops and several methods, such as colorimetric, electrochemical, fluorescence methods, etc., are currently being employed for this purpose. An example is the development of a simple and highly sensitive label-free colorimetric detection process for the cucumber green mosaic virus using AuNPs as the colorimetric probe (Wang et al. 2016). Some of the successes recorded in the application of AuNPs for plant disease detection include but are not limited to the rapid and sensitive detection of the banana bunchy top virus through AuNPs-based dot immunobinding assay (Majumder and Johari 2018); the detection of tomato yellow leaf curl virus by using unmodified AuNPs as a medium in the localized surface plasmon resonance (LSPR) process (Razmi et al. 2019); detection of the citrus tristeza virus was achieved by using a label-free AuNPs modified impedimetric biosensor (Khater et al. 2019). A similar type of visual colorimetric method with a much higher sensitivity has been developed for the chili and tomato Begomovirus respectively (Lavanya and Arun 2021).

Microbially Biosynthesized AuNPs in Managing Plant Pathogens

Available data revealed that several studies have been conducted using biosynthesized silver nanoparticles (AgNPs) to manage plant pathogens (Park et al. 2006; Jo et al. 2009; Aguilar-Mendez 2010; Jung et al. 2010; Krishnajal et al. 2012; Mishra et al. 2014; Fernandez et al. 2016; Ahmed et al. 2020; Fouda et al. 2020; Osonga et al. 2020). However, only scanty reports are available on microbial-based biosynthesized AuNPs with regard to their applications in managing plant pathogens. This is against the large volume of available data regarding applications of green (plant sources) synthesized metal nanoparticles in managing agricultural pests and diseases (Elmer and White 2018; Hossain et al. 2019; Ogunyemi et al. 2019; Ibrahim et al. 2020). To date, only a few available reports of the application of microbial-based biosynthesized AuNPs in managing agricultural plant pathogens, as presented in Table 5. One such study was carried out by (Ponmurugan (2017), where AuNPs obtained from *Trichoderma atroviride*-mediated biosynthesis were found to inhibit the mycelial growth of *Phomopsis theae*, a fungus causing canker disease in tea plants. In his report, AuNPs were applied as both wound dressing and soil applications to provide effective control of the pathogen. The other report was by Thakur et al. (2018), who utilized biosynthesized AuNPs to control root-knot nematodes of tomato (*Meloidogyne incognita*). This paucity of information has led to a lack of clear and generalizable data on the potential of microbial-based biosynthesized AuNPs in managing economically important crop diseases.

Table 5. Applications of microbial-based biosynthesized AuNPs in managing plant pathogens.

Microbial Agents Used to Produce the AuNPs	Crop Pathogen Controlled	Crop and Disease Managed by AuNPs	References
Trichoderma atroviride	*Phomopsis theae*	Tea (canker disease)	Ponmuruga (2017)
Bacillus licheniformis	*Meloidogyne incognita*	Tomato (root-knot disease)	Thakur et al. (2018)

Conclusion and Perspectives

Nanotechnology is a novel area of research that is endowed with unlimited applications in different areas due to the unique properties of nanoparticles, which makes them adaptable in critical areas of human life, such as the medical field, pharmaceutical, engineering, and agriculture. In agriculture, nanotechnology has been utilized in natural resource conservation, crop production, and management of crop pests and diseases. The approach of biological synthesis of nanoparticles using microbial agents, such as bacteria, fungi, and algae is currently receiving attention due to its low toxicity, biocompatibility, and ecologically sound nature of the approach and products obtained therefrom. Nanoparticles have displayed antibacterial and antifungal properties that if properly harnessed would prove useful in future agricultural pest and disease management. However, when compared to other metal nanoparticles, like those of silver and zinc, few studies have been conducted on the application of microbe-based biosynthesized gold nanoparticles in managing economically important plant pathogens. Furthermore, no microbe-based biosynthesized nanoparticles, especially AuNPs have been produced on a commercial basis for agricultural applications. This could be largely attributed to the fact that few field trials have been conducted in that regard compared to the large volumes of nanotechnological applications recorded in other fields, such as the biomedical and engineering industries.

This paucity of information has led to a lack of clear and generalizable data on the potential of microbial-based biosynthesized AuNPs in managing economically important crop diseases. More studies should therefore be conducted to fully elucidate the potentials of AuNPs in agricultural pest management so as to fill the currently existing knowledge gap in this regard and also improve our overall eco-friendly approach to agricultural pest management.

References

Abdel-Aziz, S.M., Prasad, R., Hamed, A.A. and Abdelraof, M. 2018. Fungal nanoparticles: a novel tool for a green biotechnology? pp. 61–87. In Fungal Nanobionics: Principles and Applications, Springer, Singapore.

Abdel-Raouf, N., Al-Enazi, N.M. and Ibraheem, I.B. 2017. Green biosynthesis of gold nanoparticles using *Galaxaura elongata* and characterization of their antibacterial activity. Arab J. Chem. 10: S3029–S3039. https://doi.org/10.1016/j.arabjc.2013.11.044.

Agnihotri, Mithila, Swanand Joshi, Ameeta Ravi Kumar, Smita Zinjarde and Sulabha Kulkarni. 2009. Biosynthesis of gold nanoparticles by the tropical marine yeast Yarrowia lipolytica NCIM 3589. Mater Lett. 63: 1231–1234. https://doi.org/10.1016/j.matlet.2009.02.042.

Aguilar-Méndez, M.A., San Martín-Martínez, E., Ortega-Arroyo, L. et al. 2011. Synthesis and characterization of silver nanoparticles: effect on phytopathogen *Colletotrichum gloesporioides*. J. Nanopart. Res. 13: 2525–2532. https://doi.org/10.1007/s11051-010-0145-6.

Ahmad, Absar, Satyajyoti Senapati, Islam Khan, M., Rajiv Kumar, Ramani, R., Srinivas, V. and Murali Sastry. 2003. Intracellular synthesis of gold nanoparticles by a novel alkalotolerant actinomycete, Rhodococcus species. Nanotechnology 14: 824. https://doi.org /10.1088/0957-4484/14/7/323.

Ahmed, Temoor, Muhammad Shahid, Muhammad Noman, Muhammad Bilal Khan Niazi, Faisal Mahmood, Irfan Manzoor, Yang Zhang et al. 2020. Silver nanoparticles synthesized by using Bacillus cereus SZT1 ameliorated the damage of bacterial leaf blight pathogen in rice. Pathog 9: 160. https://doi.org/10.3390/pathogens9030160.

Annamalai, J. and Nallamuthu, T. 2015. Characterization of biosynthesized gold nanoparticles from aqueous extract of *Chlorella vulgaris* and their anti-pathogenic properties. Appl. Nanosci. 5: 603–607. https://doi.org/10.1007/s13204-014-0353-y.

Arumugam, P. and Berchmans, S. 2011. Synthesis of gold nanoparticles: An ecofriendly approach using *Hansenula anomala*. ACS Appl. Mater Interfaces 3: 1418–1425. doi: 10.1021/am200443j.

Aswani, T., Sasi Reshmi and Suchithra, T.V. 2019. Actinomycetes: Its realm in nanotechnology. pp. 127–140. In Microbial Nanobionics, Springer, Cham.

Balagurunathan, R., Radhakrishnan, M., Rajendran, R.B. and Velmurugan, D. 2011. Biosynthesis of gold nanoparticles by actinomycete *Streptomyces viridogens* strain HM10. Indian J. Biochem. Biophys 48: 331–335.

Barabadi, H., Honary, S., Ebrahimi, P., Mohammadi, M.A., Alizadeh, A. and Naghibi, F. 2014. Microbial mediated preparation, characterization and optimization of gold nanoparticles. Braz J. Microbiol. 45: 1493–1501. https://doi.org/10.1590/S1517-83822014000400046.

Barabadi, H., Honary, S., Mohammadi, M.A., Ahmadpour, E., Rahimi, M.T., Alizadeh, A. and Saravanan, M. 2017. Green chemical synthesis of gold nanoparticles by using *Penicillium aculeatum* and their scolicidal activity against hydatid cyst protoscolices of *Echinococcus granulosus*. Environ. Sci. Pollut. Res. 24(6): 5800–5810. https://doi.org/10.1007/s11356-016-8291-8.

Bello, T.T., Coyne, D.L., Rashidifard, M. and Fourie, H. 2020. Abundance and diversity of plant-parasitic nematodes associated with watermelon in Nigeria, with focus on Meloidogyne spp. Nematology 22: 781–797. DOI: 10.1163/15685411-00003340.

Beveridge, T.J. and Murray, R.G. 1980. Sites of metal deposition in the cell wall of *Bacillus subtilis*. J. Bacteriol. 141: 876–887.

Cai, F., Li, J., Sun, J. and Ji, Y. 2011. Biosynthesis of gold nanoparticles by biosorption using *Magnetospirillum gryphiswaldense* MSR-1. J. Chem. Eng. 175: 70–75. https://doi.org/10.1016/j.cej.2011.09.041.

Castro, M.E., Cottet, L. and Castillo, A. 2014. Biosynthesis of gold nanoparticles by extracellular molecules produced by the phytopathogenic fungus *Botrytis cinerea*. Mater Lett. 115: 42–44. https://doi.org/10.1016/j.matlet.2013.10.020.

Castro-Longoria, E., Vilchis-Nestor, A.R. and Avalos-Borja, M. 2011. Biosynthesis of silver, gold and bimetallic nanoparticles using the filamentous fungus *Neurospora crassa*. Colloids Surf B Biointerfaces 83(1): 42–48. https://doi.org/10.1016/j.colsurfb.2010.10.035.

Chauhan, A., Zubair, S., Tufail, S., Sherwani, A., Sajid, M., Raman, S.C. and Owais, M. 2011. Fungus-mediated biological synthesis of gold nanoparticles: potential in detection of liver cancer. Int. J. Nanomedicine 6: 2305.

Correa-Llantén, D.N., Muñoz-Ibacache, S.A., Castro, M.E., Muñoz, P.A. and Blamey, J.M. 2013. Gold nanoparticles synthesized by *Geobacillus* sp. strain ID17 a thermophilic bacterium isolated from Deception Island, Antarctica. Microb Cell Factories 12(1): 1–6. https://doi.org/10.1186/1475-2859-12-75.

Das, S.K., Das, A.R. and Guha, A.K. 2009. Gold nanoparticles: microbial synthesis and application in water hygiene management. Langmuir 25(14): 8192–8199. https://doi.org/10.1021/la900585p.

Das, S.K., Dickinson, C., Lafir, F., Brougham, D.F. and Marsili, E. 2012. Synthesis, characterization and catalytic activity of gold nanoparticles biosynthesized with *Rhizopus oryzae* protein extract. Green Chem. 14(5): 1322–1334. DOI: 10.1039/C2GC16676C.

Elmer, W. and White, J.C. 2018. The future of nanotechnology in plant pathology. Annu Rev. Phytopathol. 56: 111–133. https://doi.org/10.1146/annurev-phyto-080417-050108.

Fabiyi, O.A., Olatunji, G.A. and Saadu, A.O. 2018. Suppression of *Heterodera sacchari* in rice with agricultural waste-silver nano particles. J. Solid Waste Technol. Manag. 44(2): 87–91. https://doi.org/10.5276/JSWTM.2018.87.

Fabiyi, O.A. and Olatunji, G.A. 2018. Application of green synthesis in nano particles preparation: *Ficus mucoso* extracts in the management of *Meloidogyne incognita* parasitizing groundnut *Arachis hypogea*. Indian J. Nematol. 48(1): 13–17.

Fabiyi, O.A., Alabi, R.O. and Ansari, R.A. 2020a. Nanoparticles' synthesis and their application in the management of phytonematodes: an overview. pp. 125–140. *In*: Ansari, R.A., Rizvi, R. and Mahmood, I. (eds.), Management of Phytonematodes: Recent Advances and Future Challenges. Singapore.

Fabiyi, O.A., Olatunji, G.A., Atolani, O. and Olawuyi, R.O. 2020b. Preparation of bio-nematicidal nanoparticles of *Eucalyptus officinalis* for the control of cyst nematode (*Heterodera sacchari*). J. Anim. Plant Sci. 30(5): 1172–1177.

Fabiyi, O.A. 2021. Sustainable management of *Meloidogyne incognita* Infecting Carrot: Green synthesis of silver nanoparticles with *Cnidoscolus aconitifolius*: (*Daucus carota*). Vegetos 34(2): 277–285. https://doi.org/10.1007/s42535-021-00216-y.

Fabiyi, O.A., Claudius-Cole, A.O., Olatunji, G.A., Abubakar, D.O. and Adejumo, O.A. 2021. Evaluation of the *in vitro* Response of *Meloidogyne incognita* to silver nano particle liquid from Agricultural wastes. Agrivita J. Agri Sci. 43(3): 524–534. https://doi.org/10.17503/agrivita.v43i3.1936.

Fernández, J.G., Fernández-Baldo, M.A., Berni, E., Camí, G., Durán, N., Raba, J. and Sanz, M.I. 2016. Production of silver nanoparticles using yeasts and evaluation of their antifungal activity against phytopathogenic fungi. Process Biochem. 51(9): 1306–1313. https://doi.org/10.1016/j.procbio.2016.05.021.

Fouda, A., Hassan, S.E.D., Abdo, A.M. and El-Gamal, M.S. 2020. Antimicrobial, antioxidant and larvicidal activities of spherical silver nanoparticles synthesized by endophytic *Streptomyces* spp. Biol. Trace Elem. Res. 195(2): 707–724. https://doi.org/10.1007/s12011-019-01883-4.

Gericke, M. and Pinches, A. 2006. Biological synthesis of metal nanoparticles. Hydrometallurgy 83: 132–140. https://doi.org/10.1016/j.hydromet.2006.03.019.

Gole, A., Dash, C., Ramakrishnan, V., Sainkar, S.R., Mandale, A.B., Rao, M. and Sastry, M. 2001. Pepsin–gold colloid conjugates: preparation, characterization, and enzymatic activity. Langmuir 17(5): 1674–1679. https://doi.org/10.1021/la001164w.

Golinska, P., Wypij, M., Ingle, A.P. et al. 2014. Biogenic synthesis of metal nanoparticles from actinomycetes: biomedical applications and cytotoxicity. Appl. Microbiol. Biotechnol. 98: 8083–8097. https://doi.org/10.1007/s00253-014-5953-7.

Gomathy, M. and Sabarinathan, K.G. 2010. Microbial mechanisms of heavy metal tolerance—a review. Agric Rev. 31(2): 133–138.

Gopal, J.V., Thenmozhi, M., Kannabiran, K., Rajakumar, G., Velayutham, K. and Rahuman, A.A. 2013. Actinobacteria mediated synthesis of gold nanoparticles using *Streptomyces* sp. VITDDK3 and its antifungal activity. Mater Lett. 93: 360–362. https://doi.org/10.1016/j.matlet.2012.11.125.

Graily-Moradi, F., Mallak, A.M. and Ghorbanpour, M. 2020. Biogenic synthesis of gold nanoparticles and their potential application in agriculture. pp. 187–204. In Biogenic Nano-Particles and their Use in Agro-ecosystems. Springer, Singapore. https://doi.org/10.1007/978-981-15-2985-6_11.

He, S., Guo, Z., Zhang, Y., Zhang, S., Wang, J. and Gu, N. 2007. Biosynthesis of gold nanoparticles using the bacteria *Rhodopseudomonas capsulata*. Mater Lett. 61(18): 3984–3987. https://doi.org/10.1016/j.matlet.2007.01.018.

Honary, S., Gharaei-Fathabad, E., Barabadi, H. and Naghibi, F. 2013. Fungus-mediated synthesis of gold nanoparticles: a novel biological approach to nanoparticle synthesis. J. Nanosci. Nanotechnol. 13(2): 1427–1430. https://doi. org/10.1166/jnn.2013.5989.

Hossain, A., Hong, X., Ibrahim, E., Li, B., Sun, G., Meng, Y. and An, Q. 2019. Green synthesis of silver nanoparticles with culture supernatant of a bacterium *Pseudomonas rhodesiae* and their antibacterial activity against soft rot pathogen *Dickeya dadantii*. Molecules 24(12): 2303. https://doi.org/10.3390/molecules24122303.

Hulkoti, N.I. and Taranath, T.C. 2014. Biosynthesis of nanoparticles using microbes—a review. Colloids Surf B Biointerfaces 121: 474–483. https://doi.org/10.1016/j.colsurfb.2014.05.027.

Husseiny, M.I., Abd El-Aziz, M., Badr, Y. and Mahmoud, M.A. 2007. Biosynthesis of gold nanoparticles using *Pseudomonas aeruginosa*. Spectrochimica Acta A Mol. Biomolr. Spectrosc. 67: 1003–1006. https://doi. org/10.1016/j.saa.2006.09.028.

Ibrahim, E., Zhang, M., Zhang, Y., Hossain, A., Qiu, W., Chen, Y. and Li, B. 2020. Green-synthesisation of silver nanoparticles using endophytic bacteria isolated from garlic and its antifungal activity against wheat Fusarium head blight pathogen *Fusarium graminearum*. Nanomaterials 10(2): 219. https://doi.org/10.3390/nano10020219.

Ikram, Saiqa. 2015. Synthesis of gold nanoparticles using plant extract: an overview. Nano Res. 1: 5.

Islam, N.U., Jalil, K., Shahid, M., Rauf, A., Muhammad, N., Khan, A. and Khan, M.A. 2019. Green synthesis and biological activities of gold nanoparticles functionalized with Salix alba. Arab J. Chem. 12(8): 2914–2925. https://doi.org/10.1016/j.arabjc.2015.06.025.

Jain, N., Bhargava, A., Majumdar, S., Tarafdar, J.C. and Panwar, J. 2011. Extracellular biosynthesis and characterization of silver nanoparticles using *Aspergillus flavus* NJP08: a mechanism perspective. Nanoscale 3(2): 635–641. DOI: 10.1039/C0NR00656D.

Jayaseelan, C., Ramkumar, R., Rahuman, A.A. and Perumal, P. 2013. Green synthesis of gold nanoparticles using seed aqueous extract of Abelmoschus esculentus and its antifungal activity. Ind. Crop Prod. 45: 423–429. https://doi.org/10.1016/j.indcrop.2012.12.019.

Jo, Y.K., Kim, B.H. and Jung, G. 2009. Antifungal activity of silver ions and nanoparticles on phytopathogenic fungi. Plant Dis. 93(10): 1037–1043. https://doi.org/10.1094/PDIS-93-10-1037.

Jung, J.H., Kim, S.W., Min, J.S., Kim, Y.J., Lamsal, K., Kim, K.S. and Lee, Y.S. 2010. The effect of nano-silver liquid against the white rot of the green onion caused by *Sclerotium cepivorum*. Mycobiology 38(1): 39–45. DOI: 10.4489/MYCO.2010.38.1.039.

Kalimuthu, K., Babu, R.S., Venkataraman, D., Bilal, M. and Gurunathan, S. 2008. Biosynthesis of silver nanocrystals by *Bacillus licheniformis*. Colloids Surf B Biointerfaces 65(1): 150–153. https://doi.org/10.1016/j. colsurfb.2008.02.018.

Kasthuri, J., Kathiravan, K. and Rajendiran, N. 2009. Phyllanthin-assisted biosynthesis of silver and gold nanoparticles: a novel biological approach. J. Nanopart. Res. 11(5): 1075–1085. https://doi.org/10.1007/s11051-008-9494-9.

Kathiraven, T., Sundaramanickam, A., Shanmugam, N. and Balasubramanian, T. 2015. Green synthesis of silver nanoparticles using marine algae Caulerpa racemosa and their antibacterial activity against some human pathogens. Appl. Nanosci. 5(4): 499–504. https://doi.org/10.1007/s13204-014-0341-2.

Khaledian, S., Nikkhah, M., Shams-bakhsh, M. and Hoseinzadeh, S. 2017. A sensitive biosensor based on gold nanoparticles to detect *Ralstonia solanacearum* in soil. J. Gen. Plant Pathol. 83(4): 231–239. https://doi. org/10.1007/s13204-014-0341-2.

Khater, M., De La Escosura-Muñiz, A., Quesada-González, D. and Merkoçi, A. 2019. Electrochemical detection of plant viruses using gold nanoparticle-modified electrodes. Anal. Chim. Acta 1046: 123–131. https://doi. org/10.1016/j.aca.2018.09.031.

Kulabhusan, P.K., Tripathi, A. and Kant, K. 2022. Gold nanoparticles and plant pathogens: an overview and prospective for biosensing in forestry. Sens. 22(3): 1259. https://doi.org/10.3390/s22031259.

Kumar, C.G., Poornachandra, Y. and Mamidyala, S.K. 2014. Green synthesis of bacterial gold nanoparticles conjugated to resveratrol as delivery vehicles. Colloids Surf B Biointerfaces 123: 311–317. https://doi.org/10.1016/j. colsurfb.2014.09.032.

Lau, H.Y., Wang, Y., Wee, E.J., Botella, J.R. and Trau, M. 2016. Field demonstration of a multiplexed point-of-care diagnostic platform for plant pathogens. Anal. Chem. 88(16): 8074–8081. https://doi.org/10.1021/ acs.analchem.6b01551.

Lavanya, R. and Arun, V. 2021. Detection of Begomovirus in chilli and tomato plants using functionalized gold nanoparticles. Sci. Rep. 11(1): 1–13. https://doi.org/10.1038/s41598-021-93615-9.

Lee, J.I., Jang, S.C., Chung, J., Choi, W.K., Hong, C., Ahn, G.R. and Chung, W.J. 2021. Colorimetric allergenic fungal spore detection using peptide-modified gold nanoparticles. Sens Actuators B Chem. 327: 128894. https://doi. org/10.1016/j.snb.2020.128894.

Lee, K.D., Nagajyothi, P.C., Sreekanth, T.V.M. and Park, S. 2015. Eco-friendly synthesis of gold nanoparticles (AuNPs) using Inonotus obliquus and their antibacterial, antioxidant and cytotoxic activities. J. Ind. Eng. Chem. 26: 67–72. https://doi.org/10.1016/j.jiec.2014.11.016.

Lei, R., Wu, P., Li, L., Huang, Q., Wang, J., Zhang, D. and Wang, X. 2021. Ultrasensitive isothermal detection of a plant pathogen by using a gold nanoparticle-enhanced microcantilever sensor. Sens Actuators B Chem. 338: 129874. https://doi.org/10.1016/j.snb.2021.129874.

Lengke, M.F., Fleet, M.E. and Southam, G. 2006. Morphology of gold nanoparticles synthesized by filamentous cyanobacteria from gold (I)−thiosulfate and gold (III)−chloride complexes. Langmuir 22(6): 2780–2787. https://doi.org/10.1021/la052652c.

Lim, H.A., Mishra, A. and Yun, S.I. 2011. Effect of pH on the extracellular synthesis of gold and silver nanoparticles by *Saccharomyces cerevisae*. J. Nanosci. Nanotechnol. 11(1): 518–522. https://doi.org/10.1166/jnn.2011.3266.

Luna-Moreno, D., Sánchez-Álvarez, A., Islas-Flores, I., Canto-Canche, B., Carrillo-Pech, M., Villarreal-Chiu, J.F. and Rodríguez-Delgado, M. 2019. Early detection of the fungal banana black Sigatoka pathogen *Pseudocercospora fijiensis* by an SPR immunosensor method. Sens 19(3): 465. https://doi.org/10.3390/s19030465.

Luo, P., Liu, Y., Xia, Y., Xu, H. and Xie, G. 2014. Aptamer biosensor for sensitive detection of toxin A of Clostridium difficile using gold nanoparticles synthesized by *Bacillus stearothermophilus*. Biosens Bioelectro. 54: 217–221. https://doi.org/10.1016/j.bios.2013.11.013.

Majumder, S. and Johari, S. 2018. Development of a gold-nanoparticle based novel dot immunobinding assay for rapid and sensitive detection of Banana bunchy top virus. J. Virol. Methods 255: 23–28. https://doi.org/10.1016/j.jviromet.2018.01.015.

Malarkodi, C., Rajeshkumar, S., Vanaja, M., Paulkumar, K., Gnanajobitha, G. and Annadurai, G. 2013. Eco-friendly synthesis and characterization of gold nanoparticles using Klebsiella pneumoniae. J. Nanaostructure Chem. 3(1): 1–7. https://doi.org/10.1186/2193-8865-3-30.

Maliszewska, I. 2013. Microbial mediated synthesis of gold nanoparticles: preparation, characterization and cytotoxicity studies. Dig J. Nanomater. Biostructures 8(3).

Manjunath, H.M., Joshi, C.G. and Raju, N.G. 2017. Biofabrication of gold nanoparticles using marine endophytic fungus–*Penicillium citrinum*. IET Nanobiotechnol. 11(1): 40–44. https://doi.org/10.1049/iet-nbt.2016.0065.

Marooufpour, N., Alizadeh, M., Hatami, M. and Lajayer, B.A. 2019. Biological synthesis of nanoparticles by different groups of bacteria. pp. 63–85. *In*: Microbial Nanobionics Springer, Cham. https://doi.org/10.1007/978-3-030-16383-9_3.

Mata, Y.N., Torres, E., Blazquez, M.L., Ballester, A., González, F.M.J.A. and Munoz, J.A. 2009. Gold (III) biosorption and bioreduction with the brown alga *Fucus vesiculosus*. J. Hazard Mater 166(2-3): 612–618. https://doi.org/10.1016/j.jhazmat.2008.11.064.

Mishra, A., Tripathy, S.K. and Yun, S.I. 2011. Bio-synthesis of gold and silver nanoparticles from *Candida guilliermondii* and their antimicrobial effect against pathogenic bacteria. J. Nanosci. Nanotechnol. 11(1): 243–248. https://doi.org/10.1166/jnn.2011.3265.

Mishra, S., Singh, B.R., Singh, A., Keswani, C., Naqvi, A.H. and Singh, H.B. 2014. Biofabricated silver nanoparticles act as a strong fungicide against *Bipolaris sorokiniana* causing spot blotch disease in wheat. Plos One 9(5): e97881. https://doi.org/10.1371/journal.pone.0097881.

Mohanpuria, P., Rana, N.K. and Yadav, S.K. 2008. Biosynthesis of nanoparticles: technological concepts and future applications. J. Nanopart. Res. 10(3): 507–517. https://doi.org/10.1007/s11051-007-9275-x.

Montes, M.O., Mayoral, A., Deepak, F.L., Parsons, J.G., Jose-Yacamán, M., Peralta-Videa, J.R. and Gardea-Torresdey, J.L. 2011. Anisotropic gold nanoparticles and gold plates biosynthesis using alfalfa extracts. J. Nanopart. Res. 13(8): 3113–3121. https://doi.org/10.1007/s11051-011-0230-5.

Mortazavi, S.M., Khatami, M., Sharifi, I., Heli, H., Kaykavousi, K., Poor, M.H.S. and Nobre, M.A.L. 2017. Bacterial biosynthesis of gold nanoparticles using Salmonella enterica subsp. enterica serovar Typhi isolated from blood and stool specimens of patients. J. Clust Sci. 28(5): 2997–3007. https://doi.org/10.1007/s10876-017-1267-0.

Muthuvel, A., Adavallan, K., Balamurugan, K. and Krishnakumar, N. 2014. Biosynthesis of gold nanoparticles using Solanum nigrum leaf extract and screening their free radical scavenging and antibacterial properties. Biomed Prev. Nutr. 4(2): 325–332. https://doi.org/10.1016/j.bionut.2014.03.004.

Namvar, F., Azizi, S., Ahmad, M.B., Shameli, K., Mohamad, R., Mahdavi, M. and Tahir, P.M. 2015. Green synthesis and characterization of gold nanoparticles using the marine macroalgae *Sargassum muticum*. Res. Chem. Intermed. 41(8): 5723–5730. https://doi.org/10.1007/s11164-014-1696-4.

Narayanan, K.B. and Sakthivel, N. 2011. Facile green synthesis of gold nanostructures by NADPH-dependent enzyme from the extract of *Sclerotium rolfsii*. Colloids Surf A Physicochem Eng. Asp 380(1-3): 156–161. https://doi.org/10.1016/j.colsurfa.2011.02.042.

Narayanan, K.B. and Sakthivel, N. 2013. Mycocrystallization of gold ions by the fungus *Cylindrocladium floridanum*. World J. Microbiol. Biotechnol. 29(11): 2207–2211. https://doi.org/10.1007/s11274-013-1379-0.

Narayanan, K.B. and Sakthivel, N. 2010. Biological synthesis of metal nanoparticles by microbes. Adv. Colloid Interface Sci. 156: 1–13. https://doi.org/10.1016/j.cis.2010.02.001.

Noruzi, M., Zare, D., Khoshnevisan, K. and Davoodi, D. 2011. Rapid green synthesis of gold nanoparticles using Rosa hybrida petal extract at room temperature. Spectrochim Acta A Mol. Biomol. Spectr. 79(5): 1461–1465. https://doi.org/10.1016/j.saa.2011.05.001.

Ogi, T., Saitoh, N., Nomura, T. and Konishi, Y. 2010. Room-temperature synthesis of gold nanoparticles and nanoplates using Shewanella algae cell extract. J. Nanopart. Res. 12(7): 2531–2539. https://doi.org/10.1007/s11051-009-9822-8.

Ogunyemi, S.O., Abdallah, Y., Zhang, M., Fouad, H., Hong, X., Ibrahim, E. and Li, B. 2019. Green synthesis of zinc oxide nanoparticles using different plant extracts and their antibacterial activity against *Xanthomonas oryzae* pv. oryzae. Artif Cells Nanomed. Biotechnol. 47(1): 341–352. https://doi.org/10.1080/21691401.2018.1557671.

Osonga, F.J., Akgul, A., Yazgan, I., Akgul, A., Eshun, G.B., Sakhaee, L. and Sadik, O.A. 2020. Size and shape-dependent antimicrobial activities of silver and gold nanoparticles: a model study as potential fungicides. Molecules 25(11): 2682. https://doi.org/10.3390/molecules25112682.

Oza, G., Pandey, S., Gupta, A., Kesarkar, R. and Sharon, M. 2012. Biosynthetic reduction of gold ions to gold nanoparticles by *Nocardia farcinica*. J. Microbiol. Biotechnol. Res. 2(4): 511–515.

Park, H.J., Kim, S.H., Kim, H.J. and Choi, S.H. 2006. A new composition of nanosized silica-silver for control of various plant diseases. Plant Pathol. J. 22(3): 295–302. https://doi.org/10.5423/PPJ.2006.22.3.295.

Paul, B., Bhuyan, B., Purkayastha, D.D., Dey, M. and Dhar, S.S. 2015. Green synthesis of gold nanoparticles using *Pogostemon benghalensis* (B) O. Ktz. leaf extract and studies of their photocatalytic activity in degradation of methylene blue. Mater Lett. 148: 37–40. https://doi.org/10.1016/j.matlet.2015.02.054.

Ponmurugan, P. 2017. Biosynthesis of silver and gold nanoparticles using *Trichoderma atroviride* for the biological control of Phomopsis canker disease in tea plants. IET Nanobiotechnol. 11(3): 261–267. https://doi.org/10.1049/iet-nbt.2016.0029.

Prasad, R., Jha, A.K. and Prasad, K. (eds.). 2018. Exploring the Realms of Nature for Nanosynthesis. Cham: Springer International Publishing.

Pimprikar, P.S., Joshi, S.S., Kumar, A.R., Zinjarde, S.S. and Kulkarni, S.K. 2009. Influence of biomass and gold salt concentration on nanoparticle synthesis by the tropical marine yeast *Yarrowia lipolytica* NCIM 3589. Colloids Surf B Biointerfaces 74(1): 309–316. https://doi.org/10.1016/j.colsurfb.2009.07.040.

Rajasree, S.R. and Suman, T.Y. 2012. Extracellular biosynthesis of gold nanoparticles using a gram negative bacterium *Pseudomonas fluorescens*. Asian Pac J. Trop. Dis. 2: S796–S799. https://doi.org/10.1016/S2222-1808(12)60267-9.

Ranjitha, V.R. and Rai, V.R. 2017. Actinomycetes mediated synthesis of gold nanoparticles from the culture supernatant of *Streptomyces griseoruber* with special reference to catalytic activity. 3 Biotech 7(5): 1–7. https://doi.org/10.1007/s13205-017-0930-3.

Razmi, A., Golestanipour, A., Nikkhah, M., Bagheri, A., Shamsbakhsh, M. and Malekzadeh-Shafaroudi, S. 2019. Localized surface plasmon resonance biosensing of tomato yellow leaf curl virus. J. Virol. Methods 267: 1–7. https://doi.org/10.1016/j.jviromet.2019.02.004.

Reddy, G.R., Morais, A.B. and Gandhi, N.N. 2013. Green synthesis, characterization and in vitro antibacterial studies of gold nanoparticles by using *Senna siamea* plant seed aqueous extract at ambient conditions. Asian J. Chem. 25(15): 8541. http://dx.doi.org/10.14233/ajchem.2013.14830.

Roy, S. and Das, T.K. 2016. Effect of biosynthesized silver nanoparticles on the growth and some biochemical parameters of *Aspergillus foetidus*. J. Environ. Chem. Eng. 4(2): 1574–1583. https://doi.org/10.1016/j.jece.2016.02.010.

Sanghi, R., Verma, P. and Puri, S. 2011. Enzymatic formation of gold nanoparticles using *Phanerochaete chrysosporium*. Adv. Chem. Engineering Sci. 1(03): 154. doi:10.4236/aces.2011.13023.

Sarkar, J., Ray, S., Chattopadhyay, D., Laskar, A. and Acharya, K. 2012. Mycogenesis of gold nanoparticles using a phytopathogen *Alternaria alternata*. Bioprocess Biosyst. Eng. 35(4): 637–643. https://doi.org/10.1007/s00449-011-0646-4.

Sastry, M., Ahmad, A., Khan, M.I. and Kumar, R. 2003. Biosynthesis of metal nanoparticles using fungi and actinomycete. Curr. Sci. 162–170.

Sawle, B.D., Salimath, B., Deshpande, R., Bedre, M.D., Prabhakar, B.K. and Venkataraman, A. 2008. Biosynthesis and stabilization of Au and Au–Ag alloy nanoparticles by fungus, *Fusarium semitectum*. Science and Technology of Advanced Materials. https://doi.org/10.1088/1468-6996/9/3/035012.

Shah, R., Oza, G., Pandey, S. and Sharon, M. 2012. Biogenic fabrication of gold nanoparticles using Halomonas salina. J. Microbiol. Biotechnol. Res. 2(4): 485–492.

Shahzad, S.A., Sajid, M.A., Khan, Z.A. and Canseco-Gonzalez, D. 2017. Gold catalysis in organic transformations: A review. Synth. Commun. 47(8): 735–755. https://doi.org/10.1080/00397911.2017.1280508.

Shankar, S.S., Ahmad, A., Pasricha, R. and Sastry, M. 2003. Bioreduction of chloroaurate ions by geranium leaves and its endophytic fungus yields gold nanoparticles of different shapes. J. Mater Chem. 13(7): 1822–1826. DOI: 10.1039/B303808B.

Shedbalkar, U., Singh, R., Wadhwani, S., Gaidhani, S. and Chopade, B.A. 2014. Microbial synthesis of gold nanoparticles: current status and future prospects. Adv. Colloid Interface Sci. 209: 40–48. https://doi.org/10.1016/j.cis.2013.12.011.

Sheikhloo, Z. and Salouti, M. 2011. Intracellular biosynthesis of gold nanoparticles by the fungus Penicillium chrysogenum. Int. J. Nanosci. Nanotechnol. 7(2): 102–105.

Shen, W., Qu, Y., Pei, X., Li, S., You, S., Wang, J. and Zhou, J. 2017. Catalytic reduction of 4-nitrophenol using gold nanoparticles biosynthesized by cell-free extracts of *Aspergillus* sp. WL-Au. J. Hazard Mater 321: 299–306. https://doi.org/10.1016/j.jhazmat.2016.07.051.

Singaravelu, G., Arockiamary, J.S., Kumar, V.G. and Govindaraju, K. 2007. A novel extracellular synthesis of monodisperse gold nanoparticles using marine alga, *Sargassum wightii* Greville. Colloids Surf B Biointerfaces 57(1): 97–101. https://doi.org/10.1016/j.colsurfb.2007.01.010.

Singh, M., Kalaivani, R., Manikandan, S., Sangeetha, N. and Kumaraguru, A.K. 2013. Facile green synthesis of variable metallic gold nanoparticle using *Padina gymnospora*, a brown marine macroalga. Appl. Nanosci. 3(2): 145–151. https://doi.org/10.1007/s13204-012-0115-7.

Singh, P., Singh, H., Kim, Y.J., Mathiyalagan, R., Wang, C. and Yang, D.C. 2016a. Extracellular synthesis of silver and gold nanoparticles by *Sporosarcina koreensis* DC4 and their biological applications. Enzyme Microb. Technol. 86: 75–83. https://doi.org/10.1016/j.enzmictec.2016.02.005.

Singh, P., Kim, Y.J., Zhang, D. and Yang, D.C. 2016b. Biological synthesis of nanoparticles from plants and microorganisms. Trends Biotechnol. 34(7): 588–599. https://doi.org/10.1016/j.tibtech.2016.02.006.

Soltani, N.M, Shahidi, B.G. and Khaleghi N. 2015. Biosynthesis of gold nanoparticles using *Streptomyces fulvissimus* isolate. Nanomed. J. 2: 153–159.

Sowani, H., Mohite, P., Munot, H., Shouche, Y., Bapat, T., Kumar, A.R. and Zinjarde, S. 2016. Green synthesis of gold and silver nanoparticles by an actinomycete Gordonia amicalis HS-11: mechanistic aspects and biological application. Process Biochem. 51(3): 374–383. https://doi.org/10.1016/j.procbio.2015.12.013.

Srinath, B.S. and Rai, V.R. 2015. Rapid biosynthesis of gold nanoparticles by Staphylococcus epidermidis: Its characterisation and catalytic activity. Mater Lett. 146: 23–25. https://doi.org/10.1016/j.matlet.2015.01.151.

Suganya, K.U., Govindaraju, K., Kumar, V.G., Dhas, T.S., Karthick, V., Singaravelu, G. and Elanchezhiyan, M. 2015. Blue green alga mediated synthesis of gold nanoparticles and its antibacterial efficacy against Gram positive organisms. Mater Sci. Eng. C 47: 351–356. https://doi.org/10.1016/j.msec.2014.11.043.

Suresh, A.K., Pelletier, D.A., Wang, W., Broich, M.L., Moon, J.W., Gu, B. and Doktycz, M.J. 2011. Biofabrication of discrete spherical gold nanoparticles using the metal-reducing bacterium *Shewanella oneidensis*. Acta Biomater 7(5): 2148–2152. https://doi.org/10.1016/j.actbio.2011.01.023.

Syed, B., Prasad, N.M. and Satish, S. 2016. Endogenic mediated synthesis of gold nanoparticles bearing bactericidal activity. J. Microsc. Ultrastruct. 4(3): 162–166. https://doi.org/10.1016/j.jmau.2016.01.004.

Tamuly, C., Hazarika, M. and Dordoloi, M. 2013. Biosynthesis of Au nanoparticles by Gymnocladus assamicus and its catalytic activity. Mater Lett. 108: 276–279. https://doi.org/10.1016/j.matlet.2013.07.020.

Tetgure, S.R., Borse, A.U., Sankapal, B.R., Garole, V.J. and Garole, D.J. 2015. Green biochemistry approach for synthesis of silver and gold nanoparticles using *Ficus racemosa* latex and their pH-dependent binding study with different amino acids using UV/Vis absorption spectroscopy. Amino Acids 47(4): 757–765. https://doi.org/10.1007/s00726-014-1906-9.

Vala, A.K. 2015. Exploration on green synthesis of gold nanoparticles by a marine-derived fungus *Aspergillus sydowii*. Environ. Prog. Sustain Energy 34(1): 194–197. https://doi.org/10.1002/ep.11949.

Vanaraj, S., Jabastin, J., Sathiskumar, S. and Preethi, K. 2017. Production and characterization of bio-AuNPs to induce synergistic effect against multidrug resistant bacterial biofilm. J. Clust. Sci. 28(1): 227–244. https://doi.org/10.1007/s10876-016-1081-0.

Vaseghi, A., Safaie, N., Bakhshinejad, B., Mohsenifar, A. and Sadeghizadeh, M. 2013. Detection of Pseudomonas syringae pathovars by thiol-linked DNA–Gold nanoparticle probes. Sens Actuators B Chem. 181: 644–651. https://doi.org/10.1016/j.snb.2013.02.018.

Venkatesan, J., Manivasagan, P., Kim, S.K., Kirthi, A.V., Marimuthu, S. and Rahuman, A.A. 2014. Marine algae-mediated synthesis of gold nanoparticles using a novel Ecklonia cava. Bioprocess Biosyst. Eng. 37(8): 1591–1597. https://doi.org/10.1007/s00449-014-1131-7.

Vetchinkina, E.P., Loshchinina, E.A., Vodolazov, I.R., Kursky, V.F., Dykman, L.A. and Nikitina, V.E. 2017. Biosynthesis of nanoparticles of metals and metalloids by basidiomycetes. Preparation of gold nanoparticles by using purified fungal phenol oxidases. Appl. Microbiol. Biotechnol. 101(3): 1047–1062. https://doi.org/10.1007/s00253-016-7893-x.

Vijayaraghavan, K., Mahadevan, A., Sathishkumar, M., Pavagadhi, S. and Balasubramanian, R. 2011. Biosynthesis of Au (0) from Au (III) via biosorption and bioreduction using brown marine alga *Turbinaria conoides*. J. Chem. Eng. 167(1): 223–227. https://doi.org/10.1016/j.cej.2010.12.027.

Waghmare, S.S., Deshmukh, A.M. and Sadowski, Z. 2014. Biosynthesis, optimization, purification and characterization of gold nanoparticles. Afri J. Microbiol. Res. 8(2): 138–146. DOI: 10.5897/AJMR10.143.

Wang, L., Shan, J., Feng, F. and Ma, Z. 2016. Novel redox species polyaniline derivative-Au/Pt as sensing platform for label-free electrochemical immunoassay of carbohydrate antigen 199. Anal. Chim. Acta 911: 108–113. https://doi.org/10.1016/j.aca.2016.01.016,

Xie, J., Lee, J.Y., Wang, D.I. and Ting, Y.P. 2007. High-yield synthesis of complex gold nanostructures in a fungal system. J. Phys. Chem. C 111(45): 16858–1686. https://doi.org/10.1021/jp0752668.

Zhang, X., Qu, Y., Shen, W., Wang, J., Li, H., Zhang, Z. and Zhou, J. 2016. Biogenic synthesis of gold nanoparticles by yeast *Magnusiomyces ingens* LH-F1 for catalytic reduction of nitrophenols. Colloids Surf A Physicochem Eng. Asp 497: 280–285. https://doi.org/10.1016/j.colsurfa.2016.02.033.

Zhao, W., Lu, J., Ma, W., Xu, C., Kuang, H. and Zhu, S. 2011. Rapid on-site detection of Acidovorax avenae subsp. citrulli by gold-labeled DNA strip sensor. Biosens Bioelectron. 26(10): 4241–4244. https://doi.org/10.1016/j.bios.2011.04.004.

Chapter 12

The Potential of Chitosan Nanoparticles to Control Plant Pathogens

Mohsen Mohamed Elsharkawy

Introduction

Agriculture and health systems suffer several problems as a result of climate change, increasing population, and decreasing agricultural land. Nanotechnology may probably alleviate these concerns by increasing food production, effective land use, and the development of new pesticides. Nanomaterials are useful in overcoming biological barriers and transporting necessary molecules to various locations in plants because of their small size. Nanoparticles made from chitosan and its derivatives, when utilized appropriately, may be invaluable in tackling concerns, such as controlling phytopathogens and improving plant health.

Chitosan's antibacterial, antioxidant, and chelating capabilities, as well as its non-toxic and biocompatible nature, have made it popular (Gedda et al. 2016). Chitosan is also an effective nanoparticle carrier because of its ability to overcome cellular barriers (Dheyab et al. 2020). Emulsion crosslinking, ionotropic gelation, emulsion-droplet coalescence, reverse micellization, and precipitation are only a few of the processes used to produce chitosan-based nanoparticles (Kashyap et al. 2015; Naskar et al. 2019). Chitosan nanoparticles (CNP) are a great way to increase the unique features of chitosan biopolymer since they have diverse physicochemical properties, such as size, surface area, and cationic nature. Considering their stability and low toxicity, as well as their convenience of preparation, CNPs are a flexible and user-friendly therapeutics agent (Asghari et al. 2016). CNP has only been linked to antifungal activity against a variety of phytopathogens (Kashyap et al. 2015). Agricultural pesticides, fungicides, and better-quality food items with a greater yield have all been thoroughly studied in the literature using chitosan-based nanoparticles (CNPs) (Kumaraswamy et al. 2018). Additionally, chitosan is a naturally found antimicrobial compound that has been proven to be effective against a wide range of bacteria and fungi (Meng et al. 2020; Kutawa et al. 2021). However, its effectiveness is strongly reliant on the kind of microorganisms. The mechanism of the antimicrobial effect of chitosan is primarily based on its physiochemical characteristics (Ke et al. 2021). Submicron dispersion of chitosan exhibited antifungal activity against the anthracnose-causing *Colletotrichum gloeosporioides*. The synthesis of cell wall-degrading enzymes in the fungus was reported to be reduced by submicron chitosan dispersions (Zahid et al. 2014, 2015). Previous investigations have demonstrated that chitosan is

Agricultural Botany Department, Faculty of Agriculture, Kafrelsheikh University, 33516 Kafr El-Sheikh, Egypt.
* Corresponding author: mohsen.abdelrahman@agr.kfs.edu.eg

efficient *in vitro* against *Xanthomonas axonopodis* pv. *poinsettiicola* (Chang et al. 2012). Chitosan's antibacterial action included membrane damage and biofilm breakdown. The antibacterial activity of chitosan may be improved by mixing it with appropriate compounds. The antibacterial capabilities of plant essential oils encapsulated in chitosan-based coatings are attracting agricultural research. These plant essential oils have antibacterial properties and were recently added to chitosan (Rabea et al. 2003). Furthermore, thymol-loaded chitosan nanoparticles (TCNPs) were reported to directly limit the development of *Xanthomonas campestris* pv. *campestris* (Xcc) by reducing the growth of biofilm formation and the generation of exopolysaccharides and xanthomonadin (Sreelatha et al. 2022).

The Interaction Between Nano Chitosan and Plant Pathogens

Henri Braconnot identified chitosan as a useful substance in 1811 (Chakraborty et al. 2020). Chitosan is a biodegradable and sustainable substance, which is also inexpensive and simple to manipulate (Park et al. 2008; Anitha et al. 2014; Khalil 2016; Malerba and Cerana 2016). In addition, chitosan has local and systemic efficacy against a wide variety of pests, including viruses, bacteria, fungi, insects, plant nematodes, and more (Abd El-Aziz and Khalil 2020; Alfy et al. 2020). The inorganic nanoparticles have the advantages of being stable, biocompatible with water, and non-toxic. Due to its biodegradable and biocompatible qualities, its antimicrobial activity, and its non-toxicity to humans and animals, chitosan has been explored for the production of microparticles and nanoparticles (Akamatsu et al. 2010; Zhou et al. 2011). It has also been observed that chitosan in the form of microparticles or nanoparticles increases the biological activity of the solution (Du et al. 2008; Huang et al. 2009). Excellent biopesticides are distinguished by their ability to connect to different biochemical interactions due to the unique and divergent qualities of chitosan nanoparticles with changed physicochemical attributes, including surface area, size, cationic capabilities, etc. (Kashyap et al. 2015; Saharan et al. 2015). Since chitosan nanoparticles are stable, simple to prepare, and have low toxicity, they are adaptable and user-friendly agents. As a consequence, nano chitosan derived from chitosan has been used in a wide range of agricultural settings, particularly for the promotion of plant development and the control of plant disease (Table 1). As a potential pesticide, nano chitosan was tested for its efficacy against *Tobacco mosaic tobamovirus* (TMV) and *Meloidogyne incognita* in greenhouse-grown tomato (Khalil et al. 2022). The findings showed that nano chitosan reduced *Meloidogyne incognita* densities by 45.89–66.61% alone or in the existence of TMV and lowered root gall densities by 10.63–67.87%. In addition, after 20 days of inoculation, the concentration of TMV on tomato leaves was reduced by 10.26% to 65.00%, and it extended up to 58.00% after 40 days of infection when *M. incognita* was present. Similarly, root-knot nematode (*M. incognita*) juveniles exposed to nano chitosan at 500 to 2,000 mg/l died at a rate of 85.2% to 97.5% after 72 hours, while egg hatching was reduced by 73% to 95.3% (Alfy et al. 2020). Under greenhouse conditions, the pot experiment showed that 2,000 mg/l of nano chitosan reduced galls, egg masses, and juveniles/250 g of soil by 90%, 78%, and 98%, respectively. The soil population was reduced by 37.67–68.50% and root galls by 70.30–83.33% when treated with chitosan of different molecular weights (Khalil and Badawy 2012). Chitosan of both high and low molecular weights reduced root galls by around 90% and 93%, respectively (El-Sayed and Mahdy 2015). The severity of the pine wilt disease (caused by *Bursaphelenchus xylophilus*) was reduced by applying chitosan with varying molecular weights (da Silva et al. 2014). Chitosan's anti-PPN (plant parasitic nematodes) effects are thought to originate from its capacity to reduce egg hatching and larval and adult viability for both root-knot and cyst nematodes, as well as from its promotion of the reduplication of chitinolytic bacteria that destroy chitin in organs of the PPNs (Westerdahl et al. 1992; Kalaiarasan et al. 2006; El-Sayed and Mahdy 2015). Chitosan exacerbated *Pochonia chlamydosporia* parasitism of *Meloidogyne javanica* eggs (El-Sayed and Mahdy 2015). Meanwhile, the PPNs are affected by the ammonia released by the nitrogen in chitosan (Fan et al. 2020). In addition to *M. incognita* (the root-knot nematode), chitosan and its derivatives elicit

Table 1. Chitosan nanoparticles used against different plant pathogens.

Nanoparticles	Pathogen	References
Chitosan nanoparticles	*Alternaria alternata, Macrophomina phaseolina, Rhizoctonia solani*	Saharan et al. (2013)
	Pyricularia grisea, Alternaria solani, Fusarium oxysporum	Sathiyabama and Parthasarathy (2016)
	Pyricularia grisea	Manikandan and Sathiyabama (2016)
	Fusarium graminearum	Kheiri et al. (2017)
	Puccinia triticina	Elsharkawy et al. (2022)
	Meloidogyne incognita	Alfy et al. (2020)
	Tobacco mosaic tobamovirus (TMV) and *Meloidogyne incognita*	Khalil et al. (2022)
Chitosan-saponin nanoparticles	*Alternaria alternata, Macrophomina phaseolina,* and *Rhizoctonia solani*	Saharan et al. (2013)
Cu-chitosan nanoparticles	*Alternaria alternata, Macrophomina phaseolina, Rhizoctonia solani*	Saharan et al. (2013)
	Fusarium solani	Vokhidova et al. (2014)
	Alternaria solani and *Fusarium oxysporum*	Saharan et al. (2015)
	Rhizoctonia solani and *Sclerotium rolfsii*	Rubina et al. (2017)
Cu(II)-chitosan Nanogel	*Fusarium graminerarium*	Brunel et al. (2013)
Oleoyl-chitosan nanoparticles	*Verticillium dahaliae*	Xing et al. (2017)
Salicylic acid-loaded chitosan nanoparticles	*Fusarium.verticillioides*	Kumaraswamy et al. (2019)
Ag-chitosan nanoparticles	*Botrytis cinerea*	Moussa et al. (2013)
	Aspergillus flavus and *Aspergillus terreus*	Mathew and Kuriakose (2013)
Silica-chitosan nanoparticles	*Phomopsis asparagi*	Cao et al. (2016)
Zataria multiflora in chitosan nanoparticles	*Botrytis cinerea*	Mohammadi et al. (2015)
Chitosan-thymol nanoparticles	*Botrytis cinerea*	Medina et al. (2019)
Chitosan-*Cymbopogon martinii*	*Fusarium graminearum*	Kalagatur et al. (2018)
Thiadiazole functionalized Chitosan derivatives	*Colletotrichum lagenarium, Phomopsis asparagi,* and *Monilinia fructicola*	Li et al. (2013)

systemic acquired resistance in plants against other diseases (Chakraborty et al. 2020). In addition, the use of chitosan or its derivatives has been shown to activate the enzymes glucanase, peroxidase, phenylalanine ammonia-lyase, chitinase, superoxide dismutase, catalase, and polyphenol oxidase in certain plant species (Burkhanova et al. 2007; Eilenberg et al. 2009; Yin et al. 2010; Chang and Kim 2012; Orzali et al. 2014; Xing et al. 2015; Li et al. 2016). In treated plants, protein concentrations, as well as several secondary metabolites that have been associated with pathogen defense, such as phenolics and phytoalexins, were elevated (Lin et al. 2005; Hadwiger 2013; Zhang et al. 2015). The negative charges on the surfaces of a variety of pathogens may be related to the positive charges on chitosan molecules, resulting in cell structural destruction (Chakraborty et al. 2020).

Viral infections may damage the host immune system in addition to affecting host cells, which can result in chronic conditions that seem to be frequent or long-lasting persistent (Boroumand et al. 2021). In order to multiply among multiple host cells, viruses have the ability to damage the host cell while releasing their RNA or DNA. It is difficult to produce antiviral drugs because of the viral life cycle. It has been proposed to use nanotechnology-based methods to treat viral infections successfully and get beyond some of the drawbacks of antiviral drugs (Boroumand et al. 2021). For instance, special characteristics of nanoparticles, such as their size and amendable surface charge, make them effective in treating viruses (Caron et al. 2010; Kumar et al. 2012). Furthermore, it is established that nanoparticles can modify biological processes (Gagliardi 2017). Applications of chitosan have an impact on viral diseases, such as *Potato virus X potexvirus, Alfalfa mosaic alfamovirus, Peanut stunt cucumovirus, Tobacco mosaic tobamovirus, Cucumber mosaic cucumovirus,* and *Tobacco necrosis alphanecrovirus* (Nagorskaya et al. 2014; Jia et al. 2016). TMV affects fruit crops in a variety of ways, including the loss of flowers or lack of blossoming, which reduces productivity (El Shafie et al. 2005). The ability of chitosan to induce a hypersensitive response in infected plants may explain its performance as an antiviral agent (Chirkov 2002). Additionally, the average frequency of polymerization, the quality of the positive charge, the level of N-deacetylation, and the nature of the molecule's chemical alterations all affect the antiviral activity of chitosan (Chakraborty et al. 2020).

Chitosan is a natural source antimicrobial compound that has been discovered to be effective against a number of bacteria and fungi (Meng et al. 2020; Kutawa et al. 2021). However, the nature of microorganisms has a significant impact on their effectiveness. Chitosan's physiochemical characteristics play a significant role in the mechanism behind its antibacterial action (Ke et al. 2021). The chitosan nanoparticles have shown to be fungistatic *in vitro* and *in vivo*, inhibiting the growth of a broad range of fungi, including *Botrytis cinerea, Aspergillus niger, Alternaria alternata, A. parasiticus,* and *Rhizopus stolonifer* (El Ghaouth et al. 1994; Bautista-Baños et al. 2006; Martínez-Camacho et al. 2010; Cota-Arriola et al. 2011). The antifungal potentials of chitosan nanoparticles against phytopathogenic fungi are related to their broad antimicrobial action (Ing et al. 2012; Saharan et al. 2013, 2015; Sathiyabama and Parthasarathy 2016). Chitosan exhibited antifungal ability against *Colletotrichum gloeosporioides,* which causes anthracnose in fresh fruits when it was used in submicron dispersion. It was reported that *C. gloeosporioides* produced fewer cell wall-degrading enzymes when exposed to submicron chitosan dispersions (Zahid et al. 2014, 2015). Furthermore, biosynthesized chitosan nanoparticles were assessed for their potential to inhibit the growth of a particularly virulent strain of the phytopathogenic fungus *Botrytis cinerea* (El-Naggar et al. 2022). Chitosan nanoparticles effectively inhibited fungal development. The effects of nano chitosan (at 25 mg/mL) on strawberry leaf gray mold severity were decreased to 3%. In contrast to lower concentrations (12.5 mg/mL and 25 mg/mL), it was discovered that the greater concentration of chitosan nanoparticles (50 mg/mL) had a negative impact on the infected area. Under controlled laboratory conditions, the antifungal and antimycotoxin activity of CMEO (*Cymbopogon martinii* essential oil) and CeCMEO-NPs (chitosan encapsulated CMEO nanoparticles) against *Fusarium graminearum* in maize grains were tested for 28 days (Kalagatur et al. 2018). Significant fungicidal and antimycotoxin actions of CMEO and Ce-CMEO-NPs have been shown in maize grains.

Reductions in fungal growth and mycotoxin (DON and ZEA) levels were detected as a function of Ce-CMEO-NPs and CMEO concentration throughout kinetic curves. The developed regression models were considered sufficient and showed excellent determination coefficients (R^2) of 0.9864 and 0.9793 (DON), 0.9694 and 0.9896 (log CFU), and 0.9873 and 0.9935 (ZEA) for Ce-CMEO-NPs and CMEO, respectively. These regression models demonstrated a dose-dependent relationship between the concentration of Ce-CMEO-NPs and CMEO and reductions in fungal growth and mycotoxins. Therefore, Ce-CMEO-NPs have shown stronger antifungal and antimycotoxin activity than CMEO. There was a significant decrease in both fungal growth and mycotoxins at 900 ppm of CMEO and 700 ppm of Ce-CMEO-NPs. The colony growth diameter of *Fusarium oxysporum* was decreased as a result of the gold-chitosan nanoparticles' antifungal effects (Figure 1). Nanoparticle concentration was shown to be inversely related to colony diameter (Lipşa et al. 2020). Germination of *Puccinia triticina* urediniospores was investigated in water-agar medium in the presence and absence of salicylic acid and chitosan nanoparticles (Elsharkawy et al. 2022). Germination rates of urediniospores were significantly reduced using chitosan nanoparticles. In comparison to the control treatment, which showed the longest latent and incubation periods across all three application methods, the chitosan nanoparticles treatment showed the best results. When compared to controls, the use of chitosan nanoparticles in three different conditions resulted in

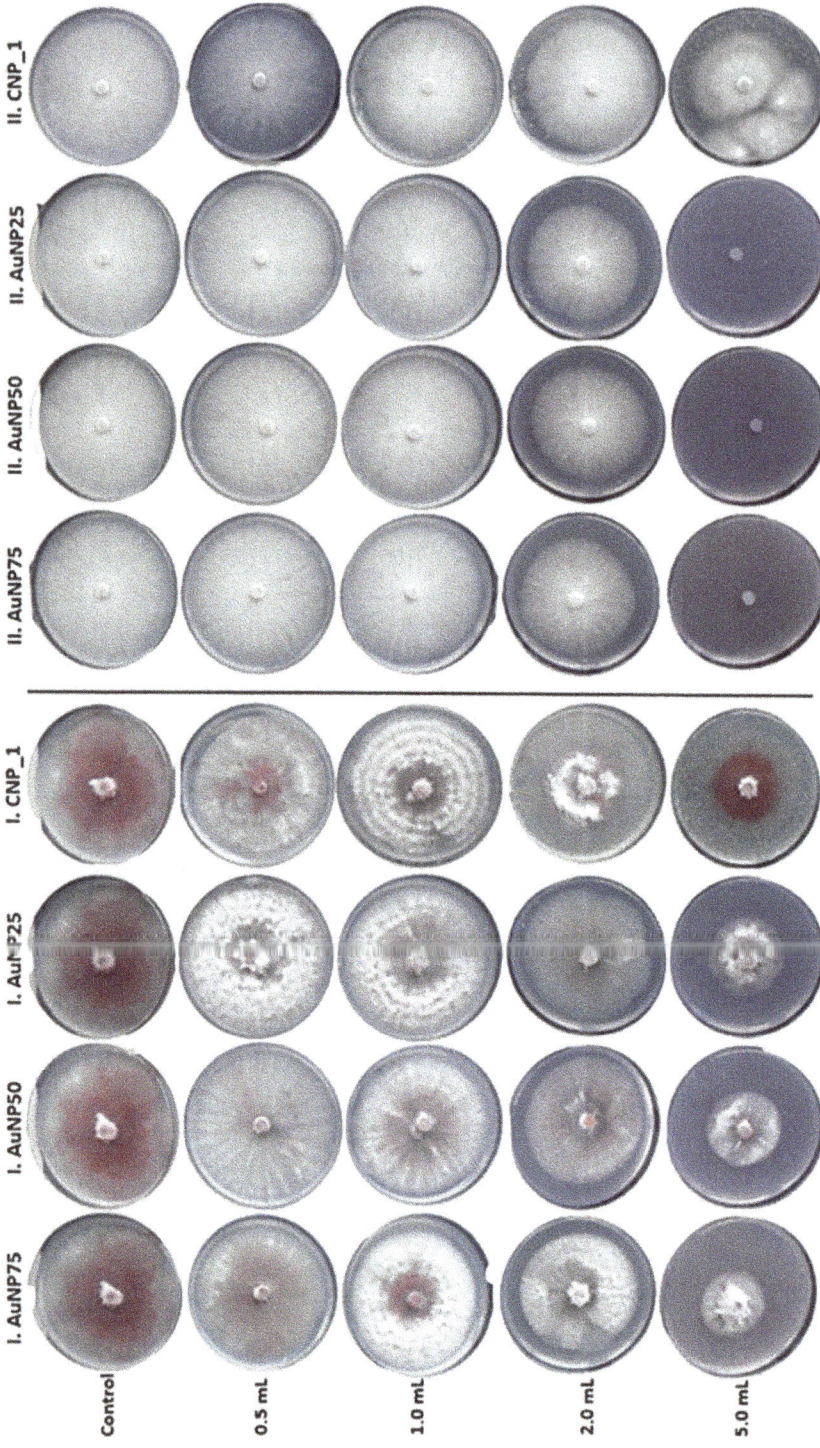

Figure 1. Effects of the interactions of different nanoparticles at different concentrations and doses on the inhibition of mycelial growth of two *Fusarium oxysporum* strains (DSM 62338 (I–F) and DSM 62360 strain (II–F)) adapted from Lipşa et al. (2020) (CC BY 4.0).

lower infection types (IT). Spraying the leaves with nanochitosan both before and after inoculation for 24 hours showed the greatest results in terms of extended incubation and latent periods, as well as reduced infection type.

Chitosan is efficient *in vitro* against *Xanthomonas axonopodis* pv. *poinsettiicola* isolated from *Euphorbia pulcherrima*. It was discovered that the antibacterial action of chitosan included membrane damage and biofilm disruption (Li et al. 2008; Chang et al. 2012). The antibacterial properties of chitosan may be improved by complexing it with the appropriate substances. A wide range of Gram-positive and Gram-negative pathogens have been observed to be sensitive to the antibacterial activity of thymol-loaded chitosan nanoparticles (TCNPs) (Hu et al. 2009). Chitosan nanoparticles and nanocapsules loaded with thyme essential oil have also been shown to have inhibitory action against a variety of foodborne bacterial pathogens (Pecarski et al. 2014). Thymol-loaded chitosan nanoparticles were manufactured and their antibacterial efficacy against *Xanthomonas campestris* pv. *campestris* (Xcc) was evaluated (Sreelatha et al. 2022). Through *in vitro* testing using liquid broth, cell viability, and live dead staining assays, the antibacterial activity of TCNPs against Xcc was determined. Additionally, it was shown that TCNPs directly inhibited the growth of Xcc by preventing the development of biofilms and the generation of xanthomonadin and exopolysaccharides. The ultrastructure results showed that TCNP-treated Xcc cells had damaged membranes, resulting in the leakage of intracellular contents. Analysis using headspace/gas chromatography (GC)-mass spectrometry (MS) revealed modifications in the volatile profile of Xcc cells treated with TCNPs. In Xcc cells cultured with TCNPs, higher levels of carbonyl components (primarily ketones) and the synthesis of novel volatile metabolites were observed.

Chitosan Nanoparticles Induced Immune Responses in Plants

Although the immunomodulatory properties of chitosan are well-known in plants, their nanoparticles have only been studied for biological applications. The capacity and mechanism of chitosan nanoparticles to stimulate and enhance plant immune responses have been examined (Chandra et al. 2015). An increase in defense enzyme activity, activation of defense-related genes, and an increase in total phenolics were all seen after chitosan nanoparticle treatment on the leaves. This resulted in a considerable increase in defense response. The effectiveness of chitosan nanoparticles as an antioxidant and defense enzyme inducer has been investigated widely. Complex gene activation cascades that code for a wide variety of proteins are involved in plants' responses to various influences. The activation of defense genes causes an increase in stress-responsive enzymes and metabolites. The main defense mechanisms include the creation and accumulation of pathogenesis-related (PR) proteins in plants in response to an invading pathogen and/or a stress environment (Ali et al. 2018). According to their functions and properties, PR-proteins have been categorized into 17 families. These families include chitinases, β-1,3-glucanases, thaumatin-like proteins, ribosome-inactivating proteins, peroxidases, defenses, thionins, oxalate oxidase, nonspecific lipid transfer proteins, and oxalate-oxidase-like proteins (Ali et al. 2018). The ribosome-inactivating protein PR-10 (PR-10) naturally inhibits fungus growth. Translation is inhibited by the tobacco isolate PR-10 in *Fusarium oxysporum*, *Erwinia amylovora*, *Pestalotia* sp., and *Pseudomonas solancearum* by releasing adenine residues from the ribosomal and non-ribosomal substrate (Kim et al. 2001). Aside from direct suppression, antimicrobial patterns cause complicated signal transduction responsibilities, including the generation of reactive oxygen species (ROS) (Alscher et al. 2002; Sharma et al. 2012). According to the transcript analysis, increased expression of defense-related genes may be linked to improved plant defense mechanisms due to chitosan nanoparticles comparative efficacy. The ability of chitosan and chitosan nanoparticles to stimulate tomato plant defensive responses against *Fusarium* wilt disease and to encourage resilient plant development in greenhouse environments.

Another explanation for plants exhibiting an increased defensive response might be due to the extracellular location of chitosan nanoparticles. The signaling molecule nitric oxide (NO), which is critical in plant defense, was also shown to elevate in concentration after exposure to chitosan nanoparticles (Chandra et al. 2015). Immunological stimulation by chitosan nanoparticles was dramatically reduced when NO generation was inhibited, suggesting that NO may play a role in this kind of immune activation. Plant tissues treated with chitosan are less likely to be infected by viruses such as *Alfalfa mosaic virus, Peanut mosaic virus, Potato mosaic virus, Tobacco mosaic virus*, and *Cucumber mosaic virus* through the mechanism of induced resistance (Kochkina et al. 1994; Chirkov 2002).

Maize defense and growth have been boosted by the use of SA-chitosan nanoparticles. Physiological and biochemical reactions have been elicited both *in vitro* and *in vivo* by SA-chitosan nanoparticles. Antioxidant defense enzyme activity (catalase, SOD, peroxidase, and others), balanced ROS, disease management, cell wall strengthening through lignin deposition, and maize plant growth were all improved (Kumaraswamy et al. 2019). Protective seed treatment and foliar application of SA-chitosan nanoparticles before flowering may improve plant innate immunity even before pathogen infection leading to development of an effective and preventive strategy to disease (Kumaraswamy et al. 2019). Wheat plants treated with chitosan nanoparticles induced *PR1, PR3,* and *PR4* expression to a much higher extent than those treated with salicylic acid (Elsharkawy et al. 2022). Specifically, transcription levels for *PR3* and *PR4* were almost five times greater in the chitosan nanoparticles treatment than in the salicylic acid treatment. At 2 days after inoculation, chitosan nanoparticles outperformed salicylic acid and a mock-inoculated control in expressing *PR1, PR3, PR4,* and *PR10*. Compared to the mock-inoculated control, chitosan nanoparticles and salicylic acid treatments resulted in roughly eight-fold greater relative expression levels of *PR2* and *PR5*.

Chitosan as a Nanocarrier for Fungicides

Research conducted *in vitro* and on infected palms has shown that the antifungal compounds hexaconazole and dazomet are effective in inhibiting the growth of *Ganoderma boninense* when used together as fungicides. However, the use of fungicides has been reported to influence plant physiology, including growth decrease, carbon metabolism, disruption of reproductive organ development, photosynthesis, and nitrogen changes (Saladin and Clément 2005; Petit et al. 2008; Petit et al. 2012; Ney et al. 2013).

Several biological properties of *Pisum sativum* plants, including anatomy, physiology, cellular damage, and cytotoxicity, were shown to be negatively affected by hexaconazole-induced stress (Shahid et al. 2018). Dazomet has been proven to have a strong phytotoxic impact on oil palm seedlings, as shown by physiological trials (Mustafa et al. 2018). Several hydrophobic fungicides may be solubilized using chitosan nanoparticles' site-specific delivery methods. Therefore, fungicides have increased bioavailability and longer circulation times (Kashyap et al. 2015). Agricultural active ingredients (such as pesticides, fertilizers, fungicides, etc.) have been transported by chitosan nanoparticles, a non-toxic biopolymer, for decades in crop protection (Agrawal and Rathore 2014; Hill and Whitham 2014; Chauhan et al. 2019). Chitosan nanoparticles may also enter plant tissue, allowing for the successful functioning of fungicides to the plant tissue's designated target (Agrawal and Rathore 2014). It is also renowned for its capacity to prevent or limit the spread of disease in plants by suppressing pathogens and enhancing the plant defense mechanisms (Pérez-de-Luque and Rubiales 2009; Ouda 2014). *Ganoderma boninense* fungicide solubility and antifungal activity were increased by chitosan encapsulation (Maluin et al. 2020). Furthermore, agronanofungicides based on chitosan showed greater antifungal efficacy against *Ganoderma boninense* when the particle size was smaller (Maluin et al. 2019a, b, c).

As a potent antimicrobial, elicitor, and growth regulator, chitosan has been studied alone and in combination with other bioactive chemicals of both organic and inorganic origins. Three subcategories of chitosan-based natural products (NMs) are shown here: (a) solely composed of natural material, (b) made from inorganic materials, and (c) made from organic materials.

Application of Nano Chitosan Alone

Recently, chitosan nanoparticles have been studied for their wide range of biological functions. They have been evaluated against a wide range of plant pathogenic fungi and have shown to be efficient in reducing fungal growth. Mycelial growth of *Rhizopus* sp. *Colletotrichum gloeosporioides* and *Colletotrichum capsici* were greatly delayed by 0.6% (w/v) of chitosan nanoparticles during *in vitro* tests (Chookhongkha et al. 2012). In comparison to chitosan, nanoparticles showed a greater ability to reduce mycelia development. Furthermore, chitosan-coated *Cicer arietinum* seeds exhibited increased vigor and extremely strong antifungal activity (Sathiyabama and Parthasarathy 2016). Ionic gelation process was used to make chitosan nanoparticles, which were tested against a variety of phytopathogenic fungi (*Alternaria alternata*, *Macrophomina phaseolina*, and *Rhizoctonia solani*) *in vitro*. *Macrophomina phaseolina* was shown to have the greatest growth inhibition (88%) at 0.1% solution. Chitosan nanoparticles decreased the radial development of *Rhizoctonia solani* at all doses (Saharan et al. 2013). In another research, anionic proteins obtained from *Penicillium oxalicum* cultures were utilized to make chitosan nanoparticles. *Alternaria solani*, *Pyricularia grisea*, and *Fusarium oxysporum* were strongly inhibited by these biologically produced chitosan nanomaterials (Sathiyabama and Parthasarathy 2016). There was a 92% inhibition rate for *Pyricularia grisea*, *Alternaria solani*, and *Fusarium oxysporum ciceri*. *Cicer arietinum* seedlings treated with these nanoparticles showed a beneficial morphological impact, including vegetative biomass, increased germination, and vigor index. The size and great permeability of biological membranes may explain effectiveness of nanoparticles (Saharan et al. 2015). These nanoparticles are more stable and effective against studied phytopathogens because of their compact size, lower PDI value, and greater zeta-potential. Rice blast disease symptoms were completely suppressed, *in vivo*, when chitosan nanoparticles produced by the ionic gelation process were applied to the leaves (Manikandan and Sathiyabama 2016). The antifungal properties of chitosan nanoparticles were tested against *Fusarium graminearum* in wheat. The inhibitory impact of these nanoparticles on this pathogen was evaluated at a variety of doses, with 5,000 ppm showing the greatest growth suppression (77.5%). Treated plants with nanoparticles showed a reduction in the AUDPC in the greenhouse experiments (Kheiri et al. 2017).

New water-soluble chitosan derivatives, such as 2-methyl-1,3,4-thiadiazole, 1,3,4-thiadiazole, and 2-phenyl-1,3,4-thiadiazole, exhibited antifungal activity against plant-threatening fungi, such as *Phomopsis asparagi*, *Monilinia fructicola*, and *Colletotrichum lagenarium* (Li et al. 2013). In the case of *C. lagenarium* growth, the inhibitory index was found to be 31.6% at a concentration of 1.0 mg/ml. In comparison to chitosan, chitosan derivatives showed superior antifungal activity. The compound 2-methyl-1,3,4-thiadiazole was shown to be the most effective among the chitosan derivatives evaluated, with inhibitory values 65–83% against the tested pathogens. The antifungal efficacy of chitosan derivatives is influenced by the hydrophobic moiety and the alkyl substituent length in thiadiazole.

Application of Metal-Based Chitosan Nanoparticles

Chitosan-based nanoparticles have been tested with metals like zinc (Zn), copper (Cu), silver (Ag), and gold (Au) since chitosan is able to chelate the metals efficiently (Choudhary et al. 2017a; Lipşa et al. 2020) (Figure 2). A wide variety of pesticides, herbicides, fungicides, and RNA-inducing compounds are carried by chitosan nanoparticles (Worrall et al. 2018).

As a classic antifungal agent, Cu has long been utilized in several commercial fungicides (Saharan et al. 2015). To control fungal disease in plants, Cu has been used to synthesize novel chitosan-based nanoparticles. Cu-chitosan nanoparticles inhibited the development of *Macrophomina phaseolina*, *Rhizoctonia solani*, and *Alternaria alternata* (Saharan et al. 2013). The synergistic action of chitosan and copper was responsible for the strong antifungal activity. Cu-chitosan nanoparticles were as effective as commercially available fungicides in controlling early blight disease (Saharan et al. 2015). These nanoparticles increased the crop's defensive responses against the *Curvularia* leaf

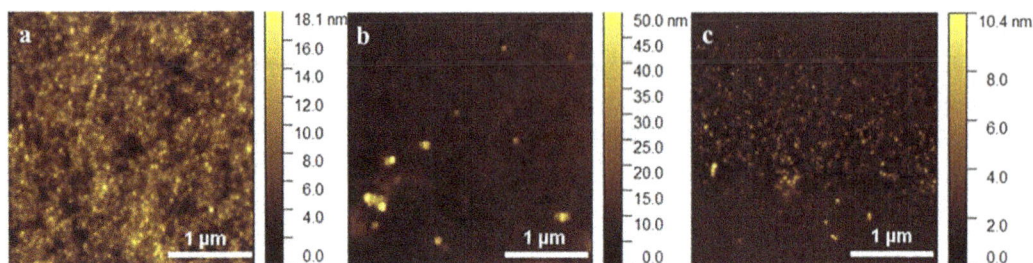

Figure 2. Atomic force microscopy (AFM) images of (a) pure chitosan, (b) gold-chitosan nanoparticles (AuNPs-chitosan), and (c) carbon nanoparticles (CNPs) adapted from Lipşa et al. (2020) (CC BY 4.0).

spot disease through the antioxidant and defensive enzymes (superoxide dismutase and peroxidase) (Choudhary et al. 2017b). In pot studies, Cu-chitosan nanoparticles considerably suppressed the disease at concentrations of 0.04%, whereas the same impact was reported in field conditions at concentrations of 0.12%. Invading fungi are effectively destroyed by the released Cu (Rubina et al. 2017). An *in vitro* antifungal study was conducted on Cu-chitosan nanoparticles and their ability to inhibit hyphal morphology and the production of sclerotia in *Rhizoctonia solani* and *Sclerotium rolfsii*. Nanoparticles proved to be effective against both pathogens in a dose-dependent approach (Rubina et al. 2017). Seed treatment and foliar spray of synthesized Zn-chitosan nanoparticles were used to test their antifungal activity in maize plants. Antifungal efficacy was shown *in vitro* by inhibiting the germination of fungal spores. The increased antioxidant enzymes and increased lignin formation balanced ROS levels generated by these nanoparticles further improved plant immunity. Zn-chitosan nanoparticles demonstrated high encapsulation effectiveness (82%) and a gradual release of Zn ions (Choudhary et al. 2017b; Kumaraswamy et al. 2018). Inhibition of *Alternaria*, *Aspergillus*, and *Rhizoctonia* species was strongest for Ag-chitosan nanoparticles. Therefore, Ag-chitosan nanoparticles might be employed in fungicides to combat seed-borne pathogens (Kaur et al. 2012).

The Application of Encapsulated Essential Oils in Chitosan Nanoparticles

Essential oils are fragrant and volatile substances derived from plants. They may be found in the stems, the leaves, and the fruits of the plant (Oussalah et al. 2006). Phenolic acids and terpenoids are some of the compounds that may be obtained from plants as essential oils (EO). Eos' antimicrobial characteristics made them popular in food production as natural agents (du Plooy et al. 2009). Many studies have demonstrated that nanoparticles functionalized with EOs have excellent antimicrobial activity due to their chemical stability and solubility, degradation of active compounds, and reduced quick evaporation. Encapsulated EOs are more effective against multidrug-resistant bacteria because of their controlled and prolonged release characteristics (Chouhan et al. 2017). Since EOs are hydrophobic, they help to separate lipids in the pathogen's cell membrane and cause the leaking of molecules and ions. EOs are active because of their composition, the functional active ingredients, and their synergistic effects. Chemical stability may be improved by using nanoencapsulation to protect bioactive substances from degradation. It may also enhance their bioactivity because of their subcellular size, which prevents them from interfering with dietary components (Donsi et al. 2011). Many researchers are interested in encapsulating chitosan with EOs because of its biocompatibility, controlled release, and biodegradability (Muzzarelli 2010; Luo et al. 2011).

Beykia et al. (2014) examined encapsulating EO of *Mentha piperita* in chitosan-cinnamic acid nanogel to improve the oil stability and antimicrobial efficacy against *Aspergillus flavus*. They discovered that the extract has outstanding antifungal activities against *A. flavus* due to its encapsulation. Under controlled conditions, the minimal inhibitory concentrations of encapsulated and free essential oil were 500 and 2,100 ppm, respectively. These results indicated the potential of chitosan-cinnamic acid nanogel as an antifungal carrier for essential oil. Luque-Alcaraz et al. (2016) created chitosan nanoparticles that encapsulated pepper essential oil with a 754 ± 7.5 nm size distribution. They evaluated the survival of *Aspergillus parasiticus* spores using varied concentrations of encapsulated essential oil of pepper plants. The survival of pathogen spores was observed to be decreased in all treatments. These findings suggest that adding essential oil into chitosan bionanocomposites is a viable option for preserving both components' antifungal characteristics while reducing essential oil volatilization and activity loss.

Chemical constituents of essential oils include humulene, geraniol caryophyllene, geranyl acetate, selinenes, linalool, limonene, etc. (Rao et al. 2005; Kakaraparthi et al. 2015). *Cymbopogon martini* EO was evaluated for its antifungal efficacy against *Fusarium graminearum* in maize (Kalagatur et al. 2018). The minimum inhibitory and fungicidal concentrations of the EO were determined to be 421.7 ppm and 618.3 ppm, respectively. A morphological alteration in vesicles, protuberance, craters, and rough surfaces was observed in treated macroconidia compared to the control. The production of ROS and lipid peroxidation increased, causing fungus death. Encapsulated nanoparticles of *Cymbopogon martini* EO with a spherical shape of between 455–480 nm in diameter have been produced. Bioactive components of EO were effectively stabilized by chitosan conjugation and successfully generated nanoparticles, according to FTIR technique. The antifungal activity of encapsulated nanoparticles was enhanced by the formation of a stable complex structure. The antifungal and antimycotoxin activity of encapsulated nanoparticles against *F. graminearum* were tested in the laboratory using maize grains. The antifungal activity was improved than the normal EO because of the regulated release from encapsulated nanoparticles of antifungal ingredients (Kalagatur et al. 2018).

Oleoyl-chitosan nanoparticles were tested for their antifungal efficacy against *Verticillium dahlia*, which has no effective treatments yet. The antifungal index of Oleoyl-chitosan nanoparticles was found to be as high as 86.81%, and the hyphae morphology and spore germination were also affected, with crumpled hyphae and spores as well as massive vacuolation of the cytoplasm, thickened cell walls, the disappearance of membrane organelles, and a separation of the cell wall. There was an inhibitory impact at all concentrations examined, which was selectively dependent on the concentration of Oleoyl-chitosan nanoparticles. Compared to the control group, the dry weight of mycelia was substantially lower at pH 5.0. Lower doses of Oleoyl-chitosan nanoparticles predominantly inhibited cell growth, but greater concentrations flocculated cells. Conidia germination and tube development are inhibited by Oleoyl-chitosan nanoparticles' antifungal properties. Furthermore, these nanoparticles with coagulant and flocculant properties might impede spore dispersion (Xing et al. 2017).

Synergistic efficacy against *M. phaseolina*, *A. alternata*, and *R. solani* was tested using chitosan–saponin nanoparticles (Saharan et al. 2013). Chitosan, saponin, and sodium tripolyphosphate interacted to form these nanoparticles using the ionic gelation process. Mycelia growth was 80.9% suppressed at 0.1% w/v concentration of chitosan-saponin nanoparticles, which also demonstrated a dose-dependent impact.

Thymol (an antifungal agent) inhibits the cell's ability to regenerate its energy (Ahmad et al. 2011). Therefore, it might be used in food storage containers as a natural antifungal agent (Mirdehghan and Valerob 2017). Adding chitosan-thymol nanoparticles generated by ionic gelation technique to quinoa protein/chitosan edible films improved their ability to extend the post-harvest life of Cyanococcus and *Piper nigrum*. All dilutions of the chitosan-thymol nanoparticles formulation showed 100% inhibition, while the normal mix of chitosan-thymol exhibited growth inhibition at higher doses only (Medina et al. 2019). Hence, chitosan-thymol nanoparticles were the only therapeutic treatment to exhibit inhibitory effects even at the lowest dosage (10%).

Inoculated rice treated with chitosan-based nanocomposite films containing thyme-peppermint, thyme-oregano, and thyme-tea tree essential oil combinations demonstrated a 51%–77% decrease in fungal growth (*Aspergillus flavus*, *A. parasiticus*, *A. niger*, and *Penicillium chrysogenum*). Bioactive film-coated rice samples remained unchanged after 12 weeks of storage considering the potential release of volatile chemicals (Hossain et al. 2019). In order to maintain the nanocomposite films' physicochemical stability and release qualities, chitosan matrix-integrated EO was necessary.

Cindi et al. (2015) used polyethylene terephthalate punnets that contained thyme oil (TO sachets) and were additionally packed with chitosan/boehmite nanocomposite lidding films. The incidence and severity of brown rot (*Monilinia laxa*) were decreased in *Prunus persica* treated fruits. People preferred treated peach fruits because the look, taste, and natural peach flavor were maintained.

One of the EO identified as a possible natural component for decreasing post-harvest loss in fruits is the extract of *Zataria multiflora* (Sajed et al. 2013). The extract was used to preserve fruit quality and prevent fungal deterioration, although they are rapidly damaged by high temperatures, light, pressure, and oxygen. For certain applications, a controlled release is necessary since they are insoluble in water (Martin et al. 2010). To maximize the advantages of utilizing EO as an antimicrobial agent, the continuous and regulated release is essential. Nanoencapsulation of the extracts in chitosan nanoparticles was used to improve the antifungal efficacy and stability of the oils against *Botrytis cinerea* in strawberries (Mohammadi et al. 2015). *In vitro* experiments also showed that the nanoparticles were released for 40 days in a regulated manner. Encapsulated oils by chitosan nanoparticles had a greater *in vitro* and *in vivo* action against *B. cinerea* compared to unmodified oils.

Conclusion

Complex compounds based on the natural biopolymer chitosan and/or on its derivative chitosan nanoparticles have provided many users with significant benefits in recent years. In particular, these compounds have improved the outcomes for crop production and health. Safe and environmentally friendly, molecules derived from chitosan are widely utilized in plants to control plant diseases. Considerations regarding particle size, concentration, and molecular weight are all critical for developing novel fungicide formulations for use in plant disease control. Therefore, if the appropriate dosage to be treated is known, a lower-concentration solution may be utilized instead of a high-concentration NP solution (which would result in lower manufacturing costs). This chapter summarizes findings from recent research on the impact of nano chitosan applications on the management of plant pathogens. Possible future approaches and unanswered questions for closing the gaps and expanding beyond the current constraints are also considered.

References

Abd El-Aziz, M.H. and Khalil, M.S. 2020. Antiviral and antinematodal potentials of chitosan: Review. Journal of Plant Science and Phytopathology 4: 055–059.

Agrawal, S. and Rathore, P. 2014. Nanotechnology pros and cons to agriculture: a review. Int. J. Curr. Microbiol. App. Sci. 3(3): 43–55.

Ahmad, A., Khan, A., Akhtar, F., Yousuf, S., Xess, I., Khan, L. and Manzoor, N. 2011. Fungicidal activity of thymol and carvacrol by disrupting ergosterol biosynthesis and membrane integrity against Candida. Eur J. Clin. Microbiol. Infect Dis. 30: 41–50.

Akamatsu, K., Kaneko, D., Sugawara, T., Kikuchi, R. and Nakao, S.I. 2010. Three preparation methods for monodispersed chitosan microspheres using the shirasu porous glass membrane emulsification technique and mechanisms of microsphere formation. Industrial and Engineering Chemistry Research 49: 3236–3241.

Alfy, H., Ghareeb, R.Y., Soltan, E. and Farag, D.A. 2020. Impact of chitosan nanoparticles as insecticide and nematicide against *Spodoptera littoralis*, *Locusta migratoria*, and *Meloidogyne incognita*. Plant Cell Biotechnology Molecular Biology 21: 126–140.

Ali, S., Ganai, B.A., Kamili, A.N., Bhat, A.A., Mir, Z.A., Bhat, J.A. et al. 2018. Pathogenesis-related proteins and peptides as promising tools for engineering plants with multiple stress tolerance. Microbiol. Res. 212–213: 29–37.

Alscher, R.G., Erturk, N. and Heath, L.S. 2002. Role of superoxide dismutases (SODs) in controlling oxidative stress in plants. J. Exp. Bot. 53: 1331–1341.

Anitha, A., Sowmya, S., Kumar, P.T.S., Deepthi, S., Chennazhi, K.P., Ehrlich, H., Tsurkan, M. and Jayakumar, R. 2014. Chitin and chitosan in selected biomedical applications. Progress in Polymer Science 39: 1644–1667.

Asghari, F., Jahanshiri, Z., Imani, M., Shams-Ghahfarokhi, M. and Razzaghi-Abyaneh, M. 2016. Chapter 10— Antifungal nanomaterials: Synthesis, properties, and applications. pp. 343–383. *In*: Nanobiomaterials in Antimicrobial Therapy—Applications of Nanobiomaterials. William Andrew Publishing: Kidlington, Oxford, UK. ISBN 978-0-323-42864-4.

Bautista-Baños, S., Hernandez-Lauzardo, A.N., Velazquez-Del Valle, M.G., Hernández-López, M., Barka, E.A., Bosquez-Molina, E. and Wilson, C. 2006. Chitosan as a potential natural compound to control pre and postharvest diseases of horticultural commodities. Crop Protection 25: 108–118.

Beykia, M., Zhaveha, S., Khalilib, S.T., Rahmani-Cheratic, T., Abollahic, A., Bayatd, M., Tabatabaeie, M. and Mohsenifar, A. 2014. Encapsulation of *Mentha piperita* essential oils in chitosan–cinnamic acid nanogel with enhanced antimicrobial activity against *Aspergillus flavus*. Ind. Crop Prod. 54: 310–319.

Boroumand, H., Badie, F., Mazaheri, S., Seyedi, Z.S., Nahand, J.S., Nejati, M., Baghi, H.B., Abbasi-Kolli, M., Badehnoosh, B., Ghandali, M., Hamblin, M.R. and Mirzaei, H. 2021. Chitosan-based nanoparticles against viral infections. Front. Cell. Infect. Microbiol. 11: 643953.

Brunel, F., Gueddari, N.E.E. and Moerschbacher, B.M. 2013. Complexation of copper (II) with chitosan nanogels: toward control of microbial growth. Carbohydr. Polym. 92: 1348–1356.

Burkhanova, G.F., Yarullina, L.G. and Maksimov, I.V. 2007. The control of wheat defense responses during infection with *Bipolaris sorokiniana* by chitooligosaccharides. Russian Journal of Plant Physiology 54: 104–110.

Cao, L., Zhang, H., Cao, C., Zhang, J., Li, F. and Huang, Q. 2016. Quaternized chitosan-capped mesoporous silica nanoparticles as nanocarriers for controlled pesticide release. NMs 6(7): 126.

Chakraborty, M., Hasanuzzaman, M., Rahman, M., Khan, M.A.R., Bhowmik, P., Mahmud, N.U., Tanveer, M. and Islam, T. 2020. Mechanism of plant growth promotion and disease suppression by chitosan biopolymer. Agriculture. https://doi.org/10.3390/agriculture10120624.

Chandra, S., Chakraborty, N., Dasgupta, A. et al. 2015. Chitosan nanoparticles: A positive modulator of innate immune responses in plants. Sci. Rep. 5: 15195. https://doi.org/10.1038/srep15195.

Chang, T. and Kim, B.S. 2012. Application of chitosan preparations for eco-friendly control of potato late blight. Research in Plant Disease 18: 338–348.

Chang, Y., McLandsborough, L. and McClements, D.J. 2012. Physical properties and antimicrobial efficacy of thyme oil nano emulsions: influence of ripening inhibitors. J. Agric. Food Chem. 60: 12056–12063. doi: 10.1021/jf304045a.

Chauhan, P., Singla, K., Rajbhar, M., Singh, A., Das, N. and Kumar, K.A. 2019. Systematic review of conventional and advanced approaches for the control of plant viruses. J. Appl. Biol. Biotechnol. 7: 89–98.

Chirkov, S.N. 2002. The antiviral activity of chitosan (review). Applied Biochemistry and Microbiology 38: 1–8.

Chookhongkha, N., Sopondilok, T. and Photchanachai, S. 2012. Effect of chitosan and chitosan nanoparticles on fungal growth and chilli seed quality. pp. 231–237. *In*: International Conference on Postharvest Pest and Disease Management in Exporting Horticultural Crops-PPDM2012 973. Acta Horticulturae, Bangkok.

Choudhary, R.C., Kumaraswamy, R.V., Kumari, S., Sharma, S.S., Pal, A., Raliya, R., Biswas, P. and Saharan, V. 2017a. Synthesis, characterization, and application of chitosan NMs loaded with zinc and copper for plant growth and protection. *In*: Prasad, R. et al. (eds.). Nanotechnology. Springer Nature Singapore Pte Ltd., Singapore.

Choudhary, R.C., Kumaraswamy, R.V., Kumari, S., Sharma, S.S., Pal, A., Raliya, R., Biswas, P. and Saharan, V. 2017b. Cu-chitosan nanoparticle boost defense responses and plant growth in maize (*Zea mays* L.). Sci. Rep. 7: 9754.

Chouhan, S., Sharma, K. and Guleria, S. 2017. Antimicrobial activity of some essential oils-present status and future perspectives. Medicines 4(3): 58.

Cindi, M.D., Shittu, T., Sivakumar, D. and Bautista-Banos, S. 2015. Chitosan boehmite-alumina nanocomposite films and thyme oil vapour control brown rot in peaches (*Prunus persica* L.) during postharvest storage. Crop Prot 72: 127–131.

Cota-Arriola, O., Cortez-Rocha, M.O., Rosas-Burgos, E.C., Burgos-Hernández, A., López-Franco, Y.L. and Plascencia-Jatomea, M. 2011. Antifungal effect of chitosan on the growth of *Aspergillus parasiticus* and production of aflatoxin B1. Polymer International 60: 937–944.

da Silva, N.M., Cardoso, A.R., Ferreira, D., Pintado, M.M.E. and Vasconcelos, M.W. 2014. Chitosan as a biocontrol agent against the pinewood nematode (*Bursaphelenchus xylophilus*). Forest Pathology 44: 420–423.

Dheyab, M.A., Aziz, A.A., Jameel, M.S., Abu Noqta, O. and Mehrdel, B. 2020. Synthesis and coating methods of biocompatible iron oxide/gold nanoparticle and nanocomposite for biomedical applications. Chin. J. Phys. 64: 305–325. doi: 10.1016/j.cjph.2019.11.014.

Donsi, F., Annunziata, M., Sessa, M. and Ferrari, G. 2011. Nanoencapsulation of essential oils to enhance their antimicrobial activity in foods. LWT-Food Sci. Technol. 44: 1908–1914.

du Plooy, W., Regnier, T. and Combrinck, S. 2009. Essential oil amended coatings as alternative to synthetic fungicides in citrus postharvest management. Postharvest Biol. Technol. 53: 117–122.

Du, W.L., Xu, Y.L., Xu, Z.R. and Fan, C.L. 2008. Preparation, characterization and antibacterial properties against *E. coli* K_{88} of chitosan nanoparticle loaded copper ions. Nanotechnology 19: 085707. doi: 10.1088/0957-4484/19/8/085707.

Eilenberg, H., Pnini-Cohen, S., Rahamim, Y., Sionov, E., Segal, E., Carmeli, S. and Zilberstein, A. 2009. Induced production of antifungal naphthoquinones in the pitchers of the carnivorous plant Nepenthes khasiana. Journal of Experimental Botany 61: 911–922.

El Ghaouth, A., Arul, J., Wilson, C. and Benhamou, N. 1994. Ultrastructural and cytochemical aspects of the effect of chitosan on decay of bell pepper fruit. Physiological and Molecular Plant Pathology 44: 417–432.

El-Naggar, N.EA., Saber, W.I.A., Zweil, A.M. and Bashir, S.I. 2022. An innovative green synthesis approach of chitosan nanoparticles and their inhibitory activity against phytopathogenic *Botrytis cinerea* on strawberry leaves. Sci. Rep. 12: 3515. https://doi.org/10.1038/s41598-022-07073-y.

El-Sayed, S.M. and Mahdy, M.E. 2015. Effect of chitosan on root-knot nematode, *Meloidogyne javanica* on tomato plants. International Journal of Chemistry and Technology Research 7: 1985–1992.

El Shafie, E., Daffalla, G., Gebre, K. and Marchoux, G. 2005. Mosaic-inducing viruses and virus like- agents infecting tomato and pepper in Sudan. International Journal of Virology 1: 28–28.

Elsharkawy, M.M., Omara, R.I., Mostafa, Y.S., Alamri, S.A., Hashem, M., Alrumman, S.A. and Ahmad, A.A. 2022. Mechanism of wheat leaf rust control using chitosan nanoparticles and salicylic acid. J. Fungi 8: 304. https://doi.org/10.3390/jof8030304.

Gedda, G., Lee, C.-Y., Lin, Y.-C. and Wu H.-F. 2016. Green synthesis of carbon dots from prawn shells for highly selective and sensitive detection of copper ions. Sens. Actuators B Chem. 224: 396–403. doi: 10.1016/j.snb.2015.09.065.

Hadwiger, L.A. 2013. Multiple effects of chitosan on plant systems: Solid science or hype. Plant Science 208: 42–49.

Hill, J. and Whitham, S. 2014. Control of virus diseases in soybeans. Adv. Virus Res. 9: 355–390.

Hossain, F., Follett, P., Salmieri, S., Vu, K.D., Fraschini, C. and Lacroix, M. 2019. Antifungal activities of combined treatments of irradiation and essential oils (EOs) encapsulated chitosan nanocomposite films in *in vitro* and *in situ* conditions. Int. J. Food Microbiol. 295: 33–40.

Huang, L., Cheng, X., Liu, C., Xing, K., Zhang, J., Sun, G., Li, X. and Chen, X. 2009. Preparation, characterization, and antibacterial activity of oleic acid-grafted chitosan oligosaccharide nanoparticles. Frontiers of Biology in China 4: 321–327.

Ing, L.Y., Zin, N.M., Sarwar, A. and Katas, H. 2012. Antifungal activity of chitosan nanoparticles and correlation with their physical properties. Int. J. Biomater. 2012: 632698.

Jia, X., Meng, Q., Zeng, H., Wang, W. and Yin, H. 2016. Chitosan oligosaccharide induces resistance to tobacco mosaic virus in *Arabidopsis* via the salicylic acid-mediated signaling pathway. Scientific Reports 6: 26144. https://doi.org/10.1038/srep26144.

Kakaraparthi, P.S., Srinivas, K.V.N.S., Kumar, J.K., Kumar, A.N., Rajput, D.K. and Anubala, S. 2015. Changes in the essential oil content and composition of palmarosa (*Cymbopogon martinii*) harvested at different stages and short intervals in two different seasons. Ind. Crops Prod. 69: 348–354.

Kalagatur, N.K., Nirmal Ghosh, O.S., Sundararaj, N. and Mudili, V. 2018. Antifungal activity of chitosan nanoparticles encapsulated with *Cymbopogon martinii* essential oil on plant pathogenic Fungi *Fusarium graminearum*. Front Pharmacol. 9: 610.

Kalaiarasan, P., Lakshmanan, P., Rajendran, G. and Samiyappan, R. 2006. Chitin and chitinolytic biocontrol agents for themanagement of root knot nematode, *Meloidogyne arenaria* in groundnut (*Arachis hypogaea* L.) cv. Co3. Indian Journal of Nematology 36: 181–186.

Kashyap, P.L., Xiang, X. and Heiden, P. 2015. Chitosan nanoparticle based delivery systems for sustainable agriculture. Int. J. Biol. Macromol. 77: 36–51.

Kaur, P., Thakur, R. and Choudhary, A. 2012. An *in vitro* study of the antifungal activity of silver/chitosan nanoformulations against important seed borne pathogens. Int. J. Sci. Technol. Res. 1(7): 83–86.

Ke, C.L., Deng, F.S., Chuang, C.Y. and Lin, C.H. 2021. Antimicrobial actions and applications of chitosan. Polymers 13: 904. doi: 10.3390/polym13060904.

Khalil, M.S. and Badawy, M.E.I. 2012. Nematicidal activity of biopolymer chitosan at different molecular weights against root-knot nematode, *Meloidogyne incognita*. Plant Protection Science 48(4): 170–178.

Khalil, M.S. 2016. Utilization of biomaterials as soil amendments and crop protection agents in integrated nematodes management. pp. 203–224. *In*: Hakeem, K.R., Akhtar, M.S. and Abdullah, S.N.A. (eds.). Plant, Soil and Microbes. 1: Interactions and Implications in Crop Science. Springer International Publishing.

Khalil, M.S., Abd El-Aziz, M.H. and Selim, R.ES. 2022. Physiological and morphological response of tomato plants to nano-chitosan used against bio-stress induced by root-knot nematode (*Meloidogyne incognita*) and *Tobacco mosaic tobamovirus* (TMV). Eur. J. Plant Pathol. 163: 799–812. https://doi.org/10.1007/s10658-022-02516-8.

Kheiri, A., Jorf, S.M., Malihipour, A., Saremi, H. and Nikkhah, M. 2017. Synthesis and characterization of chitosan nanoparticles and their effect on Fusarium head blight and oxidative activity in wheat. Int. J. Biol. Macromol. 102: 526–538.

Kim, D., Lee, D.G., Kim, K.L. and Lee, Y. 2001. Internalization of tenecin 3 by a fungal cellular process is essential for its fungicidal effect on *Candida albicans*. Eur. J. Biochem. 268: 4449–4458.

Kochkina, Z., Pospeshny, G. and Chirkov, S. 1994. Inhibition by chitosan of productive infection of T-series bacteriophages in the *Escherichia coli* culture. Microbiology 64: 211–215.

Kumaraswamy, R.V., Kumari, S., Choudhary, R.C., Pal, A., Raliya, R., Biswas, P. and Saharan, V. 2018. Engineered chitosan based nanomaterials: Bioactivities, mechanisms and perspectives in plant protection and growth. Int. J. Boil. Macromol. 113: 494–506. doi: 10.1016/j.ijbiomac.2018.02.130.

Kumaraswamy, R.V., Kumari, S., Choudhary, R.C., Sharma, S.S., Pal, A., Raliya, R. and Saharan, V. 2019. Salicylic acid functionalized chitosan nanoparticle: a sustainable biostimulant for plant. Int. J. Biol. Macromol. 123: 59–69.

Kutawa, A.B., Ahmad, K., Ali, A., Hussein, M.Z., AbdulWahab, M.A., Adamu, A. et al. 2021. Trends in nanotechnology and its potentialities to control plant pathogenic fungi: a review. Biology 10: 881. doi: 10.3390/biology10090881.

Li, B., Wang, X., Chen, R., Huangfu, W.G. and Xie, G.L. 2008. Antibacterial activity of chitosan solution against *Xanthomonas pathogenic* bacteria isolated from *Euphorbia pulcherrima*. Carbohyd. Polym. 72: 287–292. doi: 10.1016/j.carbpol.2007.08.012.

Li, P., Cao, Z., Wu, Z., Wang, X. and Li, X. 2016. The effect and action mechanisms of oligochitosan on control of stem dry rot of *Zanthoxylum bungeanum*. International Journal of Molecular Sciences 17: 1044.

Li, Q., Ren, J., Dong, F., Feng, Y., Gu, G. and Guo, Z. 2013. Synthesis and antifungal activity of thiadiazole functionalized chitosan derivatives. Carbohydr. Res. 373: 103–107.

Lin, W., Hu, X., Zhang, W., Rogers, W.J. and Cai, W. 2005. Hydrogen peroxide mediates defence responses induced by chitosans of different molecular weights in rice. Journal of Plant Physiology 162: 937–944.

Lipşa, F-D., Ursu, E-L., Ursu, C., Ulea, E. and Cazacu, A. 2020. Evaluation of the antifungal activity of gold–chitosan and carbon nanoparticles on *Fusarium oxysporum*. Agronomy 10(8): 1143. https://doi.org/10.3390/agronomy10081143.

Luo, Y., Zhang, B., Whent, M., Yu, L. and Wang, Q. 2011. Preparation and characterization of zein/chitosan complex for encapsulation of tocopherol and its *in vitro* controlled release study. Colloids Surf B: Biointerfaces 85: 145–152.

Luque-Alcaraz, A.G., Cortez-Rocha, M.O., Velázquez-Contreras, C.A., Acosta-Silva, A.L., SantacruzOrtega, H.D.C., Burgos-Hernández, A., Argüelles-Monal, W.M. and Plascencia-Jatomea, M. 2016. Enhanced antifungal effect of chitosan/pepper tree (*Schinus molle*) essential oil bionanocomposites on the viability of *Aspergillus parasiticus* spores. Journal of Nanomaterials 2016: 1–10.

Malerba, M. and Cerana, R. 2016. Chitosan effects on plant systems review. International Journal of Molecular Sciences 17: 996.

Maluin, F.N., Hussein, M.Z., Yusof, N.A., Fakurazi, S., Idris, A.S., Hilmi, N.H.Z. et al. 2019a. Preparation of chitosan–hexaconazole nanoparticles as fungicide nanodelivery system for combating *Ganoderma* disease in oil palm. Molecules 24(13): 2498.

Maluin, F.N., Hussein, M.Z., Yusof, N.A., Fakurazi, S., Idris, A.S., Hilmi, N.H.Z. et al. 2019b. A potent antifungal agent for basal stem rot disease treatment in oil palms based on chitosan-dazomet nanoparticles. Int. J. Mol. Sci. 20(9): 2247.

Maluin, F.N., Hussein, M.Z., Yusof, N.A., Fakurazi, S., Abu Seman, I., Zainol Hilmi, N.H. et al. 2019c. Enhanced fungicidal efficacy on *Ganoderma boninense* by simultaneous co-delivery of hexaconazole and dazomet from their chitosan nanoparticles. RSC Adv. 9(46): 27083–27095.

Maluin, F.N., Hussein, M.Z., Yusof, N.A., Fakurazi, S., Idris, A.S., Hilmi, N.H.Z. et al. 2020. Phytotoxicity of chitosan-based agronanofungicides in the vegetative growth of oil palm seedling. PLoS ONE 15(4): e0231315. https://doi.org/10.1371/journal.pone.0231315.

Manikandan, A. and Sathiyabama, M. 2016. Preparation of chitosan nanoparticles and its effect on detached rice leaves infected with Pyricularia grisea. Int. J. Biol. Macromol. 84: 58–61.

Martin, A., Varona, S., Navarrete, A. and Cocero, M.J. 2010. Encapsulation and co-precipitation processes with supercritical fluids: applications with essential oils. Open Chem. Eng. J. 4(1).

Martínez-Camacho, A., Cortez-Rocha, M., Ezquerra-Brauer, J., Graciano-Verdugo, A., Rodriguez-Félix, F., Castillo-Ortega, M., Yépiz-Gómez, M. and Plascencia-Jatomea, M. 2010. Chitosan composite films: Thermal, structural, mechanical and antifungal properties. Carbohydrate Polymers 82: 305–315.

Mathew, T.V. and Kuriakose, S. 2013. Photochemical and antimicrobial properties of silvernanoparticle-encapsulated chitosan functionalized with photoactive groups. Mater Sci. Eng. C 33(7): 4409–4415.

Medina, E., Caro, N., Abugoch, L., Gamboa, A., Diaz-Dosque, M. and Tapia, C. 2019. Chitosan thymol nanoparticles improve the antimicrobial effect and the water vapour barrier of chitosan-quinoa protein films. J. Food Eng. 240: 191–198.

Meng, D., Garba, B., Ren, Y., Yao, M., Xia, X., Li, M. et al. 2020. Antifungal activity of chitosan against *Aspergillus ochraceus* and its possible mechanisms of action. Int. J. Biol. Macromol. 158: 1063–1070. doi: 10.1016/j. ijbiomac.2020.04.213.

Mirdehghan, H.S. and Valerob, D. 2017. Bioactive compounds in tomato fruit and its antioxidant activity as affected by incorporation of Aloe, eugenol, and thymol in fruit package during storage. Int. J. Food Prop. 20: 798–806.

Mohammadi, A., Hashemi, M. and Hosseini, S.M. 2015. Nanoencapsulation of *Zataria multiflora* essential oil preparation and characterization with enhanced antifungal activity for controlling *Botrytis cinerea*, the causal agent of gray mould disease. Innov. Food Sci. Emerg. Technol. 28: 73–80.

Moussa, S.H., Tayel, A.A., Alsohim, A.S. and Abdallah, R.R. 2013. Botryticidal activity of nanosized silverchitosan composite and its application for the control of gray mold in strawberry. J. Food Sci. 78(10): M1589–M1594.

Mustafa, I.F., Hussein, M.Z., Seman, I.A., Hilmi, N.H.Z. and Fakurazi, S. 2018. Synthesis of dazomet-zinc/aluminum-layered double hydroxide nanocomposite and its phytotoxicity effect on oil palm seed growth. ACS Sustain. Chem. Eng. 6(12): 16064–16072.

Muzzarelli, R.A.A. 2010. Chitins and chitosans as immunoadjuvants and nonallergenic drug carriers. Mar. Drugs 8: 292–312.

Nagorskaya, V., Reunov, A., Lapshina, L., Davydova, V. and Yermak, I. 2014. Effect of chitosan on *Tobacco mosaic virus* (TMV) accumulation, hydrolase activity and morphological abnormalities of the viral particles in leaves of *N. tabacum* L. cv. samsun. Virologica Sinica 29: 250–256.

Naskar, S., Sharma, S., Kuotsu, K. and Kuotsu, K. 2019. Chitosan-based nanoparticles: An overview of biomedical applications and its preparation. J. Drug Deliv. Sci. Technol. 49: 66–81. doi: 10.1016/j.jddst.2018.10.022.

Ney, B., Bancal, M.O., Bancal, P., Bingham, I., Foulkes, J., Gouache, D. et al. 2013. Crop architecture and crop tolerance to fungal diseases and insect herbivory. Mechanisms to limit crop losses. Eur. J. Plant Pathol. 135(3): 561–580.

Orzali, L., Forni, C. and Riccioni, L. 2014. Effect of chitosan seed treatment as elicitor of resistance to *Fusarium graminearum* in wheat. Seed Science and Technology 42: 132–149.

Ouda, S.M. 2014. Antifungal activity of silver and copper nanoparticles on two plant pathogens, *Alternaria alternata* and *Botrytis cinerea*. Res. J. Microbiol. 9(1): 34–42.

Oussalah, M., Caillet, S., Saucier, L. and Lacroix, M. 2006. Inhibitory effects of selected plant essential oils on the growth of four pathogenic bacteria: *E. coli* O157:H7, *Salmonella typhimurium*, *Staphylococcus aureus* and *Listeria monocytogenes*. Food Control 18: 414–420.

Park, Y., Kim, M.H., Park, S.C., Cheong, H., Jang, M.K., Nah, J.W. and Hahm, K.S. 2008. Investigation of the antifungal activity and mechanism of action of LMWS-chitosan. Journal of Microbiology and Biotechnology 10. 1729 1731.

Petit, A.N., Fontaine, F., Clément, C. and Vaillant-Gaveau, N. 2008. Photosynthesis limitations of grapevine after treatment with the fungicide fludioxonil. J. Agric. Food Chem. 56(15): 6761–6767. pmid:18598040.

Petit, A.N., Fontaine, F., Vatsa, P., Clément, C. and Vaillant-Gaveau, N. 2012. Fungicide impacts on photosynthesis in crop plants. Photosynth. Res. 111(3): 315–326. pmid:22302592.

Pérez-de-Luque, A. and Rubiales, D. 2009. Nanotechnology for parasitic plant control. Pest Manag. Sci. 65(5): 540–545. pmid:19255973.

Rabea, E.I., Badawy, M.E.T., Stevens, C.V., Smagghe, G. and Steurbaut, W. 2003. Chitosan as antimicrobial agent: applications and mode of action. Biomacromolecules 4: 1457–1465. doi: 10.1021/bm034130m.

Rao, B.R., Kaul, P.N., Syamasundar, K.V. and Ramesh, S. 2005. Chemical profiles of primary and secondary essential oils of palmarosa (*Cymbopogon martinii* (Roxb.) Wats var. motia Burk.). Ind. Crops Prod. 21: 121–127.

Rubina, M.S., Vasil'kov, A.Y., Naumkin, A.V., Shtykova, E.V., Abramchuk, S.S., Alghuthaymi, M.A. and AbdElsalam, K.A. 2017. Synthesis and characterization of chitosan–copper nanocomposites and their fungicidal activity against two sclerotia-forming plant pathogenic fungi. J. Nanostruct. Chem. 7(3): 249–258.

Saharan, V., Mehrotra, A., Khatik, R., Rawal, P., Sharma, S.S. and Pal, A. 2013. Synthesis of chitosan based nanoparticles and their *in vitro* evaluation against phytopathogenic fungi. Int. J. Biol. Macromol. 62: 677–683.

Saharan, V., Sharma, G., Yadav, M., Choudhary, M.K., Sharma, S.S., Pal, A., Raliya, R. and Biswas, P. 2015. Synthesis and *in vitro* antifungal efficacy of cu-chitosan nanoparticles against pathogenic fungi of tomato. Int. J. Biol. Macromol. 75: 346–353.

Sajed, H., Sahebkar, A. and Iranshahi, M. 2013. Zataria multiflora Boiss. (Shirazi thyme)-an ancient condiment with modern pharmaceutical uses. J. Ethnopharmacol. 145(3): 686–698.

Saladin, G. and Clément, C. 2005. Physiological side effects of pesticides on non-target plants. Agriculture and Soil Pollution: New Research p. 53–86.

Sathiyabama, M. and Parthasarathy, R. 2016. Biological preparation of chitosan nanoparticles and its *in vitro* antifungal efficacy against some phytopathogenic fungi. Carbohydr. Polym. 151: 321–325.

Shahid, M., Ahmed, B., Zaidi, A. and Khan, M.S. 2018. Toxicity of fungicides to *Pisum sativum*: a study of oxidative damage, growth suppression, cellular death and morpho-anatomical changes. RSC Adv. 8(67): 38483–38498.

Sharma, P., Jha, A.B., Dubey, R.S. and Pessarakli, M. 2012. Reactive oxygen species, oxidative damage, and antioxidative defense mechanism in plants under stressful conditions. J. Bot. 2012: 1–26.

Sreelatha, S., Kumar, N., Si, Yin, T. and Rajani, S. 2022. Evaluating the antibacterial activity and mode of action of Thymol-Loaded chitosan nanoparticles against plant bacterial pathogen *Xanthomonas campestris* pv. *campestris*. Front. Microbiol. 2022 https://doi.org/10.3389/fmicb.2021.792737.

Vokhidova, N.R., Sattarov, M.E., Kareva, N.D. and Rashidova, S.S. 2014. Fungicide features of the nanosystems of silkworm (*Bombyx mori*) chitosan with copper ions. Microbiology 83(6): 751–753.

Westerdahl, B.B., Carlson, H.L., Grant, J., Radewald, J.D.,Welch, N., Anderson, C.A., Darso, J., Kirby, D. and Shibuya, F. 1992. Management of plant-parasitic nematodes with a chitin-urea soil amendment and other materials. Journal of Nematology 24: 669–680.

Worrall, E.A., Hamid, A., Mody, K.T., Mitter, N. and Hanu, H.R. 2018. Nanotechnology for plant disease management. Agronomy 8: 285. doi: 10.3390/agronomy8120285.

Xing, K., Zhu, X., Peng, X. and Qin, S. 2015. Chitosan antimicrobial and eliciting properties for pest control in agriculture: A review. Agronomy for Sustainable Development 35: 569–588.

Xing, K., Liu, Y., Shen, X., Zhu, X., Li, X., Miao, X. and Qin, S. 2017. Effect of O-chitosan nanoparticles on the development and membrane permeability of Verticillium dahliae. Carbohydr. Polym. 165: 334–343.

Yin, H., Zhao, X. and Du, Y. 2010. Oligochitosan: A plant diseases vaccine—a review. Carbohydrate Polymers 82: 1–8.

Zahid, N., Ali, A., Manickam, S., Siddiqui, Y., Alderson, P.G. and Maqbool, M. 2014. Efficacy of curative applications of SCD on anthracnose intensity and vegetative growth of dragon fruitplants. Crop Prot. 62: 129–134. doi: 10.1016/j.cropro.2014.04.010.

Zahid, N., Maqbool, M., Siddiqui, Y., Manickam, S. and Ali, A. 2015. Regulation of inducible enzymes and suppression of anthracnose using submicron chitosan dispersions. Sci. Hort. 193: 381–388. doi: 10.1016/j.scienta.2015.07.014.

Zhang, D., Wang, H., Hu, Y. and Liu, Y. 2015. Chitosan controls postharvest decay on cherry tomato fruit possibly via the mitogen-activated protein kinase signaling pathway. Journal of Agricultural and Food Chemistry 63(33): 7399–7404. https://doi.org/10.1021/acs.jafc.5b01566.

Zhou, H.Y., Zhou, D.J., Zhang, W.F., Jiang, L.J., Li, J.B. and Chen, X.G. 2011. Biocompatibility and characteristics of chitosan/cellulose acetate microspheres for drug delivery. Frontiers of Materials Science 5: 367–378.

Chapter 13

Biological Nano-Selenium in the Service of Plant Health

Farnoush Asghari-Paskiabi,[1,]* *Mohammad Imani*[2,3] and
Mehdi Razzaghi-Abyaneh[1]

Introduction

Selenium (Se) is called an essential poison because selenium deficiency can lead to diseases but an excessive amount of it can be toxic. In the agroecosystem, the amount of trace elements like selenium is less than or equal to 100 µg/g. At the same time, selenium is considered by many to be a significant risk to the health of the environment and living beings, as it has caused abnormalities and death of wildlife. There are several reasons contributing to environmental pollution with selenium, including sewage, fertilizers, and the frequent use of chemicals enriched with the metal. Some plants accumulate selenium up to 1% of dry weight, probably as a defense mechanism. Plants can improve selenium pollution in seleniferous areas and can also be used in Se-deficient areas to compensate for dietary selenium deficiency (El-Ramady et al. 2015).

All living aerobic organisms benefit from defense strategies to protect themselves against oxidative damage due to reactive oxygen species (ROS). Selenium as a part of the antioxidant defense process works either through the removal of free radicals or by incorporating in the chemical structure of the selenoenzyme family. The most famous enzyme is superoxide dismutase. Selenium is part of several selenoproteins, some of which play an important role in antioxidant defense activity. For example, thioredoxin reductase, a vital enzyme in selenium metabolism, both regenerates selenium compounds and regulates the intracellular redox state. Selenoprotein P is another selenoprotein that protects endothelial cells from damage caused by free radicals. The proper expression of these selenoproteins requires sufficient amounts of Se in the diet. In a population view, there is variation between individuals in terms of selenoprotein expression level due to single nucleotide polymorphism of selenoprotein genes. This diversity determines the efficiency of Se incorporation into selenoproteins (Zinicovscaia et al. 2016). It was first established in 1941 that selenium toxicity is due to its interaction with thiols (Painter 1941), which was further investigated later on. In fact, selenium sulfide is an active catalyst that oxidizes GSH, to form selenodiglutathione (GSSeSG). The toxicity of Se is due to the interaction of Se with disulfides and thiol functional

[1] Department of Mycology, Pasteur Institute of Iran, Tehran, 1316943551, Iran.
[2] Novel Drug Delivery Systems Dept., Iran Polymer and Petrochemical Institute, P.O. Box 14975-112, Tehran, Iran.
[3] Center for Nanoscience and Nanotechnology, Institute for Convergence Science & Technology, Sharif University of Technology, Tehran 14588-89694, Iran.
* Corresponding author: F_asghari@pasteur.ac.ir; asghari_393@yahoo.com

groups of selenotrisulfide-forming proteins (RSSeSR), which are similar to GSSeSG, and inhibit the activity of the enzyme (Zinicovscaia et al. 2016).

Interaction of Selenium and Plants

Selenium can be obtained from plant sludge through the process of acid treatment, typically involving the use of sulfuric acid. It is also a byproduct of lead (Pb) and copper (Cu) metals extraction; hence, there is no special mine for selenium. Selenium is continuously circulating in the environment through the land, water, and atmospheric systems, but human activities are also an important source of entering selenium in the cycles. This ends in a non-uniform distribution of selenium on the earth with varying amounts in regions, depending on different local conditions. For example, adding selenium as sodium selenite or selenate to the soil for agricultural and horticultural applications significantly affects the selenium balance in the ecosystem.

Selenium is not found naturally as elemental selenium, but most often it is found together with sulfur (S) and tellurium (Te). The stable form of Se^0 may be seen in the soil; however, due to its lack of solubility in water, it does not exist in aquatic areas at all. Gray selenium (trigonal) is the most stable form in terms of thermodynamics. At a temperature higher than 30°C, red selenium gradually turns into black selenium, and then depending on the pH and redox status of the soil, it turns into a stable gray form or oxidizes at a low speed. It has been observed that the use of plants grown in seleniferous soils causes poisoning and abnormalities. However, on the other hand, its consumption is essential in human health due to the role of selenium in glutathione peroxidase activity or even in reducing the risk of cancer. Knowing the physicochemical characteristics that regulate the selenium cycle is important due to the impact of selenium on plant, animal, and human health. The lowest level of selenium content is found in sandy soils in humid areas and the highest in calcareous and organic soils. Despite the fact of heterogeneous distribution of selenium in the environment, it is known as a rare element; its average concentration is in the range of 0.05 mg.kg^{-1} to 0.09 mg.kg^{-1}. Selenium is mostly present in the form of selenate [Se (VI)] in soil, water, and minerals. However, it is mostly in the form of selenomethionine (SeM) in plants.

The amount of selenium circulating in nature is constantly changed by various factors. Combustion of fossil fuels, forest fires, soil absorption, weathering of soils and rocks, hot springs, soil washing, groundwater movements, chemical and biological disposal, waste burning, other metals production processes and methods of watering and growth of plants in agriculture all affect the availability of this element. Considering the fact of effects of selenium on plants and microorganisms, these organisms are effective in cleaning selenium from nature in an opposite manner. The availability of selenium for plants is mostly in the form of selenite (+6) and least in the form of selenide. Plants absorb selenate (+6) much faster than selenite (+4). Selenium in water environments is affected by pH and redox reactions. Living organisms in water can create stable forms of Se^0 by reducing the oxyanions. In this way, selenium is separated from water. These processes take place in the upper layer of sediments. However, in polluted waters, there is a possibility that the regenerated selenium to be re-oxidized and re-dissolved in water, keeping the level of selenium high for years. They also accumulate in algae and other plants and enter the food cycle. Normal water usually contains less than 1 μg.L^{-1} of selenium. Evaporation of sea and ocean surface waters, as well as industrial emissions and volcanic eruptions, also cause selenium to enter the atmosphere (El-Ramady et al. 2015).

The interactions occurring between plants and nanoparticles may be physical or chemical. In chemical interactions, oxidative damage, production of ROS, lipid peroxidation, and disruption of cell membrane transport activity may occur. After entering, nanoparticles like metal ions react with carboxyl and sulfhydryl functional groups and disrupt the activity of proteins. The studies conducted on *Arabidopsis thaliana* roots in contact with TiO_2 and ZnO nanoparticles support the correlation between phytotoxicity mechanisms and the type of nanoparticles. Gene expression analyses of tomato roots and leaves exposed to carbon nanotubes (CNTs) indicated up-regulation of water channel and stress genes. The toxicity of silver nanoparticles at the early-stage soybean growth showed growth facilitation by 15 nm particles, compared to 2 nm and 50–80 nm particles (Hossain et al. 2015).

Selenium Balance in Plants, Animals, and Human Health

Selenium possesses different oxidation states like sulfur especially when it is combined with three-fold oxygen atoms (selenite, SeO_3, and Se^{4+}) or four-fold oxygen atoms (selenate, SeO_4, and Se^{6+}). Divalent selenium contributes to organic structures, such as selenocysteine or selenomethionine. Probably, the anticarcinogenic action of Se compounds, as well as their toxicity, is due to the catalytic nature of the selenide anion (RSe^-), which subsequently causes oxidative stress and induces apoptosis (Letavayová et al. 2006). In combination with sulfur, selenium increases the activity of sulfur to a great extent. As an intramolecular catalyst, the selenium in sulfide-selenium can take the place of sulfur in cyclic compounds and increase the therapeutic function of sulfur. Elemental sulfur benefits from keratolytic, parasiticidal, fungicidal, and microbicidal properties that have been known and used for years. Mechanisms of the action of sulfur are unclear; sulfur alone is biologically active and is converted to simple inorganic compounds before performing the biological activity. Interestingly, close association with selenium makes sulfur more reactive, and oxidation or reduction occurs (Mitchell et al. 1993).

The use of selenium for supplementing livestock feed has always been discussed. The appropriate amount of selenium is not known either in terms of the host's response or in terms of its consequent environmental safety. Selenium plays an important role in livestock reproduction and growth and is known as an effective agent to prevent infectious diseases. Selenium may be used by injection or orally as feed supplements for ruminants. In Finland and New Zealand with low-Se soils, fertilizer is added to Se-deficient soils to increase the amount of selenium in fodder and animal feed. Plants with high levels of selenium grow only on seleniferous soils (Rosenfield and Beath 1964). Selenium may exist as elemental, sodium selenate, or sodium selenite in food supplements. The accumulation of selenium in plants can lead to toxicity, which is called alkali disease, and it often occurs in alkaline soils having a high concentration of selenium available to plants. This high concentration has nothing to do with the use of selenium in animal feed. Usually, in such a condition, the soil must be removed due to the low yield of agriculture. There has been a suspicion that selenium might be carcinogenic, so its use in animal feed was avoided in some areas where the soil was deficient in selenium; however, this hypothesis has not been confirmed according to well-designed research yet. About 60% of the consumed selenium returns to the environment through animal feces and urine (Oldfield et al. 1994). The need of plants to selenium has been proven. Nevertheless, not many studies have been conducted regarding the possible correlation between selenium deficiency, plant diseases, or disease control by selenium. This essential trace element is completely toxic in high concentrations due to the replacement of sulfur in proteins (Wu et al. 2016).

In conclusion, it is a challenge to establish a balance between the minimum amount of selenium needed and the amount that causes toxicity. Selenium increases the bioactivity and safety of chemical insecticides. Sodium selenite and selenium sulfide can prevent the growth of some pathogens. A mixture of sodium selenite and dithane® fungicide showed an improved fungicidal effect against *Aspergillus funiculosus* and *Alternaria tenuis*. If selenium is added in minimum amounts to fungicides to control plant diseases, it may reduce environmental and human health risks. *Botrytis cinerea*, a fungus, causes gray mold decay in fresh vegetables and fruits all over the world. It is usually controlled with chemical fungicides but the residue of these fungicides circulates in the environment and causes biological risks to humans. Also, pathogen strains gradually become resistant to them however, sodium selenite can control postharvest diseases caused by *Penicillium expansum* in vegetables and fruits according to a couple of studies. In one case, *in vitro* and *in vivo* effects of sodium selenite in inhibiting *B. cinerea* and controlling gray mold disease on postharvest tomatoes were investigated. Selenite showed an inhibitory effect on germ tube elongation, spore

Table 1. Potential applications of selenium and selenium NPs in plant health.

Serial No	Target Plant	Application	Reference	Serial No	Target Plant	Application	Reference
1	*Nicotinia tabacum*	Organogenesis and root growth stimulation	Domokos-Szabolcsy et al. 2012	16	Pepper plants (*Capsicum annuum* L.)	Improvement of environmental stress resistance	Li et al. 2021
2	Pea plants (*Pisum sativum* L.)	Foliar spray	Shedeed et al. 2018	17	*Vicia faba* cv. Giza 716	Plant growth promotion and antifungal activity against *Rhizoctonia solani* RCMB 031001	Hashem et al. 2021
3	Tomatoes cv. Tiny Tim	Against parasite (*Meloidogyne* invasion)	Baycheva et al. 2018	18	Cucumber	Growth and productivity promotion under joined salinity and heat stress	Shalaby et al. 2021
4	Wheat Seedlings (*Triticum aestivum* L.)	Human diet quality and safety enhancement	Hu et al. 2018	19	Indica Rice	Cd accumulation decrease and photosynthesis improvement	Wang et al. 2021
5	Pomegranate (*Punica granatum* cv. Malase Saveh)	Growth, fruit yield, and quality	Zahedi et al. 2019	20	*Triticum aestivum* L.	Fungicidal effect against pathogenic fungi	El-Saadony et al. 2021
6	groundnut (*Arachis hypogaea* L.) cultivars (NC, Gregory and Giza 6)	Enhancement of the antioxidant defense systems and growth improvement	Hussein et al. 2019	21	Wheat plants (*Triticum aestivum* L.)	Drought stress tolerance	Sardari et al. 2022
7	Rice (*Oryza sativa* L.)	Production of agricultural Se-biofortified rice seedlings	Wang et al. 2020	22	Melon plants (*Cucumis melo* L.)	Antioxidant capacity improvement, plants insect resistance improvement, Photosynthesis enhancement and plant pathogen resistance	Kang et al. 2022
8	Rice	Rice seedlings growth promotion under Cd stress	Xu et al. 2020	23	Artemisia plants	Improving artemisinin and essential oil production	Sayed and Ahmed 2022
9	Barley plants	Barley seedling's growth promotion under salt stress	Habibi and Aleyasin 2020	24	Brassica chinensis	Plant growth and development, heavy metals mitigation, antioxidant system improvement and mineral nutrient elements improvement	Zhu et al. 2022

Table 1 contd. ...

...Table 1 contd.

Serial No	Target Plant	Application	Reference	Serial No	Target Plant	Application	Reference
10	Coffee plants	Antioxidant metabolism increase and photosynthetic pigments formation enhancement	Mateus et al. 2021	25	*Artemisia annua* L.	Essential oil accumulation	Logvinenko et al. 2022
11	*Macrotyloma uniflorum* (horse gram)	Seedling vigor enhancement	Antony et al. 2021	26	Pot marigold (*Calendula officinalis* L.)	Reducing the harmful effects of salinity stress on the growth and performance of marigolds	Shahraki et al. 2022
12	Rice	Harmful salinity mitigation, plant growth enhancement and grain yield improvement	Badawy et al. 2021	27	*Hypericum perforatum* L	Improvement of essential oils percentage and antioxidant activity	Nazari et al. 2022
13	Phaseolus vulgaris	Nutrition for antioxidant defense and salt tolerance	Rady et al. 2021	28	Wheat	Seeds germination improvement under salt stress	Ghazi et al. 2022
14	Japonica rice varieties: *Nanjing* 9108, *Jiahua* 1 and *Wuyunjing* 29	Selenium-enriched and high-yield rice varieties	Yan et al. 2021	29	Tomato	Repairing the fruit under the stress of Penthiopyrad (Pen)	Liu et al. 2022
15	Rice	Rice Cd-tolerance improvement and prevention of rice Cd-migration	Deng et al. 2021	30	Pepper plant	Improving soil qualities under Cd stress	Li et al. 2022

germination, and mycelial growth of *B. cinerea* and *P. expansum* microorganisms. ROS were found in *B. cinerea* spores treated with selenite, which indicated the presence of oxidizing molecules in the fungus under the influence of selenite.

On the other hand, selenium improves antioxidant defense and tolerance against oxidative stress in plants. It also protects the plant against biotic and abiotic stresses at low concentrations, while selenium can lead to phytotoxicity at high concentrations. In low concentrations, selenium promotes plant growth and improves fruit quality. Selenium accumulation can protect the plant against fungal infections and herbivorous invertebrates, such as aphids, spiders, and caterpillars. Selenium also makes plant leaves resistant to *Fusarium* sp. and *Alternaria brassicicola*. Selenium protects Indian mustard against fungal infections. It can protect tomatoes from *Fusarium* wilt damage. Therefore, selenium can replace chemical fungicides as an antifungal agent and prevent the growth of postharvest fungal pathogens (Wu et al. 2016). A number of different applications of selenium and selenium nanoparticles in increasing the quality of health and improving the growth and performance of plants are given in Table 1.

Selenium in nanometer size possesses more biological activity but less toxicity. Antioxidant, anticancer, anti-inflammatory, and antimicrobial properties of selenium nanoparticles have been noted. The biological activities of selenium nanoparticles can be attributed to the interactions occurring between these nanoparticles and carbonyl (C=O), amine (NH), carboxylate (COO-), and C-N functional groups frequently available in proteins structure. In nature, this element is involved in selenoenzymes structure like glutathione peroxidase and plays a critical role in the antioxidant defense function of many organisms. Also, the presence of selenium in iodothyronine deiodinase and thioredoxin reductase enzyme structure supports its important role in biological systems, i.e., detoxification and metabolism. Selenium nanoparticles obtained biologically from *Trichoderma* can control pearl millet downy disease in plants and help its growth in greenhouse conditions. The protective antimycotic and mycoparasitic role of *Trichoderma*, a root colonizing fungus, against diseases was improved in the presence of selenium nanoparticles. Also, culture filtrate, cell wall debris, and cell lysate of *Trichoderma atroviride* were able to reduce sodium selenite to selenium nanoparticles. These nanoparticles were able to inhibit *Pyricularia grisea* fungus growth, which causes blast disease in pearl millet. They also inhibited *Colletotrichum capsici, P. grisea,* and *Alternaria solani* infections affecting the leaves of chili and tomato plants. The extent of inhibition was directly related to the concentration of nanoparticles. The antifungal activity of selenium nanoparticles is found to be promising to control tomato late blight disease. Selenium nanoparticles also caused the aggregation of *Phytophthora infestans* pathogen the most dangerous pathogen of the tomato plant that may end in the loss of the entire crop (Joshi et al. 2019). Nanoparticles obtained from *Trichoderma harzianum* JF309 extract and selenium sulfide nanoparticles obtained from *Saccharomyces cerevisiae* (Asghari-Paskiabi et al. 2019) were also able to inhibit *Alternaria* (Singh et al. 2021).

Selenium Bio-Nanoparticles and Restoring Selenium Balance to Nature

According to the global calendar 2015, the issue of increasing pollution is ranked sixth in terms of importance. The World Economic Forum has announced pollution as the third most critical issue in Asia. Pollution with metalloids, which exhibit characteristics of both metals and non-metals, is classified along with heavy metals through four different paths: *ex situ* containment, *ex situ* treatment, *in situ* containment, and *in situ* treatment. Plants, fungi, and bacteria that can regenerate chemicals through the biological production of nanoparticles have become a practical solution for metalloids and heavy metals removal in an eco-friendly manner. Biorefining is a triangle whose three sides include nutrients, microorganisms, and food. Biorefining is done. Efficiently only when microorganisms can enzymatically convert pollutants into non-toxic products. Nanotechnology has increased the efficiency of refining and biorefining methods due to the smaller size of the nanoparticles produced through the process of contaminant transformation. In fact, the removal of heavy metal pollution, metalloids, and all kinds of organic and inorganic pollution with the help of nanoparticles produced by plants, fungi, and bacteria are called nanobioremediation. In addition, nanobioremediation can cause sustainability in nature by cleaning the environment from pollutants, such as microplastics, which are toxic to the environment of the marine ecosystem (Sharma and Sharma 2022). There are many cases of biological production of selenium nanoparticles as discussed elsewhere (Nikam et al. 2022) in detail.

There is a narrow border between the selenium level needed by the human body to prevent Se deficiency and promote human health and the concentration of selenium that can lead to toxicity (40–400 $\mu g.d^{-1}$). On the other hand, selenium distribution in the environment (air, soil, and water) is also diverse throughout the world. Various factors like environmental microbial collection, human activities, and natural phenomena play a role in this heterogeneous distribution pattern. Selenium ranks 76 among other elements in terms of abundance, but it ranks 145 in terms of toxicity and being hazardous. The presence of selenium in the environment is either due to natural sources or artificial ones by humans. The total amount of selenium in the soil is evaluated in five levels, the lowest of which starts from deficient meaning less than 0.125 mg.kg^{-1}, but the highest level can be more than 30 mg.kg^{-1}, which is considered excessive. The availability of selenium for plants depends on the biochemical and chemical characteristics of the soil, such as pH, the redox potential of soil organic matter, and the presence of other elements, such as phosphorus and sulfur as well as the methylation process. The use of fossil fuels and excessive use of mineral fertilizers also affect the metal composition of the soil. The type of soil vegetation, such as corn, rice, and soybeans, affects the availability of selenium and nano-selenium particles in the soil, and the cycle of soil selenium changes with the change in soil usage. Usually, plants do not have much selenium (approximately 25 μg.kg^{-1}), but there are cases where selenium content in plants gets more than 1,000 μg.kg^{-1}, which causes poisoning of animals and humans who consume them. The availability of selenium in soils depends on the fractionation of this element, which is controlled by processes, such as desorption/adsorption, precipitation/dissolution, and reduction/oxidation. Normally, selenate form (SeO_4^{2-}) in alkaline soils suffers from high pH but in acidic soils (low pH) selenite (SeO_3^{2-}) form is the predominant species. Climate changes also affect selenium distribution in soil. This may happen through atmospheric selenium sitting on the soil, or with changes in humidity and temperature, the absorption of selenium by plants will change and indirectly affect the amount of selenium in the soil. One of the issues affecting selenium levels in the soil is the methylation of selenium and volatilization to the atmosphere. About 50 to 70 percent of the global release of selenium is through the gaseous release of this element into the atmosphere. Microorganisms play an important role in the selenium cycle with nano-selenium biosynthesis (El-Ramady et al. 2022). In the same way, the biological production of selenium nanoparticles can restore the balance of selenium in the soil and cause a better quality of plant growth (Figure 1).

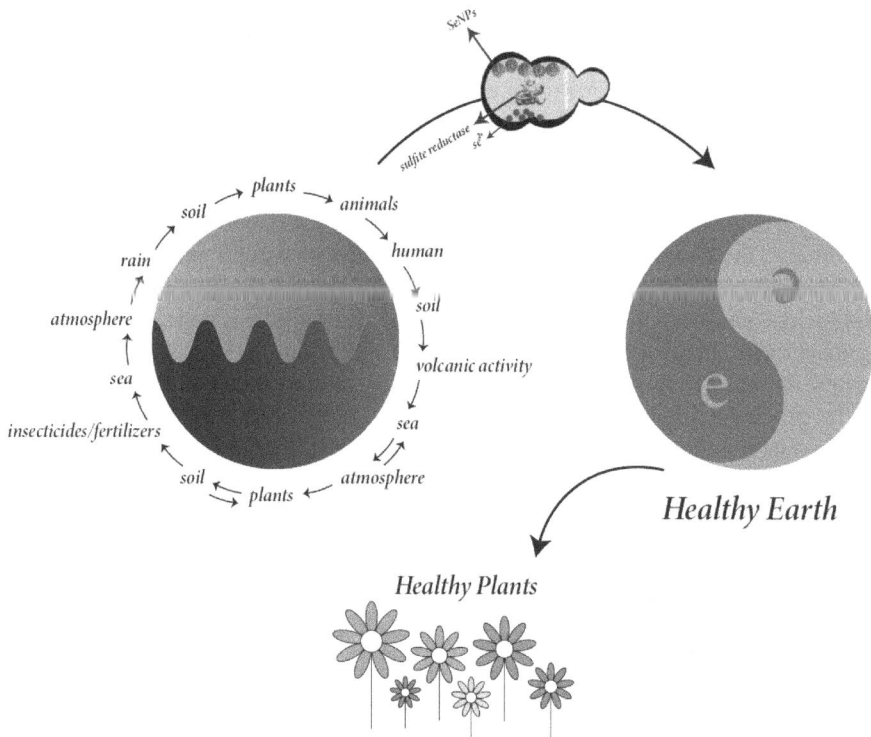

Figure 1. Green synthesized selenium nanoparticles can promise selenium balance and help the health of plants and other organisms.

Control Mechanisms of Bio-Nanoparticles Production

Biomolecules exhibit different interactions with metal ions but various factors are responsible for the biosynthesis of nanoparticles. In addition, the mechanism of intracellular and extracellular synthesis of nanoparticles in biological systems is not the same. For example, cell walls play an important role in the intracellular synthesis of nanoparticles in some microorganisms. The negatively charged cell wall interacts electrostatically with the positively charged metal ions. Enzymes in the cell wall cause the bio-reduction of metal ions into nanoparticles, and finally smaller-sized nanoparticles distribute through the cell wall. Mechanisms involved in the intracellular synthesis of nanoparticles using *Verticillium* sp. are studied and reported extensively. The biosynthesis mechanisms of nanoparticles can be classified into entrapment, bio-reduction, and synthesis steps. When the fungal cell surface comes into contact with metal ions, electrostatic interactions trap the ions. Then the enzymes present in the cell wall bio-reduce metal ions to lower oxidation states. Finally, the accumulation of the particles and synthesis of nanoparticles takes place (Mukherjee et al. 2001).

In the case of *Lactobacillus* species, it was observed that during the first stage of nanoparticle synthesis, the nucleation of metal ion clusters occurred, and then electrostatic interactions occurred between the bacterial cell wall and metal clusters leading to the formation of nanoclusters. Finally, nanoclusters that are smaller in terms of size are released from the bacterial cell wall. In actinomycetes, the reduction of metal ions occurred on the surface of mycelia along the cytoplasmic membrane leading to the formation of nanoparticles (Nair and Pradeep 2002). The mechanism of the synthesis of silver nanoparticles using lactic acid bacteria is that whenever the pH increased, more competition occurred between protons and metal ions for negatively charged binding sites (Sintubin et al. 2009).

The mechanism of extracellular synthesis of nanoparticles using microbes is also proposed based on nitrate reductase-mediated biosynthesis. Nitrate reductase enzyme secreted by fungi helped bio-reduction of metal ions and synthesis of nanoparticles (Shabani et al. 2014). Many researchers have proposed nitrate reductase for the extracellular synthesis of nanoparticles. A nitrate reductase test has been performed through the reaction of nitrite with 2,3-diaminonaphthalene. The emission spectrum showed two main fluorescence peaks at 405 nm and 490 nm, which correspond to the maximum emission of nitrite and 2,3-diaminonaphthotriazole (DAN) species, respectively. It is found that by the addition of 0.1% KNO_3 solution, the intensity of these two bands increased supporting and confirming the presence of nitrate reductase enzyme. The reductase enzyme is suggested to be responsible for the reduction of silver ions (Ag+) in silver nitrate and then the formation of silver nanoparticles (Durán et al. 2005). The researchers also used commercial nitrate reductase disks, and the white color of these disks changed to red when exposed to the fungal filtrate, indicating the presence of nitrate reductase enzyme. Therefore, the NADH-dependent reductase enzyme activity in fungi is strongly associated with the reduction of Ag^+ to Ag° (Ingle et al. 2008).

The encapsulated *Rhodoseudomonas* bacterium produced NADH cofactor and NADH-dependent enzymes. Bio-reduction of gold ions started with electron transfer from NADH by NADH-dependent reductase as an electron carrier. Then, gold ions gained electrons and were reduced to Au°, leading to the formation of gold nanoparticles (He et al. 2007). Synthesis of gold nanoparticles was also carried out using *Stenotrophomonas maltophilia* bacteria. The biosynthesis of gold nanoparticles and their stability through charge coating in *S. maltophilia* required NADPH-dependent reductase enzyme, which converted Au^{3+} to Au^0 through the electron transfer process (Nangia et al. 2009).

Bacillus cereus TAH has been investigated to produce selenium nanoparticles. The washed cells of this bacterium, as well as the supernatant, were not able to regenerate selenite and produce selenium nanoparticles. Therefore, the necessary compounds for the regeneration of selenite are secreted from bacteria only in the presence of this substance aiming for detoxification (Ghazi et al. 2022).

Tetrahymena thermophile, a protozoan was used for the bioproduction of selenium nanoparticles. Various sizes of selenium nanoparticles were obtained. The mRNA expression of glutathione synthase, an enzyme involved in GSH formation, increased during the formation of nano-selenium compared to the control group, and the possibility strengthened the hypothesis that GSH plays an important role in selenite reduction. Also, in this process, metallothionein-1 and [2Fe-2S] cluster-binding protein genes were up-regulated compared to the control group. These last two genes are associated with cysteine-rich proteins, and the amino acid cysteine provides the thiol needed by GSH. Also, total GSH content increased in the nano-selenium-producing group compared to the control group. Increased expression of the glutathione synthase gene led to increased GSH in cells. However, the amount of glutathione reduced to total glutathione in the nano-selenium-producing group had a subtle decrease compared to the control group, which could be due to the consumption of GSH during nano-selenium synthesis (Cui et al. 2016).

To answer the question, "What happens in the body of a yeast when it produces nanoparticles?" and to investigate and control the biosynthesis mechanism of the nanoparticle, glutathione metabolism genes were manipulated in the yeast *S. cerevisiae* before the intracellular synthesis of CdSe quantum dots (QDs) take place. Various aspects in the intracellular synthesis of the nanoparticles, such as prerequisites, crystallization of nanoparticles, and their encapsulation at the end stages, are all under the control of the expression of certain genes. Therefore, controlling the biosynthesis process of nanoparticles requires identifying the function of these genes. They are also useful for controlling the biological production of other nanoparticles and end to new nano-biocompounds. The metabolic pathway of glutathione is of particular importance in the biosynthesis of nano-selenium particles. This pathway is controlled through the expression of the genes responsible for the glutathione metabolism pathway. To this end, the key genes of glutathione metabolism were optimized and yeast cells became more efficient cell factories for the production of CdSe nanoparticles. The GSH1 gene in yeast encodes γ-glutamylcystein ligase (GCL), which is the first step and also the rate-determining step in the synthesis of glutathione in the cell (Scheme 1). The fluorescence of CdSe QDs was used as an intracellular probe to trace the production of these nanoparticles. By silencing this gene, the Δgsh1 mutant strain was produced. Compared to the control group, Δgsh1 mutant was unable to produce glutathione and subsequently showed significantly less fluorescence due to the intracellular production of CdSe QDs (Figure 2).

The GSH2 gene encodes GS enzyme. This protein is the second enzyme involved in the glutathione synthesis metabolic pathway. The mutant Δgsh2 in which the GSH2 gene was omitted was not able to produce glutathione, while the dipeptide γ-glutamylcysteine was accumulated in the cells. Also, a significant decrease in the fluorescence density of intracellular CdSe QDs was observed hence, γ-glutamylcystein cannot be considered a suitable substitute for glutathione in the biosynthesis of CdSe QDs. Simultaneous deletion of GSH1 and GSH2 (Δgsh1Δgsh2 mutant) was also

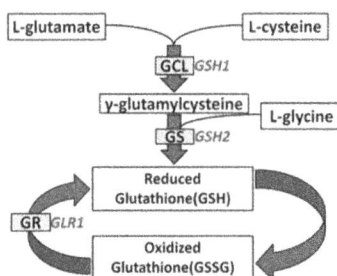

Scheme 1. Glutathione metabolism pathway in *S. cerevisiae.* The dipeptide γ-glutamylcysteine formation is catalyzed by γ-glutamylcysteine ligase (GCL) encoded by GSH1 gene. The reaction of γ-glutamylcysteine with glycine leads to glutathione production, which is catalyzed by the glutathione synthetase (GS) encoded by GSH2. The glutathione reductase (GR) is encoded by the GLR1 gene, which is responsible for the reduction of oxidized glutathione (GSSG) to reduced glutathione (GSH). Reprinted (adapted) with permission from (Li et al. 2013). Copyright 2023 American Chemical Society.

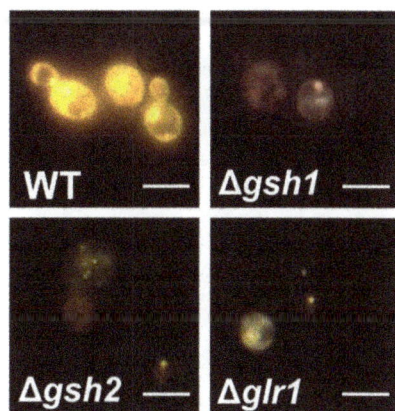

Figure 2. Fluorescence microscopic images of CdSeQDs in the wild-type group (WT), Δgsh1, Δgsh2, and Δglr1 mutants. Reprinted (adapted) with permission from (Li et al. 2013). Copyright 2023 American Chemical Society.

performed and similar results were obtained with Δgsh1 and Δgsh2, which means that glutathione synthesis is a serial metabolic process. Glutathione is oxidized when it does its antioxidant function. Oxidized glutathione (GSSG), in order to be regenerated (GSH), requires the function of the GLR1 gene. Deletion of this gene in Δgsh1 mutant cells led to the accumulation of GSSG. The fluorescence density of CdSe QDs was also decreased supporting the inherent importance of glutathione redox state in the production of nanoparticles. Production of the nanoparticles was also related to the phase of cell growth. Δgsh1 and Δgsh1Δgsh2 mutants had lower cell densities in the stationary phase. All mutants in the exponential phase grew less than the control group. For all groups, the fluorescence intensity was higher and the synthesis of QDs happened in the stationary phase compared to the exponential phase. Probably cells in this phase are more capable to respond to cadmium and selenite ions compared to the exponential phase. In all phases, the mutant cells showed lower fluorescence density compared to the control group, which again indicates the important role of glutathione in the biosynthesis of these nanoparticles. Since the first excitonic absorption peak intensity indicates the QDs content, and this peak was lower in terms of strength in the mutants compared to the control group; it turns out that the deletion of GSH1, GSH2, and GLR1 genes significantly reduces the yield of CdSe nanoparticles, which confirms the role of glutathione metabolic pathway in the production of CdSe QDs. Despite the importance of GSH2 and GLR1, significant up-regulation of these genes did not occur in the presence of cadmium supporting the unique role of the GSH1 gene in rate-limiting and regulating the amount of intracellular glutathione again, which can be a cost-effective and accurate way to biocontrol CdSe QDs nanoparticles. Also, the induction of the GSH1 gene in P_{GAL1}-GSH1 cells significantly increased the production of CdSe QDs (up to five times). Therefore, biological control of CdSe QDs production and increase in efficiency by engineering the metabolic pathway of glutathione is quite possible (Li et al. 2013).

The population of *S. cerevisiae* was investigated during the biosynthesis process of selenium sulfide nanoparticles. The synthesis of nanoparticles started after the first 100 minutes and after 240 minutes the nanoparticles completely changed the color of the culture medium. Every 100 minutes when the growth of cells was quantified as colony formation counts, in parallel to the formation of nanoparticles, a decrease in the growth of the *Saccharomyces* population was observed. During the experiments, it was observed that the production of nanoparticles was under the control of sulfite reductase enzyme activity. By increasing GSSG (versus GSH), the expression of the main genes of this enzyme, namely MET5 and MET10 was induced (Figure 3). Moreover, by considering the effect of sulfite and selenium on the expression of these genes (Asghari-Paskiabi et al. 2020), the importance of environmental factors in the functioning of the genes can be understood. In fact, in the first step, yeast cleverly took the approach of making anions insoluble in the form of nanoparticles to deal with soluble metalloids. But the nanoparticles themselves were considered to be toxic to

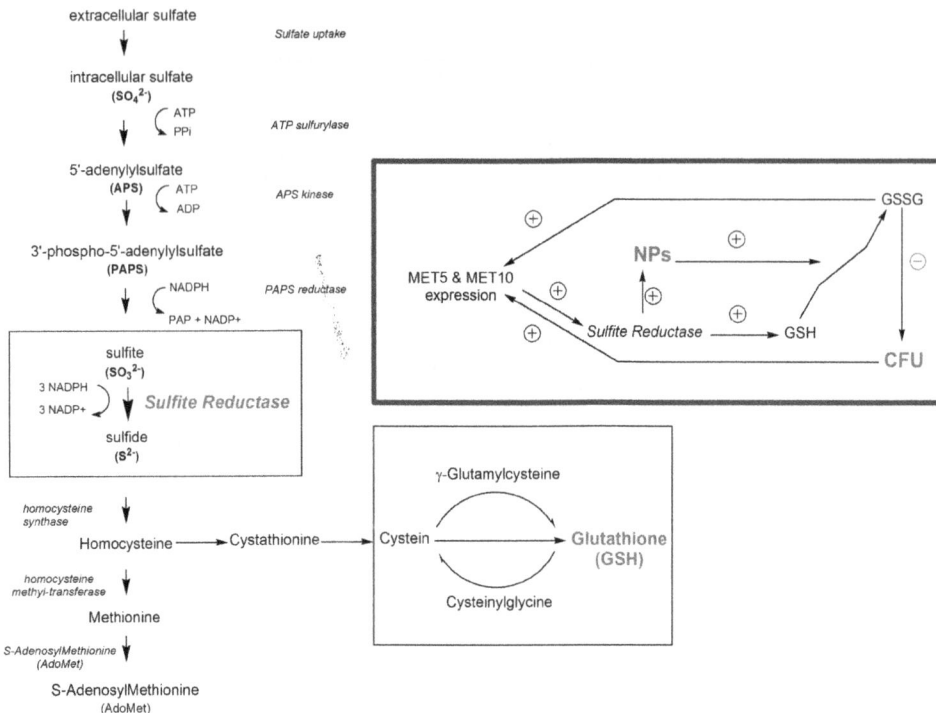

Figure 3. The correlation between NPs synthesis, sulfite reductase activity, and the expression of *MET5* and *MET10*, the main genes of sulfite reductase in sulfur pathway in *S. cerevisiae* (Asghari-Paskiabi et al. 2020).

cells. To deal with them, the cells still needed the activity of the enzyme sulfite reductase, which provides GSH for the glutathione cycle. The synthesis of sulfide and selenium nanoparticles is dependent on the glutathione cycle; in addition to the genes responsible for glutathione expression, other genes involved in this cycle, such as the important enzyme sulfite reductase, also play a role in this process, which can be considered in the biosynthesis of selenium compounds nanoparticles.

In the process of sulfate reduction to sulfide, eight electrons must be transferred, and in the enzymatic steps, two molecules of ATP and four molecules of NADPH are released. This process has three specific stages: sulfate absorption, sulfate activation, and finally sulfate reduction. The complexity of this pathway is unavoidable because sulfate cannot be reduced to sulfite in one step using NADPH. The redox potential for the direct reduction of sulfate to sulfite (−517 mV) is low compared to the same for NADPH/(NADPH+) (−320 mV). To overcome this problem, inorganic sulfate is first activated by interacting with ATP to form adenosine-5-phosphosulfate (APS), which lowers sulfate's electrochemical potential so that sulfite reduction happens more easily (Verschueren and Wilkinson 2005).

A metabolic map for sulfur metabolism in *S. cerevisiae* is depicted in Scheme 2. From the details of this map as well as the potential that genetic engineering has provided to researchers, we can hope that by determining the variables affecting performance and metabolism, such as sulfur and selenium concentrations in the environment, temperature, pH, air pressure, and the factors that make the elements enter in or leave the cycle, the whole is modeled mathematically in order to understand the achieved practical results for the control of rare and toxic selenium elements better and make biological systems predictable. Achieving such a model depends on a large amount of accurate, precise, reliable, and quantitative information. Two important events have made such a model of the metabolism of eukaryotic cells of *S. cerevisiae* possible: the first is the use of CRISPR/Cas9 to

SULFUR METABOLISM

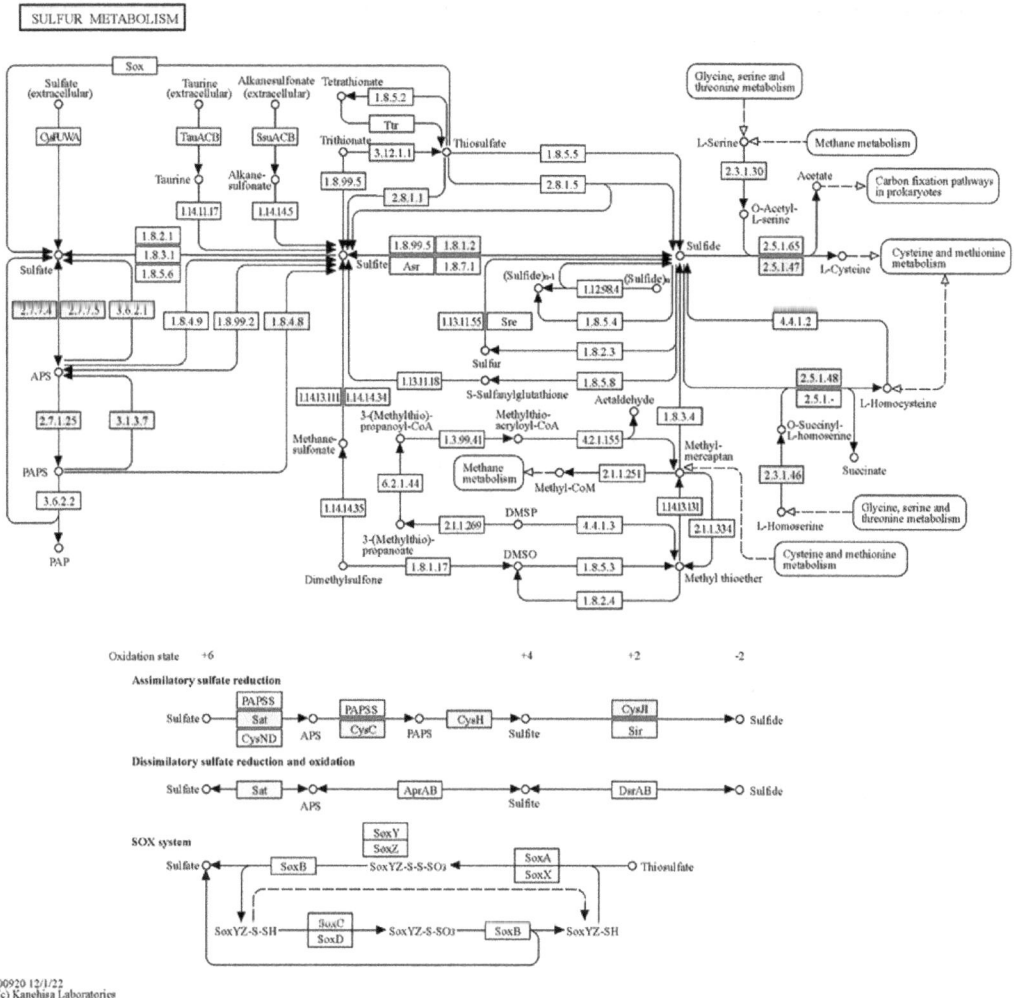

Scheme 2. Sulfur metabolic pathway in *S. cerevisiae* (KEGG, Laboratories 2022).

engineer *S. cerevisiae* genome, and the second is *S. cerevisiae* cells in which all 16 chromosomes have been designed *in vitro* (Oliver and Castrillo 2019).

Conclusions

Plants, like humans and animals, need selenium, but in a dose-dependent manner to preserve the health status at the best state optimally available. Selenium deficiency or its accumulation in the plant is directly related to the amount of selenium in the soil. Various factors affect the levels of selenium in the soil and the cycle of selenium in the environment. These factors encompass both human-induced and natural events, as well as components of the natural cycle of selenium. On the other hand, selenium in its nano-sized morphology is less toxic and more effective in terms of antioxidant and anticancer properties. The biological production of nanoparticles by living organisms is advantageous in many aspects, one of which is the removal of chemical pollutants during the biosynthesis stages. In addition, the conversion of oxyanions soluble in water, which are toxic, can be directed toward the formation of selenium nanoparticles to prevent them from entering the body of other living organisms and leaving toxicity. The biological production mechanisms involved in the formation of these nanoparticles have been investigated and even the genes controlling these syntheses can be engineered. There is hope that the balance of this metalloid can be restored to nature by metabolic engineering and biosynthesis of environmental nanoparticles.

References

Antony, D., Yadav, R. and Kalimuthu, R. 2021. Accumulation of phyto-mediated nano-CeO_2 and selenium doped CeO_2 on *Macrotyloma uniflorum* (horse gram) seed by nano-priming to enhance seedling vigor. Biocatal. Agric. Biotechnol. 31(1): 1–10.

Asghari-Paskiabi, F., Imani, M., Rafii-Tabar, H. and Razzaghi-Abyaneh, M. 2019. Physicochemical properties, antifungal activity and cytotoxicity of selenium sulfide nanoparticles green synthesized by *Saccharomyces cerevisiae*. Biochem. Biophys. Res. Commun. 516(4): 1078–1084.

Asghari-Paskiabi, F., Imani, M., Eybpoosh, S., Rafii-Tabar, H. and Razzaghi-Abyaneh, M. 2020. Population kinetics and mechanistic aspects of *Saccharomyces cerevisiae* growth in relation to selenium sulfide nanoparticle synthesis. Front. Microbiol. 11(19): 1–11.

Badawy, S.A., Zayed, B.A., Bassiouni, S.M.A., Mahdi, A.H.A., Majrashi, A., Ali, E.F. and Seleiman, M.F. 2021. Influence of nano silicon and nano selenium on root characters, growth, ion selectivity, yield, and yield components of rice (*Oryza sativa* L.) under salinity conditions. Plants 10(8): 1–8.

Baycheva, O., Samaliev, H., Udalova, Z., Trayanov, K., Zinovieva, S. and Folman, G. 2018. Selenium and its effect on plant-parasite system *Meloidogyne arenaria* - tiny tim tomatoes. Bulgarian J. Agric. Sci. 24(2): 252–258.

Cui, Y.-H., Li, L.-L., Zhou, N.-Q., Liu, J.-H., Huang, Q., Wang, H.-J., Tian, J. and Yu, H.-Q. 2016. *In vivo* synthesis of nano-selenium by *Tetrahymena thermophila* SB210. Enzyme Microb. Technol. 95(9): 185–191.

Deng, S.W., Li, P.R., Li, Y.Z., Ran, Z.X., Peng, Y.X., Yang, S.L., He, H., Zhou, K. and Yu, J. 2021. Alleviating Cd translocation and accumulation in soil-rice systems: Combination of foliar spraying of nano-Si or nano-Se and soil application of nano-humus. Soil. Use. Manag. 37(2): 319–329.

Domokos-Szabolcsy, E., Marton, L., Sztrik, A., Babka, B., Prokisch, J. and Fari, M. 2012. Accumulation of red elemental selenium nanoparticles and their biological effects in *Nicotinia tabacum*. Plant Growth Regul. 68(3): 525–531.

Durán, N., Marcato, P.D., Alves, O.L., De Souza, G.I. and Esposito, E. 2005. Mechanistic aspects of biosynthesis of silver nanoparticles by several *Fusarium oxysporum* strains. J. Nanobiotechnology 3(1): 1–7.

El-Ramady, H., A.E.-D. Omara, T. El-Sakhawy, J. Prokisch and E.C. Brevik 2022. Sources of selenium and nano-selenium in soils and plants. pp. 1–24. Mohammad Anwar Hossain, G.J.A., Zsuzsanna Kolbert, Hassan El-Ramady, Tofazzal Islam and Michela Schiavon Selenium and Nano-Selenium in Environmental Stress Management and Crop Quality Improvement. Springer.

El-Ramady, H.R., Domokos-Szabolcsy, É., Shalaby, T.A., Prokisch, J. and Fári, M. 2015. Selenium in agriculture: water, air, soil, plants, food, animals and nanoselenium. CO_2 sequestration, biofuels and depollution. J.S. Eric Lichtfouse, Didier Robert, Springer, Cham. 5: 153–232.

El-Saadony, M.T., Saad, A.M., Najjar, A.A., Alzahrani, S.O., Alkhatib, F.M., Shafi, M.E., Selem, E., Desoky, E.M., Fouda, S.E.E., El-Tahan, A.M. and Hassan, M.A.A. 2021. The use of biological selenium nanoparticles to suppress *Triticum aestivum* L. crown and root rot diseases induced by *Fusarium* species and improve yield under drought and heat stress. Saudi J. Biol. Sci. 28(8): 4461–4471.

Ghazi, A.A., El-Nahrawy, S., El-Ramady, H. and Ling, W.T. 2022. Biosynthesis of nano-selenium and its impact on germination of wheat under salt stress for sustainable production. Sustainability 14(3): 1–10.

Gholami-Shabani, M., Akbarzadeh, A., Norouzian, D., Amini, A., Gholami-Shabani, Z., Imani, A., Chiani, M., Riazi, G., Shams-Ghahfarokhi, M. and Razzaghi-Abyaneh, M. 2014. Antimicrobial activity and physical characterization of silver nanoparticles green synthesized using nitrate reductase from *Fusarium oxysporum*. Appl. Biochem. Biotechnol. 172(8): 4084–4098.

Habibi, G. and Aleyasin, Y. 2020. Green synthesis of Se nanoparticles and its effect on salt tolerance of barley plants. Int. J. Nanodimens. 11(2): 145–157.

Hashem, A.H., Abdelaziz, A.M., Askar, A.A., Fouda, H.M., Khalil, A.M.A., Abd-Elsalam, K.A. and Khaleil, M.M. 2021. *Bacillus megaterium*-mediated synthesis of selenium nanoparticles and their antifungal activity against *Rhizoctonia solani* in faba bean plants. J. Fungi. 7(3): 1–10.

He, S., Guo, Z., Zhang, Y., Zhang, S., Wang, J. and Gu, N. 2007. Biosynthesis of gold nanoparticles using the bacteria *Rhodopseudomonas capsulata*. Mater. Lett. 61(18): 3984–3987.

Hossain, Z., Mustafa, G. and Komatsu, S. 2015. Plant responses to nanoparticle stress. Int. J. Mol. Sci. 16(11): 26644–26653.

Hu, T., Li, H.F., Li, J.X., Zhao, G.S., Wu, W.L., Liu, L.P., Wang, Q. and Guo, Y.B. 2018. Absorption and bio-transformation of selenium nanoparticles by wheat seedlings (*Triticum aestivum* L.). Front. Plant Sci. 9(1): 1–8.

Hussein, H.A.A., Darwesh, O.M. and Mekki, B.B. 2019. Environmentally friendly nano-selenium to improve antioxidant system and growth of groundnut cultivars under sandy soil conditions. Biocatal. Agric. Biotechnol. 18(1): 1–9.

Ingle, A., Gade, A., Pierrat, S., Sonnichsen, C. and Rai, M. 2008. Mycosynthesis of silver nanoparticles using the fungus *Fusarium acuminatum* and its activity against some human pathogenic bacteria. Current Nanoscience 4(2): 141–144.

Joshi, S.M., De Britto, S., Jogaiah, S. and Ito, S.-i. 2019. Mycogenic selenium nanoparticles as potential new generation broad spectrum antifungal molecules. Biomolecules 9(9): 1–16.

Kang, L., Wu, Y.L., Zhang, J.B., An, Q.S., Zhou, C.R., Li, D. and Pan, C.P. 2022. Nano-selenium enhances the antioxidant capacity, organic acids and cucurbitacin B in melon (*Cucumis melo* L.) plants. Ecotoxicol. Environ. Saf. 241(10): 1–9.

KEGG. (c) Kanehisa Laboratories. Sulfur metabolism - *Saccharomyces cerevisiae* (budding yeast). Retrieved: 12.01.22 from https://www.genome.jp/pathway/sce00920.

Letavayová, L., Vlčková, V. and Brozmanová, J. 2006. Selenium: from cancer prevention to DNA damage. Toxicology 227(1-2): 1–14.

Li, D., Zhou, C.R., Ma, J.L., Wu, Y.L., Kang, L., An, Q.S., Zhang, J.B., Deng, K.L., Li, J.Q. and Pan, C.P. 2021. Nanoselenium transformation and inhibition of cadmium accumulation by regulating the lignin biosynthetic pathway and plant hormone signal transduction in pepper plants. J. Nanobiotechnology. 19(1): 1–10.

Li, D., Zhou, C.R., Wu, Y.L., An, Q.S., Zhang, J.B., Fang, Y., Li, J.Q. and Pan, C.P. 2022. Nanoselenium integrates soil-pepper plant homeostasis by recruiting rhizosphere-beneficial microbiomes and allocating signaling molecule levels under Cd stress. J. Hazard. Mater. 432(10): 1–9.

Li, Y., Cui, R., Zhang, P., Chen, B.-B., Tian, Z.-Q., Li, L., Hu, B., Pang, D.-W. and Xie, Z.-X. 2013. Mechanism-oriented controllability of intracellular quantum dots formation: the role of glutathione metabolic pathway. ACS Nano 7(3): 2240–2248.

Liu, R., Deng, Y., Zheng, M.L., Liu, Y.P., Wang, Z.K., Yu, S.M., Nie, Y.F., Zhu, W.T., Zhou, Z.Q. and Diao, J.L. 2022. Nano selenium repairs the fruit growth and flavor quality of tomato under the stress of penthiopyrad. Plant Physiol. Biochem. 184: 126–136.

Logvinenko, L., Golubkina, N., Fedotova, I., Bogachuk, M., Fedotov, M., Kataev, V., Alpatov, A., Shevchuk, O. and Caruso, G. 2022. Effect of foliar sodium selenate and nano selenium supply on biochemical characteristics, essential oil accumulation and mineral composition of Artemisia annua L. Molecules 27(23): 1–10.

Mateus, M.P.D., Tavanti, R.F.R., Tavanti, T.R., Santos, E.F., Jalal, A. and dos Reis, A.R. 2021. Selenium biofortification enhances ROS scavenge system increasing yield of coffee plants. Ecotoxicol. Environ. Saf. 209.

Mitchell, S.C., Nickson, R.M. and Waring, R.H. 1993. The biological activity of selenium sulfide. Sulfur Reports 13(2): 279–289.

Mukherjee, P., Ahmad, A., Mandal, D., Senapati, S., Sainkar, S.R., Khan, M.I., Ramani, R., Parischa, R., Ajayakumar, P. and Alam, M. 2001. Bioreduction of $AuCl_4^-$ions by the fungus, *Verticillium* sp. and surface trapping of the gold nanoparticles formed. Angew. Chem., Int. Ed. 40(19): 3585–3588.

Nair, B. and Pradeep, T. 2002. Coalescence of nanoclusters and formation of submicron crystallites assisted by *Lactobacillus* strains. Cryst. Growth Des. 2(4): 293–298.

Nangia, Y., Wangoo, N., Goyal, N., Shekhawat, G. and Suri, C. 2009. A novel bacterial isolate *Stenotrophomonas maltophilia* as living factory for synthesis of gold nanoparticles. Microb. Cell Factories 8(1): 1–7.

Nazari, M.R., Abdossi, V., Hargalani, F.Z. and Larijani, K. 2022. Antioxidant potential and essential oil properties of *Hypericum perforatum* L. assessed by application of selenite and nano-selenium. Sci. Rep. 12(1).

Nikam, P.B., Salunkhe, J.D., Minkina, T., Rajput, V.D., Kim, B.S. and Patil, S.V. 2022. A review on green synthesis and recent applications of red nano Selenium. Res. Chem. 2(100): 1–11.

Oldfield, J.E., Burau, R., Moller, G., Ohlendorf, H. and Ullrey, D. 1994. Risks and benefits of selenium in agriculture, Council for Agricultural Science and Technology Ames, IA, USA.

Oliver, S.G. and Castrillo, J.I. 2019. Yeast Systems Biology: Methods and Protocols. Humana New York, NY, Springer.

Painter, E.P. 1941. The chemistry and toxicity of selenium compounds, with special reference to the selenium problem. Chem. Rev. 28(2): 179–213.

Rady, M.M., Desoky, E.M., Ahmed, S.M., Majrashi, A., Ali, E.F., Arnaout, S. and Selem, E. 2021. Foliar nourishment with nano-selenium dioxide promotes physiology, biochemistry, antioxidant defenses, and salt tolerance in *Phaseolus vulgaris*. Plants 10(6): 1–10.

Rosenfield, I. and Beath, O. 1964. Selenium, Geobotany, Biochemistry, Toxicity and Nutrition. New York: Academic 1(1): 1–10.

Sardari, M., Rezayian, M. and Niknam, V. 2022. Comparative study for the effect of selenium and nano-selenium on wheat plants grown under drought stress. Russ. J. Plant Physiol. 69(6): 1–6.

Sayed, T.E. and Ahmed, E.S. 2022. Improving artemisinin and essential oil production from Artemisia plant through *in vivo* elicitation with gamma irradiation nano-selenium and chitosan coupled with bio-organic fertilizers. Front. Energy Res. 10(1): 1–11.

Shahraki, B., Bayat, H., Aminifard, M.H. and Atajan, F.A. 2022. Effects of foliar application of selenium and nano-selenium on growth, flowering, and antioxidant activity of pot marigold (*Calendula officinalis* L.) under salinity stress conditions. Commun. Soil Sci. Plant Anal. 53(20): 2749–2765.

Shalaby, T.A., Abd-Alkarim, E., El-Aidy, F., Hamed, E.S., Sharaf-Eldin, M., Taha, N., El-Ramady, H., Bayoumi, Y. and dos Reis, A.R. 2021. Nano-selenium, silicon and H₂O₂ boost growth and productivity of cucumber under combined salinity and heat stress. Ecotoxicol. Environ. Saf. 212(9): 1–9.

Sharma, U. and Sharma, J.G. 2022. Nanotechnology for the bioremediation of heavy metals and metalloids. Ecotoxicol. Environ. Saf. 10(5): 34–44.

Shedeed, S.I., Fawzy, Z.F. and El-Bassiony, A.M. 2018. Nano and mineral selenium foliar application effect on pea plants (*Pisum sativum* L.). Biosci. Res. 15(2): 645–654.

Singh, S., Sangwan, S., Sharma, P., Devi, P. and Moond, M. 2021. Nanotechnology for sustainable agriculture: an emerging perspective. J. Nanosci. Nanotechnol. 21(6): 3453–3465.

Sintubin, L., De Windt, W., Dick, J., Mast, J., Van Der Ha, D., Verstraete, W. and Boon, N. 2009. Lactic acid bacteria as reducing and capping agent for the fast and efficient production of silver nanoparticles. Appl. Microbiol. Biotechnol. 84(4): 741–749.

Verschueren, K.H. and Wilkinson, A. 2005. Sulfide: Biosynthesis from sulfate. eLS: 1–8.

Wang, C.R., Cheng, T.T., Liu, H.T., Zhou, F.Y., Zhang, J.F., Zhang, M., Liu, X.Y., Shi, W.J. and Cao, T. 2021. Nano-selenium controlled cadmium accumulation and improved photosynthesis in indica rice cultivated in lead and cadmium combined paddy soils. J. Environ. Sci. 103: 336–346.

Wang, K., Wang, Y.Q., Li, K., Wan, Y.N., Wang, Q., Zhuang, Z., Guo, Y.B. and Li, H.F. 2020. Uptake, translocation and biotransformation of selenium nanoparticles in rice seedlings (*Oryza sativa* L.). J. Nanobiotechnology 18(1): 1–9.

Wu, Z., Yin, X., Bañuelos, G.S., Lin, Z.-Q., Zhu, Z., Liu, Y., Yuan, L. and Li, M. 2016. Effect of selenium on control of postharvest gray mold of tomato fruit and the possible mechanisms involved. Front. Microbiol. 6(14): 1–11.

Xu, H.Z., Yan, J.P., Qin, Y., Xu, J.M., Shohag, M.J.I., Wei, Y.N. and Gu, M.H. 2020. Effect of different forms of selenium on the physiological response and the cadmium uptake by rice under cadmium stress. Int. J. Environ. Res. Public. Health. 17(19): 1–10.

Yan, J., Chen, X.J., Zhu, T.G., Zhang, Z.P. and Fan, J.B. 2021. Effects of selenium fertilizer application on yield and selenium accumulation characteristics of different japonica rice varieties. Sustainability 13(18): 1–8.

Zahedi, S.M., Hosseini, M.S., Meybodi, N.D.H. and da Silva, J.A.T. 2019. Foliar application of selenium and nano-selenium affects pomegranate (*Punica granatum* cv. Malase Saveh) fruit yield and quality. S. Afr. J. Bot. 124(12): 350–358.

Zhu, Y.Y., Dong, Y.W., Zhu, N. and Jin, H.M. 2022. Foliar application of biosynthetic nano-selenium alleviates the toxicity of Cd, Pb, and Hg in *Brassica chinensis* by inhibiting heavy metal adsorption and improving antioxidant system in plant. Ecotoxicol. Environ. Saf. 240(10): 1–10.

Zinicovscaia, I., Rudi, L., Valuta, A., Cepoi, L., Vergel, K., Frontasyeva, M.V., Safonov, A., Wells, M. and Grozdov, D. 2016. Biochemical changes in Nostoc linckia associated with selenium nanoparticles biosynthesis. Ecol. Chem. Eng. S. 23(4): 559–569.

Chapter 14

Nanomaterials for the Control Postharvest Fungal Diseases

Otniel Freitas-Silva,[1,]* *Caroline Corrêa de Souza Coelho,*[2]
Juliana Pereira Rodrigues[2] and *Daiana Ferreira Amancio*[2]

◇◇

Introduction

During transport, distribution, and storage, fruits and vegetables suffer quality loss due to postharvest physiological reactions, such as respiration, maturation, ethylene production, and senescence. These reactions can lead to water loss, softening of tissues, color change, and degradation of nutrients, which usually depend on their physiological nature (climacteric and non-climacteric fruits), chemical composition, and surface structure. At the same time, along the distribution chain fruits and vegetables can suffer injury, triggering microbial growth, and reducing the shelf life of these perishable products (Mali and Grossmann 2003; Vu et al. 2011; Thakur et al. 2018).

Although cold chain distribution is a way to minimize these reactions, this method may not be sufficient to mitigate quality losses of fruits and vegetables, prolong shelf life, and preserve sensory characteristics. Thus, the use of innovative technologies, such as nanotechnology, has been investigated to meet the needs of the market (Fakhouri et al. 2014; Rocha et al. 2019).

Various studies have investigated the use of nanomaterials as technological tools to reduce postharvest deterioration. The use of nanoparticles has grown over the last few years due to their unique properties in relation to conventional materials at micro- and macro-scales. The protective properties of nanomaterials are due to their high surface area/volume ratio and their ability to incorporate biomolecules (Akhila et al. 2022; Pushparaj et al. 2022).

Despite the many advantages of using nanomaterials as postharvest technological tools, the safety concerns of researchers and consumers are global (Sharifi et al. 2012). The migration of nanomaterials to food is a major concern, so it is important to conduct migration tests before a product reaches the market to ensure safety (Souza et al. 2019).

Still in this context, little research has been conducted on the potential toxicity of these materials, as well as on the environmental safety of their use (Pitkänen et al. 2010; Andrade et al. 2015; Menas et al. 2017). Likewise, there are few studies reporting regulatory parameters related to the use of nanomaterials. According to Souza et al. (2019), the implementation of regulatory specifications for nanomaterials is essential. In particular, particle size needs to be reliably measured

---

[1] Embrapa Agroindústria de Alimentos, Av. das Américas, 29501, 23020-470 Rio de Janeiro, Brazil.
[2] Food and Nutrition Graduate Program, Federal University of the State of Rio de Janeiro, Av. Pasteur, 296, 22290-240 Rio de Janeiro, Brazil.
* Corresponding author: otniel.freitas@embrapa.br

so that nanomaterials can be classified as safe, especially through analysis by standardized *in vitro* toxicity tests.

This chapter presents a survey of recent studies of nanomaterials for postharvest use, such as their incorporation into intelligent and active packaging for the detection and control of pathogens (Alfei et al. 2020; Chausali et al. 2022; Sagar et al. 2022) and their use as fungicides, fungistatic nanoformulations, and fungicide nanocarriers (Abd-Elsalam et al. 2019; Abdollahdokht et al. 2022). The toxicity aspects of nanomaterials, limitations, and biosafety of nanoparticles, as well as current legislation, are also discussed (Azevedo and Chasin 2003; Durán et al. 2014; Scheringer 2008; Selvaraj et al. 2018).

Nanoparticles in Innovative Packaging Technologies for Fungal Disease Control

In the coming years, the agri-food sector will face the challenge of supplying healthy products to an increasing number of people, in particular fresh fruits and vegetables. This makes understanding packaging technology fundamental (Souza and Fernando 2016; Akhila et al. 2022; Sagar et al. 2022).

Food packaging plays a key role in conservation by protecting fresh produce during the farm-to-market chain, minimizing losses, and extending the shelf life with adequate commercial quality (Nandi and Guha 2018). Concerns over food safety and nutritional and commercial quality have increased in recent years (Zhang et al. 2020; Hamad et al. 2023; Haris et al. 2023).

A sustainable alternative to the use of conventional petroleum-based plastic materials (such as polyethylene, polypropylene, and polystyrene terephthalate) is the employment of bio-based packaging from renewable resources. Demand for this solution is growing in the context of heightened concern over environmental issues. In this regard, biodegradable or biocompatible materials can be used, such as films and/or coatings for packaging fruits and vegetables. These can be composed of a wide range of materials: (i) polysaccharides such as starch, cellulose, pectin, chitosan, alginate, maltodextrin, and guar gum; (ii) lipids such as fatty acids, waxes, essential oils, and extracts; (iii) proteins such as soy proteins, gelatin, zein, etc.; or various combinations of these (Omerović et al. 2021; Sun et al. 2021; Olayil et al. 2022).

Often the terms "films" and "coatings" are used without distinction in the postharvest context. Films are produced after the evaporation of a filmogenic solution by casting. The films usually have a smooth surface and are used to wrap foods. In contrast, coatings are formed from solutions applied directly on the surface of the food, thus forming a semipermeable barrier to O_2, CO_2, and water vapor and creating a modified atmosphere between the product and the environment. Consequently, the preservation of the sensory and nutritional characteristics lasts for a longer storage time (Allesteros et al. 2018). Since applications of biopolymers as packaging materials are sometimes limited due to weak mechanical properties and high permeability to water and oxygen, biopolymers have often been incorporated into nanotechnological systems (Chausali et al. 2022).

In this respect, the use of packaging systems involving active and/or intelligent nanostructured films or coatings has shown remarkable growth along with nanotechnological advances. Thus, nanotechnology applications in the food industry, especially the incorporation of nanoparticles, improve the ability of packaging to protect food products from physical damage and/or environmental conditions, such as light, moisture, and oxygen, as well as microbiological contamination. At the same time, this packaging technology can be used to monitor the condition of the product by detecting temperature, humidity, gas, or food deterioration conditions (Chausali et al. 2022). Figure 1 describes active and intelligent packaging systems based on nanotechnology.

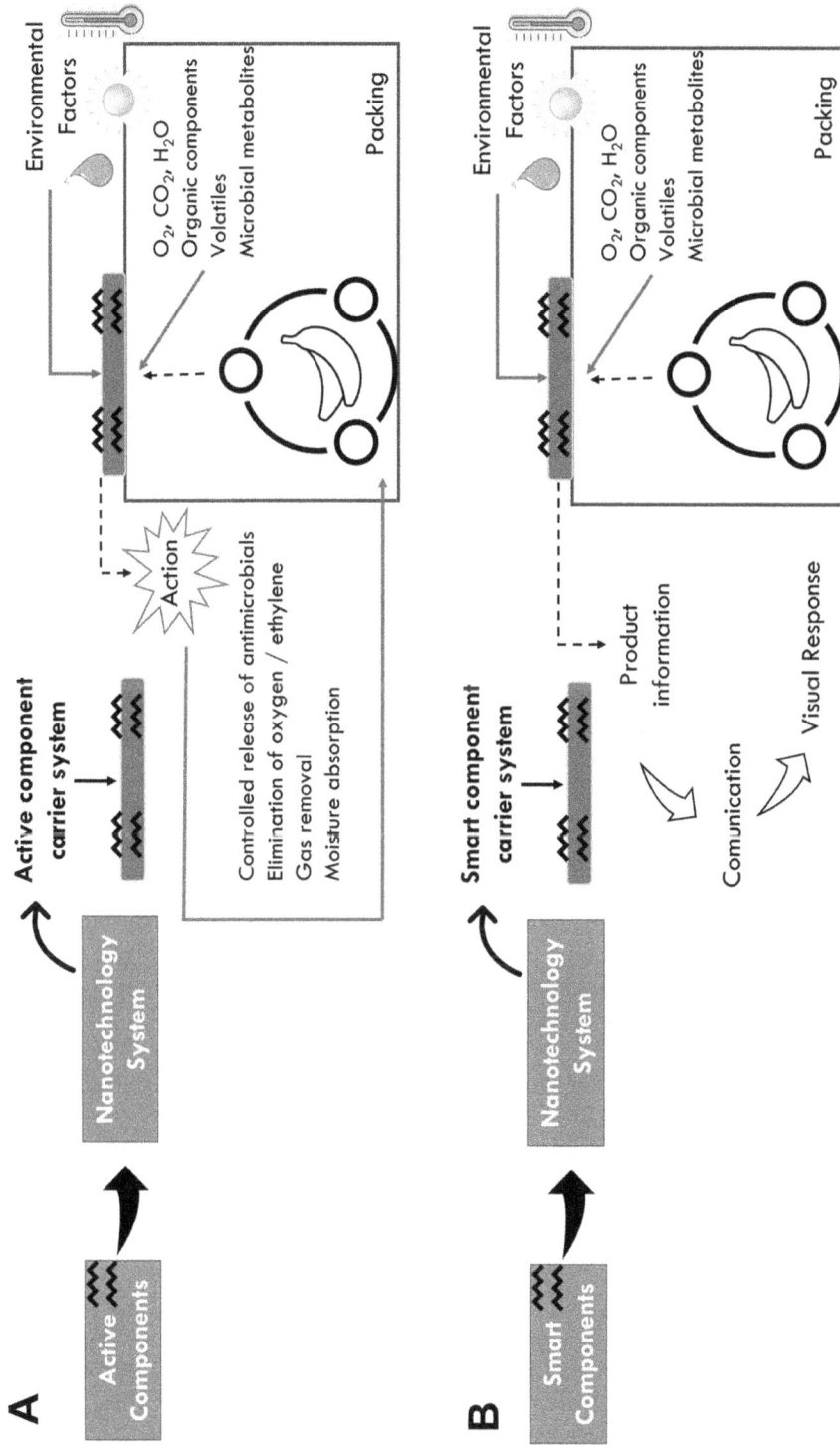

Figure 1. Applications of nanotechnology in active (A) and intelligent (B) packaging. Adapted from Mihindukulasuriya and Lim (2014).

Nanoparticles as Active Packaging Materials

Active packaging based on nanotechnology includes nanomaterials with active properties, such as metallic particles and metal oxides, bioactive compounds, antioxidants, oxygen pickups (e.g., cobalt, iron, and small organic compounds, such as ascorbic acid, ascorbates, sulfites, etc.), ethylene pickups and transporters of antioxidant agents, anti-darkening agents, and bioactive compounds. These nanomaterials are tailored to alter the headspace of the packaging to maintain the quality and/or extend the life of the product (Akhila et al. 2022; Omerović et al. 2021).

Due to the short shelf life of fruits and vegetables and their high rate of microbiological deterioration during storage, the use of packaging with active properties avoids pathogenic contamination by acting as a barrier to external agents or absorbing components that are concentrated inside the packaging that promote product deterioration, such as water and ethylene. This lengthens the useful life, making this packaging a potential market alternative (Arserim-Uçar and Çabuk 2020; Omerović et al. 2021).

In particular, inorganic nanoparticles, such as nanoclays, metal nanoparticles, and metal oxides (NPs such as Ag, Al, ZnO, Cu/CuO, TiO_2, and MgO), have attracted notable interest by improving the barrier, thermal, mechanical, and antimicrobial properties of food packaging (Alizadeh Sani et al. 2022; Mesgari et al. 2021). The incorporation of inorganic nanoparticles stands out for having excellent antimicrobial properties in food packaging systems by releasing the active ingredients in food or the surrounding atmosphere and inhibiting the growth phase of pathogens present on the surface of the packaged product. Some studies have confirmed the effectiveness of using inorganic nanoparticles for the development of active packaging due to their antimicrobial activity (Damm et al. 2008; He et al. 2011; Jamdagni et al. 2018; Akhila et al. 2022; Alizadeh Sani et al. 2022; Malandrakis et al. 2022).

The development of active packaging for postharvest protection has been investigated. The incorporation of absorbents based on Ag and cellulose NPs in minimally processed melon packages reduced by 3.0 log $CFU \cdot g^{-1}$ the growth of decaying microorganisms. Additionally, it slowed fruit senescence, resulting in remarkably lower yeast counts and a succulent appearance after 10 days of storage at 4°C (Fernández et al. 2010).

Li et al. (2017) incorporated titanium dioxide NPs in low-density polyethylene (NTLDPE) to increase the shelf life of strawberries. The NTLDPE packages showed better barrier properties, as well as forming a headspace with lower air composition of O_2 and higher CO_2 compared to LDPE. The fruits stored in the NTLDPE package presented a lower rate of decomposition and weight loss and greater firmness. In addition, the activities of antioxidant enzymes involved in the elimination of reactive oxygen species (ROS) in fruits packed with NTLDPE were significantly higher in storage, showing that the packaging may be promising for the preservation of strawberries.

Nanostructured coatings based on chitosan, alginate, and nano-ZnO were developed by Arroyo et al. (2020) and applied in guavas (*Psidium guajava* L.). After coating application, the fruits were stored for 15 days at 21 ± 1°C and 80 ± 2% RH. According to the authors, the coatings were able to prevent the appearance of rot in all guavas, confirming the high efficacy of nanoZnO against plant pathogens. The coatings with higher chitosan concentrations in the formulation protected the fruits against excessive mass loss better and delayed physical-chemical changes related to maturation.

Essential oils (EOs) are widely used in the food industry. However, despite being recognized as potential antimicrobial agents, their volatility, low water solubility, and oxidation sensitivity can limit their applications for the production of films for food packaging. In this respect, the nanoencapsulation of active OE molecules is a promising technology in which molecules are loaded in capsules with nano size (less than 100 nm). Their larger surface contact area for interaction with foods in comparison with microcapsules improves bioavailability, biological activity, solubility, and adsorption and promotes the controlled release of the active molecules (Delshadi et al. 2020).

Incorporating nanoencapsulated bioactive molecules in film packaging can increase food storage time by inhibiting the growth of pathogenic and deteriorating microorganisms, along with using lower amounts of antimicrobial agents and avoiding their direct addition to food products (Baghi et al. 2022). Recently, Robledo et al. (2018) reported the use of active edible coatings containing nanocapsules loaded with OEs (such as thymol) to prolong the shelf life of cherry tomatoes and observed the inhibition of *Botrytis cinerea*.

Yilmaz et al. (2019b) and Naserzadeh et al. (2019) encapsulated oregano and cinnamon EOs, respectively. Both obtained efficient nanoparticles with high antifungal power, even at low oil concentrations. This result was only possible due to encapsulation. The use of nanoemulsions to produce films for food packaging has also been investigated (Shen et al. 2023). Active packaging containing nanoemulsions typically requires a reinforcement material, called a matrix, and bioactive compounds are incorporated in the nanoscale. Otoni et al. (2014) reported the development of edible films with different concentrations of the antimicrobial methyl ester pectin with the incorporation of a cinnamaldehyde-based nanoemulsion having antimicrobial activity.

The antifungal effect of eugenol (EUG), carvacrol (CAR), and cinnamaldehyde (CA) nanoemulsions against *Penicillium digitatum* and the effects of nanoemulsion coating on the shelf life of citrus fruits were investigated by Yang et al. (2021). The germination of *P. digitatum* spores was significantly inhibited, and mycelial morphology was altered after treatment with the nanoemulsion. In addition, the nanoemulsion coating reduced the decomposition, weight loss, and respiratory rates.

Nanoparticles as Multifunctional Materials in Intelligent Packaging

Nanotechnology-based intelligent packaging systems have enhanced functions. Indicators and sensors manufactured using appropriate nanotechnology can interact with internal substances in the packaging headspace such as O_2, CO_2, H_2O, organic components, volatile compounds, and microbial metabolites, along with interaction with external environmental factors, such as luminosity, temperature, and humidity. As a result of this interaction, indicators and/or sensors generate a visual response, which is correlated with the state of the food product, providing useful information to consumers on product safety and quality (Chausali et al. 2022; Mihindukulasuriya and Lim 2014).

Smart packaging, in addition to providing dynamic consumer feedback on the quality of the product, supports the entire production process and distribution chain since these integration systems support decisions on when and what actions should be taken if any changes are observed due to storage time or temperature, indicating deterioration and/or contamination (Chausali et al. 2022; Mihindukulasuriya and Lim 2014; Warriner et al. 2014).

Polymeric nanoparticles have been used to produce intelligent packaging material for monitoring food freshness. In addition, they have been used in intelligent packaging as nanosensors that enable the detection of temperature, pathogenic microorganisms, organic molecules, and gases (Chen et al. 2021). When there is a change in external or internal parameters, the nanosensors in the food packaging give a signal in response to this change. Therefore, they provide visible and accurate information to consumers about the quality and freshness of food. The nanosensors are composed of receptors and transducers. The receptors are responsible for converting chemical or physical information from food packaging into energy, while transducers transform this biological or biochemical change into an electrical signal (Caon et al. 2017; Chausali et al. 2022).

In fruits and vegetables, the monitoring of freshness (change of pH or volatile compounds produced gradually during degradation, such as diacetyl, amines, ethanol, and hydrogen sulfide), ripeness (ethylene, volatile acid, etc.), and pathogenic contamination can be achieved by measurements of gases and volatile compounds in the packaging headspace using direct nanosensor labels (Sharma et al. 2017; Chen et al. 2021).

Colorimetric indication nanosensors were developed by Spricigo et al. (2017). Platforms with micro- and nanoscale based on SiO_2 and Al_2O_3 impregnated with potassium permanganate were studied to evaluate the color change to indicate ethylene removal. Nanoscale systems were more efficient in detecting the presence of oxidation by ethylene in a closed atmosphere at 45%, 60%,

75%, and 90% RH. In addition, the color changes of the nanoscale platforms, resulting from the chemical reduction of potassium permanganate, functioned as an indicator of ethylene removal, which is particularly suitable in postharvest applications.

Outlook on Action of Nanomaterials Against Fungal Diseases

The horticultural market is negatively affected by antimicrobial resistance to the fungicides, generally used in postharvest storage, leading to increasing economic losses. Some alternatives to current treatments for bacterial and fungal infection control have been investigated involving the use of nanotechnology with promising results in overcoming antimicrobial resistance. These investigations have focused on types of nanoparticles and how pathogens are affected by nanomaterials.

Nanoparticles (NPs) can be obtained from organic and inorganic substances. Organic NPs, such as proteins, carbohydrates, fats and their compounds, are usually nontoxic and are fully digestible in the human gastrointestinal tract. Some researchers have investigated the use of these NPs to increase the bioavailability of compounds (Bumbudsanpharoke and Ko 2015; Scimeca and Verron 2022).

Inorganic NPs, such as metals and metal oxides, have stability and excellent antimicrobial properties compared to organic nanomaterials (Bumbudsanpharoke and Ko 2015; Sani et al. 2022). They can interact with microbial cells, leading to cell damage, and producing secondary metabolites and oxidized cellular components (Król et al. 2017; Akhila et al. 2022).

Several studies have confirmed the use of inorganic NPs, such as silver (Ag) (Velmurugan et al. 2009; Koduru et al. 2018), gold (Au) (Mehmood et al. 2022), copper (Cu) (Kanhed et al. 2014), zinc oxide (ZnO) (Król et al. 2017; Malandrakis et al. 2022), titanium oxide (TiO_2) (Alizadeh Sani et al. 2022), magnesium oxide (MgO), iron oxide (Fe_3O_4), and silicon dioxide (SiO_2) (Omerović et al. 2021) as potential antifungal agents in the agri-food chain.

The three main pathways reported, and their synergistic effects explain the processes that lead to the antimicrobial activity of NPs: (i) direct absorption of NPs; (ii) indirect action of NPs through the formation of ROS; (iii) damage to the cell wall/membrane caused by the accumulation of NPs; and (iv) DNA damage (Król et al. 2017; Koduru et al. 2018; Omerović et al. 2021; Akhila et al. 2022) (Figure 2).

Direct contact of inorganic NPs with cell membranes causes loss of integrity, which can lead to damage to the peptidoglycan layer when in contact with bacteria, toxicity by the release of toxic metal ions, alteration of the cellular pH through proton flow pumps, generation of ROS and damage by nuclear substances, inhibiting microbiological growth (Abd Elsalam et al. 2019).

According to investigations in the food area, silver (Ag) has a high antifungal potential (Jamdagni et al. 2018; Nagaraju et al. 2020; Chen et al. 2021; Akhila et al. 2022). The release of Ag ions causes the inactivation of thiol groups in fungal cells, resulting in the rupture of membranes and cell death through the inactivation of enzymatic function and interruption of ATP synthesis. Mutations in fungal DNA, dissociation of enzymatic complexes necessary for the respiratory cycle, reduced membrane permeability, and cell death are also mechanisms of action of Ag in microorganisms (Velmurugan et al. 2009). Silver NPs (AgNPs) interact with microorganisms more effectively due to their high surface area (Jamdagni et al. 2018).

According to investigations, contact with proteins containing sulfur or thiol groups in the membranes of microorganisms causes the deregulation of their permeability and therefore the death of the microbial cells, which may explain the similarity of the antifungal mode of action of Ag NPs to that of silver ions (Koduru et al. 2018; Musa et al. 2018). In addition, other research reveals that Ag NPs can generate an imbalance in the cell wall and membrane integrity, reduce ATP levels, and bind to DNA, leading to denaturation of this molecule and prevention of cell reproduction, in addition to interfering in DNA replication and blocking the respiratory chain of microorganisms by interaction with enzymes, such as cytochrome oxidase and succinate dehydrogenase (Costa et al. 2010; Damm et al. 2008).

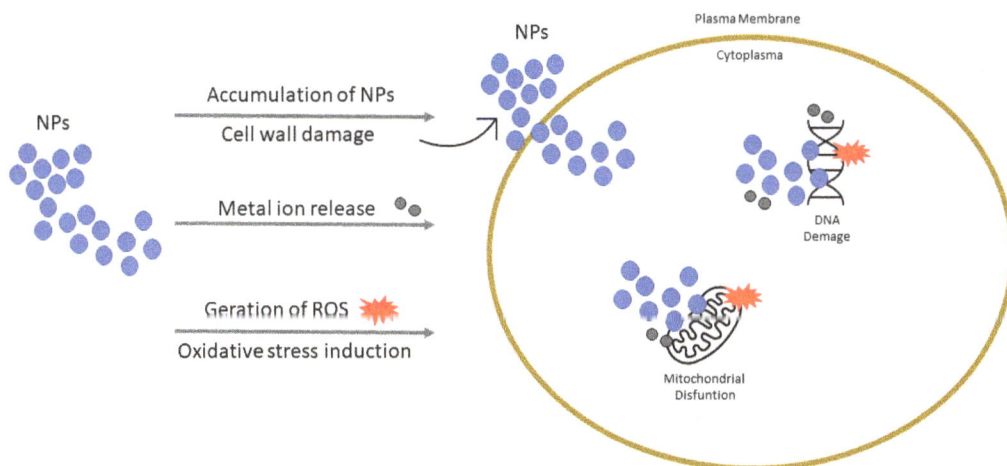

Figure 2. Potential antifungal mechanisms of nanoparticles (NPs).

The potential of AgNPs as antifungal agents was explored by Elgorban et al. (2016). Varying concentrations of NPs (0.0 mol.L^{-1}, 0.0002 mol.L^{-1}, 0.0005 mol.L^{-1}, 0.0007 mol.L^{-1}, 0.0009 mol.L^{-1}, 0.0014 mol.L^{-1}, and 0.0019 mol.L^{-1}) were tested in different groups of *Rhizoctonia solani* anastomosis (AGs) infecting cotton plants. According to the authors, all concentrations studied inhibited the fungal growth of *R. solani* AGs. Tutaj et al. (2016) developed a hybrid system synthesizing AgNPs with macrocyclic polyene amphotericin B (AmB). The AmB-Ag NPs showed high antifungal activity against the fungal species *Fusarium culmorum*, *Aspergillus niger*, and *Candida albicans*. In addition, the AmB-AgNPs were cytotoxic to the cell lines CCD841CoTr and THP-1 evaluated.

Titanium dioxide nanoparticles (TNPs) have been widely used in the food industry. They are multifunctional and have attractive characteristics due to their high stability, photocatalytic effect, UV absorbance, low toxicity, microbiological activity, and biocompatible property (Sani et al. 2022). Investigations have associated the mechanism of action of TNPs with the generation of ROS. When TNPs are exposed to ultraviolet light, ROS such as hydroxyl ions and superoxide are produced in the presence of water and oxygen species. As a result, ROS reacts with polyunsaturated phospholipids in the cell membrane of microorganisms (Mesgari et al. 2021).

Zinc (Zn) and its oxide (ZnO) are considered promising metallic nanomaterials used in investigations of nanoparticles with antimicrobial properties have been carried out in several areas, such as drugs, foods, and packaging (Król et al. 2017). Metal oxide nanoparticles are stable, have antimicrobial activity at low concentrations when compared with microparticles, and are considered nontoxic to humans (Zhang et al. 2007). Investigations have revealed the high efficiency of ZnO NPs, which can be used as effective fungicides against plant pathogens (He et al. 2011; Kairyte et al. 2013; Sharma et al. 2010).

Some studies have revealed that the size of NPs and the concentration of ZnO NPs may influence antifungal efficacy. He et al. (2011) evaluated the antifungal activity of ZnO NPs with a size of 70 ± 15 nm at different concentrations (0 mmol.L^{-1}, 3 mmol.L^{-1}, 6 mmol.L^{-1}, and 12 mmol.L^{-1}) against two postharvest pathogenic fungi (*Botrytis cinerea* and *Penicillium expansum*). Concentrations greater than 3 mmol.L^{-1} significantly inhibited the growth of both fungi. Also, as the concentration of ZnO NPs increased from 3 mmol.L^{-1} to 12 mmol.L^{-1}, and its antifungal activity increased. In addition, the fungus *P. expansum* was more sensitive to treatment with NPs than *B. cinerea*. Sharma et al. (2010) reported that 32 nm ZnO NPs showed higher antifungal activity against *Fusarium* sp. in comparison with NPs of 37 nm (20% and 7.5% respectively).

The mechanism of antifungal action *in vitro* was studied by He et al. (2011). According to the authors, SEM was used to examine the structural changes of fungal samples of *P. expansum* and *B. cinerea*. The fungal samples were inoculated in potato dextrose agar culture medium containing

12 mmol.L^{-1} of ZnO NPs and incubated at 25°C for 12 days. The researchers observed that NPs of ZnO inhibited the growth of *B. cinerea* by deforming the structure of fungal hyphae, affecting cellular functions since unusual lumps were observed on the surface of the fungal hyphae. The germination of *P. expansum conidia* was completely inhibited and its development was suppressed, which can cause the death of fungal hyphae. Fungal cell membrane rupture, which could lead to a potential decline in fungal enzyme activity, may be the cause of fungal growth inhibition (Król et al. 2017).

Nanoformulations to Control Postharvest Fungal Diseases

Agriculture has been evolving over the years to adapt to the requirements for the quantity and quality of food produced (FAO 2018, 2022). The Green Revolution that occurred in the agricultural sector, beginning in the 1960s was the result of scientific and technological advances. On the other hand, intensive food cultivation in subsequent years raised concerns about the depletion and contamination of soils and aquifers (Evenson and Gollin 2003).

It is estimated that by the year 2050, the global population will reach around 10 billion inhabitants, 20% higher than at present. This population increase will require greater food production and higher consumption of inputs and natural resources (FAO 2018). In addition to the search for higher food production, it is necessary to raise levels of quality and sustainability (FAO 2022). In a recent study by Parfitt et al. (2021), the authors highlighted that only in the primary production chain (harvest and postharvest), 1.2 billion tons of food is lost per year, partly during harvest (8.3%) and partly after harvest (7.0%).

To avoid or minimize the effects caused by intensive agriculture, several studies have been conducted to investigate the possibility of using new technologies and inputs for food production (Abdollahdokht et al. 2022; Wassermann et al. 2022; Zhang et al. 2020). Conventional fungicides are considered pesticides, with toxic chemical composition to phytopathogenic fungi, but also can be toxic to humans. Therefore, even their use in small quantities can contribute to ecosystem pollution (Abdollahdokht et al. 2022). To minimize the effects of these agrochemicals, studies of nanoformulations have been conducted, using biocompatible and biodegradable materials, to develop less polluting and more efficient fungicides (Nagaraju et al. 2020; Xu et al. 2018; Yao et al. 2018).

The development of new products should consider the need to protect the environment and human health, besides fungicidal characteristics (Smith et al. 2008). In addition, formulations for postharvest applications must meet some important requirements (Adaskaveg et al. 2021), as shown in Figure 3.

The use of commercial fungicides is still the most common method for the control of fungal diseases in the field and after harvest. As examples, for both types of application, we can mention the fungicides azoxystrobin, propiconazole, and fludioxonil among others (Adaskaveg et al. 2021).

Many studies related to the use of postharvest nanotechnology have been conducted to develop films or coatings obtained from biopolymers and incorporated with agents that will lengthen the shelf life of foods (Baldwin et al. 2011; Nešić et al. 2020). Few studies have mentioned the use of nanoformulations using commercial fungicides for direct application after harvest, but some studies of these products for application in the field can serve as a basis for further studies since many of these fungicides are also used to control postharvest diseases. The word nanofungicide is typically used to identify any particle with antifungal activity that is smaller than 1,000 nm in diameter. Nanofungicides can have new properties related to their nanoscale size (Abd-Elsalam et al. 2019).

Two of the most frequent problems related to commercial fungicides are the loss of much of the product applied by leaching or derivation, and the development of resistance in some microorganisms due to consecutive applications of the same product on the crop or the incorrect use of fungicide (Adaskaveg et al. 2021). In this sense, the use of nanotechnology in formulations plays a crucial role in the better use of products because it prevents losses and excessive use, increases efficiency,

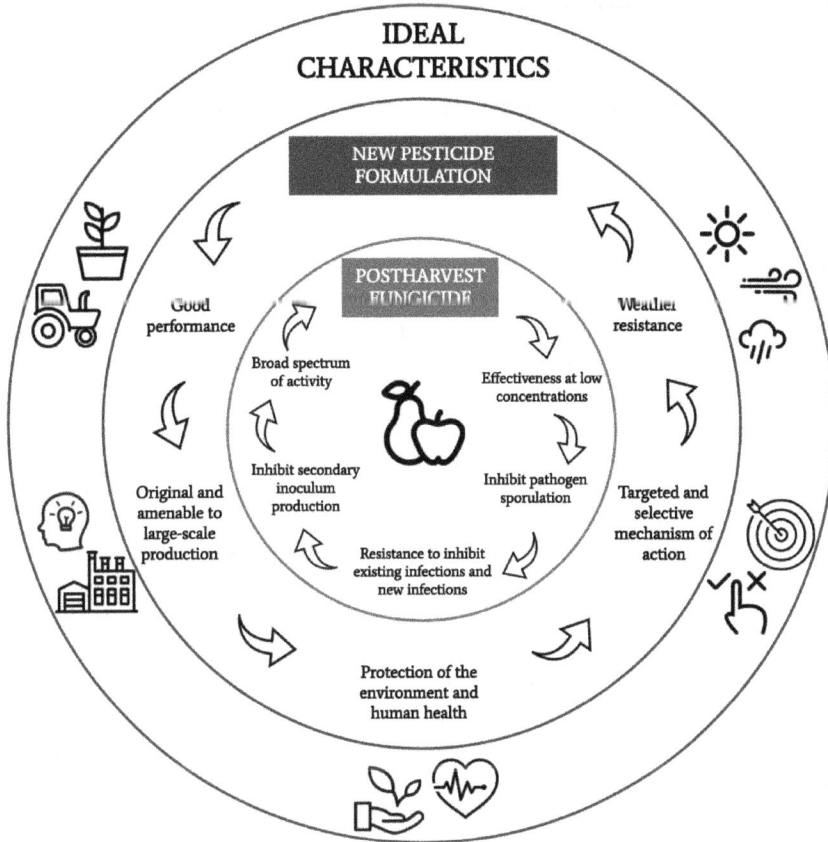

Figure 3. Characteristics for the formulation of new pesticides, adapted from Smith et al. (2008) and Adaskaveg et al. (2021).

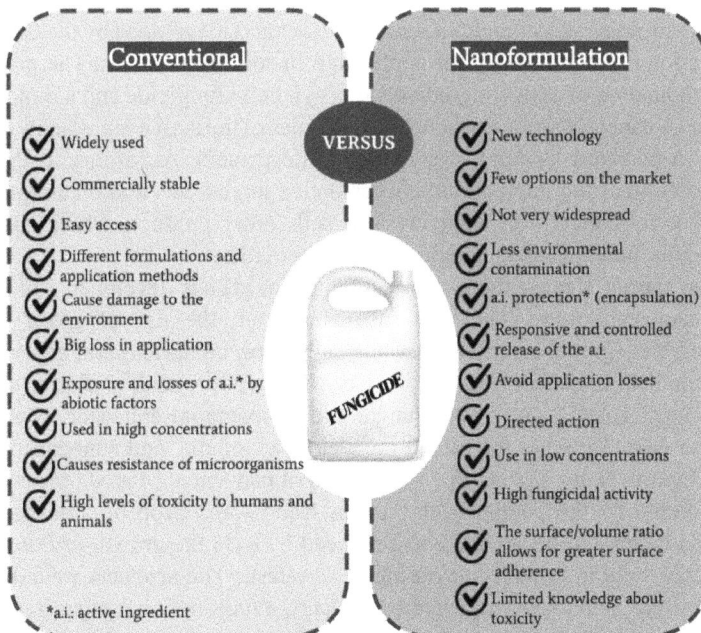

Figure 4. Comparison between the characteristics of commercial fungicides and nanoformulations.

protects the active ingredients from degradation, and reduces contamination of the environment (Tleuova et al. 2020) (see Figure 4).

The application of synthetic fungicides can contaminate ecosystems due to the accumulation of residues during and after their application. The toxicity of azoxystrobin to aquatic organisms is considered a worrisome factor due to its accumulation in groundwater and surface watercourses. Another important point about these molecules is that the incidence of UV light can form ROS, from the aromatic rings present in them, which in the presence of photons can generate such compounds (Zhang et al. 2020).

A comparison between conventionally sized and nanometric-sized azoxystrobin molecules was conducted by Zhang et al. (2020) to verify the influence of particle size on the toxicity of this active ingredient in zebrafish embryos and observe the action of UV light incidence on the toxicity of the substance. They observed that the molecules with nanometric size were more lethal to the zebrafish embryos in comparison with those of conventional size, even at lower concentrations. They attributed this to the greater bioavailability of nanoparticles, which have become more dispersive in water. Increased toxicity occurred with both particle sizes when exposed to UV light.

In order to avoid the accumulation of residues and improve the use of the fungicide sodium orthophenylphenate (SOPP), silver phosphate nanocrystals (Ag_3PO_4) with average size of 603.7 nm were combined with SOPP. The resulting formulation enhanced the antifungal activity and increased the decomposition of the fungicide residues to almost 90%, when exposed to artificial sunlight, demonstrating the high catalytic activity of these particles (Xue et al. 2016).

Postharvest treatment tests were carried out on tomato fruits inoculated with different isolates of the fungus *Alternaria alternata* (sensitive and resistant to the fungicide boscalid), using zinc oxide nanoparticles (nanoZnO) with size smaller than 100 nm and boscalid, separately and combined, with concentrations 1,000 µg.mL^{-1} nanoZnO, 500 and 1,000 µg.mL^{-1} of boscalid. The joint use of the nanoparticles and fungicide positively influenced the inhibition of the development of lesions caused by the fungus, varying between 34.21% and 82.32%, showing the strong synergistic effect between the two products (Malandrakis et al. 2022).

Current Industry

Recently the U.S. Environmental Protection Agency (EPA) approved the use of the fungicide AZterknot® (azoxystrobin + *Reynoutria sachalinensis* extract), developed by the Canadian company Vive Crop Protection (vivecrop.com), for application in soil or on leaves. The product is the first to present the combination of active ingredients of a synthetic fungicide and a biological fungicide, along with the use of nanotechnology, where the molecules of the active ingredients have nanometric size, encased by a patented nanopolymer carrier (Allosperse®). According to the company, the polymer carrier optimizes the interaction between active ingredients and other agricultural inputs, achieving better performance. The company also sells other products where nanotechnology is present, such as other fungicides, nematicides, and insecticides.

The Brazilian startup Nanoscoping (nanoscoping.com.br) has developed a line of agricultural products using nanotechnology. The products come from the use of green technology and biocompatible and biodegradable systems and are called special agricultural adjuvants. The line is composed of a repellent enhancer (Nano Agro Neem®), an antifungal, antibacterial, and repellent enhancer (Nano Agro Total®), and an antifungal and antibacterial enhancer (Nano Agro Crop®). The product Nano Agro Total has the oils of melaleuca, citronella, and nanoencapsulated neem in its composition. In turn, Nano Agro Crop is composed of nanoencapsulated oregano oil. According to the tests performed by the company, the use of Nano Agro Crop on strawberries during the postharvest period increased the shelf life and reduced by 84% the growth *of Botrytis cinerea*, the causal agent of gray mold in these fruits, during *in vitro* tests. The products are indicated for use in organic cultivation and have been inspected by Ecocert, a multinational organization that certifies sustainable agricultural practices.

The Egyptian company Bio Nano Technology (bionano-egy.com) applies methods and techniques involving nanotechnology for the production of its products. Nanocu is a fungicide and bactericide developed based on copper nanoparticles (10%) made by this company.

Nanoparticles and Nanocarriers of Different Fungicides

Nanocarriers or nanocontainers are defined as nanostructures used for encapsulation and release of active molecules (Apolinário et al. 2020; Nguyen-Tri et al. 2019). These nanostructures can originate from inorganic substances (i.e., metals, metal oxides, silica, quantum dots, clay, montmorillonite, etc.) and organic (biopolymers, lipids, proteins, carbon, and dendrimers (Kutawa et al. 2021; Nguyen-Tri et al. 2019), as identified in Table 1. The use of nanocarriers protects the active ingredient present inside from direct contact with the environment that could cause its decomposition (Nguyen-Tri et al. 2019).

One of the main interests regarding the development of new fungicides aimed at postharvest application and for use in agriculture, in general, is the manner in which the antifungal substance is released after application. This is because a slow and sustained release ensures that the fungicide will have longer contact with the fungus (Kutawa et al. 2021).

In this sense, in comparison with larger particles, nanoparticles have improved action due to easier adsorption, greater contact surface, and faster migration of compounds. Similarly, when the active ingredient is encapsulated or trapped in a nanometric-size matrix, it will be released more slowly (Ramezani et al. 2019; Zhang et al. 2016).

The chemical group strobilurin is widely used around the world for the production of commercial synthetic fungicides with systemic action. The active ingredient azoxystrobin (AZO) {methyl (E)-2-[2-[6-(2-cyanophenoxy)-pyrimidin-4-yl]oxyphenyl]-3-methoxyprop-2-enoate} is commonly used because it is a broad-spectrum fungicide belonging to this group. AZO has low solubility in aqueous media, so its indiscriminate and excess use can generate resistance of microorganisms, phytotoxicity lesions, and toxicity to non-target species, negatively affecting ecosystems (Yao et al. 2018a).

Yao et al. (2018a) encapsulated AZO in polylactic acid, a biodegradable and biocompatible polymer. It has been studied and used as a carrier for the controlled release of drugs with low stability, high degradability and toxicity, and short half-life. The encapsulation method used (emulsification-evaporation of the solvent) allowed for obtaining uniform hydrated spheres with three distinct sizes (130.9 ± 0.2 nm, 353.4 ± 6.3 nm and 3078.0 ± 336.6 nm). During the *in vitro* controlled release tests, the smallest particles had the highest initial release speed compared to the other sizes, and after 24 hours of the test had released 90.09% of the AZO. The nanoparticles with a size of 130.9 ± 0.2 nm demonstrated good physical and chemical stability during the 14 days of the storage test when submitted to three temperatures (4°C, 25°C, and 54°C) and had better wettability and adhesion in relation to larger particles. In addition, these nanoparticles provided the greatest inhibition of mycelial growth of the fungus *Colletotrichum higginsianum* Sacc *in vitro*, even with the lowest median lethal concentration of AZO (LC_{50} = 2.0386 µg.mL^{-1}), while the microparticles (3078.0 ± 336.6 nm) presented LC_{50} = 21.3 µg.mL^{-1}. The same active ingredient was encapsulated in mesoporous silica nanoparticles (NSMs) improved with carboxymethyl chitosan, which allowed a significant increase in nanocapsule loading, controlled release of the active ingredient in acid medium, and better fungicidal activity of AZO in comparison with normal sized particles (Xu et al. 2018).

To study the effect of particle size on the efficiency of the fungicide prochloraz, Zhang et al. (2016) encapsulated nano- and microparticles of this fungicide in nanoparticles developed on the basis of the copolymer methoxy polyethylene glycol-poly (lactic-co-glycolic acid) (mPEG-PLGA), with different molecular weights. The particles were obtained through the emulsion/solvent evaporation technique. The authors reported that the higher the molecular weight of the copolymer, the more difficult it became to obtain smaller particles, as well as less likely the formation of large particles with low molecular weight copolymers. When producing nanoparticles

Table 1. Use of nanocarriers for delivery of antifungal agents.

Fungicide (Code FRAC)	Nanocarrier Type	Nanocarrier Composition	Particle Size (nm)	Postharvest Species	Target Microorganism	Positive Effect	Reference
Organic							
Carbendazim (B1)	Nanocapsules	Chitosan and Pectin	70–90	-	*Fusarium oxysporum; Aspergillus parasiticus*	- Sustained release. - 100% fungal mycelial growth inhibition. - 99.2% encapsulation efficiency.	Sandhya et al. (2017)
Curcumin	Nanofibers	Zein	350	Apple	*Botrytis cinerea; Penicillium expansum*	- Nanofibers showed good stability and encapsulation efficiency (82.4–86.7%) - Encapsulation increased the inhibition of fungal mycelial growth, more effectively in *P. expansum*. - A 50% reduction in the diameter of the *P. expansum* lesion in coated fruits.	Yilmaz et al. (2016)
Prochloraz (G1)	Nanocapsules	MSN and Chitosan	340	Citrus	*Penicillium. digitatum, P. italicum,* and *G. candidum*	- Release controlled by Fickian diffusion. - Nanoparticles had excellent esterase and pH dual-responsiveness. - Toxicity of the nanoparticles to zebrafish was reduced by more than 6x compared with that of Prochloraz. - Inhibiting effect on the fungi evaluated.	Liang et al. (2018)
Azoxystrobin (C3)	Microspheres	PLA	119.4; 30.5; 2609.2 (dried)	-	*Colletotrichum higginsianum* Sacc	The smallest microspheres: - Had the fastest rate and the highest cumulative release percentage (90.09%). - Greater surface adhesion. - Caused more severe oxidative damage and had greater antagonist activity to the fungus.	Yao et al. (2018a)
Azoxystrobin (C3)	Nanosuspension	1-dodecanesulfonic acid sodium salt and PVP K30	238.1 ± 15	-	*Fusarium oxysporum*	- Increased retention volume. - Enhanced wettability and adhesion. - Had the highest antifungal activity - Reduced the defensive antioxidant capacity of the fungus.	Yao et al. (2018b)
Azoxystrobin (C3)	Nanocapsules	MSN coated with CMCS	222 (dried)	-	*Phytophthora infestans*	- Had slower release at acidic pH. - Had greater antifungal activity.	Xu et al. (2018)

						Findings	Reference
Propiconazole (G1)	Nanocapsules	PLA and PLGA	146,28	-	*Fusarium dieback*	- PLA showed higher encapsulation efficiency - Encapsulation efficiency was over 42% - The nanofungicide system enhanced the growth inhibition percentage by 5% - Good dispersivity in water - 2x longer sustained release time.	Barrera-Méndez et al. (2019)
Cinnamon (*Cinnamon zeylanicum* L.) essential oil	Nanoemulsion	Essential oil, Span 80 (Sorbitan monooleate), Tween 80 (Polysorbate 80), and Lecithin	115,33 ± 3,97	Strawberry	*Rhizopus stolonifer*; *Botrytis cinerea*	- The increase in the concentration of essential oil enabled greater antifungal action. - The nanoemulsion with the highest concentration of cinnamon essential oil (0.2%) was more efficient in controlling soft rot and gray mold than essential oil or commercial fungicide.	Naserzadeh et al. (2019)
Oregano (*Origanum vulgare* L.) essential oil	Nanoparticles	Chitosan	290–483	-	*Alternaria alternata*	- Electrospray technique developed nanoparticles with high efficiency. - High encapsulation efficiency, varying between 70.1% and 79.6%, depending on the concentration of the essential oil. - Encapsulation increased the antifungal activity of the essential oil.	Yilmaz et al. (2019)
Natamicin	Core-shell nanocapsules	Zein stabilized with CMCS	223 ± 2	Strawberry	*Botrytis cinerea*	- The increased half-life of natamycin under UV light. - Encapsulation efficiency from 68.3% to 78.4%. - Inhibition of the rate of spore germination and mycelial growth of the fungus. - Reduced the occurrence of rot and gray mold on fruits.	Lin et al. (2020)
Iron oxide	Nanoparticles	Chitosan	20–30	Peach	*Rhizopus stolonifer*	- The synthesized nanoparticles inhibited the development of soft rot in fruits. - Reduction of fruit mass loss with the use of nanoparticles.	Saqib et al. (2020)

Table 1 contd. ...

...*Table 1 contd.*

Fungicide (Code FRAC)	Nanocarrier Type	Nanocarrier Composition	Particle size (nm)	Postharvest Species	Target Microorganism	Positive Effect	Reference
Inorganic							
Extract of pomegranate fruit waste	Composite nanoparticles	Impure MT	229 (MT	Apple	*Botrytis cinerea*	- Methanolic extract had greater antifungal activity. - Greater antifungal activity *in vitro* and *in vivo* from the synergistic effect between MT and pomegranate extract.	Balooch et al. (2018)
Carbendazim, Mancozeb and Thiram (B1; M3; M3)	Nanoparticles	Silver	31.86 (green); 41.91 (chemical)	-	*Fusarium oxysporum; Penicillium expansum*	- The combined use of silver nanoparticles and fungicides enabled inhibitory action on fungal development. - Synergistic action between nanoparticles and fungicides on the fungus *F. oxysporum*.	Jamdagni et al. (2018)
Carbendazim (B1)	Nanoparticles	Silver	19–24	Mango	*Colletotrichum gloeosporioides*	- Maximum antifungal activity at low concentration (0.1%), when compared with silver nanoparticles alone and the commercial fungicide (1%). - Synergistic effect between silver nanoparticles and carbendazim.	Nagaraju et al. (2020)
Chitosan	Composite nanoparticles	Silica	48	Grape ('Italia' and 'Benitaka')	*Botrytis cinerea*	- Synergistic effect between chitosan and silica, inhibiting *in vitro* fungal growth by 100%. - Reduced the development of gray mold on fruits by more than 58% compared to the control.	Youssef et al. (2019)

FRAC: Fungicide Resistance Action Committee; PLA: polylactic acid; MSN: mesoporous silica nanoparticles; PVP:11 polyvinylpyrrolidone; PLGA: polylactic-co-glycolic acid; MT: montmorillonite; CMCS: carboxymethyl chitosan.

with the copolymer mPEG-PLGA (Pm = 45 KDa), particles were obtained with sizes between 178.2 ± 1.2 nm and 3980.0 ± 58.4 nm, depending on the proportion of copolymer and fungicide. The highest particle loading with prochloraz was verified in particles of larger size and lower proportion of copolymer/fungicide mass (5:1). An inverse result was also obtained, where smaller particles and a higher proportion of copolymer/fungicide mass (100:1) had the lowest loading. When the controlled release test was performed, the particles obtained with the mass ratio 5:1 and of different sizes (190.7 ± 2.1 nm; 708.8 ± 12.9 nm; 3980.0 ± 58.4 nm) released the fungicide in a slower and more sustained form after 24 hours in comparison with the original formula of the fungicide. However, the smaller particles with lower loading had a faster initial release in relation to the other particles, as well as better *in vitro* antifungal efficiency against the fungus *Fusarium graminearum*, attributed to its better dispersivity and higher initial release speed, allowing action on the fungus in a short time interval (14 days).

Other studies have also found higher efficiency of this same fungicide in association with nanotechnology. Liang et al. (2018) produced prochloraz nanoparticles encapsulated by mesoporous silica nanoparticles, and chitosan as a coating agent. The nanoparticles obtained had high loading power, and were responsive in an environment with acid pH and the presence of esterase, enabling the controlled release of the fungicide. In addition, prochloraz encapsulation improved its antifungal activity against diseases affecting citrus plants.

In addition to conventional fungicides, other molecules with antifungal activity have been encapsulated and presented satisfactory results in controlling the development of fungal diseases (Lin et al. 2020; Yilmaz et al. 2019).

Lin et al. (2020) encapsulated natamycin, a natural fungicide. The authors used zein reinforced with chitosan carboxymethyl to obtain nanocapsules, which served for application on strawberry fruits. Natamycin is a substance extremely sensitive to UV light, and this sensitivity decreased after encapsulation. The nanocapsules, at all concentrations, completely inhibited the development of *Botrytis cinerea* spores, and at a concentration of 10 mg.L^{-1} reduced their mycelial growth by 64.4% *in vitro*. The treatment with strawberry nanoparticles reduced the occurrences of rot and mold in the fruits and preserved them with the lowest rot rate (31.1%) for up to 8 days of storage.

Nanoparticles and Nanomaterials as Elicitors

An elicitor is a biostimulant agent capable of triggering an immune response in the plant or fruit, producing a series of modifications in metabolic processes (Juárez-Maldonado et al. 2019; Patel et al. 2020).

According to Juárez-Maldonado et al. (2019), after the action of a biostimulant agent, a series of events occurs of perception, transduction, signaling, effector action and modification of gene expression, metabolism, or cellular characteristics of the organism.

EOs are secondary metabolites formed from complex mixtures of natural and volatile compounds extracted from various parts of aromatic plants (e.g., stems, leaves, fruits, roots). EOs contain some bioactive compounds able to act as antifungal agents, antioxidants, and insecticides, among others (Robledo et al. 2018; Delshadi et al. 2020).

Yang et al. (2022) developed a nanoemulsion combining three distinct EOs in the same proportion (cinnamaldehyde, carvacrol, and eugenol) for application on 'Newhall' oranges to induce the natural resistance of fruits to the fungus *Penicillium digitatum*, the causal agent of green mold in citrus. That nanoemulsion was tested on fruits artificially inoculated at different concentrations (0.625 mg/mL; 1.25 mg/mL; 2.5 mg/mL; 5.0 mg/mL; 10.0 mg/mL). These fruits were incubated for 6 days at 25 ± 2°C and 80–90% RH. The treatment with nanoemulsion was able to inhibit the development of the fungus, with higher concentrations being more efficient. In addition, molecular analyses demonstrated the elicitor potential of the nanoemulsion, where genes encoding 5 enzymes, including peroxidase, were expressed differentially when they received treatment with the nanoemulsion. The accumulation of important amino acids, organic acids, soluble sugars, and lipids were also observed and the contents of flavonoids, total phenolics, and lignin were higher, along with increased firmness of treated fruits. These results indicated that the treatment was able to induce resistance against infection by *P. digitatum*.

Salicylic acid (SA) is a substance involved in signal transduction systems, inducing specific enzymes that catalyze biosynthetic reactions to form defense compounds, such as phenolic compounds, alkaloids, and pathogenesis-related proteins, playing a key role in plant growth, development, and defense response (Chen et al. 2006).

Pepper fruits received pre-harvest treatment with SA nanoparticles and their bulk form at concentrations of 0.7 mM and 1.4 mM, with the objective of preventing fruit infection by *Alternaria alternata*, a fungus responsible for black mold disease in this species. The nanoparticles at a concentration of 1.4 mM achieved superior results to treatment with bulk AS during *in vivo* and *in vitro* tests since they more efficiently suppressed disease severity and fungal growth, even after artificial inoculation. The genes that encode proteins linked to the defense of the pepper plants PR-1 (CaBPR1), beta-1, 3-glucanase (CaBGLU), and peroxidase (CaPO1) showed positive expression in the leaves of pepper plants treated with both forms of SA, demonstrating the elicitor effect of SA on the plants (Abdel-Rahman et al. 2021).

Silicon is a beneficial element for several plants, although it is not available in the soil. When supplied externally, its absorption can occur passively from the transpiration pathway or actively through uptake by the roots. Some studies have demonstrated the antifungal activity and other benefits associated with the external supply of silicon (Ahammed and Yang 2021).

Peng et al. (2022) applied silicon dioxide nanoparticles to ginger rhizomes to avoid their deterioration due to fungal diseases. The fungus *Fusarium solani* is responsible for causing a disease called soft rot in this species in the postharvest period. The rhizomes were immersed for 10 minutes in different concentrations of a nanoparticle suspension (50 mg, 100 mg, and 150 mg. L^{-1}), left to dry in the air, and then stored in single packs at 12°C and 85–90% RH for 30 days. After storage, the treated rhizomes were firmer and had lower mass and disease severity. Treatment with a concentration of 100 mg. L^{-1} produced the best results, increasing the activity of the antioxidant enzymes superoxide dismutase and catalase and the contents of total phenolics and flavonoids, thus reducing the accumulation of ROS and the content of malondialdehyde.

Toxicity and Leaching of Nanomaterials

The exponential growth of nanotechnology has led to the application of nanomaterials in a wide range of consumer products, such as processed foods, textiles, cosmetics, and biomedical products. The development of new food products that are more attractive, tasty, and/or easy to digest through the application of nanotechnology has been increasing significantly in the global food industry. However, new or modified products can pose risks.

Considering the lack of comprehensive information on the toxicity of nanoparticles, the precautionary principle imposes the regulation of their application to minimize the risks to human health and the environment or even their outright prohibition. Human exposure to these materials is inevitable, so an important new area of interest in the field of nanotechnology is nanotoxicology, which is the study of the toxicity of nanomaterials in contact with humans. Similarly, nano-ecotoxicology studies evaluate the effects of nanomaterials on the environment, elucidating the transfer pathways of toxic agents, as well as their interaction (Azevedo and Chasin 2003; Durán et al. 2014; Scheringer 2008; Selvaraj et al. 2018).

Studies of the toxicity of nanomaterials reported in the literature are generally focused on inorganic materials such as titanium dioxide nanoparticles (TiO$_2$ and E 171), iron oxide (E 172) applied to food, and plastic nanoparticles, which are generally absorbed by human body from environmental contamination (Chen et al. 2020; Kik et al. 2020; Voss et al. 2020). Titanium dioxide is a white dye used in food, toothpaste, and medicines. Its application is regulated by the FDA (< 1% of the product by mass) and in Europe (quantum satis), and the presence of TiO$_2$ on a nanoscale is increasingly frequent due to higher sensory quality and refraction rate, strong coverage, whiteness, and antibacterial potential. However, the smaller the size of these nanoparticles, the greater their permeability and accumulation in organs and tissues will be (Chen et al. 2020). Voss et al. (2020) identified that 50% of iron oxide pigment particles were in the nanoscale, which indicates excessive exposure and possibly deleterious effects on consumer health. Although there were no toxic effects observed on the human intestinal cell line used in the study, the authors suggested that further studies should be conducted to elucidate the cell responses of these nanoparticles.

Leaching is the movement of chemicals and water through the soil. It is mainly considered a problem associated with pest control. However, only a few investigations have been carried out on this aspect. Some articles have reported a decrease in the toxicity of fungicides loaded with nanoparticles. Some experiments of soil leaching showed that when nanoparticles were added, the fungicide release rate decreased in the soil in comparison with commercially formulated fungicides. Analysis of a formulation of nanoparticles after two weeks of use proved the effectiveness of nanoformulations over long time intervals. Kumar et al. (2017) reported a significant inhibition of fungal mycelium when polymeric nanoparticles of carbendazim were applied against *A. parasiticus* and *F. oxysporum.*

Biosafety of Nanoparticles

Various concerns have arisen due to the lack of knowledge about the interactions of materials on a nanoscale, at the molecular or physiological level, and their effects on consumer and environmental health. On the other hand, the possibility exists of applying nanotechnology in agriculture to create new fungicides and different active ingredients for the control of fungal diseases in plants.

Nanoparticles are also regularly used as nanorings to encapsulate and release active particles to form powerful formulations in the agroindustry. The use of nanoparticles as nanocarriers for fungicides has various potential benefits, including reducing losses due to leaching, improvement in the release profiles of bioactive compounds, and reducing toxicity to humans and the environment. These types of fungicide systems are an alternative for the management of fungal diseases in different plant species (Worrall et al. 2018).

Biosafety is a major concern regarding the use of nanotechnology for the management of fungal diseases. There are some uncertainties regarding the long-term impact of using nanoparticle fungicide formulations on human health and the environment (Zahid et al. 2019). Consequently, it is necessary to evaluate the uses of nanofungicide at the time of spraying by agricultural workers. By assessing the toxicity of nanoformulations, it is possible to reduce the impact on the environment and humans in relation to conventional fungicides. Nanoformulations are seen as safer and more environmentally friendly for disease control, but a high level of toxicity of nanoparticles incidentally released in the environment can cause negative effects on other microbes and humans. The toxicological impacts of nanomaterials on soil microbes and plants have been studied in general, but there are numerous gaps in knowledge about the agro-ecotoxicity of nanomaterials. Furthermore, there are numerous uncertain issues and new difficulties relating to biological impacts (Maluin et al. 2020).

NPs were found to be phytotoxic at a high concentration (3,200 ppm) when applied to pansy and cucumber plants, indicating that nanoparticles can cause negative impacts and positive impacts on the root system, which depends on plant species (such as cucumber, soybean, corn, carrot, tomato, and cabbage). Nanomaterials manufactured with TiO_2 and ZnO affected the microbial community, biomass, and diversity in the soil. To understand the potential advantages of the application of nanotechnology to agriculture, the initial step is to determine transport. In this respect, it is necessary to investigate the penetration of nanoparticles in plants and how they behave (Ge et al. 2011; Lee et al. 2012; Maluin et al. 2020; Nowack et al. 2012).

Legislation on the Use of Nanoparticles in Food

There is no specific legislation on the use of nanoparticles in food or materials in contact with food. There is still no consensus on the type of approach to conducting the safety evaluation processes of these materials. According to Yingchun et al. (2021), the damage resulting from the lack of scientific knowledge about the potential risks of nanomaterials cannot be classified as a defect of the product, since there is no flaw in its conception, but rather an absence of scientific certainty about the risks. Thus, there would be no possibility of holding the supplier liable due to the implicit adoption of the development risk theory.

Food regulation is the establishment of rules for the labeling and composition of processed foods, dissemination and advertising of foods, and maintenance of quality at all stages of the production and distribution chains (Magalhães 2017). The regulation reflects the assessment of hazards that a material, product, or service may pose to any human, animal, or component of nature. Countries, where there are large investments in research, are even more concerned with drafting laws and rules governing the development of technology.

According to StatNano (2022), nanomaterials can be found in several areas, mainly in electronics, pharmaceuticals, and cosmetics. In the last case, there are about 860 registered cosmetic products in the world, divided into 100 different types, produced by 248 companies and 29 countries. The use of nanotechnology in the food industry has promising potential in several aspects. Until April 2021, 348 products in the food area were registered in the StatNano (2022) database, divided into 50 types, produced by 138 companies in 26 countries. The products are divided into food sensors, sports nutrition, meals, packaging, and supplements, the latter two with more than 100 products each.

To date, there is no specific or unified global legislation for nanomaterials. Instead, each product is registered according to a country's regulatory agencies and evaluated individually, despite the growth of the global nano-product trade. According to the Ministry of Science, Technology and Innovations (MCTIC), Brazil already has an "ecosystem" for nanotechnology developments that covers the public and private sectors. In addition, it also has partnered with the European Union on the NANoREG Project to provide standard operating procedures for nanomaterial safety assessment (Berti et al. 2019).

The European Union has a series of specific regulations that address different aspects. In general, European Union legislation requires approval to apply nanomaterials in food based on the assessment by the European Food Safety Authority (EFSA) of potential health risks (Rasmussen et al. 2018). In the United States, the Food and Drug Administration (FDA) is the leading agency in the regulation of nanotechnology, with regulations for pharmaceuticals with nanoparticles that can assist in the validation of food products. Countries such as the United Kingdom, China, Japan, Australia, and India, among others, have also enacted regulatory measures regarding nanotechnology to ensure protection in development and human health (He et al. 2019; Ramkumar et al. 2019).

According to the FDA, products involving nanotechnology applications are not intrinsically categorized as benign or harmful. Even with the use of traditional substances, new properties can be attributed to particles in the nanometer range, which should be specifically analyzed cases (FDA 2014).

Conclusion, Opportunities, and Future Challenges

Currently, new nanotechnological products with unknown effects can be found without appropriate regulation. There are polemic discussions about the need to regulate nanotechnologies between scientists and regulatory advocates. There is a certain reluctance to buy food produced with nanotechnology due mainly to a greater perception of risks associated with new technologies than the expected benefits, and a certain degree of neophobia in relation to food technology.

The potential of nanotechnology is indisputable, however, many consumers are hesitant about accepting nanotechnology, based on a lack of knowledge and competing claims by advocates and critics, for and against its use (Suzin and Fassina 2017).

The studies by Singh et al. (2017) and Yang and Hobbs (2020) revealed that most consumers treat nanotechnology as a new and unknown subject, especially women, the elderly, and people with low schooling levels. Despite this, consumers are also curious about acquiring more information, which is important for research institutions and the food industry. Another study by Mittal et al. (2022) showed that consumers, although they have little knowledge about nanotechnology, have shown increasing interest in the use of this new technology for the development of food packaging. This indicates that consumers have an important role in consolidating these technologies through their acceptance of new products, thus determining their success or failure.

The insertion of nanometric-sized materials in formulations tailored for the prevention, control, and treatment of postharvest fungal diseases is an emerging technology that has been widely explored in recent years, as demonstrated by this review. Active ingredients contained in commercial fungicides, when used in nanoformulations, have enhanced antifungal action, even in small amounts, due to the greater contact surface between the fungus and the active molecule. In addition, when encapsulated, these ingredients can be selectively and continuously released, remaining in contact with the target microorganism for longer periods. Another benefit of encapsulation is the possibility of using active ingredients with antimicrobial potential that are sensitive to abiotic factors due to their protection when encapsulated.

Active and intelligent packaging, for use in food, developed from natural biodegradable and biocompatible polymers, incorporated with various types of nanomaterials with antifungal activity, such as EOs, metals, and metallic oxides, is a viable alternative for inhibition of postharvest diseases, by enabling targeted, prolonged, and effective action against selected microorganisms.

In the case of foods, many materials have been developed as coating films for the storage of fruits and vegetables in recent years. Evaluation of the properties of coating films prepared with various edible materials is important for their subsequent application for food storage. In recent years, nanosized particles as antimicrobial agents embedded in edible coatings have been the subject of many studies. However, further investigations involving interactions between nanoparticles and coating materials are required. In addition, further work on the effect of nanoparticle addition on the properties of coating materials should be conducted. The research results can provide information on additional improvements in physical-chemical aspects and antimicrobial properties for practical applications.

One of the major obstacles to the use of nanomaterials is the lack of unified legislation, especially when directed to their use in food, and also due to the limited knowledge regarding the interactions of nanomaterials, at the molecular or physiological level, their effects, and impacts on the health of consumers and the environment in the long term. In addition, consumers' limited knowledge about products that use nanotechnology reduces their willingness to use these products.

References

Abd-Elsalam, K.A., Al-Dhabaan, F.A., Alghuthaymi, M., Njobeh, P.B. and Almoammar, H. 2019. Nanobiofungicides: present concept and future perspectives in fungal control. pp. 315–351. In Nano-Biopesticides Today and Future Perspectives. Elsevier Inc. https://doi.org/10.1016/B978-0-12-815829-6.00014-0.

Abdel-Rahman, F.A., Khafagi, E.Y., Soliman, M.S., Shoala, T. and Ahmed, Y. 2021. Preharvest application of salicylic acid induces some resistant genes of sweet pepper against black mold disease. Eur. J. Plant Pathol. 159(4): 755–768. https://doi.org/10.1007/s10658-020-02199-z.

Abdollahdokht, D., Gao, Y., Faramarz, S., Poustforoosh, A., Abbasi, M., Asadikaram, G. and Nematollahi, M.H. 2022. Conventional agrochemicals towards nano-biopesticides: an overview on recent advances. Chem. Biol. Technol. Agric. 9(1): 1–19. https://doi.org/10.1186/s40538-021-00281-0.

Adaskaveg, J.E., Forster, H. and Chen, D. 2021. Progress on Chemical Management of Postharvest Diseases of Subtropical and Tropical Fruits James. n Postharvest Pathology. ISBN 9783030565299.

Ahammed, G.J. and Yang, Y. 2021. Mechanisms of silicon-induced fungal disease resistance in plants. Plant Physiol. Biochem. 165: 200–206. https://doi.org/10.1016/j.plaphy.2021.05.031.

Akhila, P.P., Sunooj, K.V., Navaf, M., Aaliya, B., Sudheesh, C., Sasidharan, A., Sabu, S., Mir, S.A., George, J. and Mousavi Khaneghah, A. 2022. Application of innovative packaging technologies to manage fungi and mycotoxin contamination in agricultural products: Current status, challenges, and perspectives. In Toxicon. 214: 18–29. Elsevier Ltd. https://doi.org/10.1016/j.toxicon.2022.04.017.

Alfei, S., Marengo, B. and Zuccari, G. 2020. Nanotechnology application in food packaging: A plethora of opportunities versus pending risks assessment and public concerns. Food Res. Int. 137: 109664. https://doi.org/10.1016/j.foodres.2020.109664.

Allesteros, L.F., Michelin, M., Vicente, A.A., Teixeira, J.A. and Cerqueira, M.Â. 2018. Use of lignocellulosic materials in bio-based packaging. In food applications of lignocellulosic-based packaging Materials. pp. 65–85. *In*: Lignocellulosic Materials and Their Use in Bio-based Packaging. ISBN 9783319929408.

Andrade, R., Skurtys, O. and Osorio, F. 2015. Drop impact of gelatin coating formulated with cellulose nanofibers on banana and eggplant epicarps. LWT - Food Sci. Technol. 61(2): 422–429. https://doi.org/10.1016/j.lwt.2014.12.035.

Apolinário, A.C., Salata, G.C., Bianco, A.F.R., Fukumori, C. and Lopes, L.B. 2020. Opening the pandora's box of nanomedicine: There is needed plenty of room at the bottom. Quim. Nova. 43(2): 212–225. https://doi.org/10.21577/0100-4042.20170481.

Arroyo, B.J., Bezerra, A.C., Oliveira, L.L., Arroyo, S.J., Melo, E.A. de and Santos, A.M.P. 2020. Antimicrobial active edible coating of alginate and chitosan add ZnO nanoparticles applied in guavas (*Psidium guajava* L.). Food Chem. https://doi.org/10.1016/j.foodchem.2019.125566.

Arserim-Uçar, D.K. and Çabuk, B. 2020. Emerging antibacterial and antifungal applications of nanomaterials on food products. In Nanotoxicity: Prevention and Antibacterial Applications of Nanomaterials. https://doi.org/10.1016/B978-0-12-819943-5.00027-0.

Azevedo, F.A. and Chasin, A.A.M. 2003. As bases toxicológicas da ecotoxicologia. 1ª. ed. Rima: São Carlos, SP.

Baghi, F., Gharsallaoui, A., Dumas, E. and Ghnimi, S. 2022. Advancements in biodegradable active films for food packaging: effects of nano/microcapsule incorporation. Foods. 11(5): 1–44. https://doi.org/10.3390/foods11050760.

Baldwin, E.A., Hagenmaier, R.D. and Bai, J. 2011. Edible coatings and films to improve food quality, second edition. In Edible Coatings and Films to Improve Food Quality, Second Edition. ISBN 9781420059663.

Balooch, M., Sabahi, H., Aminian, H. and Hosseini, M. 2018. Intercalation technique can turn pomegranate industrial waste into a valuable by-product. LWT - Food Sci. Technol. 98: 99–105. https://doi.org/10.1016/j.lwt.2018.08.025.

Barrera-Méndez, F., Miranda-Sánchez, D., Sánchez-Rangel, D., Bonilla-Landa, I., Rodríguez-Haas, B., Monribot-Villanueva, J.L. and Olivares-Romero, J.L. 2019. Propiconazole nanoencapsulation in biodegradable polymers to obtain pesticide-controlled delivery systems. J. Mex. Chem. Soc. 63(1): 50–60. https://doi.org/10.29356/jmcs.v63i1.564.

Berti, L.A., Mattar, D.G., Tavares, E.T., Belluci, F.S., Paiva, H.G., Estavanato, L.L., Bertotti, P.F. and Renz, S.P. 2019. Plano de ação de ciência, tecnologia e inovação para tecnologias convergentes e habilitadoras: nanotecnologia. – Brasilia: Ministério da Ciência, Tecnologia, Inovações e Comunicações. 4v. v.1; 56p.

Bumbudsanpharoke, N. and Ko, S. 2015. Nano-food packaging: an overview of market, migration research, and safety regulations. J. Food Sci. 80(3): R910–R923. https://doi.org/10.1111/1750-3841.12861.

Caon, T., Martelli, S.M. and Fakhouri, F.M. 2017. New trends in the food industry: application of nanosensors in food packaging. In Nanobiosensors. https://doi.org/10.1016/b978-0-12-804301-1.00018-7.

Chausali, N., Saxena, J. and Prasad, R. 2022. Recent trends in nanotechnology applications of bio-based packaging. J. Agric. Food Res. 7: 100257. https://doi.org/10.1016/j.jafr.2021.100257.

Chen, H., Zhang, L., Hu, Y., Zhou, C., Lan, W., Fu, H. and She, Y. 2021. Nanomaterials as optical sensors for application in rapid detection of food contaminants, quality and authenticity. In Sensors Actuators, B Chem. v. 329. https://doi.org/10.1016/j.snb.2020.129135.

Chen, J.Y., Wen, P.F., Kong, W.F., Pan, Q.H., Zhan, J.C., Li, J.M., Wan, S.B. and Huang, W.D. 2006. Effect of salicylic acid on phenylpropanoids and phenylalanine ammonia-lyase in harvested grape berries. Postharvest Biol. Technol. 40(1): 64–72. https://doi.org/10.1016/j.postharvbio.2005.12.017.

Chen, Z., Han, S., Zhou, S., Feng, H., Liu, Y. and Jia, G. 2020. Review of health safety aspects of titanium dioxide nanoparticles in food application. NanoImpact 18: 100224. https://doi.org/10.1016/j.impact.2020.100224.

Costa, C.S., Ronconi, J.V.V., Daufenbach, J.F., Gonçalves, C.L., Rezin, G.T., Streck, E.L. and Da Silva Paula, M.M. 2010. *In vitro* effects of silver nanoparticles on the mitochondrial respiratory chain. Mol. Cell. Biochem. 342(1–2): 51–56. https://doi.org/10.1007/s11010-010-0467-9.

Damm, C., Münstedt, H. and Rösch, A. 2008. The antimicrobial efficacy of polyamide 6/silver-nano- and microcomposites. Mater. Chem. Phys. 108(1): 61–66. https://doi.org/10.1016/j.matchemphys.2007.09.002.

Delshadi, R., Bahrami, A., Tafti, A.G., Barba, F.J. and Williams, L.L. 2020. Micro and nano-encapsulation of vegetable and essential oils to develop functional food products with improved nutritional profiles. Trends Food Sci. Technol. 104: 72–83. https://doi.org/10.1016/j.tifs.2020.07.004.

Durán, N., Guterres, S.S. and Alves, O.L. 2014. Nanotoxicology - Materials, Methodologies and Assessments. Springer. ISBN 9781461489924.

Elgorban, A.M., El-Samawaty, A.E.R.M., Yassin, M.A., Sayed, S.R., Adil, S.F., Elhindi, K.M., Bakri, M. and Khan, M. 2016. Antifungal silver nanoparticles: Synthesis, characterization and biological evaluation. Biotechnol. Biotechnol. Equip. 30(1): 56–62. https://doi.org/10.1080/13102818.2015.1106339.

Evenson, R.E. and Gollin, D. 2003. Assessing the impact of the Green Revolution, 1960 to 2000. Science 300(5620): 758–762. https://doi.org/10.1126/science.1078710.

Fakhouri, F.M., Casari, A.C.A., Mariano, M., Yamashita, F., Mei, L.H.I., Soldi, V. and Martelli, S.M. 2014. Effect of a gelatin-based edible coating containing cellulose nanocrystals (CNC) on the quality and nutrient retention of fresh strawberries during storage. IOP Conf. Ser. Mater. Sci. Eng. 64(1). https://doi.org/10.1088/1757-899X/64/1/012024.

FAO. 2018. Food loss and waste and the right to adequate food: Making the connection. In Right to Food Discussion Paper. Rome. 48 pp.

FAO. 2022. Perdas e desperdícios de alimentos na América Latina e no Caribe. Benítez, Raúl Osvaldo. In https://www.fao.org/americas/noticias/ver/pt/c/239394/.

FDA. 2014. Guidance for industry considering whether an FDA-regulated product involves the application of nanotechnology. Biotechnol. Law Rep. 30(5): 613–616. https://doi.org/10.1089/blr.2011.9814.

Fernández, A., Picouet, P. and Lloret, E. 2010. Cellulose-silver nanoparticle hybrid materials to control spoilage-related microflora in absorbent pads located in trays of fresh-cut melon. Int. J. Food Microbiol. 142(1–2): 222–228. https://doi.org/10.1016/j.ijfoodmicro.2010.07.001.

Ge, Y., Schimel, J.P. and Holden, P.A. 2011. Evidence for negative effects of TiO$_2$ and ZnO nanoparticles on soil bacterial communities. Environ. Sci. Technol. 45(4): 1659–1664. https://doi.org/10.1021/es103040t.

Hamad, G.M., Mehany, T., Simal-Gandara, J., Abou-Alella, S., Esua, O.J., Abdel-Wahhab, M.A. and Hafez, E.E. 2023. A review of recent innovative strategies for controlling mycotoxins in foods. Food Control. 144. https://doi.org/10.1016/j.foodcont.2022.109350.

Haris, M., Hussain, T., Mohamed, H.I., Khan, A., Ansari, M.S., Tauseef, A., Khan, A.A. and Akhtar, N. 2023. Nanotechnology—A new frontier of nano-farming in agricultural and food production and its development. Sci. Total Environ. 857: 159639. https://doi.org/10.1016/j.scitotenv.2022.159639.

He, L., Liu, Y., Mustapha, A. and Lin, M. 2011. Antifungal activity of zinc oxide nanoparticles against *Botrytis cinerea* and *Penicillium expansum*. Microbiol. Research 166(3): 207–215. https://doi.org/10.1016/j.micres.2010.03.003.

He, X., Deng, H., Aker, W.G. and Hwang, H. 2019. Regulation and Safety of Nanotechnology in the Food and Agriculture Industry. In Food Applications of Nanotechnology 1ª ed, 12p. Taylor & Francis Group. https://doi.org/10.1201/9780429297038-23.

Jamdagni, P., Rana, J.S. and Khatri, P. 2018. Comparative study of antifungal effect of green and chemically synthesised silver nanoparticles in combination with carbendazim, mancozeb, and thiram. IET Nanobiotechnology 12(8): 1102–1107. https://doi.org/10.1049/iet-nbt.2018.5087.

Juárez-Maldonado, A., Ortega-Ortíz, H., Morales-Díaz, A.B., González-Morales, S., Morelos-Moreno, Á., Cabrera-De la Fuente, M., Sandoval-Rangel, A., Cadenas-Pliego, G. and Benavides-Mendoza, A. 2019. Nanoparticles and nanomaterials as plant biostimulants. Int. J. Mol. Sci. 20(1): 1–19. https://doi.org/10.3390/ijms20010162.

Kairyte, K., Kadys, A. and Luksiene, Z. 2013. Antibacterial and antifungal activity of photoactivated ZnO nanoparticles in suspension. J. Photochem. Photobiol. B Biol. 128: 78–84. https://doi.org/10.1016/j.jphotobiol.2013.07.017.

Kanhed, P., Birla, S., Gaikwad, S., Gade, A., Seabra, A.B., Rubilar, O., Duran, N. and Rai, M. 2014. *In vitro* antifungal efficacy of copper nanoparticles against selected crop pathogenic fungi. Mater. Lett. 115: 13–17. https://doi.org/10.1016/j.matlet.2013.10.011.

Kik, K., Bukowska, B. and Sicińska, P. 2020. Polystyrene nanoparticles: Sources, occurrence in the environment, distribution in tissues, accumulation and toxicity to various organisms. Environ. Pollut. 262. https://doi.org/10.1016/j.envpol.2020.114297.

Koduru, J.R., Kailasa, S.K., Bhamore, J.R., Kim, K.H., Dutta, T. and Vellingiri, K. 2018. Phytochemical-assisted synthetic approaches for silver nanoparticles antimicrobial applications: A review. Adv. Colloid Interface Sci. 256: 326–339. https://doi.org/10.1016/j.cis.2018.03.001.

Król, A., Pomastowski, P., Rafińska, K., Railean-Plugaru, V. and Buszewski, B. 2017. Zinc oxide nanoparticles: Synthesis, antiseptic activity and toxicity mechanism. Adv. Colloid Interface Sci. 249: 37–52. https://doi.org/10.1016/j.cis.2017.07.033.

Kumar, S., Kumar, D. and Dilbaghi, N. 2017. Preparation, characterization, and bio-efficacy evaluation of controlled release carbendazim-loaded polymeric nanoparticles. Environ. Sci. Pollut. Res. 24(1): 926–937. https://doi.org/10.1007/s11356-016-7774-y.

Kutawa, A.B., Ahmad, K., Ali, A., Hussein, M.Z., Wahab, M.A.A., Adamu, A., Ismaila, A.A., Gunasena, M.T., Rahman, M.Z. and Hossain, M.I. 2021. Trends in nanotechnology and its potentialities to control plant pathogenic fungi: A review. Biology 10(9). https://doi.org/10.3390/biology10090881.

Lee, S., Kim, S., Kim, S. and Lee, I. 2012. Effects of soil-plant interactive system on response to exposure to ZnO nanoparticles. J. Microbiol. Biotechnol. 22(9): 1264–1270. https://doi.org/10.4014/jmb.1203.03004.

Li, D., Ye, Q., Jiang, L. and Luo, Z. 2017. Effects of nano-TiO$_2$-LDPE packaging on postharvest quality and antioxidant capacity of strawberry (*Fragaria ananassa* Duch.) stored at refrigeration temperature. J. Sci. Food Agric. 97(4): 1116–1123. https://doi.org/10.1002/jsfa.7837.

Liang, Y., Fan, C., Dong, H., Zhang, W., Tang, G., Yang, J., Jiang, N. and Cao, Y. 2018. Preparation of MSNs-chitosan@prochloraz nanoparticles for reducing toxicity and improving release properties of prochloraz. ACS Sustain. Chem. Eng. 6(8): 10211–10220. https://doi.org/10.1021/acssuschemeng.8b01511.

Lin, M., Fang, S., Zhao, X., Liang, X. and Wu, D. 2020. Natamycin-loaded zein nanoparticles stabilized by carboxymethyl chitosan: Evaluation of colloidal/chemical performance and application in postharvest treatments. Food Hydrocoll. 106. https://doi.org/10.1016/j.foodhyd.2020.105871.

Magalhães, R. 2017. Regulação De Alimentos No Brasil. Revista de Direito Sanitário 17(3): 113. https://doi.org/10.11606/issn.2316-9044.v17i3p113-133.

Malandrakis, A.A., Kavroulakis, N. and Chrysikopoulos, C.V. 2022. Zinc nanoparticles: Mode of action and efficacy against boscalid-resistant *Alternaria alternata* isolates. Sci. Total Environ. 829: 154638. https://doi.org/10.1016/j.scitotenv.2022.154638.

Mali, S. and Grossmann, M.V.E. 2003. Effects of Yam starch films on storability and quality of fresh strawberries (*Fragaria ananassa*). J. Agric. Food Chem. 51(24): 7005–7011. https://doi.org/10.1021/jf034241c.

Maluin, F.N., Hussein, M.Z. and Idris, A.S. 2020. An overview of the oil palm industry: Challenges and some emerging opportunities for nanotechnology development. Agronomy 10(3). https://doi.org/10.3390/agronomy10030356.

Mehmood, S., Kausar Janjua, N., Tabassum, S., Faizi, S. and Fenniri, H. 2022. Cost effective synthesis approach for green food packaging coating by gallic acid conjugated gold nanoparticles from *Caesalpinia pulcherrima* extract. Results Chem. 4: 100437. https://doi.org/10.1016/j.rechem.2022.100437.

Menas, A.L., Yanamala, N., Farcas, M.T., Russo, M., Friend, S., Fournier, P.M., Star, A., Iavicoli, I., Shurin, G.V., Vogel, U.B., Fadeel, B., Beezhold, D., Kisin, E.R. and Shvedova, A.A. 2017. Fibrillar vs crystalline nanocellulose pulmonary epithelial cell responses: Cytotoxicity or inflammation? Chemosphere 171: 671–680. https://doi.org/10.1016/j.chemosphere.2016.12.105.

Mesgari, M., Aalami, A.H. and Sahebkar, A. 2021. Antimicrobial activities of chitosan/titanium dioxide composites as a biological nanolayer for food preservation: A review. Int. J. Biol. Macromol. 176: 530–539. https://doi.org/10.1016/j.ijbiomac.2021.02.099.

Mihindukulasuriya, S.D.F. and Lim, L.T. 2014. Nanotechnology development in food packaging: A review. Trends Food Sci. Technol. 40(2): 149–167. https://doi.org/10.1016/j.tifs.2014.09.009.

Mittal, P., Saharan, A., Kapoor, R., Wilson, K. and Gautam, R.K. 2022. Nanotechnology in food packaging and its regulatory aspects. In Nanotechnology in Intelligent Food Packaging. 157 173pp. Wiley. https://doi.org/10.1002/9701119019011.ch7.

Musa, S.F., Yeat, T.S., Kamal, L.Z.M., Tabana, Y.M., Ahmed, M.A., El Ouweini, A., Lim, V., Keong, L.C. and Sandai, D. 2018. Pleurotus sajor-caju can be used to synthesize silver nanoparticles with antifungal activity against *Candida albicans*. J. Sci. Food Agric. 98(3): 1197–1207. https://doi.org/10.1002/jsfa.8573.

Nagaraju, R.S., Sriram, H.R. and Achur, R. 2020. Antifungal activity of Carbendazim-conjugated silver nanoparticles against anthracnose disease caused by *Colletotrichum gloeosporioides* in mango. J. Plant Pathol. 102(1): 39–46. https://doi.org/10.1007/s42161-019-00370-y.

Nandi, S. and Guha, P. 2018. A review on preparation and properties of cellulose nanocrystal-incorporated natural biopolymer. J. Packag. Technol. Res. 0123456789. https://doi.org/10.1007/s41783-018-0036-3.

Naserzadeh, Y., Mahmoudi, N. and Pakina, E. 2019. Antipathogenic effects of emulsion and nanoemulsion of cinnamon essential oil against *Rhizopus* rot and grey mold on strawberry fruits. Foods Raw Mater. 7(1): 210–216. https://doi.org/10.21603/2308-4057-2019-1-210-216.

Nešić, A., Cabrera-Barjas, G., Dimitrijević-Branković, S., Davidović, S., Radovanović, N. and Delattre, C. 2020. Prospect of polysaccharide-based materials as advanced food packaging. Molecules 25(1). https://doi.org/10.3390/molecules25010135.

Nguyen-Tri, P., Do, T.O., Nguyen, T.A., Le, V.T. and Assadi, A.A. 2019. Nanocontainer: An introduction. In Smart Nanocontainers: Micro and Nano Technologies. 3–6p. Elsevier. Inc. https://doi.org/10.1016/B978-0-12-816770-0.00001-0.

Nowack, B., Ranville, J.F., Diamond, S., Gallego-Urrea, J.A., Metcalfe, C., Rose, J., Horne, N., Koelmans, A.A. and Klaine, S.J. 2012. Potential scenarios for nanomaterial release and subsequent alteration in the environment. Environ. Toxicol. Chem. 31(1): 50–59. https://doi.org/10.1002/etc.726.

Olayil, R., Arumuga Prabu, V., DayaPrasad, S., Naresh, K. and Rama Sreekanth, P.S. 2022. A review on the application of bio-nanocomposites for food packaging. Mater. Today Proc. 56: 1302–1306. https://doi.org/10.1016/j.matpr.2021.11.315.

Omerović, N., Djisalov, M., Živojević, K., Mladenović, M., Vunduk, J., Milenković, I., Knežević, N., Gadjanski, I. and Vidić, J. 2021. Antimicrobial nanoparticles and biodegradable polymer composites for active food packaging applications. Compr. Rev. Food Sci. Food Saf. 20(3): 2428–2454. https://doi.org/10.1111/1541-4337.12727.

Otoni, C.G., Moura, M.R.d., Aouada, F.A., Camilloto, G.P., Cruz, R.S., Lorevice, M.V., Soares, N. de F.F. and Mattoso, L.H.C. 2014. Antimicrobial and physical-mechanical properties of pectin/papaya puree/cinnamaldehyde nanoemulsion edible composite films. Food Hydrocoll. 41: 188–194. https://doi.org/10.1016/j.foodhyd.2014.04.013.

Pabön, A., Ramirez, O., Rios, A., Löpez, E., De Las Salas, B., Cardona, F. and Blair, S. 2016. Antiplasmodial and cytotoxic activity of raw plant extracts as reported by knowledgeable Indigenous people of the amazon region (Vaupés Medio in Colombia). Planta Med. 82(8): 717–722. https://doi.org/10.1055/s-0042-104283.

Parfitt, J., Croker, T. and Brockhaus, A. 2021. Global food loss and waste in primary production: A reassessment of its scale and significance. Sustain. 13(21). https://doi.org/10.3390/su132112087.

Patel, Z.M., Mahapatra, R. and Jampala, S.S.M. 2020. Role of fungal elicitors in plant defense mechanism. In Molecular Aspects of Plant Beneficial Microbes in Agriculture. 143–158p. https://doi.org/10.1016/b978-0-12-818469-1.00012-2.

Peng, H., Hu, H., Xi, K., Zhu, X., Zhou, J., Yin, J., Guo, F., Liu, Y. and Zhu, Y. 2022. Silicon nanoparticles enhance ginger rhizomes tolerance to postharvest deterioration and resistance to *Fusarium solani*. Front. Plant Sci. 13. https://doi.org/10.3389/fpls.2022.816143.

Pitkänen, M., Honkalampi, U., Von Wright, A., Sneck, A., Hentze, H.P., Sievänen, J., Hiltunen, J. and Hellén, E.K.O. 2010. Nanofibrillar cellulose - *In vitro* study of cytotoxic and genotoxic properties. Int. Conf. Nanotechnol. For. Prod. Ind. 2010: 246–261.

Pushparaj, K., Liu, W.C., Meyyazhagan, A., Orlacchio, A., Pappusamy, M., Vadivalagan, C., Robert, A.A., Arumugam, V.A., Kamyab, H., Klemeš, J.J., Khademi, T., Mesbah, M., Chelliapan, S. and Balasubramanian, B. 2022. Nano- from nature to nurture: A comprehensive review on facets, trends, perspectives and sustainability of nanotechnology in the food sector. Energy. 240: 122732. https://doi.org/10.1016/j.energy.2021.122732.

Ramezani, M., Ramezani, F. and Gerami, M. 2019. Nanoparticles in pest incidences and plant disease control. In Nanotechnology for Agriculture: Crop Production & Protection. 233–273p. https://doi.org/10.1007/978-981-32-9374-8.

Ramkumar, C., Vishwanatha, A. and Saini, R. 2019. Regulatory aspects of nanotechnology for food industry. Nanotechnol. Appl. Dairy Sci. 2019: 169–184. https://doi.org/10.1201/9780429425370-7.

Rasmussen, K., Rauscher, H., Gottardo, S., Hoekstra, E., Schoonjans, R., Peters, R. and Aschberger, K. 2018. Regulatory status of nanotechnologies in food in the EU. In Nanomaterials for Food Applications. Elsevier Inc. https://doi.org/10.1016/B978-0-12-814130-4.00013-0.

Robledo, N., Vera, P., López, L., Yazdani-Pedram, M., Tapia, C. and Abugoch, L. 2018. Thymol nanoemulsions incorporated in quinoa protein/chitosan edible films; antifungal effect in cherry tomatoes. Food Chem. 246: 211–219. https://doi.org/10.1016/j.foodchem.2017.11.032.

Rocha, A.M. da, Santos, H.C. dos, Souza, H.M. De, Matioli, G., Barão, C.E., Pimentel, T.C., Coimbra, L.B. and Marcolino, V.A. 2019. Application of cassava starch coating prepared with stevia leaf-washing water for increasing the postharvest life of strawberries. Chem. Eng. Trans. 75: 481–486. https://doi.org/10.3303/CET1975081.

Sagar, N.A., Kumar, N., Choudhary, R., Bajpai, V.K., Cao, H., Shukla, S. and Pareek, S. 2022. Prospecting the role of nanotechnology in extending the shelf-life of fresh produce and in developing advanced packaging. Food Packag. Shelf Life. 34: 100955. https://doi.org/10.1016/j.fpsl.2022.100955.

Sandhya, Kumar, S., Kumar, D. and Dilbaghi, N. 2017. Preparation, characterization, and bio-efficacy evaluation of controlled release carbendazim-loaded polymeric nanoparticles. Environ. Sci. Pollut. Res. 24(1): 926–937. https://doi.org/10.1007/s11356-016-7774-y.

Sani, M.A., Maleki, M., Eghbaljoo-Gharehgheshlaghi, H., Khezerlou, A., Mohammadian, E., Liu, Q. and Jafari, S.M. 2022. Titanium dioxide nanoparticles as multifunctional surface-active materials for smart/active nanocomposite packaging films. Adv. Colloid Interface Sci. 300. https://doi.org/10.1016/j.cis.2021.102593.

Saqib, S., Zaman, W., Ayaz, A., Habib, S., Bahadur, S., Hussain, S., Muhammad, S. and Ullah, F. 2020. Postharvest disease inhibition in fruit by synthesis and characterization of chitosan iron oxide nanoparticles. Biocatal. Agric. Biotechnol. 28: 101729. https://doi.org/10.1016/j.bcab.2020.101729.

Scheringer, M. 2008. Environmental risks of nanomaterials. Nat. Nanotechnol. (6): 322–323. https://doi.org/10.1038/nnano.2008.145.

Scimeca, J.C. and Verron, E. 2022. Nano-engineered biomaterials: Safety matters and toxicity evaluation. Mater. Today Adv. 15. https://doi.org/10.1016/j.mtadv.2022.100260.

Selvaraj, C., Sakkiah, S., Tong, W. and Hong, H. 2018. Molecular dynamics simulations and applications in computational toxicology and nanotoxicology. Food Chem. Toxicol. 112: 495–506. https://doi.org/10.1016/j.fct.2017.08.028.

Sharifi, S., Behzadi, S., Laurent, S., Forrest, M.L., Stroeve, P. and Mahmoudi, M. 2012. Toxicity of nanomaterials. Chem. Soc. Rev. 41: 2323–2343. https://doi.org/10.1039/c1cs15188f.

Sharma, C., Dhiman, R., Rokana, N. and Panwar, H. 2017. Nanotechnology: An untapped resource for food packaging. Front. Microbiol. 8(SEP). https://doi.org/10.3389/fmicb.2017.01735.

Sharma, D., Rajput, J., Kaith, B.S., Kaur, M. and Sharma, S. 2010. Synthesis of ZnO nanoparticles and study of their antibacterial and antifungal properties. Thin Solid Films 519(3): 1224–1229. https://doi.org/10.1016/j.tsf.2010.08.073.

Shen, C., Chen, W., Li, C., Aziz, T., Cui, H. and Lin, L. 2023. Topical advances of edible coating based on the nanoemulsions encapsulated with plant essential oils for foodborne pathogen control. Food Control. 145. https://doi.org/10.1016/j.foodcont.2022.109419.

Singh, T., Shukla, S., Kumar, P., Wahla, V., Bajpai, V.K. and Rather, I.A. 2017. Application of nanotechnology in food science: perception and overview. Front. Microbiol. 8: 1501.

Smith, K., Evans, D.A. and El-Hiti, G.A. 2008. Role of modern chemistry in sustainable arable crop protection. Philos. Trans. R. Soc. B Biol. Sci. 363(1491): 623–637. https://doi.org/10.1098/rstb.2007.2174.

Souza, E., Gottschalk, L. and Freitas-Silva, O. 2019. Overview of nanocellulose in food packaging. Recent Pat. Food. Nutr. Agric. 10(i). https://doi.org/10.2174/2212798410666190715153715.

Souza, V.G.L. and Fernando, A.L. 2016. Nanoparticles in food packaging: Biodegradability and potential migration to food—A review. Food Packag. Shelf Life. 8: 63–70. https://doi.org/10.1016/j.fpsl.2016.04.001.

Spricigo, P.C., Foschini, M.M., Ribeiro, C., Corrêa, D.S. and Ferreira, M.D. 2017. Nanoscaled platforms based on SiO$_2$ and Al$_2$O$_3$ impregnated with potassium permanganate use color changes to indicate ethylene removal. Food Bioprocess Technol. 10(9): 1622–1630. https://doi.org/10.1007/s11947-017-1929-9.

STATNANO. 2022. Nano Science, Technology and Industry Information. https://statnano.com/.

Sun, X., Wu, Q., Picha, D.H., Ferguson, M.H., Ndukwe, I.E. and Azadi, P. 2021. Comparative performance of bio-based coatings formulated with cellulose, chitin, and chitosan nanomaterials suitable for fruit preservation. Carbohydr. Polym. 259: 117764. https://doi.org/10.1016/j.carbpol.2021.117764.

Thakur, R., Pristijono, P., Golding, J.B., Stathopoulos, C.E., Scarlett, C.J. and Bowyer, M. 2018. Scientia Horticulturae Development and application of rice starch based edible coating to improve the postharvest storage potential and quality of plum fruit (*Prunus salicina*). Sci. Hortic. 237: 59–66. https://doi.org/10.1016/j.scienta.2018.04.005.

Tleuova, A.B., Wielogorska, E., Talluri, V.S.S.L.P., Štěpánek, F., Elliott, C.T. and Grigoriev, D.O. 2020. Recent advances and remaining barriers to producing novel formulations of fungicides for safe and sustainable agriculture. J. Control. Release. 326: 468–481. https://doi.org/10.1016/j.jconrel.2020.07.035.

Tutaj, K., Szlazak, R., Szalapata, K., Starzyk, J., Luchowski, R., Grudzinski, W., Osinska-Jaroszuk, M., Jarosz-Wilkolazka, A., Sawstur Ciesielska, A. and Gruszecki, W.I. 2016. Amphotericin B-silver hybrid nanoparticles: Synthesis, properties and antifungal activity. Nanomedicine Nanotechnology, Biol. Med. 12(4): 1095–1103. https://doi.org/10.1016/j.nano.2015.12.378.

Velmurugan, N., Kumar, G., Han, S.S., Nahm, K.S. and Lee, Y.S. 2009. Synthesis and characterization of potential fungicidal silver nano-sized particles and chitosan membrane containing silver particles. Iran. Polym. J. 18(5): 383–392. https://doi.org/https://www.scopus.com/record/display.uri?eid=2-s2.0-67649286685&origin=inward&txGid=9309b4d411d5be04c5623af365157ac2.

Voss, L., Hsiao, I.L., Ebisch, M., Vidmar, J., Dreiack, N., Böhmert, L., Stock, V., Braeuning, A., Loeschner, K., Laux, P., Thünemann, A.F., Lampen, A. and Sieg, H. 2020. The presence of iron oxide nanoparticles in the food pigment E172. Food Chem. 327: 127000. https://doi.org/10.1016/j.foodchem.2020.127000.

Vu, K.D., Hollingsworth, R.G., Leroux, E., Salmieri, S. and Lacroix, M. 2011. Development of edible bioactive coating based on modified chitosan for increasing the shelf life of strawberries. Food Res. Int. 44: 198–203. https://doi.org/10.1016/j.foodres.2010.10.037.

Warriner, K., Reddy, S.M., Namvar, A. and Neethirajan, S. 2014. Developments in nanoparticles for use in biosensors to assess food safety and quality. Trends Food Sci. Technol. 40(2): 183–199. https://doi.org/10.1016/j.tifs.2014.07.008.

Wassermann, B., Abdelfattah, A., Cernava, T., Wicaksono, W. and Berg, G. 2022. Microbiome-based biotechnology for reducing food loss post harvest. Curr. Opin. Biotechnol. 78: 102808. https://doi.org/10.1016/j.copbio.2022.102808.

Worrall, E.A., Hamid, A., Mody, K.T., Mitter, N. and Pappu, H.R. 2018. Nanotechnology for plant disease management. Agronomy 8(12): 1–24. https://doi.org/10.3390/agronomy8120285.

Xu, C., Cao, L., Zhao, P., Zhou, Z., Cao, C., Li, F. and Huang, Q. 2018. Emulsion-based synchronous pesticide encapsulation and surface modification of mesoporous silica nanoparticles with carboxymethyl chitosan for controlled azoxystrobin release. Chem. Eng. J. 348: 244–254. https://doi.org/10.1016/j.cej.2018.05.008.

Xue, J., Zan, G., Wu, Q., Deng, B., Zhang, Y., Huang, H. and Zhang, X. 2016. Integrated nanotechnology for synergism and degradation of fungicide SOPP using micro/nano-Ag$_3$PO$_4$. Inorg. Chem. Front. 3(3): 354–364. https://doi.org/10.1039/c5qi00186b.

Yang, R., Miao, J., Shen, Y., Cai, N., Wan, C., Zou, L., Chen, C. and Chen, J. 2021. Antifungal effect of cinnamaldehyde, eugenol and carvacrol nanoemulsion against *Penicillium digitatum* and application in postharvest preservation of citrus fruit. LWT - Food Sci. Technol. 141. https://doi.org/10.1016/j.lwt.2021.110924.

Yang, R., Miao, J., Chen, X., Chen, C., Simal-Gandara, J., Chen, J. and Wan, C. 2022. Essential oils nano-emulsion confers resistance against Penicillium digitatum in "Newhall" navel orange by promoting phenylpropanoid metabolism. Ind. Crops Prod. 187(PA): 115297. https://doi.org/10.1016/j.indcrop.2022.115297.

Yang, Y. and Hobbs, J.E. 2020. Food values and heterogeneous consumer responses to nanotechnology. Can. J. Agric. Econ. 68(3): 289–313.

Yao, J., Cui, B., Zhao, X., Zhi, H., Zeng, Z., Wang, Y., Sun, C., Liu, G., Gao, J. and Cui, H. 2018a. Antagonistic effect of azoxystrobin poly (Lactic acid) microspheres with controllable particle size on colletotrichum higginsianum sacc. Nanomaterials. 8(10). https://doi.org/10.3390/nano8100857.

Yao, J., Cui, B., Zhao, X., Wang, Y., Zeng, Z., Sun, C., Yang, D., Liu, G., Gao, J. and Cui, H. 2018b. Preparation, characterization, and evaluation of azoxystrobin nanosuspension produced by wet media milling. Appl. Nanosci. 8(3): 297–307. https://doi.org/10.1007/s13204-018-0745-5.

Yilmaz, A., Bozkurt, F., Cicek, P.K., Dertli, E., Durak, M.Z. and Yilmaz, M.T. 2016. A novel antifungal surface-coating application to limit postharvest decay on coated apples: Molecular, thermal and morphological properties of electrospun zein–nanofiber mats loaded with curcumin. Innov. Food Sci. Emerg. Technol. 37: 74–83. https://doi.org/10.1016/j.ifset.2016.08.008.

Yilmaz, M.T., Yilmaz, A., Akman, P.K., Bozkurt, F., Dertli, E., Basahel, A., Al-Sasi, B., Taylan, O. and Sagdic, O. 2019. Electrospraying method for fabrication of essential oil loaded-chitosan nanoparticle delivery systems characterized by molecular, thermal, morphological and antifungal properties. Innov. Food Sci. Emerg. Technol. 52: 166–178. https://doi.org/10.1016/j.ifset.2018.12.005.

Yingchun, L., Zhenguo, W. and Zhuo, W. 2021. Total migration analysis of silver in consumer products containing nanoparticles. Asian J. Ecotoxicol. 16(6): 279–288. https://doi.org/10.7524/AJE.1673-5897.20200814005.

Youssef, K., de Oliveira, A.G., Tischer, C.A., Hussain, I. and Roberto, S.R. 2019. Synergistic effect of a novel chitosan/silica nanocomposites-based formulation against gray mold of table grapes and its possible mode of action. Int. J. Biol. Macromol. 141: 247–258. https://doi.org/10.1016/j.ijbiomac.2019.08.249.

Zahid, N., Maqbool, M., Ali, A., Siddiqui, Y. and Bhatti, Q.A. 2019. Inhibition in production of cellulolytic and pectinolytic enzymes of *Colletotrichum gloeosporioides* isolated from dragon fruit plants in response to submicron chitosan dispersions. Sci. Hortic. 243: 314–319. https://doi.org/10.1016/j.scienta.2018.08.011.

Zhang, J., Zhao, C., Liu, Y., Cao, L., Wu, Y. and Huang, Q. 2016. Size-Dependent effect of prochloraz-loaded mPEG-PLGA Micro- and nanoparticles. J. Nanosci. Nanotechnol. 16(6): 6231–6237. https://doi.org/10.1166/jnn.2016.10894.

Zhang, L., Jiang, Y., Ding, Y., Povey, M. and York, D. 2007. Investigation into the antibacterial behaviour of suspensions of ZnO nanoparticles (ZnO nanofluids). J. Nanopart. Res. 9(3): 479–489. https://doi.org/10.1007/s11051-006-9150-1.

Zhang, X., Li, B., Zhang, Z., Chen, Y. and Tian, S. 2020. Antagonistic yeasts: A promising alternative to chemical fungicides for controlling postharvest decay of fruit. Journal of Fungi 6(3): 1–15. https://doi.org/10.3390/jof6030158.

Zhang, Y., Sheedy, C., Nilsson, D. and Goss, G.G. 2020. Evaluation of interactive effects of UV light and nano encapsulation on the toxicity of azoxystrobin on zebrafish. Nanotoxicology 14(2): 232–249. https://doi.org/10.1080/17435390.2019.1690064.

Chapter 15

Inorganic Nanomaterials for Improved Abiotic Stress Tolerance in Crop Plants

Anu Kalia,[1,*] *Sreelakshmi M.V.*[2] and *Sukhjinder Kaur*[3]

Introduction

In the 21st century, the whole scientific world is focused on the agricultural impact of the climate change. The crop productivity suffers due to repeated incidence of various abiotic stresses associated with climate change, such as extreme temperatures, high CO_2, irregular precipitation, and flooding (Chen et al. 2018; Najafi et al. 2018). The stresses including heavy metal contamination and salinity resulting from the increased application of agrochemicals and inorganic pesticides pose serious environmental concerns, hindering global agricultural productivity and environmental stability (Srivastava et al. 2017; Khan et al. 2022). The crop encounters serious reactive oxygen species (ROS) bursts. Generally, the ROS content in the plant enhances during stressed conditions, and these increased levels of ROS can impair normal cell activity and lead to cell death (Waqas et al. 2019). Furthermore, exposure to various stresses adversely affects several physiological processes such as protein folding, impediment in the uptake of nutrients followed by a nutritional imbalance in the plant and eventually culminating to reduced crop productivity (Shabbir et al. 2022). The abiotic stress can also cause alterations in genetic behavior, which further induces variations in the progeny genetic profiles (Bhat et al. 2020). Different plant responds differently on exposure to various abiotic stress conditions. Some are resistant, where these will get acclimatized to that environment and survive even that stressful condition. But some plants are more susceptible to stress conditions, and lead to senescence and death of the plant. However, plants can exhibit another survival mechanism of escape, i.e., avoidance of the stresses to survive through that condition (Baweja and Kumar 2020). Among plants, abiotic stress alleviation occurs through the initiation of the defense systems in plants, which includes higher production of phytohormones, osmoprotectants, flavonoids, and polyamines (Narula et al. 2022). Even though plants exhibit a capacity to survive under stress conditions, during the process of overcoming the stress conditions, the yield attributes of the crop get affected adversely. Abiotic stress causes an annual loss in crop output with world statistics to

---

[1] Electron Microscopy and Nanoscience Laboratory, Department of Soil Science, College of Agriculture, Punjab Agricultural University, Ludhiana, Punjab, India-141004.

[2] Department of Microbiology, College of Basic Sciences and Humanities, Punjab Agricultural University, Ludhiana, Punjab, India-141004.

[3] School of Agricultural Biotechnology, College of Agriculture, Punjab Agricultural University, Ludhiana, Punjab, India-141004.

Emails: sreelakshmi-2163007@pau.edu; dr.sukhjinderkaur30@gmail.com

* Corresponding author: kaliaanu@pau.edu

exist between 51% and 82% (Oshunsanya et al. 2019). Conventional methods of protection of plants from abiotic stresses includes development of stress tolerant variety through plant breeding and biotechnological approaches besides various agronomic approaches, such as nutrient management, irrigation management, climate smart technology, and adjustment in sowing time (Hossain et al. 2021). The increased world population and proportional surge in demand for food has urged the scientific community to search for alternative sustainable methods for yield improvement under stress condition (Luqman et al. 2023).

The emergence of nanotechnology has provided sustainable options in several domains of agricultural research including the management of agricultural crops (Bala et al. 2019; Kalia et al. 2020; Kondal et al. 2021). Nanotechnology is a rapidly emerging science that delivers promising solutions to plants by ameliorating stress-related issues, thus achieving a sustainable future for crop production (Seleiman et al. 2020; Kaur et al. 2022). The already published studies indicated a promising future of nanotechnology for achieving food demands and agricultural security by furnishing novel strategies at the nano-scale (Kalia and Kaur 2019). A nanomaterial (NM) is a substance having one dimension in the 1 to 100 nanometer (nm) range, synthesized by various chemical, physical and biological methods. The NMs are broadly categorized as inorganic, organic, carbon-based, and composite nanomaterials (Majhi and Yadav 2021). For abiotic stress mitigation published reports primarily include the application of inorganic NMs. The metals and metal oxide-based NMs can help mitigate various abiotic stresses. Metal-based nanoparticles (NPs) include silver (AgNPs), gold (AuNPs), and lead nanoparticles (PbNPs). Titanium dioxide (TiO_2) NPs, silicon dioxide (SiO_2) NPs, and zinc oxide (ZnO)NPs are the most prominent metal/non-metal and their oxide-based NPs. The application of NPs as nanofertilizers results in the enhanced plant growth, thereby improving the crop productivity (Khalid et al. 2022). These NPs offer a wide surface area to volume ratio and thus improving the bioactivity and bioavailability of plants, which ultimately improves the nutrient use efficiency of nanofertilizers (Shang et al. 2019; Mittal et al. 2020; Ndaba et al. 2022). Furthermore, the controlled release of macromolecules both in a spatial and temporal manner in response to external stimuli can be observed for the NMs (Wang et al. 2020). As a consequence, the potential of nanotechnology in crop plants is gaining growing interest (Sanzari et al. 2019).

Nanoparticles in Plant Sciences

Nanoparticles are mostly applied in plants as a foliar spray where the cell penetration occurs via stomata followed by transportation inside the plant cell to various parts of plants through the apoplast/symplast pathways (Hong et al. 2021). Inside the plant cell, these NPs work as a major signaling molecule for the induction of genes involved in abiotic stress management. The extent of their impact inside the plant relies on the nature and time interval of the NP exposure. The NPs help in overcoming the stressful environment through activation of antioxidant mechanisms, increased photosynthetic efficiency, and improved signaling pathways for phytohormone synthesis (Tripathi et al. 2022).

The initial studies on NPs mainly highlighted their toxicity and adverse effects on plants at high concentrations and only a few studies focused on their beneficial effects (Sanzari et al. 2019). There is a need for research in nanotechnology for developing novel techniques targeting the unambiguous delivery of chemicals and biomolecules for the genetic modification of plants. Plant scientists in the other applied fields may also benefit from nanotechnology through the development of advanced approaches that can efficiently incorporate NPs into plants and enhance their functions, i.e., increased germination, plant growth, and resistance against biotic and abiotic stresses (Thul and Sarangi 2015; Ali et al. 2021; Sanzari et al. 2019). The NPs generally exhibit their distinct positive and deleterious effects on crop plants depending on size and interactions with the plants or microbes. Thul and Sarangi (2015) observed the transportation of NPs in *Allium cepa, Chrysanthemum,* and *Lolium* plants via roots through the vascular tissue on dipping the plant stem in the CdSe/ZnS quantum dots solutions. The affirmative effects of multiwalled Carbon Nanotubes (CNTs) on the growth and germination of tomato plants were determined (Kole et al. 2016). Thabet et al. (2021) demonstrated the biocontrol potential of nanosilica particles against crop-damaging insects. Further, the useful effects of multiwalled CNTs on numerous cereal and horticultural crops have been reported (Safdar et al. 2022). The application of NPs to reduce pathogen infestation in crops has emerged as an effective strategy, resulting in an increased yield of crops (Shang et al. 2019). The targeted and controlled release of certain cargoes can also be efficiently delivered through NMs, such as the targeted delivery of DNA in tobacco leaves through mesoporous SiNPs (Torney et al. 2007).

Nanomaterials and Abiotic Stress Tolerance

Numerous metallic and metal oxide nanoparticles, such as titanium dioxide (TiO_2), iron oxide (Fe_3O_4), zinc oxide (ZnO), silver (Ag), cobalt (Co), selenium (Se), and copper (Cu) have shown positive impacts on crop growth under abiotic stress (Adrees et al. 2020; Dimkpa et al. 2020a; Elsheery et al. 2020a; Song et al. 2021a). The synthesis of nanomaterials can be mediated through several means involving chemical, physical, and green processes and their effects on the type, size, and origin of the NPs (Sun et al. 2021). The following sub-heads describe the role of inorganic nanomaterials in increasing plant tolerance under different harsh environments.

Temperature: Extreme Stress

Cold Stress/Low-Temperature Stress

Cold stress results in the disruption of numerous physiological and metabolic functions of the cell (Petruccelli et al. 2022). Based on temperature, it is categorized as, chilling stress (4–20°C) and freezing stress (< 0°C) (Ritonga and Chen 2020). Due to reduction in temperature, the various physiological and metabolic functions in the plants are slowed down thermodynamically (Martínez-Peñalver et al. 2012). In plants, the amount of various ROS increases with the shift in temperature. The main variation occurs in the cell membrane of the plant. At low temperature membrane phase transition will occur, and this leads to structural damage of the membrane. Membrane damage successively promotes ion leakage, and collapse of the cell. Not only the cell membrane, but the other membrane systems (chloroplast and thylakoid membrane) are also affected. Therefore, the photosynthesis rate will get reduced. The overall plant growth and yield is found to be decreased under chilling condition. Cold-tolerant and sensitive plants respond differently to cold stress. Rani et al. (2021) have stated a higher electrolyte leakage index in sensitive plants than in the tolerant crop. This explains that there is a clear difference between the genetic mechanism of cold stress response in tolerant and sensitive plants. Cold-tolerant crops and TiO_2 NPs treated plants showed increased levels of transcription specific genes (*RLK, ERF,* and *VSR-6*). The TiO_2 NPs imparted improved cold stress tolerance probably through changes or modifications in the potential responsive elements involved in various mechanisms associated with metabolism, defense, specific signal transduction events, and other regulatory mechanisms (Amini et al. 2017).

Different nanoparticles have already been tested for the alleviation of cold stress in crop plants. The TiO_2, CuO, and AgNPs are the most commonly used NPs exerting adverse effects on the photosynthetic apparatus, but CeO_2 and TiO_2 help in increasing the electron flow in non-cyclic photo-phosphorylation and increase the enzyme activity of RuBisCO. Generally, cold stress is found to increase the antioxidant enzyme content, MDA, soluble sugars, proline, and other compounds but also reduces plant growth and photosynthetic content. Nanoparticles are reported to increase the amount of photosynthetic pigments if treated in low-temperature stress conditions (Hasanpour et al. 2015).

High Temperature Stress

Amid various abiotic stress, high temperature is a dominant stress limiting plant growth, metabolism, and development (Zhao et al. 2020; Wang et al. 2022a). The negative impacts of heat stress in

Table 1. Effect of inorganic nanoparticles on plant growth under cold and high-temperature stress.

Metal NP	Concentration	Method of Synthesis	Crop Tested	Temperature	Method of Application	Mechanism Behind It	Reference
Low Temperature or Cold Stress							
SiO_2	300 ppm	Mechanical	Sugarcane	6°C	Foliar spray	• Enhanced photosynthesis • Improved phytoprotection	(Elsheery et al. 2020b)
ZnO-NPs	25 mg/L, 50 mg/L, and 100 mg/L	(commercial)	Rice, Hydroponically cultured	10°C	Foliar spray	• Induced chlorophyll accumulation • Reduced oxidative stress	(Song et al. 2021b)
AgNPs	0.25 mg/dm², 1.25 mg/dm², 2.5 mg/dm²	Trisodium citrate reduction reaction 10 nm size	Green beans seeds	15–10°C	Seed treatment	• ↑ In germination rate • ↑ Germination capacity	(Prażak et al. 2020)
TiO_2 NPs	0 ppm, 2 ppm, 5 ppm, 10 ppm	7–40 nm (commercial)	Chickpea	4°C	Foliar spray	• Activation of defense mechanisms • ↓ Membrane damage • ↓ Levels of ELI and MDA	(Mohammadi et al. 2013)
TiO_2 NPs	5 mg/L	-	Chickpea	4°C	Foliar spray	• ↓ In H_2O_2 • ↑ Photosynthesis • ↑ The activity of Rubisco	(Hasanpour et al. 2015)
TiO_2 NPs	2 ppm and 5 ppm	10–25 nm (commercial)	Licorice	4°C	Foliar spray in 30 days old seedlings	• ↓ MDA and H_2O_2 • ↑ Antioxidative and osmoprotective responses	(Kardavan Ghabel and Karamian 2020a)
TiO_2 NPs +Spermine	TiO_2 NPs – 2 and 5 ppm Spermine- 1 mM	10–25 nm (commercial)	Licorice	4°C	NA	• Production of glycyrrhizin was stimulated • Alleviated the negative effects of cold stress	(Kardavan Ghabel and Karamian 2020b)
High Temperature or Heat Stress							
Se NP	10 mg/L	10–40 nm	Grain sorghum	38°C	Foliar spray	• ↑ Antioxidant enzymes activity • ↓ Concentration of signature oxidants • ↑ Levels of unsaturated phospholipids • ↑ Pollen germination percentage • ↑ Seed yield	(Djanaguiraman et al. 2018b)
Zn NPs (80 nm) and Fe NPs (50 nm)	0.25 ppm, 0.50 ppm, 0.75 ppm, 1.0 ppm, and 10 ppm	80 and 50 nm for Zn NP and FeNP respectively	Wheat	22.2–38.3°C	Foliar spray	• ↑ Antioxidant enzymes activities • ↓ Of lipid peroxidation product (MDA)	(Hassan et al. 2018)

Table 1 contd. ...

...Table 1 contd.

Metal NP	Concentration	Method of Synthesis	Crop Tested	Temperature	Method of Application	Mechanism Behind It	Reference
AgNPs	25 ppm, 50 ppm, 75 ppm, 100 ppm	AgNPs synthesis by using *M. oleifera* plant extract, 300–700 nm	Wheat	35–40°C	Along with irrigation water	• ↑ Level of osmolytes • ↑ Antioxidant and non-antioxidant enzymes • ↓ MDA, H$_2$O$_2$ and LPX • Unique plasmon-resonance optical scattering properties against heat stress • Improve plant root length, shoot length, root number, plant fresh weight and plant dry weight, leaf area, leaf number, leaf fresh weight, and leaf dry weight	(Iqbal et al. 2019)
TiO$_2$	0.05 g/L, 0.1 g/L, and 0.2 g/L	Ultrasonic vibration, 16.04 nm	Tomato	35°C	Foliar spray	• ↑ Net photosynthetic rate, conductance to H$_2$O, and transpiration rate • ↓ The minimum chlorophyll fluorescence • ↓ Relative electron transport in leaves • ↑ Regulated PS II energy dissipation • ↓ Non-regulated PS II energy dissipation	(Qi et al. 2013)
Combined ZnO$_2$ and TiO$_2$	1.5 ppm and 10 ppm	(commercial)	Wheat	32°C	Soaking of seedlings in the solution of NPs	• ↑ in more SOD and GPX activity • ↓ H$_2$O$_2$ • Stabilizing membrane by 1.5 times lowering of MDA content	(Thakur et al. 2021)

plants include water loss, yield reduction, decreased photosynthesis, oxidative stress, improper development, impaired seed germination, and alteration in phenology (Zhang et al. 2020; Wang et al. 2022b). Plants overcome heat stress by, alteration in membrane lipid composition, amount of antioxidants, increased amount of osmolytes or other metabolites, and by synthesis of different proteins (Kai and Iba 2014). Different methods are applied in crops to help them withstand heat stress, such as the application of plant growth hormones, foliar application of different micronutrients (Lamaoui et al. 2018), and application of antitranspirants (Khondoker and Kabir 2019). Now with the development of nanotechnology, we can help the plant cope with heat stress in an added economical and productive way (M. S. et al. 2021).

Metal nanoparticles of different sizes and at various concentrations have been evaluated in crop plants to study their stress-mitigating effect (Table 1). Azmat et al. (2022) studied the stress mitigation potential of ZnO-NPs in wheat crops and reported a decreasing trend of MDA content and enhanced electrolyte leakage in the treated plants, along with increased plant growth attributes. The treatment, ZnO-NPs, and *Pseudomonas* was the best treatment and showed an enhanced amount of proline and antioxidant enzymes. The ultrastructure of various plant organelles, particularly the chloroplast and mitochondria, is also affected by heat stress. According to studies, heat stress causes the outer chloroplast membrane to swell excessively (Higashi and Saito 2019), the stroma and inter-granal lamellar system to distort (Babenko and Kosakivska 2014), the thylakoids to become disorganized along with a decrease in grana stacking thickness (Salem-Fnayou et al. 2011) and a decrease in the size of the starch granules. The Si or SiNPs treatments dramatically reversed the ultrastructural abnormalities of cellular organelles brought on by heat stress, particularly in chloroplasts and the nucleus (Younis et al. 2020).

Flooding Stress

Water logging or flooding is a major concern especially now a days due to high amount of as well as frequency precipitation besides over irrigation. The details regarding the influence of nanoparticles on flooding stress tolerance in crop plants have been presented in Table 2.

Flooding leads to a chain of changes in the soil environment near roots, such as compaction of soil, changes in pH, oxygen deficiency, and accumulation of phytotoxic products. Soil composition such as mineral (45%), air (25%), organic matter (5%), and water (25%) are important for the growth of plants. In a flooded condition the air occupied in the soil pores is replaced by the water molecules, and this leads to a hypoxia condition in which the percentage of oxygen is below 21%. This makes the plant suffer, especially since the roots get immediately affected in an adverse way. The overall metabolism in the root cells will exhibit transition from aerobic to anaerobic pathway to meet the deficiency of ATP aroused due to the unavailability of oxygen (Table 2). But compared to mitochondrial respiration where the energy output is around 36 ATP/mole of glucose, the fermentative metabolic route allows only the synthesis of 2 ATP/mole of glucose. This leads to a decrease in carbohydrate reserve due to the enhanced glycolytic pathway and the increase in the accumulation of volatile fatty acid, methane, ROS, and CO_2 in the root zone. Furthermore, the plant becomes more prone to pathogen attack and the overall stress increases.

Nanoparticles, especially the inorganic NPs are exploited in studies for mitigating flooding stress. Both the dimension and concentration of the nanoparticles significantly affect the crop plants. Rezvani and Sorooshzadeh (2014) have tested the impact of the use of AgNPs in the flooding exposed saffron plants and observed that the AgNPs (40 ppm) improved the root growth and associated traits to effectively mitigate the stress. Furthermore, AgNPs can help in altering the fermentative metabolic to the normal pathways in the root during flood conditions (Mustafa et al. 2015a). AgNPs can also improve tolerance to flooding stress by upregulation of the genes involved in oxidative stress mitigation and downregulation of the ethylene signaling pathways (Kaveh et al. 2013).

Table 2. Mechanism of action of the inorganic nanoparticles on plant growth under flooding stress.

Metal NP	Concentration	Method of Synthesis	Crop Tested	Method of Application	Mechanism Involved	Reference
Al$_2$O$_3$ NPs	50 ppm	5 nm, 30–60 nm, and 135 nm	Soyabean	-	• Improved root including hypocotyl growth	(Mustafa et al. 2016)
Al$_2$O$_3$ NPs, ZnO-NPs, Ag NPs	5 ppm, 50 ppm, and 500 ppm Al$_2$O$_3$ NPs and ZnO-NPs, 0.5 ppm, 5.0 ppm, and 50 ppm of AgNPs	30–60 nm	Soyabean	Along with flooded water	• Al$_2$O$_3$ NPs at 50 ppm increased plant growth • *NmrA-like negative transcriptional regulator* was upregulated • *Flavodoxin-like quinone reductase* was down-regulated • Cell death in root including hypocotyl was less evident • Promote the growth of soybean • Regulates energy metabolism and cell death	(Mustafa et al. 2015b)
AgNPs	2 ppm	15 nm	Soyabean	Along with flooded water	• The metabolic shift was reported from fermentative to normal	(Mustafa et al. 2015a)
AgNPs	2 ppm, 15 ppm, and 50–80 ppm	15 nm	Soyabean	NA	• Increase proteins related to amino acid synthesis and waxes formation	(Hashimoto et al. 2020)
AgNPs	5 ppm	-	Soyabean	Along with flooded water	• ↑ Hypocotyl weight • ↑ Hypocotyls length • ↑ In root length • ↑ In root weight	(Hashimoto et al. 2020)
SiNPs	500 ppm	20–30 nm (purchased)	Blueberry	Foliar (500 ppm) vs. foliar and root zone application (250+250 ppm) Foliar and root zone application gives better result	• Increased accumulation of osmolytes, enzymatic antioxidants, and non-enzymatic antioxidants • Low rate of lipid peroxidation	(Iqbal et al. 2021)

Water Deficit and Chemical Drought Stress

The plants get severely affected under water-deficient conditions, which is one of the extreme environmental stresses inflicting excessive damage on the plants. The plant biomass comprises 80–95% of water which is directly involved in the various metabolic and growth processes of plants (Ahluwalia et al. 2021). Thus, drought acts as a major stress threatening food security worldwide. It is multidimensional stress dependent on several factors, such as rainfall anomalies, shifts in the pattern of monsoon, evapotranspiration, and water-holding capacity around the rhizospheric region (Seleiman et al. 2021). The adverse impacts on plant productivity and performance include inhibition of seed germination, lesser growth and efficiency of photosynthesis, altered root morphology, reduction in the synthesis of chlorophyll and membrane permeability, and increased production of ROS (Dimkpa et al. 2020a).

Plants exhibit phenotypic remodeling including lowering of leaf area as an adaptive mechanism in response to water stress. The drought stress predominantly affects the photosynthetic apparatus causing a significant decrease in the chlorophyll content. Further, the accumulation of protective pigments occurs, such as carotenoids and anthocyanine (Farooq et al. 2009; Zahedi et al. 2023). Under stressed conditions plant productivity can be improved by enhancing the rate of photosynthesis via nano-photocatalysts would result in increased absorption of light and efficiency of photosynthesis (Khatri and Rathore 2018).

Different nanomaterials are widely used to mitigate drought stress. Under drought stress conditions, the positive response of plants was reported under the application of colloidal suspension of copper and zinc nanoparticles. These responses include a decrease in the accumulation of Thiobarbituric acid reactive substances (TBRAS), higher ROS scavenging activity, a change in the content of chlorophyll a and b, and an increased accumulation of carotenoids (Singh et al. 2022). The chlorophyll synthesis and plant growth is stimulated by the nanoparticles under water stress condition by the enhancement of endogenous cytokinins content (Al-Khayri et al. 2023a). The photosynthetic pigments and the antioxidant system besides the low molecular weight osmolytes get enhanced on exposure of plant tissues with SeNPs (Zahedi et al. 2020). The contribution of inorganic NPs to the improvement of plant adaptation to drought stress has been extensively demonstrated and summarized in Table 3.

Mechanism of Stress Mitigation by Nanoparticles

Antioxidant Enzyme Regulation

The ROS produced by plants on exposure to different abiotic stress can be mitigated by ROS neutralizing properties of various antioxidant enzymes, including superoxide dismutase (sod), glutathione peroxidase (gpx), and catalase (Hasanuzzaman et al. 2020; Etesami et al. 2021). Superoxide dismutase helps in neutralizing the superoxides ($O_2^{\cdot-}$), while, the glutathione peroxidase, and catalase help in converting H_2O_2 to H_2O and O_2. The optimum level of these enzymes has to be sustained inside the plant system for proper ROS detoxification. The NPs exert several effects on crop plants, which can be short-term and/or long-lasting (Vera-Reyes et al. 2018). The published studies had reported the detrimental effects of NPs on crops via the release of ROS, which brings about severe harm to plants tissues, proteins, lipids, and DNA through the synthesis of superoxide (O_2^-), hydrogen peroxide (H_2O_2), and hydroxyl (OH^-) radicals (Ismail et al. 2014; Siddiqui et al. 2018). The ROS act as a signal molecule in the regulation of stress responses in cell (Haghighi et al. 2012). In response to ROS, the antioxidant defense mechanism of plants is activated (Vera-Reyes et al. 2018) involving scavenging of ROS-generated substances via enzymes and antioxidants (Jeevan Kumar et al. 2015). The rise in the amount of ROS indicated a direct rise in enzymatic and non-enzymatic antioxidants (flavonoids and phenols) by nanomaterials (Vera-Reyes et al. 2018). The better performance of the antioxidant system decreases the damage in cellular membranes by oxidative stress and has been reported to add certain beneficial effects such as enhanced tolerance to salinity (Kumar et al. 2021; Al-Khayri et al. 2023).

Table 3. Effect of inorganic nanoparticles on plant growth, physiology, and yield attributes of different crops under drought stress.

Crop	Nanoparticle	Concentration	Location/Conditions	Mode of Application	Effect on Crop	Reference
Wheat	Ag and Cu	CuNPs (0 mg/L, 3 mg/L, 5 mg/L, and 7 mg/L) AgNPs (0 mg/L, 10 mg/L, 20 mg/L and 30 mg/L)	Greenhouse	Hydroponically in MS medium	• Improved drought resistance and yield • Increased nutrients uptake and water retention	(Ahmed et al. 2021)
	ZnO and urea	1%	Greenhouse	Soil amendment	• Accelerated phenological development of plant • Increased grain yield and Zn uptake	(Dimkpa et al. 2020b)
	Fe	0 mg/kg, 25 mg/kg, 50 mg/kg and 100 mg/kg	Greenhouse	Soil amendment	• Improved photosynthesis, yield, and Fe concentrations in tissues • Alleviation of oxidative stress in leaves	(Adrees et al. 2020)
	Cu and Zn	-	Greenhouse	Seed treatment	• Enhanced activity of antioxidative enzymes • Reduction in the level of accumulation of thiobarbituric acid reactive substances • Stabilized the content of photosynthetic pigments and increased relative water content in leaves	(Taran et al. 2017)
	TiO$_2$	0.01%, 0.02%, and 0.03%	Field	Seed treatment	• Enhanced agronomic traits (plant height, ear weight, and number, 1,000-seed weight, yield, biomass, harvest index) • Increased gluten and starch content	(Jaberzadeh et al. 2013)
Maize	Carbon NPs	80 μg/ml	Field	Seed Priming	• Increased antioxidant defense system • Increased enzyme activities (ascorbate peroxidase, dehydro-ASC reductase, and monodehydro-ASC reductase) • Overall improved growth, physiology, and biochemistry of maize plants	(Alsherif et al. 2023)
	TiO$_2$	0 mg l^{-1}, 50 mg l^{-1}, and 100 mg l^{-1}	Field	Foliar spray	• Enhanced SOD, APX, and CAT antioxidant activities • Increased leaf proline content and grain yield	(Karvar et al. 2022)
	ZnO	100 mg L^{-1}	Greenhouse	Soil amendment	• Alleviated degradation of photosynthetic pigments • Improved stomatal movement • Higher net photosynthetic rate and enhanced water use efficiency • Enhancement in starch and sucrose biosynthesis and glycolysis metabolism in leaves	(Sun et al 2021)
	Cu, Fe, and Co	5mg/L	Field	Seed treatment	• Significant increase in germination rate and early growth • Improve enzyme activities (SOD and APX) in leaves	(Hoang et al. 2019)

Crop	Nanomaterial	Concentration	Growth condition	Application	Effects	References
Rice	ZnO + Biochar	50 mg L^{-1}	Field	Foliar application	• Significant effect on the physiological traits (chlorophyll content, relative water content, plant height, and leaf area index) • Higher yield attributes (number of panicles m^{-2}, no. of filled grain per panicle, 1,000-grain weight, biological and grain yield)	(Elshayb et al. 2022)
	CuO	0 mg L^{-1}, 2.5 mg L^{-1}, 10 mg L^{-1}, 50 mg L^{-1}, 100 mg L^{-1} and 1,000 mg L^{-1}	Axenic	Seed treatment	• Increased expression of ascorbate peroxidase and superoxide dismutase	(Costa and Sharma 2016)
Soybean	Fe, Cu, Co, and ZnO	50 mg/L (Fe, ZnO, Cu) and 0.05 mg/L (Co)	-	Seed treatment	• Improved shoot and root morphology, drought tolerance indices • Significantly enhanced expressions of the tested drought tolerance marker genes	(Linh et al. 2020)
	ZnO	0 g lit^{-1}, 0.5 g lit^{-1}, and 1 g lit^{-1}	Axenic	Seed treatment	• Improved germination, radical length, and fresh and dry weight	(Sedghi et al. 2013)
Potato	ZnO and SiO$_2$	0 ppm, 50 ppm, 100 ppm (ZnO) and 0 ppm, 25 ppm, and 50 ppm (SiO$_2$)	Field	Exogenously applied via irrigation	• Increase in leaf gas exchange, relative water contents, photosynthetic pigments, and leaf green index	(Al-Selwey et al. 2023)
	ZnO-NPs and Magnetite	0.0 ppm, 2.5 ppm, and 5.0 ppm	*In-vitro* tissue culture	Supplemented in MS medium	• Significant improvement in all morphological and harvesting parameters	(Sallam et al. 2022)

Haghighi et al. (2012) reported salinity tolerance in tomatoes by application of Si NPs and observed increased activity of both catalase and ascorbate peroxidase enzymes. Similarly, Ioannou et al. (2020) indicated a reduction in the adverse impact of salinity in *Dracocephalum moldavica* L. by enhancement in the activity of antioxidant enzymes, which further initiates the detoxification of ROS. The improved scavenging of ROS and enzymatic response in plants in regard to engineered NPs were detected in numerous plant systems (Al-Khayri et al. 2023b). Dimkpa et al. (2017) evaluated the dimension of phytotoxicity in sand-grown wheat plants amended with copper and zinc nano-oxides and found a negligible increase in catalase or peroxidase activities by treatment with zinc NPs, whereas the increase in the enzyme activity was reported in treatment with copper NPs (Vera-Reyes et al. 2018). In *Triticum aestivum,* the escalation in levels of oxidized glutathione and genes involved in detoxification was observed due to silver (Ag) NPs (Dimkpa et al. 2013). The increase in SOD and peroxidase activity because of the application of iron oxide (FeO NPs) in cadmium-stressed conditions was reported by Manzoor et al. (2021). Similarly, Djanaguiraman et al. (2018) suggested a foliar spray of SeNPs as a method of heat stress alleviation in sorghum. In their study, compared to the unsprayed control, 22, 24, 11, and 9% increase was reported for superoxide dismutase, catalase, peroxidase, and glutathione peroxidase respectively. They noted increased enzyme activities by SeNPs only under high-temperature conditions. The variation in the isozyme pattern of peroxidase was studied in heavy metal-contaminated soil with ZnO-NPs treatment in *Leucaena leucocephala* seedlings (Venkatachalam et al. 2017). In all the treatments, the four isoforms of peroxidase were detected, but the intensity of isoform 2 was high in ZnO-NPs added treatments. In heavy metal-exposed leaves modified with ZnO-NPs, the amounts of antioxidative enzymes (SOD, CAT, and POX) were enhanced. A direct correlation was identified between the nano-ZnO dosages and terminal heat stress at different doses of nano-ZnO (Kareem et al. 2022). Sarkar and Kalita (2022) have analyzed the effect of green synthesized SeNPs in salinity conditions on mustard plants. The addition of 30 mg/l SeNPs raised the activity of SOD, catalase, ascorbate peroxidase, and peroxidase by 41.20, 64.10, 63.06, and 70.43%, respectively, under the salt stress (200 mM NaCl). Many enzymes can actively interact with SeNPs which can then modify the activity of those enzymes. Further, the enzymatic antioxidant systems in mustard were triggered by the physiological and biochemical characteristics of spherical SeNPs at a specific dose under salinity stress.

Variation in Expression of Stress-Related Genes

Numerous genes attributing to the resistance to abiotic stress have been found to induce similar kinds of responses as proved through transcriptomic studies. The primary stress caused by any of the abiotic factors leads to a chain of other stress in plants. An increase in ambient CO_2 levels imparts high-temperature stress in the plant, which further leads to drought. There is an overlap between the stress responses, which is called cross-tolerance. This cross-tolerance aids in reducing the damage imparted by the other stress, followed by the primary stress. Still, studies have to be done to completely elucidate the signaling pathways and genetics behind the plants' ability to mitigate different types of stress. There are mainly three groups of environmental stress-inducible genes (Figure 1).

One group directly helps in stress mitigation through the production of various osmoprotectants, anti-freezing proteins, etc. The second one encodes for products that have a crucial role in the regulation of gene expression and signal transduction. The last one encodes transmembrane proteins that aid in the transport of different ions and water. The application of nanoparticles leads to the induction of signaling pathways related to stress, which further leads to the expression of different genes that aid in the amelioration of stress. The variation in the expression pattern of salinity-associated genes was studied in strawberries by Moradi et al. (2022). They have studied that application of silicon NPs and methyl jasmonate increased expression of salinity-associated genes (DREB, cAPX, Mn-SOD, and GST). The enzymes SOD and hormone

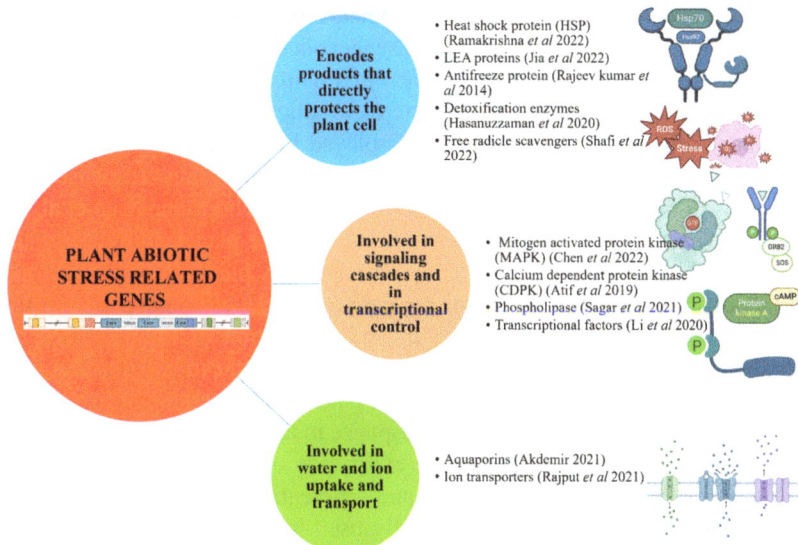

Figure 1. Classification of abiotic stress-related genes in plants.

jasmonic acid (JA) are well-known indices of salt stress. The variation in expression of SOD and JA genes under salt stress conditions by application of the chitosan-polyvinyl alcohol hydrogels (Cs-PVA) and copper (Cu) NPs were analyzed (Hernández-Hernández et al. 2018). It was observed that Cs-PVA and Cs-PVA + Cu NPs cause the SOD and JA genes to express, whether salt stress is present or not. This result strongly proved that the application of NPs stimulates a plant's antioxidant defense mechanisms.

In *Capsicum annuum,* there are reports stating the upregulation of transcription factors (bZIP1 and WRKY1) involved in the response pathways of different abiotic stresses on treatment with nano selenium (Sotoodehnia-Korani et al. 2020). The WRKY is a superfamily of transcription factors in plants and algae imparting a significant role in the growth and development of plants under stressed conditions (Li et al. 2020). Basic leucine zipper motif (bZIP) TF takes part in stress response, drought resistance, plant growth, and seed maturation. They are classified into ten subfamilies based on their structure and function (Yang et al. 2019).

In abiotically stressed plants, the plant heat stress transcription factors (HSFs) function as the regulator of transcription and control a range of critical responsive genes. The HSFs are the primary regulators of heat shock protein production. There are mainly three classes of HSFs in plants HSFA, HSFB, and HSFC. HSFA regulates the production of heat shock protein on exposure to abiotic stress. Banti et al. (2010) reported enhanced anoxic tolerance in *Arabidopsis thaliana* with the aid of HSFA2. HSFA4 confers better tolerance to oxidative and salt stress (Pérez-Salamó et al. 2014). Safari et al. (2018) tracked the variations in the expression of the HSF gene in selenium-exposed wheat plants and found the role of heat shock factor A4A as a hydrogen peroxide sensor, agent of anti-apoptosis and crosslink component with crucial signaling pathways. They reported an increase in the expression of HSFA4 F in plants. HSFA4A was upregulated by 3.4 and 9.15 times, respectively, by the application of 5 mg/L and 10 mg/L of selenium.

Increased Signaling Pathway for Phytohormones

Plant growth regulators or phytohormones are an important class of compounds aiding in plant growth and development (Hajam et al. 2017). The important phytohormones are auxin, cytokinin, gibberellins, abscisic acid, salicylic acid, ethylene, methyl jasmonate, and strigolactone (Omoarelojie et al. 2019). During stress, the plant growth regulators aid in overcoming the outcome of stress by coordinating different stress-mitigating signaling pathways and acting as a messenger molecules.

Phytohormone mediated antioxidant defense system helps the plants to overcome the abiotic stress. Notably, phytohormones operate as secondary messengers in stressed plants to activate antioxidants and scavenge ROS, reducing the detrimental effect of oxidative stress (Raza et al. 2022). Under the abiotic stress condition, hormones like brassinosteroids, salicylic acid, strigolactone, and ethylene impart a stimulatory role, and an inhibitory role is played by cytokinin, gibberellic acid, abscisic acid (ABA), auxin, and JA. The ABA is involved in various kinds of stress response signaling alone or in combination with other hormones and is also called a stress hormone (Luo et al. 2023).

Nanoparticles play the role of regulatory molecules inside plants and can modulate various biochemical and physiological processes. The NPs can help plants to withstand abiotic stress conditions by induction of various hormonal regulations. There are reports stating the induction of various phytohormone synthesis pathways by the application of nanoparticles. Azhar et al. (2021) had notified the effect of SiO_2 nanoparticles on the biosynthesis of cytokinin in *Arabidopsis thaliana* by transcriptomic studies. They analyzed the pattern of expression of two genes taking part in cytokinins regulation (ARR7 and ARR15) in response to temperature stress (Li et al. 2021). The treatment with SiO_2 nanoparticles increased the expression of both of the above genes. A higher amount of cytokinins helps in plant cell division and differentiation.

JA is a plant signaling molecule that is widely used in the stress response mechanism. The physiological response to JA includes the control of opening and closing of stomata, induction of antioxidant defense system, and buildup of amount of osmoprotectants. Jiang et al. (2017) analyzed various effects of TiO_2 nanoparticles on wheat under normal and elevated CO_2 levels. As per their report, an increase in the level of JA content is proportional to the dose of TiO_2 nanoparticles under elevated CO_2 conditions, but no effect was shown under normal CO_2 levels. In the climate change scenario, this shows that TiO_2 nanoparticles can be used in the future to mitigate abiotic stresses. The development of an appropriate crosstalk mechanism between nanoparticles and the manufacture of phytohormones is still in its initial stages.

Conclusions

The rising prevalence of diverse abiotic stresses owing to increasing population, scarcity of natural resources and climate change has adversely affected the crop yield. Realizing, the concern of abiotic stresses at the global level, the scientific community has concentrated on the ability of nanotechnology in raising agricultural yield under stressed conditions. The advancements in nanotechnology have helped to improve the resistance of plants against numerous stresses prevailing in the environment, especially the most widespread and commonly experienced water and temperature stress. The application of different nanoparticles in crops has emerged as a novel approach for induction of abiotic stress tolerance.

References

Adrees, M., Khan, Z.S., Ali, S., Hafeez, M., Khalid, S., ur Rehman, M.Z., Hussain, A., Hussain, K., Shahid Chatha, S.A. and Rizwan, M. 2020. Simultaneous mitigation of cadmium and drought stress in wheat by soil application of iron nanoparticles. Chemosphere 238: 124681. doi: 10.1016/j.chemosphere.2019.124681.

Ahluwalia, O., Singh, P.C. and Bhatia, R. 2021. A review on drought stress in plants: Implications, mitigation and the role of plant growth promoting rhizobacteria. Resour. Environ. Sustain. 5: 100032. doi: 10.1016/j.resenv.2021.100032.

Ahmed, F., Javed, B., Razzaq, A. and Mashwani, Z. 2021. Applications of copper and silver nanoparticles on wheat plants to induce drought tolerance and increase yield. IET Nanobiotechnology 15: 68–78. doi: 10.1049/nbt2.12002.

Akdemir, H. 2021. Evaluation of transcription factor and aquaporin gene expressions in response to Al_2O_3 and ZnO nanoparticles during barley germination. Plant Physiology and Biochemistry 166: 466–476. https://doi.org/10.1016/j.plaphy.2021.06.018.

Al-Khayri, J.M., Rashmi, R., Surya Ulhas, R., Sudheer, W.N., Banadka, A., Nagella, P., Aldaej, M.I., Rezk, A.A.-S., Shehata, W.F. and Almaghasla, M.I. 2023a. The role of nanoparticles in response of plants to abiotic stress at physiological, biochemical, and molecular levels. Plants 12: 292. doi: 10.3390/plants12020292.

Al-Khayri, J.M., Rashmi, R., Surya Ulhas, R., Sudheer, W.N., Banadka, A., Nagella, P., Aldaej, M.I., Rezk, A.A.-S., Shehata, W.F. and Almaghasla, M.I. 2023b. The role of nanoparticles in response of plants to abiotic stress at physiological, biochemical, and molecular levels. Plants 12: 292. doi: 10.3390/plants12020292.

Al-Selwey, W.A., Alsadon, A.A., Ibrahim, A.A., Labis, J.P. and Seleiman, M.F. 2023. Effects of zinc oxide and silicon dioxide nanoparticles on physiological, yield, and water use efficiency traits of potato grown under water deficit. Plants 12: 218. doi: 10.3390/plants12010218.

Ali, S.S., Al-Tohamy, R., Koutra, E., Moawad, M.S., Kornaros, M., Mustafa, A.M., Mahmoud, Y.A.-G., Badr, A., Osman, M.E.H., Elsamahy, T., Jiao, H. and Sun, J. 2021. Nanobiotechnological advancements in agriculture and food industry: Applications, nanotoxicity, and future perspectives. Sci. Total Environ. 792: 148359. doi: 10.1016/j.scitotenv.2021.148359.

Alsherif, E.A., Almaghrabi, O., Elazzazy, A.M., Abdel-Mawgoud, M., Beemster, G.T.S. and AbdElgawad, H. 2023. Carbon nanoparticles improve the effect of compost and arbuscular mycorrhizal fungi in drought-stressed corn cultivation. Plant Physiol. Biochem. 194: 29–40. doi: 10.1016/j.plaphy.2022.11.005.

Amini, S., Maali-Amiri, R., Mohammadi, R. and Kazemi-Shahandashti, S.-S. 2017. cDNA-AFLP analysis of transcripts induced in chickpea plants by TiO_2 nanoparticles during cold stress. Plant Physiol. Biochem 111: 39–49. doi: 10.1016/j.plaphy.2016.11.011.

Atif, R.M., Shahid, L., Waqas, M., Ali, B., Rashid, M.A.R., Azeem, F. and Chung, G. 2019. Insights on calcium-dependent protein kinases (CPKs) signaling for abiotic stress tolerance in plants. International Journal of Molecular Sciences 20(21): 5298. https://doi.org/10.3390/ijms20215298.

Azhar, B.J., Noor, A., Zulfiqar, A., Zeenat, A., Ahmad, S., Chishti, I., Abbas, Z. and Shakeel, S.N. 2021. Effect of ZnO, SiO_2 and composite nanoparticles on *Arabidopsis thaliana* and involvement of ethylene and cytokinin signaling pathways. Pakistan J. Bot. 53. doi: 10.30848/PJB2021-2(40).

Azmat, A., Tanveer, Y., Yasmin, H., Hassan, M.N., Shahzad, A., Reddy, M. and Ahmad, A. 2022. Coactive role of zinc oxide nanoparticles and plant growth promoting rhizobacteria for mitigation of synchronized effects of heat and drought stress in wheat plants. Chemosphere 297: 133982. doi: 10.1016/j.chemosphere.2022.133982.

Babenko, L. and Kosakivska, I. 2014. Effect of temperature stresses on pigment content, lipoxygenase activity and cell ultrastructure of winter wheat seedlings. Genet Plant 4: 117–125.

Bala, R., Kalia, A. and Dhaliwal, S.S. 2019. Evaluation of efficacy of ZnO nanoparticles as remedial zinc nanofertilizer for rice. J. Soil Sci. Plant. Nutr. 19: 379–389. doi: 10.1007/s42729-019-00040-z.

Banti, V., Mafessoni, F., Loreti, E., Alpi, A. and Perata, P. 2010. The heat-inducible transcription factor HsfA2 enhances anoxia tolerance in Arabidopsis. Plant Physiol. 152: 1471–1483. doi: 10.1104/pp.109.149815.

Baweja, P. and Kumar, G. 2020. Abiotic stress in plants: an overview. pp. 1–15. *In*: Plant Stress Biology. Springer Singapore, Singapore.

Bhat, M.A., Kumar, V., Bhat, M.A., Wani, I.A., Dar, F.L., Farooq, I., Bhatti, F., Koser, R., Rahman, S. and Jan, A.T. 2020. Mechanistic insights of the interaction of plant growth-promoting rhizobacteria (PGPR) with plant roots toward enhancing plant productivity by alleviating salinity stress. Front Microbiol. 11. doi: 10.3389/fmicb.2020.01952.

Chen, Y., Zhang, Z. and Tao, F. 2018. Impacts of climate change and climate extremes on major crops productivity in China at a global warming of 1.5 and 2.0°C. Earth Syst. Dyn. 9: 543–562. doi: 10.5194/esd-9-543-2018.

Chen, Y., Chen, L., Sun, X., Kou, S., Liu, T., Dong, J. and Song, B. 2022. The mitogen-activated protein kinase kinase MKK2 positively regulates constitutive cold resistance in the potato. Environmental and Experimental Botany 194: 104702. https://doi.org/10.1016/j.envexpbot.2021.104702.

Costa, M.V.J. and Sharma, P.K. 2016. Effect of copper oxide nanoparticles on growth, morphology, photosynthesis, and antioxidant response in *Oryza sativa*. Photosynthetica 54: 110–119. doi: 10.1007/s11099-015-0167-5.

Dimkpa, C.O., McLean, J.E., Martineau, N., Britt, D.W., Haverkamp, R. and Anderson, A.J. 2013. Silver nanoparticles disrupt wheat (*Triticum aestivum* L.) growth in a sand matrix. Environ. Sci. Technol. 47: 1082–1090. doi: 10.1021/es302973y.

Dimkpa, C.O., Bindraban, P.S., Fugice, J., Agyin-Birikorang, S., Singh, U. and Hellums, D. 2017. Composite micronutrient nanoparticles and salts decrease drought stress in soybean. Agron Sustain Dev. 37: 1–13. doi: 10.1007/s13593-016-0412-8.

Dimkpa, C.O., Andrews, J., Fugice, J., Singh, U., Bindraban, P.S., Elmer, W.H., Gardea-Torresdey, J.L. and White, J.C. 2020a. Facile coating of urea with low-dose ZnO nanoparticles promotes wheat performance and enhances Zn uptake under drought stress. Front Plant Sci. 11. doi: 10.3389/fpls.2020.00168.

Dimkpa, C.O., Andrews, J., Fugice, J., Singh, U., Bindraban, P.S., Elmer, W.H., Gardea-Torresdey, J.L. and White, J.C. 2020b. Facile coating of urea with low-dose ZnO nanoparticles promotes wheat performance and enhances Zn uptake under drought stress. Front Plant Sci. 11. doi: 10.3389/fpls.2020.00168.

Djanaguiraman, M., Belliraj, N., Bossmann, S.H. and Prasad, P.V.V. 2018a. High-temperature stress alleviation by selenium nanoparticle treatment in grain sorghum. ACS Omega 3: 2479–2491. doi: 10.1021/acsomega.7b01934.

Djanaguiraman, M., Belliraj, N., Bossmann, S.H. and Prasad, P.V.V. 2018b. High-temperature stress alleviation by selenium nanoparticle treatment in grain sorghum. ACS Omega 3: 2479–2491. doi: 10.1021/acsomega.7b01934.

Elshayb, O.M., Nada, A.M., Sadek, A.H., Ismail, S.H., Shami, A., Alharbi, B.M., Alhammad, B.A. and Seleiman, M.F. 2022. The integrative effects of biochar and ZnO nanoparticles for enhancing rice productivity and water use efficiency under irrigation deficit conditions. Plants 11: 1416. doi: 10.3390/plants11111416.

Elsheery, N.I., Helaly, M.N., El-Hoseiny, H.M. and Alam-Eldein, S.M. 2020a. Zinc oxide and silicone nanoparticles to improve the resistance mechanism and annual productivity of salt-stressed mango trees. Agronomy 10: 558. doi: 10.3390/agronomy10040558.

Elsheery, N.I., Sunoj, V.S.J., Wen, Y., Zhu, J.J., Muralidharan, G. and Cao, K.F. 2020b. Foliar application of nanoparticles mitigates the chilling effect on photosynthesis and photoprotection in sugarcane. Plant Physiol. Biochem. 149: 50–60. doi: 10.1016/j.plaphy.2020.01.035.

Etesami, H., Fatemi, H. and Rizwan, M. 2021. Interactions of nanoparticles and salinity stress at physiological, biochemical and molecular levels in plants: A review. Ecotoxicol Environ. Saf. 225: 112769. doi: 10.1016/j.ecoenv.2021.112769.

Farooq, M., Wahid, A., Kobayashi, N., Fujita, D. and Basra, S.M.A. 2009. Plant drought stress: effects, mechanisms and management. pp. 153–188. *In*: Sustainable Agriculture. Springer Netherlands, Dordrecht.

Haghighi, M., Afifipour, Z. and Mozafarian, M. 2012. The effect of N-Si on tomato seed germination under salinity levels. J. Biol. Environ. Sci. 6: 87–90.

Hajam, M.A., Hassan, G.I., Bhat, T.A., Bhat, I.A., Asif, M., Ejaz, A., Wani, M.A. and Khan, I.F. 2017. Understanding plant growth regulators, their interplay: For nursery establishment in fruits. Int. J. Chem. Stud. 5: 905–910.

Hasanpour, H., Maali-Amir, R. and Zeinali, H. 2015. Effect of TiO$_2$ nanoparticles on metabolic limitations to photosynthesis under cold in chickpea. Russ J. Plant Physiol. 62: 779–787. doi: 10.1134/S1021443715060096.

Hasanuzzaman, M., Bhuyan, M.H.M.B., Parvin, K., Bhuiyan, T.F., Anee, T.I., Nahar, K., Hossen, M.S., Zulfiqar, F., Alam, M.M. and Fujita, M. 2020. Regulation of ROS metabolism in plants under environmental stress: a review of recent experimental evidence. Int. J. Mol. Sci. 21: 8695. doi: 10.3390/ijms21228695.

Hashimoto, T., Mustafa, G., Nishiuchi, T. and Komatsu, S. 2020. Comparative analysis of the effect of inorganic and organic chemicals with silver nanoparticles on soybean under flooding stress. Int. J. Mol. Sci. 21: 1300. doi: 10.3390/ijms21041300.

Hassan, N.S., Salah, El Din, T.A., Hendawey, M.H., Borai, I.H. and Mahdi, A.A. 2018. Magnetite and zinc oxide nanoparticles alleviated heat stress in wheat plants. Curr. Nanomater. 3: 32–43. doi: 10.2174/2405461503666180619160923.

Hernández-Hernández, H., Juárez-Maldonado, A., Benavides-Mendoza, A., Ortega-Ortiz, H., Cadenas-Pliego, G., Sánchez-Aspeytia, D. and González-Morales, S. 2018. Chitosan-PVA and copper nanoparticles improve growth and overexpress the SOD and JA genes in tomato plants under salt stress. Agronomy 8: 175. doi: 10.3390/agronomy8090175.

Higashi, Y. and Saito, K. 2019. Lipidomic studies of membrane glycerolipids in plant leaves under heat stress. Prog. Lipid Res. 75: 100990. doi: 10.1016/j.plipres.2019.100990.

Hoang, S.A., Nguyen, L.Q., Nguyen, N.H., Tran, C.Q., Nguyen, D.V., Le, N.T., Ha, C.V., Vu, Q.N. and Phan, C.M. 2019. Metal nanoparticles as effective promotors for Maize production. Sci. Rep. 9: 13925. doi: 10.1038/s41598-019-50265-2.

Hong, J., Wang, C., Wagner, D.C., Gardea-Torresdey, J.L., He, F. and Rico, C.M. 2021. Foliar application of nanoparticles: mechanisms of absorption, transfer, and multiple impacts. Environ. Sci. Nano. 8: 1196–1210. doi: 10.1039/D0EN01129K.

Hossain, A., Skalicky, M., Brestic, M., Maitra, S., Ashraful Alam, M., Syed, M.A., Hossain, J., Sarkar, S., Saha, S., Bhadra, P., Shankar, T., Bhatt, R., Kumar Chaki, A., EL Sabagh, A. and Islam, T. 2021. Consequences and mitigation strategies of abiotic stresses in wheat (*Triticum aestivum* L.) under the changing climate. Agronomy 11: 241. doi: 10.3390/agronomy11020241.

Ioannou, A., Gohari, G., Papaphilippou, P., Panahirad, S., Akbari, A., Dadpour, M.R., Krasia-Christoforou, T. and Fotopoulos, V. 2020. Advanced nanomaterials in agriculture under a changing climate: The way to the future? Environ. Exp. Bot. 176: 104048. doi: 10.1016/j.envexpbot.2020.104048.

Iqbal, M., Raja, N.I., Mashwani, Z., Wattoo, F.H., Hussain, M., Ejaz, M. and Saira, H. 2019. Assessment of AgNPs exposure on physiological and biochemical changes and antioxidative defence system in wheat (*Triticum aestivum* L.) under heat stress. IET Nanobiotechnology 13: 230–236. doi: 10.1049/iet-nbt.2018.5041.

Iqbal, Z., Sarkhosh, A., Balal, R.M., Rauf, S., Khan, N., Altaf, M.A., Camara-Zapata, J.M., Garcia-Sanchez, F. and Shahid, M.A. 2021. Silicon nanoparticles mitigate hypoxia-induced oxidative damage by improving antioxidants activities and concentration of osmolytes in Southern Highbush Blueberry plants. Agronomy 11: 2143. doi: 10.3390/agronomy11112143.

Ismail, A., Takeda, S. and Nick, P. 2014. Life and death under salt stress: same players, different timing Journal of Experimental Botany 65(12): 2963–2979. https://doi.org/10.1093/jxb/eru159.

Jaberzadeh, A., Moaveni, P., Moghadam, H.R. and Zahedi, H. 2013. Influence of bulk and nanoparticles titanium foliar application on some agronomic traits, seed gluten and starch contents of wheat subjected to water deficit stress. Not Bot. Horti. Agrobot Cluj-Napoca 41: 201. doi: 10.15835/nbha4119093.

Jeevan Kumar, S.P., Rajendra Prasad, S., Banerjee, R. and Thammineni, C. 2015. Seed birth to death: dual functions of reactive oxygen species in seed physiology. Ann. Bot. 116: 663–668. doi: 10.1093/aob/mcv098.

Jia, C., Guo, B., Wang, B., Li, X., Yang, T., Li, N. and Yu, Q. 2022. The LEA gene family in tomato and its wild relatives: genome-wide identification, structural characterization, expression profiling, and role of SlLEA6 in drought stress. BMC Plant Biology 22(1): 596. https://link.springer.com/article/10.1186/s12870-022-03953-7.

Jiang, F., Shen, Y., Ma, C., Zhang, X., Cao, W. and Rui, Y. 2017. Effects of TiO_2 nanoparticles on wheat (Triticum aestivum L.) seedlings cultivated under super-elevated and normal CO_2 conditions. PLoS One 12: e0178088. doi: 10.1371/journal.pone.0178088.

Kai, H. and Iba, K. 2014. Temperature Stress in Plants. *In*: eLS. Wiley.

Kalia, A. and Kaur, H. 2019. Nanofertilizers: An innovation towards new generation fertilizers for improved nutrient use efficacy (NUE) and environmental sustainability. pp. 45–61. *In*: Bhoop, B., Katare, O. and Souto, E. (eds.). Emerging Trends in NanoBioMedicine. Taylor & Francis (CRC Press), Boca Raton, FL, USA, Boca Raton, FL, USA.

Kalia, A., Abd-Elsalam, K.A. and Kuca, K. 2020. Zinc-based nanomaterials for diagnosis and management of plant diseases: Ecological safety and future prospects. J. Fungi 6. doi: 10.3390/jof6040222.

Kardavan Ghabel, V. and Karamian, R. 2020a. Effects of TiO_2 nanoparticles and spermine on antioxidant responses of Glycyrrhiza glabra L. to cold stress. Acta Bot. Croat 79: 137–147. doi: 10.37427/botcro-2020-025.

Kardavan Ghabel, V. and Karamian, R. 2020b. Effects of TiO_2 nanoparticles and spermine on antioxidant responses of Glycyrrhiza glabra L. to cold stress. Acta Bot. Croat 79: 137–147. doi: 10.37427/botcro-2020-025.

Kareem, H.A., Saleem, M.F., Saleem, S., Rather, S.A., Wani, S.H., Siddiqui, M.H., Alamri, S., Kumar, R., Gaikwad, N.B., Guo, Z., Niu, J. and Wang, Q. 2022. Zinc oxide nanoparticles interplay with physiological and biochemical attributes in terminal heat stress alleviation in mungbean (*Vigna radiata* L.). Front Plant Sci. 13. doi: 10.3389/fpls.2022.842349.

Karvar, M., Azari, A., Rahimi, A., Maddah-Hosseini, S. and Ahmadi-Lahijani, M.J. 2022. Titanium dioxide nanoparticles (TiO_2-NPs) enhance drought tolerance and grain yield of sweet corn (Zea mays L.) under deficit irrigation regimes. Acta Physiol. Plant 44: 14. doi: 10.1007/s11738-021-03349-4.

Kaur, H., Kaur, J., Kalia, A. and Kuca, K. 2022. The Janus face of nanomaterials: physiological responses as inducers of stress or promoters of plant growth? pp. 395–426. *In*: Plant and Nanoparticles. Springer Nature Singapore, Singapore.

Kaveh, R., Li, Y.-S., Ranjbar, S., Tehrani, R., Brueck, C.L. and Van Aken, B. 2013. Changes in *Arabidopsis thaliana* gene expression in response to silver nanoparticles and silver ions. Environ. Sci. Technol. 47: 10637–10644. doi: 10.1021/es402209w.

Khalid, M.F., Iqbal Khan, R., Jawaid, M.Z., Shafqat, W., Hussain, S., Ahmed, T., Rizwan, M., Ercisli, S., Pop, O.L. and Alina Marc, R. 2022. Nanoparticles: The plant saviour under abiotic stresses. Nanomaterials 12: 3915. doi: 10.3390/nano12213915.

Khan, F., Pandey, P. and Upadhyay, T.K. 2022. Applications of nanotechnology-based agrochemicals in food security and sustainable agriculture: an overview. Agriculture 12: 1672. doi: 10.3390/agriculture12101672.

Khatri, K. and Rathore, M.S. 2018. Plant nanobionics and its applications for developing plants with improved photosynthetic capacity. *In*: Photosynthesis—From Its Evolution to Future Improvements in Photosynthetic Efficiency Using Nanomaterials. InTech.

Khondoker, R. and Kabir, M.H. 2019. Influence of anti-transpirant and cycocel on growth and flowering of tuberose under different moisture regimes. Int. J. Nat. Soc. Sci. 6: 27–43.

Kole, C., Kumar, D.S. and Khodakovskaya, M.V. 2016. Plant nanotechnology: Principles and practices. Plant Nanotechnol Princ. Pract. 1–383. doi: 10.1007/978-3-319-42154-4.

Kondal, R., Kalia, A., Krejcar, O., Kuca, K., Sharma, S.P., Luthra, K., Dheri, G.S., Vikal, Y., Taggar, M.S., Abd-Elsalam, K.A. and Gomes, C.L. 2021. Chitosan-urea nanocomposite for improved fertilizer applications: The effect on the soil enzymatic activities and microflora dynamics in N cycle of potatoes (Solanum tuberosum L.). Polymers (Basel) 13. doi: 10.3390/polym13172887.

Kumar, V., Srivastava, A., Wani, S.H., Shriram, V. and Penna, S. 2021. Transcriptional and post-transcriptional mechanisms regulating salt tolerance in plants. Physiol. Plant 173: 1291–1294. doi: 10.1111/ppl.13592.

Lamaoui, M., Jemo, M., Datla, R. and Bekkaoui, F. 2018. Heat and drought stresses in crops and approaches for their mitigation. Front Chem. 6: 1–14. doi: 10.3389/fchem.2018.00026.

Li, S.-M., Zheng, H.-X., Zhang, X.-S. and Sui, N. 2021. Cytokinins as central regulators during plant growth and stress response. Plant Cell Rep. 40: 271–282. doi: 10.1007/s00299-020-02612-1.

Li, W., Pang, S., Lu, Z. and Jin, B. 2020. Function and mechanism of WRKY transcription factors in abiotic stress responses of plants. Plants 9: 1515. doi: 10.3390/plants9111515.

Linh, T.M., Mai, N.C., Hoe, P.T., Lien, L.Q., Ban, N.K., Hien, L.T.T., Chau, N.H. and Van, N.T. 2020. Metal-based nanoparticles enhance drought tolerance in soybean. J. Nanomater. 2020: 1–13. doi: 10.1155/2020/4056563.

Luo, X., Xu, J., Zheng, C., Yang, Y., Wang, L., Zhang, R., Ren, X., Wei, S., Aziz, U., Du, J., Liu, W., Tan, W. and Shu, K. 2023. Abscisic acid inhibits primary root growth by impairing ABI4-mediated cell cycle and auxin biosynthesis. Plant Physiol. 191: 265–279. doi: 10.1093/plphys/kiac407.

Luqman, M., Mahmood, F. and Al-Ansari, T. 2023. Supporting sustainable global food security through a novel decentralised offshore floating greenhouse. Energy Convers Manag. 277: 116577. doi: 10.1016/j.enconman.2022.116577.

M. S., A., Sridharan, K., Puthur, J.T. and Dhankher, O.P. 2021. Priming with nanoscale materials for boosting abiotic stress tolerance in crop plants. J. Agric Food Chem. 69: 10017–10035. doi: 10.1021/acs.jafc.1c03673.

Majhi, K.C. and Yadav, M. 2021. Synthesis of inorganic nanomaterials using carbohydrates. pp. 109–135. *In*: Green Sustainable Process for Chemical and Environmental Engineering and Science. Elsevier.

Manzoor, N., Ahmed, T., Noman, M., Shahid, M., Nazir, M.M., Ali, L., Alnusaire, T.S., Li, B., Schulin, R. and Wang, G. 2021. Iron oxide nanoparticles ameliorated the cadmium and salinity stresses in wheat plants, facilitating photosynthetic pigments and restricting cadmium uptake. Sci. Total Environ. 769: 145221. doi: 10.1016/j.scitotenv.2021.145221.

Martínez-Peñalver, A., Graña, E., Reigosa, M.J. and Sánchez-Moreiras, A.M. 2012. Early photosynthetic response of *Arabidopsis thaliana* to temperature and salt stress conditions. Russ. J. Plant Physiol. 59: 640–647. doi: 10.1134/S1021443712030119.

Mittal, D., Kaur, G., Singh, P., Yadav, K. and Ali, S.A. 2020. Nanoparticle-based sustainable agriculture and food science: recent advances and future outlook. Front Nanotechnol. 2. doi: 10.3389/fnano.2020.579954.

Mohammadi, R., Maali-Amiri, R. and Abbasi, A. 2013. Effect of TiO_2 nanoparticles on chickpea response to cold stress. Biol. Trace Elem. Res. 152: 403–410. doi: 10.1007/s12011-013-9631-x.

Moradi, P., Vafaee, Y., Mozafari, A.A. and Tahir, N.A. 2022. Silicon nanoparticles and methyl jasmonate improve physiological response and increase expression of stress-related genes in strawberry cv. Paros under salinity stress. Silicon 14: 10559–10569. doi: 10.1007/s12633-022-01791-8.

Mustafa, G., Sakata, K., Hossain, Z. and Komatsu, S. 2015a. Proteomic study on the effects of silver nanoparticles on soybean under flooding stress. J. Proteomics 122: 100–118. doi: 10.1016/j.jprot.2015.03.030.

Mustafa, G., Sakata, K. and Komatsu, S. 2015b. Proteomic analysis of flooded soybean root exposed to aluminum oxide nanoparticles. J. Proteomics 128: 280–297. doi: 10.1016/j.jprot.2015.08.010.

Mustafa, G., Sakata, K. and Komatsu, S. 2016. Proteomic analysis of soybean root exposed to varying sizes of silver nanoparticles under flooding stress. J. Proteomics 148: 113–125. doi: 10.1016/j.jprot.2016.07.027.

Najafi, E., Devineni, N., Khanbilvardi, R.M. and Kogan, F. 2018. Understanding the changes in global crop yields through changes in climate and technology. Earth's Futur. 6: 410–427. doi: 10.1002/2017EF000690.

Narula, S., Chaudhry, S. and Sidhu, G.P.S. 2022. Ameliorating abiotic stress tolerance in crop plants by metabolic engineering. pp. 25–59. *In*: Metabolic Engineering in Plants. Springer Nature Singapore, Singapore.

Ndaba, B., Roopnarain, A., Rama, H. and Maaza, M. 2022. Biosynthesized metallic nanoparticles as fertilizers: An emerging precision agriculture strategy. J. Integr. Agric. 21: 1225–1242. doi: 10.1016/S2095-3119(21)63751-6.

Omoarelojie, L.O., Kulkarni, M.G., Finnie, J.F. and Van Staden, J. 2019. Strigolactones and their crosstalk with other phytohormones. Ann. Bot. 124: 749–767. doi: 10.1093/aob/mcz100.

Oshunsanya, S.O., Nwosu, N.J. and Li, Y. 2019. Abiotic stress in agricultural crops under climatic conditions. pp. 71–100. *In*: Sustainable Agriculture, Forest and Environmental Management. Springer Singapore, Singapore.

Pérez-Salamó, I., Papdi, C., Rigó, G., Zsigmond, L., Vilela, B., Lumbreras, V., Nagy, I., Horváth, B., Domoki, M., Darula, Z., Medzihradszky, K., Bögre, L., Koncz, C. and Szabados, L. 2014. The heat shock factor A4A confers salt tolerance and is regulated by oxidative stress and the mitogen-activated protein kinases MPK3 and MPK6. Plant Physiol. 165: 319–334. doi: 10.1104/pp.114.237891.

Petruccelli, R., Bartolini, G., Ganino, T., Zelasco, S., Lombardo, L., Perri, E., Durante, M. and Bernardi, R. 2022. Cold stress, freezing adaptation, varietal susceptibility of Olea europaea L.: A review. Plants 11: 1367. doi: 10.3390/plants11101367.

Prażak, R., Święciło, A., Krzepiłko, A., Michałek, S. and Arczewska, M. 2020. Impact of Ag nanoparticles on seed germination and seedling growth of green beans in normal and chill temperatures. Agriculture 10: 312. doi: 10.3390/agriculture10080312.

Qi, M., Liu, Y. and Li, T. 2013. Nano-TiO$_2$ improve the photosynthesis of tomato leaves under mild heat stress. Biol. Trace Elem. Res. 156: 323–328. doi: 10.1007/s12011-013-9833-2.

Rajeev Kumar, S., Kiruba, R., Balamurugan, S., Cardoso, H.G., Birgit, A.S., Zakwan, A. and Sathishkumar, R. 2014. Carrot antifreeze protein enhances chilling tolerance in transgenic tomato. Acta Physiologiae Plantarum 36: 21–27. https://link.springer.com/article/10.1007/s11738-013-1383-x.

Rajput, V.D., Minkina, T., Kumari, A., Singh, V.K., Verma, K.K., Mandzhieva, S. and Keswani, C. 2021. Coping with the challenges of abiotic stress in plants: New dimensions in the field application of nanoparticles. Plants 10(6): 1221. https://doi.org/10.3390/plants10061221.

Ramakrishna, G., Singh, A., Kaur, P., Yadav, S.S., Sharma, S. and Gaikwad, K. 2022. Genome wide identification and characterization of small heat shock protein gene family in pigeonpea and their expression profiling during abiotic stress conditions. International Journal of Biological Macromolecules 197: 88–102. https://doi.org/10.1016/j.ijbiomac.2021.12.016.

Rani, A., Kiran, A., Sharma, K.D., Prasad, P.V.V., Jha, U.C., Siddique, K.H.M. and Nayyar, H. 2021. Cold tolerance during the reproductive phase in chickpea (Cicer arietinum L.) is associated with superior cold acclimation ability involving antioxidants and cryoprotective solutes in anthers and ovules. Antioxidants 10: 1693. doi: 10.3390/antiox10111693.

Raza, A., Salehi, H., Rahman, M.A., Zahid, Z., Madadkar Haghjou, M., Najafi-Kakavand, S., Charagh, S., Osman, H.S., Albaqami, M., Zhuang, Y., Siddique, K.H.M. and Zhuang, W. 2022. Plant hormones and neurotransmitter interactions mediate antioxidant defenses under induced oxidative stress in plants. Front Plant Sci. 13. doi: 10.3389/fpls.2022.961872.

Rezvani, N. and Sorooshzadeh, A. 2014. Effect of nano-silver on root and bud growth of saffron in flooding stress condition. Int. J. Biol. Biomol. Agric Food Biotechnol. Eng. 2: 91–104.

Ritonga, F.N. and Chen, S. 2020. Physiological and molecular mechanism involved in cold stress tolerance in plants. Plants 9: 560. doi: 10.3390/plants9050560.

Safari, M., Oraghi Ardebili, Z. and Iranbakhsh, A. 2018. Selenium nano-particle induced alterations in expression patterns of heat shock factor A4A (HSFA4A), and high molecular weight glutenin subunit 1Bx (Glu-1Bx) and enhanced nitrate reductase activity in wheat (*Triticum aestivum* L.). Acta Physiologiae Plantarum 40: 1–8. https://link.springer.com/article/10.1007/s11738-018-2694-8.

Safdar, M., Kim, W., Park, S., Gwon, Y., Kim, Y.-O. and Kim, J. 2022. Engineering plants with carbon nanotubes: a sustainable agriculture approach. J. Nanobiotechnology 20: 275. doi: 10.1186/s12951-022-01483-w.

Sagar, S. and Singh, A. 2021. Emerging role of phospholipase C mediated lipid signaling in abiotic stress tolerance and development in plants. Plant Cell Reports 40(11): 2123–2133. https://link.springer.com/article/10.1007/s00299-021-02713-5.

Salem-Fnayou, A. Ben, Bouamama, B., Ghorbel, A. and Mliki, A. 2011. Investigations on the leaf anatomy and ultrastructure of grapevine (Vitis vinifera) under heat stress. Microsc. Res. Tech. 74: 756–762. doi: 10.1002/jemt.20955.

Sallam, A.R., Mahdi, A.A. and Farroh, K.Y. 2022. Improving drought stress tolerance in potatos (Solanum tuberosum L.) using magnetite and zinc oxide nanoparticles. Plant Cell Biotechnol. Mol. Biol. 1–16. doi: 10.56557/pcbmb/2022/v23i37-387886.

Sanzari, I., Leone, A. and Ambrosone, A. 2019. Nanotechnology in plant science: to make a long story short. Front Bioeng. Biotechnol. 7: 1–12. doi: 10.3389/fbioe.2019.00120.

Sarkar, R.D. and Kalita, M.C. 2022. Green synthesized Se nanoparticle-mediated alleviation of salt stress in field mustard, TS-36 variety. J. Biotechnol. 359: 95–107. doi: 10.1016/j.jbiotec.2022.09.013.

Sedghi, M., Hadi, M. and Toluie, S.G. 2013. Effect of nano zinc oxide on the germination parameters of soybean seeds under drought stress. Ser. Biol. XVI: 73–78.

Seleiman, M.F., Al-Suhaibani, N., Ali, N., Akmal, M., Alotaibi, M., Refay, Y., Dindaroglu, T., Abdul-Wajid, H.H. and Battaglia, M.L. 2021. Drought stress impacts on plants and different approaches to alleviate its adverse effects. Plants 10: 259. doi: 10.3390/plants10020259.

Seleiman, M.F., Almutairi, K.F., Alotaibi, M., Shami, A., Alhammad, B.A. and Battaglia, M.L. 2020. Nano-fertilization as an emerging fertilization technique: why can modern agriculture benefit from its use? Plants 10: 2. doi: 10.3390/plants10010002.

Shabbir, R., Singhal, R.K., Mishra, U.N., Chauhan, J., Javed, T., Hussain, S., Kumar, S., Anuragi, H., Lal, D. and Chen, P. 2022. Combined abiotic stresses: challenges and potential for crop improvement. Agronomy 12: 2795. doi: 10.3390/agronomy12112795.

Shafi, A., Hassan, F. and Khanday, F.A. 2022. Reactive Oxygen and Nitrogen Species: Oxidative Damage and Antioxidative Defense Mechanism in Plants under Abiotic Stress. pp. 71–99. In Plant Abiotic Stress Physiology. Apple Academic Press. https://books.google.co.in/books?hl=en&lr=&id=tDVVEAAAQBAJ&oi=fnd&pg=PA71& dq=Shafi,+A.,+Hassan,+F.,+%26+Khanday,+F.+A.+(2022).+Reactive+Oxygen+and+Nitrogen+Species: +Oxidative+Damage+and+Antioxidative+ Defense+Mechanism+in+Plants+under+Abiotic+Stress.+In+Pla.

Shang, Y., Kamrul Hasan, M., Ahammed, G.J., Li, M., Yin, H. and Zhou, J. 2019. Applications of nanotechnology in plant growth and crop protection: A review. Molecules 24. doi: 10.3390/molecules24142558.

Siddiqui, Z.A., Khan, A., Khan, M.R. and Abd-Allah, E.F. 2018. Effects of zinc oxide nanoparticles (zno nps) and some plant pathogens on the growth and nodulation of lentil (lens culinaris medik.). Acta Phytopathol. Entomol. Hungarica 53: 195–212. doi: 10.1556/038.53.2018.012.

Singh, A., Sharma, R., Rawat, S., Singh, A.K., Rajput, V.D., Fedorov, Y., Minkina, T. and Chaplygin, V. 2022. Nanomaterial-plant interaction: Views on the pros and cons. pp. 47–68. *In*: Toxicity of Nanoparticles in Plants. Elsevier.

Song, X.-P., Verma, K.K., Tian, D.-D., Zhang, X.-Q., Liang, Y.-J., Huang, X., Li, C.-N. and Li, Y.-R. 2021a. Exploration of silicon functions to integrate with biotic stress tolerance and crop improvement. Biol. Res. 54: 19. doi: 10.1186/s40659-021-00344-4.

Song, Y., Jiang, M., Zhang, H. and Li, R. 2021b. Zinc oxide nanoparticles alleviate chilling stress in rice (*Oryza Sativa* L.) by regulating antioxidative system and chilling response transcription factors. Molecules 26: 2196. doi: 10.3390/molecules26082196.

Sotoodehnia-Korani, S., Iranbakhsh, A., Ebadi, M., Majd, A. and Oraghi Ardebili, Z. 2020. Selenium nanoparticles induced variations in growth, morphology, anatomy, biochemistry, gene expression, and epigenetic DNA methylation in Capsicum annuum; an *in vitro* study. Environ. Pollut. 265: 114727. doi: 10.1016/j. envpol.2020.114727.

Srivastava, V., Sarkar, A., Singh, S., Singh, P., de Araujo, A.S.F. and Singh, R.P. 2017. Agroecological responses of heavy metal pollution with special emphasis on soil health and plant performances. Front Environ. Sci. 5. doi: 10.3389/fenvs.2017.00064.

Sun, L., Song, F., Zhu, X., Liu, S., Liu, F., Wang, Y. and Li, X. 2021. Nano-ZnO alleviates drought stress via modulating the plant water use and carbohydrate metabolism in maize. Arch. Agron. Soil Sci. 67: 245–259. doi: 10.1080/03650340.2020.1723003.

Taran, N., Storozhenko, V., Svietlova, N., Batsmanova, L., Shvartau, V. and Kovalenko, M. 2017. Effect of zinc and copper nanoparticles on drought resistance of wheat seedlings. Nanoscale Res. Lett. 12: 60. doi: 10.1186/ s11671-017-1839-9.

Thabet, A.F., Boraei, H.A., Galal, O.A., El-Samahy, M.F.M., Mousa, K.A., Zhang, Y.Z., Tuda, M., Helmy, E.A., Wen, J. and Nozaki, T. 2021. Silica nanoparticles as pesticide against insects of different feeding types and their non-target attraction of predators. Sci. Rep. 11: 14484. https://doi.org/10.1038/s41598-021-93518-9.

Thakur, S., Asthir, B., Kaur, G., Kalia, A. and Sharma, A. 2021. Zinc oxide and titanium dioxide nanoparticles influence heat stress tolerance mediated by antioxidant defense system in wheat. Cereal Res. Commun. doi. 10.1007/s42976-021-00190-w.

Thul, S.T. and Sarangi, B.K. 2015. Nanotechnology and Plant Sciences: Nanoparticles and Their Impact on Plants. Springer International Publishing, Cham.

Torney, F., Trewyn, B.G., Lin, V.S.-Y. and Wang, K. 2007. Mesoporous silica nanoparticles deliver DNA and chemicals into plants. Nat. Nanotechnol. 2: 295–300. doi: 10.1038/nnano.2007.108.

Tripathi, D., Singh, M. and Pandey-Rai, S. 2022. Crosstalk of nanoparticles and phytohormones regulate plant growth and metabolism under abiotic and biotic stress. Plant Stress 6: 100107. doi: 10.1016/j.stress.2022.100107.

Venkatachalam, P., Jayaraj, M., Manikandan, R., Geetha, N., Rene, E.R., Sharma, N.C. and Sahi, S.V. 2017. Zinc oxide nanoparticles (ZnONPs) alleviate heavy metal-induced toxicity in Leucaena leucocephala seedlings: A physiochemical analysis. Plant Physiol. Biochem. 110: 59–69. doi: 10.1016/j.plaphy.2016.08.022.

Vera-Reyes, Edgar, V.-N., H. L-SR, B., M.-A. 2018. Effects of Nanoparticles on Germination, Growth, and Plant Crop Development I.

Wang, J., Wu, X., Shen, P., Wang, J., Shen, Y., Shen, Y., Webster, T.J. and Deng, J. 2020. Applications of inorganic nanomaterials in photothermal therapy based on combinational cancer treatment. Int. J. Nanomedicine 15: 1903–1914. doi: 10.2147/IJN.S239751.

Wang, X., Shi, X., Zhang, R., Zhang, K., Shao, L., Xu, T., Li, D., Zhang, D., Zhang, J. and Xia, Y. 2022a. Impact of summer heat stress inducing physiological and biochemical responses in herbaceous peony cultivars (Paeonia lactiflora Pall.) from different latitudes. Ind. Crops Prod. 184: 115000. doi: 10.1016/j.indcrop.2022.115000.

Wang, X., Shi, X., Zhang, R., Zhang, K., Shao, L., Xu, T., Li, D., Zhang, D., Zhang, J. and Xia, Y. 2022b. Impact of summer heat stress inducing physiological and biochemical responses in herbaceous peony cultivars (Paeonia lactiflora Pall.) from different latitudes. Ind. Crops Prod. 184: 115000. doi: 10.1016/j.indcrop.2022.115000.

Waqas, M.A., Kaya, C., Riaz, A., Farooq, M., Nawaz, I., Wilkes, A. and Li, Y. 2019. Potential mechanisms of abiotic stress tolerance in crop plants induced by thiourea. Front Plant Sci. 10. doi: 10.3389/fpls.2019.01336.

Yang, S., Xu, K., Chen, S., Li, T., Xia, H., Chen, L., Liu, H. and Luo, L. 2019. A stress-responsive bZIP transcription factor OsbZIP62 improves drought and oxidative tolerance in rice. BMC Plant Biol. 19: 260. doi: 10.1186/s12870-019-1872-1.

Younis, A.A., Khattab, H. and Emam, M.M. 2020. Impacts of silicon and silicon nanoparticles on leaf ultrastructure and TaPIP1 and TaNIP2 gene expressions in heat stressed wheat seedlings. Biol. Plant 64: 343–352. doi: 10.32615/bp.2020.030.

Zahedi, S.M., Moharrami, F., Sarikhani, S. and Padervand, M. 2020. Selenium and silica nanostructure-based recovery of strawberry plants subjected to drought stress. Sci. Rep. 10: 17672. doi: 10.1038/s41598-020-74273-9.

Zahedi, S.M., Hosseini, M.S., Fahadi Hoveizeh, N., Kadkhodaei, S. and Vaculík, M. 2023. Comparative morphological, physiological and molecular analyses of drought-stressed strawberry plants affected by SiO_2 and SiO_2-NPs foliar spray. Sci. Hortic. (Amsterdam) 309: 111686. doi: 10.1016/j.scienta.2022.111686.

Zhang, H., Zhao, Y. and Zhu, J.-K. 2020. Thriving under Stress: How plants balance growth and the stress response. Dev Cell 55: 529–543. doi: 10.1016/j.devcel.2020.10.012.

Zhao, J., Lu, Z., Wang, L. and Jin, B. 2020. Plant responses to heat stress: physiology, transcription, noncoding rnas, and epigenetics. Int. J. Mol. Sci. 22: 117. doi: 10.3390/ijms22010117.

Chapter 16

Role of Plant-Derived Extracellular Nanovesicles in Triggering Innate Immune Response Against Plant Pathogens

Vandana Sharma,[1] *Victor Samuel,*[1] *Chenicheri K Keerthana,*[2,3] *Praveen Archana,*[2]
Shifana Chembothumparambil Sadiq,[2] *Mundanattu Swetha,*[2,3]
Tennyson Prakash Rayginia,[2,3] *Sreekumar Usha Devi Aiswarya,*[2,4]
Smitha Vadakkeveettil Bava,[4] *Manikandan Mohan,*[5,6]
Johnson Retnaraj Samuel Selvan Christyraj,[7] *Jaison Arivalagan,*[8]
Sam Aldrin Chandran[9] and *Kalimuthu Kalishwaralal*[2,]*

Introduction

Raising awareness of the adverse health effects in response to chemical pesticides on humans urged the scientific community to develop eco-friendly crop protection strategies in recent times (Ekström et al. 2011). However, understanding the molecular mechanism of plant immunity and pathogen inhibition would lead to the selection of pathogen-resistant crops, which would eventually reduce our dependence on chemical pesticides for crop protection. Over the past two decades, several studies decoded the existence of plant innate immunity that acts as the first line of defense against their pathogens (Chisholm et al. 2006). It has been suggested that the immune response is triggered by the recognition of pathogen-associated molecular patterns (PAMPs) by pattern recognition receptors (PRRs), leading to the activation of pattern-triggered immunity (PTI) (Nürnberger et al. 2004; Chisholm et al. 2006). It is shown that extracellular vesicles (EVs) are secreted during such immune responses; however, nothing is known about their secretion, contents, and function.

---

[1] Rajiv Gandhi Centre for Biotechnology, Thiruvananthapuram, 695014, Kerala, India.
[2] Division of Cancer Research, Rajiv Gandhi Centre for Biotechnology, Thiruvananthapuram, 695014, Kerala, India.
[3] Department of Biotechnology, University of Kerala, Thiruvanathapuram, 695011, Kerala, India.
[4] Department of Biotechnology, University of Calicut, Malappuram, India.
[5] College of Pharmacy, University of Georgia, Athens, GA, USA.
[6] VAXIGEN International Research Center Private Limited, India.
[7] Regeneration and Stem Cell Biology Lab, Centre for Molecular and Nanomedical Sciences, International Research Centre, Sathyabama Institute of Science and Technology, Chennai, 600119, Tamil Nadu, India.
[8] Department of Chemistry, Molecular Biosciences and Proteomics Center of Excellence, Northwestern University, Evanston, IL, 60208, USA.
[9] School of Chemical and Biotechnology, SASTRA University, Thanjavur, 613 401 India.
* Corresponding author: kalimuthu@rgcb.res.in

EVs are considered natural nanoparticles, which are secreted by prokaryotic and eukaryotic cells. Based on their origin and size, EVs are classified as exosomes, microvesicles, and apoptosis-derived vesicles (Woith et al. 2009; Badierah et al. 2021). Edible plants, such as ginger, strawberry, lemon, grapefruit, tomato, blueberry, etc., secrete exosomes (Kalarikkal et al. 2021). Plant exosomes contain mRNA, ncRNA, miRNAs, proteins, polyphenols, and glycolipids (Figure 1) (Chen et al. 2022). These exosomes serve as extracellular messengers and help in mediating cell-cell communication via cargo transfer to the recipient cells (Akuma et al. 2019; van Niel et al. 2022). One common method for isolating and purifying them is by collecting apoplastic wash fluid (AWF) from sources, such as leaves, seeds, or pollen germination media, as described by Huang et al. (2021).

Recently, it was discovered that plant multivesicular bodies (MVBs) excrete exosome-like vesicles for cell-to-cell communication and in the regulation of the immune system in response to plant-pathogen invasion (Suharta et al. 2021). Plant-derived exosomes have been reported to contain specialized and characteristic miRNA and small RNA, which possess an inherent biological function in plant host-pathogen interactions (Woith et al. 2019). The biocompatibility and stability of the plant-derived exosomes make them suitable candidates for drug delivery compared to synthetic nanoparticles (Dad et al. 2021). Given these advantages, exosomes derived from different plant cells find a wide range of applications. The use of plant-based miRNA, i.e., small RNA for developing disease-resistant varieties of crops is one such example. The role of plant-derived EVs in the plant-pathogen interaction is a budding area of research with immense application potential (Hansen et al. 2018). In this book chapter, we focus on the isolation and characterization of plant-derived exosome methods and highlight recent literature on plant EV signaling and the ability of plant pathogens to suppress PTI as a key virulence approach. We also discuss the current applications of plant-derived exosomes for therapeutic purposes as well as the prospects of exosome technology.

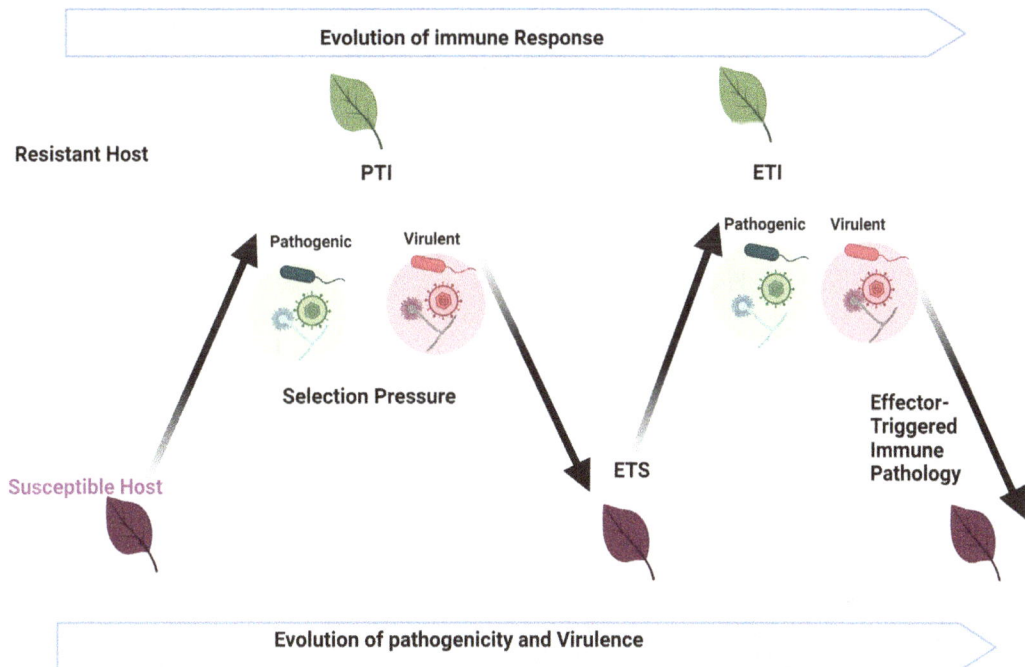

Figure 1. The schematic representation of the Zig-Zag model of the evolution of plant immune response (adapted from Jones and Dangl (2006) and created using BioRender.com).

Biogenesis, Isolation, and Characterization of Plant-Derived Exosome-Like Nanoparticles

In 1967, MVBs were first recognized in plant cells. It has been suggested that the origin of exosomes in plant cells involves the fusion of multivesicular bodies (MVBs) with the plasma membrane, which leads to the release of exosomes into the extracellular space (An et al. 2007). Recently, transmission electron microscopy (TEM) studies on barley leaf cells infected by fungus have verified and confirmed this idea. EVs were first discovered from the plasma membranes of soybean protoplasts (Banerjee et al. 2009). There are previous reports on the isolation of EV-like small vesicles (50 nm to 300 nm) from *Arabidopsis thaliana* apoplastic fluid. These EV-like small vesicles enhance the production of stress response proteins upon exposure to biotic and abiotic stress elements (Rutter et al. 2017). Plants infected with pathogenic viruses/bacteria/parasites cause an increase in the amount of EV secretion, which implies the vital role of EVs in basal immunity and stress response in plants. Thus, exploiting the immunomodulatory properties of EVs for developing pathogen-resistant varieties of crops would be an interesting area of research (Cai et al. 2021).

Previously, some report developed different methods of exosome isolation from biological fluids or cell supernatant were studied (Kalishwaralal et al. 2019; Kalimuthu et al. 2019). However, no consensus could be established as to the preferred method since none of them guarantees 100% separation of exosomes from other types of vesicles (Soares Martins et al. 2018). Commonly used methods for the isolation of plant-derived exosomes like nanoparticles include ultracentrifugation, size-exclusion chromatography, precipitation, and immunoaffinity-based isolation. Of all these methods, ultracentrifugation is widely used, and it facilitates obtaining relatively pure exosomes. However, this long technique needs the tenure of sophisticated equipment (Sidhom et al. 2020). In addition, the low yield of exosomes in isolation from certain biological fluids (urine, serum, and plasma) has been a requirement of high-speed centrifugation for a long period (Kalimuthu et al. 2009). The poor reproducibility of the isolation protocols is another major concern associated with this technique (Li et al. 2017; Kalimuthu et al. 2019). There are other methods of isolating exosomes, such as filtration, size-exclusion chromatography, immunoaffinity isolation, the use of microfluidic devices, density gradient isolation, as well as the use of commercial polymer-based reagents (Li et al. 2017; Kalimuthu et al. 2019). Size-exclusion chromatography allows the production of pure exosomes; however, the exosomes obtained are always diluted and these procedures are relatively long and time-consuming (Li et al. 2017; Sidhom et al. 2020). Commercially available solutions based on the use of polymers provide an easy and rapid method of isolation of exosomes by low-speed centrifugation. This method does not require any specialized equipment, and it yields a good quantity of exosomes; however, sometimes this process can interfere with downstream analysis (Li et al. 2017; Kalishwaralal et al. 2019; Sidhom et al. 2020). Of late, researchers are working on strategies to adapt the existing exosome purification techniques using "lab-on-a-chip" technologies, which require less sample consumption per analysis, enable a reduction in costs compared to that of the benchtop equipment, and at the same time allow automation of analyzes for better performance and greater precision (Chiriacò et al. 2018).

Exosomes can be visualized and characterized by different physical or biochemical techniques. The TEM is the only technique that enables direct observation of the morphology and size of exosomes (Kalishwaralal et al. 2019). Exosomes can be simply observed after fixation on colored and contrasted with heavy metals. However, this technique can also be combined with the use of antibodies coupled to gold nanoparticles, which facilitates the detection of specific markers (immuno-TEM) (Pascucci et al. 2021). Other electron microscopy techniques, such as scanning electron microscopy, sample preparation for which is much easier than TEM (cryo-TEM), which limits the alteration of the structure of exosomes due to the use of dehydrating chemicals or atomic force microscopy can be used to characterize the three-dimensional structure of exosomes (Pascucci et al. 2021).

Exosomes can be analyzed quantitatively and qualitatively by flow cytometry; however, this technique requires the adsorption of exosomes on particles carrying specific antibodies, fluorescent drugs, or antibodies to specific proteins known to be enriched on exosomes so that they are distinguished from background noise (Huda et al. 2021; Hartjes et al. 2019). The size and distribution of vesicles may be determined by individual particle tracking (or nanoparticle tracking analysis). This technique can also be used to determine the concentration and size of vesicles in a sample; however, the presence of other elements in the exosome suspension, such as protein aggregates, may skew the results (Hartjes et al. 2019; Pascucci et al. 2021). A similar technique, the analysis of the dynamic light scattering, allows the calculation of the average particle size in a sample but requires working on a sample very pure (Hartjes et al. 2019; Pascucci et al. 2021). The biochemical characterization of exosomes can be carried out using different techniques, such as western blotting or immunoaffinity analysis (Kalishwaralal et al. 2019).

There are very few reports pertaining to the protein content of EVs. However, the syntaxin PEN1 is associated with extracellular membranous material, suggesting that it may be packaged into exosomes (Rutter et al. 2017). Recently, Doyle et al. (2019) reported that the surface proteins of vesicles derived from citrus and successfully identified several common marker proteins, such as patellin-3-like proteins and clathrin heavy chain proteins. Analysis of exosomal protein markers using western blotting is a quick and easy technique, but it requires a large sample volume. These markers can also be analyzed by ELISA (enzyme-linked immunosorbent assay), an easy-to-use and highly specific technique, which can be miniaturized on a chip (ExoChip), or by an alternative method using secondary antibodies carrying a fluorochrome (FLISA, fluorophore-linked immunosorbent assay) (Doyle et al. 2009; Hartjes et al. 2019). Finally, the protein profile of exosomes can be determined by mass spectrometry, their lipid profile can be elucidated by a thin layer or gas-liquid chromatography, and mass spectrometry (Skotland et al. 2017). Next-generation sequencing or quantitative real-time polymerase chain reaction (qRT-PCR) can be used to analyze the nucleic acid content of EVs derived from plants (Adlerz et al. 2020b). In the coming years, studies aimed at evaluating common surface proteins present in all plant-derived exosome-like nanoparticles need to be carried out for the identification of exosome-specific biomarkers. Such novel biomarkers would enable investigators to better identify and characterize plant-derived exosome-like nanoparticles.

Exosome-Mediated Innate Immune Response Against the Virus, Bacteria, and Fungus in Plants

Owing to continuous exposure to a diverse variety of pathogens and pests, plants have developed diverse mechanisms of innate immunity. It is divided into two branches: one branch detects and responds to features common to different microbes and the other branch responds to pathogen virulence factors. This is called the PTI/ETI model. It is based on the principle of perception of signals of danger (Andolfo and Ercolano 2015). (1) PTI is the mechanism that utilizes PAMPs or danger-associated molecular patterns (DAMPs) that are recognized by PRRs present on the host. PAMPs are highly conserved molecules present in a class of microbes that are essential for survival and thus indispensable (Eder and Cosio 1994). (2) Effector-triggered immunity (ETI) has evolved as a stronger mechanism to counter pathogen virulence by recognizing the direct and indirect effectors through nucleotide-binding and oligomerization domain (NOD)-like receptors (NLRs) (Figure 2). In 2006, the "Zig-Zag model" proposed by Jones and Dangl quantified the amplitude of resistance or susceptibility to PTI and ETI mechanisms (Jones and Dangl 2006).

There is compelling evidence regarding the presence of effector-triggered susceptibility. The presence of pathogen effector proteins alters/attacks the PTI and ETI in the host plants. Bacteria secrete these effector molecules through the type-III secretion system and interfere with host immunity, but they are also a target of ETI and can impart resistance to plants once detected. In 2001, Dangl and Jones proposed the guard model of plant-pathogen interactions. According to this

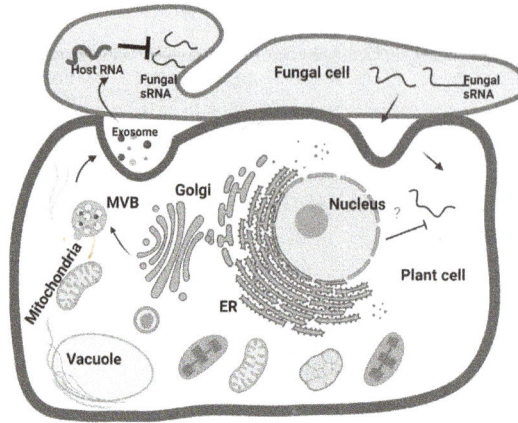

Figure 2. Host-pathogen interaction: Role of plant-derived EVs in silencing pathogenic nucleic acids (Created using Biorender.com).

model, the effector proteins have evolved from pathogens for targeting host proteins and enhancing virulence (Yuan et al. 2021).

In animal systems, exosomes were found to aid in the acceleration or inhibition of the infection process. Possible connections among host cells and host-pathogen cells were observed. Most studies on plant exosomes have been done using Transmission Electron Microscopy (TEM) (Zhang et al. 2018). Through these studies, MVBs have been observed during pathogen attacks and observed at different stages of their fusion with the plasma membrane near papillae, which is an extracellular structure for the prevention of pathogen entry. Furthermore, the presence of small vesicles in the papillae matrix and the accumulation of GTPases near the site of infection were observed (Ruano et al. 2019). The GTPases were speculated to promote the fusion of EVs with the plasma membrane. Taken together, these observations were suggestive of the role of exosomes as key players in the plant defense system in response to pathogen attacks (Hansen et al. 2018). In 2009, Cheng et al. found that 50% of extracellular proteins were secreted without signal peptides needed for secretion through the secretory pathway. These proteins were found to possess antimicrobial activity and are known to be released in response to pathogens. In the context of host-pathogen interactions, EVs can be utilized by pathogens to modulate host immune responses and by plants to resist microbial infections (Regente et al. 2017).

Recent reports have highlighted the role of EVs in the context of plant-microbe interactions and plant defense. Interestingly, plant-derived exosomes seemingly have varied functions in the regulation of gene expression in response to oxidative stress and function as a protective compartment and mediator of defense response against pathogens (Regente et al. 2017). In plants, exocytosis occurs via two major pathways; these are the conventional protein secretion pathway and the unconventional protein secretion pathway. It is believed that the unconventional protein secretion (UPS) pathway works through the formation of EVs (Wang et al. 2016; U. Stotz et al. 2022).

Defense Against Fungi

With recent advances in the proteomics and transcriptomics analysis of EV composition, it is becoming increasingly evident that plants can inhibit fungal growth and virulence by transferring their defense protein and siRNA cargo packed inside exosomes into fungal cells (Cai et al. 2018; Hill et al. 2020) (Figure 3).

The presence of exosome-like vesicles in the extrahaustorial matrix of the Arabidopsis-Golovinomycesorontii (Powdery Mildew) pathosystem was first reported (Micali et al. 2011). Schweizer's research group demonstrated the impact of host-induced gene silencing (HIGS)

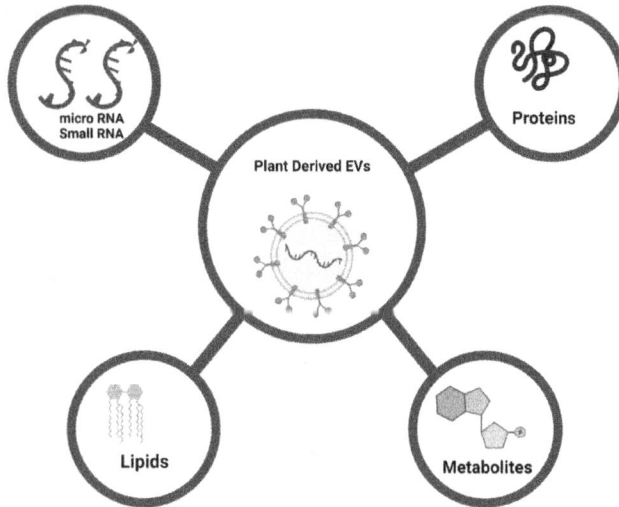

Figure 3. General composition of plant-derived extracellular nanovesicles (Created using Biorender.com).

on the growth of the powdery mildew fungus *Blumeria graminis* on barley and wheat. This involves the transfer of sRNAs from the plant to the fungus, which targets and silences fungal transcripts via RNA interference (Nowara et al. 2010).

In a separate study, researchers attempted to isolate EVs from the apoplastic fraction of sunflower plants and performed proteomics analysis to identify their cargo. Their findings demonstrated that plant EVs contain approximately 240 different proteins, many of which are related to plant defense mechanisms. These include pathogenesis-related (PR) proteins, such as chitinases II (PR-4), thaumatins (PR-5), proteinase inhibitors (PR-6), peroxidases (PR-9), and lipid transfer proteins (PR-14), as well as dirigent protein-disease resistance, PMR5, and Gnk2 antifungal protein, GDSL lipase acyl hydrolases, lectins, and germin-like proteins (Regente et al. 2017).

Furthermore, the same group reported that the isolated plant EVs contain antifungal properties against the severe fungal pathogen *Sclerotinia sclerotiorum*, which causes white molds in a wide range of hosts and causes sclerotinia stem root. It was observed that when these fungal spores were incubated with labeled EVs, they were internalized by membrane fusion and there was about a 60–80% loss in the viability of fungal cells (Regente et al. 2017).

De Palma et al. (2020) pioneered the isolation of EVs released as root exudates from hydroponically grown tomato plants. Through proteomics analysis, they reported that 100 different proteins were present in tomato EVs, among which 23 proteins were defense-related. These EVs strongly inhibited the spore germination of soil-borne fungal pathogens like *Fusarium oxysporum* even at low EV dosage (1.5×10^{10} EVs) and air-borne fungal pathogens like *Botrytis cinerea* and *Alternaria alternata* at high EV dosage (6.0×10^{10} EVs). The authors also suggest the possibility of RNAi-mediated gene silencing in this spore inhibition (De Palma et al. 2020). The sRNA-mediated silencing of fungal virulence using *Arabidopsis - Botrytis cinerea* pathosystem is another interesting domain of research. *B. cinerea* is a necrotrophic fungus affecting a range of horticultural crops, especially grapevine, causing bunch rot or gray mold, which incurs a huge economic loss. Profiling of sRNA from purified protoplast of fungal pathogen revealed 42 Arabidopsis sRNA (Weiberg et al. 2103). The same pattern was observed in the sRNA profiling from EVs in the apoplastic region, and they are protected inside vesicular structures as revealed by the nuclease protection assay. These sRNAs do not follow concentration gradients because several low-abundance sRNAs (~ 25 out of 42) in the host are selectively transferred into the pathogens. These sRNAs silence several virulence genes inside the pathogen with a bias toward genes involved in the vesicular trafficking pathway (Cai et al. 2018).

This selective packaging of sRNA cargo occurs inside the protective TET8-positive exosome-like EVs. Studies by He et al. revealed three important RNA-binding proteins (RBPs) in this selective loading process (He et al. 2021): (1) AGO1 (argonaute protein 1), which binds preferentially to siRNAs of 20–22 nucleotides (nts) long with 5'-U; (2) DEAD-box RNA helicases (RHs), which bind to sRNAs having secondary structures and rearrange them in an energy-dependent manner; (3) annexins (ANNs), which non-specifically bind to sRNA and stabilize them inside the EVs.

A comparative analysis of the sRNA profiles in total sRNA, EV-sRNA, and EV-depleted apoplastic RNA sources revealed that EVs are enriched with unusually short sRNAs of 10–17 nts, which were termed "tiny RNAs (tyRNAs)." Their genomic origin corresponds primarily to the regions of miRNAs, and transposable elements, and they were also expected to be degradation products of mRNAs, primary miRNAs, siRNAs, tasiRNAs, or hcRNAs. The exact function of tyRNAs is not yet determined but they are expected to act as "small activating RNAs (saRNAs)," which are sRNAs involved in the activation of specific gene expression by mostly binding to the UTR regions (Baldrich et al. 2019).

All the above results point toward the defensive role of exosome-like EVs during a fungal infection in plants. Both the pathogen and the host can transfer their cargo packed inside these vesicles as a means of cross-kingdom communication (Cai et al. 2021). Consequently, the host immune system gets activated by recognizing the PAMPs and starts retaliating by secreting defense proteins and sRNAs targeting invading pathogens, safely packed inside the EVs (Wang et al. 2015). However, the biogenesis of these EVs is diverse and the mechanism of targeting the pathogen and means of crossing the cell wall are not well understood. Future studies are required to gain a deeper understanding of these processes.

Defense Against Bacteria

Information about plant defense against bacterial pathogens utilizing EVs is rudimentary, and this necessitates further scientific exploration in this direction. The first observations of bacterial outer membrane vesicles (OMVs) were around 60 years ago when these were observed in cell-free supernatants from pathogenic *Vibrio cholerae* Pacini culture that had toxic effects on human cells (Cai et al. 2021). Electron microscopy images were further published that showed various stages of budding from the bacterial cell walls. Structurally, OMVs are secretions of gram-negative bacteria, whereas secretions of Gram-positive bacteria are known as "microvesicles." Thus, OMVs are similar in composition to Gram-negative cell envelopes (Liu et al. 2018). OMVs possess a diverse set of contents ranging from peptidoglycans, outer membrane proteins, lipopolysaccharides, phospholipids, and soluble proteins to enzymes and nucleic acids, such as DNA and RNA. These compositions differ based on bacterial species and growth conditions (Kaparakis-Liaskos and Ferrero 2015).

The role of OMVs in eliciting mammalian immune responses is already known and well-established. For example, OMVs released from cells infected with *Mycobacterium tuberculosis* can trigger TLR2 in uninfected macrophages and consequently result in cytokine responses. But recent evidence suggests the role of phytopathogenic bacterial species that play a critical role in plant immune modulation by virtue of the various types of effector molecules present in these species (Wang et al. 2019).

McMillan et al. (2021) have reported that OMVs released from *Pseudomonas syringae* and *P. fluorescens* have a role in the activation of plant defense mechanisms. In the study, *A. thaliana* leaves were treated with OMVs followed by exposure to *Pst* bacteria. They also demonstrated that OMV induces various hormone-dependent and hormone-independent plant immune responses (McMillan et al. 2021).

Park et al. (2021) studied the ultrastructure of the phytoplasma-infected jujube plant and determined the presence of MVBs in the phloem. These MVBs were observed to be associated with the cell wall. They also observed a positive correlation between increased MVB production and ER-stress-dependent exosome release due to disruption of homeostasis. Further studies are required for characterizing the contents of MVBs and determining their potential role in phytoplasma infection (Park et al. 2021).

Inne et al. (2017) reported the possibility of immune-modulatory roles of plant EVs against bacterial infection through proteomic analysis. *Pseudomonas syringae* pv. *tomato* is a Gram-negative bacterial pathogen of tomato and Arabidopsis plants. Its pathogenesis has two phases: (1) initially, it infects the plant leaf surface locally and survives epiphytically; (2) upon favorable conditions, it enters the plant tissue through the natural openings like stomata and multiplies in the apoplastic region (Xin and He 2013). A proteomic analysis of EVs isolated from the apoplastic fluid of *P. syringae*-infected *Arabidopsis thaliana* leaves revealed that the EV proteome is enriched with biotic and abiotic stress response proteins, and there is an enhancement in the EV secretion upon bacterial infection or salicylic acid (SA) treatment. However, there is little change in the proteome in response to infection (Rutter and Innes 2017).

Defense Against Virus

Plants have evolved various specialized defense mechanisms against viruses. These strategies are controlled by intricate interconnections between biochemical, cellular, and genetic factors. There are various surveillance pathways mediated via RNA and proteins (Pumplin et al. 2013). Nonsense-mediated decay (NMD) is one of the pathways for the antiviral defense that leads to the disruption of aberrant mRNA transcripts. Viruses are known to produce such transcripts due to genome restrictions. Another known strategy is natural recessive resistance, which is produced because of mutations induced in recessive genes that are important for viral replication (Musidlak et al. 2017). There are also different protein-mediated pathways present with known functions in viral defense (Mandadi et al. 2013). These are pathogenesis-related (PR) proteins, which are produced in response to pathogenic attacks and environmental stresses. Apart from PR proteins, a ribosome-inactivating protein (RIP) also inactivates ribosomes and suppresses the translation of viral proteins. Certain RNA-binding proteins that interact with target mRNAs altering the plant physiology also act as defense proteins. RISC-mediated RNA silencing is the most known and widely studied method of antiviral defense (Musidlak et al. 2017).

Research studies have established that EVs play a crucial role in viral infection by affecting viral entry, transmission, and immune evasion. EVs act as mediators between uninfected and infected cells, and their biogenesis pathways overlap with those of viruses, making them interconnected (Cui et al. 2020; U. Stotz et al. 2022). EVs released from infected cells can either elicit an antiviral response or enhance viral infection by transferring viral elements and activating antiviral mechanisms in various cell types. While exosomes containing viral genomes can promote viral spread by transferring them to susceptible cells, exosomes containing viral proteins and nucleic acids can stimulate immune responses, particularly in myeloid cells (Saad et al. 2021).

The information related to plant EVs in response to viral defense is scarce, except for a study showing the release of rice dwarf virus from insect vector cells, involving exosome production from MVBs. Turnip Mosaic Virus (TuMV) generates quasi-organelle structures, namely, "viral factories." These vesicles move from infected cells to neighboring healthy cells and were observed to be present in xylem vessels. Recently based on the research done by Movahed et al. (2019), it was shown that TuMV components released into the extracellular space of infected leaves were associated with EVs. They observed that abundant MVBs were released from *Nicotiana benthamiana* leaves. These MVBs released intraluminal vesicles that contained viral RNA. In MVBs of Arabidopsis leaves, they observed the presence of viral proteins and host factors that were responsible for the induction of an immune response (Movahed et al. 2019).

However, further investigation on the types of immune response mediators and the exact mechanism of action of EVs in plant defense against viral infections ought to be carried out to gain in-depth knowledge regarding the role of EVs against viral pathogens.

Development of Exosomedatabase and Its Application in Artificial Intelligence

Artificial intelligence (AI) based applications in the field of medicine reached $4 billion in 2019 (Pifer 2020). We propose that developing a futuristic AI-based plant exosome database would be a novel idea for extrapolating the benefits of exosome technology for the benefit of mankind. The development of a "plant exosome database" can be rendered possible by taking the help of next-generation sequencing techniques. A compilation of all miRNAs, sRNAs, and mRNA sequences associated with exosomes would help in further studies directed at elucidating the molecular events underlying plant-pathogen interactions. This can help the researchers to gain a better understanding of the molecular cascade, which triggers the immune response against plant pathogens.

Conclusion and Future Directions

Despite their promising potential and wide range of applications, the major limitation associated with plant-based exosomes is the lack of reproducibility. Studies focusing on developing an optimized isolation protocol or technique would be of immense benefit. Besides, our proposal of an AI-based plant exosome miRNA database would be a novel approach for studying exosome-based regulation of gene expression in plants and may be of immense value for aiding in the development of disease-resistant varieties of plants and an overall improvement in crop production. Traditional Indian agricultural practices include the use of "panchagavya," which is organic manure or biofertilizer mainly consisting of five natural substances namely, cow dung, cow urine, milk, ghee, and curd. It functions as an organic growth stimulant (Chandra et al. 2019). Panchagavya contains plant growth regulatory substances such as IAA, GA, cytokinins, various essential plant nutrients, effective microorganisms like lactic acid bacteria, and biofertilizers like acetobacter, etc. (Vallimayil et al. 2012; Yu et al. 2020). This natural mixture has been proven to promote plant growth, resist pest attack, improve the nutrient profile, fertility status, and water retention capacity of the soil, and enhance the population of beneficial microorganisms. Panchagavya treatment elicits holistic benefits on the crops and soil and hence represents a sustainable agricultural regimen. It is interesting to note that the ingredients of panchagavya may be rich in exosome content since they comprise body fluids derived from cows. Whether the exosomes present in this mixture are responsible for conferring protection against pathogen attack remains unknown to date. If exosomes are the underlying reason either partly or fully, for the efficacy of panchgavya treatment, then what would be the mechanism of action by which it confers disease resistance to plants? The inter-relationship between budding exosome technology and traditional agricultural science would be a novel research area that is worth exploring since it would bridge the gap between ancient and modern scientific approaches.

Acknowledgment

Dr. Kalimuthu Kalishwaralal, MK Bhan Young Researcher Fellowship for 2020–2021 (Ref: No, HRD-12/4/2020-AFS-DBT) awarded by DBT, India. We thank Dr. Ramakrishnan Muthuswamy, Associate Professor, Nanjing Forestry University, China, for the help rendered in image generation using BioRender software.

References

Adlerz, K., Patel, D., Rowley, J., Ng, K. and Ahsan, T. 2020. Strategies for scalable manufacturing and translation of MSC-derived extracellular vesicles. Stem Cell Res. 48: 101978.

Akuma, P., Okagu, O.D. and Udenigwe, C.C. 2019. Naturally occurring exosome vesicles as potential delivery vehicle for bioactive compounds. Front. Sustain. Food Syst. 3: 23.

An, Q., van Bel, A.J. and Hückelhoven, R. 2007. Do plant cells secrete exosomes derived from multivesicular bodies? Plant signal. Behav. 2(1): 4–7.

Andolfo, G. and Ercolano, M.R. 2015. Plant innate immunity multicomponent model. Front. Plant Sci. 6: 987.

Badierah, R.A., Uversky, V.N. and Redwan, E.M. 2021. Dancing with Trojan horses: an interplay between the extracellular vesicles and viruses. J. Biomol. Struct. Dyn. 39(8): 3034–3060.

Baldrich, P., Rutter, B.D., Karimi, H.Z., Podicheti, R., Meyers, B.C. and Innes, R.W. 2019. Plant extracellular vesicles contain diverse small RNA species and are enriched in 10- to 17-nucleotide "tiny" RNAs. The Plant Cell 31(2): 315–324.

Banerjee, K., Pramanik, P., Maity, A., Joshi, D.C., Wani, S.H. and Krishnan, P. 2019. Methods of using nanomaterials to plant systems and their delivery to plants (Mode of entry, uptake, translocation, accumulation, biotransformation and barriers). pp. 123–152. In Advances in Phytonanotechnology. Academic Press.

Cai, Q., Qiao, L., Wang, M., He, B., Lin, F.M., Palmquist, J., Huang, S.D. and Jin, H. 2018. Plants send small RNAs in extracellular vesicles to fungal pathogen to silence virulence genes. Science 360(6393): 1126–1129.

Cai, Q., He, B., Wang, S., Fletcher, S., Niu, D., Mitter, N., Birch, P.R. and Jin, H. 2021. Message in a bubble: shuttling small RNAs and proteins between cells and interacting organisms using extracellular vesicles. Annu. Rev. Plant. Biol. 72: 497.

Chandra, M.S., Naresh, R.K., Lavanya, N., Varsha, N., Wasim, S. and Navsare, R.I. 2019. Production and potential of ancient liquid organics panchagavya and kunapajala to improve soil health and crop productivity: A review. J. Pharmacogn. Phytochem. 8(6): 702–713.

Chen, N., Sun, J., Zhu, Z., Cribbs, A.P. and Xiao, B. 2022. Edible plant-derived nanotherapeutics and nanocarriers: recent progress and future directions. Expert Opin. Drug Deliv. 19(4): 409–419.

Cheng, F., Blackburn, K., Lin, Y., Goshe, M.B. and Williamson, J.D. 2009. Absolute protein quantification by LC/MSE for global analysis of salicylic acid-induced plant protein secretion responses. J. Proteome Res. 8: 82–93. https://doi.org/10.1021/pr800649s.

Chiriacò, M.S., Bianco, M., Nigro, A., Primiceri, E., Ferrara, F., Romano, A., Quattrini, A., Furlan, R., Arima, V. and Maruccio, G. 2018. Lab-on-chip for exosomes and microvesicles detection and characterization. Sensors 18(10): 3175.

Chisholm, S.T., Coaker, G., Day, B. and Staskawicz, B.J. 2006. Host-microbe interactions. shaping the evolution of the plant immune response. Cell 124(4): 803–814.

Cui, Y., Gao, J., He, Y. and Jiang, L. 2020. Plant extracellular vesicles. Protoplasma 257(1): 3–12.

Dad, H.A., Gu, T.W., Zhu, A.Q., Huang, L.Q. and Peng, L.H. 2021. Plant exosome-like nanovesicles: emerging therapeutics and drug delivery nanoplatforms. Molecular Therapy 29(1): 13–31.

De Palma, M., Ambrosone, A., Leone, A., Del Gaudio, P., Ruocco, M., Turiák, L., Bokka, R., Fiume, I., Tucci, M. and Pocsfalvi, G. 2020. Plant roots release small extracellular vesicles with antifungal activity. Plants 9(12): 1777.

Doyle, L.M. and Wang, M.Z. 2019. Overview of extracellular vesicles, their origin, composition, purpose, and methods for exosome isolation and analysis. Cells 8(7): 727.

Eber, J. and Cosio, E.G. 1994. Elicitors of plant defence responses. Int. Rev. Cytol. 148: 1–36.

Ekström, G. and Ekbom, B. 2011. Pest control in agro-ecosystems: an ecological approach. Crit. Rev. Plant Sci. 30(1-2): 74–94.

Hansen, L.L. and Nielsen, M.E. 2018. Plant exosomes: using an unconventional exit to prevent pathogen entry? J. Exp. Bot. 69(1): 59–68.

He, B., Cai, Q., Qiao, L., Huang, C.Y., Wang, S., Miao, W., Ha, T., Wang, Y. and Jin, H. 2021. RNA-binding proteins contribute to small RNA loading in plant extracellular vesicles. Nat. Plants 7(3): 342–352.

Huang, Y., Wang, S., Cai, Q. and Jin, H. 2021. Effective methods for isolation and purification of extracellular vesicles from plants. J. Integr. Plant Biol. 63(12): 2020–2030.

Jones, J.D. and Dangl, J.L. 2006. The plant immune system. Nature 444(7117): 323–329.

Kalarikkal, S.P. and Sundaram, G.M. 2021. Inter-kingdom regulation of human transcriptome by dietary microRNAs: Emerging bioactives from edible plants to treat human diseases? Trends in Food Sci. Technol. 118: 723–734.

Kalimuthu, K., Kwon, W.Y. and Park, K.S. 2019. A simple approach for rapid and cost-effective quantification of extracellular vesicles using a fluorescence polarization technique. J. Biol. Eng. 13(1): 1–7.

Kalishwaralal, K., Kwon, W.Y. and Park, K.S. 2019. Exosomes for non-invasive cancer monitoring. Biotechnol. J. 14(1): 1800430.

Kaparakis-Liaskos, M. and Ferrero, R.L. 2015. Immune modulation by bacterial outer membrane vesicles. Nat. Rev. Immunol. 15(6): 375–387.

Liu, Y., Defourny, K.A., Smid, E.J. and Abee, T. 2018. Gram-positive bacterial extracellular vesicles and their impact on health and disease. Front. Microbiol. 9: 1502.

Madison, M.N. and Okeoma, C.M. 2015. Exosomes: implications in HIV-1 pathogenesis. Viruses 7: 4093–4118.

Mandadi, K.K. and Scholthof, K.B.G. 2013. Plant immune responses against viruses: how does a virus cause disease? The plant cell 25(5): 1489–1505.

McMillan, H.M., Zebell, S.G., Ristaino, J.B., Dong, X. and Kuehn, M.J. 2021. Protective plant immune responses are elicited by bacterial outer membrane vesicles. Cell Rep. 34(3): 108645.

Micali, C.O., Neumann, U., Grunewald, D., Panstruga, R. and O'Connell, R. 2011. Biogenesis of a specialized plant-fungal interface during host cell internalization of Golovinomycesorontiihaustoria. Cell. Microbiol. 13(2): 210–226.

Ming, M., Weiberg, A. and Hailing, J. 2015. Pathogen small RNAs: a new class of effectors for pathogen attacks. Mol. Plant Pathol. 16(3): 219–223.

Movahed, N., Cabanillas, D.G., Wan, J., Vali, H., Laliberté, J.F. and Zheng, H. 2019. Turnip mosaic virus components are released into the extracellular space by vesicles in infected leaves. Plant Physiol. 180(3): 1375–1388.

Musidlak, O., Nawrot, R. and Goździcka-Józefiak, A. 2017. Which plant proteins are involved in antiviral defense? Review on *in vivo* and *in vitro* activities of selected plant proteins against viruses. Int. J. Mol. Sci. 18(11): 2300.

Nowara, D., Gay, A., Lacomme, C., Shaw, J., Ridout, C., Douchkov, D., Hensel, G., Kumlehn, J. and Schweizer, P. 2010. HIGS: host-induced gene silencing in the obligate biotrophic fungal pathogen Blumeriagraminis. The Plant Cell 22(9): 3130–3141.

Nürnberger, T., Brunner, F., Kemmerling, B. and Piater, L. 2004. Innate immunity in plants and animals: striking similarities and obvious differences. Immunol. Rev. 198(1): 249–266.

Park, J., Kim, H.J., Huh, Y.H. and Kim, K.W. 2021. Ultrastructure of phytoplasma-infected jujube leaves with witches' broom disease. Micron. 148: 103108.

Pascucci, L. and Scattini, G. 2021. Imaging extracelluar vesicles by transmission electron microscopy: coping with technical hurdles and morphological interpretation. Biochimi. Biophys. Acta –Gen. Sub. 1865(4): 129648.

Pifer, R. 2020. Health AI Startups Netted a Record $4B in Funding Last Year. Healthcare Dive.

Pumplin, N. and Voinnet, O. 2013. RNA silencing suppression by plant pathogens: defence, counter-defence and counter-counter-defence. Nat. Rev. Microbiol. 11(11): 745–760.

Regente, M., Pinedo, M., San Clemente, H., Balliau, T., Jamet, E. and De La Canal, L. 2017. Plant extracellular vesicles are incorporated by a fungal pathogen and inhibit its growth. J. Exp. Bot. 68(20): 5485–5495.

Ruano, G. and Scheuring, D. 2020. Plant cells under attack: Unconventional endomembrane trafficking during plant defense. Plants 9(3): 389.

Rutter, B.D. and Innes, R.W. 2017. Extracellular vesicles isolated from the leaf apoplast carry stress-response proteins. Plant Physiol. 173(1): 728–741.

Saad, M.H., Badierah, R., Redwan, E.M. and El-Fakharany, E.M. 2021. A comprehensive insight into the role of exosomes in viral infection. dual faces bearing different functions. Pharmaceutics 13(9). 1405.

Sidhom, K., Obi, P.O. and Saleem, A. 2020. A review of exosomal isolation methods: is size exclusion chromatography the best option? Int. J. Mol. Sci. 21(18): 6466.

Skotland, T., Sandvig, K. and Llorente, A. 2017. Lipids in exosomes: Current knowledge and the way forward. Progress in Lipid Research 66: 30–41. https://doi.org/10.1016/j.plipres.2017.03.001.

Soares Martins, T., Catita, J., Martins Rosa, I., AB da Cruz e Silva, O. and Henriques, A.G. 2018. Exosome isolation from distinct biofluids using precipitation and column-based approaches. PloS One 13(6): e0198820.

Suharta, S., Barlian, A., Hidajah, A.C., Notobroto, H.B., Ana, I.D., Indariani, S., Wungu, T.D.K. and Wijaya, C.H. 2021. Plant-derived exosome-like nanoparticles: A concise review on its extraction methods, content, bioactivities, and potential as functional food ingredient. J. Food Sci. 86(7): 2838–2850.

U. Stotz, H., Brotherton, D. and Inal, J. 2022. Communication is key: extracellular vesicles as mediators of infection and defence during host-microbe interactions in animals and plants. FEMS Microbiol. Rev. 46(1): fuab044.

Vallimayil, J. and Sekar, R. 2012. Investigation on the effect of panchagavya on sounthernsunnhemp mosaic virus (SSMV) infected plant systems. Glob. J. Environ. Res. 6(2): 75–79.

vanNiel, G., Carter, D.R., Clayton, A., Lambert, D.W., Raposo, G. and Vader, P. 2022. Challenges and directions in studying cell-cell communication by extracellular vesicles. Nat. Rev. Mol. Cell Biol. 23(5): 369–382.

Wang, H., Zhuang, X., Wang, X., Law, A.H.Y., Zhao, T., Du, S., Loy, M.M. and Jiang, L. 2016. A distinct pathway for polar exocytosis in plant cell wall formation. Plant Physiol. 172(2):1003–1018.

Wang, J., Wang, Y., Tang, L. and Garcia, R.C. 2019. Extracellular vesicles in mycobacterial infections: their potential as molecule transfer vectors. Front. Immunol. 10: 1929.

Weiberg, A., Wang, M., Lin, F.M., Zhao, H., Zhang, Z., Kaloshian, I., Huang, H.D. and Jin, H. 2013. Fungal small RNAs suppress plant immunity by hijacking host RNA interference pathways. Science 342(6154): 118–123.

Woith, E., Fuhrmann, G. and Melzig, M.F. 2019. Extracellular vesicles—connecting kingdoms. Int. J. Mol. Sci. 20(22): 5695.

Xin, X.F. and He, S.Y. 2013. Pseudomonas syringae pv. tomato DC3000: a model pathogen for probing disease susceptibility and hormone signaling in plants. Annu. Rev. of Phytopathol. 51: 473–498.

Yu, L., Deng, Z., Liu, L., Zhang, W. and Wang, C. 2020. Plant-derived nanovesicles: a novel form of nanomedicine. Front. Bioeng. Biotechnol. 8: 584391.

Yuan, M., Ngou, B.P.M., Ding, P. and Xin, X.F. 2021. PTI-ETI crosstalk: an integrative view of plant immunity. Curr. Opin. Plant Biol. 62: 102030.

Zhang, W., Jiang, X., Bao, J., Wang, Y., Liu, H. and Tang, L. 2018. Exosomes in pathogen infections: a bridge to deliver molecules and link functions. Front. Immunol. 9: 90.

Chapter 17

Carbon Nanomaterials as Alternative to Control Plant Pathogens

Yolanda González-García,[3] *Gregorio Cadenas-Pliego*[2] and
Antonio Juárez-Maldonado[1,]*

Introduction

Carbon is one of the most abundant elements in the biosphere and is found in three states of hybridization: sp, sp[2], and sp[3]. This diversity in bonding can create numerous possibilities for the formation of carbon nanomaterials (CNMs), which can vary in shape, size, and dimension. These variations include 0D structures, which are hollow spherical structures (such as fullerene and carbon nanodots), 1D structures, such as carbon nanotubes carbon, 2D as graphene sheets, and 3D as graphite (Gabris and Ping 2021).

CNMs have proven to be of great importance in recent times due to their excellent structural, thermophysical, and chemical properties. In addition to having great stability, they have the potential to be derived from natural organic raw materials and therefore can be considered environmentally friendly (Goswami et al. 2021).

CNMs have the characteristic that they can be easily absorbed by plant cells, which has a positive impact on plant growth and development (Ghorbanpour and Hadian 2015). Among the most outstanding aspects of its application is the high potential for the elimination of pesticides in the water (Dehghani et al. 2019), its ability to remove heavy metals from soil and water (Fiyadh et al. 2019), and its antifungal and bactericidal effects. In agriculture, they have been applied to stimulate seed germination (Verma et al. 2019); they also act as growth regulators and stimulate the antioxidant activity of plants (Ghorbanpour and Hadian 2015).

CNM can generate toxicity in plants; however, it is well known that the extent of the toxicity depends mainly on the concentration applied and the type of CNM. Furthermore, the exposure time and the plant species used also influence the outcome (Wang et al. 2019).

It is known that as the world population increases, so does the demand for healthy and quality food (Wagle et al. 2022). However, global food security, especially of plants, is continually threatened by climate change, increased environmental pollution, and the co-evolution of plant pests and pathogens (Omayio and Ndombi 2022).

[1] Department of Botany, Autonomous Agrarian University Antonio Narro, Saltillo, 25315, Mexico.
[2] Research Center in Applied Chemistry, Saltillo, 25294, Mexico.
[3] National Institute of Forestry, Agricultural and Livestock Research (INIFAP); Northwest Regional Research Center, Todos Santos Experimental Field, La Paz 23070, Mexico.
* Corresponding author: antonio.juarez@uaaan.edu.mx

Invasive plant pathogens, such as viruses, fungi, oomycetes, and bacteria are often unwanted shocking agents that can potentially cause various plant diseases and decrease crop productivity (Cardoso et al. 2022). These can be classified as biotrophic or necrotrophic according to their different infection strategies (Li et al. 2019). Biotrophic pathogens first penetrate epidermal cells and multiply in intercellular spaces by feeding on living host tissue, most biotrophic pathogens are host-specific (Jaswal et al. 2020). While necrotrophic pathogens kill plant cells using toxic metabolites to later feed on their remains. Most of these infect a wide range of hosts (Li et al. 2019).

Several aspects influence the occurrence of damage by diseases in plants; however, it is necessary that there are adequate environmental conditions, the presence and viability of the pathogen, and the variety of the crop that is susceptible to generating damage (Wei et al. 2022). Biotic stress caused by pest and disease damage constantly puts crop productivity at risk, causing up to 25% of production losses (Pandey et al. 2020).

Although there are numerous chemical products that are commercially available, their prolonged exposure generates multi-resistance in pathogens; in addition, they can cause considerable risk to the environment and human health. In the current scenario, it has become a challenge to develop substitute pesticides to control phytopathogens resistant to multiple agrochemicals or to induce resistance in plants, so that they can defend themselves against some stressful agents (Pandey et al. 2020).

In this chapter, the use of CNMs for the control of pathogens is proposed as an alternative, in addition to describing how they impact plants from their physiology to defense mechanisms.

Biotic Stress in Crop Plants

Biotic stress caused by pathogenic agents puts the productivity of a crop at constant risk since it causes substantial losses in agriculture (Pandey et al. 2020). Recently, it has been documented that worldwide, there are about 137 pests and pathogens that cause qualitative and quantitative losses in the five main plant crops that are potato, soybean, rice, corn, and wheat, inducing yield losses ranging from 10% and 40% (Al-zaban et al. 2022).

Of the estimated annual global economic losses of crops in general, 50% is generated by viruses, transmitted mainly assisted (vectorized) by insects, which transmit approximately 70% of the total number of viruses that infect plants or are unassisted (not vectorized) (Farooq et al. 2021).

For their part, phytopathogenic fungi affect 25% of the world's agricultural crops, causing losses of up to 10% of production (López-Seijas et al. 2019). Many of these fungi survive in the soil for long periods (Soliman et al. 2022). In addition, some have evolved from being endophytes to become pathogens since they can live in the seeds and systemically colonize the host plant without any sign of disease, only to cause symptoms when they reach the leaves and fruits and thus making it difficult to control (Collinge et al. 2022). Furthermore, some can behave as opportunistic pathogens and infect not only plants but also humans and animals with compromised immunity, therefore causing lethality (Nag et al. 2022).

There are more than 100 species of bacteria responsible for multiple plant diseases worldwide, which generate economic losses of up to 40% of agricultural production (Martinez-Gil and Ramos 2018). Most of these are difficult to treat as they produce biofilm-like aggregations in xylem vessels, on the root and leaf surfaces of host plants, but not in intercellular spaces, contributing to virulence through xylem blockage, as well as increased resistance to plant antimicrobial compounds and increased colonization of the host plant (Mori et al. 2016).

In each of its phenological stages, plants are susceptible to the attack of some pathogenic agent, from germination, development, growth, and flowering. Some of them can become harmful until the post-harvest stage (Khamsaw et al. 2022), generating a negative impact on the growth, productivity, quality, and yield of the plants (Maglangit et al. 2021).

Synthetic pesticides and agricultural antibiotics are the most common way to protect crops against pests and diseases (Qi et al. 2022). In recent decades, its use at a global level has increased dramatically, initially inducing multi-resistance in plant pathogens (Pandey et al. 2020) and generating an increasing risk of exposure for all organisms and the environment (Cech et al. 2023).

Abuse of these can have destructive consequences as their residues do not remain restricted to the application site but can drift to surrounding areas due to evaporation of spray droplets, air movement, and volatilization during the application or subsequently by leaching and soil erosion, water runoff, or wind transport (Córdoba Gamboa et al. 2020). In addition, most of the pesticides manufactured are not biodegradable, so they accumulate causing contamination of the soil and groundwater (Saravanan 2022). Also, fruits can be a source of pesticides for humans through ingestion (Philippe et al. 2021), increasing the risk of diseases such as different types of cancer, asthma, and congenital malformations (Qi et al. 2022). Pesticide abuse is extraordinarily serious, with approximately 700,000 people dying directly or indirectly each year from pesticides (Chen et al. 2022).

Therefore, effective pathogen management is a major challenge in modern agriculture, where control efficacy, cost affordability, environmental safety, toxicity to non-target organisms, and sustainability of the production system are important factors to consider (Gwiazdowska et al. 2022; Schiavi et al. 2022).

Nanotechnology

Nanotechnology is a developing science related to the engineering of nanometric particles composed of materials of inorganic (metals and metal oxides), organic (natural products), and combined nature (He et al. 2019). This science is distinguished as the study of operative matter at the molecular and atomic scale, which allows the creation of innovative nanomaterials and devices with sizes in the range of 1 nm to 100 nm (Saleem et al. 2021). The use of these materials has great potential since it is multidisciplinary, ranging from mechanics to medical applications without ruling out agriculture (Saleem et al. 2021). Due to this, global demand for nanoparticles was valued at US$ 25,373.92 million in 2020 and is projected to increase by 20% by 2028 (Francis et al. 2022).

Nanotechnology is considered to have great potential in the different areas used since materials behave more efficiently at the nanoscale, mainly due to their high surface/volume ratio and their excellent chemical, physical and biological properties (Vishwanath and Negi 2021).

Nanoparticle synthesis can generally be achieved through a wide variety of strategies that can be classified as top-down methods, i.e., the fragmentation of large molecules into smaller ones, followed by the conversion of these smaller molecules into suitable nanoparticles, and bottom-up method in which atomic-sized particles are assembled to produce nanoparticles (Weldegebrieal 2020).

The size of the nanoparticles and their mode of synthesis is decisive for their application since the smaller the size, the wider the branch of uses (Francis et al. 2022). In particular, the polydispersity index (PDI) is considered a critical factor when performing the synthesis since it determines the heterogeneity of the size of the nanoparticles in a solution, which varies from 0 to 1, where 0 indicates a monodisperse solution, while that the highest values indicate a great heterogeneity of size of the particles in solution (Francis et al. 2022).

Carbon Nanomaterials (CNMs)

With the emergence of nanotechnology, the synthesis of CNM has evolved significantly in the last two decades, allowing the creation of materials that can vary in shape, size, and dimension. It started with the discovery of fullerene (Mukherjee et al. 2016) to which have been added graphene oxide (GO), carbon nanotubes (CNT), carbon nanohorns (CNH), nanodiamonds (ND), and carbon quantum dots (CQD) (Khorsandi et al. 2021) (Figure 1).

Fullerene in its most common structure (C_{60}) is an icosahedron made up of 20 hexagons and 12 pentagons. The next most common fullerene is C_{70}, which is more ovoid than completely spherical; however other fullerenes with 72, 76, 84, and 108 carbon atoms have been developed

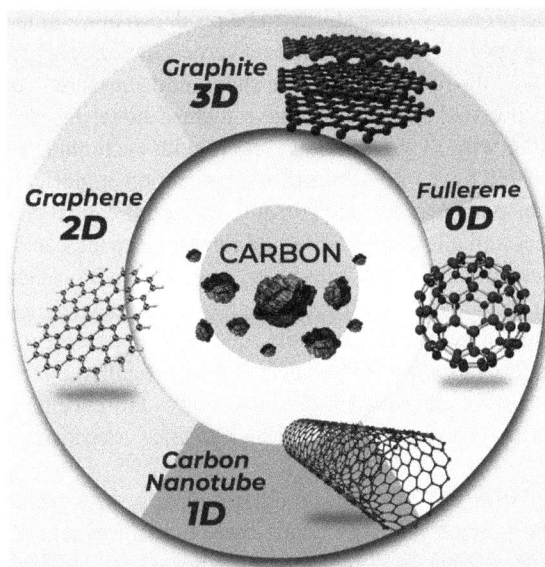

Figure 1. Different structures of carbon-based nanomaterials considering different dimensions. 0D: all dimensions at nanoscale. 1D: two dimensions at nanoscale. 2D: one dimension at nanoscale. 3D: no nanoscale dimension.

(Hamblin 2018). Its use in agriculture has important practical limitations due to the lack of solubility in water, its high tendency to aggregation, and its low biocompatibility. It can also be more toxic than the other CNM since it is extremely sensitive and unstable to the environment (Cantelli et al. 2022).

Graphene is a two-dimensional planar structure monolayer of carbon atoms with sp^2 bonds arranged in a lattice resembling a honeycomb, with a theoretical specific surface area of 2,630 m^2 g^{-1}. Graphene is a semiconductor and shows remarkable electron mobility at room temperature, in addition to exhibiting extremely high thermal conductivity (Papageorgiou et al. 2017). Monolayer graphene is a single-layer material, but there are other different forms so it is also possible to find few-layer graphene (two to five sheets stacked on top of each other) and multilayer graphene (five to ten sheets), both of which are commonly called simply graphene. When more than ten layers of graphene are stacked, they form graphite (Morales-Flórez and Domínguez-Rodríguez 2022).

Carbon nanotubes, which can be single-walled, made up of a single atomic layer of graphene rolled into a tube with a diameter of up to 30 nm and a length of up to 6 mm or multi-walled when two or more layers of graphene are concentrically coiled, with an arrangement of concentric cylinders of individual graphene layers, approximately 42 nm apart and with an external diameter of up to 140 nm (Mallakpour and Soltanian 2016). The coiling direction of the graphene layers can be defined by a vector in the hexagonal lattice and defines their alignment, also known as chirality (Morales-Flórez and Domínguez-Rodríguez 2022). Furthermore, they can be easily functionalized by attaching different chemical groups, such as hydroxyl or carboxyl groups and many others, which makes them promising for use in agriculture (Zemtsova et al. 2022).

Carbon nanohorns are closed structures of sp^2-bonded carbon atoms, typically 2 nm to 5 nm in diameter and 40 nm to 50 nm in length, and differ from other carbon allotropes by their cone-shaped tips (Karousis et al. 2016). Due to its elongated shape, its structure is similar to single-walled carbon nanotubes with which they share a common chemistry. CNH stand out from other allotropes of carbon due to their unique geometry, their ability to be produced at room temperature, high thermal and chemical stability, high porosity, and roughness (Verde-Gómez et al. 2022).

Nanodiamonds are usually round or oval in shape and measure around 5 nm in diameter. Their small size and narrow size distribution make them a novel research topic; however, they have to be highly pure otherwise they tend to aggregate with each other and form large aggregate structures (Chipaux et al. 2018). Its possibilities of application in agriculture are due to its high biocompatibility, great transport capacity, and versatile surface chemical properties that improve the union with other molecules and allow their sustained release (Zupančič and Veranič 2022).

CQD are derivatives of single- or few-layer graphene with a lateral size of less than 100 nm, generated when the lateral dimensions of graphene sheets are reduced to the nanoscale (Yan et al. 2018). Due to their zero dimensionality, quantum confinement, and edge effects, GQD exhibit extraordinary properties such as large surface area, excellent water solubility, controllable photoluminescence, high biocompatibility, low cytotoxicity, and excellent photostability, making them extremely desirable in the fields of materials, physics, chemistry, biology among others (Abbas et al. 2022).

The unique characteristics of CNM can be attributed to several factors such as its structure, as it has a layer of carbon atoms with sp^2 bonds that result in excellent electrical conductivity and the ability to form charge transfer complexes on exposure to electron donor groups (Asadian et al. 2019). Furthermore, their unparalleled biocompatibility, wide surface area, and simple organic functionalization are the reasons why CNM act as laudable carriers of multiple elements, such as antibodies, enzymes, inorganic nanomaterials, redox molecules, and more (Karimi-Maleh et al. 2022), which is why its use is extensive in numerous areas.

Derived from the above, doubts have been raised about the potential toxicity of CNM for humans and the environment. The physicochemical characteristics of these materials are believed to be key determinants of the interaction of CNT with living organisms and thus determine their toxicity (Lanone et al. 2013). In most cases, the toxicity and biocompatibility of CNM can be predicted based on dose or concentration, in addition to the size of the CNM (diameter and length) and geometric structure. Various studies have shown that elongated materials such as MWCNT can be more toxic than spherical molecules (Chen and Li 2022).

Recent advances in nanotechnology have also made it possible to convert liquid, solid, and even gaseous biomass residues into CNM with minimal or no hazardous consequences through methods such as microwave-assisted pyrolysis, hydrothermal carbonization, molten salt process, ball milling method, and pyrolysis single pass in the inert atmosphere among others (Tiwari et al. 2022).

Given the current situation, our society would be better off if we could effectively recycle and valorize selected waste materials for CNM synthesis; for instance, using agricultural waste as initial raw material for CNM is potentially profitable providing a solution to environmental burdens and the same time, reduces the negative impact of waste pollution (Samaddar et al. 2018).

Applications of Carbon Nanomaterials in Agriculture

The application of CNM in the agricultural industry to increase the productivity of land and crops, especially in suboptimal situations, began in the early 21st century; however, it remains a relatively unexplored area (Ioannou et al. 2020). In the last decade, they have been developed and applied to monitor crop health, promote growth, improve fertilizer and pesticide efficiency, manage and control pests and diseases, and mitigate abiotic stress (Verma et al. 2019). Compared to other nanomaterials, especially those of inorganic nature, carbon-based nanomaterials show lower environmental toxicity as well as higher biocompatibility and degradation due to their non-toxic carbon structure (Mukherjee et al. 2016).

The most common forms of application of CNM in plants are by embedding seeds, in a foliar way, directly in the soil or substrate, in the nutrient solution, or by dosing the culture medium when studied *in vitro* (González-García et al. 2019a; Kumar et al. 2018). Thus, a CNM-plant

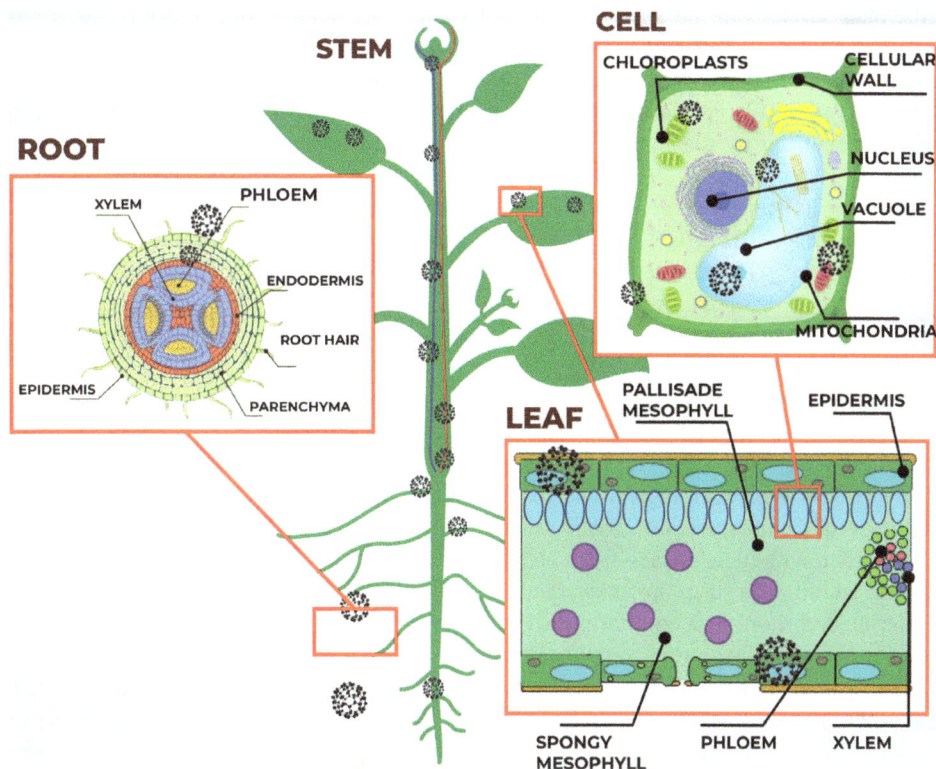

Figure 2. CNM transport pathways in plants.

interaction arises that creates a series of changes at the physiological, biochemical, and genetic levels in the selected plant species, depending on the properties of the CNM such as size, type, shape, charge, chemical composition, reactivity, surface coverage, dose and exposure time (Verma et al. 2019; Wang et al. 2019). The efficacy of a particular type of CNM varies from plant to plant, and the beneficial, null, or adverse outcome of CNM-plant interactions generally depends on the concentration (Tripathi et al. 2011).

Like other types of nanomaterials, when CNM interacts with the plant, three main types of transport can be identified (Figure 2). These include macroscale transport, which considers movement through plant organ structures, such as the leaf cuticle and the vasculature of the plant; microscale transport, which considers the passage through the cell wall; lastly, molecular transport, which considers the crossing of the lipid membranes of the cells and the organelles, where the physicochemical properties of the CNM play an important role (Hubbard et al. 2020).

The routes of entry to the plant of the CNM when applied by the foliar route are stomatal and cuticular. The absorption by the stomata can be limited by their size, the stomatal density in the leaves of the plants, and the opening cycles. However, the larger amount of cuticle area in leaves compared to the stomatal area makes cuticular uptake potentially a more efficient pathway for CNM entry (Avellan et al. 2019).

After crossing the cuticle, the nanomaterials pass into the epidermis and mesophyll tissues, where they interact with these tissues and their structures for subsequent translocation from the shoot to the root, which is achieved by the vascular systems of the plant (Hubbard et al. 2020).

Through the root, the CNM can penetrate the cell walls, and the cytoplasmic membrane and cross the Casparian band to enter the vascular system and translocate to the other organs of the plant (He et al. 2021; Jordan et al. 2018). After entering the plant, penetrating the cell wall and plasma membranes of the epidermal layer, and entering the interior of the vascular tissues, CNM enter the plant along with the absorption of water and other solutes, then move to the stems and subsequently to the leaves through the process of transpiration (Tripathi et al. 2017). The transport of nanomaterials in the plant can be through the apoplast route, that is through the cell wall, where the transport of larger particles (~ 200 nm) is favored. Or by the symplast pathway, through cell to cell, mediated by plasmodesmata; this pathway favors the transport of smaller particles (< 50 nm) (Schwab et al. 2016). Another route of transport of CNM is through the xylem and phloem (Milewska-Hendel et al. 2017).

After entering the cell, CNM adheres to various organelles where they begin to transform metabolic processes, increasing ROS production and modifying biochemical processes in the cell (Miao et al. 2009; Tripathi et al. 2017). They are also transported to the nucleus by passive diffusion (Ahmed et al. 2021), which allows the expression of different genes to be promoted (Yan et al. 2013).

In microscale transport, when CNM interact with the cell wall, they generate a large surface free energy derived from the exposure of electrons at the interface of the material and the surrounding medium, in addition to a high surface/volume ratio (Syahirah et al. 2021). When nanomaterials come into contact with a biological environment, such as the epidermis or the internal fluids of plants, they retain components of organic matter or biomolecules from the environment, forming a protein corona with high surface-free energy (Juárez-Maldonado et al. 2019). This can maintain a long-term equilibrium state known as a hard corona, determined by biomolecules attached to the nanomaterial, or a short-term equilibrium state known as a soft corona (Docter et al. 2015).

The formation of the protein corona largely determines the distribution, cellular internalization, fate, and effects of CNM (Prakash and Deswal 2020). The positive charge of the CNM positively influences the absorption of these and other ions and biomolecules due to the favorable electrostatic interactions with the negatively charged cell membrane; however, some negatively charged nanomaterials are also absorbed, although in smaller measures (Jackson et al. 2017). Therefore, the initial interactions do not limit the interaction with anionic or cationic components of the cell walls through positively and negatively charged groups located on the surface of the corona (Prakash and Deswal 2020).

On the other hand, the electrical potential of the membrane maintains functional ion transport channels, integral proteins, and specific transporters that favor the internalization of CNM. In addition, the absorption of these is also related to adhesion forces such as Van der Waals forces, hydrophobic forces, and ligand-receptor binding, among others (Zhao and Stenzel 2018).

Once inside the cell, a cell signaling process begins that modifies the membrane potential and integral proteins for energy transduction and production of elicitors, metabolites, and gene expression, followed by the transformation of nanomaterials into ionic forms and their internalization in the cytoplasm (Juárez-Maldonado et al. 2021).

Impact of Carbon Nanomaterials on Plants

Some types of CNM have gained importance due to their potential use in regulating plant growth. For the most part, research reports have shown both a beneficial and an adverse influence on plants, depending on CNM types, concentration, exposure time, plant species, and growing conditions in which they are applied (Verma et al. 2019) (Table 1).

Some research reports that CNM can penetrate the seed coat and improve plant germination and growth (Khodakovskaya et al. 2009). This is because they can activate water channel proteins; for example, the tomato aquaporin channel (*LeAqp1*), which is directly related to germination efficiency, can also increase the absorption of water and nutrients, positively impacting most of the physiological processes of plants (Ratnikova et al. 2015; Villagarcia et al. 2012). However, the penetration of CNM into seeds could be restricted in some plant species due to the morphology and composition of the seed coat that protects the embryo (Ratnikova et al. 2015).

Table 1. Impact of carbon nanomaterials on plants.

CNM	Concentration and Route of Application	Plant Species	Observed Effect	Reference
CNT and GP	50 µg mL^{-1}, 100 µg mL^{-1}, 250 µg mL^{-1}, 500 µg mL^{-1} and 1,000 µg mL^{-1}/Seed priming	*Sorghum bicolor* and *Panicum virgatum*	Increased germination and seedling growth	Pandey et al. (2018)
GO	0.5 mg mL^{-1} and 0.10 mg mL^{-1}/Foliar	*Daucus carota*	Increased the content of carotenoids in the root	Siddiqui et al. (2019)
MWCNT	10 µg mL^{-1}, 20 µg mL^{-1}, 30 µg mL^{-1} and 40 µg mL^{-1}/ Seed priming	*Lycopersicum esculentum*	Increased germination and water absorption	Khodakovskaya et al. (2009)
MWCNT	50 µg mL^{-1}, 100 µg mL^{-1}, and 200 µg mL^{-1}/Seed priming	*Hordeum vulgare, Glycine max, Zea mays*	Increased germination and water absorption	Lahiani et al. (2013)
MWCNT	70 µg mL^{-1}, 80 µg mL^{-1}, and 90 µg mL^{-1}/Seed priming	*Oriza sativa*	Increased biomass and photosynthetic activity	Joshi et al. (2018)
GP	0.1 g L^{-1}, 0.2 g L^{-1} and 0.3 g L^{-1}/Foliar	*Capsicum annuum* and *Solanum melongena*	Increased fruit yield and induced lipid peroxidation	Younes et al. (2019)
CNT and C$_{60}$	200 mg L^{-1}/Foliar	*Nicotiana benthamiana*	Increased the content of chlorophylls, silicic acid, abscisic acid, SOD, POD, and CAT enzymes	Adeel et al. (2021)
C$_{60}$	25 mg L^{-1}, 250 mg L^{-1}, 500 mg L^{-1} and 1,000 mg L^{-1}/ Foliar	*Tanacetum parthenium*	Increased parthenolide content	Ahmadi et al. (2020)
C$_{60}$	10 nM, 40 nM, 80 nM, and 120 nM/seed priming	*Triticum aestivum*	Increased the content of vitamin C, free amino acids, and proline	Shafiq et al. (2021)
GO	50 and 100 mg L^{-1}/Foliar	*Ocimum basilicum*	Increased the content of phenols estragole, methyl chavicol, germacrene, and linalool	Ganjavi et al. (2021)
MWCNT	50 mg L^{-1}, 100 mg L^{-1}, 250 mg L^{-1}, 500 mg L^{-1} and 1,000 mg L^{-1}/Foliar	*Calendula officinalis*	Increased content of reduced glutathione and ascorbic acid	Sharifi et al. (2021)
MWCNT	50, 100 and 150 mg L^{-1}/ Added to MS culture medium	*Catharanthus roseus*	Increased the content of carotenoids, phenols, and alkaloids	Ghasempour et al. (2019)
MWCNT	25, 50, 75 and 100 mg L^{-1}/ Added to MS culture medium	*Salvia nemorosa*	Increased the content of rosmarinic acid, salvianolic acid, ferulic acid, and cinnamic acid	Heydari et al. (2020)
MWCNT	25 µg mL^{-1}, 50 µg mL^{-1}, 100 µg mL^{-1} and 250 µg mL^{-1}/Added to culture medium B5	*Satureja khuzestanica*	Increased the content of phenols, flavonoids, rosmarinic acid, and caffeic acid	Ghorbanpour and Hadian (2015)
MWCNT	250 mg L^{-1}/Added to MS culture medium	*Thymus daenensis* celak	Increased the content of phenols and flavonoids	Samadi et al. (2020)

Table 1 contd. ...

...Table 1 contd.

CNM	Concentration and Route of Application	Plant Species	Observed Effect	Reference
Carbon dots (CDs)	20%/coating film	*Litchi chinensis*	Increased vitamin C content and DPPH antioxidant capacity	Mei et al. (2022)
Carbon dots functionalized with trehalose CNPT and glucose CNPG	20 mg L^{-1}/seed priming	*Vigna radiata*	Increased germination and content of Chlorophyll *a* and *b*, carotenoids, total sugar, reducing sugar, and glycine betaine	Sarkar et al. (2022)
Carbon nanoparticles	200 mg kg^{-1}	*Zea mays*	Increased the efficiency of N, P, and K	Zhao et al. (2021)
MWCNT	50, 100 and 200 mg L^{-1}/ Added to MS culture medium	*Saccharum* spp.	Increased shoot development and total chlorophyll content	Sorcia-Morales et al. (2021)
SWCNT	10 mg L^{-1} and 40 mg L^{-1}/ Foliar	*Dianthus caryophyllus*	Promoted flower diameter and extended maximum flower opening period, increased vase life	Ahmadi-Majd et al. (2022)

In addition, CNM act as inducers in the regulation of plant growth since they activate the biosynthesis of some hormones such as indoleacetic acid and abscisic acid (Patel et al. 2019; Tripathi et al. 2011). They also promote the expression of marker genes for cell division (*CycB*) and cell wall elongation (*NtLRX1*) that directly influence growth (Khodakovskaya et al. 2012). Likewise, they increase the activity of antioxidant enzymes such as superoxide dismutase (SOD), guaiacol peroxidase (POD), catalase (CAT), ascorbate peroxidase (APX), and glutathione peroxidase (GPX), in addition to non-enzymatic antioxidants, which results in more vigorous plants and healthy (González-García et al. 2019b; Patel et al. 2019).

López-Vargas et al. (2020) showed that by soaking tomato seeds in a solution of 1,000 mg L^{-1} of MWCNT and GP, the germination percentage decreased compared to untreated seeds. On the contrary, the application of 100 mg L^{-1} of these same CNM increased the root biomass of plants.

Pandey et al. (2018) reported an increase in the rate of seed germination, in addition to an increase in the length of shoots and roots as well as the fresh biomass of sorghum (Sorghum bicolor) and millet (*Panicum virgatum*) incubated in MS culture medium (Murashige and Skoog) supplemented with 50 mg L^{-1} and 200 mg L^{-1} of GP and MWCNT, respectively.

Lahiani et al. (2013) showed that dosing the MS culture medium with 25 μg mL^{-1}, 50 μg mL^{-1}, and 100 μg mL^{-1} of MWCNT considerably improved the germination percentage, in addition to increasing the aerial and root biomass in barley plants (*Hordeum vulgare*), soybean (*Glycine max*), and maize (*Zea mays*). In addition, they demonstrated the penetration and internalization of MWCNT in the three species using Raman spectroscopy and TEM microscopy.

It has also been shown that CNM influences photosynthetic processes since they are good electron donors and acceptors; by joining on the surface of the chloroplast, they can transfer electrons and accelerate their transfer in the light reaction of photosynthesis (Li et al. 2020).

CNM can increase the activity of photosystem I and the enzyme ribulose bisphosphate carboxylase oxygenase (RUBISCO) (Wang et al. 2018). In addition, within the chloroplast, they can induce the production of chlorophylls and carotenoids and can also act as a carbon source that facilitates carbon fixation and increases the speed of electron transport, thus inducing an improvement in photosynthesis (Lahiani et al. 2018).

Lahiani et al. (2018) reported that foliar application of 100 mg L^{-1} of MWCNT increased the efficiency of photosynthesis in maize plants. Fan et al. (2018) reported that supplementing the

MS culture medium with 50 mg L^{-1} of MWCNT increased the photosynthetic rate of *Arabidopsis thaliana* plants, as well as the number of lateral roots. Rahmani et al. (2020) demonstrated that the foliar application of 50 mg L^{-1} of CNT improved the photosynthetic capacity of *Salvia verticillata* leaves by increasing the levels of photosynthetic pigments.

Antimicrobial Activity of Carbon Nanomaterials

CNM are suggested as an alternative for the control of pathogens because they have a great antimicrobial activity attributed to several factors such as physical damage that is induced through direct contact of CNM with the pathogen (Maksimova 2019), where the surface/volume ratio of the CNM causes a link with the cell wall or membrane of microorganisms, exerting physical damage on these (Khan et al. 2016). This damage produces lysis of the cell membrane, lipid degradation, release of cytoplasmic material, and leakage of electrolytes and enzymes, which generates a reduction in the growth and germination of spores, stimulating a decrease in the population or viability of microorganisms (Siddiqui et al. 2019; Wang et al. 2013). There are some studies that demonstrate the antimicrobial potential of CNM *in vitro*. Siddiqui et al. (2019) showed that the growth of *Alternaria dauci* and *Fusarium solani* can be reduced by 36% and 33% respectively, by adding 10 mg mL^{-1} of GO to the culture medium of potato dextrose agar (PDA). Zhao et al. (2016) reported the antimicrobial activity of GO against the pathogens *Staphylococcus aureus* and *Candida albicans* by supplementing Luria-Bertani (LB) and Sabouraud gentamicin chloramphenicol (SGC) culture media with 200 µg mL^{-1} of the nanomaterial. Chang et al. (2016) reported the antimicrobial activity of GO against *Escherichia coli* and *Staphylococcus aureus* by adding 250 mg L^{-1} to the LB culture medium.

Another antibacterial mechanism of CNM is attributed to their laminar structures, which can cover the bacterial surface and decrease their viability, reproduction, and proliferation (Ji et al. 2016).

After the effect of the CNM on the microorganisms derived from direct contact, when entering the plant cell, the CNM adhere to the various organelles where they begin to modify the metabolic processes by increasing the production of reactive oxygen species (ROS) and affecting biochemical processes in the cell (Miao et al. 2009; Tripathi et al. 2017). This in turn induces the activation of the antioxidant defense system that includes enzymatic and non-enzymatic compounds (Czarnocka and Karpiński 2018). Some of these with antimicrobial characteristics, such as phenolic compounds (Kumar and Goel 2019). The accumulation of enzymes related to defense and metabolites are key for the activation of plant defense genes and the translation of PR proteins, chitinases, and β-1,3-glucanases that generate greater resistance to biotic stress (Chun and Chandrasekaran 2019). In addition to promoting lignification that improves protection against different pathogens by strengthening the barrier of the plant cell wall (Chandra et al. 2015).

MWCNT has been applied through the MS culture medium (50–150 mg L^{-1}) in *Catharanthus roseus* plants where the content of carotenoids, phenols, and alkaloids increased (Ghasempour et al. 2019). In *Thymus daenensis* celak, the content of phenols and flavonoids increased when 250 mg L^{-1} of MWCNT was added to the MS culture medium (Samadi et al. 2020). The addition of MWCNT-COOH in the MS culture medium (25–100 mg L^{-1}) increased the content of rosmarinic acid, salvianolic acid B, ferulic acid, and cinnamic acid in *Salvia nemorosa* (Heydari et al. 2020). While the addition of MWCNT to the B5 culture medium (25–250 µg mL^{-1}) induced the increase of phenols, flavonoids, rosmarinic acid, and caffeic acid in *Satureja khuzestanica* (Ghorbanpour and Hadian 2015).

The foliar application of fullerene C$_{60}$ (25–1,000 mg L^{-1}) increased the content of parthenolides in *Tanacetum parthenium* (Ahmadi et al. 2020). On the other hand, the application of fullerene C$_{60}$-OH$_{20}$ through seed priming (10–120 nM) in wheat seeds (*Triticum aestivum*) induced an increase in vitamin C (Shafiq et al. 2021). For its part, the foliar application of GO (50 mg L^{-1} and

Table 2. Impact of carbon nanomaterials on plant pathogens.

CNM	Concentration and Route of Application	Plant Species	Plant Pathogen	Reference
CNT and C_{60}	200 mg L^{-1} Foliar	*Nicotiana benthamiana*	Inhibited the spread of tobacco mosaic virus (TMV)	Adeel et al. (2021)
GO	0.05 mg mL^{-1} and 0.10 mg mL^{-1} Foliar	*Daucus carota*	Decreased the spread of *Alternaria dauci* and *Fusarium solani*	Siddiqui et al. (2019)
CNT	50 mg L^{-1} and 100 mg L^{-1} Foliar	*Solanum lycopersicum*	Decreased the incidence and severity of *Alternaria solani*	González-García et al. (2021)
fullerene (C_{60})	100 mg L^{-1} Foliar	*Nicotiana benthamiana*	Suppressed the growth of viral symptoms of Cucurbit chlorotic yellows virus (CCYV)	Al-zaban et al. (2022)
MWCNTs or (C_{60})	50 mg L^{-1} and 200 mg L^{-1} Foliar	*Nicotiana benthamiana*	Inhibited the spread of turnip mosaic virus (TuMV)	Hao et al. (2018)
CNT and GO	100 mg L^{-1} Foliar	*Solanum lycopersicum*	Decreased the incidence and severity of *Fusarium oxysporum*	González-García et al. (2022)
MWCNT and GO	200 mg L^{-1} Foliar	*Rosa rugosa*	Inhibited the development of *Podosphaera pannosa*	Hao et al. (2019)
GO	50 mg L^{-1} Foliar	*Brassica oleracea*	Decreased the incidence and severity of *Alternaria brassicicola*	Saleem et al. (2021)

100 mg L^{-1}) in *Ocimum basilicum* plants induced the increase of phenols estragole, methyl chavicol, germacrene D, and linalool (Ganjavi et al. 2021).

Tolerance to biotic stress can also be induced with the use of CNM since they can serve as carriers for pesticides or other active ingredients (Majeed et al. 2020). The encapsulation of pesticides changes their original characteristics in the form of pesticide nanoparticles and with this increases the contact area with pathogens, in addition to a gradual release (Sarlak et al. 2014). Likewise, they can serve as nanofertilizers since they have the capacity to release nutrients at the specific point of demand and generate a healthier and more stress-tolerant plant (Rafique et al. 2017).

Nanomaterials alone have the potential to be applied directly to plant seeds, foliage, or roots to protect against pests and diseases caused by insects, bacteria, fungi, and viruses (Worrall et al. 2018) (Table 2).

Carbon nanomaterials have been shown to increase tolerance to various diseases. The foliar application of 200 mg L^{-1} of C_{60} and CNT in *Nicotiana benthamiana* plants inhibited the replication of the tobacco mosaic virus (TMV), in addition to limiting its dissemination to the apical tissues. The immunity of the plant was also improved, the photosynthetic yield was increased, as well as the activity of antioxidant enzymes and phytohormones related to defense (Adeel et al. 2021). Hao et al. (2019) demonstrated that foliar application of 200 mg L^{-1} of MWCNT and GO inhibited the development of the pathogen *Podosphaera pannosa*, which caused powdery mildew in roses (*Rosa rugosa* Thunb). Saleem et al. (2021) demonstrated the decrease in incidence and severity of *Alternaria brassicicola* in *Brassica oleracea* plants with a foliar application of 50 mg L^{-1} of GO.

Conclusions

CNMs are potentially applicable products in the field as an alternative to traditional pesticides. These can act directly on pathogens, partially or totally by inhibiting them or inducing the generation of defense metabolites in the plant.

CNMs are easily acquired, have a low cost, are used in small quantities, and have low toxicity. They can also be obtained from vegetable waste, such as agricultural and/or forest residues, which contributes to reducing the progressive and cumulative deterioration of the environment.

However, it is necessary to carry out more studies that consider biochemical, micromorphological, and genetic processes, as well as other relevant eco-physiological processes, to elucidate how CNM induces tolerance or resistance to pathogens. At present, this technology is not yet available for agricultural production; however, with the increasing research carried out on this topic, it is possible that in the future there will be commercial products based on nanomaterials.

References

Abbas, A., Liang, Q., Abbas, S., Liaqat, M., Rubab, S. and Tabish, T. 2022. Eco-friendly sustainable synthesis of graphene quantum dots from biowaste as a highly selective sensor. Nanomaterials 20: 1–13.

Adeel, M., Farooq, T., White, J.C., Hao, Y., He, Z. and Rui, Y. 2021. Carbon-based nanomaterials suppress tobacco mosaic virus (TMV) infection and induce resistance in *Nicotiana benthamiana*. J. Hazard. Mater. 404: 124–167.

Ahmadi-Majd, M., Mousavi-Fard, S., Rezaei Nejad, A. and Fanourakis, D. 2022. Carbon nanotubes in the holding solution stimulate flower opening and prolong vase life in carnation. Chem. Biol. Technol. Agric. 9: 15–31.

Ahmadi, S.Z., Ghorbanpour, M., Aghaee, A. and Hadian, J. 2020. Deciphering morpho-physiological and phytochemical attributes of *Tanacetum parthenium* L. plants exposed to C_{60} fullerene and salicylic acid. Chemosphere 259: 127406.

Ahmed, S., Gao, X., Jahan, M.A., Adams, M., Wu, N. and Kovinich, N. 2021. Nanoparticle-based genetic transformation of *Cannabis sativa*. J. Biotechnol. 326: 48–51.

Al-zaban, M.I., Alhag, S.K., Dablool, A.S., Ahmed, A.E., Alghamdi, S., Ali, B., Al-saeed, F.A., Saleem, M.H. and Poczai, P. 2022. Manufactured nano-objects confer viral protection against cucurbit chlorotic yellows virus (CCYV) infecting *Nicotiana benthamiana*. Microorganisms 10: 1837.

Asadian, E., Ghalkhani, M. and Shahrokhian, S. 2019. Electrochemical sensing based on carbon nanoparticles: A review. Sensors Actuators B Chem. 293: 183–209.

Avellan, A., Yun, J., Zhang, Y., Spielman-Sun, E., Unrine, J.M., Thieme, J., Li, J., Lombi, E., Bland, G. and Lowry, G.V. 2019. Nanoparticle size and coating chemistry control foliar uptake pathways, translocation, and leaf-to-rhizosphere transport in wheat. ACS Nano. 13: 5291–5305.

Cantelli, A., Malferrari, M., Mattioli, E.J., Marconi, A., Mirra, G., Soldà, A., Marforio, T.D., Zerbetto, F., Rapino, S., Di Giosia, M. and Calvaresi, M. 2022. Enhanced uptake and phototoxicity of C_{60}@albumin hybrids by folate bioconjugation. Nanomaterials 12: 3501.

Cardoso, R.M., Pereira, T.S., Facure, M.H.M., dos Santos, D.M., Mercante, L.A., Mattoso, L.H.C. and Correa, D.S. 2022. Current progress in plant pathogen detection enabled by nanomaterials-based (bio) sensors. Sensors and Actuators Reports 4: 100068.

Cech, R., Zaller, J.G., Lyssimachou, A., Clausing, P., Hertoge, K. and Linhart, C. 2023. Pesticide drift mitigation measures appear to reduce contamination of non-agricultural areas, but hazards to humans and the environment remain. Sci. Total Environ. 854: 158814.

Chandra, S., Chakraborty, N., Dasgupta, A., Sarkar, J., Panda, K. and Acharya, K. 2015. Chitosan nanoparticles: A positive modulator of innate immune responses in plants. Sci. Rep. 5: 1–14.

Chang, Y., Gong, J., Zeng, G., Ou, X., Song, B., Guo, M., Zhang, J. and Liu, H. 2016. Antimicrobial behavior comparison and antimicrobial mechanism of silver coated carbon nanocomposite. Process Saf. Environ. Prot. 10: 596–605.

Chen, R., Peng, X., Song, Y. and Du, Y. 2022. A paper-based electrochemical sensor based on PtNP/COF TFPB – DHzDS @ rGO for sensitive detection of furazolidone. Biosensors 10: 904.

Chen, Y. and Li, X. 2022. The utilization of carbon-based nanomaterials in bone tissue regeneration and engineering: Respective featured applications and future prospects. Med. Nov. Technol. Devices 16: 100168.

Chipaux, M., van der Laan, K.J., Hemelaar, S.R., Hasani, M., Zheng, T. and Schirhagl, R. 2018. Nanodiamonds and their applications in cells. Small. 14: 1704263.

Chun, S.-C. and Chandrasekaran, M. 2019. Chitosan and chitosan nanoparticles induced expression of pathogenesis-related proteins genes enhances biotic stress tolerance in tomato. Int. J. Biol. Macromol. 125: 948–954.

Collinge, D.B., Jensen, B. and Jørgensen, H.J. 2022. Fungal endophytes in plants and their relationship to plant disease. Curr. Opin. Microbiol. 69: 102177.

Córdoba Gamboa, L., Solano Diaz, K., Ruepert, C. and van Wendel de Joode, B. 2020. Passive monitoring techniques to evaluate environmental pesticide exposure: Results from the Infant's environmental health study (ISA). Environ. Res. 184: 109243.

Czarnocka, W. and Karpiński, S. 2018. Friend or foe? Reactive oxygen species production, scavenging and signaling in plant response to environmental stresses. Free Radic. Biol. Med. 122: 4–20.

Dehghani, M.H., Kamalian, S., Shayeghi, M., Yousefi, M., Heidarinejad, Z., Agarwal, S. and Gupta, V.K. 2019. High-performance removal of diazinon pesticide from water using multi-walled carbon nanotubes. Microchem. 145: 486–491.

Docter, D., Westmeier, D., Markiewicz, M., Stolte, S. and Knauer, S.K. 2015. The nanoparticle biomolecule corona : lessons learned—challenge accepted ? Chem. Soc. Rev. 9: 6094–6121.

Fan, X., Xu, J., Lavoie, M., Peijnenburg, W.J.G.M., Zhu, Y., Lu, T., Fu, Z., Zhu, T. and Qian, H. 2018. Multiwall carbon nanotubes modulate paraquat toxicity in *Arabidopsis thaliana*. Environ. Pollut. 233: 633–641.

Farooq, T., Adeel, M., He, Z., Umar, M., Shakoor, N., da Silva, W., Elmer, W., White, J.C. and Rui, Y. 2021. Nanotechnology and plant viruses: An emerging disease management approach for resistant pathogens. ACS Nano. 15: 6030–6037.

Fiyadh, S.S., AlSaadi, M.A., Jaafar, W.Z., AlOmar, M.K., Fayaed, S.S., Mohd, N.S., Hin, L.S. and El-Shafie, A. 2019. Review on heavy metal adsorption processes by carbon nanotubes. J. Clean. Prod. 230: 783–793.

Francis, D.V., Aiswarya, T. and Gokhale, T. 2022. Optimization of the incubation parameters for biogenic synthesis of WO_3 nanoparticles using Taguchi method. Heliyon. 8: 10640.

Gabris, M.A. and Ping, J. 2021. Carbon nanomaterial-based nanogenerators for harvesting energy from environment. Nano Energy. 90: 106494.

Ganjavi, A.S., Oraei, M., Gohari, G., Akbari, A. and Faramarzi, A. 2021. Glycine betaine functionalized graphene oxide as a new engineering nanoparticle lessens salt stress impacts in sweet basil (*Ocimum basilicum* L.). Plant Physiol. Biochem. 162: 14–26.

Ghasempour, M., Iranbakhsh, A., Ebadi, M. and Oraghi Ardebili, Z. 2019. Multi-walled carbon nanotubes improved growth, anatomy, physiology, secondary metabolism, and callus performance in *Catharanthus roseus*: an *in vitro* study. Biotech. 9: 404.

Ghorbanpour, M. and Hadian, J. 2015. Multi-walled carbon nanotubes stimulate callus induction, secondary metabolites biosynthesis and antioxidant capacity in medicinal plant *Satureja khuzestanica* grown *in vitro*. Carbon. 94: 749–759.

González-García, Y., López-Vargas, E.R., Cadenas-Pliego, G., Benavides-Mendoza, A., González-Morales, S., Robledo-Olivo, A., Alpuche-Solís, Á.G. and Juárez-Maldonado, A. 2019. Impact of carbon nanomaterials on the antioxidant system of tomato seedlings. Int. J. Mol. Sci. 20: 11–35.

González-García, Y., Cadenas-Pliego, G., Alpuche-Solís, Á.G., Cabrera, R.I. and Juárez-Maldonado, A. 2021. Carbon nanotubes decrease the negative impact of *Alternaria solani* in tomato crop. Nanomaterials 11: 1–15.

González-García, Y., Cadenas-Pliego, G., Alpuche-Solís, Á.G., Cabrera, R.I. and Juárez-Maldonado, A. 2022. Effect of carbon-based nanomaterials on *Fusarium* wilt in tomato. Sci. Hortic. 23: 291.

Goswami, A.D., Trivedi, D.H., Jadhav, N.L. and Pinjari, D.V. 2021. Sustainable and green synthesis of carbon nanomaterials: A review. J. Environ. Chem. Eng. 9: 106118.

Gwiazdowska, D., Marchwińska, K., Juś, K., Uwineza, P.A., Gwiazdowski, R., Waśkiewicz, A. and Kierzek, R. 2022. The concentration-dependent effects of essential oils on the growth of *Fusarium graminearum* and mycotoxins biosynthesis in wheat and maize grain. Appl. Sci. 12: 473.

Hamblin, M.R. 2018. Fullerenes as photosensitizers in photodynamic therapy: pros and cons. Photochem. Photobiol. Sci. 17: 1515–1533.

Hao, Y., Yuan, W., Ma, C., White, J.C., Zhang, Z., Adeel, M., Zhou, T., Rui, Y. and Xing, B. 2018. Engineered nanomaterials suppress Turnip mosaic virus infection in tobacco (*Nicotiana benthamiana*). Environ. Sci. Nano. 5: 1685–1693.

Hao, Y., Fang, P., Ma, C., White, J.C., Xiang, Z., Wang, H., Zhang, Z., Rui, Y. and Xing, B. 2019. Engineered nanomaterials inhibit *Podosphaera pannosa* infection on rose leaves by regulating phytohormones. Environ. Res. 170: 1–6.

He, A., Jiang, J., Ding, J. and Sheng, G.D. 2021. Blocking effect of fullerene nanoparticles (nC_{60}) on the plant cell structure and its phytotoxicity. Chemosphere 278: 130474.

He, X., Deng, H. and Hwang, H. 2018. The current application of nanotechnology in food and agriculture. J. Food Drug Anal. 27: 1–21.

Heydari, H.R., Chamani, E. and Esmaielpour, B. 2020. Cell line selection through gamma irradiation combined with multi-walled carbon nanotubes elicitation enhanced phenolic compounds accumulation in *Salvia nemorosa* cell culture. Plant Cell, Tissue Organ Cult. 142: 353–367.

Hubbard, J.D., Lui, A. and Landry, M.P. 2020. Multiscale and multidisciplinary approach to understanding nanoparticle transport in plants. Curr. Opin. Chem. Eng. 30: 135–143.

Ioannou, A., Gohari, G., Papaphilippou, P., Panahirad, S., Akbari, A., Reza, M., Krasia-christoforou, T. and Fotopoulos, V. 2020. Advanced nanomaterials in agriculture under a changing climate: The way to the future? Environ. Exp. Bot. 176: 104048.

Jackson, T.C., Patani, B.O. and Israel, M.B. 2017. Nanomaterials and cell interactions: A Review. J. Biomater. Nanobiotechnol. 8, 220–228.

Jaswal, R., Kiran, K., Rajarammohan, S., Dubey, H., Singh, P.K., Sharma, Y., Deshmukh, R., Sonah, H., Gupta, N. and Sharma, T.R. 2020. Effector biology of biotrophic plant fungal pathogens: Current advances and future prospects. Microbiol. Res. 241: 126567.

Ji, H., Sun, H. and Qu, X. 2016. Antibacterial applications of graphene-based nanomaterials: Recent achievements and challenges. Adv. Drug Deliv. Rev. 105: 176–189.

Jordan, J.T., Singh, K.P. and Cañas-Carrell, J.E. 2018. Carbon-based nanomaterials elicit changes in physiology, gene expression, and epigenetics in exposed plants: A review. Curr. Opin. Environ. Sci. Heal. 6: 29–35.

Joshi, A., Kaur, S., Singh, P., Dharamvir, K., Nayyar, H. and Verma, G. 2018. Tracking multi-walled carbon nanotubes inside oat (*Avena sativa* L.) plants and assessing their effect on growth, yield, and mammalian (human) cell viability. Appl. Nanosci. 8: 1399–1414.

Juárez-Maldonado, A., Ortega-Ortiz, H., González-Morales, S., Morelos-Moreno, Á., Cabrera-de la Fuente, M., Sandoval-Rangel, A., Cadenas-Pliego, G. and Benavides-Mendoza, A. 2019. Nanoparticles and nanomaterials as plant biostimulants. Int. J. Mol. Sci. 20: 1–19.

Juárez-Maldonado, A., Tortella, G., Rubilar, O., Fincheira, P. and Benavides-Mendoza, A. 2021. Biostimulation and toxicity: the magnitude of the impact of nanomaterials in microorganisms and plants. J. Adv. Res. 31: 113–126.

Karimi-Maleh, H., Beitollahi, H., Senthil Kumar, P., Tajik, S., Mohammadzadeh Jahani, P., Karimi, F., Karaman, C., Vasseghian, Y., Baghayeri, M., Rouhi, J., Show, P.L., Rajendran, S., Fu, L. and Zare, N. 2022. Recent advances in carbon nanomaterials-based electrochemical sensors for food azo dyes detection. Food Chem. Toxicol. 164: 112961.

Karousis, N., Suarez-Martinez, I., Ewels, C.P. and Tagmatarchis, N. 2016. Structure, properties, functionalization, and applications of carbon nanohorns. Chem. Rev. 116: 4850–4883.

Khamsaw, P., Sangta, J., Chaiwan, P., Rachtanapun, P., Sirilun, S., Sringarm, K., Thanakkasaranee, S. and Sommano, S.R. 2022. Bio-circular perspective of citrus fruit loss caused by pathogens: Occurrences, active ingredient recovery and applications. Horticulturae 8: 748.

Khan, A.A.P., Khan, A., Rahman, M.M., Asiri, A.M. and Oves, M. 2016. Lead sensors development and antimicrobial activities based on graphene oxide/carbon nanotube/poly (O-toluidine) nanocomposite. Int. J. Biol. Macromol. 89: 198–205.

Khodakovskaya, M., Dervishi, E., Mahmood, M., Xu, Y., Li, Z., Watanabe, F. and Biris, A.S. 2009. Carbon nanotubes are able to penetrate plant seed coat and dramatically affect seed germination and plant growth. ACS Nano 3: 3221–3227.

Khodakovskaya, M.V., de Silva, K., Biris, A.S., Dervishi, E. and Villagarcia, H. 2012. Carbon nanotubes induce growth enhancement of tobacco ells. ACS Nano. 3: 3221–3227.

Khorsandi, Z., Borjian-Boroujeni, M., Yekani, R. and Varma, R.S. 2021. Carbon nanomaterials with chitosan: A winning combination for drug delivery systems. J. Drug Deliv. Sci. Technol. 66: 102847.

Kumar, A., Singh, A., Panigrahy, M., Sahoo, P.K. and Panigrahi, K.C.S. 2018. Carbon nanoparticles influence photomorphogenesis and flowering time in *Arabidopsis thaliana*. Plant Cell Rep. 37: 901–912.

Kumar, N. and Goel, N. 2019. Phenolic acids: Natural versatile molecules with promising therapeutic applications. Biotechnol. Reports 24: 370.

Lahiani, M.H., Dervishi, E., Chen, J., Nima, Z., Gaume, A., Biris, A.S. and Khodakovskaya, M.V. 2013. Impact of carbon nanotube exposure to seeds of valuable crops. ACS Appl. Mater. Interfaces 5: 7965–7973.

Lahiani, M.H., Nima, Z.A., Villagarcia, H., Biris, A.S. and Khodakovskaya, M.V. 2018. Assessment of effects of the long-term exposure of agricultural crops to carbon nanotubes. J. Agric. Food Chem. 66: 6654–6662.

Lanone, S., Andujar, P., Kermanizadeh, A. and Boczkowski, J. 2013. Determinants of carbon nanotube toxicity. Adv. Drug Deliv. Rev. 65: 2063–2069.

Li, N., Han, X., Feng, D., Yuan, D. and Huang, L.-J. 2019. Signaling crosstalk between salicylic acid and ethylene/jasmonate in plant defense: Do we understand what they are whispering? Int. J. Mol. Sci. 20: 671.

Li, Y., Xu, X., Wu, Y., Zhuang, J., Zhang, X., Zhang, H., Lei, B., Hu, C. and Liu, Y. 2020. A review on the effects of carbon dots in plant systems. Mater. Chem. Front. 4: 437–448.

López-Seijas, J., García-Fraga, B., da Silva, A.F. and Sieiro, C. 2019. Wine lactic acid bacteria with antimicrobial activity as potential biocontrol agents against *Fusarium oxysporum* f. sp. lycopersici. Agronomy 10: 31.

López-Vargas, E.R., González-García, Y., Pérez-Álvarez, M., Cadenas-Pliego, G., González-Morales, S., Benavides-Mendoza, A., Cabrera, R.I. and Juárez-Maldonado, A. 2020. Seed priming with carbon nanomaterials to modify the germination, growth, and antioxidant status of tomato seedlings. Agronomy 10: 1–22.

Maglangit, F., Yu, Y. and Deng, H. 2021. Bacterial pathogens: threat or treat (a review on bioactive natural products from bacterial pathogens). Nat. Prod. Rep. 38: 782–821.

Majeed, N., Panigrahi, K.C.S., Sukla, L.B., John, R. and Panigrahy, M. 2020. Application of carbon nanomaterials in plant biotechnology. Mater. Today Proc. 30: 340–345.

Maksimova, Y.G. 2019. Microorganisms and Carbon Nanotubes: Interaction and Applications (Review). Appl. Biochem. Microbiol. 55: 1–12.

Mallakpour, S. and Soltanian, S. 2016. Surface functionalization of carbon nanotubes: fabrication and applications. RSC Adv. 6: 116–135.

Martinez-Gil, M. and Ramos, C. 2018. Role of cyclic di-GMP in the bacterial virulence and evasion of the plant immunity. Curr. Issues Mol. Biol. 25: 199–222.

Mei, S., Fu, B., Su, X., Chen, H., Lin, H., Zheng, Z., Dai, C. and Yang, D.-P. 2022. Developing silk sericin-based and carbon dots reinforced bio-nanocomposite films and potential application to litchi fruit. LWT 164: 113630.

Miao, A.-J., Schwehr, K.A., Xu, C., Zhang, S.-J., Luo, Z., Quigg, A. and Santschi, P.H. 2009. The algal toxicity of silver engineered nanoparticles and detoxification by exopolymeric substances. Environ. Pollut. 157: 3034–3041.

Milewska-Hendel, A., Zubko, M., Karcz, J., Stróz, D. and Kurczyńska, E. 2017. Fate of neutral-charged gold nanoparticles in the roots of the *Hordeum vulgare* L. cultivar Karat. Sci. Rep. 7: 1–13.

Morales-Flórez, V. and Domínguez-Rodríguez, A. 2022. Mechanical properties of ceramics reinforced with allotropic forms of carbon. Prog. Mater. Sci. 128: 100966.

Mori, Y., Inoue, K., Ikeda, K., Nakayashiki, H., Higashimoto, C., Ohnishi, K., Kiba, A. and Hikichi, Y. 2016. The vascular plant-pathogenic bacterium *Ralstonia solanacearum* produces biofilms required for its virulence on the surfaces of tomato cells adjacent to intercellular spaces. Mol. Plant Pathol. 17: 890–902.

Mukherjee, A., Majumdar, S., Servin, A.D., Pagano, L., Dhankher, O.P. and White, J.C. 2016. Carbon nanomaterials in agriculture: A Critical review. Front. Plant Sci. 7: 172.

Nag, P., Paul, S., Shriti, S. and Das, S. 2022. Defence response in plants and animals against a common fungal pathogen, *Fusarium oxysporum*. Curr. Res. Microb. Sci. 3: 100135.

Omayio, D.O. and Ndombi, S.T. 2022. Tool for determining levels and classifying; host plant resistance, tolerance to stress, vigour and pathogen virulence in plants. Sci. African. 16: 1218.

Pandey, K., Lahiani, M.H., Hicks, V.K., Hudson, M.K., Green, J. and Khodakovskaya, M. 2018. Effects of carbon-based nanomaterials on seed germination, biomass accumulation and salt stress response of bioenergy crops. PLoS One 8: 1–17.

Pandey, S., Giri, V.P., Tripathi, A., Kumari, M., Narayan, S., Bhattacharya, A., Srivastava, S. and Mishra, A. 2020. Early blight disease management by herbal nanoemulsion in *Solanum lycopersicum* with bio-protective manner. Ind. Crops Prod. 150: 112421.

Papageorgiou, D.G., Kinloch, I.A. and Young, R.J. 2017. Mechanical properties of graphene and graphene-based nanocomposites. Prog. Mater. Sci. 90: 75–127.

Philippe, V., Neveen, A., Marwa, A. and Ahmad Basel, A.-Y. 2021. Occurrence of pesticide residues in fruits and vegetables for the eastern Mediterranean region and potential impact on public health. Food Control. 119: 107457.

Prakash, S. and Deswal, R. 2020. Analysis of temporally evolved nanoparticle-protein corona highlighted the potential ability of gold nanoparticles to stably interact with proteins and influence the major biochemical pathways in *Brassica juncea*. Plant Physiol. Biochem. 146: 143–156.

Qi, Y., Wang, M., Zhang, B., Liu, Y., Fan, J., Wang, Z., Song, L., Mohamed Abdul, P. and Zhang, H. 2022. Effects of natural rheum tanguticum on the cell wall integrity of resistant phytopathogenic *Pectobacterium carotovorum* subsp. Carotovorum. Molecules 27: 5291.

Rafique, M., Sadaf, I., Rafique, M.S. and Tahir, M.B. 2017. A review on green synthesis of silver nanoparticles and their applications. Artif. Cells, Nanomedicine Biotechnol. 45: 1272–1291.

Rahmani, N., Radjabian, T. and Soltani, B.M. 2020. Impacts of foliar exposure to multi-walled carbon nanotubes on physiological and molecular traits of *Salvia verticillata* L., as a medicinal plant. Plant Physiol. Biochem. 150: 27–38.

Ratnikova, T.A., Podila, R., Rao, A.M. and Taylor, A.G. 2015. Tomato seed coat Permeability to selected carbon nanomaterials and enhancement of germination and seedling growth. Sci. World J. 1: 61–75.

Saleem, H., Zaidi, S.J. and Center, N.A.A. 2021. Recent advancements in the nanomaterial application in concrete and its ecological impact. Materials 14: 6387.

Saleem, S., Bytešníková, Z., Richtera, L. and Pokluda, R. 2021. The effects of *Serendipita indica* and guanidine-modified nanomaterial on growth and development of cabbage seedlings and black spot infestation. Agriculture 11: 1295.

Samaddar, P., Ok, Y.S., Kim, K.-H., Kwon, E.E. and Tsang, D.C.W. 2018. Synthesis of nanomaterials from various wastes and their new age applications. J. Clean. Prod. 197: 1190–1209.

Samadi, S., Saharkhiz, M.J., Azizi, M., Samiei, L. and Ghorbanpour, M. 2020. Multi-walled carbon nanotubes stimulate growth, redox reactions and biosynthesis of antioxidant metabolites in *Thymus daenensis* celak. in vitro. Chemosphere 249: 126069.

Saravanan, G. 2022. Plants and phytochemical activity as botanical pesticides for sustainable agricultural crop production. J. Agric. Food Res. 9: 100345.

Sarkar, M.M., Pradhan, N., Subba, R., Saha, P. and Roy, S. 2022. Sugar-terminated carbon-nanodots stimulate osmolyte accumulation and ROS detoxification for the alleviation of salinity stress in *Vigna radiata*. Sci. Rep. 12: 17567.

Sarlak, N., Taherifar, A. and Salehi, F. 2014. Synthesis of nanopesticides by encapsulating pesticide nanoparticles using functionalized carbon nanotubes and application of new nanocomposite for plant disease treatment. J. Agric. Food Chem. 62: 4833–4838.

Schiavi, D., Francesconi, S., Taddei, A.R., Fortunati, E. and Balestra, G.M. 2022. Exploring cellulose nanocrystals obtained from olive tree wastes as sustainable crop protection tool against bacterial diseases. Sci. Rep. 12: 6149.

Schwab, F., Zhai, G., Kern, M., Turner, A., Schnoor, J.L. and Wiesner, M.R. 2016. Barriers, pathways and processes for uptake, translocation and accumulation of nanomaterials in plants—Critical review. Nanotoxicology 10: 257–278.

Shafiq, F., Iqbal, M., Ali, M. and Ashraf, M.A. 2021. Fullerenol regulates oxidative stress and tissue ionic homeostasis in spring wheat to improve net-primary productivity under salt-stress. Ecotoxicol. Environ. Saf. 211: 111901.

Sharifi, P., Bidabadi, S.S., Zaid, A. and Abdel Latef, A.A.H. 2021. Efficacy of multi-walled carbon nanotubes in regulating growth performance, total glutathione and redox state of *Calendula officinalis* L. cultivated on Pb and Cd polluted soil. Ecotoxicol. Environ. Saf. 213: 112051.

Siddiqui, Z.A., Parveen, A., Ahmad, L. and Hashem, A. 2019. Effects of graphene oxide and zinc oxide nanoparticles on growth, chlorophyll, carotenoids, proline contents and diseases of carrot. Sci. Hortic. 249: 374–382.

Soliman, S.A., Khaleil, M.M. and Metwally, R.A. 2022. Evaluation of the antifungal activity of *Bacillus amyloliquefaciens* and *B. velezensis* and characterization of the bioactive secondary metabolites produced against plant pathogenic fungi. Biology 11: 1390.

Sorcia-Morales, M., Gómez-Merino, F.C., Sánchez-Segura, L., Spinoso-Castillo, J.L. and Bello-Bello, J.J. 2021. Multi-walled carbon nanotubes improved development during *in vitro* multiplication of sugarcane (*Saccharum* spp.) in a semi-automated bioreactor. Plants 10: 2015.

Syahirah Kamarudin, N., Jusoh, R., Dina Setiabudi, H., Fateha Sukor, N. and Haslinda Shariffuddin, J. 2021. Potential nanomaterials application in wastewater treatment: Physical, chemical and biological approaches. Mater. Today Proc. 42: 107–114.

Tiwari, S.K., Bystrzejewski, M., De Adhikari, A., Huczko, A. and Wang, N. 2022. Methods for the conversion of biomass waste into value-added carbon nanomaterials: Recent progress and applications. Prog. Energy Combust. Sci. 92: 101023.

Tripathi, D.K., Tripathi, A., Shweta, Singh, S., Singh, Y., Vishwakarma, K., Yadav, G., Sharma, S., Singh, V.K., Mishra, R.K., Upadhyay, R.G., Dubey, N.K., Lee, Y. and Chauhan, D.K. 2017. Uptake, accumulation and toxicity of silver nanoparticle in autotrophic plants, and heterotrophic microbes: A concentric review. Front. Microbiol. 8: 1–16.

Tripathi, S., Sonkar, S.K. and Sarkar, S. 2011. Growth stimulation of gram (*Cicer arietinum*) plant by water soluble carbon nanotubes. Nanoscale 3: 1176.

Verde-Gómez, Y., Montiel-Macías, E., Valenzuela-Muñiz, A.M., Alonso-Lemus, I., Miki-Yoshida, M., Zaghib, K., Brodusch, N. and Gauvin, R. 2022. Structural study of sulfur-added carbon nanohorns. Materials 15: 3412.

Verma, S.K., Das, A.K., Gantait, S., Kumar, V. and Gurel, E. 2019. Applications of carbon nanomaterials in the plant system: A perspective view on the pros and cons. Sci. Total Environ. 667: 485–499.

Villagarcia, H., Dervishi, E., De Silva, K., Biris, A.S. and Khodakovskaya, M.V. 2012. Surface chemistry of carbon nanotubes impacts the growth and expression of water channel protein in tomato plants. Small. 8: 2328–2334.

Vishwanath, R. and Negi, B. 2021. Conventional and green methods of synthesis of silver nanoparticles and their antimicrobial properties. Curr. Res. Green Sustain. Chem. 4: 100205.

Wagle, S.A., R, H., Varadarajan, V. and Kotecha, K. 2022. A new compact method based on a convolutional neural network for classification and validation of tomato plant disease. Electronics 11: 2994.

Wang, H., Zhang, M., Song, Y., Li, H., Huang, H., Shao, M., Liu, Y. and Kang, Z. 2018. Carbon dots promote the growth and photosynthesis of mung bean sprouts. Carbon. 136: 94–102.

Wang, Q., Li, C., Wang, Y. and Que, X. 2019. Phytotoxicity of graphene family nanomaterials and its mechanisms: A review. Front. Chem. 7: 292.

Wang, X., Liu, X. and Han, H. 2013. Evaluation of antibacterial effects of carbon nanomaterials against copper-resistant *Ralstonia solanacearum*. Colloids surfaces. Biointerfaces 103: 136–142.

Wei, S.J., Al Riza, D.F. and Nugroho, H. 2022. Comparative study on the performance of deep learning implementation in the edge computing: Case study on the plant leaf disease identification. J. Agric. Food Res. 10: 100389.

Weldegebrieal, G.K. 2020. Synthesis method, antibacterial and photocatalytic activity of ZnO nanoparticles for azo dyes in wastewater treatment: A review. Inorg. Chem. Commun. 120: 108140.

Worrall, E., Hamid, A., Mody, K., Mitter, N. and Pappu, H. 2018. Nanotechnology for plant disease management. Agronomy 8: 1–24.

Yan, S., Zhao, L., Li, H., Zhang, Q., Tan, J., Huang, M., He, S. and Li, L. 2013. Single-walled carbon nanotubes selectively influence maize root tissue development accompanied by the change in the related gene expression. J. Hazard. Mater. 3: 246–247.

Yan, Y., Chen, J., Li, N., Tian, J., Li, K., Jiang, J., Liu, J., Tian, Q. and Chen, P. 2018. Systematic bandgap engineering of graphene quantum dots and applications for photocatalytic water splitting and CO_2 reduction. ACS Nano. 12: 3523–3532.

Younes, N.A., Dawood, M.F.A. and Wardany, A.A. 2019. Biosafety assessment of graphene nanosheets on leaf ultrastructure, physiological and yield traits of *Capsicum annuum* L. and *Solanum melongena* L. Chemosphere 228: 318–327.

Zemtsova, E.G., Arbenin, A.Y., Sidorov, Y.V., Morozov, N.F., Korusenko, P.M., Semenov, B.N. and Smirnov, V.M. 2022. The use of carbon-containing compounds to prepare functional and structural composite materials: A review. Appl. Sci. 12: 9945.

Zhao, J. and M.H. Stenzel. 2018. Entry of nanoparticles into cells: The importance of nanoparticle properties. Polym. Chem. 9: 259–272.

Zhao, F., Xin, X., Cao, Y., Su, D., Ji, P., Zhu, Z. and He, Z. 2021. Use of carbon nanoparticles to improve soil fertility, crop growth and nutrient uptake by corn (*Zea mays* L.). Nanomaterials 11: 2717.

Zupančič, D. and Veranič, P. 2022. Nanodiamonds as possible tools for improved management of bladder cancer and bacterial cystitis. Int. J. Mol. Sci. 23: 8183.

Chapter 18

Antiviral Activity of Nanoparticles in Plants

Mohsen Mohamed Elsharkawy

Introduction

The world's population has grown rapidly over the last years, and it is predicted to reach more than 8 billion by the year 2030. This makes it challenging for scientists to achieve the highest agricultural output. As a result, the use of synthetic agrochemicals has multiplied dramatically in order to meet the expanding need for food. Nearly all agricultural chemicals are run off into the environment and contaminate agricultural crops; this amounts to an estimated 4.6 million tons worldwide. Pesticides are a class of agrochemicals that have long been employed to protect crops against the severe impact of phytopathogens. Almost all crops are affected by plant diseases, which forces farmers to overuse greater and greater pesticides in order to maximize agricultural yields (Zhang et al. 2011a; Zhang 2018). As a result of pesticide usage, crop losses have decreased by 40% (Zhang et al. 2011a, b). Pesticide use is on the increase all over the world. Agrochemicals are important in agriculture, but their non-degradable and harmful nature has resulted in an irreparable negative impact on the environment as a result of careless and injudicious usage (Kumaraswamy et al. 2018). Almost all pesticides are not completely absorbed by plants and instead leak into the soil and groundwater, where they ultimately accumulate in living beings as well (Dietz and Herth 2011; Marutescu et al. 2017). Pesticide usage across the world has also led to the extinction of many species of plants and animals (Kumar et al. 2013). Pesticide usage has also resulted in several human and animal diseases, reduced human fertility, and lowered cognitive scores in recent years (Chen et al. 2004). The rise of agrochemical-resistant plant diseases has also become a significant concern (Xing et al. 2017). As a result, either novel agrochemicals have been produced or greater dosages of the old ones have been employed, raising costs and hastening the emergence of new plant diseases.

Plant diseases and pests generate large qualitative and quantitative yield losses, making attempts to maintain food security more complicated (Savary et al. 2019). According to recent research, diseases, and pests are linked to yield and quality losses in five important worldwide crops (maize, wheat, rice, potato, and soybean), with production losses ranging from 10% to 50%. Plant viruses pose a severe danger to global food security by launching an attack on crop yields throughout the globe (Kang et al. 2005; Soosaar et al. 2005). There are about 900 different types of plant viruses that may infect more than 700 different crop species (Cai et al. 2019; Chauhan et al. 2019). Phytoviruses of agricultural crops, which account for roughly half of all known plant diseases and cause major crop losses globally, have an estimated yearly economic effect of over \$30 billion (Sastry and Zitter 2014; Bernardo et al. 2018). According to the stage of infection, the strain of the virus, environment,

Agricultural Botany Department, Faculty of Agriculture, Kafrelsheikh University, 33516 Kafr El-Sheikh, Egypt.
Email: mohsen.abdelrahman@agr.kfs.edu.eg

host type, speed of replication, and viral concentration, a virus may have a major influence on a plant (Hill and Whitham 2014). Viral infections may produce symptoms, such as drying of leaves, mottling, distortion of fruit, stunting, necrosis, and even death of plants under favorable conditions (Cai et al. 2019). The fundamental issue in virus control is that there is no cure for viruses and the only alternatives are prevention or avoidance of contamination (Almasi and Almasi 2018; Xiang et al. 2020). Among the different aided ways of viral transmission in the environment, insect vectors play a relatively major role, spreading more than 70% of all plant-infecting viruses (Hogenhout et al. 2008; Jones and Naidu 2019). Even though plant viruses have simple RNA or DNA genomes, they are hard to prevent and among the most effective agricultural pathogens due to factors, such as dynamic genome diversity, genome structure, transmission by a wide range of vectors, rapid evolution, and wide adaptation ability to various climatic conditions (Lauring et al. 2013; French and Holmes 2020). Plant viral outbreaks need the use of long-term and sustainable management approaches. The advent of nanoscience and the use of nanotechnological instruments has increased expectations for the delivery of next-generation agrochemicals that are safe for the environment and powerful at low dosages. A wide variety of nanostructured materials, including nanometals, carbon nanomaterials, metallic oxides, metalloids, dendrimers, and functionalized liposomes are available as engineered nanoparticles (ENPs) (Elmer and White 2018). It is possible to divide these materials into metallic and non-metallic nanoparticles (Yih and Al-Fandi 2006).

The significance of nanoscale materials is attained from their intrinsic features (optically active, surface area, reactivity toughness, chemically reactive, and mechanically strong). Applied nanobiotechnology includes a wide range of sectors, such as agriculture and medicine as well as biotechnology at the nanoscale (Asghari et al. 2016). Biosensors, biological separation, molecular imaging, and cancer treatment are all examples of nanoscale uses in biology (Morais et al. 2014). Since nanoparticles were effective for *in vitro* nanomedicine, there has been a great deal of interest in their usage in agro-nanotechnology (Tripathi et al. 2018). Nanostructured materials might be used to create next-generation insecticides that can be used in a controlled way when necessary. Crop production may benefit from its usage because it reduces the use of pesticides, water, and nutrient management. Nanomaterials improve the absorption of nutrients during fertilization and increase yield via pest and nutrient control (Prasad et al. 2017). Nanomaterials have gained attention as agrichemicals because of their huge surface area to volume ratio, lower size, tunability, high reactivity, and amenability to coating. Nanoscale materials are emerging as new chemicals for the treatment of bacterial, fungal, and viral infections in animals and plants, respectively (Fernando et al. 2018; Parveen et al. 2018; Cai et al. 2019). In the last decade, a rising number of studies have shown that plant pathology scholars are more interested in the antiviral effect of ENPs (Eugene and Zholobak 2016; El-Sawy et al. 2017b; El-Dougdoug et al. 2018; Cai et al. 2019; Elsharkawy and Derbalah 2019). Inorganic and organic bioactive substances that have yet to be investigated may be combined with nanotechnology to create new nano-based products for use in agriculture, particularly in the fight against viral disease. These materials have shown benefits over traditional plant disease control systems, and they have crucial features that result in potential inputs, increased efficiency, and less eco-toxicity (Fu et al. 2020). Nanoparticles have been examined for their potential as nanofertilizers, biostimulants, nanocarriers, and antimicrobial agents. However, the mechanisms by which nanoparticles interact with plant viruses have yet to be fully described (Elmer and White 2018; Fu et al. 2020).

Virus Diseases Control

Crop losses have been exacerbated as a result of the virus infection because of variations in cultural practices and climate changes (Jones 2016). Virus infections have grown in frequency and importance, necessitating the adoption of proactive management strategies to keep them under control. Insect vector populations may be reduced by the application of chemical compounds or biological control in the greenhouses because of viruses' need for insect vectors for survival, transmission, and proliferation (Sarwar 2020; Ziegler-Graff 2020). Other measures include the removal of viral sources, interference with a vector's impact by modifying the attraction of insects to colors and interfering with the transmission process via the application of mineral oils (Sarwar 2020). Conventional and innovative procedures have also been used in the laboratory to control viral infections in crops (Chauhan 2019). Cryotherapy, thermal therapy, and chemotherapy are among the most common treatments used in traditional control methods. Tissue cultures and a decrease in the initial inoculum restrict the use of these procedures, which have significant drawbacks. Advanced techniques include RNA silencing, transgenic plants, cross-protection, protein-protein interaction, and gene pyramiding (Chauhan 2019). Instead of only preventing or delaying infection, the virus may be eradicated from plant tissues using the earlier method, although this procedure interferes with vectors, necessitates the use of attenuated strains and resistant or tolerant cultivars, and is very costly (Bragard et al. 2013; Maruthi et al. 2019; Pechinger et al. 2019).

Antivirals may be used to cure various animal diseases, and they can also be used to treat viruses in plant tissue cultures (Wang et al. 2018). *Potato leafroll virus* (PLRV), *Grapevine vitivirus*, and *Sugarcane mosaic virus* (SCMV) are among the pathogens that have been successfully treated (Panattoni et al. 2008; Singh et al. 2015). The drawbacks of chemical antivirals include a lack of efficacy *in vitro*, a lack of specificity for viral crop diseases, and the fact that they all have distinct mechanisms of action (Chauhan et al. 2019). Sophisticated approaches include RNA interference (RNAi)-mediated response, which uses silencing through dsRNA. The development of resistant cultivars or a cross-protection approach, which consists of a systematic infection with a second virus to promote resistance to the target virus, is also used to manage viral infections in plants. Since resistant varieties are the most cost-efficient and beneficial way to minimize crop losses, using them is a significant strategy; nevertheless, the creation of highly effective and long-lasting crop cultivars that are resistant to viruses is necessary (Zhao et al. 2020). A further drawback of gene pyramiding is that it has an epistatic effect and significant costs since it entails the accumulation of several genes (Chauhan et al. 2019). The antiviral activities of metallic nanoparticles are not fully known; however, the available research might give evidence of the involved processes. Metallic nanoparticles have been shown to have antiviral activity both *in vitro* and *in vivo* with a variety of plant species (Table 1). This activity is effective against positive-and negative-sense single-stranded RNA viruses. The antiviral activity of ZnONPs and SiO_2NPs against *Tobacco mosaic virus* (TMV) was investigated and documented (Cai et al. 2019). The findings of this research point to the direct inactivation of TMV by metallic nanoparticles owing to the contact with envelope glycoproteins. This interaction causes direct damage to TMV shell proteins, which in turn causes TMV aggregation and even breakage. TMV particles that have been treated with Fe_3O_4NPs still exhibit aggregation and fracture. In addition, direct interactions between Fe_3O_4NPs and TMV particles result in higher particle sizes (Cai et al. 2020). AgNPs also have the ability to attach to the coat proteins of viruses, including the *Potato virus Y* and *Tomato mosaic* virus (El-Dougdoug et al. 2018a). AuNPs caused damage to the virus-like particles (VLPs) of the *Barley yellow dwarf virus-PAV* (Alkubaisi and Aref 2017). Using transmission electron microscopy (TEM), the researchers observed deteriorated VLPs decorated with AuNPs, in addition to destroyed and vanished particles. *In vivo* experiments on the efficiency of metallic nanoparticles against viral infections in the Solanaceae, Fabaceae, Asteraceae, Cucurbitaceae, and Poaceae families have also been performed (Elbeshehy et al. 2015; Elsharkawy and Mousa 2015; Eugene and Zholobak 2016; Elazzazy et al. 2017; El-shazly et al. 2017; Elsharkawy and Derbalah 2019). The use of metallic nanoparticles in plant cultivation disrupts the normal viral life cycle in a variety of ways, which ultimately prevents virus replication.

In addition, nanoparticles have an impact on disorders that are caused by viruses. *Nicotiana tabacum* and *Datura stramonium* leaf segments infected with TMV are penetrated via the vascular system by CeO_2NPs, which suppress viral reproduction (Eugene and Zholobak 2016). Reduced *Tomato bushy stunt virus* concentration and disease severity are two benefits of using graphene-based silver nanoparticles in lettuce (Elazzazy et al. 2017). There is antiviral action in both foliar spray and soil treatment of nanoparticles. Soil treatment with SiO_2 nanoparticles resulted in a considerable

Table 1. Antiviral activity of different nanoparticles.

Nanoparticle	Host	Virus	References
Cerium oxide nanoparticles	Datura and tobacco	Tobacco mosaic virus (TMV)	Eugene et al. 2016
Graphene oxide-silver nanoparticles	Lettuce	Tomato bushy stunt virus (TBSV)	Elazzazy et al. 2017
Gold nanoparticles	Barley	Barley yellow mosaic virus (BaYMV)	Aref et al. 2012
Iron oxide nanoparticles	Tobacco	Turnip mosaic virus (TuMV)	Hao et al. 2018
Nickel oxide nanoparticles	Cucumber	Cucumber mosaic virus (CMV)	Derbalah and Elsharkawy 2019
Silver nanoparticles	Cluster beans	Sunhemp rosette virus (SHRV)	Jain 2014
	Faba bean	Bean yellow mosaic virus (BYMV)	Elbeshehy 2015
	Potato	Potato virus Y (PVY)	El-shazly et al. 2017
	Tomato	Tomato mosaic virus (ToMV)	El-Dougdoug et al. 2018
	Tomato	Potato virus Y (PVY)	El-Dougdoug et al. 2018
Schiff base nanosilver	Tobacco	Tobacco mosaic virus (TMV)	Wang et al. 2016
Titanium dioxide nanoparticles	Faba bean	Broad bean stain virus (BBSV)	Elsharkawy and Derbalah 2019
	Tobacco	Turnip mosaic virus (TuMV)	Hao et al. 2018
Copper oxide nanoparticles	Squash	Zucchini yellow mosaic virus (ZYMV)	Derbalah et al. 2022
Silicon dioxide nanoparticles	Cucumber	Papaya ringspot virus (PRSV)	Elsharkawy and Mousa 2015
	Tomato	Tomato yellow leaf curl virus (TYLCV)	El-Sawy et al. 2017
	Tobacco	Tobacco mosaic virus (TMV)	Cai et al. 2019
Iron oxide nanoparticles	Tobacco	Turnip mosaic virus (TuMV)	Hao et al. 2018
	Tobacco	Tobacco mosaic virus (TMV)	Cai et al. 2019
Zinc oxide nanoparticles	Tobacco	Tobacco mosaic virus (TMV)	Cai et al. 2019
	Eggplant	Cucumber mosaic virus (CMV)	El-Sawy et al. 2017

decrease in disease symptoms in cucumber plants infected with *Papaya ringspot virus* (Elsharkawy and Mousa 2015). After the viral infection has occurred, the antiviral effects and responses of the plants vary depending on the time of the treatment application. *Bacillus persicus*, *B. pumilus*, and *B. licheniformis* bacteria generated AgNPs had positive effects on the *Bean yellow mosaic virus* (BYMV) (Elbeshehy et al. 2015). To avoid the harmful side effects of virus, a post-infection treatment was applied 24 hours after the initial infection was performed. AgNP-treated plants infected with BYMV concurrently displayed modest symptoms, while a 72-hours pre-infection treatment had little influence on viral titer or disease severity. AgNPs sprayed on *Chenopodium amaranticolor* plants had the greatest antiviral efficacy against the *Tomato spot wilt virus*. There was no difference in the infection rate between plants treated after inoculation and those that were sprayed before inoculation (Shafie et al. 2018). On the other hand, weak infection was exhibited in the plants sprayed after inoculation, whereas the plants treated before inoculation had a weak inhibitory effect (El-shazly et al. 2017). In contrast, spraying AgNPs on *Cymopsis tetragonoloba* leaves completely suppressed the disease and inactivated *Sunhemp rosette virus* propagation (Jain 2014). Additional investigations revealed antiviral activity when nanoparticles were treated prior to inoculation. Cucumber and tomato plants treated with nickel oxide nanoparticles and silica nanoparticles one day before *Tomato yellow leaf curl virus* (TYLCV) inoculation were found to have fewer severe viral symptoms (El-Sawy et al. 2017a). Antiviral interaction between AgNPs and salicylic acid (SA) applied before 3 and 7 days of virus inoculation was found to be more effective than either AgNPs or SA used alone against *Tomato bushy stunt virus* (El-shazly et al. 2017). Recent research showed that metallic nanoparticles stimulate the plant's defense mechanisms, leading to a reduction in virus concentration (Hao et al. 2018; Cai et al. 2019; Derbalah and Elsharkawy 2019; Cai et al. 2020). Squash plants treated with copper oxide nanostructures (CONS) either as a soil drench or a foliar spray showed a substantial decrease in *Zucchini yellow mosaic virus* (ZYMV) symptoms, including severe mosaic and distorted leaves, 14 days after infection, compared to the untreated control (Derbalah et al. 2022). Treatment of squash plants with CONS significantly decreased ZYMV concentrations at 1, 2, and 3 weeks post-infection, in comparison to the untreated control. Soil drenching squash with CONS reduced ZYMV severity by 85%, 85%, and 75% after 1, 2, and 3 weeks of inoculation, respectively, whereas applying CONS as a foliar spray reduced ZYMV severity by 76%, 86%, and 88% after 1, 2, and 3 weeks of inoculation, respectively (Derbalah et al. 2022). Compared to untreated control plants, which displayed significant mosaic symptoms with tiny malformed leaves at 14 days after inoculation, titanium dioxide nanostructures (TDNS)-treated faba bean plants showed a remarkable decrease in BBSV (*Broad bean stain virus*) symptoms (Elsharkawy and Derbalah 2018). Two weeks after BBSV infection, faba bean plants treated with titanium dioxide exhibited much less disease severity than untreated plants. In addition, the foliar spray application of the TDNS resulted in a higher reduction in disease severity in the faba bean plants compared to the soil-drenched plants. Compared to untreated plants, TDNS-treated faba bean plants drastically decreased BBSV accumulation 2 weeks after inoculation. The titer of BBSV on faba bean plants treated with TDNS through foliar spray was less than on plants treated with soil drench (Table 1). Fe_2O_3 or TiO_2NPs-treated tobacco demonstrated antiviral effects, reduced the quantity of *Turnip mosaic virus* (TuMV) protein and activated defensive pathways (Hao et al. 2018). As a result of the nanoparticles attaching to *Tomato mosaic virus* (ToMV) and *Potato virus Y* (PVY), the disease severity and viral concentration were decreased at 7 days before inoculation with ToMV and PVY (El-Dougdoug et al. 2018). Plant leaves treated for 12 days with Fe_3O_4NP demonstrated great resistance to TMV, which is even more powerful than the commonly used lentinan (Cai et al. 2020). Several compounds with metals, such as chitosan Schiff base nanosilver, greatly decreased the number of lesions on TMV-infected tobacco leaves, generating TMV resistance and increasing immunity (Wang et al. 2016). Consequently, scientists are looking to inorganic and organic materials for the synthesis of nanomaterials (NMs) with biodegradability, biocompatibility, extensive biological activity, and ecological safety properties (Shukla et al. 2013; Kashyap et al. 2015). Chitosan—a hetero-aminopolysaccharide that may be readily produced from shrimp waste, crab shells, and the cell wall of the fungus—is now being investigated in this regard (Malerba and Cerana 2016). The smaller size, increased surface area, and cationic nature of chitosan nanoparticles make them ideal for biological applications. Furthermore, because of the abundance of functional groups in their structures, they provide ideal blending materials for a wide range of organic and inorganic compounds (Choudhary et al. 2019a, b). Chitosan nanoparticles may be produced using chitosan alone or in combination with other inorganic and organic compounds. Improved effectiveness and less environmental impact might be achieved by using chitosan-based nanocomposites, which are designed to release active chemicals slowly, systemically, targeted, and protected (Saharan et al. 2015; Choudhary et al. 2017a, b). Nanochitosan-functionalized with other substances might eventually lead to precision farming at a lower cost and provide an intelligent biochitosan nanoagri-input.

Interaction Mechanisms Between Nanoparticles and Plant Viruses

As previously stated, the use of nanoparticles to control plant viruses is a new strategy with little understanding of the underlying pathways. Researchers have found that nanoparticles can be used as antivirals in plants because they inhibit not only the replication of viruses but also activate the plant defense mechanisms that lead to the plant's immune response and growth response in which antioxidants, plant hormones, and resistance genes are involved. Hence, we summarize an in-depth review of the known and possible mechanisms of nanoparticle-virus interactions. In addition to emphasizing the role of nanoparticles against virus-transmitting insect vectors, we discuss the challenges that may restrict the effective implementation of nanoparticles against these pathogens and how nanomaterials may solve these obstacles to become a practical and efficient plant virus management method.

Gold nanoparticles (AuNPs) were shown to have antiviral activity *in vitro* against the *Barley yellow mosaic virus*. AuNPs treatment caused the total dissociation of virus particles (Aref et al. 2012). Although this finding depends on *in vitro* experiments, the effectiveness of AuNPs in a growing plant is yet unconfirmed. Metal nanoparticles and metal oxides have been extensively studied for their ability to combat plant viruses, as well as for their potential benefits for plant health (Elsharkawy and Derbalah 2019; Derbalah et al. 2022).

Direct Interactions Between Viral Particles and Nanoparticles

The antiviral mechanisms of nanoparticles have recently been studied in various recent research. Before viral infection, carbon-based nanoparticles suppressed TMV multiplication in *Nicotiana benthamiana* and prevented systemic transmission to uninfected tissues (Adeel et al. 2021). Antiviral action was not conducted on the basis of concurrent infection or even recovery following infection. Similar research found that SiO_2 nanoparticles and zinc oxide nanoparticles were effective against TMV *in vitro* (Cai et al. 2019). The direct antiviral activity of TDNS against BBSV was tested by spraying a mixture of TDNS and BBSV on faba bean plants (Elsharkawy and Derbalah 2018). Compared to untreated plants, sprayed faba bean plants showed a considerable decrease in BBSV severity. Also, compared to untreated plants, TDNS-treated faba bean plants had a considerably lower viral titer as assessed by ELISA. TMV aggregation and fast deactivation of viral particles were attributed to structural disruption caused by the interaction of nanoparticles with viral capsid proteins. Similarly, TuMV particles responded to treatment with Fe_2O_3 nanoparticles (Hao et al. 2018). They asserted that phytohormonal activity was a major factor in the activation of the immune system and the subsequent antiviral activity of nanoparticles. However, the actual process by which phytohormones were activated is still unknown. Furthermore, the absence of ionic and bulk limitations on the nanoparticles affects our comprehension of the nanoparticles' mechanistic effect. PVY and ToMV coat proteins were shown to be bindable to silver nanoparticles in virus-infected plants (El-Dougdoug et al. 2018). Moreover, foliar application of AgNPs (at 50 mg/L) on tomato plants cultivated in greenhouses decreased ToMV-induced lesions. The antioxidant enzymes (polyphenol oxidase and peroxidase) and total soluble protein levels improved the potential of treatment (El-Dougdoug et al. 2018).

However, challenges about AgNPs-induced differential phytotoxic effects were not fully addressed. Another research found that AgNPs stimulated plant development even after viral infection, despite the fact that they could not combat the virus (Adeel et al. 2021). AuNPs were shown to degrade and disintegrate viral particles, resulting in a decrease in total viral abundance (Alkubaisi and Aref 2017). There are just a few research that has shown promising outcomes; however, overall the number of studies published lacks robustness, and mechanistic explanations are either elusive or simply hypothesized in certain circumstances. Using non-nanoscale controls would help mechanistic understanding and compare biochemical, morpho-physiological, and molecular responses of nanoparticles-treated plants before and after viral infection. It is also conceivable that *in vitro*-treated viral particles may respond differently to the same treatment in a living plant. The temporal dynamics of infection and efficient treatment methods are still unexplored. To maximize the potential of nanoparticles-based phytovirus control techniques, all of these variables must be recognized at a molecular level. No research has examined these interactions across the whole life cycle of any host.

Indirect Interactions Between Viral Particles and Nanoparticles

Nanoparticles may develop indirect interactions with plant pathogenic viruses, in addition to direct nanoparticle-plant virus interactions (Figure 1). Plant growth and development may be stimulated via improved photosynthetic processes or enhanced defensive responses (Hao et al. 2018; Adeel et al. 2021). The application of nanoparticles has been linked to an increase in phytohormonal levels [abscisic acid (ABA), SA, zeatin riboside (ZR), jasmonic acid (JA), and brassinosteroids (BR)], antioxidant production, reactive oxygen species (ROS), and increased expression of genes involved in JA, ABA and SA mediated defense signaling pathways (Hao et al. 2018; Cai et al. 2019; Adeel et al. 2021). A precise and well-timed concentration of nanoparticles may also have a good effect on the growth and development of plants (Rastogi et al. 2017; Elmer and White 2018; Adeel et al. 2021). Treatment with the nanoparticles might conceivably have therapeutic benefits on plants against viral infection. However, the dosage strategy and the duration of antiviral activity are still unknown.

Increases in ROS are a common stress response in plants. These ROS help to restrict pathogen entry and spread, as well as trigger systemic and local defensive responses, such as activation of pathogenesis-related (PR) genes (Cai et al. 2019). Plants produce ROS in response to a wide range of biotic and abiotic stimuli. The plant's antioxidant system counteracts the negative effects of oxidants by generating a defense mechanism. It is superoxide dismutase, the first line of defense, that breaks down oxygen molecules into hydrogen peroxide and water (Soares et al. 2018; Tan et al. 2018). Other enzymes that make up the antioxidant system include catalase (CAT), guiacol peroxidase, and ascorbate peroxidase (Tan et al. 2018). Nanoparticle treatment has been proven to suppress oxidative stress and activate the antioxidant system in different crops. Hydrogen peroxide was aggregated by SiO_2 and ZnO nanoparticles, even when the virus was not found. However, Fe_3O_4, ZnO, and SiO_2 nanoparticles increased the activity of CAT and peroxidase enzymes (Cai et al. 2019; Cai et al. 2020). The presence of the virus induced the expression of several enzymes. One day after PRSV inoculation, SiO_2 nanoparticles-treated cucumber plants expressed *peroxidase* and *phenylalanine ammonia-lyase* genes (Elsharkawy and Mousa 2015). After 4 days of infection with cucumber mosaic virus (CMV), plants treated with NiO nanoparticles showed an increase in *peroxidase* gene expression (Derbalah and Elsharkawy 2019). There was a considerable rise in the activity of POD and PPO in tomato plants treated with Ag nanoparticles and inoculated with ToMV or PVY (El-Dougdoug et al. 2018).

Resistance in plants is mediated by a complicated system of defensive mechanisms using JA, SA, and ET pathways to induce appropriate responses. During various plant hormone signals, crosstalk

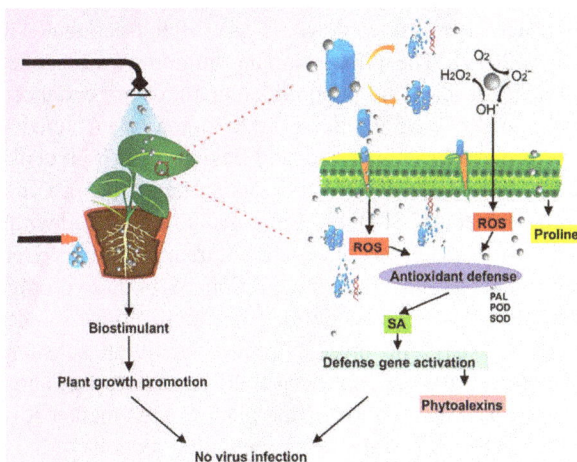

Figure 1. Antiviral mechanism of metal and metal oxide nanoparticles in plants adapted from Vargas-Hernandez et al. (2022) (CC BY 4.0).

mediates the balance between plant defense and growth. Various kinds of stress affect different hormonal pathways in different ways; for example, nanoparticles affect plant hormone balance (Rastogi et al. 2017). Transcript levels of genes associated with SA and JA signaling are increased in *Arabidopsis thaliana* after exposure to CuO nanoparticles (Landa et al. 2017). ZR, abscisic acid, and Brassinosteroid phytohormone levels are affected by TMV-infected tobacco containing Fe_2O_3 and TiO_2 nanoparticles (Hao et al. 2018). Tobacco plants infected with TMV and treated with Fe_2O_3 and TiO_2 nanoparticles showed increased levels of the phytohormones abscisic acid, zeratin riboside, and Brassinosteroid (Hao et al. 2018). Increased zeatin ribose levels were seen in TuMV-infected *N. benthamiana* leaves that were treated with TiO_2 nanoparticles. The proline content of diseased tomato plants treated with Ag nanoparticles was also increased (El-Dougdoug et al. 2018).

PR proteins are both components of a plant's innate immune system and diagnostic molecular markers of defensive signaling pathways. PR proteins play an important role in protecting plants against pathogens. Uninfected *Nicotiana benthamiana* plants that were treated with SiO_2 and ZnO nanoparticles had an increase in the transcription of SA-responsive *PR1* and *PR2* genes (Cai et al. 2019), and the same result was observed by the influence of Fe_3O_4 nanoparticles (Cai et al. 2020). Cucumber treated with SiO_2 nanoparticles also induces PR1 transcription when infected with PRSV (1 dpi) (Elsharkawy and Mousa 2015). Two days after BBSV inoculation, a pathogenesis-related gene (*PR1*) was expressed in faba bean plants that had been treated with TDNS by foliar spray or soil drench (Elsharkawy and Derbalah 2018). The resistance gene expression in faba bean plants was not significantly different between spray or soil drench treatments of TDNS. During CMV infection, cucumber plants similarly increased PR1 after 2 and 4 days (Eugene et al. 2016). Whether applied as a foliar spray or a soil drench, CONS, and SA caused an increase in the transcription levels of PR genes [peroxidase (POX)] and catalase (CAT) in squash plants (Derbalah et al. 2022). After one day of inoculation with ZYMV, squash plant *CAT*, and *POX* gene expressions were significantly higher than control and remained raised at two days post-inoculation.

Several investigations have been conducted on the influence of metallic nanoparticles on plant secondary metabolism; however, the mechanism is still unclear. ROS generation caused by nanoparticle contact could potentially interact with secondary plant metabolism (Marslin et al. 2017). The enzyme phenylalanine ammonia-lyase serves as a link between different metabolisms in plants. Non-oxidative phenylalanine deamination is catalyzed by phenylalanine ammonia-lyase, and trans-cinnamate is the end product. The phenylpropanoid pathway also contributes to SA production. It is a SA/JA-inducible enzyme that is also generated in response to abiotic and biotic stress. Plants infected with PRSV showed an increase in JA-inducible phenylalanine ammonia-lyase after treatment with SiO_2 nanoparticles (Elsharkawy and Mousa 2015). Similar results were obtained in cucumbers infected with CMV after treatment with NiO nanoparticles (Derbalah and Elsharkawy 2019). The phenylalanine ammonia-lyase gene in *Brassica rapa* is upregulated by the CuO nanoparticles and phenolic compounds are enhanced in the plant (Chung et al. 2019). The application of ZnO nanoparticles led to increased PAL enzyme activity in tobacco plants and the production of phenolic compounds and flavonoids (Tirani et al. 2019). A foliar spray treatment of TiO_2 and FeO_3 nanoparticles resulted in a significant increase in fresh and dry weights of TuMV-infected tobacco (Hao et al. 2018). PVY-infected tubers treated with Ag nanoparticles had better quality attributes than infected plants that were not treated (El-shazly et al. 2017).

The cheap cost, excellent stability, and integration of both hydrophilic and hydrophobic chemicals make nanoparticles ideal for a wide range of applications. Since new or re-emerging viruses develop on a regular basis, the present trend of antiviral substances seems promising (Galdiero et al. 2011). It is essential to take into consideration the dosage and the time of treatment, as well as monitor the phenological condition of the plant to see whether it is showing the desired effects. Nanoparticles should also be established in the edible sections of the plant as well as at the nutraceutical level so that they are not harmful and may also have a favorable impact on customers. Nanoparticles can accumulate in fruits and seeds despite the fact that there are no studies about the length of time that nanoparticles stay in the plants. Nanoparticles take a long time to biodegrade and they can cause damage to potentially vulnerable biota. The development of an application management system to reduce the potential for toxicity in the involved biological communities should be considered an important responsibility.

Utilization of Nanoparticles as Nanocarriers for Antiviral Protection

The use of nanoparticles as delivery methods for nucleic acids to control plant virus activity via RNA interference is an interesting research field. Anti-parasitic and anti-pathogenic nucleic acids may be prevented from entering cells through dsRNA molecules, which regulate mRNA stability and translation (Hannon 2002; Setten et al. 2019). Double hydroxide bioclay nanosheets (nontoxic and biodegradable) were shown to be an excellent delivery method for dsRNA. An array of dsRNA molecules aggregate in plant cells, causing RNA interference (RNAi). Additionally, suppression of the CMV and *Pepper mild mottle virus* (PMMoV) was demonstrated to be efficient for at least 20 days (Mitter et al. 2017). Recently, the same team revealed that this method might also protect plants from virus vectors (Worrall et al. 2019). *Bean common mosaic virus* (BCMV) transmission by aphids was prevented at 5 days after topical treatment of dsRNA molecules using bioclay on tobacco and cowpea. More research is necessary to clarify virus resistance to this technique, as well as the dose and timing. Moreover, the carrier's fate must be understood and recognized. Foliar application of naked dsRNA was difficult because of their quick uptake by plant regulatory systems and the need to balance with degradation by external variables. In addition to demonstrating stable dsRNA expression (up to 30 days after application) and the breakdown of layered double hydroxide, the aforementioned approach also indicates the decomposition of this compound. Plant-virus-vector systems in which insect species transmit many potentially harmful viruses to host plants have not yet been evaluated for this strategy despite the great positive results obtained by this approach. Additionally, it is critical to understand how dsRNA delivery may modify the plant response to insect vectors (Farooq et al. 2021). Finally, the kinetics of control and viral reaction, as well as the dose in the strategy, are unclear, and effectiveness has not been thoroughly shown under field conditions.

Detection of Plant Viruses by Nanoparticles

To identify viral nucleic acids or proteins, nanoparticles may be coupled with a variety of probes (Banerjee and Jaiswal 2018; Farooq et al. 2021). Many nanoscale-biosensors have been designed to detect human pathogenic viruses, and these nanomaterials include silica, graphene oxide, gold and silver, carbon nanotubes and quantum dots, as well as zinc oxide and magnetic nanoparticles (Mokhtarzadeh et al. 2017). Nanoscale treatment and identification of human and animal viruses would be particularly beneficial since the work is better established there and methodologies are informative because of the general commonality of all viral life cycles (Alsteens et al. 2017; Singh et al. 2017; Farooq et al. 2021). As a consequence, the field of phytovirus detection may benefit from this approach. Using the unique optical characteristics and localized surface plasmon (LSP) of gold nanoparticles, a colorimetric nanobiosensing system for the detection of TYLCV in infected plants was developed (Razmi et al. 2019). An immunochromatographic lateral flow assay (ICA) was also established, which uses a sandwich immunoassay protocol to detect grapevine leafroll-associated virus 3 in grape leaf extracts in just 10 minutes (Byzova et al. 2018). Raman spectroscopy (RS), a nondestructive, fast, and low-cost approach that creates a sample's chemical fingerprint, has recently been effectively used to distinguish between healthy and virus-infected plants (Mandrile et al. 2019). The development of viral detection nanobiosensors has the potential to revolutionize the field of disease control. Another option for phytovirus detection testing is the use of nanoparticles-based particle bombardment devices employing "gene guns" (Abubakar et al. 2019). Using this technology, temporary or stable gene expression systems may be established in order to explore the interactions between plants and viruses at the cellular level (Acanda et al. 2019).

To improve diagnostic platforms, researchers should concentrate on isolating nanomaterials with adjustable characteristics that improve biosensor affinity, selectivity, and effectiveness.

Challenges for Effective Management of Plant Viruses by Nanoparticles

Although the antiviral activity of nanoparticles offers enormous promises, there are numerous substantial difficulties that may impede its development and use in the future. Adequate nanoscale materials with desirable qualities and behaviors will be synthesized and selected in this process. The function of the synthesis technique and the relevance of particle attributes, such as size, crystal structure, morphology, and other physiochemical parameters should be established and examined (Stankic et al. 2016). The size and charge of nanoparticles have the greatest impact on their ability to enter cells and interact with biomolecules. The cellular uptake efficiency of nanoparticles is greatly affected by their shape and surface charge; positive surface charges are associated with more effective cellular uptake (Zhang 2018). The movement of nanoparticles from the leaves to the roots and back again has been reported after penetration. An important consideration in determining how to effectively distribute and transport nanoparticles inside plants is the plants' unique morphology and physiology. Furthermore, nanoparticles' size and coating chemical composition have been shown to influence their absorption routes in planta translocation and leaf-to-rhizosphere transfer of nanoparticles (Su et al. 2019). Details concerning engineered nanoparticles' absorption, delivery, and movement in plants were studied (Avellan et al. 2019; Lv et al. 2019). To assure the safety and stability of the viral control approach, relevant end goals must be included, including an evaluation of the consequences on humans (Abdal Dayem et al. 2017; Wang et al. 2017). The time and spacing of nanoparticle application is also critical to enhance control. Treatment of nanoparticles following viral infection is ineffective in the majority of cases (Cai et al. 2019; Adeel et al. 2021). Even with these biostimulatory effects, the prophylactic application of nanoparticles not only allows direct contact control procedures but also promotes general plant growth and development via biostimulatory impacts. Additional study is required to determine the most effective kinetics of treatments and possible retreatment, as well as the optimization of dose/concentration. Moreover, there is a fundamental lack of knowledge on the influence of environmental variables on effectiveness. However, *in vitro* studies have shown that nanoparticles may degrade viral particles, but the existing evidence is inadequate to establish the specific antiviral function of nanoparticles *in vivo*.

Future Research on Antiviral Nanoparticles

Nanophytovirology is a potential strategy for long-term crop protection against viruses. *In vitro* studies have contributed to the great majority of the research on nanoparticles' potential antiviral activity. Furthermore, despite the fact that DNA phytoviruses are less common and less characterized than RNA phytoviruses, they continue to pose a danger to world agricultural production. There are several reports on agricultural losses in Africa, the United States, Pakistan, and India as a result of geminivirus infections (Moffat 1999; Briddon and Markham 2000; Malathi et al. 2003). Furthermore, the influence of nanomaterials at the interface of virus vector plant interactions is unclear (Farooq et al. 2021). Future research should focus on getting a greater understanding of the precise mechanisms that control virus-mediated plant infection. Given the efficiency of most nanoparticles as a protective treatment, research should concentrate on the first stages of the disease. As a result, a diverse collection of plant biologists, pathologists, and agricultural scientists would collaborate to design, develop, and apply nano-enabled antiviral techniques to combat these critical infections. Measures should be focused on developing and implementing sustainable, environmentally friendly, long-term, and multifunctional antiviral crop protection and forward-thinking efforts to attain global food security.

Conclusions

After reviewing the progress achieved in the last ten years, it is clear that the use and manufacture of nanomaterials have advanced significantly. There is a significant possibility that the use of nanoparticles might be an effective solution to the issue of plant viruses. However, much more may be accomplished by developing better methods for producing and applying nanomaterials against phytoviruses. It is vital to building exact complementing techniques in order to make the technology usable without posing any dangers to either the environment or customers. Additionally, it is necessary to know dosages, the time of the plant's phenological development at which applications should be made, and the particular nanoparticle forms that give the best benefits. Taken together, we may make a conclusion that nanomaterials are useful for preventing and treating phytovirus infections. Nanoencapsulation may be used to develop nanopesticide formulations for plant virus management in agriculture.

References

Abdal Dayem, A., Hossain, M.K., Lee, S.B., Kim, K., Saha, S.K., Yang, G.-M., Choi, H.Y. and Cho, S.-G. 2017. The role of reactive oxygen species (Ros) in the biological activities of metallic nanoparticles. Int. J. Mol. Sci. 18(1): 120.

Abubakar, A.L., Abarshi, M.M. and Maruthi, M.N. 2019. Testing the infectivity of a begomovirus by particle bombardment method using a gene gun. Nigerian Journal of Biotechnology 35(2): 58−65.

Acanda, Y., Wang, C. and Levy, A. 2019. Gene expression in citrus plant cells using helios gene gun system for particle bombardment. pp. 219−228. *In*: Catara, A.F., Bar-Joseph, M., Licciardello, G. (eds.). Citrus Tristeza Virus: Methods and Protocols. Springer: New York.

Adeel, M., Farooq, T., White, J.C., Hao, Y., He, Z. and Rui, Y. 2021. Carbon-based nanomaterials suppress tobacco mosaic virus (TMV) infection and induce resistance in *Nicotiana benthamiana*. J. Hazard. Mater. 404: 124167.

Alkubaisi, N.A. and Aref, N.M.A. 2017. Dispersed gold nanoparticles potentially ruin gold barley yellow dwarf virus and eliminate virus infectivity hazards. Appl. Nanosci. 7: 31–40.

Almasi, M.A. and Almasi, G. 2018. Colorimetric immunocapture loop mediated isothermal amplification assay for detection of Impatiens necrotic spot virus (INSV) by GineFinderTM dye. Eur. J. Plant Pathol. 150: 533–538.

Alsteens, D., Newton, R., Schubert, R., Martinez-Martin, D., Delguste, M., Roska, B. and Müller, D.J. 2017. Nanomechanical mapping of first binding steps of a virus to animal cells. Nat. Nanotechnol. 12(2): 177−183.

Aref, N., Alkubaisi, N., Marraiki, N. and Hindi, A. 2012. Multi-functional effects of gold nano-particles inducing plant virus resistance crops. In Proceedings of the 5th Annual World Congress of Industrial Biotechnology-2012, Xi'an, China.

Asghari, F., Jahanshiri, Z., Imani, M., Shams-Ghahfarokhi, M. and Razzaghi-Abyaneh, M. 2016. Chapter 10—Antifungal nanomaterials: Synthesis, properties, and applications. pp. 343–383. In Nanobiomaterials in Antimicrobial Therapy—Applications of Nanobiomaterials; William Andrew Publishing: Kidlington, Oxford, UK. ISBN 978-0-323-42864-4.

Avellan, A., Yun, J., Zhang, Y., Spielman-Sun, E., Unrine, J.M., Thieme, J., Li, J., Lombi, E., Bland, G. and Lowry, G.V. 2019. Nanoparticle size and coating chemistry control foliar uptake pathways, translocation, and leaf to rhizosphere transport in wheat. ACS Nano 13(5): 5291−5305.

Banerjee, R. and Jaiswal, A. 2018. Recent advances in nanoparticle based lateral flow immunoassay as a point of care diagnostic tool for infectious agents and diseases. Analyst 143(9): 1970−1996.

Bernardo, P., Charles-Dominique, T., Barakat, M., Ortet, P., Fernandez, E., Filloux, D., Hartnady, P., Rebelo, T.A., Cousins, S.R., Mesleard, F., Cohez, D., Yavercovski, N., Varsani, A., Harkins, G.W., Peterschmitt, M., Malmstrom, C.M., Martin, D.P. and Roumagnac, P. 2018. Geometagenomics illuminates the impact of agriculture on the distribution and prevalence of plant viruses at the ecosystem scale. ISME J. 12(1): 173−184.

Bragard, C., Caciagli, P., Lemaire, O., López-Moya, J.J., MacFarlane, S., Susi, P. and Torrance, L. 2013. Status and prospects of plant viruses control through interference vector transmission. Annu. Rev. Phytopathol. 51: 177–201.

Briddon, R.W. and Markham, P.G. 2000. Cotton leaf curl virus disease. Virus Res. 71(1-2): 151–9.

Byzova, N.A., Vinogradova, S.V., Porotikova, E.V., Terekhova, U.D., Zherdev, A.V. and Dzantiev, B.B. 2018. Lateral flow immunoassay for rapid detection of grapevine leafroll-associated virus. Biosensors (Basel) 8(4): 111.

Cai, L., Liu, C., Fan, G., Liu, C. and Sun, X. 2019. Preventing viral disease by ZnO NPs through directly deactivating TMV and activating plant immunity in *Nicotiana benthamiana*. Environmental Science: Nano 6(12): 3653−3669.

Cai, L., Cai, L., Jia, H., Liu, C., Wang, D. and Sun, X. 2020. Foliar exposure of Fe_3O_4 nanoparticles on *Nicotiana benthamiana*: Evidence for nanoparticles uptake, plant growth promoter and defense response elicitor against plant virus. J. Hazard. Mater. 393: 122415.

Chauhan, P., Singla, K., Rajbhar, M., Singh, A., Das, N. and Kumar, K. 2019. A systematic review of conventional and advanced approaches for the control of plant viruses. J. Appl. Biol. Biotechnol. 7: 89–98.

Chen, J.P., Lin, G. and Zhou, B.S. 2004. Correlation between pesticides exposure and morbidity and mortality of breast cancer. Chin. J. Public Health-Shenyang 20: 289–290.

Choudhary, R.C., Kumaraswamy, R.V., Kumari, S., Sharma, S.S., Pal, A., Raliya, R., Biswas, P. and Saharan, V. 2017a. Synthesis, characterization, and application of chitosan NMs loaded with zinc and copper for plant growth and protection. *In*: Prasad R. et al. (eds.). Nanotechnology. Springer Nature Singapore Pte Ltd., Singapore.

Choudhary, R.C., Kumaraswamy, R.V., Kumari, S., Sharma, S.S., Pal, A., Raliya, R., Biswas, P. and Saharan, V. 2017b. Cu-chitosan nanoparticle boost defense responses and plant growth in maize (Zea mays L.). Sci. Rep. 7: 9754.

Choudhary, R.C., Kumaraswamy, R.V., Kumari, S., Sharma, S.S., Pal, A., Raliya, R. and Saharan, V. 2019a. Zinc encapsulated chitosan nanoparticle to promote maize crop yield. Int. J. Biol. Macromol. 127: 126–135.

Choudhary, R.C., Kumari, S., Kumaraswamy, R.V., Pal, A., Raliya, R., Biswas, P. and Saharan, V. 2019b. Characterization methods for chitosan-based NMs. pp. 103–116. Plant Nanobionics. Springer, Cham.

Chung, I.-M., Rekha, K., Venkidasamy, B. and Thiruvengadam, M. 2019. Effect of copper oxide nanoparticles on the physiology, bioactive molecules, and transcriptional changes in Brassica rapa ssp. rapa Seedlings. Water Air Soil Pollut. 230: 48.

Derbalah, A. and Elsharkawy, M. 2019. A new strategy to control Cucumber mosaic virus using fabricated NiO-nanostructures. J. Biotechnol. 2019: 134–141.

Derbalah, A., Abdelsalam, I., Behiry, S.I., Abdelkhalek, A., Abdelfatah, M., Ismail, S. and Elsharkawy, M.M. 2022. Copper oxide nanostructures as a potential method for control of zucchini yellow mosaic virus in squash. Pest Management Science 78: 3587–3595.

Dietz, K.J. and Herth, S. 2011. Plant nanotoxicology. Trends Plant Sci. 16(11): 582–589.

Elazzazy, A., Elbeshehy, E. and Betiha, M. 2017. *In vitro* assessment of activity of graphene silver composite sheets against multidrug-resistant bacteria and Tomato Bushy Stunt Virus. Trop. J. Pharm. Res. 16: 2705–2711.

Elbeshehy, E.K.F., Elazzazy, A.M. and Aggelis, G. 2015. Silver nanoparticles synthesis mediated by new isolates of *Bacillus* spp., nanoparticle characterization and their activity against Bean Yellow Mosaic Virus and human pathogens. Front. Microbiol. 6: 453.

El-Dougdoug, N.K., Bondok, A.M. and El-Dougdoug, K.A. 2018. Evaluation of silver nanoparticles as antiviral agent against ToMV and PVY in tomato plants. Middle East J. Appl. Sci. 08: 100–111.

Elmer, W. and White, J.C. 2018. The future of nanotechnology in plant pathology. Annu. Rev. Phytopathol. 56(1): 111–133.

El-Sawy, M., Elsharkawy, M., Abass, J. and Kasem, M. 2017a. Antiviral activity of 2-nitromethyl phenol, zinc nanoparticles and seaweed extract against cucumber mosaic virus (CMV) in eggplant. Journal of Virology and Antiviral Research 06. DOI: 10.4172/2324- 8955.1000173.

El-Sawy, M., Elsharkawy, M., Abass, J. and Hagag, E. 2017b. Inhibition of tomato yellow leaf curl virus by *Zingiber officinale* and *Mentha longifolia* extracts and silica nanoparticles and its reflection on tomato growth and yield. Int. J. Antivir. Antiretrovir. 1: 1–6.

Elsharkawy, M.M. and Mousa, K.M. 2015. Induction of systemic resistance against Papaya ring spot virus (PRSV) and its vector *Myzus persicae* by *Penicillium simplicissimum* GP17-2 and silica (SiO_2) nanopowder. Int. J. Pest Manag. 61: 353–358.

Elsharkawy, M.M. and Derbalah, A. 2019. Antiviral activity of titanium dioxide nanostructures as a control strategy for Broad bean stain virus in faba bean. Pest Manage. Sci. 75(3): 828–834.

El-shazly, M., Attia, Y., Kabil, F., Anis, E. and Hazman, M. 2017. Inhibitory effects of salicylic acid and silver nanoparticles on Potato Virus Y-infected potato plants in Egypt. Middle East J. Agric. Res. 6: 835–848.

Eugene, K. and Zholobak, N. 2016. Antiviral activity of cerium dioxide nanoparticles on tobacco mosaic virus model. In Proceedings of the Topical Issues of New Drugs Development; NUPH: Kharkiv, Ukraine, 21 April; Vol. 355, (1).

Farooq, T., Adeel, M., He, Z., Umar, M., Shakoor, N., da Silva, W., Elmer, W., White, J.C. and Rui, Y. 2021. Nanotechnology and plant viruses: An emerging disease management approach for resistant pathogens. ACS Nano 15(4): 6030–6037.

Fernando, S.S.N., Gunasekara, C. and Holton, J. 2018. Antimicrobial Nanoparticles: Applications and mechanisms of action. SRI Lankan J. Infect. Dis. 8: 2.

French, R.K. and Holmes, E.C. 2020. An ecosystems perspective on virus evolution and emergence. Trends Microbiol. 28(3): 165–175.

Fu, L., Wang, Z., Dhankher, O.P. and Xing, B. 2020. Nanotechnology as a new sustainable approach for controlling crop diseases and increasing agricultural production. J. Exp. Bot. 71(2): 507–519.

Galdiero, S., Falanga, A., Vitiello, M., Cantisani, M., Marra, V. and Galdiero, M. 2011. Silver nanoparticles as potential antiviral agents. Molecules 16: 8894–8918.

Goldbach, R. 1998. All you wanted to know about plant virus control. Trends Plant Sci. 3(12): 490.

Hannon, G.J. 2002. RNA interference. Nature 418(6894): 244–251.

Hao, Y., Yuan, W., Ma, C., White, J., Zhang, Z., Adeel, M., Zhou, T., Yukui, R. and Xing, B. 2018. Engineered nanomaterials suppress Turnip mosaic virus infection in tobacco (*Nicotiana benthamiana*). Environ. Sci. Nano 5: 1685–1693.

Hill, J. and Whitham, S. 2014. Control of virus diseases in soybeans. Adv. Virus Res. 90: 355–390.

Hogenhout, S.A., Ammar, E.-D., Whitfield, A.E. and Redinbaugh, M.G. 2008. Insect vector interactions with persistently transmitted viruses. Annu. Rev. Phytopathol. 46(1): 327–359.

Jain, D. 2014. Green synthesis of silver nanoparticles and their application in plant virus inhibition. J. Mycol. Plant Pathol. 44: 21–24.

Jones, R.A.C. 2016. Future Scenarios for Plant Virus Pathogens as Climate Change Progresses. pp. 87–147. In Advances in Virus Research; Elsevier Science: Amsterdam, The Netherlands, Volume 95.

Jones, R.A.C. and Naidu, R.A. 2019. Global dimensions of plant virus Diseases: current status and future perspectives. Annu. Rev. Virol. 6(1): 387–409.

Kang, B.C., Yeam, I. and Jahn, M.M. 2005. Genetics of plant virus resistance. Annu. Rev. Phytopathol. 43: 581–621.

Kashyap, P.L., Xiang, X. and Heiden, P. 2015. Chitosan nanoparticle based delivery systems for sustainable agriculture. Int. J. Biol. Macromol. 77: 36–51.

Kumaraswamy, R.V., Kumari, S., Choudhary, R.C., Pal, A., Raliya, R., Biswas, P. and Saharan, V. 2018. Engineered chitosan based nanomaterials: bioactivities, mechanisms and perspectives in plant protection and growth. Int. J. Biol. Macromol. 113: 494–506.

Kumar, J.N., Bora, A., Kumar, R.N., Amb, M.K. and Khan, S. 2013. Toxicity analysis of pesticides on cyanobacterial species by 16S rDNA molecular characterization. Proc. Int. Acad. Ecol. Environ. Sci. 3(2): 101.

Landa, P., Dytrych, P., Prerostova, S., Petrova, S., Vankova, R. and Vanek, T. 2017. Transcriptomic response of *Arabidopsis thaliana* exposed to CuO nanoparticles, bulk material, and ionic copper. Environ. Sci. Technol. 51: 10814–10824.

Lauring, A.S., Frydman, J. and Andino, R. 2013. The role of mutational robustness in RNA virus evolution. Nat. Rev. Microbiol. 11(5): 327–36.

Lv, J., Christie, P. and Zhang, S. 2019. Uptake, translocation, and transformation of metal-based nanoparticles in plants: recent advances and methodological challenges. Environmental Science: Nano 6(1): 41–59.

Malathi, V.G., Radhakrishnan, G. and Varma, A. 2003. Cotton. pp. 743–754. In: Loebenstein, G. and Thottappilly, G. (eds.). Virus and Virus-Like Diseases of Major Crops in Developing Countries. Springer: Netherlands: Dordrecht.

Malerba, M. and Cerana, R. 2016. Chitosan effects on plant systems. Int. J. Mol. Sci. 17(7): 996. doi: 10.3390/ijms17070996.

Mandrile, L., Rotunno, S., Miozzi, L., Vaira, A.M., Giovannozzi, A.M., Rossi, A.M. and Noris, E. 2019. Nondestructive raman spectroscopy as a tool for early detection and discrimination of the infection of tomato plants by two economically important viruses. Anal. Chem. 91(14): 9025–9031.

Marslin, G., Sheeba, C.J. and Franklin, G. 2017. Nanoparticles alter secondary metabolism in plants via ROS burst. Front. Plant Sci. 8: 832.

Marutescu, L., Popa, M., Saviuc, C., Lazar, V. and Chifiriuc, M.C. 2017. Botanical pesticides with virucidal, bactericidal, and fungicidal activity. pp. 311–335. In: New Pesticides and Soil Sensors. Academic Press, London.

Maruthi, M.N., Whitfield, E.C., Otti, G., Tumwegamire, S., Kanju, E., Legg, J.P., Mkamilo, G., Kawuki, R., Benesi, I., Zacarias, A. et al. 2019. A method for generating virus-free cassava plants to combat viral disease epidemics in Africa. Physiol. Mol. Plant Pathol. 105: 77–87.

Mitter, N., Worrall, E., Robinson, K., Li, P., Jain, R., Taochy, C., Fletcher, S., Carroll, B., Lu, G. and Xu, Z. 2017. Clay nanosheets for topical delivery of Rnai for sustained protection against plant viruses. Nature Plants 3: 16207.

Moffat, A.S. 1999. Geminiviruses emerge as serious crop threat. Science 286(5446): 1835–1835.

Mokhtarzadeh, A., Eivazzadeh-Keihan, R., Pashazadeh, P., Hejazi, M., Gharaatifar, N., Hasanzadeh, M., Baradaran, B. and de la Guardia, M. 2017. Nanomaterial-based biosensors for detection of pathogenic virus. TrAC, Trends Anal. Chem. 97: 445–457.

Morais, M., Martins, V., Steffens, D., Pranke, P. and Costa, J.A. 2014. Biological applications of nanobiotechnology. J. Nanosci. Nanotechnol. 14: 1007–1017.

Panattoni, A., D'Anna, F., Cristani, C. and Triolo, E. 2008. Grapevine vitivirus A eradication in Vitis vinifera explants by antiviral drugs and thermotherapy. J. Virol. Methods 146: 129–135.

Parveen, S., Wani, A.H., Shah, M., Devi, H., Bhat, M. and Koka, J. 2018. Preparation, characterization and antifungal activity of iron oxide nanoparticles. Microb. Pathog. 115.

Pechinger, K., Chooi, K.M., MacDiarmid, R.M., Harper, S.J. and Ziebell, H. 2019. A new era for mild strain cross-protection. Viruses 11: 670.

Prasad, R., Bhattacharyya, A. and Nguyen, Q.D. 2017. Nanotechnology in sustainable agriculture: recent developments, challenges, and perspectives. Front. Microbiol. 8: 1014.

Rastogi, A., Zivcak, M., Sytar, O., Kalaji, H.M., He, X., Mbarki, S. and Brestic, M. 2017. Impact of metal and metal oxide nanoparticles on plant: a critical review. Front. Chem. 5: 78.

Razmi, A., Golestanipour, A., Nikkhah, M., Bagheri, A., Shamsbakhsh, M. and Malekzadeh-Shafaroudi, S. 2019. Localized surface plasmon resonance biosensing of tomato yellow leaf curl virus. J. Virol. Methods 267: 1–7.

Saharan, V., Sharma, G., Yadav, M., Choudhary, M.K., Sharma, S.S., Pal, A., Raliya, R. and Biswas, P. 2015. Synthesis and *in vitro* antifungal efficacy of cu-chitosan nanoparticles against pathogenic fungi of tomato. Int. J. Biol. Macromol. 75: 346–353.

Sarwar, M. 2020. Chapter 27—Insects as transport devices of plant viruses. pp. 381–402. In Applied Plant Virology: Advances, Detection, and Antiviral Strategies. Academic Press: Cambridge, MA, USA. ISBN 978-0-12-818654-1.

Sastry, K.S. and Zitter, T.A. 2014. Management of virus and viroid diseases of crops in the tropics. pp. 149–480. *In*: Sastry, K.S. and A. Zitter, T. (eds.). Plant Virus and Viroid Diseases in the Tropics: Vol. 2: Epidemiology and Management. Springer: Dordrecht, Netherlands.

Savary, S., Willocquet, L., Pethybridge, S.J., Esker, P., McRoberts, N. and Nelson, A. 2019. The Global burden of pathogens and pests on major food crops. Nature Ecology and Evolution 3(3): 430–439.

Setten, R.L., Rossi, J.J. and Han, S.-p. 2019. The current state and future directions of RNAi-based therapeutics. Nat. Rev. Drug Discovery 18(6): 421–446.

Shafie, R.M., Salama, A.M. and Farroh, K.Y. 2018. Silver nanoparticles activity against Tomato spotted wilt virus. Middle East J. Agric. Res. 7: 1251–1267.

Shukla, S.K., Mishra, A.K., Arotiba, O.A. and Mamba, B.B. 2013. Chitosan-based NMs: a state-of-the-art review. Int. J. Biol. Macromol. 59: 46–58.

Singh, B. 2015. Effect of antiviral chemicals on *in vitro* regeneration response and production of PLRV-free plants of potato. J. Crop Sci. Biotechnol. 18: 341–348.

Singh, L., Kruger, H.G., Maguire, G.E.M., Govender, T. and Parboosing, R. 2017. The role of nanotechnology in the treatment of viral infections. Ther. Adv. Infect. Dis. 4(4): 105–131.

Soares, C., Pereira, R. and Fidalgo, F. 2018. Metal-based nanomaterials and oxidative stress in plants: current aspects and overview. pp. 197–227. *In*: Faisal, M., Saquib, Q., Alatar, A. and Al-Khedhairy, A. (eds.). Phytotoxicity of Nanoparticles. Springer International Publishing: Cham, Switzerland. ISBN 978-3-319-76707-9.

Soosaar, J.L., Burch-Smith, T.M. and Dinesh-Kumar, S.P. 2005. Mechanisms of plant resistance to viruses. Nat. Rev. Microbiol. 3(10): 789–98.

Stankic, S., Suman, S., Haque, F. and Vidic, J. 2016. Pure and multi metal oxide nanoparticles: synthesis, antibacterial and cytotoxic properties. J. Nanobiotechnol. 14(1): 73.

Su, Y., Ashworth, V., Kim, C., Adeleye, A.S., Rolshausen, P., Roper, C., White, J. and Jassby, D. 2019. Delivery, uptake, fate, and transport of engineered nanoparticles in plants: A critical review and data analysis. Environmental Science: Nano 6(8): 2311–2331.

Tan, B.L., Norhaizan, M.E., Liew, W.-P.-P. and Sulaiman Rahman, H. 2018. Antioxidant and oxidative stress: a mutual interplay in age-related diseases. Front. Pharmacol. 9: 1162.

Tirani, M.M., Haghjou, M.M. and Ismaili, A. 2019. Hydroponic grown tobacco plants respond to zinc oxide nanoparticles and bulk exposures by morphological, physiological and anatomical adjustments. Funct. Plant Biol. 46: 360–375.

Tripathi, M., Kumar, S., Kumar, A., Tripathi, P. and Kumar, S. 2018. Agro-nanotechnology: A Future technology for sustainable agriculture. Int. J. Curr. Microbiol. Appl. Sci. 7: 196–200.

Vargas-Hernandez, M., Macias-Bobadilla, I., Guevara-Gonzalez, R.G., Rico-Garcia, E., Ocampo-Velazquez, R.V., Avila-Juarez, L. and Torres-Pacheco, I. 2020. Nanoparticles as potential antivirals in agriculture. Agriculture 10(10): 444. https://doi.org/10.3390/agriculture10100444.

Wang, M.-R., Cui, Z.-H., Li, J.-W., Hao, X.-Y., Zhao, L. and Wang, Q.-C. 2018. *In vitro* thermotherapy-based methods for plant virus eradication. Plant Methods 14: 87.

Wang, Y., Sun, C., Xu, C., Wang, Z., Zhao, M., Wang, C., Liu, L. and Chen, F. 2016. Preliminary experiments on nano-silver against tobacco mosaic virus and its mechanism. Tob. Sci. Technol. 49: 22–30.

Worrall, E.A., Bravo-Cazar, A., Nilon, A.T., Fletcher, S.J., Robinson, K.E., Carr, J.P. and Mitter, N. 2019. Exogenous application of RNAi-inducing double-stranded RNA inhibits aphid-mediated transmission of a plant virus. Front. Plant Sci. 10: 265.

Xiang, Y., Nie, X., Bernardy, M., Liu, J., Su, L., Bhagwat, B., Dickison, V., Holmes, J., Grose, J.M. and Creelman, A.C. 2020. Genetic diversity of strawberry mild yellow edge virus from eastern Canada. Arch. Virol. 165: 923–935.

Xing, K., Liu, Y., Shen, X., Zhu, X., Li, X., Miao, X. and Qin, S. 2017. Effect of O-chitosan nanoparticles on the development and membrane permeability of *Verticillium dahliae*. Carbohydr. Polym. 165: 334–343.

Yih, T.C. and Al-Fandi, M. 2006. Engineered nanoparticles as precise drug delivery systems. J. Cell. Biochem. 97: 1184–1190.

Zhang, W., Jiang, F. and Ou, J. 2011a. Global pesticide consumption and pollution: with China as a focus. Proc. Int. Acad. Ecol. and Environ. Sci. 1(2): 125.

Zhang, W.J., van der Werf, W. and Pang, Y. 2011b. A simulation model for vegetable-insect pest-insect nucleopolyhedrovirus epidemic system. J. Environ. Entomol. 33(3): 283–301.

Zhang, W. 2018. A long-term trend of cancer-induced deaths in European countries. Network 3(1–2): 1–9.

Zhao, Y., Yang, X., Zhou, G. and Zhang, T. 2020. Engineering plant virus resistance: From RNA silencing to genome editing strategies. Plant Biotechnol. J. 18: 328–336.

Ziegler-Graff, V. 2020. Molecular insights into host and vector manipulation by plant viruses. Viruses 12: 263.

Chapter 19

Nanomaterials for Quarantine Fungi and Bacteria Plant Pathogens

Graciela Avila-Quezada,[1,*] *Mahendra Rai,*[2,3] *Cynthia Alvarez-Alvarez*[1] and
Damaris L. Ojeda Barrios[1]

Introduction

The application of pesticides has restricted all kinds of plant pathogens for safer food production. Government campaigns focused on quarantining plant pathogens with an aim to restrict the spread of pathogens and their vectors since there is no or limited cure for the infection (USDA 2020). The losses caused by quarantined pathogens cause strong economic damage in the world and a great loss of food. Various regulated fungi and bacteria affect a wide range of plants. These pathogens are quarantined because they are difficult to eliminate and have a tremendous capacity for survival. Unfortunately, for some of them, there is no effective treatment to control (Avila-Quezada et al. 2018).

Various plant diseases caused by bacteria and fungi are devastating to agriculture. Disease in plants refers to the manifestation of distinctive symptoms due to pathogens that can cause severity of up to 100%. An epidemic of quarantined pathogens implies rapid actions to delimit a control belt, where infected plants are eradicated within a radius. This area is treated and monitored (Strona et al. 2020) for the international restrictions and serious economic losses that these diseases represent (Avila-Quezada and Rai 2022).

Thus, the search for new alternatives for disease management is fundamental. Although conventional pesticide formulations are effective in controlling pathogens and ensuring food production, they have many disadvantages commonly related to toxicity, in addition to their low water solubility (Hazra et al. 2017). Thus, they must be gradually replaced by effective non-polluting products. Among non-contaminating pathogen control products are biocontrol agents, which can be effective, but they have the disadvantage of slow action. In addition, it is necessary to explore alternative methods to control pathogens and their vectors.

Thus, nanotechnology (NT) is a tool with direct application in the management of quarantine-type diseases. In agriculture, it is applied in the nutrition and health of plants with innovations, such as controlled administration of nutrients, active ingredients, and DNA, RNA, or

---

[1] Universidad Autonoma de Chihuahua, Facultad de Ciencias Agrotecnologicas, Escorza 900, col. Centro, Chihuahua 31000, Mexico.

[2] SGB Amravati University, Department of Biotechnology, Nanobiotechnology Lab., Amravati-444 602, Maharashtra, India.

[3] Nicolaus Copernicus University, Department of Microbiology, Toruń, 87-100 Poland.

* Corresponding author: gdavila@uach.mx

proteins in nanosensors for diagnosis (Pramanik et al. 2020; Avila-Quezada et al. 2021). Besides, nanophytopathology is a new concept that refers to the application of nanotechnology to combat plant diseases (Avila-Quezada et al. 2023a).

The prefix "nano" refers to one-millionth of a millimeter. This nanometric size of a metal, due to the high surface-to-volume ratio, influence its physical properties. These properties include the antimicrobial nature of nanomaterials. For instance, metal nanoparticles adhere to bacteria, penetrate the cell and generate reactive oxygen species (ROS), causing lipid peroxidation, triggering damage to proteins, DNA, and membrane, ultimately leading to cell death (Avila-Quezada and Espino-Solis 2019; Liao et al. 2019).

Given the urgent need to maintain low populations of phytopathogens that are so aggressive with crops, various diagnostic and control strategies can be designed taking advantage of the benefits of currently available nanomaterials. Several NMs have been applied to counter the negative effects of quarantined pathogens. Generally, to control these pathogens, large amounts of agrochemicals are used conventionally, which could gradually be replaced by low volumes of NP.

The global trend is the gradual reduction of synthetic products for the control of phytopathogens. In this sense, all research efforts should be directed toward the search for new methods to integrate into the integrated management of pathogens. Based on the properties of NMs and with the experience of previous research, this chapter deals with the diagnosis and management of pathogens using various nanomaterials, which are the new generation of materials to fight against pathogens in order to safeguard food for the growing population.

Current Quarantine Plant Pathogens, the Devastation of Crops, and the Cost of Applying Quarantine

The European and Mediterranean Plant Protection Organization (EPPO) provides lists of pests that are recommended for regulation as quarantine pests. EPPO first approved a list of quarantined pests and pathogens in 1975. In 2022, it lists 14 bacteria and 37 fungi (EPPO 2022a).

A "quarantine pest" is one of potential economic importance to the area where it is not present. If the pest is already present, it is not widespread and is controlled by the country's authorities. While "regulated non-quarantine pest" is one whose presence influences the proposed use for crops with economically unacceptable repercussions, which is regulated in the territory of the importing country (ISPM 2018).

For example, musaceous wilt caused by the fungus *Fusarium oxysporum* f. sp. *cubense* race 1 (Foc R1) was responsible for the death of bananas in the American continent in the 1950s and 1960s, driving almost the total disappearance of commercial plantations, as well as an economic impact on exports of US \$2.3 billion. Moreover, *Fusarium oxysporum* f. sp. *cubense* Tropical Race 4 (Foc R4T) was originally from Southeast Asia in the early 90s and has caused millionaire losses. This is more destructive than Foc R1 and is considered one of the ten most aggressive pathogens for agriculture worldwide. It severely affects varieties of the Cavendish subgroup, as well as other bananas (*Musa* AAA) and Bluggoe-type cooking bananas (*Musa* AAB, and ABB).

Another example is coffee rust, caused by the fungus *Hemileia vastatrix*, which first appeared in 1867 in Sri Lanka (formerly called Ceylon, the world's leading coffee-producing region), where it caused serious damage to coffee plantations. Sri Lanka exports almost 100 million pounds of coffee a year, and as a result of the rust and devastation, producers began to plant tea instead of coffee. Consequently, the tradition of drinking coffee changed to tea consumption, especially among the British. The fungus *H. vastatrix* was responsible for the devastation (Arneson 2011).

Wheat stem rust race Ug99, caused by *Puccinia graminis* f. sp. *tritici*, was first reported in Uganda. "Ug" denotes Uganda, while 99 for the year it appeared (Pretorius et al. 2000). Rust spores are easily spread by the wind, reaching remote places and posing a phytosanitary risk for food production in the world. Thus, this rust was detected in South Africa in 2000, Kenya in 2001, and Ethiopia in 2003. Later, it arrived in Yemen and Sudan in 2006, Iran in 2007, Tanzania in 2009, Ethiopia and Zimbabwe in 2010, Eritrea in 2012, and Rwanda and Egypt in 2014 (RustTracker org 2022).

The seriousness of the problem is that 80% of the wheat varieties in the world are susceptible to race 99 (Senasica 2019c); in addition, the economic losses can be from 70% to 100% (RustTracker org 2022). The rapid spread and aggressiveness of emerging pathogens alert countries to quarantine them and seek rapid control measures.

Another disease caused by the quarantined pathogen *Tilletia indica* is the Karnal bunt of wheat. It bears this name because the pathogen appeared for the first time in the city of Karnal, India in 1931. From there it spread to Asia, Africa, Mexico, and the United States. Although the yield losses are not high, even a 1% severity affects grain quality (EPPO 2022b). Thus, the economic impact is significant due to the cost of quarantine measures. For instance, in the USA, these expenses were 35 million dollars from 1996 to 2001 (Vocke et al. 2002).

Among the most destructive quarantined bacteria in the plant is *Ralstonia solanacearum* species complex. Specifically, *Ralstonia solanacearum* Phylotype II causes the disease known as banana moko that can destroy complete banana plantations (Álvarez et al. 2015). Moko disease has affected bananas and plantains in large areas of Central and South America, affecting subsistence agriculture. Yield losses of up to 74% are reported in Guyana (Senasica 2019a). Moko has also affected the subsistence economy in Indonesia and the Philippines (EPPO 2021).

Besides, Pierce's disease is caused by *Xylella fastidiosa* subsp. *fastidiosa*, which was responsible for the loss of more than 30 million dollars in the almost 500 ha vineyards in California, USA, between the years 1994 and 2000 (Senasica 2019b). The bacterium affects not only vines (EPPO 2022c) but also plants like almonds, alfalfa, coffee, and avocado, among others (Senasica 2019b). Given this host diversity, control measures should be strengthened to keep crop areas free of *Xylella fastidiosa* subsp. *fastidiosa*.

Another bacterium *Xanthomonas citri* causes citrus canker, a disease that results in economic losses throughout the world due to reduced fruit production and restrictions on export markets coupled with the costs of quarantine measures. The bacterium affects the entire tree leading to death.

The pathogen is dispersed between regions and countries through the movement of diseased citrus fruit and seedlings (CABI 2021), the reason why the bacterium is in quarantine in many countries.

The entry of quarantined pathogens into new territories brings devastating economic and food security consequences. The previous examples of plant pathogens show that bacteria and fungi will reach the last corner of the globe and adapt to a diversity of conditions. Being new pathogens in a new territory, plants will be susceptible and cause serious food losses.

Therefore, the best option to avoid the impact of pathogens on food production is the application of various control strategies, such as integrated management and exclusion. NMs are a viable option to integrate into control strategies and for a rapid and accurate diagnosis.

Nanomaterials (NM) Used for the Quarantine Pathogens

Among the most surveillance bacteria worldwide due to the economic damage they cause to crops, include *Xanthomonas citri, X. campestris* pv. *musacearum, Ralstonia solanacearum* race 2, *Xylella fastidiosa* subsp. *fastidiosa, X. fastidiosa,* and *X. fastidiosa* subsp. *pauca* (EPPO 2022a).

Nanomaterials are important powerful tools for preventing the adhesion and development of bacteria. Hyaluronan/chitosan nanofilms are efficient since natural polymers, such as sodium alginate, carboxymethylcellulose, starch, chitosan, and pectin trap and release antimicrobial substances due to their bioadhesive nature (Croisier and Jérôme 2013). Chitosan is an efficient polymer for the administration of molecules due to its antibacterial properties and activation of the plant defense system (Deepmala et al. 2014).

Moreover, Khan et al. (2022) demonstrated the antibacterial activity of silicon dioxide and titanium dioxide nanoparticles in reducing the development of *Ralstonia solanacearum* in eggplant.

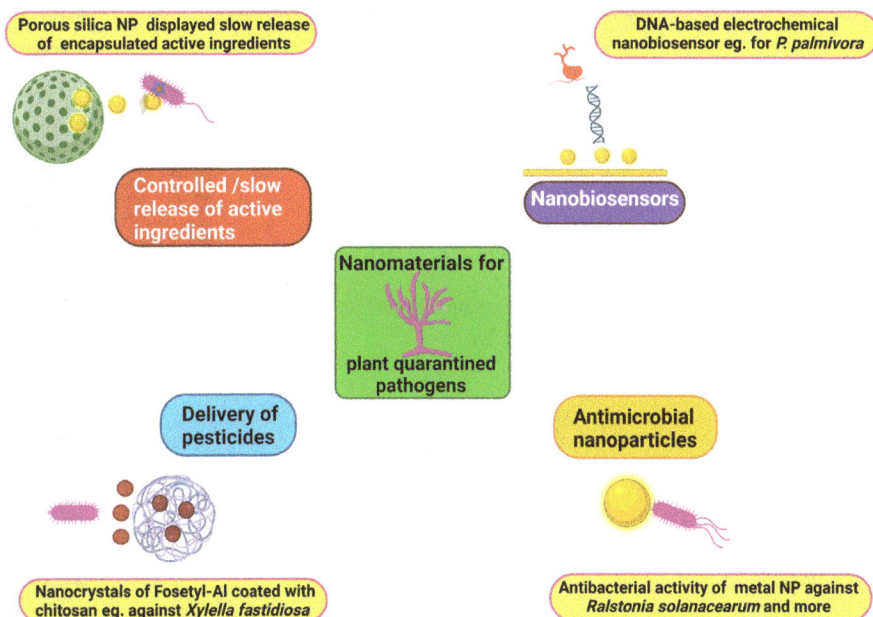

Figure 1. Some nanomaterials for controlling or for diagnostic of bacteria and fungi plant pathogens.

In another study, Varympopi et al. (2022) tested copper nanoparticles against *Xanthomonas campestris* pv. *vesicatoria* in tomato plants which were effective to reduce the disease. Successful cases of antimicrobial activity with the use of NPs have been reported by various authors. For example, Dzimitrowicz et al. (2018) reported the antibacterial activity of fructose-stabilized AgNPs synthesized by atmospheric pressure glow discharge toward the quarantine pathogens *Erwinia amylovora, Clavibacter michiganensis, Ralstonia solanacearum, Xanthomonas campestris* pv. *campestris*, and *Dickeya solani*.

The nanomaterials can also be used for the diagnosis of quarantined pathogens (Figure 1). For this, nanobiosensors are used that are based on the recognition of the sample and the translation of the signal. Nanobiosensors can be optical, thermal, electrochemical, and piezoelectric. For instance, Fluorescence Resonance Energy Transfer (FRET) nanosensors are based on fluorescence resonance energy with indicator fluorophores (Kwak et al. 2017), where the target is DNA or antibodies (Ambrin et al. 2019). These have been used to diagnose *Candidatus* Phytoplasma aurantifolia (Rad et al. 2012).

On the other hand, electrochemical nanosensors (ECN) are those that emit an electrical signal when faced with a biological or chemical sample (Perdomo et al. 2021). These have three different electrodes with a nanomaterial surface (Barry and O'Riordan 2015) and use instruments such as an amperemeter, potentiometer, voltammeter, and conductometer (Muniandy et al. 2019). They have the advantage that it is possible to use them for the diagnosis of plant pathogens in the field (Srivastava et al. 2018). They have been used to diagnose *Pseudomonas syringae* (Lau et al. 2017).

Another device is the Surface-Enhanced Raman Scattering (SERS), which has the surfaces of metallic nanoparticles and its mode of action is by enhancing the Raman scattering signals of the adsorbed analyte through laser excitation (Kahraman et al. 2017). Piezoelectric nanosensors function by translating mechanical energy, ultrasonic vibrations, and hydraulic energy from the biological sample into an electrical signal. The principle of these devices makes them very useful due to their sensitivity to diagnosing pathogens in the laboratory and the field.

These new nanodevices offered by nanotechnology for the diagnosis of plant pathogens are valuable in food production and even more so when dealing with quarantined phytopathogens. Some successful examples are nanobiosensors based on visual colorimetry which let a rapid and

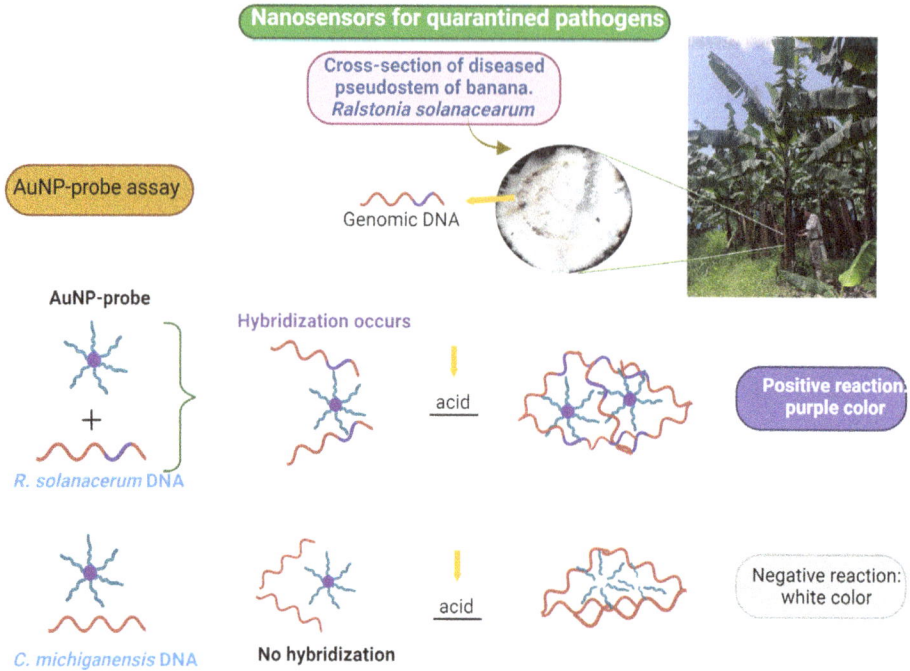

Figure 2. AuNPs probes are stabilized when *Ralstonia solanacearum* genomic DNA is added to the sensor, thus aggregation is prevented after the addition of acid and the "purple" color reaction is a positive indicator of *Ralstonia solanacearum* genomic DNA. Contrary, in the presence of genomic DNA from other bacteria, the AuNP-probes lose stability, and tend to aggregate, no purple color is observed indicating the absence of target DNA; this is a negative reaction.

accurate diagnostic for quarantined bacteruim. For instance, Khaledian et al. (2017) developed a nanobiosensor based on gold nanoparticles to detect unamplified genomic DNA of *Ralstonia solanacearum*. The device of gold NPs operates with a probe of single-stranded oligonucleotides that detect the bacterium genomic DNA.

The optical properties of gold nanoparticles (AuNPs) are because of the resonant excitation of their conduction electrons or localized surface plasmon resonance (LSPR). AuNPs selectively absorb and scatter photons, then the interaction of these nanoparticles with biological compounds or DNA generates a change in frequency. This is how aggregation-based or colorimetric sensors can be fabricated because the colloid nanoparticles change color when their size changes.

The AuNPs-oligo probe hybridizes with the complementary DNA (there is no aggregation) and in an acidic medium, a color change is observed. The solution then turns purple when it does not have the complementary DNA (Draz and Shafiee 2018). This method can be designed for the detection of bacteria such as the causal agent of banana Moko *Ralstonia solanacearum* race 2 (Figure 2).

Other diagnostic methods are based on electrochemical transduction, which was applied for *Xylella fastidiosa* subsp. *pauca*, allows on-chip detection of the bacterium from plant leaves with this method. This lab-on-chip developed by Chiriacò et al. (2018) includes a microfluidic module.

Moreover, nanotechnology can reduce time and take more efficient steps in the diagnosis with the PCR technique. Shoala and Abd-El-Aziz (2019) extracted a DNA template from *Ralstonia solanacearum* using the combination of magnetic AgNPs and Fe_2O_3NPs of sizes less than 23 nm. With this technique, they achieved an increased yield in DNA extraction and PCR efficiency even with a low concentration of template bacterial DNA. PCR performance is due to rapid heat transfer in the presence of metallic NPs, which can reduce the number of PCR cycles.

Other NMs have been successful in controlling *in vivo* quarantined bacteria, such as *Xanthomonas citri* subsp. *citri*, which causes citrus canker. Graham et al. (2016) evaluated the NMs formulated with zinc oxide: Zinkicide SG4 with a two-dimensional structure in the form of a plate

Table 1. Some examples of successful NMs to combat quarantined phytopathogenic bacteria.

Pathogen	NM	Size	Effect *In Vitro* or Inhibition	Effect *In Vivo*	Reference
	MgONPs	50–100 nm	MIC 200 µg/mL MBC 250 µg/mL. Physical injury to the cell membranes. Decreased motility and biofilm formation.	Reducing the bacterial wilt index in plants in the greenhouse	Cai et al. 2018
	AgNPs	< 40 nm	Inhibition zone of diffusion disk, MIC (30 µg mL^{-1}) and MBC (40 µg mL^{-1}),	-	Tortella et al. 2019
Ralstonia solanacearum	Fe_2O_3NPs	56–350 nm	*R. solanacearum* was inhibited by 6 mg/mL of Fe_2O_3NPs.	The bacterial wilt severity was reduced with 6% w/v of Fe_2O_3NPs in planta	Alam et al. 2019
	TiO_2NPs and SiO_2NPs	SiO_2 NPs, 5–15 nm. TiO_2NPs, < 150 nm	Inhibition zone. SiO_2NPs 0.62 mm. TiO_2NPs 0.51 mm. Both NP fragmented the mycelium and the conidia		Khan et al. 2022
Xanthomonas citri	Silver and copper nanoparticles	-	Inhibition zone of AgNPs + CuNP: 21.06 mm. AgNP: 18.26 mm. CuNPs: 15.27 mm.	Disease severity of citrus canker under field conditions. control 49%. AgNPs 40%. CuNPs 43%. AgNPs + CuNPs 36%.	Atiq et al. 2022
Xanthomonas citri subsp. *citri*	Zinkicide SG4 (Bidimensional structure). Zinkicide SG6 (Similar structure as a gel composed of interconnected particles	espesor zinkicide SG4: ~ 10 nm. zinkicide SG6:4-6 nm		Both NM of zinc and cuprous oxide/zinc oxide reduced the incidence of fruit canker from 63% (control) to 7.0 and 9.0% (NM, with a better effect than zinc oxide. cuprous and cuprous oxide/zinc oxide. The fruits did not show signs of phytotoxicity due to the sprayed NM	Graham et al. 2016
Xylella fastidiosa subsp. *fastidiosa* and *Xylella fastidiosa* subsp. *pauca*	Fosetyl-Al nanocrystal coated with chitosan		Inhibited planktonic growth 6 days after treatment at 100 µg/mL, for both bacterial species		Baldassarre et al. 2020

MIC: minimum inhibitory concentration
MBC: minimum bactericidal concentration

and Zinkicide SG6 in particles. Both products were sprayed on sweet orange foliage and fruit of grapefruit, after inoculation with *Xanthomonas citri* subsp. *citri* in the intercellular space. Both NM reduced the development of citrus canker, surpassing the effect of commercial bactericides based on copper and zinc oxide due to the translaminar movement of Zinkicide NMs.

Some other studies are focused on delivering the active ingredient encapsulated in nanochitosan. For example, Baldassarre et al. (2020) tested the systemic fungicide fosetyl-Al formulated as nanocrystals coated with chitosan (CH-nanoFos) against *Xylella fastidiosa* subsp. *pauca* and *X. fastidiosa* subsp. *fastidiosa* and against biofilm formation. CH-nanoFos inhibited planktonic growth 6 days after treatment at 100 µg/mL, for both bacterial species. Control treatments did not inhibit growth or biofilm synthesis, even one of them is nano size: fosetyl-Al nanocrystals. Then, 15 days later (and after the third administration), the growth of *Xylella fastidiosa* subsp. *pauca* was inhibited from low concentrations (10 µg/mL), while *X. fastidiosa* subsp. *fastidiosa* was not affected. The formation of both bacteria biofilms, after 15 days of treatment, was inhibited only at 100 µg/mL. More examples of NMs applied to quarantined bacteria are cited in Table 1. These types of evidence suggest that nanomaterials must be administered repeatedly in the crop cycle to obtain effective control of the pathogen.

NMs Used for the Quarantine Pathogenic Fungi

Among the most economically important fungal pathogens in plants are *Guignardia citricarpa*, *Guignardia bidwellii*, *Hemileia vastatrix*, Polyphagous beetle-borer complex (*Euwallacea* sp.-*Fusarium euwallaceae*), *Fusarium oxysporum* f. sp. *cubense* Tropical Race 4, *Fusarium guttiforme*, *Moniliophthora perniciosa*, *Phakopsora euvitis*, *Phakopsora pachyrhizi*, *Puccinia graminis* f. sp. *tritici* race Ug99, *Pseudocercospora angolensis*, *Phytophthora palmivora*, *Phytophthora ramorum*, and *Tilletia indica* (EPPO 2022).

In the search for control options for pathogenic fungi, bioreduction of metal ions has been successfully achieved with the use of fungi to obtain metallic nanoparticles (Avila-Quezada et al. 2023b). Several NPs have been applied experimentally as fungicides in plants. As an example, CuNPs have been used against the complex beetle/*Fusarium euwallaceae*. Cu-NPs inhibited fungal growth by 60% and damaged cell morphology, even at very low concentrations (0.1 mg/mL) (Cruz et al. 2021). Moreover, Barrera-Méndez et al. (2019) nanoencapsulated the fungicide propiconazole using polylactic acid polymer and poly(lactic-co-glycolic) acid copolymer as carrier materials against *Fusarium euwallaceae*. In this study, the nanoencapsulated fungicide increases the percentage of growth inhibition of the *Fusarium solani* fungus by 5% compared to the fungicide alone. The release was more prolonged with the nanoencapsulated than the commercial fungicide, which gives it advantages over the commercial product since it increases the availability of the active ingredient.

Recent studies by Caetano et al. (2022) with the pathogenic fungus *Hemileia vastatrix*, the causal organism of coffee rust, demonstrated *in vitro* antifungal activity of 100% at 1,000 µl l^{-1} with nanoencapsulated essential oils of *Eucalyptus citriodora, E. camaldulensis*, and *E. grandis*. It also reduced the severity of the disease by up to 90% with the nanoencapsulated *E. grandis* essential oil.

In the case of *Fusarium oxysporum* f. sp. *cubense* Tropical Race 4 (FocTR4), which is resistant to most fungicides, the active ingredient of the fungicide captan and the lipophilic cations could be nanoencapsulated since, according to Cannon et al. (2022), both chemical compounds suppress the pathogen.

Hydroxyapatite/α-silver vanadate (HA/α-AgVO$_3$) nanocomposites: HA nanorods deposited on α-AgVO$_3$ microrods, inhibit the growth of *Fusarium guttiforme* (da Silva et al. 2020), which cause pineapple fusariosis. Moreover, Cu-chitosan nanoparticles (CuChNp) sprayed inhibited wheat stem rust by *Puccinia graminis* Pers. f. sp. *tritici* in Egyptian wheat genotypes studied by Omar et al. (2021).

Nanomaterials have also been tested for the control of oomycete species. Natsir et al (2021) demonstrated that TiO_2 nanoparticles doped with Cu (Cu/TiO_2) in concentrations less than 2.5% had antifungal activity against the quarantined oomycete *Phytophthora palmivora* with inhibition percentages of up to 75%.

It is well-documented that NPs damage the fungal cell. Metal NPs break the cell wall and cytoplasmic membrane of fungal cells. Cellular content leaks out and the cell becomes distorted (Chwalibog et al. 2010).

A wide variety of nanoparticles have been used for the diagnosis of phytopathogenic fungi and indirectly to reduce the production of mycotoxins in food. In addition, a range of nanosensors has also been used for the diagnosis of phytopathogenic fungi (Kashyap et al. 2019).

Nanotechnology is transforming diagnostic methods and will also transform plant pathogen management methods. NM as a treatment for crop diseases continues to be investigated prior to commercialization and regulatory issuance due to its potential toxicity to humans. Since the diagnosis of phytopathogens is based on internationally agreed protocols published by the EPPO, these instances would regulate the use of NMs for quarantine pathogens.

Conclusion

Several plant pathogens including bacteria and fungi destroy a broad range of plants. Such pathogens are quarantined as they cannot be eradicated. Unfortunately, some of the pathogens have no potential treatment options in the traditional way of control. In this context, nanotechnology has emerged as a potential tool for early diagnosis of the disease using nanobiosensor, the use of nanomaterials, such as the application of different nanoparticles, and their combinations. The development of novel biosensors for the fast detection of the disease has made it possible to treat the disease at an early stage. In addition, such tools are sensitive, portable, and time-saving. The slow and controlled delivery of chemicals will minimize the toxicity to humans and the environment. Nanotechnology has huge potential to diagnose and control plant pathogens on a large scale. Lastly, there is a greater need to study the plant-nanoparticles interaction for a better understanding of plant fitness for sustainable crop production.

References

Alam, T., Khan, R.A.A., Ali, A., Sher, H., Ullah, Z. and Ali, M. 2019. Biogenic synthesis of iron oxide nanoparticles via *Skimmia laureola* and their antibacterial efficacy against bacterial wilt pathogen *Ralstonia solanacearum*. Mater Sci. Eng. C. 98: 101–108.

Álvarez, E., Pantoja, A., Gañán, L. and Ceballos, G. 2015. Current status of Moko disease and the Caribbean, and options for managing them. Centro Internacional de Agricultura Tropical (CIAT), Food and Agriculture Organization of the United Nations (FAO), 40 p. CIAT publication No. 404. https://www.fao.org/3/i3400e/i3400e.pdf.

Ambrin, G., Ahmad, M., Alqarawi, A.A., Hashem, A., Abd-Allah, E.F. and Ahmad, A. 2019. Conversion of cytochrome P450 2D6 of human into a FRET-based tool for real-time monitoring of Ajmalicine in living cells. Front Bioeng. Biotechnol. 7: 375. https://doi.org/10.3389/fbioe.2019.00375.

Arneson, P.A. 2011. Coffee rust. The Plant Health Instructor. https://www.apsnet.org/edcenter/disandpath/fungalbasidio/pdlessons/Pages/CoffeeRust.aspx https://doi.org/10.1094/PHI-I-2000-0718-02.

Atiq, M., Mazhar, H.M.R., Rajput, N.A., Ahmad, U., Hameed, A. and Lodhi, A. 2022. Green synthesis of silver and copper nanoparticles from leaves of *Eucalyptus globulus* and assessment of its antibacterial potential towards *Xanthomonas citri* pv. *citri* causing citrus canker. Appl. Ecol. Environ. Res. 20(3): 2205–2213. http://dx.doi.org/10.15666/aeer/2003_22052213.

Avila-Quezada, G.D. and Espino-Solis, G.P. 2019. Silver nanoparticles offer effective control of pathogenic bacteria in a wide range of food products. pp. 203–211. *In*: Kirmusaoglu, S. and Bhardwaj, S.B. (eds.). Pathogenic Bacteria. IntechOpen Croatia. http://dx.doi.org/10.5772/intechopen.89403.

Avila-Quezada, G., Rai, M., Orduño-Cruz, N., Rivas-Valencia, P., Golinska, P., Muñoz-Castellanos, L., Mercado-Meza, D. and Saenz-Hidalgo, H. 2023a. Nanophytopathology: A new and emerging science. pp. X–x.

In: Avila-Quezada, G. and Rai, M. (eds.). Nanophytopathology. CRC Press Taylor & Francis, USA. http://dx.doi.org/10.

Avila-Quezada, G., Rai, M., Orduño-Cruz, N., Mercado-Meza, D.Y. and Sáenz-Hidalgo, H.K. 2023b. Strategic role of myconanotechnology in agriculture for control of fungal pathogens. pp. 171–192. *In*: Rai, M. and Golinska, P. (eds.). Mycosynthesis of Nanomaterials Perspectives and Challenges. CRC Press Taylor & Francis, USA. http://dx.doi.org/10.1201/9781003327356.

Avila-Quezada, G.D., Esquivel, J.F., Silva-Rojas, H.V., Leyva-Mir, S.G., Garcia-Avila, C., Quezada-Salinas, A., Noriega-Orozco, L., Rivas-Valencia, P., Ojeda-Barrios, D. and Melgoza-Castillo, A. 2018. Emerging plant diseases under a changing climate scenario: Threats to our global food supply. Emir. J. Food Agric. 30(6): 443–450. https://doi.org/10.9755/ejfa.2018.v30.i6.1715.

Avila-Quezada, G.D., Golinska, P. and Rai, M. 2021. Engineered nanomaterials in plant diseases: can we combat phytopathogens? Appl. Microbiol. Biotechnol. 106: 117–129. https://doi.org/10.1007/s00253-021-11725-w.

Avila-Quezada, G.D. and Rai, M. 2022. Diseases of fruits, tubers, and seeds caused by *Phoma* sensu lato species complex. pp. 57–64. *In*: Rai, M., Zimowska, B., Kovics, G.J. (eds.). *Phoma*: Diversity, Taxonomy, Bioactivities, and Nanotechnology. Springer, Cham. Switzerland. https://doi.org/10.1007/978-3-030-81218-8_4.

Baldassarre, F., Tatulli, G., Vergaro, V., Mariano, S., Scala, V., Nobile, C., Pucci, N., Dini, L., Loreti, S. and Ciccarella, G. 2020. Sonication-assisted production of Fosetyl-Al nanocrystals: Investigation of human toxicity and in vitro antibacterial efficacy against *Xylella Fastidiosa*. Nanomaterials 10(6): 1174. https://doi.org/10.3390/nano10061174.

Barrera-Méndez, F., Miranda-Sánchez, D., Sánchez-Rangel, D., Bonilla-Landa, I., Rodríguez-Haas, B., Monribot-Villanueva, J.L. and Olivares-Romero, J.L. 2019. Propiconazole nanoencapsulation in biodegradable polymers to obtain pesticide-controlled delivery systems. J. Mex Chem. Soc. 63(1): 50–60. https://doi.org/10.29356/jmcs.v63i1.564.

Barry, S. and O'Riordan, A. 2015. Electrochemical nanosensors: advances and applications. Rep. Electrochem. 2016(6): 1–14. https://pdfs.semanticscholar.org/f21e/f4d938f905479b61a1184d94a6c24e8b6ede.pdf.

CABI. 2021. Datasheet. *Xanthomonas citri* (citrus canker). Crop Protection Compendium. Global Module. CAB International. UK. https://www.cabi.org/isc/datasheet/56921.

Caetano, A.R.S., Cardoso, M.G., Resende, M.L.V., Chalfuon, S.M., Martins, M.A., Gomes, H.G., Andrade, M.E.R., Brandao, R.M., Campolina, G.A., Nelson D.L. and de Oliveira, J.E. 2022. Antifungal activity of poly (ε-caprolactone) nanoparticles incorporated with *Eucalyptus* essential oils against *Hemileia vastatrix*. Lett. Appl. Microbiol. 75(4): 1028–1041. https://doi.org/10.1111/lam.13782.

Cai, L., Chen, J., Liu, Z., Wang, H., Yang, H. and Ding, W. 2018. Magnesium oxide nanoparticles: effective agricultural antibacterial agent against *Ralstonia solanacearum*. Front Microbiol. 9: 790. https://doi.org/10.3389/fmicb.2018.00790.

Cannon, S., Kay, W., Kilaru, S., Schuster, M., Gurr, S.J. and Steinberg, G. 2022. Multi-site fungicides suppress banana Panama disease, caused by *Fusarium oxysporum* f. sp. *cubense* Tropical Race 4. PLoS Pathog. 18(10): e1010860. https://doi.org/10.1371/journal.ppat.1010860.

Chiriacò, M.S., Luvisi, A., Primiceri, E., Sabella, E., De Bellis, L. and Maruccio, G. 2018. Development of a lab-on-a-chip method for rapid assay of *Xylella fastidiosa* subsp. *pauca* strain CoDiRO. Sci. Rep. 8: 7376. https://doi.org/10.1038/s41598-018-25747-4.

Chwalibog, A., Sawosz, E., Hotowy, A., Szeliga, J., Mitura, S., Mitura, K., Grodzik, M., Orlowski, P. and Sokolowska, A. 2010. Visualization of interaction between inorganic nanoparticles and bacteria or fungi. Int. J. Nanomed. 5: 1085–1094. https://doi.org/10.2147/IJN.S13532.

Croisier, F. and Jérôme, C. 2013. Chitosan-based biomaterials for tissue engineering. Eur. Polym. J. 49(4): 780–792. https://doi.org/10.1016/j.eurpolymj.2012.12.009.

Cruz, L.F., Cruz, J.C., Carrillo, D., Mtz-Enriquez, A.I., Lamelas, A., Ibarra-Juarez, L.A. and Pariona, N. 2021. *In-vitro* evaluation of copper nanoparticles as a potential control agent against the fungal symbionts of the invasive ambrosia beetle *Euwallacea fornicatus*. Crop Protec. 143: 105564. https://doi.org/10.1016/j.cropro.2021.105564.

da Silva, J.S., Machado, T.R., Trench, A.B., Silva, A.D., Teodoro, V., Vieira, P.C., Martins, T.A. and Longo, E. 2020. Enhanced photocatalytic and antifungal activity of hydroxyapatite/α-AgVO$_3$ composites. Mater Chem. Phys. 252: 123294. https://doi.org/10.1016/j.matchemphys.2020.123294.

Deepmala, K., Hemantaranjan, A., Bharti, S. and Nishant Bhanu, A. 2014. A future perspective in crop protection: Chitosan and its oligosaccharides. Adv. Plants Agric. Res. 1(1): 1–8. http://dx.doi.org/10.15406/apar.2014.01.00006.

Draz, M.S. and Shafiee, H. 2018. Applications of gold nanoparticles in virus detection. Theranostics 8(7): 1985–2017. http://dx.doi.org/10.7150/thno.23856.

Dzimitrowicz, A., Motyka-Pomagruk, A., Cyganowski, P., Babinska, W., Terefinko, D., Jamroz, P., Lojkowska, E., Pohl, P. and Sledz, W. 2018. Antibacterial activity of fructose-stabilized silver nanoparticles produced by direct current atmospheric pressure glow discharge towards quarantine pests. Nanomaterials 8(10): 751. https://doi.org/10.3390/nano8100751.

EPPO. 2022a. EPPO A1 and A2 lists of pests recommended for regulation as quarantine pests. https://www.eppo.int/media/uploaded_images/RESOURCES/eppo_standards/pm1/pm1-002-31-en_A1A2_2022.pdf.

EPPO. 2022b. *Tilletia indica* datasheet. https://gd.eppo.int/taxon/NEOVIN/datasheet.

EPPO. 2022c. *Xylella fastidiosa* subsp. *fastidiosa* (XYLEFF). EPPO Global database. https://gd.eppo.int/taxon/XYLEFF/distribution.

EPPO. 2021. EPPO Datasheet: *Ralstonia solanacearum* species complex. https://gd.eppo.int/taxon/RALSSO/datasheet.

Graham, J.H., Johnson, E.G., Myers, M.E., Young, M., Rajasekaran, P., Das, S. and Santra, S. 2016. Potential of nanoformulated zinc oxide for control of citrus canker on grapefruit trees. Plant Dis. 100(12): 2442–2447. https://doi.org/10.1094/pdis-05-16-0598-re.

Hazra, D.K., Karmakar, R., Poi Rajlakshmi, Bhattacharya, S. and Mondal, S. 2017. Recent advances in pesticide formulations for eco-friendly and sustainable vegetable pest management: A review. Arch. Agric. Environm. Sci. 2(3): 232–237. https://www.aesacademy.org/journal/volume2/issue3/AAES-02-03-017.pdf.

ISPM 5, 2018. International standards for phytosanitary measures. Glossary of phytosanitary terms. International Plant Protection Convention. https://www.ippc.int/static/media/files/publication/en/2018/06/ISPM_05_2018_En_Glossary_2018-05-20_PostCPM13_R9GJ0UK.pdf.

Kahraman, M., Mullen, E.R., Korkmaz, A. and Wachsmann-Hogiu, S. 2017. Fundamentals and applications of SERS-based bioanalytical sensing. Nanophotonics. https://doi.org/10.1515/nanoph-2016-0174.

Kashyap, P.L., Kumar, S., Jasrotia, P., Singh, D.P. and Singh, G.P. 2019. Nanosensors for plant disease diagnosis: current understanding and future perspectives. *In*: Pudake, R., Chauhan, N. and Kole, C. (eds.). Nanoscience for Sustainable Agriculture. Springer, Cham. https://doi.org/10.1007/978-3-319-97852-9_9.

Khaledian, S., Nikkhah, M., Shams-bakhsh, M. and Hoseinzadeh, S. 2017. A sensitive biosensor based on gold nanoparticles to detect *Ralstonia solanacearum* in soil. J. Gen. Plant Pathol. 83: 231–239. https://doi.org/10.1007/s10327-017-0721-z.

Khan, M., Siddiqui, Z.A., Parveen, A., Khan, A.A., Moon, I.S. and Alam, M. 2022. Elucidating the role of silicon dioxide and titanium dioxide nanoparticles in mitigating the disease of the eggplant caused by *Phomopsis vexans*, *Ralstonia solanacearum*, and root-knot nematode *Meloidogyne incognita*. Nanotechnol. Rev. 11(1): 1606–1619. https://doi.org/10.1515/ntrev-2022-0097.

Kwak, S-Y., Wong, M.H., Lew, T.T.S., Bisker, G., Lee, M.A., Kaplan, A., Dong, J., Liu, A.T., Koman, V.B., Sinclair, R., Hamann, C. and Strano, M.S. 2017. Nanosensor technology applied to living plant systems. Ann. Rev. Anal. Chem. 10: 113–140. https://doi.org/10.1146/annurev-anchem-061516-045310.

Lau, H.Y., Wu, H., Wee, E.J.H., Trau, M., Wang, Y. and Botellab, J.R. 2017. Specific and sensitive isothermal electrochemical biosensor for plant pathogen DNA detection with colloidal gold nanoparticles as probes. Sci. Rep. 7: 38896. https://doi.org/10.1038/srep38896.

Liao, C., Li, Y. and Tjong, S.C. 2019. Bactericidal and cytotoxic properties of silver nanoparticles. Int. J. Mol. Sci. 20(2): 449. https://doi.org/10.3390/ijms20020449.

Muniandy, S., Teh, S.J., Thong, K.L., Thiha, A., Dinshaw, I.J., Lai, C.W., Ibrahim, F. and Leo, B.F. 2019. Carbon nanomaterial-based electrochemical biosensors for foodborne bacterial detection. Crit. Rev. Anal. Chem. 49(6): 510–533. https://doi.org/10.1080/10408347.2018.1561243.

Natsir, M., Maulidiyah, M., Watoni, A.H., Arif, J., Sari, A., Salim, L. O.A., Sarjuna, S., Irwan I. and Nurdin, M. 2021. Synthesis and charcterization of Cu-doped TiO_2 (Cu/TiO_2) nanoparticle as antifungal *Phytophthora palmivora*, in J. Physics: Conference Series (Vol. 1899, No. 1, p. 012039). IOP Publishing. https://doi.org/10.1088/1742-6596/1899/1/012039.

Omar, H.S., Al Mutery, A., Osman, N.H., Reyad, N.E.H.A. and Abou-Zeid, M.A. 2021. Genetic diversity, antifungal evaluation and molecular docking studies of Cu-chitosan nanoparticles as prospective stem rust inhibitor candidates among some Egyptian wheat genotypes. PloS One 16(11): e0257959. https://doi.org/10.1371/journal.pone.0257959.

Perdomo, S.A., Marmolejo-Tejada, J.M. and Jaramillo-Botero, A. 2021. Bio-Nanosensors: fundamentals and recent applications. J. Electrochem. Soc. 168(10): 107506. https://doi.org/10.1149/1945-7111/ac2972.

Pramanik, P., Krishnan, P., Maity, A., Mridha, N., Mukherjee, A. and Rai, V. 2020. Application of nanotechnology in agricultura. pp. 317–348. *In*: Dasgupta, N., Ranjan, S. and Lichtfouse, E. (eds.). Environmental Nanotechnology Volume 4. Environmental Chemistry for a Sustainable World. Springer, Cham. Switzerland. https://doi.org/10.1007/978-3-030-26668-4_9.

Pretorius Z.A., Singh, R.P., Wagoire, W.E. and Payne, T.S. 2000. Detection of virulence to wheat stem rust resistance gene Sr31 in *Puccinia graminis.* f. sp. *tritici* in Uganda. Plant Dis. 84(2): 203. https://doi.org/10.1094/PDIS.2000.84.2.203B.

Rad, F., Mohsenifar, A., Tabatabaei, M., Safarnejad, M.R., Shahryari, F., Safarpour, H., Foroutan, A., Mardi, M., Davoudi, D. and Fotokian, M. 2012. Detection of *Candidatus* Phytoplasma aurantifolia with a quantum dots FRET-based biosensor. J. Plant Pathol. 94(3): 525–534. http://www.jstor.org/stable/45156279.

RustTracker org, 2022. A global wheat rust monitoring system. Ug99 (Race TTKSK). https://rusttracker.cimmyt.org/?page_id=260.

Senasica. 2019a. Moko del platano. *Ralstonia solanacearum* race 2 Smith. Code EPPO: PSDMS2. Aviso público de riesgo y situación actual. https://prod.senasica.gob.mx/SIRVEF/ContenidoPublico/Avisos%20y%20alertas/Avisos%20publicos/Aviso%20p%C3%BAblico%20Moko%20del%20pl%C3%A1tano.pdf.

Senasica. 2019b. Enfermedad de Pierce (*Xylella fastidiosa* subsp. *fastidiosa*). Dirección General de Sanidad Vegetal-Programa de Vigilancia Epidemiológica Fitosanitaria. Cd. De México. Ficha Técnica No. 26. 14 p. https://prod.senasica.gob.mx/SIRVEF/ContenidoPublico/Fichas%20tecnicas/Ficha%20T%C3%A9cnica%20de%20enfermedad%20de%20Pierce.pdf.

Senasica. 2019c. Roya negra del tallo del trigo raza Ug99 (*Puccinia graminis* f.sp. *tritici* raza Ug99). Dirección General de Sanidad Vegetal-Sistema Nacional de Vigilancia Epidemiológica Fitosanitaria. Cd. De México. Ficha Técnica No. 25. 18 p. https://prod.senasica.gob.mx/SIRVEF/ContenidoPublico/Fichas%20tecnicas/Ficha%20T%C3%A9cnica%20de%20Roya%20del%20tallo%20del%20trigo%20Ug99.pdf.

Shoala, T. and Abd-El-Aziz, R.M. 2019. Silver and magnetic iron oxide nanoparticles-assisted PCR for the phytopathogenic bacteria *Ralstonia solanacearum.* J. Plant Protec. Pathol. 10(9): 471–476. https://doi.org/10.21608/JPPP.2019.63659.

Srivastava, A.K., Dev, A. and Karmakar, S. 2018. Nanosensors and nanobiosensors in food and agriculture. Environ. Chem. Lett. 16(1): 161–182. https://doi.org/10.1007/s10311-017-0674-7.

Strona, G., Castellano, C., Fattorini, S., Ponti, L., Gutierrez, A.P. and Beck, P.S. 2020. Small world in the real world: Long distance dispersal governs epidemic dynamics in agricultural landscapes. Epidemics 30: 100384. https://doi.org/10.1016/j.epidem.2020.100384.

Tortella, G., Navas, M., Parada, M., Durán, N., Seabra, A.B., Hoffmann, N. and Rubilar O. 2019. Synthesis of silver nanoparticles using extract of weeds and optimized by response surface methodology to the control of soil pathogenic bacteria *Ralstonia solanacearum.* J. Soil Sci. Plant Nutr. 19: 148–156. https://doi.org/10.1007/s42729-019-00021-2.

USDA. 2020. Plant Protection and Quarantine: Helping U.S. Agriculture Thrive-Across the Country and Around the World. https://www.aphis.usda.gov/publications/plant_health/report-ppq-2020.pdf.

Varympopi, A., Dimopoulou, A., Papafotis, D., Avramidis, P., Sarris, I., Karamanidou, T. et al. 2022. Antibacterial activity of copper nanoparticles against *Xanthomonas campestris* pv. *vesicatoria* in tomato plants. Int. J. Mol. Sci. 23(8): 4080. https://doi.org/10.3390/ijms23084080.

Vocke, G., Allen, E.W. and Price, J.M. 2002. Economic Analysis of Ending the Issuance of Karnal Bunt Phytosanitary Wheat Export Certificates. Economic Research Service/USDA. Economic Analysis of Ending the Issuance of Karnal Bunt Phytosanitary Wheat Export Certificates. https://www.aphis.usda.gov/plant_health/plant_pest_info/kb/downloads/2000ERSEcoAnalysisStopPhytos.pdf.

Chapter 20

Recent Advances in Using Nanotechnology in the Management of Soilborne Plant Pathogens

Mohamed A. Mosa,[1,2] *Dina S. S. Ibrahim,*[3,*] *Mohamed A. M. El-Tabakh*[4] and *Sozan E. El-Abeid*[1,2]

Introduction

Although more than 18 distinct nanoparticles (NPs) of carbon nanomaterials and single elements have been reported to alter disease and/or plant infections, silver (Ag), copper (Cu), and zinc (Zn) were the only metals that have received great attention so far. Some NPs work instantly as antibacterial agents, while others affect the nutritional state of the host, thus activating defensive systems. Some NPs of Ag and Cu can be directly harmful to microorganisms, whereas NPs of Mn, Si, B, Cu, and Zn appear to operate as fertilizers in host defense (Elmer et al. 2018). As food production demands rise in response to warmer weather, nanoparticles would reduce the increased hurdles in disease control, as a result of which active metals and other chemical inputs are reduced. In addition to being necessary for cell wall formation, copper is essential for oxidative stress defense, carbon and nitrogen metabolism, photosynthesis, and mitochondrial respiration (Ashraf et al. 2021).

According to Maroufpour et al. (2020), numerous NPs have been investigated for their potential use in agriculture, including nano-zinc oxide (ZnO), nano-silica (SiO_2), nanosilver (Ag), nano-titanium dioxide (TiO_2), nano-copper (Cu), carbon nanotubes, and nano-aluminum. In addition, nano-copper (Cu), carbon nanotubes, and nano-aluminum were reported for their potential in agriculture (Baazaoui et al. 2021).

Numerous of these encounters lead to the oxidation of the spores' surface chemicals, which ultimately results in cell death (Lin and Xing 2007). Recent research was critically reviewed in this study to establish the extent to which nanomaterials affect commercial items. There is currently no systematic research in the literature examining the efficacy and environmental impact of nano-agroproducts as antimicrobials, particularly soil microorganisms. This is a serious knowledge

[1] Nanotechnology and Advanced Nanomaterials Laboratory (NANML), Plant Pathology Research Institute, Agricultural Research Center, Giza 12619, Egypt.
[2] Mycology and Disease Survey Research Department, Plant Pathology Research Institute, Agricultural Research Center, Giza 12619, Egypt.
[3] Nematology and Biotechnology Department of Nematodes Diseases and Central Lab of Biotechnology, Plant Pathology Research Institute, Agriculture Research Center, 9 Cairo University St., Giza 12619, Egypt
[4] Zoology Department, Faculty of Science, Al-Azhar University, Cairo, Egypt
* Corresponding author: Dina.Serag@arc.sci.eg

gap, thus further research will be required to adequately analyze the new advantages and hazards that agrochemical nanostructures imply for current products.

The aim of this chapter is to examine the role of nanoparticles as a highly efficient antimicrobial agent, as well as their mode of action. Furthermore, we will focus on commercial products and the use of nanopesticides to have nanoparticle photocatalytic and antimicrobial mechanisms. Also, the potential and difficulties of using nanopesticides in the future to control plant pests and diseases.

Comparison Between Different Nanoparticles Movement and Effect on Plant Infection

Silver-NPs Compounds

Recently, nanoparticles had attention as an applicable pesticide, with some afraid of their future toxic effect and this uncontrollable and insensible distribution. Nevertheless, silver nanoparticles have optical and antimicrobial characteristics. Also, nano-Ag can be used as an antifungal and antibacterial agent through their reduction of rapid growth or prevention of microbes growth. Importantly, these properties can be boosted when combined within the composite.

Silver acts as a powerful antibacterial, antifungal, as well as antinematode. Silver was one of the most popular and frequently reported materials responsible for a wide variety of uses. Inversely, it appears to be playing a powerful antimicrobial role in the soil in the literature that has tested the strength of silver for use as an alternative pesticide. The controversy over its uses in agriculture still needs to be controlled. Various techniques are used to synthesize the NPs, which then affect the shape, size, and efficiency. Zake et al. (2022) mentioned that the silver nanoparticles produced by the bioagent as *Trichoderma harzianum* (bio-synthesized) greatly reduced soilborne diseases (*Fusarium, Rhizoctonia, Macrophomina*) in cotton seedlings.

Antimicrobial activity of silver hydrogen peroxide (SHP) against selected fungi (*Fusarium solani, Pythium aphanidermatum*) and bacteria (*Ralstonia solanacearum*) *in vitro* was observed (Mahesha et al. 2021). Because of their size-dependent characteristics, the surface-to-volume ratio is very high, as well as distinct physiochemical characteristics, nanoparticle-based antibacterial medicines have evolved as a multidisciplinary subject integrating medicine, biology, material science, and chemistry.

Chen et al. (2016) used electrostatic self-assembly at the interface to construct a studded GO-AgNPs nanocomposite and assessed for its antifungal activity against the phytopathogen *Fusarium graminearum in vitro* and *in vivo* for the first time. When compared to pure AgNPs and GO suspension, the GO-AgNPs nanocomposite demonstrated a roughly 3-fold and a 7-fold increase in inhibitory efficiency, respectively. Hyphae and the spores were destroyed; this might be related to an antibacterial process as a result of the exceptional synergistic impact of GO-AgNPs, generating physical harm and the formation of chemically reactive oxygen species. All of the results point to the GO-AgNPs nanocomposite created as a potential material for the development of new antibacterial agents against pathogenic fungi or bacteria.

Le et al. (2019) tested the AgNPs as antifungal agents against *Phytophthora capsici* in composite with chitosan and utilized this composite as nano-antibiotic materials as more efficient and promising materials.

Matei et al. (2018) used AgNPs against *Phytophthora cinnamomic*. When the AgNPs were formulated with polyphenols, such as garlic acid, ferulic acid, curcumin, and silymarin, and prepared in an aqueous solution or a solution DES (urea/chloride/glycerin), the inhibition was more pronounced in DES solution than any other solutions used the reduction was more pronounced in the DES solution than in any other solutions used to reduce *Phytophthora cinnamomi* growth.

The effect of AgNPs on *Fusarium solani* is investigated by using a transmission electron microscope to observe the role of the AgNPs on fungal hyphae changes and the distribution of the element in the cell wall, membranes, and cytoplasm (Shen et al. 2020). According to an electron-microscopic examination, the treated cells had blurred cell walls and contracting membranes. AgNPs were detected on the membrane of the cell. According to Gordienko et al. (2019) AgNPs altered the integrity, structure, and activity of the membrane enzymes of the cell. Furthermore, AgNPs were determined to be small enough to pass the bacterium and cause cell death. The researchers also observed that virtually all of the DEGs encoding carbohydrate, nucleotide, fatty acid, and amino acid metabolism were down-regulated in the exposed group during the 6–12 hours period but recovered after 24 hours. Furthermore, their research found a reduction in the hexokinase gene, a critical enzyme in glucose metabolism. Many genes, particularly isocitrate dehydrogenase, which is involved in the citrate cycle (TCA cycle), were differentially activated after being exposed to AgNPs for 24 hours. There was also a decrease in the glycolysis-gluconeogenesis pathways that create ATPs. As a second messenger, manufacturing reactive oxygen species (ROS) is crucial in several signaling pathway events. Excessive ROS can accelerate apoptosis and potentially cause cell death. The rapid response to a typical metal and nanoparticle stress response increased ROS production (Tian et al. 2018).

In the genetic study, the genes cysC, cysl, cysH, ssuD, and MET17, which are involved in sulfur metabolism, were interestingly enhanced when exposed to AgNPs after 6–12 hours (Mars 1966; Rosen 2002; Ahsan et al. 2008; Shen et al. 2020).

When AgNPs decreased the *Fusarium solani* species complex, the abundance of HSP proteins increased dramatically. The role of PI3K and MAPK, which are engaged in a variety of cellular processes, needs to be investigated further. The genes that code after being exposed to AgNPs, and the proteins for both the "homologous recombination pathway" and "nucleotide excision repair pathway" increased in *F. solani* cells (Shen et al. 2020).

Gupta et al. (2018) studied the spectrum of activity, bio-applicability, and eco-sensitivity of silver nanoparticles. These silver nanoparticles have a phyto-stimulatory impact on the rice plants (*Oryza sativa* L. cv. Swarna), seedlings development, and germination. *In vitro* experiments showed no stress of AgNPs on rice seedlings. Very low ROS levels, as well as a lower amount of H_2O_2 content and lipid peroxidation, were observed in the promotion of rice seedling development. Increased vegetative growth, root length, and biomass for rice seedlings were also observed. Increased chlorophyll content and phenolic determination refer to a non-toxic impression and a low ROS level, while also decreasing peroxidation of the lipid and H_2O_2 amount. All of these indicators suggest no toxic stress of AgNPs on rice seedlings. For detecting and proving to induce antioxidant enzyme activity changes like an increase in catalase (CAT) and ascorbate peroxidase (APX), all of these indicators that enzymes of the ascorbate cycle are significantly stimulated glutathione reductase (GR), in anticipation of oxidative damage and no changes were recorded in superoxide dismutase (SOD). With regard to changes at the genetic level, the expression of CAT and APX gene increased up-regulated in seedling exposure to AgNPs since decreased the level of the expression of CuZnSOD gene in parallel even though the AgNPs concentration.

Moreover, Martinez-Gutierrez et al. (2010) indicated that plants may absorb AgNPs that have been placed into the surroundings; this might have ramifications for both the environment and humans silver accumulation in plant tissues can have a detrimental effect on the membrane transporters and vascular tissues that move water and other nutrients. At concentrations higher than 5%, silver nanoparticles' cytotoxic impact on THP-1 cells by employing a "comet assay" test to gauge the DNA damage was significantly increased at a concentration of over 5 µg/ml. The first generation of silver nanoparticles is an excellent candidate for use as antimicrobials against bacteria. The experiment's findings demonstrated that nanoparticle exposure to THP-1 cells did not result in any DNA damage. In this study, authors demonstrated that antibacterial properties of AgNPs was significantly enhanced when the size of AgNPs decreased from 29 nm to 20–25 nm (Zapór 2016; Noori et al. 2020).

Silver nanoparticles (AgNPs) influence genotoxicity by influencing cellular absorption, which determines a genotoxic cell response. Mutagenicity, clastogenicity, and DNA strand-break damage were all evaluated. AgNPs of all diameters (10 nm, 20 nm, 50 nm, and 100 nm) were shown to be mutagenic in bacteria, as was silver nitrate ($AgNO_3$). TEM found no AgNP within the bacterial cells, indicating that these bacteria lack the ability to consume AgNP 10 nm or larger. The data suggest that silver ions are the primary and maybe only cause of genotoxicity. The flow cytometry-based micronucleus test and the Comet assay were used to assess clastogenicity and intermediate DNA damage in two mammalian white blood cell lines: E6-1 Jurkat Clone and THP-1. 2015 (Butler et al. 2015).

The transfer of AgNPs into plants can occur via different pathways. Through the tiny holes in the cell wall known as plasmodesmata, AgNPs may enter the plants. Plants may be exposed to dissolved and nanoscale forms of silver because of this. To safeguard crops and prevent tainted food from reaching humans, it is essential to comprehend the destiny, localization, and behavior of AgNP in plant tissues. Plasmodesmata have a size of 50 nm on average (Reagan et al. 2018). Apoplastic or symplastic pathways might then be used by particles to penetrate cells or go to vascular tissues (Tripathi et al. 2017; Yan and Chen 2019). AgNPs may go from the xylem to the plant's aerial parts (Yin et al. 2012; Pérez-de-Luque 2017; Cocozza et al. 2019). As exposed to all types of silver, the membrane transporters potassium transporter, H+-ATPase, and sulfate transporter were increased by 23.50%, 52.09%, and 7.6%, respectively compared to the control group. These changes impact plant cell electrochemical potential, water and nutrient dynamics, and plant development (Noori et al. 2020).

Titanium NPs

TiO_2 has been employed in pesticide formulations as a plant growth stimulant or catalyst; however, dissolved Ti is unlikely to be an active component in these formulations. Regardless of soil Ph, the bulk of TiO_2 NP in soil solutions remains in particulate form. The filamentous fungus *Aspergillus niger* may accumulate bioavailable TiO_2 NPs. Approximately 7.5 percent of TiO_2 NPs were conserved in fungal mycelia, suggesting that some soil fungi may influence the geographical distribution of this type of nanoparticle in soils. The pH was also found to have a major influence on the stability of TiO_2 NPs in soils. Their dissolving rate improves significantly as soil acidity increases (Šebesta et al. 2020).

Aslam et al. (2021), TiO_2NPs properties can also be described as eco-friendly biocide properties (Aslam et al. 2021). In particular, antimicrobial characteristics of metal oxide NPs, have been intensively explored due to their positive influence, with encouraging results published. When TiO_2 nanoparticles interact with microbial cells, they start the production of ROS, which can efficiently kill bacteria by compromising the integrity of their cell walls, mostly by phospholipid oxidation, which lowers adhesion and alters ion balance. It also inhibits respiratory cytosolic enzymes within the cytosol and alters macromolecule forms. It reduces phosphate consumption and cellular contact across the cell.

In 2020, research conducted by Irshad et al., synthetic TiO-NPs were evaluated for their ability to combat plant diseases, especially wheat rust (*Ustilago tritici*). Less than 15 nm was typical of TiO_2 NPs size made using chemical and green methods. Both sol-gel and green-prepared TiO_2-NPs exhibit effective antifungal effects encountered in *U. tritici*, but green-prepared NPs had the highest antifungal activity. This means that a green approach may be used to produce TiO_2NPs on a wide scale while presenting less risk.

Zhang et al. (2013) indicated that TiON/PdO nanocomposite was prepared using a palladium-activated visible light as a high-efficiency photocatalytic. This composite is used as complete disinfection of *Fusarium graminearum* macroconidia causes head blight in wheat, and produces various mycotoxins. These nanoparticles adsorbed on the surface of macroconidia then caused membrane and cell wall damage to the existence of ROS species. Satti et al. (2022) investigated TiO_2NPs on *Puccinia striiformis* and discovered that they were beneficial in triggering metabolic modifications to alleviate stress in plants. Titanium nanoparticles were also shown to increase phytohormone production, including gibberellins (GA) and indole acetic acid (IAA) while decreasing ABA synthesis. These hormones regulate cellular activities such as development, germination, and the response of the plant to drought stress. Also improved was the chlorophyll concentration; all of the following parameters were calculated: relative water content, membrane stability index, and osmolyte concentration (proline and sugar). The combination application TiO_2NPs increases superoxide dismutase, peroxidase, and catalase enzymes (Kamal and Mogazy 2021; Mustafa et al. 2021). The effects of titanium dioxide on *A. thaliana* tissue transcriptomes and productivity were explored using genetic analysis. Nano-titanium induced more differentially expressed genes in rosette leaves and roots, including photosynthesis-related genes. Using map-man analyses, whereas metabolic pathways in both tissues were typically increased in nano-titania, these nanoparticles enriched ontology areas such as defensive responses to pathogens, reactions to endogenous stimuli, and responses to stress (abiotic and biotic); also, nano-titanium activated many photosynthesis-related genes (Tumburu et al. 2017).

The cytotoxicity of titanium dioxide nanoparticles at various concentrations was investigated in the bronchial epithelial cell line culture for humans at 5 g/ml, 10 g/ml, 20 g/ml, and 40 g/ml. Cell death was produced by nanoparticle exposure, as was the activation of oxidative stress-related genes, such as heme glutathione-S-transferase, thioredoxin reductase, oxygenase-1, catalase, and a hypoxia-inducible gene. The p38 mitogen-activated protein kinase (MAPK) pathway and/or extracellular signaling were used to create IL-8 (ERK). Titanium dioxide nanoparticles seemed to infiltrate the cytoplasm and accumulate in the nucleus' peri-region, perhaps encouraging direct interactions between the particles and cellular components, resulting in negative biological consequences. Additionally, Park et al. (2008) discussed that TiO_2, ZnO-NPs, and their combination substantially increased chromosomal aberrations (CAs) and decreased the mixture in *A. cepa* root cells. When treated meristematic cells were put in water for recovery, the number of aberrant cells decreased. The interactive factor analysis of the mixture's effects revealed antagonism. This discovery, which demonstrates the ability of investigated NPs to cause a somatic cell mutation, has public and environmental health implications (Fadoju et al. 2020).

Titanium translocation on diverse plant species revealed that *R. crispus*, in both 10 mg L^{-1} and 30 mg L^{-1} inclusion of nanoparticles, demonstrated substantial Ti translocation. Plants may move ionic Ti from the soil to the branches (Jacob et al. 2013).

AgNPs and TiNPs produced mitochondrial hypertrophy and increased ROS. Long-term NP exposure maintained high ROS levels while diminishing the endogenous antioxidant system. Changes in mitochondrial function, specifically an uncoupling of the oxidation of the phosphorylation membrane system, have been linked to NP cytotoxicity. Both worked synergistically, with the negative effect on mitochondrial redox status being larger when both were present (Pereira et al. 2018).

ZnO-NPs

Isah and Garba (2022) demonstrated zinc oxide nanoparticles (ZnO-NPs) were tested for antimicrobial activity against date fruit pathogens. Antibacterial activity increases in direct proportion to the concentration of ZnO. The increasing concentration causes bacterial cell death, inhibits the function of mitochondria, increases leakage of lactate dehydrogenase, and changes the cell shape. The influence of size and concentration was explored, and the larger surface area of ZnO allowed for strong antibacterial action. Several factors, including bacterial type and ZnO-NPs size, affect the size of the bacteria's inhibitory zone.

Zinc is a mineral that is essential for the growth and development of plants because it plays a role as a cofactor in more than 300 different proteins, some of which include zinc finger proteins and RNA and DNA polymerases. Oxidoreductase, catalase, hydrolases, lyases, isomerize, and ligases are only a few of the enzyme types that contain it. In addition to its primary involvement in membrane integrity and the alleviation of oxidative stress, it also functions as a hormone regulator (tryptophan synthesis and their role in IAA), and signal transmission through mitogen-activated protein kinases. It considers the structural or catalytic unit component or regulating activity in addition to its primary involvement in membrane integrity and the alleviation of oxidative stress (Lopez Millan et al. 2005; Hansch and Mendel 2009; Gupta et al. 2016). The primary factor controlling the distribution of zinc in the soil is the pH level, which affects how soluble zinc is in the soil solution. Zinc is more likely to bind to cation exchange sites on soil components when the pH of the soil is higher. This results in a decrease in the amount of zinc available in the soil.

For each unit rise in soil pH between 5.5 and 7.0, the Zn content in the soil drops noticeably by 30 to 45 times. Particularly in soils with low levels of soluble organic matter, zinc solubility and the ratio of Zn^{2+} to an organic-Zn complex ligand rises as pH levels get lower. Another physical aspect is soil moisture that promotes plant roots to absorb zinc by diffusion (Gupta et al. 2016).

Cu NPs

Cu^{1+} and Cu^{2+} are the two oxidation states of copper, and monovalent copper is unstable. Because of its capacity to accelerate the generation of free radicals, which can harm proteins, DNA, and other macromolecules, copper is potentially hazardous due to this feature. To stop copper from developing in a toxic form, the overwhelming majority of copper ions bound via protein scavenging, such as metallothioneins, right after intake. Since dioxygen molecules have a strong affinity for copper, many oxidases use copper as their catalytic metal. Copper is found in chloroplasts and is involved in photosynthetic activity, making up more than half of the copper in plants. Copper deficiency is most noticeable in immature leaves and reproductive organs, followed by stunted development and light green leaves that wither quickly (Hansch and Mendel 2009). Copper nanoparticles inhibited bacteria more than just fungus; they also showed a greater inhibitory zone in *Escherichia coli* (26 mm) than *Candida albicans* (23 mm) (Ramyadevi et al. 2012). Furthermore, all CuNPs demonstrated considerable Gram-positive antibacterial activity (*Listeria monocytogenes*) and Gram-negative antibacterial activity (*Escherichia coli*) pathogens found in food (Sankarand Rhim 2014). Nano-copper-treated plants dramatically improved soil macronutrients compared to bulk copper and carbendazim-treated plants. Furthermore, there was a minor alteration in microbial population dynamics in plants treated with nano-copper. A positive finding showed nano-copper might be an effective new fungicide for the treatment of tea plants that suffer from red root rot (Ponmurugan et al. 2016; Pham et al. 2016; Viet et al. 2016).

Akturk et al. (2020) reported that capped soluble starch Cu NPs (CuS NPs) and sodium alginate capped Cu NPs (CuA NPs) were originally made utilizing copper sulfate, hydrazine, soluble starch, and sodium alginate as reduction, oxidation, and capping agents, respectively. The findings revealed that CuS NPs outperform CuA NPs in terms of antifungal activity and stability in the open environment. CuS NPs are thus regarded as having the ability to be utilized in fields as well as biotechnology and packaging food.

Copper oxide nanoparticles (CuO NPs) have been the subject of numerous investigations due to their fascinating properties and potential applications in various fields. Here are some key points about the role of copper oxide nanoparticles:

- *Optical properties*: Copper nanoparticles (Cu NPs) exhibit a phenomenon called Localized Surface Plasmon Resonance (LSPR), which is influenced by their size, shape, and density. The optical characteristics of copper NPs have been extensively studied for potential applications in catalysts, biosensors, optoelectronic devices, and optical devices (Maliki et al. 2022) .

- *Anticancer and antibacterial properties*: Copper oxide nanoparticles have shown promise as potential therapeutic agents for the treatment of cancer and bacterial infections. Studies have highlighted their remarkable anticancer and antibacterial properties (Maliki et al. 2022). However, it's worth noting that most approaches are still in the research phase, and additional work is required to address challenges and bring copper nanosystems loaded with anticancer and antibacterial agents to clinical applications (Chavali and Nikolova 2019).

Iron NPs

Iron is a metal that undergoes redox reactions involved in nitrogen absorption, photosynthesis, hormone production (ethylene, gibberellic acid, and jasmonic acid), mitochondrial respiration, ROS formation in addition to scavenging, osmoprotectant, and pathogen defense. The chloroplasts contain up to 80% of the cellular iron and play a fundamental role in photosynthesis (Hansch and Mendel 2009).

Antifungal activity of Fe_2O_3 nanoparticles was evaluated by suppressing spore germination and defining fungal pathogen inhibition zones induced by varying the concentrations of iron oxide nanoparticles in growth media at concentrations of 0.063 and 0.016 mg/ml, which gave it anti-mycotic efficacy against different fungi investigated (Parveen et al. 2018). Iron oxide-based NPs have the benefit, even as they discover particular localization to deliver their load, which is critical in the research of nanoparticle delivery in plants. Furthermore, iron oxide nanoparticles are ecologically friendly, cheaply accessible or produced, magnetically sensitive, redox reactive, and biooperable. Iron is a microelement component of chlorophyll and is involved in various physiological functions. Furthermore, the phytotoxicity profile of nanoparticles was tested using root elongation tests and seed germination, which revealed NPs have no deleterious influence on plant physiology (González-Melendi et al. 2008; Zhu et al. 2008). As a result, iron oxides are reasonably safe for delivering nanoparticles to plants. As a result, an attempt was undertaken to investigate the influence of Fe_2O_3 NPs on seed growth germination and growth transferred across plant tissue. Fe_2O_3 NPs accumulate preferentially in root hairs, root tips, nodal zones, and the plant's middle zone. Reduced ferric to ferrous iron is due to the presence of many phytoconstituents in the plant's body. The growth of *Solanum lycopersicum* plants cultivated with NPs has no detrimental impact (ca. toxicity) (Shankramma et al. 2016). When chitosan-Fe NPs concentration was boosted, antifungal activity *in vitro* and *in vivo* significantly improved. The overall findings showed that organometallic $CH-Fe_2O_3$ NPs had a high level of synergistic antifungal activity against *Rhizopus oryzae* and recommended using $CH-Fe_2O_3$ NPs against additional phytopathological diseases because of their biodegradability (Saqib et al. 2019).

Chitosan NPs

Chitosan is a polymeric polysaccharide with antibacterial characteristics and can activate plant defense responses. It has the ability to inhibit the oomycete disease *Phytophthora nicotianae* (Breda de Haan), which affects different plants such as tomato crops. This polymer had no effect on colony shape or diameter, but it did lower the number of oospores at the lowest dose tested. The zoospores created cysts that could germinate and were thus unable to infect the roots of tomato plants. But the foliar treatment of 0.1 and 1 g/L before inoculation generated host resistance, resulting in higher tomato crop resistance (Morales-Rodriguez et al. 2012).

Common chitosan-induced damages in different fungal species, *Fusarium oxysporum* f. sp. *lycopersici, Rhizopus stolonifer, Pythium aphanidermatum, Alternaria alternata, Botrytis cinerea*, and *Penicillium expansum*, include mycelial aggregation, excessive branching, swelling of the cell wall, hyphae size reduction, vacuolation, and protoplasm disintegration. Moreover, chitosan has been shown to damage the vacuole system and plasmalemasomes of *Phytophthora capsici* hyphae. Plasmalemasome structures are involved in cell wall synthesis, suggesting that chitosan could retard mycelial growth by interfering with cell wall biogenesis. Differences in the antifungal effectiveness of chitosan are most probably connected with the fungal species or to the physical-chemical properties of the polymer (González-Peña Fundora et al. 2022). Chitosan foliar spraying resulted in small necrotic lesions on treated tomato leaves, notably on the axial leaf surfaces. Superoxide anion (O_2) formation was measured after 12–36 hours of nitro-tetrazolium staining, and H_2O_2 synthesis was measured using 3,3-diaminbenzidine-4HCl labeling. Peroxidase activity increased over time in treated tomato leaf tissues. The amounts of free and total salicylic acid in treated foliar leaves were significantly greater than in untreated leaves. Furthermore, treated and untreated tomato top leaves accumulated disease-related proteins such as ß-1,3-glucanase, chitinase, and PR 14. Chitosan research on tomato late blight disease revealed two effects: (a) direct interference in *Phytophthora infestans* embryonic stages and (b) lesion development, which leads to disease resistance (Atia et al. 2005; Huang et al. 2021). Chitosan is a non-toxic, biodegradable polymer with versatile applications. ChNPs have been proven to be effective in controlling plant diseases. The activity of ChNPs against selected plant pathogens *R. solani, F. oxysporum, Colletotrichum acutatum*, and *Phytophthora infestans* was studied. The antioxidant activity of the ChNPs was also analyzed. ChNPs had good antifungal activity against all selected pathogens compared to Amphotericin B (Divia et al. 2018).

Nanoparticles as Antimicrobials and Possible Modes of Action

Compared to the root and foliar treatments, the foliar treatment had a larger concentration of Ag in the stem, and the poplar plant had a higher concentration than oak and pine Cocozza et al. (2019). Poplars, but not oaks or pines, had their aboveground biomass and stem length decreased by foliar application of Ag-NPs. Oak's leaves accumulated H_2O_2 after foliar treatment, and poplar leaves increased O_2 after both foliar and root treatments. NPs invade the tree stem via the leaves more quickly than the roots. Compared to the root and foliar treatment had a large concentration of Ag nanoparticles in the stem, and the types of plants differed in translocation of the element as translocation of Ag in poplar was higher than in oak and pine tree.

The fungal cell wall is essential for maintaining cell homeostasis. This cell wall is composed of a strong, tensile scaffold and robust that accounts for 40% of the cell volume. This system is supported by a plethora of proteins and carbs. The fungal cell wall's outer layer comprises mannosylated glycoproteins with N- and O-linked oligosaccharides modified. The inner layer is made up of chitin and glucan, with -(1-3)-glucan taking into consideration 50–60% of the dry weight. Fungi can be classified into three major groups: Ascomycetes, Basidiomycetes, and Deuteromycetes, according to their morphologies (Kurtzman et al. 1992). There are more than 1 million species belonging to these three classes that have been identified so far, and most of them can cause serious diseases for animals and humans, including plant infection (Goh and Kohn 2003; Rennie and Denning 2005). Stress conditions such as hypoxia and nutrient starvation activate the masking of a major immunostimulatory PAMP at the cell surface, which is regulated by micronutrients iron, manganese, and zinc. The combination of nutrient deprivation and inhibition of chitin synthesis resulted in the loss of viability and impaired morphology but not increased β-glucan exposure at its cell surface. This suggests that the latter might be triggered as an alternative mechanism for cell protection during stress conditions that disrupt endocytosis and vesicular traffickings, such as nutrient starvation or lack of copper. Advances in understanding fungal cell wall structure and structure have paved the way for novel antifungal pathogen targeting. Antifungal nanoparticles (AgNPs) might be critical in overcoming this resistance. Cell wall breakdown is induced by AgNPs, degradation of surface protein, nucleic acid damage, ROS and free radical buildup, and proton pump obstruction. The antifungal action of nanoparticles is related to their small size to big surface ratio. Plants are good natural sources for metallic nanoparticles due to their unique properties like high reducibility, non-toxic nature, biocompatibility, low cost, availability, and eco-friendliness.

Recent advances in research on the composition, structure, and significance of fungal cell walls in drug resistance have opened the door to novel targets against fungal diseases, and as a clearer understanding of the process through which antifungal resistance was developing from. AgNPs could be significant in eliminating such resistance. AgNPs induce surface protein degradation, cell wall disintegration, proton pump obstruction, and nucleic acid damage by producing and accumulating free radicals and ROS. It has been proposed that AgNPs cause silver ion buildup, which inhibits respiration via intracellular ion efflux and thereby affects the electron transport system (Du et al. 2012). The antifungal action of nanoparticles is due to their small size to a big surface ratio. AgNPs of a smaller size can readily pass across cell borders. The toxicity of AgNPs is ascribed in part to the formation of ROS, which causes apoptosis. It has been proposed that the *in vitro* toxicity of AgNPs is caused by either the combined action of Ag ions and AgNPs or by their independent effects (Beer et al. 2012; Cronholm et al. 2013; Kim and Ryu 2013). In the future, greater research into the precise mechanisms and modes of action of AgNPs is required (Mansoor et al. 2021).

Chwalibog et al. (2010) demonstrated that diamond and metal nanoparticles exhibited various effects on bacteria and fungi. Nano-D is dielectrics and shows a positive zeta potential, being very different from the membrane potential of microorganisms. They uniformly surrounded the microorganisms, without causing visible damage and destruction to the cells. In the case of *Candida albicans*, glucan, and chitin are cellular components that confer rigidity to the overall cell wall structure. In the outermost cell wall layer, proteins and mannoproteins appear to predominate, and nano-D could possibly bind them closely, without visible cell destruction. We could observe a clear link between zeta potential and the electron structure of the nanoparticles and their interaction with microorganisms. All metal nanoparticles with negative zeta potential, less than that of bacterial cells, showed cell-damaging properties. Silver showed self-organization properties with the cells, disintegrating the cell walls and releasing a substance (probably cytoplasm) outside the cell. Nano-Ag showed high antibacterial activity against both types of bacteria, including highly multiresistant strains, such as *S. aureus*. This suggests nano-Ag may exert a fungicidal activity by destroying cell membrane integrity. Other results 30 have also shown that fungal cells show significant damage, characterized by the formation of a pit in their cell walls and pores in their plasma membrane. Nano-Au showed different effects on bacteria and fungi cells compared with nano-Ag and nano-Pt. The arrangement of nano-AU with microorganisms did not create a system of self-organization. However, a "noncontact" interaction between nanoparticles and microorganisms caused damage to fungal cells. This observation could be an essential guideline for the selection of applications of nanoparticles as antimicrobial compounds and/or carriers of active medical substances. Nano-Ag, nano-Au, and nano-Pt (metal nanoparticles) are harmful to bacteria and fungi. Nano-Ag attaches specifically to the microbial cell wall, evokes the release of a substance from the microorganisms, and binds to it. Nano-Au acts in a "noncontact" way, stimulating biofilm production and aggregating within this biofilm.

Effect of Nanoparticles on the Pathogenic Microbes

Numerous nanoparticles (NPs) have been investigated in terms of their potential uses in agriculture, including nano-silica (SiO_2), nano-zinc oxide (ZnO), nanosilver (Ag), nano-copper (Cu), nano-titanium dioxide (TiO_2), nano-aluminum, and carbon nanotubes. The major purpose (75%) is plant protection, followed by fertilization and UV protection, with around half of the NP components designed as active ingredients and the other half as supplements (Gogos et al. 2012). Because nanoforms' physicochemical characteristics differ greatly from those of bulk forms, it is critical to investigate how NPs influence microorganisms to be able to restrict the use of this technology to phytopathogen-based plant protection.

NPs may affect microbial activity because they are too tiny, even more minute than a viral particle, and incredibly sensitive (Khan and Rizvi 2014). Because of its excellent antibacterial efficiency toward fungus, bacteria, and viruses, silver nanoparticles (AgNPs) have been demonstrated to become the most effective. Numerous nanostructures have been studied, including zinc, copper, and titanium (Gu et al. 2003), magnesium, gold, alginate, and silver (Alvarez-Puebla et al. 2004; Lead and Wilkinson 2006; Gong et al. 2007). Much research has been conducted to investigate the potential interactions between NPs and biomolecules found in living organisms. Because bacteria and nanoparticles have different charges, NPs act as an electromagnetic absorber between the microbe and the NPs, allowing the NPs to connect to the cell membrane. Many of these interactions promote the oxidation of germ surface chemicals, which leads to cell death (Lin and Xing 2007a).

Table 1. Some important plant pathogenic bacteria controlled with nanoparticles.

Bacterial Species	Nanoparticles Type	Activity	Reference
Pantoea agglomerans	Ag	Growth suppression was 27 mm	Mohammad and El-Rahman (2015)
Ralstonia solanacearum	Ag	Zone of inhibition was 19.66 mm	
Erwinia amylovora	Ag	Growth inhibition was 16.66 mm	
Pseudomonas lachrymans	Ag	Inhibition was 13 mm	
Agrobacterium tumefaciens	Ag	Zone of inhibition was 9 mm	
Erwinia carotovora	Ag	No effect was observed	
Xanthomonas arboricola	Ag	Zone of inhibition was between 10 mm and 11.6 mm	Ghadamgahi et al. (2014)
Pseudomonas syringae	Ag	Zone of inhibition was between 11 mm and 15 mm	
Erwinia carotovora	Ag	Growth suppression was 20 mm	Al-Askar et al. (2013)
Pectobacterium wasabiae	Ag	Growth inhibition was 18 mm	
Dickeya dianthicola	Ag	Inhibition was 18 mm	
Dickeya chrysanthemi	Ag	Zone of inhibition was 10 mm	
Xanthomonas campestris pv. *vesicatoria*	Silica-Ag	Growth of bacteria stopped completely at a concentration of 100 mg/L	Park et al. (2006)
Pseudomonas syringae	Silica-Ag	Growth of bacteria stopped completely at a concentration of 100 mg/L	
Pseudomonas syringae	Cu NPs	Growth was completely inhibited at 200 mg/L	Banik and Pérez-de-Luque (2017)
Ralstonia solanacearum	MgO	Treated plants exhibited little inhibition of bacterial wilt, but when roots were drenched with a MgO NP suspension prior to inoculation with the pathogen, the incidence of disease was significantly decreased	Imada et al. (2016)
Pseudomonas syringae	ZnO	Growth inhibition was between 23 mm and 24 mm	Ghadamgahi et al. (2014)
Xanthomonas arboricola	ZnO	Growth suppression (only at 200 mg/L) with a value of 8 mm	
Xanthomonas axonopodis	Cu	Nanoparticles exhibited inhibitory properties at all concentrations	Mondal and Mani (2012)
Agrobacterium tumefaciens	CuCO₃	The growth of bacteria stopped completely at a concentration of 250 μg/ml	Ataee et al. (2011)

Effect of Nanoparticles on Plant-Pathogenic Bacteria

Some NPs influence the growth of bacteria tolerance to stress, susceptibility of plants to bacterial infection, and interactions between related plants and bacteria (Degrassi et al. 2012). Multiple investigations on the capabilities of antibacterial of NPs have indeed been carried out, and as a result, numerous publications revealing NPs' remarkable capacity to battle infections resistant to conventional antibiotics have been published (Elechiguerra et al. 2005). Many nanomaterials, particularly carbon nanotubes (Kang et al. 2007; Liu et al. 2009), iron-based NPs (Hu et al. 2010), silver (Sondi and Salopek-Sondi 2004), graphene-based nanomaterials (Hu et al. 2010), titanium oxide, zinc, and copper NPs have indeed been found to be toxic to pure bacteria cultures (Ge et al. 2011; Xie et al. 2011). However, researchers disagree on how NPs alter bacteria's secondary metabolites. Shape and oxidative stress are crucial determinants of antibacterial activity, according to a study on zinc nanoparticles (Raghupathi et al. 2011; Dwivedi et al. 2014). Several chemicals, including SiO_2, ZnO, and inorganic TiO_2, have been shown to be lethal to bacteria. The toxicity of these chemicals increases significantly in the presence of light (Adams et al. 2006).

The most sensitive microbial activity in the cycle of nitrogen is nitrification, yet the exposed to a restricted span of sublethal AgNP concentrations can potentially have minor stimulatory effects. When AgNP concentrations reach inhibitory levels, nitrification activity declines first, followed by other nitrogen-cycling processes (Yang et al. 2013). The antibacterial effects of AgNPs were caused by changes in the cytoplasm in the wall of the cell (Dhas et al. 2014; Lysakowska et al. 2015); permeability of membranes and respiration (Morones et al. 2005; Manjumeena et al. 2014); morphology (Tamayo et al. 2014); disjunction of the cell wall's cytoplasmic membrane (Shameli et al. 2012; Wang et al. 2014). To live, bacteria are assumed to need an enzyme to extract oxygen; however, silver ions block the enzyme, suffocating and killing the bacteria (Alghuthaymi et al. 2015). The deterring effect of NPs on many types of bacteria may be due to plasma membrane or damage to bacterial enzymes. Bacterial cells die due to cytoplasmic material spilling into the damaged metabolic pathways around them (Li et al. 2006).

Effect of Nanoparticles on Plant-Pathogenic Fungi

According to certain research, AgNPs have an antifungal impact on several harmful fungi (Kim et al. 2009). With the addition of AgNPs, incubated spore germination and mycelial development are significantly reduced (Morones et al. 2005). The influence of prevention in NPs may occur from the extracellular enzymes and metabolites are released, which may work as a survival factor when pressured by highly dangerous compounds and temperature variations, as demonstrated by the fungus *Trichoderma* Reese and many others (Pérez-de-Luque and Rubiales 2009). Many fungi, including *Fusarium* species, wood-rotting fungi, and other phytopathogenic fungi, are resistant to a range of AgNPs. Previous research has demonstrated that inhibiting fungal enzymes and toxins can give antifungal action (Bhainsa and D'Souza 2006; Khabat et al. 2011). Min et al. (2009) discovered the antifungal activity of AgNPs toward the sclerotium-forming phytopathogens *Rhizoctonia solani*, *Sclerotinia sclerotiorum*, and *S. minor.*

Data showed that NPs dramatically reduced growth, sclerotial germination, and fungal growth. Furthermore, it was hypothesized that the silvers with a few nanometers in diameter might have a range of attributes as a consequence of morphological, structural changes, and physiological (Baek and An 2011). Table 2 lists the number of NPs that affect bacteria. The information from the microscope showed that hyphae treated with AgNP had severe damage to their walls, which resulted in the hyphae plasmolysis (Min et al. 2009). According to a recent study, AgNPs were found to interfere with many transport processes, including ion efflux (Morones et al. 2005). Silver ions may quickly accumulate as a result of ion efflux system failure, inhibiting cellular functions at lower concentrations, including respiration and metabolism, by reacting to molecules. Additionally, the reaction between oxygen and silver ions can result in ROS, which is bad for cells and can destroy proteins, lipids, and fatty acids (Hwang et al. 2008).

Numerous fungi have demonstrated high efficacy at deterrence at a concentration of AgNPs of 100 mg/l. Generally, the level of inhibition rose as AgNP concentration did. This increase is a result of the solution's ability to penetrate and adhere to fungus hyphae at densities above those necessary to kill plant-pathogenic fungi. After being exposed to Ag, the DNA loses its capacity to replicate, which results in the deactivation of the production of ribosomal subunit proteins. In addition, ATP production requires a few additional cellular proteins and enzymes. Ag^+ is expected to have the most influence on enzymes that bind to the membrane, such as those found in the mitochondrial membrane (Kim et al. 2012). Since technological development has made it more affordable to produce nano-sized silver particles, their utilization for their antibacterial qualities has gained wider acceptance. The antifungal action of NPs and silver ions has a significant promising role in the management of spore-producing fungal plant diseases. Plant-pathogenic fungi, *Magnaporthe grisea*, and *Bipolaris sorokiniana*, have been the subject of investigations into the antifungal activity of various types of NPs and silver (Jo et al. 2009). Findings of the experiments show that the utilization

Table 2. Activity of different types of nanoparticles against some phytopathogenic fungi.

Fungal Species	Nanoparticles Type	Activity	Reference
Macrophomina phaseolina	Ag	Growth inhibition was between 31% and 71%	Bahrami-Teimoori et al. (2017)
Alternaria alternata	Ag	Growth suppression was between 9% and 57%	
Fusarium oxysporum	Ag	Inhibition was between 9% and 53%	
Sclerotinia homoeocarpa	Ag	Growth inhibition was observed at 25–100 µg/ml; this value was between 63% and 67% in the sensitive strain and 64–81% in the resistant strain	Li et al. (2017)
Penicillium digitatum	Ag	Growth suppression was between 53% and 78%	Abkhoo and Panjehkeh (2016)
Alternaria citri	Ag	Growth inhibition was between 57% and 81%	
Alternaria alternata	Ag	Inhibition was between 62% and 83%	
Alternaria solani	Ag	Significant growth suppression was noticed	
Fusarium oxysporum	Ag	Colony formation after 5 hours for 1,000 and 5,000 mg/L was approximately 85% and 24%, respectively	
Colletotrichum gloeosporioides	Ag	Nanoparticles exhibited good antifungal activity	Shanmugam et al. (2015)
Aspergillus flavus	Ag	Decrease in the frequency of the fungal population in the treated soil	Ali et al. (2015)
Aspergillus fumigatus	Ag	Completely inhibited the fungal population in the treated soil	
Aspergillus japonicas	Ag	Decreased the frequency of the fungal population in the treated soil	
Aspergillus niger	Ag	Decreased the frequency of the fungal population in the treated soil	
Cladosporium cladosporioides	Ag	Completely inhibited the fungal population in the treated soil	
Epicoccum nigrum	Ag	Decreased the frequency of the fungal population in the treated soil	
Penicillium funiculosum	Ag	Decreased the frequency of the fungal population in the treated soil	
Penicillium duclauxii	Ag	Decreased frequency of the fungal population in the treated soil.	
Rhizopus stolonifer	Ag	Decreased the frequency of the fungal population in the treated soil	
Trichoderma harzianum	Ag	Decreased the frequency of the fungal population in the treated Soil	
Fusarium oxysporum	Ag	Decreased the frequency of the fungal population in the treated soil; however, inhibition was approximately 55–78%	
Penicillium expansum	Ag	Inhibition was approximately 65–85%	
Fusarium graminearum	Ag	Growth suppression was between 57% and 78%	
Fusarium solani	Ag	Decreased the frequency of the fungal population in the treated soil, and growth inhibition was approximately 78–91%	

Table 2 contd. ...

...Table 2 contd.

Fungal Species	Nanoparticles Type	Activity	Reference
Macrophomina phaseolina	Ag	Growth was inhibited after 3 days upon exposure to all concentrations	Mahdizadeh et al. (2015)
Alternaria alternata	Ag	Growth inhibition was 26 mm	Al-Askar et al. (2013)
Fusarium oxysporum	Ag	Growth suppression was 20 mm	
Aspergillus flavus	Ag	Growth inhibition was 18 mm	
Cladosporium cladosporioides	Ag	At a concentration of 50 mg/L, growth inhibition was 90%	Pulit et al. (2013)
Aspergillus niger	Ag	At a concentration of 50 mg/L, growth inhibition was 70%	
Bipolaris sorokiniana	Ag	Nanoparticles showed good control	Jo et al. (2009)
Magnaporthe grisea	Ag	Nanoparticles showed good control	
Fusarium culmorum	Ag	Significant reduction in mycelial growth was observed for spores incubated with silver nanoparticles	Kasprowicz et al. (2010)
Magnaporthe grisea	Silica-Ag	Growth inhibition was between 1.4% and 100%	Park et al. (2006)
Botrytis cinerea	Silica-Ag	Growth suppression was between 2.6% and 100%	
Colletotrichum gloeosporioides	Silica-Ag	Growth inhibition was between 11.8% and 100%	
Pythium ultimum	Silica-Ag	Growth suppression was between 15.5% and 100%	
Rhizoctonia solani	Silica-Ag	Growth inhibition was between 54.8% and 100%	
Fusarium solani	S	Smaller particles revealed greater growth-inhibiting efficiency	Rao and Paria (2013)
Venturia inaequalis	S	Smaller particles exhibited greater growth-inhibiting efficiency	
Fusarium oxysporum	MgO	Suppression of spore germination was between 34% and 88%	Wani and Shah (2012)
Alternaria alternata	MgO	Inhibition of spore germination was between 56% and 91%	
Rhizopus stolonifer	MgO	Inhibition of spore germination was between 57% and 91%	
Mucor plumbeus	MgO	Suppression of spore germination was between 59% and 94%	
Alternaria alternata	ZnO	Inhibition of spore germination was between 30% and 79%	
Rhizopus stolonifer	ZnO	Inhibition of spore germination was between 44% and 77%	
Mucor plumbeus	ZnO	Suppression of spore germination was between 32% and 69%	
Fusarium oxysporum	ZnO	Inhibition of spore germination was between 22% and 58%	
Fusarium oxysporum	ZnO	Growth inhibition was 61–91%	He et al. (2011)
Botrytis cinerea	ZnO	Inhibition was between 63% and 80%	

Table 2 contd. ...

...Table 2 contd.

Fungal Species	Nanoparticles Type	Activity	Reference
Phytophthora cinnamomi	Cu	Growth suppression was approximately 60%	Banik and Pérez-de-Luque (2017)
Alternaria alternata	Cu	Growth inhibition was between 23% and 100%	
Sclerotinia homoeocarpa	ZnO	Significant growth inhibition was observed at 200–400 µg/ml; this amount was 47–67% in the sensitive strain and 61–76% in the resistant strain	Li et al. (2017)
Fusarium oxysporum f. sp. *lycopersici*	Ni	The percentage inhibition of mycelia growth was between 39% and 60%	Ahmed et al. (2016)
Fusarium oxysporum f. sp. *lactucae*	Ni	Growth inhibition was between 52% and 59%	Ahmed et al. (2016)
F. oxysporum f. sp. *radicis-lycopersici*	rGO-CuO	Disease severity was reduced by 98 and 95% in tomato plants treated with 1 and 100 mg/L rGO-CuO NPs	El-Abeid et al. (2020)
Alternaria solani	Ag	Pathogen growth inhibition was around 90–100%	Mohamed (2015)
Alternaria solani	Ag	Pathogen growth inhibition was around 95.6%	Mohamed et al. (2018)
Alternaria solani	Ag	Pathogen growth inhibition was around 95.6%	Abdel hafez et al. (2017)
F. oxysporum f. sp. *radicis-lycopersici*	MSNs	Pathogen growth inhibition was around 98%	Mosa et al. (2022)

time greatly affects the action of silver and that uses for the prevention of AgNPs are more effective before the isolation of fungi enters and colonizes the plant tissue (Kim et al. 2012).

Commercial Products and Uses of Nanopesticides

The protection solutions for the plant with nanoparticle usage (nano-enabled pesticides) promise many advantages over traditional pesticides, including improved species targeting, reduced application rates, increased efficacy, and improved safety for the environment (Walker et al. 2017). The most researched substance for reducing insect and disease impact on crops is nanosilver. The development of the manufacture of silver nanomaterials, notably silver nanoformulation, was creating them, and they appear to have been on the market for a while. To improve the antibacterial and biosafety properties of silver nanoformulations, a variety of methods are constantly being researched. Alstasan Silvox, a nanosilver hydrogen peroxide product, is one of the most recent offerings from Chemtex Speciality Limited Company. The solution was touted as a "biodegradable disinfectant" that was both safe and ecologically beneficial (https://www.chemtexlimited.com/search.html?ss=Alstasan+Silvox).

It was produced for broad-spectrum antibiotics/bactericides, fungicides, virucides, and nematicides as well, though. Additionally, the product can be used to prevent and manage several agricultural diseases, such as powdery mildew, bacterial blights, and root-knot nematodes. The solution is appropriate for both cleaning and disinfection because of its nonfoaming qualities. It is also effective against a wide range of pathogens, environmentally friendly, biodegradable, colorless, and odorless. Unlike chlorine, bromine, formalin, and aldehydes, it is noncarcinogenic and nonmutagenic, with no chance of microbial resistance, nonpolluting and nonstaining, and does not emit any hazardous fumes.

This silver peroxide kills biofilms, is a cold sterilant, has great thermostability, preserves the ease of handling and dose, is free of carcinogenic aldehydes, and has no efficacy gaps. It is also nonflammable, which reduces the risks associated with shipping and storage. This silver peroxide has minimal contact times and provides long-lasting disinfection while leaving no harmful residues. Up to a concentration of 25 mg/l, the product is said to be safe for human consumption. This solution is a clear liquid that may be used in four ways: dipping, spraying, filling, soaking, and cleaning-in-place (CIP).

This product's nanosilver possesses the stabilizing, activating, and oligo-dynamic properties of hydrogen peroxide, a naturally occurring antimicrobial. Because nanoformulations of metals, such as silver, have a broad spectrum of antimicrobial activity and have been widely applied as an active disinfecting and sterilizing agent in a variety of fields, including water management, food service storage, pharmaceutical, cosmetics, treatment of various human diseases, and agriculture, it appears that their product is safe and cannot be classified as an agro-nano.

In this product, nanosilver was utilized as a stabilizer, activator and to enhance hydrogen peroxide's oligodynamic activity. Given all the aforementioned characteristics, this nanosilver product looks to be appealing and appropriate for its safe, efficient, and successful agricultural applications. According to Khot et al. (2012) and Sangeetha et al. (2017), the development of nanomaterials with intense fluid-solid phase dispersion and interaction, well-understood toxicodynamics and toxic kinetics, soil and environmental biodegradability, less toxicity, and greater photogenerativity, and smart and stable utilization in agriculture would be beneficial and ideal for their effective and efficient application fields.

Although the length of time or expiry of the long-term effect of the nanosilver (in the product) was unclear, Because of the increased reactivity of nanosilver from the continuous application, a bioaccumulation risk assessment is required that nanosilver product contains broad-spectrum nanopesticides that are effective and efficient against plant pests and diseases. This product, on the other hand, showed that producing a nanoformulation including nanometal as a pesticidal component or as a formula product stabilizer and activator is a step toward the development of metal-based nano pesticides. Nano chemical pesticides are another sort of nanopesticide. The creation of nanochemical pesticides (or nanopesticides), which contain minuscule chemical poisons, was one of the earliest nano industrial uses (Kuzma and Verhage 2006). BASF, Bayer Crop Science, Monsanto, and Syngenta are among the leading agrochemical businesses creating nano-based insecticides. However, because the effects of nanoparticles on the environment and human health are still largely unknown and unpredictable, smart pesticide marketing is currently restricted, particularly by environmental groups and risk assessors opposed to their introduction or potential risks associated with nanoscale materials (Annex E 2018, http://www.nanotechproject.org/). Table 3 summarizes some of the most recent nano pesticides launches.

Table 3. Nano-based products on the market (Annex E 2018, https://nanotechproject.org/).

Company	Product	Mechanism
Syngenta	PRIMO Maxx and KARATE ZEON	Inhibit neural system
Nano Green	Nano Green	Attack respiratory apparatus
Agro Nano-technology Crop	Nano Gro	Mimics stress conditions, increasing crop activity, and yield

Prospects and Challenges of Nano Pesticides Formulation and Application in Plant Pest and Disease Management

Through the use of nano pesticides formulations that combine (1) nanoparticles (metals, agrochemicals, and EOs), as an innovative pesticidal ingredient that is more active than conventional pesticides and (2) nanostructured materials (bioactive agents, agrochemicals, metals, etc.) nanomaterials, as a sustained release mediator, provide plant pest and disease pathogen-targeted applications in plant protection. Pesticides, agrochemicals, and volatile oils are all toxic (such as citronella grass, clove, eucalyptus, thyme, camphor, and Monarda species), nanostructured materials (such as silver and copper), and bioactive substances are all potential pesticide component targets (microorganisms, bio-materials, microbial products). The challenge with essential oil, pesticidal metal, agrochemical, and bioactive agent ingredients, as well as their application, was solved by nanostructures and nanoencapsulation. The antibacterial and insecticidal effects of bulk metals are minimal. Compared to bulk silver, silver nanoparticles with a wide surface area and a high surface atom fraction have a better antibacterial impact (Patel et al. 2014).

Pesticides made from chemicals cannot control nematodes or other deadly disease pathogens like viruses and bacteria since nematode populations continue to grow months after nematicidal treatment (Thakur and Shirkot 2017). As a competing strategy, nanocarrier-based formulations with controlled release technology have evolved with the promise of resolving issues associated with using some agrochemicals while minimizing potential negative consequences for others (Han et al. 2009). In comparison to chemical pesticides, small, designed structures that have pesticidal properties have demonstrated delayed deterioration and regulated active component release over a long period (Chhipa 2017). Such nanopesticide delivery techniques enable the delaying or control of pesticide delivery to the target species (Singh et al. 2015).

Due to their huge droplet size and lack of suitably sized spraying equipment at the farmer level, EOs are unstable during pesticidal action and have low vapor pressure when utilized as fumigant agents on a wide scale. Water solubility and low vapor pressure issues might be resolved by nanoemulsions or nanoencapsulation of essential oils, respectively. The nanoformulations could increase the biological agent's stability and against desiccation of biomaterials, UV radiation inactivation, and heat stressors. The creation of intelligent delivery systems utilizing nano-biomaterials may enable the regulated use of biologically active molecules (equivalent to agrochemicals). The nanocomposite should break down in plants and grow swiftly in soil but slowly in plants in a way to sustain the level of residue below the key limits allowed in accordance with food safety regulations (Khan and Rizvi 2014). In plants, such nanomaterials have demonstrated long-term regulated release of active compounds and gradual disintegration.

Because of their strong adherence to bacterial and fungal cell surfaces, well-dispersed and stabilized AgNP solutions are increasingly being utilized to treat plant diseases (Aziz et al. 2016). This promotes the growth of healthy plants. Because of their tailored approach, metal nanoparticles have a lot of potential for controlling pest populations; Nanoparticles allow for increased uptake of active components with a lower chemical concentration employed since they quickly permeate through cell walls and generate effect at maximum (Thakur and Shirkot 2017).

One usage of nanotechnology that has a significant potential for use in agriculture and plant pest control is the direct application of nanoparticles to the soil, seeds, or leaves to protect plants against various forms of pathogen invasion by inhibiting them, as chemical treatments do (Thakur and Shirkot 2017). The silver nanoparticle is now employed in planter soils and hydroponic systems to get rid of undesirable microbes. Because the particle size is also reduced by this nanoformulation technique, utilizing nano pesticides will promote efficient and productive spraying and lessen splash losses, which are frequently seen when using conventional pesticides and EO sprays.

Silver is also a remarkable plant growth promoter (Oldenburg 2017). Compared to conventional chemical antibacterial agents that cause multidrug resistance, silver nanoparticles have advantages (Guzman et al. 2009). fungicidal, bactericidal, nematicidal, virucidal, and insecticidal properties were revealed using silver metallic nanoparticles (Park et al. 2006; Abbassy et al. 2017; El-Shazly et al. 2017; Sharma et al. 2017), nanosilver-based compounds offer promising broad-spectrum pesticides. Such nanomaterials are also free of harmful compounds, making them safer to use and compatible with living things. AgNP may also offer the advantage of controlling various plant pathogens causing complicated diseases that frequently manifest in the field. These results could be the result of its non-specific method of action, which is linked to numerous cellular systems. AgNP is, therefore, a broad-spectrum antibacterial drug capable of inhibiting bacteria and fungi that cause plant disease (Park et al. 2006). For instance, AgNP is likely to have antifungal properties on many fungus infections connected with roots (e.g., *Rhizoctonia solani* and *Gaeumannomyces graminis*). Because of some defense against extra stress generated by these other infections, pathogens that have been treated with AgNPs may become more resistant to the harm caused by root-knot nematodes (Taha 2016). While compared to conventional pesticides, nano pesticides appear to be safer and more effective than the perspective of silver nanoparticles advantage. However, the Biosafety to Ecosystems Debate has been addressed in terms of its field application. Because of the increasing usage of silver nanoparticles and their wide geographic distribution, a significant number of nanoparticles are being released into the environment (Nam et al. 2014). Pesticide and nanopesticide biodegradation are crucial for removing hazardous substances from the ecosystem. To reduce the negative effects of harmful nanoparticles, soil, and water remediation measures should be taken after the use of nanoparticles made of silver with agrochemicals (as is typically done with conventional chemical pesticides). Pesticide-contaminated soil and water are typically treated utilizing photochemical reactions, oxidation, biodegradation, and phytoremediation. According to certain research, chemical and biological nanoparticles play an important function in crop protection, such as pesticide removal and irrigation water filtration (Patil et al. 2016). Because of its extensive use in environmental remediation and wastewater purification, iron nanoparticles are a significant nanomaterial.

Using nanosized zerovalent iron, pesticides like chlorpyrifos and atrazine can be broken down. Likewise, some use a TiO_2-doped catalyst; the pesticide may also be broken down photocatalytically, direct spraying of Fe_2O_3 or the addition of Fe_2O_3 into pesticide formulations (Sasson et al. 2007). Natural processes use iron oxide as an adsorbent and catalyst and artificial contaminants (Rajabi et al. 2012). A possible bio-remedial therapy is gradually growing in acceptance. Several types of plants have shown the capacity to absorb hazardous substances, including heavy metals. Citronella grass, an essential oil plant, demonstrated aptitude in phytoremediation for cleaning up soil from metal contamination (Handique and Handique 2009). Practical acceptance of bioactive agent-based and essential oils nanopesticides as efficient and secure insecticides exist. To control pests and alleviate diseases, particularly storage pests, EOs-based nano pesticides are regarded as an effective, safe, and environmentally friendly alternative to synthetic pesticides (Williams 2002; Khan and Rizvi 2014). Because of the abundance of active ingredients, such as volatile substances, such oils provide a broad spectrum of pesticidal and insecticidal actions against a wide range of pests and disease pathogens, with varying locations and mechanisms of toxic action ranging from extremely fumigant and contact toxicity against crop pests and storage pests to extremely fumigant and contact toxicity against crop pests and storage pests. Chitosan is one of the most promising future candidates for nanopesticides formulation as alternative safe and effective pesticidal active ingredients and/or the nanostructure for controlled release delivery system of pesticidal ingredients due to its unique properties such as abundance, biocompatibility, biodegradability, hydrophilicity, safety, and nontoxicity. These biologically active chemicals have several applications, including antibacterial properties (Du et al. 2009), nano fertilizers (Corradini et al. 2010), antifungal activities (Saharan et al. 2013), and plant growth promotion activity (Van et al. 2013). The commercialization of nanomaterials for agricultural applications (such as metal and agrochemicals) requires significant work and calls for appropriate protection requirements, testing priorities, risk assessments, and worldwide regulatory guidelines (Chen and Yada 2011). Otherwise, future research and industry will face challenges and opportunities in the development and marketing of nanopesticides based on bioactive agents and essential oils as environmentally friendly insecticides. Plant-based pesticides are highly sought after since they are environmentally friendly, safe, and enjoyable to be used (Maia and Moore 2011). For usage against pests in the home and garden, essential oils from the flavor and fragrance industries are readily available and well-liked by customers. These oils also work well in agricultural settings for organic food production, and the legalization of IPM has hastened the commercialization of essential oil-based pesticides. These oils also work well in agricultural settings for organic food production, and the legalization of IPM has hastened the commercialization of essential oil-based pesticides.

However, developing efficient, affordable, and scalable nanoemulsions for applying nanopesticides will take much work. Different outcomes can be seen depending on the essential oil distillation process or technology used, the plant's ripening phase, the growing area, and other factors (Ozdemir and Gozel 2017). The biocompatibility and biodegradability of nanoparticles made from natural materials (silica, chitosan, clay) and biopolymers are advantages in nanoformulations.

Conclusion and Future Strategies

Nanoparticles have unique effects, as they interfere in major biological and physiological processes in plants and living organisms; many articles track them inside fungus and plant cells and outcomes, whether at the level of plants or humans, their distribution, and their impact on public health.

To add safer products that have smart qualities in terms of specialization, they are introduced into the environment in very small quantities and are more effective, selective, and secure than conventional pesticides or fungicides. These products are used in organic farming to increase yields of fruit, vegetables, and flowers, reduce losses due to fungi and bacteria and increase the overall quality of fruits, vegetables, and flowers by reducing pests and diseases that affect crop yields and the production of fruits, vegetables, and flowers in organic agriculture. This is more profitable and environmentally friendly by making the best use of natural resources and energy, preserving biodiversity, and conserving soil fertility and water quality while maintaining good soil structure, water retention, air circulation, and control of weeds, insects, rodents, birds, etc.

Metals have always been present in the earth's crust, even before life began. They are natural soil ingredients, although in limited amounts. The metal and metal oxide business in agriculture is a major substantial sector; metal/metal oxide nanoparticles produced will have increased from 58,000 tons in 2002 to roughly 58,000 tons per year between 2011 and 2020. This finding demonstrates the significance of metal/metal oxide nanoparticles in several domains, particularly in agriculture. Metals/metal oxide nanoparticles have certain unique qualities that help them to stand out in nanotechnology. These properties include magnetic, photocatalytic, fluoresce, and controls in size that allow them to be used in various applications. Using nanopesticide formulations that combine (1) nanoparticles (metals, agrochemicals, and essential oils) as a novel pesticidal component that is more active than traditional pesticides and (2) nanostructured materials (bioactive agents, agrochemicals, metals, etc.). As a sustained release mediator, nanomaterials enable plant pest and disease-pathogen-targeted uses in plant protection. Bulk metals have few antibacterial and insecticidal properties. Nanostructures and nanoencapsulation solved the problem of essential oil, pesticidal metal, agrochemical, and bioactive agent components and their application.

References

Abbassy, M.A., Abdel-Rasoul, M.A., Nassar, A.M.K. and Soliman, B.S.M. 2017. Nematicidal activity of silver nanoparticles of botanical products against root-knot nematode, *Meloidogyne incognita*. 1–18, https://doi.org/1 0.1080/03235408.2017.1405608.

Abdel-Aziz, S.M., Prasad, R., Hamed, A.A. and Abdelraof, M. 2018. Fungal nanoparticles: A novel tool for a green biotechnology? pp. 61–87. *In*: Prasad, R., Kumar, V., Kumar, M. and Wang, S. (eds.). Fungal Nanobionics: Principles and Applications. Springer Singapore.

Abdel-Hafez, S.I., Nafady, N.A., Abdel-Rahim, I.R., Shaltout, A.M., Daròs, J.A. and Mohamed, M.A. 2017. Biosynthesis of silver nanoparticles using the compound curvularin isolated from the endophytic fungus Epicoccum nigrum: characterization and antifungal activity. J. Pharm. and Appl. Che. 3(2): 135–146.

Abdelmalek, G.A. and Salaheldin, T.A. 2016. Silver nanoparticles as a potent fungicide for citrus phytopathogenic fungi. Nanomed. Res. 3: 00065.

Abkhoo, J. and Panjehkeh, N. 2016. Evaluation of antifungal activity of silver nanoparticles on *Fusarium oxysporum*. Int. J. Infect 4: e 41126.

Adams, L.K., Lyon, D.Y. and Alvarez, P.J. 2006. Comparative eco-toxicity of nanoscale TiO_2, SiO_2, and ZnO water suspensions. Water Res. 40: 3527–3532.

Ahsan, N., Lee, D.G., Alam, I., Kim, P.J., Lee, J.J., Ahn, Y.O. and Lee, B.H. 2008. Comparative proteomic study of arsenic-induced differentially expressed proteins in rice roots reveals glutathione plays a central role during As stress. Proteomics 8(17): 3561–3576.

Ahmed, A.I., Yadav, D.R. and Lee, Y.S. 2016. Applications of nickel nanoparticles for control of *Fusarium* wilt on lettuce and tomato. Int. J. Innov. Res. Sci. Eng. Technol. 5: 7378–7385.

Akturk, A., Güler, F.K., Taygun, M.E., Goller, G. and Küçükbayrak, S. 2020. Synthesis and antifungal activity of soluble starch and sodium alginate capped copper nanoparticles. Mater. Res. Express 6(12): 1250g3.

Al-Askar, A., Hafez, E., Kabeil, S. and Meghad, A. 2013. Bioproduction of silver-nano particles by *Fusarium oxysporum* and their antimicrobial activity against some plant pathogenic bacteria and fungi. Life Sci. J. 10: 2470–2475.

Alghuthaymi, M.A., Almoammar, H., Rai, M., Said-Galiev, E. and Abd-Elsalam, K.A. 2015. Myconanoparticles: synthesis and their role in phytopathogens management. Biotechnol. Equip 29: 221–236.

Ali, S.M., Yousef, N.M. and Nafady, N.A. 2015. Application of biosynthesized silver nanoparticles for the control of land snail Eobania vermiculata and some plant pathogenic fungi. J. Nanomater. Article ID 218904.

Alvarez-Puebla, R., Dos Santos, J.D. and Aroca, R. 2004. Surface-enhanced Raman scattering for ultrasensitive chemical analysis of 1 and 2-naphthalenethiols. Analyst 129: 1251–1256.

Ashraf, H., Anjum, T., Riaz, S., Ahmad, I.S., Irudayaraj, J., Javed, S. and Naseem, S. 2021. Inhibition mechanism of green-synthesized copper oxide nanoparticles from Cassia fistula towards *Fusarium oxysporum* by boosting growth and defense response in tomatoes. Environ. Sci.: Nano 8(6): 1729–1748.

Aslam, M., Abdullah, A.Z. and Rafatullah, M. 2021. Recent development in the green synthesis of titanium dioxide nanoparticles using plant-based biomolecules for environmental and antimicrobial applications. J. Ind. Eng. Chem. 98: 1–16.

Ataee, R., Derakhshanpour, J., Mehrabi, T.A. and Eydi, A. 2011 Antibacterial effect of calcium carbonate nanoparticles on *Agrobacterium tumefaciens*. J. Mil. Med. 13: 65–70.

Atia, M.M.M., Buchenauer, H., Aly, A.Z. and Abou-Zaid, M.I. 2005. Antifungal activity of chitosan against *Phytophthora infestans* and activation of defence mechanisms in tomato to late blight. Biol. Agric. Hortic. 23(2): 175–197.

Baazaoui, N., Sghaier-Hammami, B., Hammami, S.B., Khefacha, R., Chaari, S., Elleuch, L. and Abdelly, C. 2021. A handbook guide to better use of nanoparticles in plants. Commun. Soil. Sci. Plant Anal. 52(4): 287–321.

Baek, Y.W. and An, Y.J. 2011. Microbial toxicity of metal oxide nanoparticles (CuO, NiO, ZnO, and Sb_2O_3) to *Escherichia coli, Bacillus subtilis*, and *Streptococcus aureus*. Sci. Total Environ. 409: 1603–1608.

Bahrami-Teimoori, B., Nikparast, Y., Hojatianfar, M., Akhlaghi, M., Ghorbani, R. and Pourianfar, H.R. 2017. Characterisation and antifungal activity of silver nanoparticles biologically synthesised by *Amaranthus retroflexus* leaf extract. J. Exp. Nanosci. 12: 129–139.

Banik, S. and Pérez-de-Luque, A. 2017. *In vitro* effects of copper nanoparticles on plant pathogens, beneficial microbes and crop plants. Span J. Agric Res. 15: e1005.

Bhainsa, K.C. and D'Souza, S. 2006. Extracellular biosynthesis of silver nanoparticles using the fungus *Aspergillus fumigatus*. Colloids Surf B: Biointerfaces 47: 160–164.

Butler, K.S., Peeler, D.J., Casey, B.J., Dair, B.J. and Elespuru, R.K. 2015. Silver nanoparticles: correlating nanoparticle size and cellular uptake with genotoxicity. Mutagenesis 30(4): 577–591.

Chavali, M.S. and Nikolova, M.P. 2019. Metal oxide nanoparticles and their applications in nanotechnology. SN Appl. Sci. 1: 607. https://doi.org/10.1007/s42452-019-0592-3.

Chen, H. and Yada, R. 2011. Nanotechnologies in agriculture: new tools for sustainable development. Trends Food Sci. Technol. 22: 585–594.

Chen, J., Sun, L., Cheng, Y., Lu, Z., Shao, K., Li, T. and Han, H. 2016. Graphene oxide-silver nanocomposite: novel agricultural antifungal agent against *Fusarium graminearum* for crop disease prevention. ACS Appl. Mater, 8(36): 24057–24070.

Chhipa, H. 2017. Nanopesticide: current status and future possibilities. Agri Res. Tech: Open Access J. 5(1). https://doi.org/10.19080/ARTOAJ.2017.05.555651.

Chwalibog, A., Sawosz, E., Hotowy, A., Szeliga, J., Mitura, S., Mitura, K. and Sokolowska, A. 2010. Visualization of interaction between inorganic nanoparticles and bacteria or fungi. Int. J. Nanomed. 5: 1085.

Cocozza, C., Perone, A., Giordano, C., Salvatici, M.C., Pignattelli, S., Raio, A. and Cherubini, P. 2019. Silver nanoparticles enter the tree stem faster through leaves than through roots. Tree Physiol. 39(7): 1251–1261.

Corradini, E., de Moura, M.R. and Mattoso, L.H.C. 2010. A preliminary study of the incorporation of NPK fertilizer into chitosan nanoparticles. Express Polym. Lett. 4(8): 509–515.

Degrassi, G., Bertani, I., Devescovi, G., Fabrizi, A., Gatti, A. and Venturi, V. 2012. Response of plant-bacteria interaction models to nanoparticles. EQA-Int. J. Environment Qual. 8: 39–50.

Dhas, S.P., John, S.P., Mukherjee, A. and Chandrasekaran, N. 2014. Autocatalytic growth of biofunctionalized antibacterial silver nanoparticles. Biotechnol. Appl. Biochem. 61: 322–332.

Dimkpa, C.O., Zeng, J., McLean, J.E., Britt, D.W., Zhan, J. and Anderson, A.J. 2012. Production of indole-3-acetic acid via the indole-3-acetamide pathway in the plant-beneficial bacterium *Pseudomonas chlororaphis* O6 is inhibited by ZnO nanoparticles but enhanced by CuO nanoparticles. Appl. Environ. Microbiol. 78: 1404–1410.

Divya, K., Smitha, V. and Jisha, M.S. 2018. Antifungal, antioxidant and cytotoxic activities of chitosan nanoparticles and its use as an edible coating on vegetables. Int. J. Biol. Macromol. 114: 572–577.

Du, W.L., Niu, S.S., Xu, Y.L., Xu, Z.R. and Fan, C.L. 2009. Antibacterial activity of chitosan tripolyphosphate nanoparticles loaded with various metal ions. Carbohydr. Polym. 75: 385–389.

Dwivedi, S., Wahab, R., Khan, F., Mishra, Y.K., Musarrat, J. and Al-Khedhairy, A.A. 2014. Reactive oxygen species mediated bacterial biofilm inhibition via zinc oxide nanoparticles and their statistical determination. PLoS One 9: e111289.

El-Abeid, S.E., Ahmed, Y., Daròs, J.A. and Mohamed, M.A. 2020. Reduced graphene oxide nanosheet-decorated copper oxide nanoparticles: A potent antifungal nanocomposite against fusarium root rot and wilt diseases of tomato and pepper plants. Nanomater. 10(5): 1001.

Elechiguerra, J.L., Burt, J.L., Morones, J.R., Camacho-Bragado, A., Gao, X., Lara, H.H. et al. 2005. Interaction of silver nanoparticles with HIV-1. J. Nanobiotechnol. 3: 6.

Elmer, W., Ma, C. and White, J. 2018. Nanoparticles for plant disease management. Curr. Opin. Environ. Sci. Health. 6: 66–70.

El-Sayed, A.A., Khalil, A.M., El-Shahat, M., Khaireldin, N.Y. and Rabie, S.T. 2016. Antimicrobial activity of PVC-pyrazolone-silver nanocomposites. J. Macromol. Sci. Part A 53(6): 346–353.

El-Shazly, M.A., Attia, Y.A., Kabil, F.F., Anis, E. and Hazman, M. 2017. Inhibitory effects of salicylic acid and silver nanoparticles on potato virus Y-infected potato plants in Egypt. Middle East J. Agric. Res. 6: 835–848.

Fadoju, O.M., Osinowo, O.A., Ogunsuyi, O.I., Oyeyemi, I.T., Alabi, O.A., Alimba, C.G. and Bakare, A.A. 2020. Interaction of titanium dioxide and zinc oxide nanoparticles induced cytogenotoxicity in Allium cepa. The Nucleus 63(2): 159–166.

Gajjar, P., Pettee, B., Britt, D.W., Huang, W., Johnson, W.P. and Anderson, A.J. 2009. Antimicrobial activities of commercial nanoparticles against an environmental soil microbe, *Pseudomonas putida* KT2440. J. Biol. Eng. 3: 9.

Ge, Y., Schimel, J.P. and Holden, P.A. 2011. Evidence for negative effects of TiO_2 and ZnO nanoparticles on soil bacterial communities. Environ. Sci. Technol. 45: 1659–1664.

Ghadamgahi, F., Mehraban, S., Atash, M. and Shahidi, B.G. 2014. Comparison of inhibitory effects of silver and zinc oxide nanoparticles on the growth of plant pathogenic bacteria. Int. J. Adv. Biol. Biomed. Res. 2: 1163–1167.

Gogos, A., Knauer, K. and Bucheli, T.D. 2012. Supporting information: nanomaterials in plant protection and fertilization: current state, foreseen applications, and research priorities. J. Agric. Food Chem. 60(39): 9781–9792.

Gong, P., Li, H., He, X., Wang, K., Hu, J., Tan, W. et al. 2007. Preparation and antibacterial activity of Fe_3O_4Ag nanoparticles. Nanotechnology 18: 285604.

González-Peña Fundora, D., Falcón-Rodríguez, A.B., Costales Menendez, D., Foroud, N.A., Vaillant Flores, D., Aispuro-Hernández, E. and Martínez-Téllez, M.Á. 2022. Chitosan induces tomato basal resistance against *Phytophthora nicotianae* and inhibits pathogen development. Can. J. Plant Pathol. 44(3): 400–414.

Gordienko, M.G., Palchikova, V.V., Kalenov, S.V., Belov, A.A., Lyasnikova, V.N., Poberezhniy, D.Y. and Skladnev, D.A. 2019. Antimicrobial activity of silver salt and silver nanoparticles in different forms against microorganisms of different taxonomic groups. J. Hazard. Mater. 378: 120754.

Gu, H., Ho, P., Tong, E., Wang, L. and Xu, B. 2003. Presenting vancomycin on nanoparticles to enhance antimicrobial activities. Nano Lett. 3: 1261–1263.

Gunawan, C., Teoh, W.Y., Marquis, C.P. and Amal, R. 2011. Cytotoxic origin of copper (II) oxide nanoparticles: comparative studies with micron-sized particles, leachate, and metal salts. ACS Nano 5: 7214–7225.

Gupta, S.D., Agarwal, A. and Pradhan, S. 2018. Phytostimulatory effect of silver nanoparticles (AgNPs) on rice seedling growth: An insight from antioxidative enzyme activities and gene expression patterns. Ecotoxicol. Environ. Saf. 161: 624–633.

Gupta, N., Ram, H. and Kumar, B. 2016. Mechanism of Zinc absorption in plants: uptake, transport, translocation and accumulation. Rev. Environ. Sci. Biotechnol. 15(1): 89–109.

Han, X., Chen, S. and Hu, X. 2009. Controlled-release fertilizer encapsulated by starch/polyvinyl alcohol coating. Desalination 240: 21–26.

Handique, G.K. and Handique, A.K. 2009. Proline accumulation in lemongrass (*Cymbopogon flexuosus* Stapf.) due to heavy metal stress. J. Environ. Biol. 30(2): 299–302.

Hänsch, R. and Mendel, R.R. 2009. Physiological functions of mineral micronutrients (cu, Zn, Mn, Fe, Ni, Mo, B, cl). Curr. Opin. Plant Biol. 12(3): 259–266.

He, L., Liu, Y., Mustapha, A. and Lin, M. 2011. Antifungal activity of zinc oxide nanoparticles against *Botrytis cinerea* and *Penicillium expansum*. Microbiol. Res. 166: 207–215.

Hu, C.M.J., Aryal, S. and Zhang, L. 2010. Nanoparticle-assisted combination therapies for effective cancer treatment. Ther Deliv. 1: 323–334.

Huang, X., You, Z., Luo, Y., Yang, C., Ren, J., Liu, Y. and Ren, M. 2021. Antifungal activity of chitosan against *Phytophthora infestans*, the pathogen of potato late blight. Int. J. Biol. Macromol. 166: 1365–1376.

Hwang, E.T., Lee, J.H., Chae, Y.J., Kim, Y.S., Kim, B.C., Sang, B.I. et al. 2008. Analysis of the toxic mode of action of silver nanoparticles using stress-specific bioluminescent bacteria. Small 4: 746–750.

Imada, K., Sakai, S., Kajihara, H., Tanaka, S. and Ito, S. 2016. Magnesium oxide nanoparticles induce systemic resistance in tomato against bacterial wilt disease. Plant Pathol. 65: 551–560.

Irshad, M.A., Nawaz, R., ur Rehman, M.Z., Imran, M., Ahmad, J., Ahmad, S. and Ali, S. 2020. Synthesis and characterization of titanium dioxide nanoparticles by chemical and green methods and their antifungal activities against wheat rust. Chemosphere 258: 127352.

Isah, U.I. and Garba, M.D. 2022. Antimicrobial evaluation of zinc oxide nanoparticles compared with date fruit extracts. Proc. Natl. Acad. Sci. India - Phys. Sci. 92(3): 311–318.

Jacob, D.L., Borchardt, J.D., Navaratnam, L., Otte, M.L. and Bezbaruah, A.N. 2013. Uptake and translocation of Ti from nanoparticles in crops and wetland plants. Int. J. Phytoremediation 15(2): 142–153.

Jain, J., Arora, S., Rajwade, J.M., Omray, P., Khandelwal, S. and Paknikar, K.M. 2009. Silver nanoparticles in therapeutics: development of an antimicrobial gel formulation for topical use. Mol. Pharm. 6: 1388–1401.

Jo, Y.K., Kim, B.H. and Jung, G. 2009. Antifungal activity of silver ions and nanoparticles on phytopathogenic fungi. Plant Dis. 93: 1037–1043.

Jones, N., Ray, B., Ranjit, K.T. and Manna, A.C. 2008. Antibacterial activity of ZnO nanoparticle suspensions on a broad spectrum of microorganisms. FEMS Microbiol. Lett. 279: 71–76.

Kamal, R. and Mogazy, A.M. 2021. Effect of doping on TiO_2 nanoparticles characteristics: Studying of fertilizing effect on cowpea plant growth and yield. J. Soil Sci. Plant Nutr. 1–13.

Kang, S., Pinault, M., Pfefferle, L.D. and Elimelech, M. 2007. Single-walled carbon nanotubes exhibit strong antimicrobial activity. Langmuir 23: 8670–8673.

Kasprowicz, M.J., Kozioł, M. and Gorczyca, A. 2010. The effect of silver nanoparticles on phytopathogenic spores of *Fusarium culmorum*. Can J. Microbiol. 56: 247–253.

Khabat, V., Mansoori, G.A. and Karimi, S. 2011. Biosynthesis of silver nanoparticles by fungus *Trichoderma Reesei*. Insciences J. 1: 65–79.

Khan, M.R. and Rizvi, T.F. 2014. Nanotechnology: scope and application in plant disease management. Plant Pathol. J. 13: 214–231.

Khot, L.R., Shankaran, S., Maja, J.M., Ehsani, R. and Schuster, E.W. 2012. Applications of nanomaterials in agricultural production and crop protection: a review. Crop Prot. 35: 64–70.

Kim, S.W., Kim, K.S., Lamsal, K., Kim, Y.J., Kim, S.B., Jung, M. et al. 2009. An *in vitro* study of the antifungal effect of silver nanoparticles on oak wilt pathogen *Raffaelea* sp. J. Microbiol. Biotechnol. 19: 760–764.

Kim, S.W., Jung, J.H., Lamsal, K., Kim, Y.S., Min, J.S. and Lee, Y.S. 2012. Antifungal effects of silver nanoparticles (AgNPs) against various plant pathogenic fungi. Mycobiology 40: 53–58.

Kuzma, J. and Verhage, P. 2006. Nanotechnology in agriculture and food production: anticipated applications. http://www.nanotechproject.org/file_download/files/PEN4_AgFood.

Latif, U., Al-Rubeaan, K. and Saeb, A.T. 2015. A review on antimicrobial chitosan-silver nanocomposites: a roadmap toward pathogen targeted synthesis. International Int. J. Polym. Mater. Polym. 64(9): 448–458.

Le, V.T., Bach, L.G., Pham, T.T., Le, N.T.T., Ngoc, U.T.P., Tran, D.H.N. and Nguyen, D.H. 2019. Synthesis and antifungal activity of chitosan-silver nanocomposite synergize fungicide against Phytophthora capsici. J. Macromol. Sci. A. 56(6): 522–528.

Lead, J.R. and Wilkinson, K.J. 2006. Aquatic colloids and nanoparticles: current knowledge and future trends. Environ. Chem. 3: 159–171.

Li, J., Sang, H., Guo, H., Popko, J.T., He, L., White, J.C. et al. 2017. Antifungal mechanisms of ZnO and Ag nanoparticles to *Sclerotinia homoeocarpa*. Nanotechnology 28: 155101.

Li, Y., Leung, P., Yao, L., Song, Q. and Newton, E. 2006. Antimicrobial effect of surgical masks coated with nanoparticles. J. Hosp. Infect. 62: 58–63.

Lin, D. and Xing, B. 2007. Phytotoxicity of nanoparticles: inhibition of seed germination and root growth. Environ. Pollut. 150: 243–250.

Liu, S., Wei, L., Hao, L., Fang, N., Chang, M.W., Xu, R. et al. 2009. Sharper and faster "nano darts" kill more bacteria: a study of antibacterial activity of individually dispersed pristine single-walled carbon nanotube. ACS Nano 3: 3891–3902.

López-Millán, A.F., Ellis, D.R. and Grusak, M.A. 2005. Effect of zinc and manganese supply on the activities of superoxide dismutase and carbonic anhydrase in Medicago truncatula wild type and raz mutant plants. Plant Science 168(4): 1015–1022.

Lysakowska, M.E., Ciebiada-Adamiec, A., Klimek, L. and Sienkiewicz, M. 2015. The activity of silver nanoparticles (Axonnite) on clinical and environmental strains of *Acinetobacter* spp. Burns 41: 364–371.

Mahdizadeh, V., Safaie, N. and Khelghatibana, F. 2015. Evaluation of antifungal activity of silver nanoparticles against some phytopathogenic fungi and *Trichoderma harzianum*. J. Crop Protect. 4: 291–300.

Mahesha, H.S., Vinay, J.U., Ravikumar, M.R., Visweswarashastry, S., Keerthi, M.C., Halli, H.M. and Elansary, H.O. 2021. Colloidal silver hydrogen peroxide: new generation molecule for management of phytopathogens. Horticulturae 7(12): 573.

Maia, M.F. and Moore, S.J. 2011. Plant-based insect repellents: a review of their efficacy, development and testing. Malar J. 10(Suppl 1): S11. https://www.ncbi.nlm.nih.gov/pmc/articles/PMC3059459/. https://doi.org/10.1186/1475-2875-10-S1-S11.

Maliki, M., Ifijen, I.H., Ikhuoria, E.U. et al. 2022. Copper nanoparticles and their oxides: optical, anticancer and antibacterial properties. Int. Nano Lett. 12: 379–398. https://doi.org/10.1007/s40089-022-00380-2.

Manjumeena, R., Duraibabu, D., Sudha, J. and Kalaichelvan, P. 2014. Biogenic nanosilver incorporated reverse osmosis membrane for antibacterial and antifungal activities against selected pathogenic strains: an enhanced eco-friendly water disinfection approach. J. Environ. Sci. Health A 49: 1125–1133.

Mansoor, S., Zahoor, I., Baba, T.R., Padder, S.A., Bhat, Z.A., Koul, A.M. and Jiang, L. 2021. Fabrication of silver nanoparticles against fungal pathogens. Front. Nanotechnol. 3: 679358. doi: 10.3389/fnano.

Maroufpour, N., Mousavi, M., Abbasi, M. and Ghorbanpour, M. 2020. Biogenic nanoparticles as novel sustainable approach for plant protection. pp. 161–172. *In*: Biogenic Nano-Particles and their Use in Agro-ecosystems. Springer, Singapore.

Marrs, K.A. 1996. The functions and regulation of glutathione S-transferases in plants. Annu. Rev. Plant Biol. 47(1): 127–158.

Martinez-Gutierrez, F., Olive, P.L., Banuelos, A., Orrantia, E., Nino, N., Sanchez, E.M. and Av-Gay, Y. 2010. Synthesis, characterization, and evaluation of antimicrobial and cytotoxic effect of silver and titanium nanoparticles. Nanomedicine: Nanotechnology, Biology and Medicine 6(5): 681–688.

Matei, P.M., Martín-Gil, J., Michaela Iacomi, B., Pérez-Lebeña, E., Barrio-Arredondo, M.T. and Martín-Ramos, P. 2018. Silver nanoparticles and polyphenol inclusion compounds composites for *Phytophthora cinnamomi* mycelial growth inhibition. Antibiotics 7(3): 76.

Min, J.S., Kim, K.S., Kim, S.W., Jung, J.H., Lamsal, K., Kim, S.B. et al. 2009. Effects of colloidal silver nanoparticles on sclerotium-forming phytopathogenic fungi. Plant Pathol. J. 25: 376–380.

Mohammad, T.G. and El-Rahman, A.A. 2015. Environmentally friendly synthesis of silver nanoparticles using *Moringa oleifera* (Lam') leaf extract and their antibacterial activity against some important pathogenic bacteria. Mycopath 13: 1–6.

Mohamed, A.M. 2015. One-step functionalization of silver Nanoparticles using the orsellinic acid compound isolated from the endophytic fungus Epicoccum Nigrum: characterization and antifungal activity. Int. J. Nano Chem. 1(3): 103–110.

Mohamed, M.A., Hussein, H.A. and Ali, A.A.M. 2018. Antifungal activity of different size controlled stable silver nanoparticles biosynthesized by the endophytic fungus *Aspergillus terreus*. J. Phytopathology Pest Manag. 5(2): 88–107.

Mondal, K.K. and Mani, C. 2012. Investigation of the antibacterial properties of nanocopper against *Xanthomonas axonopodis* pv. punicae, the incitant of pomegranate bacterial blight. Ann. Microbiol. 62: 889–893.

Morales-Rodríguez, C., Picón-Toro, J., Palo, C., Palo, E.J., García, Á. and Rodríguez-Molina, C. 2012. *In vitro* inhibition of mycelial growth of *Phytophthora nicotianae* Breda de Haan from different hosts by Brassicaceae species. Effect of the developmental stage of the biofumigant plants. Pest Management Science 68(9): 1317–1322.

Morones, J.R., Elechiguerra, J.L., Camacho, A., Holt, K., Kouri, J.B., Ramírez, J.T. et al. 2005. The bactericidal effect of silver nanoparticles. Nanotechnology 16: 2346–2353.

Mosa, M.A., El-Abeid, S.E., Khalifa, M.M.A., Elsharouny, T.H., El-Baz, S.M. and Ahmed, A.Y. 2022. Smart pH-responsive system based on hybrid mesoporous silica nanoparticles for delivery of fungicide to control Fusarium crown and root rot in tomato. J. Plant Pathol. 1–14.

Mustafa, H., Ilyas, N., Akhtar, N., Raja, N.I., Zainab, T., Shah, T. and Ahmad, P. 2021. Biosynthesis and characterization of titanium dioxide nanoparticles and its effects along with calcium phosphate on physicochemical attributes of wheat under drought stress. Ecotoxicol. Environ. Saf. 223: 112519.

Nam, D., Lee, B., Eom, I., Kim, P. and Yeo, M. 2014. Uptake and bioaccumulation of titanium- and silver-nanoparticles in aquatic ecosystems. Mol. Cell Toxicol. 10: 9–17.

Noori, A., Ngo, A., Gutierrez, P., Theberge, S. and White, J.C. 2020. Silver nanoparticle detection and accumulation in tomato (Lycopersicon esculentum). J. Nanoparticle Res. 22(6): 1–16.

Oldenburg, S.J. 2017. Silver nanoparticles: properties and applications. Merck KGaA, Darmstadt, Germany and/or its affiliates. 14 pp. https://www.sigmaaldrich.com/technical-documents/articles/materials-science/nanomaterials/silver-nanoparticles.html.

Ozdemir, E. and Gozel, U. 2017. Efficiency of some plant essential oils on root-knot nematode *Meloidogyne incognita*. J. Agricul. Sci. Technol. A 7(3): 178–183. https://doi. org/10.17265/2161-6256/2017.03.005.

Paret, M.L., Vallad, G.E., Averett, D.R., Jones, J.B. and Olson, S.M. 2013. Photocatalysis: effect of light-activated nanoscale formulations of TiO_2 on *Xanthomonas perforans* and control of bacterial spot of tomato. Phytopathol. 103(3): 228–236.

Park, E.J., Yi, J., Chung, K.H., Ryu, D.Y., Choi, J. and Park, K. 2008. Oxidative stress and apoptosis induced by titanium dioxide nanoparticles in cultured BEAS-2B cells. Toxicology Letters 180(3): 222–229.

Park, H.J., Kim, S.H., Kim, H.J. and Choi, S.H. 2006. A new composition of nanosized silica-silver for control of various plant diseases. Plant Pathol. J. 22: 295–302.

Parveen, S., Wani, A.H., Shah, M.A., Devi, H.S., Bhat, M.Y. and Koka, J.A. 2018. Preparation, characterization and antifungal activity of iron oxide nanoparticles. Microb. Pathog. 115: 287–292.

Patel, N., Desai, P., Patel, N., Jha, A. and Gautam, H.K. 2014. Agronanotechnology for plant fungal disease management: a review. Int. J. Curr. Microbiol. App. Sci. 3: 71–84.

Patil, S.S., Shedbalkar, U.U., Truskewycz, A., Chopade, B.A. and Ball, A.S. 2016. Nanoparticles for environmental clean-up: a review of potential risks and emerging solutions. Environ. Technol. Innov. 5: 10–21.

Pereira, L.C., Pazin, M., Franco-Bernardes, M.F., da Cunha Martins Jr, A., Barcelos, G.R.M., Pereira, M.C. and Dorta, D.J. 2018. A perspective of mitochondrial dysfunction in rats treated with silver and titanium nanoparticles (AgNPs and TiNPs). J. Trace Elem. Med. Biol. 47: 63–69.

Pérez-de-Luque, A. and Rubiales, D. 2009. Nanotechnology for parasitic plant control. Pest Manag. Sci. 65: 540–545.

Pham, N.D., Duong, M.M., Le, M.V. and Hoang, H.A. 2019. Preparation and characterization of antifungal colloidal copper nanoparticles and their antifungal activity against *Fusarium oxysporum and Phytophthora capsici*. Comptes Rendus Chimie 22(11-12): 786–793.

Ponmurugan, P., Manjukarunambika, K., Elango, V. and Gnanamangai, B.M. 2016. Antifungal activity of biosynthesised copper nanoparticles evaluated against red root-rot disease in tea plants. J. Exp. Nanosci. 11(13): 1019–1031.

Pulit, J., Banach, M., Szczygłowska, R. and Bryk, M. 2013. Nanosilver against fungi. Silver nanoparticles as an effective biocidal factor. Acta Biochim. Pol. 60: 795–798.

Raghupathi, K.R., Koodali, R.T. and Manna, A.C. 2011. Size-dependent bacterial growth inhibition and mechanism of antibacterial activity of zinc oxide nanoparticles. Langmuir 27: 4020–4028.

Rajabi, F., Karimi, N., Saidi, M.R., Primo, A., Varma, R.S. and Luque, R. 2012. Unprecedented selective oxidation of styrene derivatives using a supported iron oxide nanocatalyst in aqueous medium. Adv. Synth. Catal. 453(9): 1707–1711. https://doi.org/10.1002/adsc.201100630.

Ramyadevi, J., Jeyasubramanian, K., Marikani, A., Rajakumar, G. and Rahuman, A.A. 2012. Synthesis and antimicrobial activity of copper nanoparticles. Materials Letters 71: 114–116.

Rao, K.J. and Paria, S. 2013. Use of sulfur nanoparticles as a green pesticide on *Fusarium solani* and *Venturia inaequalis* phytopathogens. RSC Adv. 3: 10471–10478.

Rosen, B.P. 2002. Transport and detoxification systems for transition metals, heavy metals and metalloids in eukaryotic and prokaryotic microbes. Comp. Biochem. Physiol. Part A Mol. Integr. Physiol. 133(3): 689–693.

Saharan, V., Mehrotra, A., Khatik, R., Rawal, P., Sharma, S.S. and Pal, A. 2013. Synthesis of chitosan based nanoparticles and their *in vitro* evaluation against phytopathogenic fungi. Int. J. Biol. Macromol. 62: 677–683.

Saqib, S., Zaman, W., Ullah, F., Majeed, I., Ayaz, A. and Hussain Munis, M.F. 2019. Organometallic assembling of chitosan-Iron oxide nanoparticles with their antifungal evaluation against *Rhizopus oryzae*. Appl. Organomet. Chem. 33(11): e5190.

Sangeetha, J., Thangadurai, D., Hospet, R., Purushotham, P., Karekalammanavar, G., Mundaragi, A.C., David, M., Shinge, M.R., Thimmappa, S.C., Prasad, R. and Harish, E.R. 2017. Agricultural nanotechnology: concepts, benefits, and risks. pp. 1–18. *In*: Nanotechnology an Agricultural Paradigm. Springer, Singapore.

Sasson, Y., Levy-Ruso, G., Toledano, O. and Ishaaya, I. 2007. Nanosuspension: emerging novel agrochemical formulations. pp. 1–32. *In*: Isaaya, I., Nauen, R. and Horowitz, A.R. (eds.). Insecticides Design Using Advanced Technologies. Springer, Dordrecht.

Satti, S.H., Raja, N.I., Ikram, M., Oraby, H.F., Mashwani, Z.U.R., Mohamed, A.H. and Omar, A.A. 2022. Plant-based titanium dioxide nanoparticles trigger biochemical and proteome modifications in *Triticum aestivum* L. under biotic stress of Puccinia striiformis. Molecules 27(13): 4274.

Šebesta, M., Nemček, L., Urík, M., Kolenčík, M., Bujdoš, M., Hagarová, I. and Matúš, P. 2020. Distribution of TiO_2 nanoparticles in acidic and alkaline soil and their accumulation by *Aspergillus niger*. Agronomy 10(11): 1833.

Shameli, K., Ahmad, M.B., Jazayeri, S.D., Shabanzadeh, P., Sangpour, P., Jahangirian, H. et al. 2012. Investigation of antibacterial properties silver nanoparticles prepared via green method. Chem. Cent. J. 6: 73.

Shankar, S. and Rhim, J.W. 2014. Effect of copper salts and reducing agents on characteristics and antimicrobial activity of copper nanoparticles. Materials Letters 132: 307–311.

Shankramma, K., Yallappa, S., Shivanna, M.B. and Manjanna, J. 2016. Fe_2O_3 magnetic nanoparticles to enhance S. lycopersicum (tomato) plant growth and their biomineralization. Applied Nanoscience 6(7): 983–990.

Shanmugam, C., Gunasekaran, D., Duraisamy, N., Nagappan, R. and Krishnan, K. 2015 Bioactive bile salt-capped silver nanoparticles activity against destructive plant pathogenic fungi through *in vitro* system. RSC Adv. 5: 71174–71182.

Sharma, H., Dhirta, B. and Shirkot, P. 2017. Evaluation of biogenic iron nanoformulations to control *Meloidogyne incognita* in okra. Int. J. Chem. Stud. 5(5): 1278–1284.

Shen, T., Wang, Q., Li, C., Zhou, B., Li, Y. and Liu, Y. 2020. Transcriptome sequencing analysis reveals silver nanoparticles antifungal molecular mechanism of the soil fungi Fusarium solani species complex. J. Hazard. Mater. 388: 122063.

Singh, S., Singh, B.K., Yadav, S.M. and Gupta, A.K. 2015. Applications of nanotechnology in agricultural and their role in disease management. Res. J. Nanosci. Nanotech. 5: 1–5.

Sondi, I. and Salopek-Sondi, B. 2004. Silver nanoparticles as antimicrobial agent: a case study on *E. coli* as a model for Gram-negative bacteria. J. Colloid Interface Sci. 275:177–182

Taha, E.H. 2016. Nematicidal effects of silver nanoparticles on root-knot nematodes (*Meloidogyne incognita*) in laboratory and screenhouse. J. Plant Prot. Path Mansoura Univ 7(5): 333–337.

Thakur, R.K. and Shirkot, P. 2017. Potential of biogold nanoparticles to control plant pathogenic nematodes. J. Bioanal. Biomed. 9(4): 220–222. https://doi.org/10.4172/1948-593X.1000182.

Tumburu, L., Andersen, C.P., Rygiewicz, P.T. and Reichman, J.R. 2017. Molecular and physiological responses to titanium dioxide and cerium oxide nanoparticles in Arabidopsis. Environ. Toxicol. Chem. 36(1): 71–82.

Van, S.N., Minh, H.D. and Anh, D.N. 2013. Study on chitosan nanoparticles on biophysical characteristics and growth of Robusta coffee in greenhouse. Biocatal. Agric Biotechnol. 2(4): 289–294.

Viet, P.V., Nguyen, H.T., Cao, T.M. and Hieu, L.V. 2016. Fusarium antifungal activities of copper nanoparticles synthesized by a chemical reduction method. J. Nanomater.

Walker, G.W., Kookana, R.S., Smith, N.E., Kah, M., Doolette, C.L., Reeves, P.T., Lovell, W., Anderson, D.J., Turney, T.W. and Navarro, D.A. 2017. Ecological risk assessment of nano-enabled pesticides: a perspective on problem formulation. J. Agric Food Chem. 66(26): 6480–6486. https://doi.org/10.1021/acs.jafc.7b02373.

Wang, C., Huang, X., Deng, W., Chang, C., Hang, R. and Tang, B. 2014. A nano-silver composite based on the ion-exchange response for the intelligent antibacterial applications. Mater Sci. Eng. C 41: 134–141.

Wani, A. and Shah, M. 2012. A unique and profound effect of MgO and ZnO nanoparticles on some plant pathogenic fungi. J. Appl. Pharmaceut. Sci. 2: 40–44.

Williams, D. 2002. Medical technology: how small we can go? Med. Device Tech 4: 7–9.

Xie, Y., He, Y., Irwin, P.L., Jin, T. and Shi, X. 2011. Antibacterial activity and mechanism of action of zinc oxide nanoparticles against *Campylobacter jejuni*. Appl. Environ. Microbiol. 77: 2325–2331.

Yang, Y., Wang, J., Xiu, Z. and Alvarez, P.J. 2013. Impacts of silver nanoparticles on cellular and transcriptional activity of nitrogen-cycling bacteria. Environ. Toxicol. Chem. 32: 1488–1494.

Zaki, S.A., Ouf, S.A., Abd-Elsalam, K.A., Asran, A.A., Hassan, M.M., Kalia, A. and Albarakaty, F.M. 2022. Trichogenic silver-based nanoparticles for suppression of fungi involved in damping-off of cotton seedlings. Microorganisms 10(2): 344.

Zhang, J., Liu, Y., Li, Q., Zhang, X. and Shang, J. K. (2013). Antifungal activity and mechanism of palladium-modified nitrogen-doped titanium oxide photocatalyst on agricultural pathogenic fungi *Fusarium graminearum*. ACS Applied Materials & Interfaces 5(21): 10953–10959. https://doi.org/10.1021/am4031196.

Chapter 21

Selenium, Sulfur, and Tellurium Chalcogen-Containing Nanostructures

Synthesis, Properties, and Practical Applications in Agrochemistry and Phytopathology

Alla I. Perfileva[1] and *Konstantin V. Krutovsky*[2,3,4,5,6,]*

Introduction

Chalcogens in nature are most often found in the composition of the ore, which are sulfides, pyrites, oxides, and selenides (Yang et al. 2022). Chalcogens include non-metals and metals, including such biologically important ones as sulfur (S), selenium (Se), and tellurium (Te), promising for use in agriculture as nanoparticles (NPs) and nanocomposites (NCs). According to their electronic structure, chalcogens are *p*-elements. There are six electrons in their outer energy contour. To complete the *p*-orbital, two more electrons are needed, therefore, in compounds, chalcogens exhibit the properties of an oxidizing agent. With an increase in the energy levels in a group, the bond with external electrons weakens; therefore, Te is a reducing agent (Chivers 2005).

S is an element of the main subgroup A of group VI of the 3rd period of Mendeleev's periodic table. At room temperature, elemental S is a bright yellow solid crystalline substance. It forms over 30 solid allotropes, more than any other element. α-S occurs in nature and forms insoluble, chemically stable lemon-yellow crystals; β-S exists in the form of white crystalline plates or a plastic, rubbery, unstable, brown-green substance. There are 23 known isotopes of S, four of which are stable: ^{32}S, ^{33}S, ^{34}S, and ^{36}S (Skabara 2004).

[1] Laboratory of Plant-Microbe Interactions, Siberian Institute of Plant Physiology and Biochemistry, Siberian Branch of the Russian Academy of Sciences, 664033 Irkutsk, Russia.

[2] Department of Forest Genetics and Forest Tree Breeding, Faculty of Forest Sciences and Forest Ecology, Georg-August University of Göttingen, Büsgenweg 2, 37077 Göttingen, Germany.

[3] Center for Integrated Breeding Research (CiBreed), Georg-August University of Göttingen, Albrecht-Thaer-Weg 3, 37075 Göttingen, Germany.

[4] Laboratory of Population Genetics, N.I. Vavilov Institute of General Genetics, Russian Academy of Sciences, Gubkin Str. 3, 119333 Moscow, Russia.

[5] Genome Research and Education Center, Laboratory of Forest Genomics, Department of Genomics and Bioinformatics, Institute of Fundamental Biology and Biotechnology, Siberian Federal University, 660036 Krasnoyarsk, Russia.

[6] Scientific and Methodological Center, G.F. Morozov Voronezh State University of Forestry and Technologies, 8 Timiryazeva Str., 394036 Voronezh, Russia.

* Corresponding author: konstantin.krutovsky@forst.uni-goettingen.de

Se is an element of group VI of the 4th period of Mendeleev's periodic table. Se was discovered in 1817 by J. Berzelius (Boud 2011). Se has the following valency in chemical compounds: 4+, 6+, and 2+. Six isotopes are known for Se: ^{74}Se, ^{76}Se, ^{77}Se, ^{78}Se, ^{80}Se, and ^{82}Se (Pyrzynska 2002). It occurs in nature as selenides SeO_4^{2-}, SeO_3^{2-} and Se^{2-}, which can be reduced to the atomic state. Se has several allotropic modifications in the free state: red, gray, and black Se (Atkins et al. 2010; Zhu et al. 2019). In nature, Se can be found in various organic forms, as well as gaseous (dimethyl selenide and dimethyl diselenide) and in compounds with amino acids selenocysteine (SeCys) and SeMet (SeMet) (Pyrzynska 2002).

Te is a metalloid present in nature as a soluble oxyanion in a fourfold oxidation state: $^{-2}$(H_2Te), $^{+2}$(TeO_2^{2-}), $^{+4}$(TeO_3^{2-}), and $^{+6}$(TeO_4^{2-}). Elemental Te^0 has the chemical properties of a non-metal or metalloid and the physical properties of a metal (Vávrová et al. 2021). Pure Te is a brittle, silvery-white substance with a metallic sheen. In thin layers, it has a red-brown color under the light; in pairs, it is golden yellow. Te is soluble in alkalis, nitric and sulfuric acids. With water, metallic Te begins to react at 100°C. Te is oxidized in the air as a powder, even at room temperature, forming TeO_2 oxide. When heated in air, Te burns out, forming a solid substance with less volatility than Te itself, TeO_2 (Carapella 1981).

Physiological and Biochemical Role of Sulfur, Selenium, and Tellurium in Plants

S is a macronutrient needed for plants and is contained in 90% of plant proteins. It is involved in the formation of chlorophyll and the synthesis of oils and is a part of glutathione, coenzyme-A, vitamins, amino acids methionine, cysteine, and cystine. In plants, S is found in the form of organic and mineral compounds composing up to 0.2–1% of dry weight (Narayan et al. 2022). It stimulates the germination of plant seeds (Mondal et al. 2022) and is actively involved in the regulation of the redox status of the cell (Mukwevho et al. 2014; Francioso et al. 2020). Its content in the cell is relatively small compared to the amount of nitrogen or phosphorus, but it is irreplaceable. Plant families such as legumes, cabbages, nightshades, hazeweeds, umbrellas, Asteraceae, and lilies consume the most S (Hell et al. 2010).

Se plays an important role for plants. In 1957, this element was assigned by K. Schwarz and S. Foltz to the group of microelements necessary for both plants and animals (Vikhreva et al. 2012). It is involved in a number of redox reactions as a cofactor (Zoidis et al. 2018; White 2018). It takes part in the formation of chlorophyll (Sali et al. 2018), the synthesis of tricarboxylic acids, and the metabolism of long-chain fatty acids (Gupta and Gupta 2017; Trippe and Pilon-Smits 2021). Se has an antagonistic effect on the absorption and transport of heavy metals (Feng et al. 2021) and increases resistance to water stress and salt and drought tolerance. In addition, it is involved in the synthesis of tocopherols, tocotrienols, and ubiquinones (Nessel and Gupta 2020). The direct conversion of Se compounds in the plant cell occurs in chloroplasts (synthesis of SeCys from selenide) and cytoplasm (synthesis of SeMet and SeCys) (van Hoewyk 2013; Trippe and Pilon-Smits 2021).

Se is an essential trace element for living organisms and is beneficial in a narrow range of low concentrations but toxic at high doses (Lanctot et al. 2017; Li et al. 2019). Both excess and deficiency of Se in the nutrient medium adversely affect the growth and development of plants. With an excess of Se, chlorosis, necrosis, and slow growth are observed (Aslam et al. 1990; Kolbert et al. 2019). The Se content in plants averages 0.0001 wt % and depends on soil characteristics, growing climatic conditions, plant development phase, and biological characteristics of the species (White 2016).

Te can be toxic even at very low concentrations (1 μg ml^{-1}) (Vaigankar et al. 2018). The maximum allowable concentration of Te for humans varies for various compounds: 0.007–0.01 mg/m³ in air and 0.001–0.01 mg/L in water. Recently, the conversion of tellurite to black elemental Te, including extracellular accumulation, volatilization, and methylation, has aroused the interest of researchers (Huang et al. 2016). The physiological role of Te for plants, animals, and humans is less studied than Se. It is known that Te is less toxic than Se, and Te compounds in the body are quickly reduced to elemental Te, which in turn combines with organic substances. Te is not an essential trace element. It is toxic to humans and does not occur naturally in any of the known biomolecules (Kaur et al. 2013). However, organic Te compounds exhibit a wide range of biological properties, including antioxidant ones (You et al. 2003).

Thus, despite the fact that Se and Te are very similar to S in chemical properties, these elements have different meanings in terms of their physiological roles in the cell. Thus, S plays an important role in the life of plants, animals, and humans; it is included in almost all proteins, in sulfur-containing amino acids, cysteine, and methionine, and also in vitamin B1 and the hormone insulin. Se is an important trace element for a living cell, a coenzyme in many enzymes. Te is not found in known biomolecules, but it is detected in living organisms. At the same time, all the discussed chalcogens have pronounced antioxidant activity, which is extremely important for the resistance of organisms to stress factors.

Selenium Accumulating Plants

According to their ability to accumulate Se in their tissues, plants can be divided into non-accumulators (contain less than 100 mg Se per kg of dry tissue), accumulators (contain 100–1,000 mg/kg when grown in the Se-enriched soil), and hyperaccumulators (1,000–15,000 mg/kg) (Pilon-Smits 2019). Accumulator plants are able to concentrate Se by actively absorbing it from the soil (White 2016). An example of a hyperaccumulator plant is *Astragalus racemosus* Pursh from the Fabaceae family. Hyperaccumulator plants actively absorb Se from the soil and transfer it along the xylem to the aboveground organs, where Se is evaporated from the leaf surface (Trippe and Pilon-Smits 2021). Se can also be excreted in plant root exudates (Pilon-Smits 2019).

In hyperaccumulator species, most of the Se is in the form of methylselenocysteine due to the high activity of the SeCys methylation process, which is considered one of the resistance mechanisms providing plant tolerance to high Se concentrations. Further, methylselenocysteine can be converted to volatile dimethyl diselenide (Lima et al. 2018). In addition, accumulator plants are able to accumulate Se-methylselenocysteine and γ-glutamine-Se-methylselenocysteine in cell vacuoles (White 2016; Schiavon and Pilon-Smits 2017). Non-hyperaccumulating species tend to have a slower selenate assimilation rate than hyperaccumulating plants, accumulating relatively more inorganic Se (Schiavon and Pilon-Smits 2017).

Other differences between hyperaccumulators and non-hyperaccumulators are that the former move Se more in the xylem from root to shoot and in the phloem from leaves to reproductive organs, and often accumulate Se in specialized tissues that create leaf pubescence (White 2016; Lima et al. 2018; Schiavon and Pilon-Smits 2017; Pilon-Smits 2019). The high content of Se in plant tissues performs a protective function. Se can accumulate in plants to levels that are toxic to bacteria, fungi, and animals. In addition, Se is able to be metabolized in plant tissues to gaseous forms, acting as a repellant. In this regard, it was proposed to use Se and its compounds as insecticides to treat agricultural plants (Mechora 2019).

Thus, the brief analysis of the role of Se for plants shows that it is essential for any plant organism. Se is involved in a number of essential cellular and organismal processes, but its content in the tissues of different plants is not the same and depends on the species.

Sulfur, Selenium, and Tellurium in Soil and Their Absorption by Plants

S is the tenth most abundant element by mass in the universe and the fifth most abundant on Earth. Although sometimes S can be found in pure natural form, it is usually found in the nature in the form of sulfide and sulfate minerals. Large reserves of native S have been found in Central Asia, Poland, Mexico, and southern Italy (Sosa-Torres et al. 2020). Some of the natural S is of volcanic origin, and some were formed from sulfates due to the sulfate-reducing bacteria activity (Kushkevych et al. 2021).

Agricultural plant species contain different amounts of S in their body, so the need for it is different. The largest amount of S is found in leaves and seeds, while the smallest amount is found in stems and the root system. Plant families that consume S the most are legumes, cabbages, nightshades, hazeweeds, umbrellas, asteraceae, and lilies. Cereal crops are not so sensitive to S deficiency. All agricultural plants that need S are divided into three groups: (1) sensitive (cabbage, garlic, onion, turnip, mustard, and especially rapeseed), which consume 40–80 kg of S per hectare; (2) moderately demanded (legumes, maize, and beets) consuming 20–40 kg of S per hectare, and (3) slightly sensitive (potatoes, grasses, and cereals) consuming 10–25 kg of S per hectare.

The root system consumes S from the soil in the form of SO_4 ions, and a small part of S is absorbed from the air by the leaves in the oxidized form (Bloem et al. 2015). The amount of S consumption varies depending on the plant species and development phase. For example, rapeseed needs S mostly during flowering and pod formation, while wheat needs S mostly during the tillering and milky ripeness of grains. Maize needs S throughout its growth period (Bouranis et al. 2020).

S in the organic form is mostly found in the soil (in plant residues and humus), up to 80–90%, but is inaccessible to plants. Only about 10–20% of S occurs in the mineral form. In order for the element to be well absorbed by plants, sulfur-containing organic substances must be mineralized to sulfates. Microorganisms carry out this process, but it is very slow, so some plants that are characterized by intensive growth can experience sulfur starvation (Fuentes-Lara et al. 2019).

The availability and distribution of Se in soils depend on its acidity and soil organic matter content, redox conditions, soil microbiota activity, soil structure, temperature, and moisture (Sager 2006; Ullah et al. 2019). It is shown that Se is concentrated mainly in the humus and illuvial soil layers. The content of Se in the upper layer (0–20 cm) is mainly determined by the presence of humus. Se is present in soil solution, usually in the form of selenate (Se VI), selenite (Se IV), and organic forms (SeCys and SeMet). Elemental Se and selenide may also occur depending on the redox potential of the soil (Terry et al. 2000). Selenate dominates in aerobic soils with neutral pH, while selenite dominates at lower pH and redox (Ullah et al. 2019).

Plant roots can absorb selenate, selenite, or other organic Se compounds from the soil but cannot absorb colloidal selenides. Selenate is transported across the plasma membrane with the participation of special carriers. Selenite is thought to be transported by phosphate transporters (Ullah et al. 2019) or enters plant cells passively via diffusion (Terry et al. 2000). Studies showed that at low acidity, selenite could enter root cells through the aquaporin channel (Trippe and Pilon-Smits 2021). At high doses, selenites and selenates are toxic to plants. Both of these forms are rapidly absorbed and incorporated into plant metabolism. For example, selenate is distributed through the xylem from the roots to other parts of the plant. The transformation of inorganic forms into SeCys occurs in chloroplasts; therefore, the highest content of selenates is observed in leaves. Further, SeCys is transformed into SeMet and other volatile forms (Trippe and Pilon-Smits 2021). Organic Se compounds then move through the phloem to the roots and other organs of the plant. Selenite is easily converted to SeMet, accumulating mainly in the roots (Mechora 2019).

Se content in soils varies widely around the world and can range from 10 to 1000 µg/kg or more. Soils rich in Se originate from Cretaceous shale rock. Such soils are found in North America, Great Britain, India, Pakistan, and Australia (Winkel et al. 2015). In the last century, the amount of Se in the soil has decreased significantly. In many regions of the world, there is a deficiency of Se in the soil, resulting in its deficiency in plants. For example, studies of the Se content in cow's milk, grass, and rabbit liver have shown that about 75% of organisms living in Poland are Se deficient (Zachara and Pilecki 2000; Dębski et al. 2001). Fertilizers containing Se do not have a pronounced effect due to nitrates, chlorides, and phosphates, which bind Se into insoluble compounds.

The earth's crust contains 1×10^{-6}% of Te by mass (0.01 g/t). The industrial source of Te is the sludge from the electrolytic treatment of copper (Cu), molybdenum (Mo), and lead (Pb) ores. There are more than 100 Te minerals, such as altaite PbTe, hessite Ag_2Te, sylvanite $AgAuTe_4$, calaverite $AuTe_2$, tetradymite Bi_2Te_2S, Te ocher TeO_2. Native Te also occurs together with Se and S. For example, Japanese Te S ore contains 0.17% of Te and 0.06% of Se (Carapella 1971; Kudryavtsev 1974).

Te concentrations in soil vary widely from 0.008 mg/kg to 0.03 mg/kg, and in some places, that are naturally enriched in Te compounds can reach 0.166 and even 0.5 mg/kg (Jabłonska-Czapla et al. 2021). As a result of human industrial activities, Te-containing compounds can occur as environmental pollutants. These compounds are most often found as tellurides in Cu, gold (Au), and silver (Ag) ores. In areas adjacent to Au mines, Te concentration reaches extreme values up to 14.8 ppm. Te is also a by-product of the electrolytic refining of Cu (Vávrová et al. 2021). Tellurite is highly toxic to living organisms. Elemental Te^0 has been classified as non-toxic (Chasteen et al. 2009) compared to soluble Te oxyanions (Taylor 1999). Te compounds can be divided into three groups: (1) inorganic tellurides, (2) Te-containing complex structures, and (3) tellurides (Cooper 1971). Te inorganic compounds occur in various oxidation states from –II to +VI, namely –II (H_2Te, hydrogen telluride), 0 (elemental Te^0), +II (TeO, Te monoxide), +IV (TeO_3^{2-}, tellurite), and +VI (TeO_4^{2-}, tellurate). These oxides form tellurous (H_2TeO_3) and telluric (H_2TeO_4) acids and their salts are known as tellurites TeO_3^{2-} and tellurates TeO_4^{2-}. The TeO_2^{2-} anion in the +II oxidation state also exists (Vávrová et al. 2021).

From the abovementioned, we can summarize that S and Se are widely distributed in the soil over a wide range. In the soil, chalcogens are found in the form of various organic and inorganic compounds. These substances are absorbed by the plant roots with the participation of special carriers or by means of diffusion. Then, they are transferred through the xylem to the aboveground organs. Most transformations of Se-containing substances occurred in the chloroplast. S compounds are actively involved in the synthesis of proteins and amino acids. These substances are distributed throughout the body, evaporate from the surface of the leaf or accumulate in the roots.

Advantages of Sulfur, Selenium, and Tellurium Nanoparticles

Due to climate change, about 40% of crops are subjected to abiotic and biotic stress every year, causing large yield losses (Raza et al. 2019; Rivero et al. 2022). The global demand for food is growing, but at the same time, the environmental requirements for agriculture are being tightened. Nanotechnology is an advanced and popular technology widely used in the fields of electronics, aircraft and automotive, construction, biomedicine, pharmacology, etc., but its use in agriculture is still limited despite its huge potential. In particular, nanotechnologies can help to solve agricultural problems related to crop disease control by reducing the use of chemicals, such as herbicides, pesticides, and fungicides. The use of these toxic chemicals harms humans, animals, and the environment. Therefore, the use of NPs as fungicides/bactericides or nanofertilizers, due to their small size and high reactivity surface area, increases the effectiveness of plant disease control. Due to their advantages, the use of NP-based fungicides has been gradually increasing in recent years compared to conventional chemical fungicides (Kallol Das et al. 2021). For example, the lack of S is compensated by the application of fertilizers, which are not always beneficial to the environment and may exhibit toxicity, while S NPs are safe for use in agriculture and can also be used to improve plants and control phytopathogens. For example, a stronger effect of S NPs compared to elemental S and sulfur-containing salts (sodium thiosulfate and sodium metabisulfite) against bacteria (*Escherichia coli* and *Staphylococcus aureus*) and fungi (*Aspergillus flavus, Candida albicans*) was shown in the study of Kim et al. (2020). S NPs showed stronger activity against bacteria than against fungi. Among the microorganisms tested, *Escherichia coli* (Gram-negative) was the most susceptible to S NPs, followed by *Staphylococcus aureus* (Gram-positive), *Candida albicans* (yeast), and *Aspergillus flavus* (molds). Scanning electron micrographs of microorganisms treated with S NPs showed different patterns of cell destruction depending on the type of microorganisms (Kim et al. 2020). It has been shown that S NPs can be used to treat microbial infections and potentially solve the problem of antibiotic resistance by replacing them since they have a broad spectrum of antimicrobial activity (Rai et al. 2016).

The problem of Se deficiency in food products is one of the widely discussed topics at present. Se biofortification (Newman et al. 2019; Surai and Kochish 2020) and biotransformation (Hawrylak-Nowak 2013; Moreno-Martin et al. 2019) of plants are actively used to combat Se deficiency. In addition, to solve this problem, the possibility of treating plants using Se NPs and their compounds is also being considered. Compared to total inorganic and organic Se, Se NPs demonstrated better bioavailability, increased biological activity, and reduced toxicity (Wang et al. 2013; Zsiros et al. 2019). For example, it was found that the use of selenate at a concentration of 10 mg/L had a negative effect on the photosynthetic activity and growth of tobacco *Nicotiana tabacum* L., while the treatment of plants with Se in the form of NPs, even at higher concentrations (100 mg/L) had no negative effects (Zsiros et al. 2019). The effectiveness of NPs depends on their size, shape, and ability to aggregate and precipitate. Se is toxic to organisms at elevated levels because inorganic forms of Se can cause oxidative stress. It can also be incorporated into proteins, replacing S and leading to the formation of SeCys and SeMet, and disrupting the functional properties of proteins (Stadtman 1990).

Te NPs are promising due to their excellent biocompatibility (Medina et al. 2018) and antimicrobial, antioxidant, and anticancer activities (Medina et al. 2019; Vahidi et al. 2021). Using some microorganisms, such as *Bacillus cereus, Erythromonas ursincola, Shewanella fridigimarina,* and *Rhodobacter capsulatus*, due to their reducing enzymes, it is possible to convert metalloids into less toxic elemental forms and increase the Se content and Te bioavailability (Dhanjal and Cameotra 2010; Varvarova et al. 2021). Moreover, these same microorganisms can be used for soil bioremediation due to their ability to reduce biologically the cations of these metalloids. In the future, they can be used to produce Se and Te nanomaterials.

The characterization of these NPs is carried out using a variety of routine laboratory methods, including the study of their shape, size, porosity, surface chemistry, crystallinity, and dispersion pattern. Among the widely used methods for characterizing NPs are: UV spectroscopy, luminescence spectroscopy (LS), scanning electron microscopy - energy dispersive X-ray spectroscopy (SEM-EDX), transmission electron microscopy (TEM), Fourier transform infrared spectroscopy (FTIR), and X-ray diffraction (XRD).

XRD confirms the presence of NPs and determines their lattice structure, crystallinity, and crystallite size. FTIR is an efficient technique that provides reproducible analyses used to reveal the presence of functional groups on the NP surface. These groups may be involved in the reduction of the metal ions and/or the NP capping that ensures colloidal stability. In addition to determining the surface charge (z potential) of the NPs, dynamic light scattering (DLS) provides the NP hydrodynamic diameter and good insight into their stability/aggregation by measuring their Brownian motion. Atomic force microscopy (AFM) provides quantitative information about the length, width, height, morphology, and surface texture of NPs through a tridimensional visualization (Zambonino et al. 2021).

Thus, chalcogens in the form of NPs are more advantageous compounds in comparison with organic and inorganic chalcogen-containing substances. This is achieved due to the reduced toxicity of NPs and increased biological activity. Biogenic NPs have promising applications in medicine, biosensors, and environmental remediation.

There are various ways to obtain chalcogen NPs: chemical and physical synthesis and obtaining biogenic NPs using living organisms, an example of the so-called "green chemistry."

Chemical and Physical Synthesis

One of the simple chemical methods for obtaining S NPs is the synthesis using sodium thiosulfate and hydrochloric acid (Shankar et al. 2018). The synthesis and characterization of water-dispersible polymeric NPs with S content greater than 75% obtained from interfacial polymerization between 1,2,3-trichloropropane and sodium polysulfide in water has been reported (Lim et al. 2015). Among the existing methods for synthesizing materials, the sonochemical method is a highly efficient and convenient production method. Thus, NC from copper sulfide (CuS-covellite) and S-doped reduced graphene oxide (S-rGO) was synthesized (Karikalan et al. 2017). Ultrathin S NPs with 10–20 nm in diameter were synthesized using the membrane deposition method (Chen et al. 2013).

Physical synthesis methods include pulsed laser ablation, microwave synthesis, hydrothermal treatment, and vapor deposition. Physical methods have their advantages; for example, the size of nanoclusters can be controlled by laser parameters, such as fluence, wavelength, and pulse duration, as well as ambient gas conditions such as pressure and flow parameters (Marine et al. 2000). In addition, sputtering and laser ablation maintain material stoichiometry (Singh and Narayan 1990). Synthesis by laser ablation is carried out using a liquid phase in deionized water. The result is a colloidal solution with spherical particles of various sizes (Guisbiers et al. 2017). Laser ablation is popular due to the ease of obtaining stable NPs and the absence of chemical contamination (Guisbiers et al. 2014).

S NPs can also be synthesized physically (Saikia and Lens 2020). An ultrasonication-promoted strategy was proposed to synthesize luminescent S-dots, which reduced the synthesis time from the commonly used five days to several hours. The as-synthesized S-dots showed high photostability and low cytotoxicity and were then successfully applied for cellular imaging (Zhang et al. 2019).

The chemical synthesis of Se NPs is carried out by combining various chemicals. Organic (for example, sodium bis(phenylethyl)diselenophosphinate (Papkina et al. 2015; Perfileva et al. 2018) and inorganic (Se oxide) Se precursor compounds) are used as a precursor. As a rule, chemical reduction and stabilization of the obtained product are carried out by various chemical agents in the course of reactions.

The synthesis of Se NPs by laser ablation in various liquid monomers is also of interest. In this case, isodecyl acrylate, carboxyethyl acrylate, and ethylene glycol phenyl acrylate are used as condensing liquids for laser ablation (Galiová et al. 2014). Microwave synthesis is also actively used for the synthesis of Se NPs based on heating an aqueous solution of Se salt by microwaves. The method is sensitive to reaction time, irradiation power, and the type of substances used (Hou et al. 2011; Panahi-Kalamuei et al. 2014). The synthesis of Se NPs on nanotubes using an autoclave is also known among physical methods (Xi et al. 2006).

Se NPs are highly unstable and readily convert to inactive forms as they easily form aggregates in aqueous suspensions (Hosnedlova et al. 2018; Zhu et al. 2019; Li et al. 2020). Therefore, in order to increase the stability of NPs during preparation, storage, and use, they can be packaged in various polysaccharides because their molecular structure contains reactive amino, hydroxyl, or carboxyl groups that affect the formation, stabilization, and growth of NPs. Such packaging has a number of advantages: high biocompatibility, biodegradability, and active hydroxyl groups. For NCs, polysaccharides derived from various sources such as mushrooms, fruits, trees, medicinal plants, and synthetic polymers are used. The most commonly used polysaccharides are chitosan and glucans derived from mushrooms. Usually, to obtain Se compounds with such polymers, the selenium acid solution is reduced with ascorbic acid in the presence of polysaccharides (Górska et al. 2021).

Chemically synthesized NPs can be packed into polymeric polysaccharide matrixes during the synthesis; for example, Se NCs can be obtained based on polysaccharide matrixes. Se NPs can be obtained from acetic, oxalic, and gallic acids (Dwivedi et al. 2011). Known is the synthesis of hybrid NPs based on Se NPs and zinc (Zn) selenide, stabilized with water-soluble polymers—polyvinylpyrrolidone and polymethacrylic acid. Depending on the mass ratio of the initial elements, NPs of different morphology (irregularly shaped spheres and micelles containing nuclei) and different sizes, 30–135 nm, can be formed. At the same time, the nature of the polysaccharide is the determining factor in the formation of Se-containing NPs and optimization of their parameters (Plucinski et al. 2021). For example, in the chemical synthesis of Se, NPs are synthesized using sodium borohydride as a reducing agent and gum as a stabilizer. The size of NPs varied from 44.4 nm to 200 nm with the average size of 105.6 nm. It has been shown that such NPs exhibited high activity of radial absorption (Kora 2018). An example of the chemical production of Se NPs is ionic liquid-induced synthesis with sodium selenosulfide as Se precursor in the presence of a polyvinyl alcohol stabilizer, which can produce spherical Se NPs in the size range of 76–150 nm (Langi et al. 2010).

The study of the absorption spectrum of UV–Vis Se NPs in the range of 200–400 nm makes it possible to obtain a peak indicating the synthesis of Se NPs. In addition, one of the characteristics of NPs is the Z-mean size and zeta potential of NPs, which indicate the size and stability of the bond between NPs. FTIR of Se NPs detects the presence of different chemical groups in the test substance; for example, two peaks at 3,427.88 cm^{-1} and 1,638.83 cm^{-1} corresponding to –OH and –NH groups, which indicate the presence of carboxylic and amide groups, respectively (Joshi et al. 2019). XRD allows to explore the crystal structure of NPs (Giannini et al. 2016). SEM-EDS and HR-TEM provide information on surface morphology, elemental composition, and sizes of NPs (Zhou and Greer 2016; Anjum 2016).

Various chemical methods for the synthesis of Te NPs are known. There is a simple method for synthesizing various Te NPs, including nanorods and nanospheres, by chemical reduction at low temperatures. The synthesis is carried out using potassium borohydride (KBH_4) and hydrazine hydrate ($N_2H_4 \cdot H_2O$) as reducing agents (Panahi-Kalamuei et al. 2014). Te NPs were synthesized by a simple sonochemical method by reduction of $TeCl_4$ to Te under ultrasonic irradiation in methanol. When the reaction was carried out in an alkaline medium, TeO_2 NPs were obtained (Arab et al. 2017). Synthesized cadmium telluride NPs (CdTe NPs) can be stabilized with thioglycolic acid (TGA), 1-thioglycerol (TGC), and L-cysteine (L-C) (Jo et al. 2021). The ratio of the stabilizer to Cd^{2+} was 1 to 2.4, respectively. The resulting particles had a crystalline structure. The average sizes of synthesized NPs were 4.2 nm, 4.1 nm, and 3.7 nm (Jo et al. 2021). A method for the synthesis of flower-like Te NPs has been described by Wang et al. (2010). It is based on the reduction of tellurite precursor by-products formed during the decomposition of sulforaphane at elevated temperatures in an aqueous medium. These particles and other organic molecules present in the reaction mixture are adsorbed on the surface of the Te nuclei and control the further growth of Te in the form of nanoflowers. The average particle size of the nanoflower was 112 nm, and they consisted of smaller domains, the size of which was approximately 30 nm in diameter. The domains were crystalline and consisted of trigonal Te (Krug et al. 2020).

Fernández-Lodeiro et al. (2017) described the synthesis of novel multi-crystalline NPs and $PtTe_2$ of various sizes through an annealing process using novel nanostructured organometallic Pt-Te NPs as a single precursor source. This precursor was obtained in a single reaction step using Ph_2Te_2 and H_2PtCl_6, and its size could be successfully controlled in the nanoscale range. The sizes of $PtTe_2$ crystallites formed using large spheres have been estimated in the range of 2.5–6.5 nm (Fernández-Lodeiro et al. 2017). A synthesis of Te NPs was developed as a result of the reaction for obtaining trialkylphosphane tellurides formed by dissolving Te in pre-dried commercial trihexyltetradecylphosphonium chloride ($P_{66614}Cl$) at high temperatures with common polar protic solvents (for example, water, alcohols or amides). Highly homogeneous Te nano- and microstructures with various sizes and morphologies, including three-dimensional (3D) Te fusiform assemblies and 3D aloe-like Te microarchitectures, were obtained (Zhang et al. 2020).

Physicochemical methods for synthesizing NPs are not always environmentally friendly and can be expensive. Therefore, the use of synthesis methods based on the use of living organisms is gaining popularity. Such synthesis usually consists of a single stage and is environmentally friendly and more cost-effective because it requires less reaction time and reagent use (Shoeibi et al. 2017). The synthesis of Se NPs is carried out due to the reducing enzymes present in various species, such as bacteria, fungi, algae and plants (Husen and Siddiqi 2014). When comparing the biological activity of biogenic and chemically synthesized Se NPs, it was found that the antimicrobial activity of biogenic NPs was much higher (Cremonini et al. 2016). The authors attribute this effect to the biological activity of the matrixes themselves, in which NPs, for example, chitosan, were packaged.

Synthesis Using Plants

The biosynthesis of Se NPs from plant extracts is the most preferable among all other biosynthesis methods due to its low cost and no need for special equipment and conditions. S NPs are obtained in a biogenic way based on sodium thiosulfate and an aqueous extract of basil leaves (Ragab and Saad-Allah 2020). Biogenic S NPs (BioS NPs) can be obtained from pomegranate (*Punica granatum*) peel extract using a simple biogenic technique (Salem et al. 2016). These NPs have shown antimicrobial and wound-healing activity (Samrat et al. 2021). S NPs were also synthesized from *Catharanthus roseus* extract using sodium sulfide (Paralikar, Rai 2017). Their size was 480 ± 39.6 nm, and they had high antifungal activity with minimal inhibitory concentration (MIC) equaled 1.56 mg/ml (Chantongsri et al. 2021). Paralikar and Rai (2017) showed that S NPs derived from sodium polysulfide and *Azadirachta indica, Catharanthus roseus, Mangifera indica,* and *Polyalthia longifolia* plant extracts had a particle size of 70–80 nm and were effective against *E. coli* and *S. aureus*. Khairan et al. (2019) reported that *Allium sativum* extract could be used to synthesize S NPs that had antifungal activity against *Candida albicans*. A detailed review of the synthesis of S NPs using plants is described in Ghotekar et al. (2020).

Se NPs can also be obtained from plant extracts. For example, Se NPs with a particle size of 1–3 nm in diameter were obtained by microwave heating an extract of the cocoa bean shell (*Theobroma cacao* L.) (Mellinas et al. 2019). They were characterized by high antioxidant properties (Mellinas et al. 2019). Plants from the laurel (Lauraceae) and rue (Rutaceae) families can also be used for the biosynthesis of Se NPs. For example, Se NPs can be obtained using a sodium thiosulfate-based bark extract of *Cinnamomum zeylanicum* (Lauraceae) (Najafi et al. 2020). The synthesis of Se NPs using an aqueous extract of the berries of the tropical plant *Murraya koenigii* (Rutaceae) has been described in Yazhiniprabha and Vaseeharan (2019). Se NPs with 50 and 150 nm in size were spherical and exhibited antibacterial activity against Gram-positive (*Enterococcus faecalis, Streptococcus*) and Gram-negative (*Shigella sonnei* and *Pseudomonas aeruginosa*) bacteria at concentrations of 40 μg/mL and 50 μg/mL. They also had an anti-biofilm effect on these types of bacteria. These Se NPs were characterized by low cytotoxicity (Yazhiniprabha and Vaseeharan 2019). Se NPs were also synthesized from the tree of the rue family *Clausena dentate* with sizes ranging from 46 nm to 78 nm (Sowndarya et al. 2017).

There are also studies that demonstrated the possibility for "green synthesis" of Se NPs using fruit trees and shrubs. For example, Se NPs with a mean size of 113 nm were obtained from an extract of hawthorn berries, *Crataegi fructus* (Cui et al. 2018). Zhang et al. (2018) presented a method for the synthesis of Se NPs based on polysaccharides of common wolfberry (also known as goji berry), *Lycium barbarum*. Such Se NPs were characterized by high antioxidant activity due to the active absorption of free radicals and protection of cells from apoptosis (Zhang et al. 2018). Very small Se NPs with 3–18 nm in size can be obtained using dried cultivated grape *Vitis vinifera* (Sharma et al. 2014). Colloidal Se NPs with ~ 60–80 nm in diameter and fluorescent properties were synthesized based on an aqueous extract from the lemon (*Citrus limon*) leaves (Prasad et al. 2013). An aqueous extract of amla fruit or Indian gooseberry (*Emblica officinalis*) is rich in various secondary metabolites and was suitable for the synthesis of amorphous Se NPs with 15–40 nm in size and a wide spectrum of antibacterial and fungicidal activity (Gunti et al. 2019).

Se NPs also can be biosynthesized using aquatic plants. For example, Se NPs were isolated from *Spirulina platensis* filtrate after sonication of its biomass (Abbas et al. 2020). They had a spherical shape with an average size of 79 ± 44 nm and antimicrobial activity against Gram-negative bacteria and yeast *Candida albicans*. It has been shown that Se NPs with an average diameter of 28 nm can be recovered by crude extraction from *Caulerpa taxifolia* with Se (Men et al. 2009). Plants capable of accumulating Se in their tissues, such as *Astragalus*, were also used for the synthesis of Se NPs (Meng et al. 2018).

Te is not important for plant metabolism and is toxic in most cases (Tanaka et al. 2020). Despite this, it has been documented that some plants have the ability to metabolize Te and convert it into telluroamino acids (Anan et al. 2013) and organotellurium (Hartwig et al. 2022). For instance, common garlic (*A. sativum*) can assimilate chalcogens Te and S and produce Te-methyltellurocysteine (MeTeCys) and S-methyltellurosulfide metabolites, respectively. The size of garlic-based Te NPs was 40–55 nm (Tanaka et al. 2020). In some cases, Te NPs synthesized by plants can look like spheres, rods, and plates. For example, the synthesis of Te NPs in the form of rods, consisting of highly crystalline elemental Te, was demonstrated in root tissues and garlic cloves of *A. sativum* (Tanaka et al. 2020).

Extracts of orange, lemon, and lime have been used as reducing agents for the green synthesis of Te NPs using microwaves. Te NPs showed uniform size distribution, rod, and cubic shapes. They had important antibacterial activity against Gram-negative and positive bacteria in a concentration range from 5 μg/mL to 50 μg/mL over a 24-hour time period (Medina et al. 2019).

Te NPs were synthesized from cell lyalizate of the marine planktonic diatom *Haloferax alexandrinus* (Alvares and Furtado 2021). The synthesized Te nanorods exhibited antibiofilm activity against the *P. aeruginosa* pathogen. At a dose of 50 μg/mL, these Te nanorods showed a 75.03% reduction of biofilms *in vitro* (Alvares and Furtado 2021). Thus, plants belonging to various genera, including fruit, shrubs, tropical, medicinal, and aquatic species, have the ability to synthesize chalcogen NPs.

Synthesis Using Fungi

The synthesis of NPs using fungi, myconanotechnology, is an emerging branch of nanotechnology that has the potential to reduce the use of agrochemicals in agriculture (Rai et al. 2009). The bioreduction of metal oxides and chalcogens to their elemental form in fungi is catalyzed by extracellular exoenzymes and metabolites secreted by fungi, which form the basis of NPs mycosynthesis (fungal fermentation). Through biodegradation, toxic elements are converted into non-toxic NPs (Moghaddam et al. 2015).

The synthesis of S NPs as nanocrystals using fungi was reviewed in Yanchatuña Aguayo et al. (2022). According to this review, CdS NPs can be synthesized by various fungal species, such as *Aspergillus niger*, *F. oxysporum*, *Pleurotus ostreatus*, *S. cerevisiae*, and *Trichoderma harzianum*; ZnS NPs are formed in fungi *Aspergillus flavus*, *F. oxysporum*, and *S. cerevisiae*; PbS NPs are formed in *Rhodosporidium diobovatum*, *A. flavus*, and *Torulopsis* spp.; Ag_2S NPs are formed in *P. ostreatus*; CuS NPs in *F. oxysporum*.

There are many studies on the synthesis of Se NPs by fungi. Hybrid biocomposites of elemental Se, potentially antimicrobial, represent a particular interest (Khiralla and El-Deeb 2015). Se NPs with an average particle size of 55 nm were synthesized using the fungus *Aspergillus oryzae*, fermented *Lupinus albus* aqueous extract, and gamma radiation to hinder the growth of some multidrug-resistant bacteria and pathogenic fungi (Mosallam et al. 2018).

A technique for obtaining Se NPs based on extracellular metabolites of higher macro basidiomycete fungi *Ganoderma lucidum*, *Grifola umbellata*, *Laetiporus sulphureus*, *Lentinula edodes*, and *Pleurotus ostreatus* with the formation of elemental Se biocomposites *in vivo* by bioreduction of the organoselenium substrate has been developed (Tsivileva and Perfileva 2017). The conditions for Se transformations in fungal cultures under the influence of diacetophenonyl selenide were established and optimized by Tsivileva and Perfileva (2017). The resulting Se nanobiocomposites had antibacterial and antibiofilm effects against the phytopathogenic bacterium *Clavibacter sepedonicus* (Perfileva et al. 2016, 2018a, b). The maximum efficiency of biocomposites based on extracellular metabolites of *L. edodes* and *G. lucidum* was found; these Se nanobiocomposites reduced the biofilm formation of *C. sepedonicus* (Perfileva et al. 2018a,b). The influence of Se biocomposites obtained using deep cultures of macrobasidiomycetes *Ganoderma applantum*, *G. cattienensis*, *G. colossus G. lucidum*, *G. neojaponicum*, and *G. valesiacum* on phytopathogenic bacteria *Micrococcus luteus*, *Pectobacterium atrosepticum*, *Pectobacterium carotovorum* sp. *carotovorum*, *Pseudomonas fluorescens*, *Pseudomonas viridiflava*, and

Xanthomonas campestris was studied in Perfileva et al. (2017a). Oxopropyl-4-hydroxychromenones were used as components of fungal nutrient media. Bacteriostatic and bactericidal activity of Se-containing and Se-free substances of fungal origin against phytopathogenic bacteria was studied by determining the number of colony-forming units, diffusion into agar, and measuring the optical density of a bacterial suspension. Composites based on *G. valesiacum* 120702 isolates with $S(NO_2)$ showed the maximum antibacterial activity against *X. campestris* B-610. A high antimicrobial effect of *G. lucidum* 1315 and *G. colossus* SIE1301 with $S(NO_2)$ against *X. campestris* B-610 and *P. fluorescens* EL-2.1, respectively, was found by Perfileva et al. (2017a).

The influence of biocomposites obtained from mushrooms on potato plant growth and tuber germination was studied in two potato cultivars: Lukyanovsky and Lugovskoi (Tsivileva and Perfileva 2022). Biocomposites based on *Grifola umbellata* demonstrated the strongest positive effect on the number of leaves and plant height in both cultivars without a negative effect on the biomass of the vegetative part. Treatment of the potato tubers with Se NC obtained from *Gr. umbellata* also significantly increased germ length. Potato plants exposed to Se-bio-composite obtained from *Ganoderma lucidum* SIE1303 experienced an increase in the potato vegetative biomass by up to 55% versus the control. Biocomposites based on *P. ostreatus* promoted the potato root biomass increase in the Lugovskoi cultivar by up to 79% versus the control. The phytostimulating ability of mushroom-based Se-containing biocomposites and their anti-phytopathogenic activity testify in favor of the bifunctional mode of action of these Se-biopreparations. The application of stimulatory green Se NCs for growth enhancement could be used to increase crop yield (Tsivileva and Perfileva 2022).

The fungi *Aureobasidium pullulans*, *Mortierella humilis*, *Trichoderma harzianum*, and *Phoma glomerata* can be used to produce Se NPs during their growth on a Se-containing medium at a concentration of 1 mM (Liang et al. 2019). It has also been demonstrated that Se NPs obtained using the fungus *Trichoderma* suppressed the growth, sporulation, and zoospore viability of *Sclerospora graminicol,* causing downy mildew disease in pearl millet and improved its growth in greenhouse conditions (Nandini et al. 2017). The authors suggested that Se NPs induced systemic and localized plant resistance to a number of biotic stress factors. Se NPs obtained using the fungus *Trichoderma atroviride* were spherical with size from 60.48 nm to 123.16 nm and demonstrated high fungicidal activity against the fungi *Pyricularia grisea*, *Colletotrichum capsici*, and *Alternaria solani* (Joshi et al. 2019).

The baker's yeast *Saccharomyces cerevisiae* was also used to synthesize Se NPs from inorganic Se, and these Se NPs ranged in size from a few nm to 750 nm (Álvarez-Fernández García et al. 2020; Faramarzi et al. 2020). Asghari-Paskiabi et al. (2019) described the synthesis of Se sulfide NPs by baker's yeast with the Se NP size ranging from 6 nm to 153 nm and a pronounced fungicidal activity against fungi of the genera *Aspergillus, Candida,* and *Alternaria*. The yeast *Nematospora coryli* was also used to synthesize Se NPs of 50–250 nm in size (Rasouli 2019). It was shown that extracts of the mycelium of the basidiomycetes *Agaricus bisporus* and *A. arvensis* and their filtrates in immersed and solid media were able to reduce ions of Se compounds, forming Se^0 NPs (Loshchinina et al. 2018). Se nanospheres obtained from the cell-free culture filtrate of *A. bisporus* and *A. arvensis* had a diameter of 100–250 nm and 150–550 nm, respectively (Loshchinina et al. 2018).

Extracellular synthesis of Se NPs with a size of 100–250 nm was demonstrated using the culture supernatant of *Streptomyces griseoruber*, a representative of actinomycetes, isolated from the soil (Ranjitha and Ravishankar 2018). A genetically modified strain of the methylotrophic yeast *Pichia pastoris* was used for the biosorption of Ag and Se and the production of their stable NPs with 70–180 nm in size (Elahian et al. 2017). The synthesis of small Se NPs (4–12.7 nm) using the fungus *Penicillium expansum* ATTC 36200 was described in Hashem et al. (2021).

Fungi synthesizing Se NPs from their precursors are able to participate in important geochemical processes and carry out bioremediation. Using strains of the environmentally widespread fungi *Ascomycete* spp., *Paraconiothyrium sporulosum,* and *Stagonospora* spp., Rosenfeld et al. (2020)

observed that the aerobic bioreduction of Se compounds to NPs, which occurs simultaneously with the opposite process of redox biomineralization of mycogenic Mn(II) yields stable Se(0) NPs and organoselenium compounds. However, mycogenic Mn oxides rapidly oxidize volatile Se products, turning them back into soluble forms. Given their abundance in natural systems, biogenic Mn oxides likely play an important role in the Se biogeochemistry (Rosenfeld et al. 2020). The simultaneous removal of phenol and selenite from polluted wastewater has been investigated using the method of co-cultivation of the fungus *Phanerochaete chrysosporium* and the bacterium *Delftia lacustris*. Separately grown fungal and bacterial biomass was then cultured together (as a suspended co-culture) and incubated with various concentrations of phenol (0–1,200 mg/L) and selenite (10 mg/L). Selenite ions were biologically reduced to Se(0) NPs (with a diameter of 3 nm) with simultaneous reduction of phenol down to 800 mg/L. When a fungus and a bacterium were co-cultivated, the bacterium grew as a biofilm on the fungus (Chakraborty et al. 2019).

It was shown that the fungi *Aureobasidium pullulans, Mortierella humilis, Trichoderma harzianum,* and *Phoma glomerata* are capable of producing Se and Te NPs during growth on Se- and Te-containing nutrient media at concentrations of 1 mM. At the same time, extensive deposition of elemental Se and Te on the surface of the fungi was observed as the red and black colors, respectively, on the surface of the hyphae (Liang et al. 2019). The fungus *Aureobasidium pullulans* has a similar ability; it is able to synthesize monodisperse Se NPs (45–90 nm) based on selenite and polydisperse Te NPs (5–65 nm) based on tellurite (Nwoko et al. 2021). *Aspergillus welwitschiae* is able to reduce potassium tellurite (K_2TeO_3) to elemental oval and spherical Te NPs with 60.8 nm in size. The resulting Te NPs showed antibacterial activity against *E. coli* and *S. aureus* at 25 mg/ml concentration (Abo Elsoud et al. 2018).

Biogenic Te NPs have been successfully generated using potassium tellurite (K_2TeO_3 $3H_2O$) and extracellular enzymes and biomolecules secreted from *Penicillium chrysogenum* at room temperature. The average hydrodynamic diameter of Te NPs was about 50.16 nm. The authors believe that *P. chrysogenum* could be a potential nanofactory for the production of Te NPs due to several advantages, including a non-pathogenic organism, fast growth rate, and high elemental ion reduction ability, as well as ease and economical handling of biomass (Barabadi et al. 2019).

The presented review demonstrates the ability of various types of fungi, including basidiomycetes, molds, and yeasts, to form Se NPs. The variety of fungal-reducing enzymes, in parallel with the synthesis of Se NPs, makes it possible to use them for soil bioremediation, particularly in areas contaminated with heavy metals.

Synthesis Using Bacteria

Information on the synthesis of S NPs using bacterial cells is insufficient. However, energy dispersive X-ray showed that the NPs extracted from the antimony-transforming bacterium *Serratia marcescens* isolated from the Caspian Sea in northern Iran represent aggregated antimony and S atoms (Bahrami et al. 2012). A review article by Yanchatuña Aguayo et al. (2022) is devoted to the biogenic S-based chalcogenide nanocrystals. According to this review, *Clostridium thermoaceticum* and *Klebsiella pneumoniae* can synthesize cadmium sulfide NPs (CdS NPs), and the latter can synthesize Zn sulfide NPs (ZnS NPs). Various strains of *E. coli* can also synthesize CdS nanocrystals both intracellularly and extracellularly. Sulfate-reducing bacteria *Desulfovibrio caledoiensis* can produce CdS NPs with a 40–80 nm diameter. The same bacteria can synthesize ZnS and lead sulfide (PbS) NPs in the shape of 2–5 nm long spheres. *Clostridiaceae* spp. can also synthesize spheric and cubic PbS NPs. The bacterium *Desulfovibrio desulfuricans* can produce ZnS NPs. A mixture of sulfate-reducing bacteria *Desulfovibrio* spp., *Clostridiaceae* spp., *Proteiniphilum* spp., *Geotoga* spp., and *Sphaerochaeta* spp. was used to obtain photoluminescent ZnS NPs with 6.5 nm in size. Bismuth sulfide (Bi_2S_3) NPs were obtained using *Clostridiaceae* spp. (Yanchatuña Aguayo et al. 2022).

Microorganisms play an important role in the transport and transformation of Se in the environment, thereby influencing the accumulation of Se in plants. Microbial selenite reductases can convert the water-soluble oxyanion SeO_4^{2-} into SeO_3^{2-} and further into insoluble elemental Se without charge (Se^0) in a two-step reduction reaction (Wadhwani et al. 2016). As a rule, soil bacteria have the ability to reduce Se oxides to their NPs. There are many examples of Se NPs synthesis by bacteria. For instance, in China, two strains of *Lysinibacillus xylanilyticus* and *L. macrolides* were isolated from soil rich in Se, which can reduce selenite at a concentration of 1 mmol/L to NPs of elemental Se under aerobic conditions in 36 hours (Zhang et al. 2019). The *Pseudomonas moraviensis* subsp. *stanleyae* strain isolated from the rhizosphere of the Se hyperaccumulator plant, the perennial shrub *Stanleya pinnata*, is able to tolerate lethal concentrations of SeO_2^{2-} in liquid culture and synthesize Se NPs (Ni et al. 2015). Cultures of *Duganella* spp. and *Agrobacterium* spp. isolated from arable soil can convert water-soluble selenite into Se NPs with 140–200 nm in size under aerobic conditions (Bajaj et al. 2012). When growing the aerobic bacterium *Rhodococcus aetherivorans* on a nutrient medium containing selenite, the bacterium can produce Se NPs and Se nanorods. At the same time, Se NPs were stable, polydisperse, and non-aggregated (Presentato et al. 2018). Biogenic Se NPs with a diameter of 160–250 nm were obtained by reducing selenite with the bacterium *Azospirillum thiophilum* (VKM strain B-2513) (Tugarova et al. 2018). It was shown that the culture of the bacterium *Acinetobacter* spp. SW30 can synthesize amorphous nanospheres of 78 nm in size at a sodium selenite concentration of 1.5 mM and crystalline nanorods at a Na_2SeO_3 concentration above 2 mM. NPs with an average size of 79 nm were found in the supernatant (Wadhwani et al. 2016). The aerobic soil bacterium *Comamonas testosteroni* S44 reduces Se(VI)/Se(IV) to less toxic elemental Se NPs (Tan et al. 2018). A new strain of nitrate and selenite-reducing bacteria *Bacillus oryziterrae* sp. nov. isolated from the soil of rice fields in Dehong, China, can produce Se NPs more uniform in size than those from chemical processing and thus have better application potential (Bao et al. 2016). The strain of *Bacillus cereus* CC-1 can not only reduce selenite and selenate to Se NPs but also synthesize several types of metallic selenide NPs with the simultaneous addition of metal ions (Pb^{2+}, Ag^+, and Bi^{3+}) and selenite (Che et al. 2019). This strain is expected to be used for the production of biocompatible photothermal and thermoelectric nanomaterials (Che et al. 2019). Two halotolerant strains of *Bacillus megaterium* (BSB6 and BSB12) isolated from saline mangrove habitat without Se contamination were able to reduce Se^{+IV} to elemental Se even in the presence of salt in high concentrations (Mishra et al. 2011).

Bacteria possessing selenite reductases are of interest for use in bioremediation. In southwest China, a strain of *Streptomyces* sp. ES2-5 isolated from soil from Se mining sites can reduce selenite and selenate to less toxic Se^0 with the formation of Se NPs with sizes of 50–500 nm (Tan et al. 2016). It was suggested to use this strain for Se bioremediation (Tan et al. 2016). In addition, it was proposed to use such microorganisms not only on lands contaminated with selenates and selenites but also for the restoration of soils contaminated with other hazardous substances, such as mercury (Wang et al. 2017). For instance, the bacterium *Citrobacter freundii* Y9 demonstrated a high ability to reduce selenite. It can synthesize Se NPs under both aerobic and anaerobic conditions. With this microbe, up to 50% of the elemental mercury (Hg^0) in contaminated soil was converted into insoluble mercury selenide (HgSe) under anaerobic and aerobic conditions using biogenic Se NPs (Wang et al. 2017). The addition of sodium dodecylsulfonate enhanced Hg^0 remediation, probably due to the release of intracellular Se NPs from bacterial cells for Hg fixation. The reaction product after remediation was identified as non-reactive HgSe, which was formed due to the fusion of Se NPs and Hg0. The researchers concluded that the biosynthesis of Se NPs under both aerobic and anaerobic conditions provides a versatile and cost-effective approach to the remediation of mercury-contaminated soils, in which the redox potential often fluctuates dramatically (Wang et al. 2017).

Different types of bacteria have been used for the biosynthesis of Se NPs, such as the species of phylum *Proteobacteria* (*Escherichia coli, Ralstonia eutropha, Enterobacter cloacae, Pseudomonas aeruginosa, Klebsiella pneumoniae, Pantoea agglomerans, Zooglea ramigera, Rhodopseudomonas palustris, Shewanella* spp., *Azoarcus* spp., *Burkholderia fungorum,* and *Stenotrophomonas maltophilia*), *Firmicutes* (*Lactobacillus casei, Lactobacillus acidophilus, Lactobacillus helveticus, Enterococcus faecalis, Streptococcus thermophilus, Staphylococcus carnosus, Bacillus* spp., *Bacillus subtilis, Bacillus mycoides,* and *Bacillus licheniformis*), *Actinobacteria* (*Streptomyces* spp. and *Bifidobacterium*), and *Cyanobacteria* (Hosnedlova et al. 2019).

For the bacterial synthesis of Te NPs, K_2TeO_3, or Na_2TeO_3 precursors are usually used because they are the least toxic. Te has different oxidation states: telluride (Te^{2-}), tellurite (TeO_3^{2-}), and tellurate (TeO_4^{2-}). The biotransformation of Te NPs from nanospheres to nanorods using *Escherichia coli* has been shown by Gómez-Gómez et al. (2020). Te NPs were also synthesized by the anaerobic haloalkaliphilic bacteria *Bacillus selenitireducens*. After 30 days of incubation of these bacteria on a medium with 0.6 mM Te (IV), Te nanorods first formed, then larger compounds of Te NPs in the form of rosettes (Wang et al. 2019a).

Resting (non-growing) cells of *Rhodococcus aetherivorans* have demonstrated the ability to produce Te-based NPs and nanorods through bioconversion of dangerous and toxic to living organisms anion - tellurite TeO_3^{2-}, depending on the initial concentration of the oxyanion and the time of cell incubation. Initially, Te NPs appeared in the cytoplasm of *R. aetherivorans* cells as spherical NPs, which turned into NPs as the exposure time increased. This observation confirmed the existence of an intracellular mechanism for the assembly and growth of Te NPs. Te NPs produced by *R. aetherivorans* had a mean length of > 700 nm, almost twice as long as those observed in other studies. In addition, biogenic Te nanorods exhibit a regular single-crystal structure (Presentato et al. 2018). The bacterial strain *Ochrobactrum* sp. isolated from the waste of burned arsenopyrites as a residue of sulfuric acid production near Scarlino (Tuscany, Italy) was analyzed for its ability to efficiently bioreduce selenite (SeO_3^{2-}) and tellurite (TeO_3^{2-}) chalcogen oxyanions to their respective elemental forms (Se^0 and Te^0) under aerobic conditions with the formation of Se and Te NPs (Zonaro et al. 2017). The isolate can bioconvert 2 mM SeO_3^{2-} and 0.5 mM TeO_3^{2-} into the corresponding Se^0 and Te^0 in 48 and 120 hours, respectively (Zonaro et al. 2017). A bacterial strain, *Shewanella baltica*, isolated from the mouth of the Zuari River, Goa, India, demonstrated bioremediation, a complete reduction of 2 mM toxic tellurite to elemental Te during the late stationary phase (Vaigankar et al. 2018). These NPs have demonstrated photocatalytic and antibiotic film activity as well as genotoxicity (Vaigankar et al. 2018). The photosynthetic bacterium *Rhodobacter capsulatus* is able to convert toxic oxyanion tellurite (TeO_3^{2-}) into Te NPs in a culture medium (Borghese et al. 2020).

From the above studies, one can conclude that soil microorganisms are very capable of forming Se NPs in their metabolism. They are diverse in morphology and their physiological and biochemical characteristics. Among them are such bacteria as Gram-positive aerobes or facultative anaerobes p. *Bacillus*, Gram-negative aerobes *Agrobacterium*, Gram-positive aerobes p. *Rhodococcus*, and Gram-negative facultative anaerobes p. *Citrobacter*. All these bacteria possess selenite reductases, which allow them to synthesize NPs. Various types of bacteria isolated from the environment (soil and water) are capable of synthesizing Te NPs, including *Rhodobacter capsulatus, Shewanella baltica, Ochrobactrum* spp., *Rhodococcus aetherivorans, Bacillus selenitireducens* and *Escherichia coli*.

Cytotoxic Effect of Sulfur, Selenium, and Tellurium Nanoparticles on Microbial Cells and Phytopathogens

The following mechanisms of antimicrobial activity of NPs are known: generation of reactive oxygen species (ROS), interaction with the cell wall (its destruction and change in permeability), inhibition of protein and DNA synthesis, and influence on gene expression (Filipović et al. 2021; Godoy-Gallardo et al. 2022). At the same time, an important property of NPs is their ability to simultaneously act in several directions, which prevents the formation of microbial resistance to them. When it comes to metal-based NPs, antimicrobial activity is very often attributed to ROS products (hydroxyl radicals, superoxide anions, and hydrogen peroxide). These types of ROS can additionally inhibit DNA replication and amino acid synthesis and also induce membrane damage in bacterial cells (Hemeg 2017).

S NPs are known to have a broad spectrum of antimicrobial activity and therefore can be used to treat microbial infections and potentially address the problem of antibiotic resistance (Rai et al. 2017). Orthorhombic (spherical; ~ 10 nm) and monoclinic (cylindrical; ~ 50 nm) S NPs were synthesized and investigated for their effect on total lipid content and *A. niger* desaturase enzymes by Choudhury et al. (2012). Both S NPs significantly reduced total lipids in treated fungal isolates with significant down-regulation of the expression of various desaturase enzymes (linoleoyl-CoA desaturase, stearoyl-CoA-9 desaturase, and phosphatidylcholine desaturase). The unusually high accumulation of lipid-depleted saturated fatty acids can be considered one of the main causes of fungistatic mediated by S NPs (Choudhury et al. 2012). S/Se NPs on the surface of rGO graphene oxide had antibacterial properties against antibiotic-resistant Gram-positive pathogens *S. aureus* and *Enterococcus faecalis* (> 90% growth inhibition noted at 200 μg/mL). Microscopy studies have shown that composite rGO NPs can be deposited on the surface of a bacterial cell, leading to membrane disruption and oxidative stress (Niranjan et al. 2022). S NPs interfere with mitochondrial enzymes involved in cellular respiration and oxidative phosphorylation of Gram-positive (*S. aureus*) and multidrug-resistant Gram-negative (*E. coli, P. aeruginosa, Klebsiella pneumoniae,* and *Acinetobacter*) bacteria, causing a reduction in their growth (Tran and Webster 2011; Shankar et al. 2018; Paralikar et al. 2019).

The biological activity of Se depends on its chemical form and structure. Elemental Se is insoluble and was previously thought to be biologically inert. Therefore, the dominant forms of Se used in biomedical research were SeMet, SeCys, methylselenocysteine, and sodium selenite. With the development of nanotechnology, a new form of Se has appeared in the form of NPs, which has the advantages of reduced toxicity, bioavailability, and efficiency. The shape, size, and surface structure of NPs are important characteristics that determine the interaction of nanomaterials with biological objects. In addition, it is necessary to take into account the narrow boundary between the stimulating and toxic effects of Se. However, it is known that Se NPs obtained in different ways, i.e., using various stabilizers and reducing agents, exhibited different antimicrobial activity and cytotoxicity.

The mechanism of cytotoxicity of Se NPs toward bacteria is not fully understood. The zeta potential of NPs plays an important role in the interaction of NPs with bacteria. Due to the presence of a layer of negatively charged lipopolysaccharides, Gram-negative bacteria have a greater negative charge on the cell surface than Gram-positive bacteria. It is believed that strong repulsive forces exist between Se NPs and strongly negatively charged Gram-negative bacteria such as *E. coli*. For Gram-positive bacteria with a lower (or neutral) surface net charge, such as *S. aureus*, the interaction between Se NPs and bacteria can occur as well. As a result, NPs are able to attach to the bacterial cell wall, changing its membrane permeability and negatively affecting cell division and viability. High concentrations of negatively charged Se NPs can induce significant antibacterial activity due to molecular crowding (Bisht et al. 2022). In Gram-positive bacteria, this effect can be explained by a significant difference in the composition of the bacterial wall, including abundant pores and a thin layer of peptidoglycan. Many authors have previously reported similar results (Tran and Webster 2013; Guisbiers et al. 2016; Tran et al. 2016).

It was found in our studies (Perfileva et al. 2017b, 2018c; Graskova et al. 2019), that Se NCs in natural polymer matrixes have a negative effect on the phytopathogenic Gram-positive bacterium *Clavibacter sepedonicus*. Figure 1 shows this bacterium incubated for 24 hours with arabinogalactan-based Se NCs (Se/AG NCs) and in control (modified Figure 4 in Perfileva et al. 2017b).

Toxic Se NPs are densely packed into a polymeric polysaccharide matrix and can be released from it only when the polysaccharide is cleaved by bacterial exoenzymes. In the case of polysaccharide cleavage, toxic Se NPs are released from Se NCs; due to the negative charge of the cell wall, they attach to bacteria. As a result, there is a disruption of the transmembrane potential

Control   **Se/AG NCs**

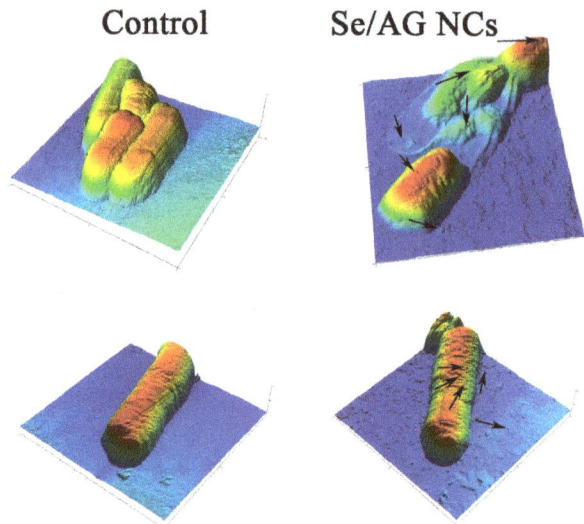

Figure 1. Images of bacteria *Clavibacter sepedonicus* Ac1405 incubated for 24 hours with arabinogalactan-based Se NCs (Se/AG NCs) and in control obtained using an SMM 2000 scanning probe microscope (SPM) (Zelenograd, Russia). Black arrows indicate selenium nanoparticles (modified Fig. 4 in Perfileva et al. 2017b).

of the cell due to oxidative stress, which leads to damage to the cell integrity and its rupture. The schematic interaction of Se NCs with a bacterial cell is presented in Figure 2.

Inhibition of biofilm formation and viability of resistant strains of microorganisms under the influence of Se NPs has been shown in several studies (Shakibaie et al. 2015; Prateeksha et al. 2017; Cremonini et al. 2018; Lin et al. 2021; Filipović et al. 2021; Godoy-Gallardo et al. 2022). *S. aureus* biofilm formation decreased under the influence of Se NPs, reaching a significant level of inhibition at 2 µg/mL (41%) and 4 µg/mL (58%) (Shakibaie et al. 2015).

Te is toxic to living beings (Vávrová et al. 2021). The mechanism of toxicity of thiol-binding metals (loids) is based on the interaction and subsequent inhibition of the essential thiol groups of enzymes and proteins (Ulricha and Jakoba 2019). The similarity of the physical and electrochemical properties of Te with Se and S leads to their substitution in proteins (Moroder 2005). Erroneous incorporation of the resulting Te-cysteine and/or Te-methionine into the protein structure leads to a change in the activity or inactivity of the protein. When TeO_3^{2-} enters the cell in *E. coli*, the transmembrane proton gradient is disrupted, regardless of the level of resistance. This effect is accompanied by the inhibition of ATP synthesis, which leads to the depletion of intracellular ATP stores during aerobic growth. The same damaging effect of Te compounds was found in the synthesis of proteins in relation to proteins containing amino acids with reduced thiol groups of both low- and highly-resistant microbes. Two highly resistant bacteria, namely *Erythromonas ursincola* and *Erythromicrobium ramosum*, differ from other microbes in that both show increased protein and ATP synthesis in the presence of Te compounds (Vávrová et al. 2021). The mechanism for this increase is unclear. Se and Te NPs exhibit antimicrobial activity and disrupt biofilm formation. These NPs effectively inhibited the growth of *E. coli, P. aeruginosa*, and *S. aureus* (Zonaro et al. 2015).

Hybrid NCs of Te and lignin (TeLig NPs) have shown strong bactericidal effects against Gram-negative bacteria *E. coli* and *P. aeruginosa*. Studies of the antimicrobial mechanism of action have shown that the TeLig NPs are able to disrupt the membranes of bacterial models and generate ROS in Gram-negative bacteria (Morena et al. 2021). Te nanorods biosynthesized from cell lysates of the marine planktonic diatom *Haloferax alexandrinus* GUSF-1 were active against *P. aeruginosa*

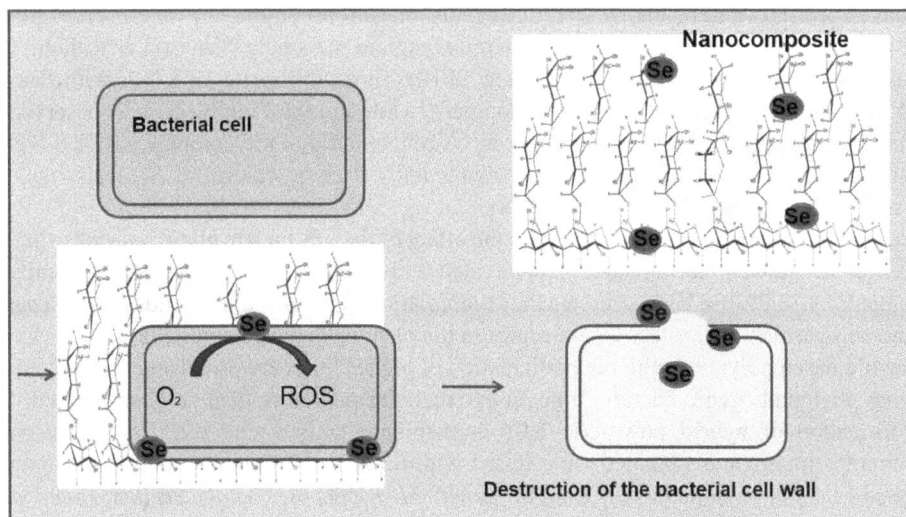

Figure 2. Schematic interaction of selenium (Se) nanocomposites with a bacterial cell.

biofilms (Alvares and Furtado 2021). Te NPs reduced the size of biofilm produced by *S. aureus* and *E. coli* bacteria. At the same time, the interaction between Te NPs and bacterial communities led to the transformation of spherical Te NPs into Te nanorods (Gómez-Gómez et al. 2020). The antibiofilm effect of Se and Te NPs was explained by the disruption of quorum sensing between cells (Gómez-Gómez et al. 2019).

Small-size S NPs (~ 35 nm) have been found to be very effective in preventing the growth of the fungi *Fusarium solani* (isolated from an infected tomato leaf and responsible for *Alternaria* and *Fusarium* wilt) and *Venturia inaequalis* (causing apple scab). The authors attribute the observed fungicidal effect to the deposition of S NPs on the fungal cell wall and its subsequent damage. A likely imbalance in cell wall structure is also confirmed by a Biuret assay test (Rao et al. 2013). The stable and uniform S NPs with a medium diameter of 35–45 nm had a positive effect on the growth and development of sunflowers (Subramanian et al. 2021). Plants fertilized with S NPs had 11–12% higher dry matter content, 15% higher seed yield, and 14.7% higher oil content than control plants. Data have also shown that S NPs increase S availability in soils (Subramanian et al. 2021). The fungicidal effect of silver sulfide nanoparticles (Ag$_2$S NPs) synthesized using ultrasound was shown on phytopathogenic fungi *Fusarium verticillioides, Bipolaris oryzae, Ustilago hordei,* and *Uromyces viciafabia* in Sidhu et al. (2021). Under the influence of Ag$_2$S NPs, the germination of fungal spores and the growth of their hyphae slowed down. At the same time, the fungicidal effect of Ag$_2$S NPs exceeded the effect of standard fungicides Carboxin, Bavistin, and Captan (Sidhu et al. 2021). S NPs increased tomato resistance to the fungal pathogen *F. oxysporum* f. sp. *lycopersici* (Cao et al. 2021). Foliar application and seed treatment with S NPs (30–100 mg/L; 30 nm and 100 nm) suppressed pathogen infection in tomatoes in a concentration and size-dependent manner in a greenhouse experiment (Cao et al. 2021). The disease control efficiency of S NPs was 1.43 times higher than that of the commercial pesticide chemexazole. It was found that S NPs activated the salicylic acid-dependent systemic acquired resistance pathway in tomato shoots and roots with the help of salicylic acid, followed by activation of the expression of genes associated with the pathogenesis and synthesis of antioxidants.

In addition, TEM imaging showed that S NPs were found in tomato stem tissues and directly inactivated pathogens *in vivo*. Oxidative stress in tomato shoots and roots, root plasma membrane damage, and pathogen growth in the stem were significantly reduced with S NPs (Cao et al. 2021).

S NPs have been shown to be highly effective against the soft rot of ginger *F. oxysporum* (Athawale et al. 2018). S NPs were compared with commercial fungicides, namely "Bavistin", "Ridomil Gold", "Sunflex" and "Streptocycline" (Athawale et al. 2018). The combination of S NPs with "Bavistin" showed maximum inhibition of fungal development, while minimal inhibition was observed in its combination with "Ridomil Gold." Thus, it can be concluded that the combination of S NPs with "Bavistin" can be used for effective and environmentally friendly control of *F. oxysporum*, which causes soft rot in ginger (Athawale et al. 2018).

The recent study by Joshi et al. (2021) on the effect of Se NPs on late blight resistance in tomato *S. lycopérsicum* showed that infected plants treated with Se NPs had higher viability compared to control plants by 72.9%. Se NPs promoted the accumulation of lignin, callose, and hydrogen peroxide in tomato cells, which are protective molecules in the plant cells (Joshi et al. 2021). Se NPs capped with triazole-based polymers with the participation of higher fungi are promising nanomaterials for combating phytopathogenic bacteria. Fungal extracellular polymers made a positive contribution to the formation of hybrid polymeric NCs containing elements and their biological activity. Sixty biopreparations thus obtained were tested with Se NPs. Element-containing biocomposites suppressed bacterial phytopathogens from the genera *Micrococcus, Pectobacterium, Pseudomonas,* and *Xanthomonas* (Tsivileva et al. 2021).

Se NPs biologically synthesized with *Lactobacillus acidophilus* ML14 (BioSe NPs) had broad fungicidal activity (El-Saadony et al. 2020). BioSe NPs removed 88% and 92% of DPPH and ABTS radicals, respectively, and successfully inhibited the growth of *Fusarium culmorum, Fusarium graminearum, F. culmorum,* and *F. graminearum* in the range of 20–40 μg/mL in a study by El-Saadony et al. (2020). This biological activity was associated with the small size of BioSe NPs and the content of phenols in their suspension. Under greenhouse conditions, wheat supplemented with BioSe NPs (100 μg/ml) has been shown to significantly reduce the incidence of root rot diseases by 75% and significantly improve plant growth, grain quantity, and quality by 5–40% (El-Saadony et al. 2020). Mycogenic Se NPs synthesized from *Trichoderma atroviride* and applied *in vitro* to chili and tomato leaves at concentrations of 50 ppm and 100 ppm have demonstrated excellent antifungal properties against *Pyricularia grisea, Colletotrichum capsici,* and *Alternaria solani* (Shreya et al. 2019).

Effect of Sulfur, Selenium, and Tellurium Nanoparticles on Plant Growth and Development

The information available from the published studies indicates a positive effect of S NPs on the growth and development of plants. S NPs, which were obtained biogenically from sodium thiosulfate and an aqueous extract of basil leaves, have been shown to reduce the detrimental effects of NaCl salinity on wheat by reducing oxidative stress and maintaining membrane fluidity, activating antioxidant status, and inducing ionic homeostasis (Saad-Allah et al. 2020). All these processes led to the acceleration of plant growth and development under the influence of S NPs (Saad-Allah et al. 2020). An increased percentage of germination, seedling length, seedling vigor, and an increase in dry weight of soybean seedlings after soaking them for 5 minutes in a solution containing S NPs was found by Kavanashree et al. (2020). S NPs obtained by "green" synthesis using *Melia azedarach* leaves aqueous extract and citric acid stimulated the growth of the biomass of the roots and aboveground parts of *Cucurbita pepo* plants under field-experimental conditions (Salem et al. 2016).

A decrease in Mn toxicity was demonstrated with the use of S NPs, whose medium doses enhanced the growth of *Helianthus annuus* seedlings and increased the content of photosynthetic pigments and mineral nutrients (Ragab and Saad-Allah 2020). Application of S NPs reduced Mn uptake and increased S metabolism by increasing cysteine levels. Similarly, S NPs increased the water content of seedlings and eliminated physiological drought by increasing the content of osmolytes such as amino acids and proline. The authors believe that S NPs synthesized with *Ocimum basilicum* leaf extract can reduce the harmful effects of Mn stress (Ragab and Saad-Allah 2020). Treatment of lettuce plants with 1 mg/ml S NPs improved the growth and photosynthesis parameters of lettuce plants compared to the control (Najafi et al. 2020). Some physiological parameters, such as levels of proline, glycine, betaine, and soluble sugars, as well as some phytochemical parameters, such as anthocyanins, total phenol, flavonoids, and tannins, were improved after treatment of plants with the same concentration of S NPs. At the same time, the activity of antioxidant enzymes (catalase, ascorbate peroxidase, and polyphenol oxidase) and the level of stress markers, the content of malondialdehyde and H_2O_2 were reduced in lettuce plants treated with S NPs (Najafi et al. 2020).

Over the past 5 years, a lot of research has been done to study the effect of Se NPs on the growth and development of plants, as well as their ability to resist stresses of various natures. Published data on the action of Se NPs indicate mostly their positive effect on plants. For example, exogenous spraying by Se NPs has been shown to increase the antioxidant potential of basil (*Ocimum basilicum* L.) (Ardebili et al. 2015), enhance the growth of tobacco (*N. tabacum* L.) (Jiang et al. 2015) and peanut (*Arachis hypogaea* L.) (Hussein et al. 2019). The effect of Se NPs on *Hordeum vulgare* L. seeds' germination was studied by Siddiqui et al. (2021). Nano-selenium dioxide increased the yield and intensity of plant growth and enhanced salt tolerance in *Phaseolus vulgaris* growing in a field experiment on saline soils (Rady et al. 2021). It was suggested that an increase in the growth of higher plants treated by Se NCs occurs due to an increase in the productivity of photosynthesis (Rady et al. 2021). It was shown that Se NPs synthesized using *Allium sativum* L. clove as a reducing, capping, and stabilizing agent at a concentration of 75 mg/L increased the amount of chlorophyll and carotenoids, the membrane stability index, the number of total soluble sugars, and the activity of antioxidant enzymes: superoxide dismutase (SOD), peroxidase (POD), and catalase (CAT), as well as total flavonoid content in mandarin plants (Ikram et al. 2022). Significant reductions in hydrogen peroxide (H_2O_2), malondialdehyde (MDA), and proline (PRO) were noted in "Kinnow" mandarin plants infected with *Liberibacter* bacteria causing citrus greening or huanglongbing (HLB) disease compared to untreated diseased tangerine plants (Ikram et al. 2022).

It was found that in the seedlings of the grain crop *Chenopōdium quīnoa* germinated from seeds treated with Se NPs, the germination parameters and the activity of antioxidant enzymes, as well as the content of proline and protein, were increased compared to the control (Gholami et al. 2022). Under salt stress conditions, the highest activity of catalase (CAT), superoxide dismutase (SOD), and ascorbate peroxidase (APC) enzymes was observed after treatment of seeds with Se NPs at concentrations of 4.5 mg L^{-1}, 6.0 mg L^{-1}, and 4.5 mg L^{-1}, respectively. Se NPs concentrations above 3 mg/L and sodium selenate concentrations of 3 mg/L resulted in the maximum accumulation of photosynthetic pigments under controlled (non-stress) conditions. This study showed that seed treatment with Se NPs can reduce the harmful effects of drought on quinoa by altering germination and biochemical properties (Gholami et al. 2022). The effect of Se NPs obtained by laser ablation of Se in water using a ytterbium fiber laser was tested on vegetables (Gudkov et al. 2020). The eggplant growth on the soil with the Se NPs at a concentration of 10 μg/kg of leaf plate surface area was twice as high as compared to the eggplant growth in untreated soil (Gudkov et al. 2020). A similar result was obtained with tomato plants in the same study. The surface area of the leaf blade of cucumber plants grown using Se NPs was 50% higher compared to the control. In addition, resistance to hyperthermia increased in plants under the influence of Se NPs (Gudkov et al. 2020).

Spraying cherry tomato plants *Solanum lycopersicum* L. var. *cerasiforme* with an aqueous solution containing Se NPs led to an increase in the length of the shoot and root, the wet and dry mass of the shoot and root, the content of chlorophyll and Zn elements in the leaves, and the level of activity of antioxidant enzymes: catalase and superoxide dismutase (Neysanian et al. 2021). It was concluded that treatment with Se NPs at a concentration of 4 mg/L increased the tolerance of tomato plants to drought and stimulated their growth (Neysanian et al. 2021).

It was assumed that an increase in the growth of higher plants under the influence of Se NPs occurs due to an increase in the productivity of photosynthesis (Feng et al. 2015). A change in the fatty acid profile of lipids in plant cells under the influence of Se NPs was also shown (Hussein et al. 2019). In addition, it was found that Se NPs affect the activity of antioxidant enzymes in various plant organs—nitrate reductase in leaves and peroxidase in roots (Babajani et al. 2019).

It is believed that the effect of Se NPs on plants depends on their size (Bano et al. 2021). Thus, it was found that Se NPs of different sizes (50 nm, 100 nm, and 150 nm) are absorbed and transported differently in plants (Joudeh and Linke 2022). Studies have shown that the intensity of absorption and distribution of Se NPs in the body of wheat *Triticum aestivum* L. and rice *Oryza sativa* L. plants depend on the size of Se NPs and the acidity of the medium (Wang et al. 2019b). The highest Se content in wheat shoots was observed upon treatment with Se NPs with a size of 50 nm after 24 hours and 72 hours, respectively. In addition, the Se transfer coefficient in wheat upon treatment with Se NPs with a size of 50 nm was twice as higher as upon treatment with a size of 100 nm and 150 nm. Se content in rice roots treated for 24 h with Se NPs with a size of 50 nm increased by 11% and 41% compared to those treated with Se NPs with a size of 100 nm and 150 nm, respectively. Se content in rice sprouts and Se transfer coefficient peaked with Se NPs with the size of 50 nm. In addition, it was found that the absorption of Se by plants is affected by the acidity of the medium. In particular, the amount of Se taken up by the roots of wheat treated with Se NPs was greatest at pH 6 after 24 hours, which was 89% higher than the amount of Se in wheat treated with selenite. In addition, the highest Se transfer coefficient was noted at pH 4 in wheat (Wang et al. 2019b).

The effect of Se NPs on plants also depends on the concentration of Se NPs. For example, the effect of various concentrations of Se NPs (0.5 mg/L, 1 mg/L, 10 mg/L, and 30 mg/L) on the growth, genetic, and biochemical characteristics of the *Capsicum annuum* plants was studied by Sotoodehnia-Korani et al. (2020). It was shown that the introduction of Se NPs into the plant cultivation medium caused changes in morphology and growth depending on the dose of Se NPs. They showed growth-stimulating effects at low doses while causing severe toxicity and disturbances in leaf and root development at 10 mg/L and 30 mg/L. It was found that the toxicity of Se NPs was associated with DNA hypermethylation. Treatment with Se NPs at a Se concentration of 0.5 or 1 mg/L led to a significant induction of nitrate reductase activity and an increase in the proline concentration. There was a change in the activity of peroxidase and CAT, as well as a decrease in the activity of phenylalanine ammonia-lyase and the concentration of soluble phenols. The toxicity of Se NPs is also associated with the inhibition of xylem tissue differentiation (Sotoodehnia-Korani et al. 2020).

A dose-dependent effect of Se NPs on the biomass of medicinal plants was found by (Ghasemian et al. 2021) *Melissa officinalis* plants were treated with Se NPs with two concentrations of 10 mg/L and 50 mg/L, respectively. When plants were treated with Se NPs at a concentration of 10 mg/L, a sharp increase in biomass, activation of lateral buds, and stimulation of the development of lateral roots were observed. However, at a concentration of 50 mg/L, Se NPs reduced plant growth by 45.5% compared to the control, having a strong toxic effect on the plant (Ghasemian et al. 2021).

Studies have been carried out on the effect of Se NPs complexes with solid polymer NC matrixes on the viability of the phytopathogenic bacterium *C. sepedonicus* (Papkina et al. 2015b; Perfileva et al. 2018c; Graskova et al. 2019) and the phytopathogenic fungus *Phytophthora cactorum* (Perfileva et al. 2021a), potato plants *in vitro* (Papkina et al. 2015a; Perfileva et al. 2017b, 2018c, 2019, 2021b; Lesnichaya et al. 2022) and soybean, pea, and potato seed germination (Nurminsky et al. 2020). Arabinogalactan (Sosedova et al. 2018), starch (Perfileva et al. 2020), and carrageenan (Nozhkina et al. 2019) were chosen as biopolymers for Se NCs. These Se NCs reduced the viability of the phytopathogenic bacterium *C. sepedonicus* and the phytopathogenic fungus *P. cactorum* (Perfileva et al. 2021a). Some of them stimulated the growth and development of potatoes *in vitro*, and the germination of soybean, pea, and potato seeds increased the productivity and quality of the potato crop in a field experiment (Perfileva et al. 2021a). There was no accumulation of Se NPs in plant tissues after their treatment with Se NCs (Perfileva et al. 2019; Nozhkina et al. 2019). The studied Se NCs did not inhibit the viability of useful soil microorganisms *Acinetobacter guillouiae, Rhodococcus erythropolis,* and *Pseudomonas oryzihabitans* (Perfileva et al. 2021a). The data on the biological activity of Se NCs published in these studies indicate that Se NCs are promising and relatively safe for the practical use of them as agents for the recovery of plants from phytopathogens.

When studying the effects of Se NPs, it was shown that Se NPs could function as plant development promoters, improving their antioxidant defense system and therefore their ability to tolerate stress (Hussein et al. 2019). Se NPs affected cellular processes; for example, they regulated the activity of antioxidant enzymes and affected the photosynthetic apparatus (Azimi et al. 2021). It was shown that Se NPs significantly reduced the content of heavy metals in rice grains grown on technogenically polluted soil (Wang et al. 2021). Spraying plants with a solution of Se NPs improved the growth, increased the yield of rice, radish, and corn, and accelerated the growth of lettuce plants (El-Ramady et al. 2020). It was found that Se NPs not only enhance the resistance of tomato plants to salt (Morales-Espinoza et al. 2019) and biotic stress (caused, for example, by *Alternaria solani* or nematodes) but also increase their productivity (Quiterio-Gutiérrez et al. 2019). The increased resistance of plants to stress is explained by the induction of the activity of the enzymes superoxide dismutase, ascorbate peroxidase, glutathione peroxidase, phenylalanine ammonia-lyase in leaves, and glutathione peroxidase in the tomato fruits. In addition, the content of chlorophylls a and b was increased in the leaves, and the amount of vitamin C, glutathione, phenols, and flavonoids increased in the tomato fruits. Treatment with Se NPs increased the activity of lipoxygenase, phenylalanine lyase, β-1,3-glucanase, and superoxide dismutase enzymes in tomato tissues (Joshi et al. 2021).

Salt stress tolerance and increased yields have been observed when strawberry (*Fragaria moschata*) and wheat plants were sprayed with Se NPs (Zahedi et al. 2019). The resulting effect was explained by a decrease in the level of lipid peroxidation (LPO), an increase in the activity of antioxidant enzymes, superoxide dismutase, and an increase in the content of proline in plant tissues. In addition, an increase in the quality and nutritional properties of strawberries has been noted due to an increase in the content of organic acids, such as malic, citric, and succinic acids, and sugars, such as glucose, fructose, and sucrose, in the berries of plants treated with Se NPs (Zahedi et al. 2019). There is no sufficient information in the literature about the effect of Te NPs on plants.

Conclusions

This review of the influence of S and Se NPs indicates their positive effect on the viability and resistance of plants to stress. S and Se NPs activate antioxidant and other protective enzymes in plant tissues, intensifying the most important cellular processes.

Chalcogens include a number of elements that differ in the degree of necessity for plants. S is an essential macronutrient that is included in the composition of proteins, enzyme cofactors, vitamins, and other biologically active substances. Se is a microelement necessary for the plant organism. It is involved in the redox reactions of the cell and in the synthesis of compounds important for plants (chlorophyll, vitamins, and fatty acids). Se increases the body's resistance to stress factors of various natures. Plant roots can only take up certain chalcogen compounds. Selenate is transported to the root tissue through the plasma membrane with the participation of special carriers. Selenite is transported by phosphate transporters or enters plant cells passively by diffusion. Then, the Se compounds are transported through the xylem to the aboveground plant organs, which contain many chloroplasts in which inorganic Se is converted into SeCys, SeMet, and volatile dimethyl diselenide. After that, organic Se compounds move through the phloem to the roots, where they can accumulate, and to other plant organs (Zhou et al. 2020).

The Se content in plants averages 0.0001 mg% (by weight). According to the ability to accumulate Se in their tissues, plants are divided into non-accumulators, accumulators, and hyperaccumulators. Increased content of Se in plant tissues can perform a protective function for them. The metabolic pathways of Se hyperaccumulator plants are due to the active uptake of selenate from the soil, followed by its transportation through the xylem to the aboveground organs. Some of the Se compounds are deposited in vacuoles, while the remaining portion undergoes a series of transformations. It is first converted into methyl-selenocysteine and then into volatile dimethyl diselenide, which evaporates from the surface of the leaf and acts as a repellant.

The availability and distribution of Se in soils depend on many factors (composition and structure of the soil, climatic conditions, etc.). The content of Se in soils around the world varies considerably and can range from 10 to 1,000 µg/kg or more. However, Se is useful for living organisms in a narrow range of concentrations, it is an essential trace element, and its compounds are toxic at high doses. High concentrations of Se are dangerous because Se can replace S in some important proteins, leading to their inactivation. In the last century, Se and S in soils have decreased significantly. In many regions of the world, there is a deficiency of these elements in soils and, as a result, a deficiency in plants. Therefore, there is a lack of Se in food. To solve this problem, biofortification of Se plants, biotransformation, and the possibility of treating plants with Se NPs and its compounds are used. NPs are of interest for their increased bioavailability, biological activity, and reduced toxicity. There are various ways to obtain NP: physical synthesis, chemical synthesis, and synthesis using living beings. Physical synthesis methods include pulsed laser ablation, microwave synthesis, hydrothermal treatment, and vapor deposition. The chemical synthesis of chalcogen NPs is carried out by combining various chemicals, where organic and inorganic compounds are used as a precursor. Synthesis of NPs with the help of bacteria, fungi, and plants is carried out due to the participation of various reducing enzymes. Among plants, berry, and fruit shrubs, aquatic plants have this ability. Bacteria are predominantly representatives of soil microflora.

Most of the available published data indicate a positive effect of nanochalcogens on the viability of plants and their resistance to biotic and abiotic stress factors. The effect of NPs on plants depends on the size (smaller particles are more effective) and concentration of NPs that the plants are exposed to. The stimulating effect of Se and S NPs is explained by an increase in the productivity of photosynthesis, a change in the fatty acid profile of lipids, a decrease in the level of LPO, an increase in the content of proline and organic acids in plant tissues, as well as an increase in the activity of antioxidant enzymes in various plant organs—nitrate reductase, lipoxygenase, phenylalanine lyase, peroxidase, and superoxide dismutase in plant cells under the influence of Se NPs.

This review demonstrates that fungicides based on S and Se NPs can be useful for protecting important crops, such as tomatoes, potatoes, apple trees, grapes, etc., from various diseases, mainly for "organic" farming. Nanoproducts in the form of nanofertilizers and nanopesticides have smart delivery mechanisms and controlled release of active ingredients, thus minimizing the release of polluted effluents into the environment.

Thus, S and Se NPs are promising agents for healing and stimulating the growth of cultivated plants. We hope that the presented critical analysis and systematization of the accumulated published information in this area will help to promote research in this promising area of agricultural science and technology.

References

Abbas, H.S., Abou Baker, D.H. and Ahmed, E.A. 2020. Cytotoxicity and antimicrobial efficiency of selenium nanoparticles biosynthesized by *Spirulina* platensis. Arch. Microbiol. 203(2): 523–532. doi: 10.1007/s00203-020-02042-3.

Abo Elsoud, M.M., Al-Hagar, O.E.A., Abdelkhalek, E.S. and Sidkey, N.M. 2018. Synthesis and investigations on tellurium myconanoparticles. Biotechnol Rep (Amst). 18: e00247. doi: 10.1016/j.btre.2018.e00247.

Alvares, J.J. and Furtado, I.J. 2021. Anti-*Pseudomonas aeruginosa* biofilm activity of tellurium nanorods biosynthesized by cell lysate of *Haloferax alexandrinus* GUSF-1 (KF796625). Biometals 34(5): 1007–1016. doi: 10.1007/s10534-021-00323-y.

Anan, Y., Yoshida, M., Hasegawa, S., Katai, R., Tokumoto, M., Ouerdane, L., Lobinski, R. and Ogra, Y. 2013. Speciation and identification of tellurium-containing metabolites in garlic, *Allium sativum*. Metallomics 5: 1215–1224. doi: 10.1039/c3mt00108c.

Anjum, D.H. 2016. Characterization of nanomaterials with transmission electron microscopy. IOP Conf Ser: Mater Sci. Eng. 146(1): 012001. doi: 10.1088/1757-899X/146/1/012001.

Arab, F., Mousavi-Kamazani, M. and Salavati-Niasari, M. 2017. Facile sonochemical synthesis of tellurium and tellurium dioxide nanoparticles: Reducing Te(IV) to Te via ultrasonic irradiation in methanol. Ultrason Sonochem. 37: 335–343. doi: 10.1016/j.ultsonch.2017.01.026.

Ardebili, Z.O., Ardebili, N.O., Jalili, S. and Safiallah, S. 2015. The modified qualities of basil plants by selenium and/ or ascorbic acid. Turk. J. Bot. 39: 401–407. doi: 10.3906/bot-1404-20.

Asghari-Paskiabi, F., Imani, M., Rafii-Tabar, H. and Razzaghi-Abyaneh, M. 2019. Physicochemical properties, antifungal activity and cytotoxicity of selenium sulfide nanoparticles green synthesized by *Saccharomyces cerevisiae*. Biochem. Biophys. Res. Commun. 516(4): 1078–1084. doi: 10.1016/j.bbrc.2019.07.007.

Aslam, C.J., Harbit, K.B. and Huaker, R.C. 1990. Comparative effects of selenite and selenate on nitrate assimilation in barley seedling. Plant Cell Environ. 13: 773–782.

Atkins, P.W., Overton, T., Rourke, Jonathan, Weller, M. and Armstrong, F.A. 2010. Shriver & Atkins' Inorganic Chemistry. Oxford: Oxford University Press. 824.

Azimi, F., Oraei, M., Gohari, G., Panahirad, S. and Farmarzi, A. 2021. Chitosan-selenium nanoparticles (Cs-Se NPs) modulate the photosynthesis parameters, antioxidant enzymes activities and essential oils in *Dracocephalum moldavica* L. under cadmium toxicity stress. Plant Physiol. Biochem. 167: 257–268. doi: 10.1016/j.plaphy.2021.08.013.

Bahrami, K., Nazari, P., Sepehrizadeh, Z., Zarea, B. and Shahverdi, A.R. 2012. Microbial synthesis of antimony sulfide nanoparticles and their characterization. Ann. Microbiol. 62: 1419–1425. doi 10.1007/s13213-011-0392-5.

Bano, I., Skalickova, S., Sajjad, H., Skladanka, J. and Horky, P. 2021. Uses of selenium nanoparticles in the plant production. Agronomy 11: 2229. doi: 10.3390/agronomy11112229.

Bao, P., Xiao. K.Q., Wang, H.J., Xu, H., Xu, P.P., Jia, Y., Häggblom, M.M. and Zhu, Y.G. 2016. Characterization and potential applications of a selenium nanoparticle producing and nitrate-reducing bacterium *Bacillus oryziterrae* sp. nov. Sci. Rep. 6: 34054. doi: 10.1038/srep34054.

Barabadi, H., Kobarfard, F. and Vahidi, H. 2018. Biosynthesis and characterization of biogenic tellurium nanoparticles by using *Penicillium chrysogenum* PTCC 5031: A novel approach in gold biotechnology. Iran J. Pharm. Res. 17(2): 87–97. doi:10.22037/IJPR.2018.2360.

Bisht, N., Phalswal, P. and Khanna, P.K. 2022. Selenium nanoparticles: a review on synthesis and biomedical applications. Mater Adv. 3: 1415–1431. doi: 10.1039/D1MA00639H.

Bloem, E., Haneklaus, S. and Schnug, E. 2015. Milestones in plant sulfur research on sulfur-induced-resistance (SIR) in Europe. Front Plant Sci. 5: 779. doi: 10.3389/fpls.2014.00779.

Borghese, R., Malferrari, M., Brucale, M., Ortolani, L., Franchini, M., Rapino, S., Borsetti, F. and Zannoni, D. 2020. Structural and electrochemical characterization of lawsone-dependent production of tellurium-metal nanoprecipitates by photosynthetic cells of *Rhodobacter capsulatus*. Bioelectrochemistry 133: 107456. doi: 10.1016/j.bioelechem.2020.107456.

Boud, R. 2011. Selenium stories. Nat. Chem. 3: 570. doi: 10.1038/nchem.1076.

Bouranis, D.L., Malagoli, M., Avice, J.-C. and Bloem, E. 2020. Advances in plant sulfur research. Plants 9: 256. doi: 10.3390/plants9020256.

Cao, X., Wang, C., Luo, X., Yue, L., White, J.C., Elmer, W., Dhankher, O.P., Wang, Z. and Xing, B. 2021. Elemental sulfur nanoparticles enhance disease resistance in tomatoes. ACS Nano 15(7): 11817–11827. doi: 10.1021/acsnano.1c02917.

Carapella, S.C. 1971. "History and occurrence of tellurium", Tellurium. Van Nostrand Reinhold Co. 1–49.

Chakraborty, S., Rene, E.R. and Lens, P.N.L. 2019. Reduction of selenite to elemental Se(0) with simultaneous degradation of phenol by co-cultures of *Phanerochaete chrysosporium* and *Delftia lacustris*. J. Microbiol. 57(9): 738–747. doi: 10.1007/s12275-019-9042-6.

Chantongsri, A., Phuektes, P., Borlace, G.N. and Aiemsaard, J. 2021. Antifungal activity of green sulfur nanoparticles synthesized using *Catharanthus roseus* extract against *Microsporum canis*. Thai J. Vet. Med. 51(4): 705–713. doi: 10.14456/tjvm.2021.85.

Chasteen, T.G., Fuentes, D.E., Tantaleán, J. and Vásquez, C.C. 2009. Tellurite: history, oxidative stress, and molecular mechanisms of resistance. FEMS Microbiol. Rev. 33: 820–832. doi: 10.1111/j.1574-6976.2009.00177.x.

Che, L., Xu, W., Zhan, J., Zhang, L., Liu, L. and Zhou, H. 2019. Complete genome sequence of *Bacillus cereus* cc-1, a novel marine selenate/selenite reducing bacterium producing metallic selenides nanomaterials. Curr. Microbiol. 76(1): 78–85. doi: 10.1007/s00284-018-1587-9.

Chen, H., Dong, W., Ge, J., Wang, C., Wu, X., Lu, W. and Chen, L. 2013. Ultrafine sulfur nanoparticles in conducting polymer shell as cathode materials for high performance lithium. Sulfur batteries. Sci. Rep. 3: 1910. doi: 10.1038/srep01910.

Chivers, T. 2005. A Guide to Chalcogen-Nitrogen Chemistry, World Scientific Publishing Co.: Singapore. 340. doi: 10.1142/5701.

Choudhury, R.S., Ghosh, M. and Goswami, A. 2012. Inhibitory effects of sulfur nanoparticles on membrane lipids of *Aspergillus niger*: a novel route of fungistasis. Curr. Microbiol. 65: 91–97. doi: 10.1007/s00284-012-0130-7.

Cooper, W.C. 1971. Tellurium; Van Nostrand Reinhold Company: New York. USA, 34.

Cowgill, U.M. 1988. The tellurium content of vegetation. Biol. Trace Elem. Res. 17: 43–67. doi: 10.1007/BF02795446.

Cremonini, E., Zonaro, E., Donini, M., Lampis, S., Boaretti, M. and Dusi, S. 2016. Biogenic selenium nanoparticles: characterization, antimicrobial activity and effects on human dendritic cells and fibroblasts. Microb. Biotechnol. 9(6): 758–771. doi: 10.1111/1751-7915.12374.

Cremonini, E., Boaretti, M., Vandecandelaere, I., Zonaro, E., Coenye, T., Lleo, M.M., Lampis, S. and Vallini, G. 2018. Biogenic selenium nanoparticles synthesized by *Stenotrophomonas maltophilia* SeITE02 loose antibacterial and antibiofilm efficacy as a result of the progressive alteration of their organic coating layer. Microb. Biotechnol. 11(6): 1037–1047. doi: 10.1111/1751-7915.13260.

Cruz, D.M., Mi, G. and Webster, T.J. 2018. Synthesis and characterization of biogenic selenium nanoparticles with antimicrobial properties made by *Staphylococcus aureus*, methicillin-resistant *Staphylococcus aureus* (MRSA), *Escherichia coli*, and *Pseudomonas aeruginosa*. J. Biomed. Mater Res. A. 106: 1400–1412. doi: 10.1002/jbm.a.36347.

Cruz, D.M., Tien-Street, W., Zhang, B., Huang, X., Crua, A.V., Nieto-Argüello, A., Cholula-Díaz, J.L., Martínez, L., Huttel, Y., Ujué González, M., García-Martín, J.M. and Webster, T.J. 2019. Citric juice-mediated synthesis of tellurium nanoparticles with antimicrobial and anticancer properties. Green Chem. 21(8): 1982–1988. doi: 10.1039/c9gc00131j.

Cui, D., Liang, T., Sun, L., Meng. L., Yang, C., Wang, L., Liang, T. and Li, Q. 2018. Green synthesis of selenium nanoparticles with extract of hawthorn fruit induced HepG$_2$ cells apoptosis. Pharm. Biol. 56(1): 528–534. doi: 10.1080/13880209.2018.1510974.

Das, K., Jhan, P.K., Das, S.C., Aminuzzaman, F.M. and Ayim, B.Y. 2021. Nanotechnology: Past, present and future prospects in crop protection. pp. 211–226. Chapter 11 in Technology in Agriculture, edited by Fiaz Ahmad, Muhammad Sultan. London: IntechOpen, 2021. doi: 10.5772/intechopen.98703.

Dębski, B., Zachara, B. and Wąsowicz, W. 2001. An attempt to evaluate the level of selenium in Poland and its influence on the healthiness of people and animals. Folia Univ Agric Stetin Zootech. 224: 31–38 [in Polish].

Dhanjal, S. and Cameotra, S.S. 2010. Aerobic biogenesis of selenium nanospheres by *Bacillus cereus* isolated from coalmine soil. Microb Cell Fact. 9: 52. doi: 10.1186/1475-2859-9-52.

Dwivedi, C., Shah, C.P., Singh, K.M., Kumar, M. and Bajaj, P.N. 2011. An organic acid-induced synthesis and characterization of selenium nanoparticles. J. Nanotechnol. 651971. doi: 10.1155/2011/651971.

Elahian, F., Reiisi, S., Shahidi, A. and Mirzaei, S.A. 2017. High-throughput bioaccumulation, biotransformation, and production of silver and selenium nanoparticles using genetically engineered *Pichia pastoris*. Nanomedicine 13(3): 853–861. doi: 10.1016/j.nano.2016.10.009.

El-Ramady, H., Faizy, S.E.D., Abdalla, N., Taha, H., Domokos-Szabolcsy, É., Fari, M. and Brevik, E.C. 2020. Selenium and nano-selenium biofortification for human health: Opportunities and challenges. Soil Syst. 4: 57. doi: 10.3390/soilsystems4030057.

El-Saadony, M.T., Saad, A.M., Najjar, A.A., Alzahrani, S.O., Alkhatib, F.M., Shafi, M.E., Selem, E., Desoky, E.-S.M., Fouda, S.E.E., El-Tahan, A.M. and Hassan, M.A.A. 2020. The use of biological selenium nanoparticles to suppress *Triticum aestivum* L. crown and root rot diseases induced by *Fusarium* species and improve yield under drought and heat stress. Saudi J. Biol. Sci. 28(8): 4461–4471. doi: 10.1016/j.sjbs.2021.04.043.

Faramarzi, S., Anzabi, Y. and Jafarizadeh-Malmiri, H. 2020. Nanobiotechnology approach in intracellular selenium nanoparticle synthesis using *Saccharomyces cerevisiae* - fabrication and characterization. Arch. Microbiol. 202(5): 1203–1209. doi: 10.1007/s00203-020-01831-0.

Feng, R., Wang, L., Yang, J., Zhao, P., Zhu, Y., Li, Y., Yu, Y., Liu, H., Rensing, C., Wu, Z., Ni, R. and Zheng, S. 2021. Underlying mechanisms responsible for restriction of uptake and translocation of heavy metals (metalloids) by selenium via root application in plants. J. Hazard Mater. 15: 402. doi: 10.1016/j.jhazmat.2020.123570.

Feng, T., Chen, S., Gao, D., Liu, G., Bai, H., Li, A., Peng, L. and Ren, Z. 2015. Selenium improves photosynthesis and protects photosystem II in pear (*Pyrus bretschneideri*), grape (*Vitis vinifera*), and peach (*Prunus persica*). Photosynthetica 53: 609–612. doi: 10.1007/s11099-015-0118-1.

Fernández-Lodeiro, J., Rodríguez-Gónzalez, B., Novio, F., Fernández-Lodeiro, A., Ruiz-Molina, D., Capelo, J.L., Santos, A.A.D. and Lodeiro, C. 2017. Synthesis and characterization of PtTe(2) multi-crystallite nanoparticles using organotellurium nanocomposites. Sci. Rep. 7(1): 9889. doi: 10.1038/s41598-017-10239-8.

Filipović, N., Ušjak, D., Milenković, M.T., Zheng, K., Liverani, L., Boccaccini, A.R. and Stevanović, M.M. 2021. Comparative study of the antimicrobial activity of selenium nanoparticles with different surface chemistry and structure. Front Bioeng. Biotechnol. 8: 1591. doi: 10.3389/fbioe.2020.624621.

Francioso, A., Conrado, A.B., Mosca, L. and Fontana, M. 2020. Chemistry and biochemistry of sulfur natural compounds: key intermediates of metabolism and redox biology. Oxid. Med. Cell Longev. 8294158. doi: 10.1155/2020/8294158.

Fuentes-Lara, L.O., Medrano-Macías, J., Pérez-Labrada, F., Rivas-Martínez, E.N., García-Enciso, E.L., González-Morales, S., Juárez-Maldonado, A., Rincón-Sánchez, F. and Benavides-Mendoza, A. 2019. From elemental sulfur to hydrogen sulfide in agricultural soils and plants. Molecules 24: 2282. doi: 10.3390/molecules24122282.

Galiová, M.V., Kanický, V. and Havliš, J. 2014. Laser ablation inductively coupled plasma mass spectrometry as a tool in biological sciences. pp. 313–348. *In*: Natural Products Analysis (John Wiley & Sons, Inc.). doi: 10.1002/9781118876015.ch9.

Ghasemian, S., Masoudian, N., Nematpour, F.S. and Afshar A.S. 2021. Selenium nanoparticles stimulate growth, physiology, and gene expression to alleviate salt stress in *Melissa officinalis*. Biologia 76: 2879–2888. doi: 10.1007/s11756-021-00854-2.

Gholami, S., Dehaghi, M.A., Rezazadeh, A. and Naji, A.M. 2022. Seed germination and physiological responses of quinoa to selenium priming under drought stress. Bragantia. 81: e0722. doi: 10.1590/1678-4499.20210183.

Ghotekar, S., Pagar T., Pansambal, S. and Oza, R. 2020. A review on green synthesis of sulfur nanoparticles via plant extract, characterization and its applications. Ad J. Chem. B. 2(3): 128–143. doi: 10.33945/SAMI/AJCB.2020.3.5.

Giannini, C., Ladisa, M., Altamura, D., Siliqi, D., Sibillano, T. and De Caro, L. 2016. X-ray diffraction: A powerful technique for the multiple-length-scale structural analysis of nanomaterials. Crystals 6: 87. doi: 10.3390/cryst6080087.

Godoy-Gallardo, M., Ulrich, E., Delgado, L.M., de Roo Puente, Y.J.D., Hoyos-Nogués, M., Gil, F.J. and Pereza, R.A. 2022. Antibacterial approaches in tissue engineering using metal ions and nanoparticles: From mechanisms to applications. Bioact. Mater. 6(12): 4470–4490. doi: 10.1016/j.bioactmat.2021.04.033.

Gómez-Gómez, B., Arregui, L., Serrano, S., Santos, A., Pérez-Corona, T. and Madrid, Y. 2019. Selenium and tellurium-based nanoparticles as interfering factors in quorum sensing-regulated processes: violacein production and bacterial biofilm formation. Metallomics 11(6): 1104–1114. doi: 10.1039/c9mt00044e.

Gómez-Gómez, B., Sanz-Landaluce, J., Pérez-Corona, M.T. and Madrid, Y. 2020. Fate and effect of in-house synthesized tellurium based nanoparticles on bacterial biofilm biomass and architecture. Challenges for nanoparticles characterization in living systems. Sci. Total Environ. 719: 137501. doi: 10.1016/j.scitotenv.2020.137501.

Górska, S., Maksymiuk, A. and Turło, J. 2021. Selenium-containing polysaccharides—structural diversity, biosynthesis, chemical modifications and biological activity. Appl. Sci. 11(8): 3717. doi: 10.3390/app11083717.

Graskova, I.A., Perfileva, A.I., Nozhkina, O.A., Dyakova, A.V., Nurminsky, V.N., Klimenkov, I.V., Sudakov, N.P., Borodina, T.M., Aleksandrova, G.P., Lesnichaya, M.V., Sukhov, B.G. and Trofimov, B.A. 2019. The effect of nanoscale selenium on the causative agent of ring rot and potato *in vitro*. Khimiya Rastitel'nogo Syr'ya. 3: 345–354. doi: 10.14258/jcprm.2019034794 (in Russian with English Abstract).

Gudkov, S.V., Shafeev, G.A., Glinushkin, A.P., Shkirin, A.V., Barmina, E.V., Rakov, I.I., Simakin, A.V., Kislov, A.V., Astashev, M.E., Vodeneev, V.A. and Kalinitchenko, V.P. 2020. Production and use of selenium nanoparticles as fertilizers. ACS Omega 5: 17767–17774. doi: 10.1021/acsomega.0c02448.

Guisbiers, G., Wang, Q., Khachatryan, E., Arellano-Jimenez, M.J., Webster, T.J., Larese-Casanova, P. and Nash, K.L. 2014. Anti-bacterial selenium nanoparticles produced by UV/VIS/NIR pulsed nanosecond laser ablation in liquids. Laser Phys. Lett. 12(1): 016003. doi: 10.1088/1612-2011/12/1/016003.

Guisbiers, G., Wang, Q., Khachatryan, E., Mimun, L., Mendoza-Cruz. R., Larese-Casanova, P., Webster, T. and Nash, K. 2016. Inhibition of *E. coli* and *S. aureus* with selenium nanoparticles synthesized by pulsed laser ablation in deionized water. Int. J. Nanomed. 11: 3731–3736. doi: 10.2147/IJN.S106289.

Guisbiers, G., Lara, H.H., Mendoza-Cruz, R., Naranjo, G., Vincent, B.A., Peralta, X.G. and Nash, K.L. 2017. Inhibition of *Candida albicans* biofilm by pure selenium nanoparticles synthesized by pulsed laser ablation in liquids. Nanomedicine 13(3): 1095–1103. doi: 10.1016/j.nano.2016.10.011.

Gunti, L., Dass, R.S. and Kalagatur, N.K. 2019. Phytofabrication of selenium nanoparticles from emblica officinalis fruit extract and exploring its biopotential applications: antioxidant, antimicrobial, and biocompatibility. Front Microbiol. 10: 931. doi: 10.3389/fmicb.2019.00931.

Gupta, M. and Gupta, S. 2017. An overview of selenium uptake, metabolism, and toxicity in plants. Front Plant Sci. 11(7): 2074. doi: 10.3389/fpls.2016.02074.

Hartwig, D., Jacob, R.G., Lenardão, E.J., Nascimento, J.E.R., Abenante, L., Soares, L.K. and Schiesser, C.H. 2022. Semisynthetic bioactive organoselenium and organotellurium compounds. pp. 253–289. *In*: Organochalcogen Compounds. Synthesis, Catalysis and New Protocols with Greener Perspectives Advances in Green and Sustainable Chemistry. (Elsevier). doi: 10.1016/B978-0-12-819449-2.00003-3.

Hasanuzzaman, M., Bhuyan, M.H.M.B., Raza, A., Hawrylak-Nowak, B., Matraszek-Gawron, R., Nahar, K. and Fujita, M. 2020. Selenium toxicity in plants and environment: biogeochemistry and remediation possibilities. Plants. 9(12): 1711. doi: 10.3390/plants9121711.

Hashem, A.H., Khalil, A.M.A., Reyad, A.M. and Salem, S.S. 2021. Biomedical applications of mycosynthesized selenium nanoparticles using *Penicillium expansum* ATTC 36200. Biol. Trace Elem. Res. 199: 3998–4008. doi: 10.1007/s12011-020-02506-z.

Hawrylak-Nowak, B. 2013. Comparative effects of selenite and selenate on growth and selenium accumulation in lettuce plants under hydroponic conditions. Plant Growth Reg. 70: 149–157. doi: 10.1007/s10725-013-9788-5.

Hell, R., Khan, M.S. and Wirtz, M. 2010. Cellular biology of sulfur and its functions in plants. *In*: Cell biology of metals and nutrients. Plant Cell Monographs, 17 (Berlin: Springer, Heidelberg). doi:10.1007/978-3-642-10613-2_11.

Hosnedlova, B., Kepinska, M., Skalickova, S., Fernandez, C., Ruttkay-Nedecky, B., Peng, Q., Baron, M., Melcova, M., Opatrilova, R., Zidkova, J., Bjørklund, G., Sochor, J. and Kizek, R. 2018. Nano-selenium and its nanomedicine applications: a critical review. Int. J. Nanomedicine 13: 2107–2128. doi: 10.2147/IJN.S157541.

Hou, J.Y., Ai, S.Y. and Shi, W.J. 2011. Preparation and characterization of nanoSe/silk fibroin colloids. Chem. Res. Chin. Univ. 27(1): 158–160. doi: 10.1016/j.colsurfb.2015.10.044.

Huang, W., Wu, H., Li, X. and Chen, T. 2016. Facile one-pot synthesis of tellurium nanorods as antioxidant and anticancer agents. Chem. Asian J. 11: 2301–2311. doi: 10.1002/asia.201600757.

Husen, A. and Siddiqi, K.S. 2014. Plants and microbes assisted selenium nanoparticles: characterization and application. J. Nanobiotechnology 12: 28. doi: 10.1186/s12951-014-0028-6.

Hussein, H.A., Darwesh, O.M. and Mekki, B.B. 2019. Environmentally friendly nano-selenium to improve antioxidant system and growth of groundnut cultivars under sandy soil conditions. Biocatal. Agric Biotechnol. 18: 101080. doi: 10.1016/j.bcab.2019.101080.

Hussein, H.A., Darwesh, O.M., Mekki, B.B. and El-Hallouty, S.M. 2019. Evaluation of cytotoxicity, biochemical profile and yield components of groundnut plants treated with nano-selenium. Biotechnol. Rep. (Amst) 12(24): 1–7. doi: 10.1016/j.btre.2019.e00377.

Ibers, J. 2009. Tellurium in a twist. Nature Chem. 1: 508. doi: 10.1038/nchem.350.

Ikram, M., Raja, N.I., Mashwani, Z.-U.-R., Omar, A.A., Mohamed, A.H., Satti, S.H. and Zohra, E. 2022. Phytogenic selenium nanoparticles elicited the physiological, biochemical, and antioxidant defense system amelioration of huanglongbing-infected 'kinnow' mandarin plants. Nanomaterials 12: 356. doi: 10.3390/nano12030356.

Jabłonska-Czapla, M. and Grygoyc, K. 2021. Development of a tellurium speciation study using ic-icp-ms on soil samples taken from an area associated with the storage, processing, and recovery of electrowaste. Molecules 26: 2651. doi: 10.3390/molecules26092651.

Jiang, C., Zu, C., Shen, J., Shao, F. and Li, T. 2015. Effects of selenium on the growth and photosynthetic characteristics of flue-cured tobacco (*Nicotiana tabacum* L.). Acta Soc. Bot. Pol. 84: 71–77. doi: 10.5586/asbp.2015.006.

Jo, I., Kang, J.W. and Kim, K.S. 2021. Synthesis of cadmium telluride nanoparticles using thioglycolic acid, thioglycerol, and L-cysteine. J. Nanosci. Nanotechnol. 21(7): 4073–4076. doi: 10.1166/jnn.2021.19182.

Joshi, S.M., de Britto, S., Jogaiah, S. and Ito, S. 2019. Mycogenic selenium nanoparticles as potential new generation broad spectrum antifungal molecules. Biomolecules 9: 419. doi: 10.3390/biom9090419.

Joshi, S.M., de Britto, S. and Jogaiah, S. 2021. Myco-engineered selenium nanoparticles elicit resistance against tomato late blight disease by regulating differential expression of cellular, biochemical and defense responsive genes. J. Biotechnol. 325: 196–206. doi: 10.1016/j.jbiotec.2020.10.023.

Joudeh, N. and Linke, D. 2022. Nanoparticle classification, physicochemical properties, characterization, and applications: a comprehensive review for biologists. J. Nanobiotechnology 20: 262. doi: 10.1186/s12951-022-01477-8.

Karikalan, N., Karthik, R., Chen, S.-M., Karuppiah, C. and Elangovan, A. 2017. Sonochemical synthesis of sulfur doped reduced graphene oxide supported cus nanoparticles for the non-enzymatic glucose sensor applications. Sci. Rep. 7: 2494. doi: 10.1016/j.ultsonch.2020.105043.

Kaur, M., Rob, A., Caton-Williams, J. and Huang, Z. 2013. Biochemistry of nucleic acids functionalized with sulfur, selenium, and tellurium: Roles of the single-atom substitution *In*: Biochalcogen Chemistry: The Biological Chemistry of Sulfur, Selenium, and Tellurium (ACS Symposium Series) 1152: 89–126.

Kavanashree, K., Jahagirdar, S., Patil, M.S., Kambrekar, D.N., Basavaraja, G.T. and Krishnaraj, P.U. 2020. Influence of different nanoformulations on soybean seed quality parameters. Int. J. Chem. Stud. 8(5): 1073–1077. doi: 10.22271/chemi.2020.v8.i5o.10439.

Khairan, K. and Zahraturriaz, J.Z. 2019. Green synthesis of sulphur nanoparticles using aqueous garlic extract (*Allium sativum*). Rasayan J. Chem. 12: 50–57.

Khiralla, G.M. and El-Deeb, B.A. 2015. Antimicrobial and antibiofilm effects of selenium nanoparticles on some foodborne pathogens. LWT-Food Sci. and Technol. 63(2): 1001–1007. doi: 10.1016/j.lwt.2015.03.086.

Kim, Y.H., Kim, G.H., Yoon, K.S., Shankar, S. and Rhim, J.-W. 2020. Comparative antibacterial and antifungal activities of sulfur nanoparticles capped with chitosan. Microb. Pathog. 144: 104178. doi: 10.1016/j.micpath.2020.104178.

Kolbert, Z., Molnár, Á., Feigl, G. and van Hoewyk, D. 2019. Plant selenium toxicity: Proteome in the crosshairs. J. Plant Physiol. 232: 291–300. doi: 10.1016/j.jplph.2018.11.003.

Kora, A.J. 2018. Tree gum stabilised selenium nanoparticles: characterisation and antioxidant activity. IET Nanobiotechnol. 12(5): 658–662. doi: 10.1049/iet-nbt.2017.0310.

Krug, P., Wiktorska, K., Kaczyńska, K., Ofiara, K., Szterk, A., Kuśmierz, B. and Mazur, M. 2020. Sulforaphane-assisted preparation of tellurium flower-like nanoparticles. Nanotechnology 31(5): 055603. doi: 10.1088/1361-6528/ab4e38.

Kudryavtsev, A.A. 1974. Tellurium compounds. pp. 83–132. *In*: The Chemistry and Technology of Selenium and Tellurium (Collet's Ltd., London and Wellingbrough).

Kushkevych, I., Kováˇrová, A., Dordevic, D., Gaine, J., Kollar, P., Víťezová, M. and Rittmann, S.K.-M.R. 2021. Distribution of sulfate-reducing bacteria in the environment: cryopreservation techniques and their potential storage application. Processes 9: 1843. doi: 10.3390/pr9101843.

Lanctot, C.M., Cresswell, T., Callaghan, P.D. and Melvin, S.D. 2017. Bioaccumulation and biodistribution of selenium in metamorphosing tadpoles. Environ. Sci. Technol. 51: 5764–5916. doi: 10.1021/acs.est.7b00300.

Langi, B., Shah, C., Singh, K., Chaskar, A., Kumar, M.S. and Bajaj, P.N. 2010. Ionic liquid-induced synthesis of selenium nanoparticles. Mater Res. Bull. 45(6): 668–671. doi: 10.1016/j.materresbull.2010.03.005.

Lesnichaya, M.V., Perfileva, A.I., Gazizova, A.V. and Graskova, I.A. 2022. Synthesis, toxicity evaluation and determination of possible mechanisms of antimicrobial effect of arabinogalactane-capped selenium nanoparticles. J. Trace Elem. Med. Biol. 69: 126904. doi: 10.1016/j.jtemb.2021.126904.

Li, J., Shen, B., Nie, S., Duan, Z. and Chen, K. 2019. A combination of selenium and polysaccharides: Promising therapeutic potential. Carbohydr. Polym. 206: 163–173. doi: 10.1016/j.carbpol.2018.10.088.

Li, Q., Gao, Y. and Yang, A. 2020. Sulfur homeostasis in plants. Int. J. Mol. Sci. 21: 8926. doi:10.3390/ijms21238926.

Liang, X., Perez, M.A.M., Nwoko, K.C., Egbers, P., Feldmann, J., Csetenyi, L. and Gadd, G.M. 2019. Fungal formation of selenium and tellurium nanoparticles. Appl. Microbiol. Biotechnol. 103(17): 7241–7259. doi: 10.1007/s00253-019-09995-6.

Lim, J., Jung, U., Joe, W.T., Kim, E.T., Pyun, J. and Char, K. 2015. High sulfur content polymer nanoparticles obtained from interfacial polymerization of sodium polysulfide and 1,2,3-trichloropropane in water. Macromol. Rapid Commun. 36(11): 1103–1107. doi: 10.1002/marc.201500006.

Lima, L.W., Pilon-Smits, E.A.H. and Schiavon, M. 2018. Mechanisms of selenium hyperaccumulation in plants: A survey of molecular, biochemical and ecological cues. Biochim. Biophys. Acta 1862(11): 2343–2353. doi: 10.1016/j.bbagen.2018.03.028.

Lin, W., Zhang, J., Xu, J.-F. and Pi, J. 2021. The advancing of selenium nanoparticles against infectious diseases. Front Pharmacol. 12: 682284. doi: 10.3389/fphar.2021.682284.

Loshchinina, E.A., Vetchinkina, E.P., Kupryashina, M.A., Kursky, V.F. and Nikitina, V.E. 2018. Nanoparticles synthesis by *Agaricus* soil basidiomycetes. J. Biosci. Bioeng. 126(1): 44–52. doi: 10.1016/j.jbiosc.2018.02.002.

Marine, W., Patrone, L., Luk'yanchuk, B. and Sentis, M. 2000. Strategy of nanocluster and nanostructure synthesis by conventional pulsed laser ablation. Appl. Surf. Sci. 154: 345–352. doi: 10.1016/S0169-4332(99)00450-X.

Mechora, Š. 2019. Selenium as a protective agent against pests: A review. Plants 8(8): E262. doi: 10.3390/plants8080262.

Mellinas C., Jiménez A. and Garrigós M.D.C. 2019. Microwave-assisted green synthesis and antioxidant activity of selenium nanoparticles using *Theobroma Cacao* L. bean shell extract. Molecules 24(22): E4048. doi: 10.3390/molecules24224048.

Men, X.Y., Xu, W.G., Zhu, X. and Ma, W.C. 2009. Extraction, selenium-nanoparticle preparation and anti-virus bioactivity determination of polysaccharides from *Caulerpa taxifolia*. Zhong Yao Cai (J. Chin. Med. Mater.). 32(12): 1891–1894. PMID: 20432908 (in Chinese with English Abstract).

Meng, Y., Zhang, Y., Jia, N., Qiao, H., Zhu, M., Meng, Q., Lu, Q. and Zu, Y. 2018. Synthesis and evaluation of a novel water-soluble high Se-enriched *Astragalus* polysaccharide nanoparticles. Int. J. Biol. Macromol. 118(Pt B): 1438–1448. doi: 10.1016/j.ijbiomac.2018.06.153.

Mishra, R.R., Prajapati, S., Das, J., Dangar, T.K., Das, N. and Thatoi, H. 2011. Reduction of selenite to red elemental selenium by moderately halotolerant *Bacillus megaterium* strains isolated from *Bhitarkanika mangrove* soil and characterization of reduced product. Chemosphere 84(9): 1231–1237. doi: 10.1016/j.chemosphere.2011.05.025.

Moghaddam, A.B., Namvar, F., Moniri, M., Md. Tahir, P., Azizi, S. and Mohamad, R. 2015. Nanoparticles biosynthesized by fungi and yeast: A review of their preparation, properties, and medical applications. Molecules 20(9): 16540–16565. doi: 10.3390/molecules200916540.

Mondal, S., Pramanik, K., Panda, D., Dutta, D., Karmakar, S. and Bose, B. 2022. Sulfur in seeds: an overview. Plants 11: 450. doi: 10.3390/plants11030450.

Morales-Espinoza, M.C., Cadenas-Pliego, G., Pérez-Alvarez, M., Hernández-Fuentes, A.D., Cabrera de la Fuente, M., Benavides-Mendoza, A., Valdés-Reyna, J. and Juárez-Maldonado, A.S. 2019. Nanoparticles induce changes in the growth, antioxidant responses, and fruit quality of tomato developed under NaCl stress. Molecules 24(17): E3030. doi: 10.3390/molecules24173030.

Morena, A.G., Bassegoda, A., Hoyo, J. and Tzanov, T. 2021. Hybrid tellurium-lignin nanoparticles with enhanced antibacterial properties. ACS Appl. Mater. Interfaces 13(13): 14885–14893. doi: 10.1021/acsami.0c22301.

Moreno-Martin, G., Sanz-Landaluze, J., León-Gonzalez, M.E. and Madrid, Y. 2019. *In-vivo* solid phase microextraction for quantitative analysis of volatile organoselenium compounds in plants. Anal. Chim. Acta 12: 72–80. doi: 10.1016/j.aca.2019.06.061.

Moroder, L. 2005. Isoteric replacement of sulfur with other chalcogens in peptides and proteins. J. Pept. Sci. 11: 187–214. doi: 10.1002/psc.654.

Mosallam, F.M., El-Sayyad, G.S., Fathy, R.M. and El-Batal, A.I. 2018. Biomolecules-mediated synthesis of selenium nanoparticles using *Aspergillus oryzae* fermented Lupin extract and gamma radiation for hindering the growth of some multidrug-resistant bacteria and pathogenic fungi. Microb. Pathog. 122: 108–116. doi: 10.1016/j.micpath.2018.06.013.

Mukwevho, E., Ferreira, Z. and Ayeleso, A. 2014. Potential role of sulfur-containing antioxidant systems in highly oxidative environments. Molecules 19(12): 19376–19389. doi: 10.3390/molecules191219376.

Najafi, S., Razavi, S.M., Khoshkam, M. and Asadi, A. 2022. Effects of green synthesis of sulfur nanoparticles from *Cinnamomum zeylanicum* barks on physiological and biochemical factors of Lettuce (*Lactuca sativa*). Physiol. Mol. Biol. Plants 26(42). doi: 10.1007/s12298-020-00793-3.

Nandini, B., Hariprasad, P., Prakash, H.S., Shetty, H.S. and Geetha, N. 2017. Trichogenic-selenium nanoparticles enhance disease suppressive ability of *Trichoderma against* downy mildew disease caused by *Sclerospora graminicola* in pearl millet. Sci. Rep. 7: 2612. doi: 10.1038/s41598-017-02737-6.

Narayan, O.P., Kumar, P., Yadav, B., Dua, M. and Johri, A.K. 2022. Sulfur nutrition and its role in plant growth and development. Plant Signal Behav. doi: 10.1080/15592324.2022.2030082.

Nessel, T.A. and Gupta, V. 2020. Selenium. Treasure Island (FL): StatPearls Publishing. 32491483.

Newman, R., Waterland, N., Moon, Y. and Tou, J.C. 2019. Selenium biofortification of agricultural crops and effects on plant nutrients and bioactive compounds important for human health and disease prevention—a Review. Plant Foods Hum. Nutr. 74(4): 449–460. doi: 10.1007/s11130-019-00769-z.

Neysanian, M., Iranbakhsh, A., Ahmadvand, R., Oraghi Ardebili, Z. and Ebadi, M. 2021. Investigation of the effect of selenium nanoparticles on drought stress tolerance in cherry tomato plant (*Solanum Lycopersicum* L. var. *Cerasiforme*). Iran J. Plant Biotechnol. 16(3): 27–41. https://www.sid.ir/en/journal/ViewPaper.aspx?id=893935.

Ni, T.W., Staicu, L.C., Nemeth, R.S., Schwartz, C.L., Crawford, D., Sellgman, J.D., Hunter, W.J., Pilon-Smits, E.A. and Ackerson, C.J. 2015. Progress toward clonable inorganic nanoparticles. Nanoscale 7(41): 17320–17327. doi: 10.1039/c5nr04097c.

Niranjan, R., Zafar, S., Lochab, B. and Priyadarshini, R. 2022. Synthesis and characterization of sulfur and sulfur-selenium nanoparticles loaded on reduced graphene oxide and their antibacterial activity against Gram-positive pathogens. Nanomaterials 12: 191. doi: 10.3390/nano12020191.

Nozhkina, O.A., Perfileva, A.I., Graskova, I.A., Djyakova, A.V., Nurminsky, V.N., Klimenkov, I.V., Ganenko, T.V., Borodina, T.N., Aleksandrova, G.P., Sukhov, B.G. and Trofimov, B.A. 2019. The biological activity of a selenium nanocomposite encapsulated with carrageenan macromolecules regarding the ring rot pathogen and potato plants. Nanotechnologies Russ. 14(5–6): 255–262. doi: 10.1134/S1995078019030091.

Nurminsky, V.N., Perfileva, A.I., Kapustina, I.S., Graskova, I.A., Sukhov, B.G. and Trofimov, B.A. 2020. Growth-stimulating activity of natural polymer-based nanocomposites of selenium during the germination of cultivated plant seeds. Dokl Biochem. 495: 296–299. doi: 10.1134/S1607672920060113.

Nwoko, K.C., Liang, X., Perez, M.A., Krupp, E., Gadd, G.M. and Feldmann, J. 2021. Characterisation of selenium and tellurium nanoparticles produced by *Aureobasidium pullulans* using a multi-method approach. J. Chromatogr. A. 1642: 462022. doi: 10.1016/j.chroma.2021.462022.

Panahi-Kalamuei, M., Mohandes, F., Mousavi-Kamazani, M., Salavati-Niasaria, M., Fereshteh, Z. and Fathide, M. 2014. Tellurium nanostructures: Simple chemical reduction synthesis, characterization and photovoltaic measurements. Mater Sci. Semicond. 27: 1028–1035. doi: 10.1016/j.mssp.2014.09.015.

Panahi-Kalamuei, M., Salavati-Niasari, M. and Hosseinpour-Mashkani, S.M. 2014. Facile microwave synthesis, characterization, and solar cell application of selenium nanoparticles. J. Alloys Compd. 617: 627–632. doi: 10.1016/j.jallcom.2014.07.174.

Papkina, A.V., Perfileva, A.I., Zhivet'yev, M.A., Borovskii, G.B., Graskova, I.A., Klimenkov, I.V., Lesnichaya, M.V., Sukhov, B.G. and Trofimov, B.A. 2015a. Complex effects of selenium-arabinogalactan nanocomposite on both phytopathogen *Clavibacter michiganensis* subsp. *sepedonicus* and potato plants. Nanotechnologies Russ. 10(5–6): 484–491. doi: 10.1134/S1995078015030131.

Papkina, A.V., Perfileva, A.I., Zhivetev, M.A., Borovskiy, G.B., Graskova, I.A., Lesnichaya, M.V., Klimenkov, I.V., Sukhov, B.G. and Trofimov, B.A. 2015b. Effect of selenium and arabinogalactan nanocomposite on viability of the phytopathogen *Clavibacter michiganensis* subsp. *sepedonicus*. Dokl Biol Sci. 461(1): 89–91. doi: 10.1134/S001249661501010X.

Paralikar, P. and Rai, M. 2017. Bio inspired synthesis of sulphur nanoparticles using leaf extract of four medicinal plants with special reference to their antibacterial activity. IET Nanobiotechnol. 12: 25–31. doi: 10.1049/iet-nbt.2017.0079.

Paralikar, P., Ingle, A.P., Tiwari, V., Golinska, P., Dahm, H. and Rai, M. 2019. Evaluation of antibacterial efficacy of sulfur nanoparticles alone and in combination with antibiotics against multidrug-resistant uropathogenic bacteria. J. Environ. Sci. Health Part A. 54: 381–390. doi: 10.1080/10934529.2018.1558892.

Perfileva, A.I., Tsivileva, O.M. and Koftin, O.V. 2016. Growth behavior of phytopathogen *Clavibacter michiganensis* ssp. *sepedonicus* treated with selenium biocomposites of mushroom origin. J. Stress Physiol. Biochem. 12(1): 13–20.

Perfileva, A.I., Tsivileva, O.M., Koftin, O.V., Ibragimova, D.N. and Fedotova, O.V. 2017a. Effect of selenium-containing biocomposites based on ganoderma mushroom isolates grown in the presence of oxopropyl-4-hydroxycoumarins, on bacterial phytopathogens. Microbiology 86(2): 183–191. doi: 10.1134/S0026261717020163.

Perfileva, A.I., Moty'leva, S.M., Klimenkov, I.V., Graskova, I.A., Skhov, B.G. and Trofimov, B.A. 2017b. Development of antimicrobial nano-selenium biocomposite for protecting potatoes from bacterial phytopathogens. Nanotechnologies Russ. 12(9–10): 553–558. doi: 10.1134/S1995078017050093.

Perfileva, A.I., Tsivileva, O.M., Drevko, Ya.B., Ibragimova, D.N. and Koftin, O.V. 2018a. Effect of selenium-containing biocomposites from medicinal mushrooms on the potato ring rot causative agent. Dokl Biol. Sci. 479(4): 472–475. doi: 10.1134/S0012496618020072.

Perfileva, A.I., Tsivileva, O.M., Koftin, O.V., Anis'kov, A.A. and Ibragimova, D.N. 2018b. Selenium-containing nanobiocomposites of fungal origin reduce the viability and biofilm formation of the bacterial phytopathogen *Clavibacter michiganensis* subsp. *sepedonicus*. Nanotechnologies Russ. 13(5–6): 268–276. doi: 10.1134/S1995078018030126.

Perfileva, A.I., Nozhkina, O.A., Graskova, I.A., Sidorov, A.V., Lesnichaya, M.V., Aleksandrova, G.P., Dolmaa, G., Klimenkov, I.V. and Sukhov, B.G. 2018c. Synthesis of selenium and silver nanobiocomposites and their influence on phytopathogenic bacterium *Clavibacter michiganensis* subsp. *sepedonicus*. Russ. Chem. Bull. 67(1): 157–163. doi: 10.1007/s11172-018-2052-4.

Perfileva, A.I., Nozhkina, O.A., Graskova, I.A., Dyakova, A.V., Pavlova, A.G., Aleksandrova, G.P., Klimenkov, I.V., Sukhov, B.G. and Trofimov, B.A. 2019. Selenium nanocomposites having polysaccharid matrices stimulate growth of potato plants *in vitro* infected with ring rot pathogen. Dokl Biol. Sci. 489: 184–188. doi: 10.31857/S0869-56524893325-330.

Perfileva, A.I., Nozhkina, O.A., Tretyakova, M.S., Graskova, I.A., Klimenkov, I.V., Sudakov, N.P., Alexandrova, G.P. and Sukhov, B.G. 2020. Biological activity and safety for the environment of selenium nanoparticles encapsulated in starch macromolecules. Nanotechnologies Russ. 15(1): 96–104. doi:10.1134/S1995078020010152.

Perfileva, A.I., Tsivileva, O.M., Nozhkina, O.A., Karepova. M.S., Ganenko, T.V., Sukhov, B.G. and Krutovsky, K.V. 2021a. Effect of natural polysaccharide matrix-based selenium nanocomposites on *Phytophthora cactorum* and rhizospheric microorganisms. Nanomaterials 11: 2274. doi: 10.3390/nano11092274.

Perfileva, A.I., Nozhkina, O.A., Ganenko, T.V., Graskova, I.A., Sukhov, B.G., Artem'ev, A.V., Trofimov, B.A. and Krutovsky, K.V. 2021b. Selenium nanocomposites in natural matrices as potato recovery agent. Int. J. Mol. Sci. 22: 4576. doi: 10.3390/ijms22094576.

Pilon-Smits, E.A.H. 2019. On the ecology of selenium accumulation in plants. Plants 8(7): E197. doi: 10.3390/plants8070197.

Plucinski, A., Lyu, Z. and Schmidt, B.V.K.J. 2021. Polysaccharide nanoparticles: from fabrication to applications. J. Mater Chem. B. 9(35): 7030–7062. doi: 10.1039/d1tb00628b.

Prasad, K.S., Patel, H., Patel, T., Patel, K. and Selvaraj, K. 2013. Biosynthesis of Se nanoparticles and its effect on UV-induced DNA damage. Colloids Surf B Biointerfaces 103: 261–266. doi: 10.1016/j.colsurfb.2012.10.029.

Prateeksha, Singh B.R., Shoeb, M., Sharma, S., Naqvi, A.H., Gupta, V.K. and Brahma, N. 2017. Singh scaffold of selenium nanovectors and honey phytochemicals for inhibition of *Pseudomonas aeruginosa* quorum sensing and biofilm formation. Front Cell Infect. Microbiol. 7: 93. doi: 10.3389/fcimb.2017.00093.

Presentato, A., Piacenza, E., Anikovskiy, M., Cappelletti, M., Zannoni, D. and Turner, R.J. 2018. Biosynthesis of selenium-nanoparticles and - nanorods as a product of selenite bioconversion by the aerobic bacterium *Rhodococcus aetherivorans* BCP1. New Biotechnol. 41: 1–8. doi: 10.1016/j.nbt.2017.11.002.

Pyrzynska, K. 2002. Determination of selenium species in environmental samples. Mikrochim Acta. 140: 55–62. doi:10.15224/978-1-63248-055-2-14.

Quiterio-Gutiérrez, T., Ortega-Ortiz, H., Cadenas-Pliego, G., Hernández-Fuentes, A.D., Sandoval-Rangel, A., Benavides-Mendoza, A., Cabrera-de la Fuente, M. and Juárez-Maldonado, A. 2019. The application of selenium and copper nanoparticles modifies the biochemical responses of tomato plants under stress by *Alternaria solani*. Int. J. Mol. Sci. 20(8): E1950. doi: 10.3390/ijms20081950.

Rady, M.M., Desoky, E.M., Ahmed, S.M., Majrashi, A., Ali, E.F., Arnaout, S.M.A.I. and Selem, E. 2021. Foliar nourishment with nano-selenium dioxide promotes physiology, biochemistry, antioxidant defenses, and salt tolerance in *Phaseolus vulgaris*. Plants 10(6): 1189. doi: 10.3390/plants10061189.

Ragab, G. and Saad-Allah, K. 2020. Green synthesis of sulfur nanoparticles using *Ocimum basilicum* leaves and its prospective effect on manganese-stressed *Helianthus annuus* (L.) seedlings. Ecotoxicol. Environ. Saf. 191: 110242. doi: 10.1016/j.ecoenv.2020.110242.

Rai, M., Yadav, A., Bridge, P. and Gade, A. 2009. Myconanotechnology: a new and emerging science. pp. 258–267. *In*: Rai, M. and Bridge, P. (eds.). Applied Mycology. CABI, UK.

Rai, M., Ingle, A.P. and Paralikar, P. 2016. Sulfur and sulfur nanoparticles as potential antimicrobials: from traditional medicine to nanomedicine. Expert Rev. Anti Infect. Ther. 14(10): 969–978. doi: 10.1080/14787210.2016.1221340.

Rai, M., Pandit, R., Paralikar, P., Shende, S., Gaikwad, S., Ingle, A.P. and Gupta, I. 2017. Nanoparticles as therapeutic agent for treatment of bacterial infections. pp. 191–208. *In*: Essential Oils and Nanotechnology for Treatment of Microbial Diseases. CRC Press.

Ranjitha, V.R. and Ravishankar, V.R. 2018. Extracellular synthesis of selenium nanoparticles from an *Actinomycetes Streptomyces* griseoruber and evaluation of its cytotoxicity on HT-29 cell line. Pharm. Nanotechnol. 6(1): 61–68. doi: 10.2174/2211738505666171113141010.

Rao, K. and Paria, S. 2013. Use of sulfur nanoparticles as a green pesticide on *Fusarium solani* and *Venturia inaequalis* phytopathogens. RSC Adv. 3(26): 10471–10478. doi: 10.1039/c3ra40500a.

Rasouli, M. 2019. Biosynthesis of selenium nanoparticles using yeast *Nematospora coryli* and examination of their anti-candida and anti-oxidant activities. IET Nanobiotechnol. 13(2): 214–218. doi: 10.1049/iet-nbt.2018.5187.

Raza, A., Razzaq, A., Mehmood, S.S., Zou, X., Zhang, X., Lv, Y. and Xu, J. 2019. Impact of climate change on crops adaptation and strategies to tackle its outcome: A review. Plants 8(2): 34. doi: 10.3390/plants8020034.

Rivero, R.M., Mittler, R., Blumwald, E. and Zandalinas, S.I. 2022. Developing climate-resilient crops: improving plant tolerance to stress combination. Plant J. 109(2): 373–389. doi: 10.1111/tpj.15483.

Rosenfeld, C.E., Sabuda, M.C., Hinkle, M.A.G., James, B.R. and Santelli, C.M. 2020. A fungal-mediated cryptic selenium cycle linked to manganese biogeochemistry. Environ. Sci. Technol. 54(6): 3570–3580. doi: 10.1021/acs.est.9b06022.

Saad-Allah, K. and Ragab, G. 2020. Sulfur nanoparticles mediated improvement of salt tolerance in wheat relates to decreasing oxidative stress and regulating metabolic activity. Physiol. Mol. Biol. Plants 26(11): 2209–2223. doi:10.1007/s12298-020-00899-8.

Sager, M. 2006. Selenium in agriculture, food, and nutrition. Pure Appl. Chem. 78: 111–133. doi:10.1351/pac200678010111.

Saikia, S. and Lens, P.N.L. 2020. Synthesis and application of sulfur nanoparticles. pp. 445–475. *In*: Environmental Technologies to Treat Sulfur Pollution: Principles and Engineering (Canada IWA Publishing). doi: 10.2166/9781789060966_0445.

Salem, N., Albanna, L., Awwad, A., Ibrahim, Q. and Abdeen, A. 2016. Green synthesis of nano-sized sulfur and its effect on plant growth. J. Agric. Sci. 8(1): 188–194. doi: 10.5539/jas.v8n1p188.

Sali, A., Zeka, D., Fetahu, S., Rusinovci, I. and Kaul, H. 2018. Selenium supply affects chlorophyll concentration and biomass production of maize (*Zea mays* L.). Die Bodenkultur: Journal of Land Management, Food and Environment 69(4): 249–255. doi: 10.2478/boku-2018-0021.

Samrat, K., Chandraprabha, M.N., Hari Krishna, R., Sharath, R. and Harish, B.G. 2021. Biogenic synthesis of nano-sulfur using *Punica granatum* fruit peel extract with enhanced antimicrobial activities for accelerating wound healing. Nano Futures 5(4): 045003. doi: 10.1088/2399-1984/ac279b.

Schiavon, M. and Pilon-Smits, E.A. 2017. The fascinating facets of plant selenium accumulation - biochemistry, physiology, evolution and ecology. New Phytol. 213(4): 1582–1596. doi: 10.1111/nph.14378.

Shakibaie, M., Forootanfar, H., Golkari, Y., Mohammadi-Khorsand, T. and Shakibaie, M.R. 2015. Anti-biofilm activity of biogenic selenium nanoparticles and selenium dioxide against clinical isolates of *Staphylococcus aureus, Pseudomonas aeruginosa* and *Proteus mirabilis*. J. Trace Elem. Med. Biol. 29: 235–41. doi: 10.1016/j. jtemb.2014.07.020.

Shankar, S., Pangeni, R., Park, J.W. and Rhim, J.-W. 2018. Preparation of sulfur nanoparticles and their antibacterial activity and cytotoxic effect. Mater Sci. Eng. C. 92: 508–517. doi: 10.1016/j.msec.2018.07.015.

Sharma, G., Sharma, A.R., Bhavesh, R., Park, J., Ganbold, B., Nam, J.S. and Lee, S.S. 2014. Biomolecule-mediated synthesis of selenium nanoparticles using dried *Vitis vinifera* (raisin) extract. Molecules 19(3): 2761–2770. doi: 10.3390/molecules19032761.

Shoeibi, S., Mozdziak, P. and Golkar-Narenji, A. 2017. Biogenesis of selenium nanoparticles using green chemistry. Top Curr. Chem. (Cham) 375(6): 88. doi: 10.1007/s41061-017-0176-x.

Siddiqui, S.A., Blinov, A.V., Serov, A.V., Gvozdenko, A.A., Kravtsov, A.A., Nagdalian, A.A., Raffa, V.V., Maglakelidze, D.G., Blinova, A.A. and Kobina, A.V. 2021. Effect of selenium nanoparticles on germination of hordéum vulgáre barley seeds. Coatings 11(7): 862. doi: 10.3390/coatings11070862.

Sidhu, A., Sethi, G., Bala, A. and Ahuja, R. 2021. Evaluation of the myco-toxicities of silver sulfide nanoparticles against phytopathogenic fungi. Agric. Res. J. 58(5): 821–827. doi: 10.5958/2395-146X.2021.00117.4.

Singh, R.K. and Narayan, J. 1990. Pulsed-laser evaporation technique for deposition of thin films: physics and theoretical model. Phys. Rev. B. 41(13): 8843–8859. doi: 10.1103/PhysRevB.41.8843.

Skabara, P.J. 2004. Oxygen, sulfur, selenium and tellurium. Annu. Rep. Prog. Chem. Sect. A: Inorg. Chem. 100: 113–129. doi: 10.1039/b311782k.

Sosedova, L.M., Rukavishnikov, V.S., Sukhov, B.G., Borovsky, G.B., Titov, E.A., Novikov, M.A., Vokina, V.A., Yakimova, N., Lesnichay, M.V., Konkova, T.V., Borovskaya, M.K., Graskova, I.A., Perfileva, A.I. and Trofimov, B.A. 2018. Synthesis of chalcogen-containing nanocomposites of selenium and tellurium with arabinogalactan and a study of their toxic and antimicrobial properties. Nanotechnologies Russ. 13(5-6): 290–294. doi:10.1134/S1995078018030175.

Sotoodehnia-Korani, S., Iranbakhsh, A., Ebadi, M., Majd, A. and Oraghi Ardebili, Z. 2020. Selenium nanoparticles induced variations in growth, morphology, anatomy, biochemistry, gene expression, and epigenetic DNA methylation in *Capsicum annuum;* an *in vitro* study. Environ. Pollut. 265(Pt B): 114727. doi: 10.1016/j. envpol.2020.114727.

Sowndarya, P., Ramkumar, G. and Shivakumar, M.S. 2017. Green synthesis of selenium nanoparticles conjugated *Clausena dentata* plant leaf extract and their insecticidal potential against mosquito vectors. Artif. Cells Nanomed. Biotechnol. 45(8): 1490–1495. doi: 10.1080/21691401.2016.1252383.

Stadtman, T.C. 1990. Selenium biochemistry. Annu. Rev. Biochem. 59: 111–127. doi: 10.1146/annurev. bi.59.070190.000551.

Subramanian, K.S., Rajeswary, R., Yuvaray, M., Pradeep, D., Guna, M. and Yoganathan, G. 2021. Synthesis and characterization of nano-sulfur and its impact on growth, yield, and quality of sunflower (*Helianthus annuus* L.). Commun. Soil Sci. Plant Anal. doi: 10.1080/00103624.2022.2072867.

Surai, P.F. and Kochish, I.I. 2020. Food for thought: nano-selenium in poultry nutrition and health. Anim. Health Res. Rev. 23: 1–5. doi: 10.1017/S1466252320000183.

Tan, Y., Yao, R., Wang, R., Wang, D., Wang, G. and Zheng, S. 2016. Reduction of selenite to Se(0) nanoparticles by filamentous bacterium *Streptomyces* sp. ES2-5 isolated from a selenium mining soil. Microb. Cell Fact. 15(1): 157. doi: 10.1186/s12934-016-0554-z.

Tan, Y., Wang, Y., Wang, Y., Xu, D., Huang, Y., Wang, D., Wang, G., Rensing, C. and Zheng, S. 2018. Novel mechanisms of selenate and selenite reduction in the obligate aerobic bacterium *Comamonas testosteroni* S44. J. Hazard Mater. 359: 129–138. doi: 10.1016/j.jhazmat.2018.07.014.

Tanaka, Y.K., Takada, S., Kumagai, K., Kobayashi, K., Hokura, A. and Ogra, Y. 2020. Elucidation of tellurium biogenic nanoparticles in garlic, *Allium sativum*, by inductively coupled plasma-mass spectrometry. J. Trace Elem. Med. Biol. 62: 126628. doi: 10.1016/j.jtemb.2020.126628.

Tarrahi, R., Khataee, A., Movafeghi, A., Rezanejad, F. and Gohari, G. 2017. Toxicological implications of selenium nanoparticles with different coatings along with Se$_{4+}$ on *Lemna minor.* Chemosphere 181: 655–665. doi: 10.1016/j.chemosphere.2017.04.142.

Taylor, D.E. 1999. Bacterial tellurite resistance. Trends Microbiol. 7: 111–115.

Terry, N., Zayed, A.M., de Souza, M.P. and Tarun, A.S. 2000. Selenium in greater plants. Annu. Rev. Plant Physiol. 51: 401–432. doi: 10.1146/annurev.arplant.51.1.401.

Torres, M.E.S., Morales, A.R., Solano-Peralta, A. and Kroneck, P.M. 2020. Sulfur, the versatile non-metal. Met. Ions Life Sci. 20: 19–50. doi:10.1515/9783110589757-008.

Tran, P.A. and Webster, T.J. 2011. Selenium nanoparticles inhibit *Staphylococcus aureus* growth. Int. J. Nanomed. 6: 1553. doi: 10.2147/IJN.S21729.

Tran, P.A. and Webster, T.J. 2013. Antimicrobial selenium nanoparticle coatings on polymeric medical devices. Nanotechnology 24(15): 155101. doi: 10.1088/0957-4484/24/15/155101.

Tran, P.A., O'Brien-Simpson, N., Reynolds, E.C., Pantarat, N., Biswas, D.P. and O'Connor, A.J. 2016. Low cytotoxic trace element selenium nanoparticles and their differential antimicrobial properties against *S. aureus* and *E. coli*. Nanotechnology 27(4): 045101. doi: 10.1088/0957-4484/27/4/045101.

Trippe, R.C. and Pilon-Smits, E.A.H. 2021. Selenium transport and metabolism in plants: Phytoremediation and biofortification implications. J. Hazard Mater. 404(Pt B): 124178. doi: 10.1016/j.jhazmat.2020.124178.

Tsivileva, O.M. and Perfileva, A.I. 2017. Selenium compounds biotransformed by mushrooms: not only dietary sources, but also toxicity mediators. Curr. Nutr. Food Sci. 13(2): 82–96. doi: 10.2174/1573401313666170117 144547.

Tsivileva, O.M., Perfileva, A.I., Ivanova, A.A., Pozdnyakov, A.S. and Prozorova, G.F. 2021. The effect of selenium- or metal-nanoparticles incorporated nanocomposites of vinyl triazole based polymers on fungal growth and bactericidal properties. J. Polym. Environ. 29: 1287–1297. doi:10.1007/s10924-020-01963-w.

Tsivileva, O.M. and Perfileva, A.I. 2022. Mushroom-derived novel selenium nanocomposites' effects on potato plant growth and tuber germination. Molecules 27(14): 4438. doi: 10.3390/molecules27144438.

Tugarova, A.V., Mamchenkova, P.V., Dyatlova, Y.A. and Kamnev, A.A. 2018. FTIR and Raman spectroscopic studies of selenium nanoparticles synthesised by the bacterium *Azospirillum thiophilum*. Spectrochim Acta A Mol. Biomol. Spectrosc. 192: 458–463. doi: 10.1016/j.saa.2017.11.050.

Ullah, H., Liu, G., Yousaf, B., Ali, M.U., Irshad, S., Abbas, Q. and Ahmad, R. 2019. A comprehensive review on environmental transformation of selenium: recent advances and research perspectives. Environ. Geochem. Health. 41(2): 1003–1035. doi: 10.1007/s10653-018-0195-8.

Ulricha, K. and Jakoba, U. 2019. The role of thiols in antioxidant systems. Free Radic. Biol. Med. 140: 14–27. doi: 10.1016/j.freeradbiomed.2019.05.035.

Vahidi, H., Kobarfard, F., Alizadeh, A., Saravanan, M. and Barabadi, H. 2021. Green nanotechnology-based tellurium nanoparticles: Exploration of their antioxidant, antibacterial, antifungal and cytotoxic potentials against cancerous and normal cells compared to potassium tellurite. Inorg. Chem. Commun. 124: 108385. doi: 10.1016/j.inoche.2020.108385.

Vaigankar, D.C., Dubey, S.K., Mujawar, S.Y. and D'Costa, A.S.K.S. 2018. Tellurite biotransformation and detoxification by *Shewanella baltica* with simultaneous synthesis of tellurium nanorods exhibiting photo-catalytic and anti-biofilm activity. Ecotoxicol. Environ. Saf. 165: 516–526. doi: 10.1016/j.ecoenv.2018.08.111.

van Hoewyk, D. 2013. A tale of two toxicities: malformed selenoproteins and oxidative stress both contribute to selenium stress in plants. Ann. Bot. 112(6): 965–972. doi: 10.1093/aob/mct163.

Vávrová, S., Struhárňanská, E., Turňa, J. and Stuchlík, S. 2021. Tellurium: A rare element with influence on prokaryotic and eukaryotic biological systems. Int. J. Mol. Sci. 22(11): 5924. doi: 10.3390/ijms22115924.

Wadhwani, S.A., Shedbalkar, U.U., Singh, R. and Chopade, B.A. 2016. Biogenic selenium nanoparticles: current status and future prospects. Appl. Microbiol. Biotechnol. 100(6): 2555–2566. doi: 10.1007/s00253-016-7300-7.

Wadhwani, S.A., Gorain, M., Banerjee, P., Shedbalkar, U.U., Singh, R., Kundu, G.C. and Chopade, B.A. 2017. Green synthesis of selenium nanoparticles using Acinetobacter sp. SW30: optimization, characterization and its anticancer activity in breast cancer cells. Int. J. Nanomedicine 12: 6841–6855. doi: 10.2147/IJN.S139212.

Wang, C., Cheng, T., Liu, H., Zhou, F., Zhang, J., Zhang, M., Liu. X., Shi, W. and Cao, T. 2021. Nano-selenium controlled cadmium accumulation and improved photosynthesis in indica rice cultivated in lead and cadmium combined paddy soils. Res. J. Environ. Sci. 103: 336–346. doi: 10.1016/j.jes.2020.11.005.

Wang, K., Zhang, X., Kislyakov, I.M., Dong, N., Zhang, S., Wang, G., Fan, J., Zou, X., Du, J., Leng, Y., Zhao, Q., Wu, K., Chen, J., Baesman, S.M., Liao, K.S., Maharjan, S., Zhang, H., Zhang, L., Curran, S.A., Oremland, R.S., Blau, W.J. and Wang, J. 2019a. Bacterially synthesized tellurium nanostructures for broadband ultrafast nonlinear optical applications. Nat. Commun. 10(1): 3985. doi: 10.1038/s41467-019-11898-z.

Wang, S., Guan, W., Ma, D., Chen, X., Wan, L., Huang, S. and Wang, J. 2010. Synthesis, characterization and optical properties of flower-like tellurium. CrystEngComm. 12(1): 166–171. doi:10.1039/B905053C.

Wang, X., Zhang, D., Pan, X., Lee, D.J., Al-Misned, F.A., Mortuza, M.G. and Gadd, G.M. 2017. Aerobic and anaerobic biosynthesis of nano-selenium for remediation of mercury contaminated soil. Chemosphere 170: 266–273. doi: 10.1016/j.chemosphere.2016.12.020.

Wang, Y., Yan, X. and Fu, L. 2013. Effect of selenium nanoparticles with different sizes in primary cultured intestinal epithelial cells of crucian carp, *Carassius auratus gibelio.* Int. J. Nanomed. 8: 4007–4013. doi: 10.2147/IJN. S43691.

Wang, Y.Q., Zhu, L.N., Li, K., Wang, Q., Wang, K., Guo, Y.B. and Li, H.F. 2019b. Absorption and transportation of selenium nanoparticles in wheat and rice. Huan Jing Ke Xue 40(10): 4654–4660. doi: 10.13227/j.hjkx.201904048 (in Chinese).

White, P.J. 2016. Selenium accumulation by plants. Ann. Bot. 117(2): 217–235. doi: 10.1093/aob/mcv180.

White, P.J. 2018. Selenium metabolism in plants. Biochim. Biophys. Acta 1862(11): 2333–2342. doi: 10.1016/j.bbagen.2018.05.006.

Winkel, L., Vriens, B., Jones, G.D., Schneider, L.S., Pilon-Smits, E.A.H. and Bañuelos, G.S. 2015. Selenium cycling across soil-plant-atmosphere interfaces: A critical review. Nutrients 7: 4199–4239. doi: 10.3390/nu7064199.

Xi, G., Xiong, K., Zhao, Q., Zhang, R., Zhang, H. and Qian, Y. 2006. Nucleation-dissolutionrecrystallization: A new growth mechanism for t-selenium nanotubes. Cryst. Growth Des. 6: 577–582. doi: 10.1021/cg050444c.

Yanchatuña Aguayo, O.P., Mouheb, L., Villota Revelo, K., Vásquez-Ucho, P.A., Pawar, P.P., Rahman, A., Jeffryes, C., Terencio, T. and Dahoumane, S.A. 2022. Biogenic sulfur-based chalcogenide nanocrystals: methods of fabrication, mechanistic aspects, and bio-applications. Molecules 27(2): 458. doi: 10.3390/molecules27020458.

Yang, H., Yang, X., Ning, Z., Kwon, S.Y., Li, M.-L., Tack, F.M.G., Kwon, E.E., Rinklebe, J. and Yin, R. 2022. The beneficial and hazardous effects of selenium on the health of the soil-plant-human system: An overview. J. Hazard Mater. 422: 126876. doi: 10.1016/j.jhazmat.2021.126876.

Yazhiniprabha, M. and Vaseeharan, B. 2019. *In vitro* and *in vivo* toxicity assessment of selenium nanoparticles with significant larvicidal and bacteriostatic properties. Mater Sci. Eng. C Mater Biol. Appl. 103: 109763. doi: 10.1016/j.msec.2019.109763.

You, Y., Ahsan, K. and Detty, M.R. 2003. Mechanistic Studies of the tellurium(ii)/tellurium(iv) redox cycle in thiol peroxidase-like reactions of diorganotellurides in methanol. J. Am. Chem. Soc. 125: 4918–4927. doi: 10.1021/ja029590m.

Yu, Y., Liu, Y., Luo, S. and Peng X. 2003. Effects of selenium on soybean chloroplast ultra-structure and microelement content of soybean leaves under continuous cropping stress. Ying Yong Sheng Tai Xue Bao (Chin. J. App. Ecol.) 14(4): 573–576. PMID: 12920905.

Yu, S., Zhang, W., Liu, W., Zhu, W., Guo R., Wang, Y., Zhang, D. and Wang, J. 2015. The inhibitory effect of selenium nanoparticles on protein glycation *in vitro*. Nanotechnology 26: 145703. doi: 10.1088/0957-4484/26/14/145703.

Zachara, B.A. and Pilecki, A. 2000. Selenium concentration in the milk of breast-feeding mothers and its geographic distribution. Environ. Health Persp. 10: 1043–1046. doi: 10.1289/ehp.001081043.

Zahedi, S.M., Abdelrahman, M., Hosseini, M.S., Hoveizeh, N.F. and Tran, L.P. 2019. Alleviation of the effect of salinity on growth and yield of strawberry by foliar spray of selenium-nanoparticles. Environ. Pollut. 253: 246–258. doi: 10.1016/j.envpol.2019.04.078.

Zambonino, M.C., Quizhpe, E.M., Jaramillo, F.E., Rahman, A., Santiago, Vispo, N., Jeffryes, C. and Dahoumane, S.A. 2021. Green synthesis of selenium and tellurium nanoparticles: Current trends, biological properties and biomedical applications. Int. J. Mol. Sci. 22(3): 989. doi: 10.3390/ijms22030989.

Zhang, C., Zhang, P., Ji, X., Wang, H., Kuang, H., Cao, W., Pan, M., Shi, Y.-E. and Wang, Z. 2019. Ultrasonication-promoted synthesis of luminescent sulfur nano-dots for cellular imaging applications. Chem. Commun. (Camb). 55(86): 13004–13007. doi: 10.1039/c9cc06586e.

Zhang, J., Wang, Y., Shao, Z., Li, J., Zan, S., Zhou, S. and Yang, R. 2019. Two selenium tolerant *Lysinibacillus* sp. strains are capable of reducing selenite to elemental Se efficiently under aerobic conditions. J. Environ. Sci. (China) 77: 238–249. doi: 10.1016/j.jes.2018.08.002.

Zhang, T., Doert, T., Schwedtmann, K., Weigand, J.J. and Ruck, M. 2020. Facile synthesis of tellurium nano- and microstructures by trace HCl in ionic liquids. J. Chem. Soc, Dalton Trans. 49: 1891–1896. doi: 10.1039/C9DT04604F.

Zhang, W., Zhang, J., Ding, D., Zhang, L., Muehlmann, L.A., Deng, S.E., Wang, X., Li, W. and Zhang, W. 2018. Synthesis and antioxidant properties of *Lycium barbarum* polysaccharides capped selenium nanoparticles using tea extract. Artif. Cells Nanomed. Biotechnol. 46(7): 1463–1470. doi: 10.1080/21691401.2017.1373657.

Zhou, X., Yang, J., Kronzucker, H.J. and Shi, W. 2020. Selenium biofortification and interaction with other elements in plants: A review. Front Plant Sci. 11: 586421. doi: 10.3389/fpls.2020.586421.

Zhou, W. and Greer H.F. 2016. What can electron microscopy tell us beyond crystal structures? Eur. J. Inorg. Chem. 7: 941–950. doi: 10.1002/ejic.201501342.

Zhu, M., Niu, G. and Tang, J. 2019. Elemental Se: Fundamentals and its optoelectronic applications. J. Mater. Chem. C. 7: 2199–2206. doi: 10.1039/C8TC05873C.

Zoidis, E., Seremelis, I., Kontopoulos, N. and Danezis, G.P. 2018. Selenium-dependent antioxidant enzymes: actions and properties of selenoproteins. Antioxidants 7(5): 66. doi: 10.3390/antiox7050066.

Zonaro, E., Lampis, S., Turner, R.J., Qazi, S.J.S. and Vallini, G. 2015. Biogenic selenium and tellurium nanoparticles synthesized by environmental microbial isolates efficaciously inhibit bacterial planktonic cultures and biofilms. Front Microbiol. 584. doi: 10.3389/fmicb.2015.00584.

Zonaro, E., Piacenza, E., Presentato, A., Monti, F., Dell', A.R., Lampis, S. and Vallini, G. 2017. *Ochrobactrum* sp. MPV1 from a dump of roasted pyrites can be exploited as bacterial catalyst for the biogenesis of selenium and tellurium nanoparticles. Microb. Cell Fact. 16(1): 215. doi: 10.1186/s12934-017-0826-2.

Zsiros, O., Nagy, V., Párducz, Á., Nagy, G., Ünnep, R., El-Ramady, H., Prokisch, J., Lisztes-Szabó, Z., Fári, M., Csajbók, J., Tóth, S., Garab, G. and Domokos-Szabolcsy, É. 2019. Effects of selenate and red Se-nanoparticles on the photosynthetic apparatus of *Nicotiana tabacum*. Photosynth. Res. 139: 449–460. doi: 10.1007/s11120-018-0599-4.

Chapter 22

Toxicity of Nanomaterials to Plants

Joanna Trzcińska-Wencel,[1] *Patrycja Golińska*[1,*] and *Mahendra Rai*[1,2]

Introduction

As nanomaterial production and applications in industry, medicine, agriculture, and environment protection sectors continuously expand, nanomaterial residues are present in the environment and generate concerns regarding their potentially adverse impact on non-target organisms (Montes et al. 2017). The diversity of nanomaterials (NMs) includes inorganic nanoparticles (metal and metal oxide), carbon-based nanomaterials (graphene, graphene oxide, fullerenes, and carbon nanotubes), other organic nanostructures (liposomes and polymer-based) and a wide range of combinations (nanocomposites). The potential phytotoxicity of NMs in plants at different levels is shown in Figure 1. Their activity is associated with unique characteristics (extremely small size, high surface area to volume ratio, variety of shapes, coating agents, and many others) (Dhyani et al. 2022). Nevertheless, their toxic effect is determined by the exposure conditions and the nature of the plants. A number of techniques and methods are used for toxicity assessment and the fate of NMs in plant organisms. Identification of interactions at the cellular, subcellular, and molecular

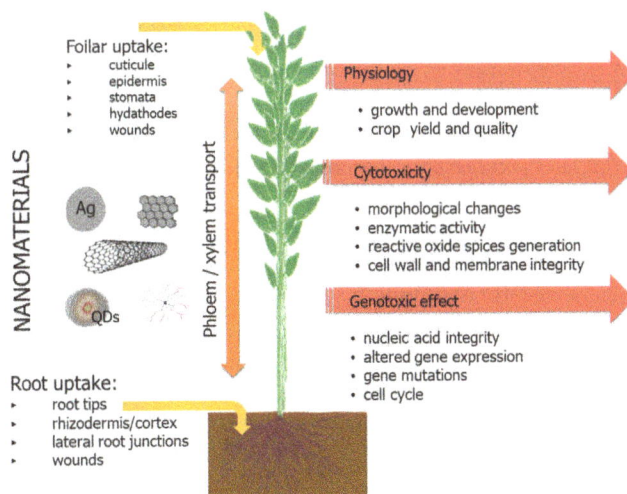

Figure 1. Uptake, transport, and potential toxic effect of various nanomaterials on plants.

[1] Faculty of Biological and Veterinary Sciences, Nicolaus Copernicus University, Toruń, Poland.
[2] Nanobiotechnology Research Lab., Department of Biotechnology, Sant Gadge Baba Amravati University, Amravati, India.
* Corresponding author: golinska@umk.pl

levels provides a particular understanding of toxicity to specific tissues and organisms, which facilitates the advancement of the safety of NM use (Barabadi et al. 2019). In some conditions, NMs negatively affect the life cycle of plants and alter seed germination, seedling growth, and organs and impair the efficiency of processes, such as photosynthesis, flowering, and yield (Ruttkay-Nedecky et al. 2017). However, the genomics and proteomics studies, combined with biochemical analyses, confirm a range of plant responses to the NMs, indicating substances and processes that mitigate NMs-induced stress (Fen et al. 2022).

Diversity of Nanomaterials Used for Plant Protection

Nanomaterials have found applications for monitoring and detecting plant diseases and also as plant protection products with antimicrobial, antifungal, and anti-insecticide properties. In addition, some of NMs in the form of nano-fertilizers or hydrogels, are used for improving soil quality and facilitating nutrient availability for plants (Dhyani et al. 2022).

Nanomaterials for Biosensing of Plant Pathogens

It is essential to detect plant pathogens before disease symptoms appear. To ensure proper prevention, diagnosis, and management of plant diseases and pests, different types of developments involving nanotechnology products, such as metal nanoparticles, carbon nanomaterials, hybrid nanocomposites, etc., are used (Babu et al. 2022). Nanoparticles (NPs) can be used for pathogen detection and as diagnostic tools to mark specific substances synthesized during plant infections (Anand and Panigrahi 2021; Sellappan et al. 2022). Due to the use of nanomaterials with outstanding chemical and physical properties, it is possible to transform and modify conventional molecular tests into modern and high-performance diagnostic tools that can be directly used in agricultural fields and outside laboratories (Li et al. 2020). Studies on applications of metal nanoparticles and carbon nanomaterials to plant pathogen biosensors have been presented in Table 1. Two types of nanomaterials are increasingly being considered for use in biosensors, namely gold nanoparticles and carbon nanotubes. Gold nanoparticles (AuNPs) demonstrate a variety of electrical, optical, and catalytical properties as well as the ability to bind molecules, such as nucleic acids, peptides, proteins, and polymers. These make them ideal for application in biosensing devices (Yadav et al. 2020). Thus, Zhan and coworkers (2018) have developed AuNPs-based biosensor for the detection of *Phytophthora infestans* with a detection threshold of 0.1 pg μL^{-1} of pathogen gDNA. In another study, AuNPs were used by Lei et al. (2021) to produce a dynamic microcantilever biosensor for the rapid detection of *Leptosphaeria maculans* responsible for severe rapeseed yield losses. Moreover, Yang and coworkers (2015) summarized the advantages of a carbon nanotube-based biosensors, pointing out the high sensitivity and fast reaction rate, the ability to immobilize enzymes without the loss of their biological activity, the low effect of contamination, and long-term stability. The high potential of carbon nanotubes (CNTs) for use in biosensors is attributed to their properties, including large surface area ratio and hollow tube or capacity to mediate rapid electron transfer kinetics. For example, semiconducting single-walled carbon nanotubes with anti-SDE1 have been used in a chemoreceptive biosensor for the early detection of Huanglongbing (HLB) in citrus trees. The device demonstrated improved sensitivity compared to the standard ELISA test, the minimum required concentration for assessment was in a range from 3 nM to 2.6 µM, while the detection limit for ELISA ranged between 1.4 nM to 140 nM (Tran et al. 2020).

Nanomaterials for Controlling Plant Diseases

In addition, the antimicrobial activity of nanomaterials has been widely studied against important plant pathogens, indicating their use for pest and disease control, as shown in Table 2. In the number of nanomaterials antimicrobial activity of carbon nanomaterials (Lipsa et al. 2020; El-Abeid et al. 2020), polymeric nanostructures (Oh et al. 2019; Vanti et al. 2020; Ahmed et al. 2021a) and

Table 1. Nanomaterials used for the detection of plant pathogens.

Type of Nanomaterial	Biosensor Format	Plant	Target	References
AuNPs	Lateral flow biosensor	Potato (*Solanum tuberosum*)	DNA of *Phytophthora infestans*, *Alternaria solani*, and *Rhizoctonia solani*	Zhan et al. 2018
AuNPs	Dynamic microcantilever biosensor	Oilseed rape (*Brassica oleracea*)	DNA of *Leptosphaeria maculans*	Lei et al., 2021
AuNPs	Integrated RPA and an AuNP probe	Tomato and genus *Solanum*	DNA of tomato yellow leaf curl virus	Wang and Yang 2019
CNPs	Fluorometric-based nanobiosensor	Citrus	Citrus tristeza virus	Shojaei et al. 2016
rGO and ZnONPs	Electrochemical biosensor	Oil palm	Chemical markers for early detection of Basal stem rot disease (induced by *Ganoderma boninense*)	Rahmat et al. 2020
MWCNTs and CuNPs	Electrochemical biosensor	Cotton	DNA viruses that belong to the genera *Begomovirus* of the family *Geminiviridae*	Tahir et al. 2018
MWCNTs	Chemical sensors based on CNTs	Strawberry (*Fragaria x ananassa*)	Volatile organic compounds (VOC) of *Aspergillus* sp. *Rhizopus* sp.	Greenshields et al. 2016
SWNTs	Electrical gas sensor with SWNTs functionalized with ssDNA	Citrus trees	Biomarkers of citrus huanglongbing disease caused by *Candidatus Liberibacter* (ethyl hexanol, linalool, tetradecene, and phenylacetaldehyde)	Wang et al. 2019
SWNTs	SWNT-based chemiresistive biosensor	Citrus trees	Label-free detection and quantification of protein-based biomarker SDE1 for citrus greening	Tran et al. 2020

AuNPs, gold nanoparticles; CNPs, carbon nanoparticles; rGO, reduced graphene oxide; ZnONPs, zinc oxide nanoparticles; SWCNTs, single-walled carbon nanotubes; MWCNTs, multi-walled carbon nanotubes.

nanoparticles based on silver, magnesium, zinc, and copper (Cai et al. 2018; Pham et al. 2019; Win et al. 2020; El-Batal et al. 2020; Irshad et al. 2021) were studied. Khan et al. (2021) reported a comparative investigation of the antiparasitic and antimicrobial activity of silver nanoparticles (AgNPs) on nematode *Meloidogyne incognita*, bacterium *Ralstonia solanacearum*, and fungus *Fusarium oxysporum*. Their results proved that biologically synthesized rectangular AgNPs with sizes ranging between 55–70 nm at a concentration of 100 µg/mL exhibited antinematode, antifungal, and antibacterial activity. These findings imply the potential of bio-AgNPs for application for controlling plant pathogens (Khan et al. 2021). Furthermore, El-Abeid et al. (2020) have identified a potential mechanism of action for the reduced graphene oxide-copper nanoparticle (rGO-CuONP) nanocomposite against *Fusarium oxysporum,* a causal agent of wilt in tomato and pepper. Microscopic observation illustrated strong damage and the appearance of numerous cavities in the cell wall surface in hyphae, macro- and microconidia, and chlamydospores. This action may be due to the non-specific interaction of rGO with the cell wall and/or the effect of CuONPs on biomolecules present on the cell wall surface including phosphorus or sulfur in their structure (El-Abeid et al. 2020).

Interestingly, several reports also provide insight into the simultaneous antimicrobial and pathogen resistance-enhancing effects in plants of nanomaterials based on iron oxide, copper, magnesium, and zinc (Bilesky-Jose et al. 2021; Haggag and Eid 2022; Abdelkhalek and Al-Askar 2020; Shende et al. 2021). Therefore, chemically prepared sulfur nanoparticles (SNPs) were evaluated for enhancement of the immune response of *Fusarium oxysporum*-infected tomato.

Table 2. Biological activity of nanomaterials in plant protection against pathogens.

Nanomaterial	Synthesis Method	Average Size/ Zeta Potential	Activity	References
α-Fe$_2$O$_3$ nanoparticles	Biological from *Trichoderma harzianum*	207 nm/–13 mV	Enhancement of biological activity of *Trichoderma* against *Sclerotinia sclerotiorum*	Bilesky-Jose et al. 2021
AgNPs	Biological from strawberry waste	55 nm	Inhibitory effect on nematode, *Meloidogyne incognita*, bacterium, *Ralstonia solanacearum* and fungus, *Fusarium oxysporum*	Khan et al. 2021
SiO$_2$-NPs	Chemical	76.7 nm	Enhancement of immune response by activation of salicylic acid-dependent defense pathway in *Arabidopsis thaliana* against *Pseudomonas syringae pv.* tomato.	El-Shetehy et al. 2021
Sulfur nanoparticles (SNPs)	Chemical	30 nm/–25.7 mV	Inhibition of *Fusarium oxysporum* f. sp. *Lycopersici* growth and activation of systemic acquired resistance (SAR) in tomatoes	Cao et al. 2021
CuNPs	Biological from *Aspergillus flavus*	8 nm/–26 mV	Antifungal activity against *Aspergillus niger, Fusarium oxysporum,* and *Alternaria alternata*	Shende et al. 2021
ZnONPs	Biological from leaf extract of *Mentha spicata*	74.68 nm	Induction of SAR after foliar treatment of tomato during Tobacco mosaic virus (TMV) infection	Abdelkhalek and Al-Askar 2020
Ag@FeO-NPs@Chitosan composite	Biological from *Streptomyces aureofaciens*	< 50 nm	Inhibition of the growth and spores germination of soilborne pathogens	Haggag and Eid 2022
CuONPs	Biological from *Penicilium chrysogenum*	9.70 nm	Antimicrobial activity against *Fusarium oxysporum* and *Ralstonia solanacearum*	El-Batal et al. 2020
MgONPs	Biological from strawberry	100 nm	Antinematode activity against *Meloidogyne incognita*	Khan et al. 2022
Chitosan guar nanoparticle (CGNP)	Ionic gelation method	122 nm/–30 mV	Inhibition of rice pathogens *Pyricularia grisea* and *Xanthomonas oryzae*	Sathiyabama and Muthukumar 2020

Foliar spraying resulted in 1.43-fold higher disease suppression than commercial fungicide hymexazol; after SNPs treatment activation of the stress response (increased content and activity of disease-related enzymes and antioxidants) was observed (Cao et al. 2021).

How do Nanomaterials Interact With Plants?

In fact, nanomaterials are able to overcome physiological barriers and penetrate plant tissues, negatively affecting plant functions by inducing oxidative stress, damaging cell walls, membranes, and organelles, or interfering with biomolecules crucial to metabolic processes. However, the defense mechanisms of plants against various biotic and abiotic stresses depend considerably on the plant species, growth stage, and other factors. However, the exact mechanism of plant response to nanomaterials is not fully understood (Dev et al. 2018). Nanomaterials might induce oxidative stress and impair the synthesis of molecules essential for plant growth and development. Accordingly, studies mainly focus on the antioxidant system, gene expression, and synthesis of proteins or hormones responsible for alleviating the effects of stress (Ghafari et al. 2020).

Antioxidative System Activity in Response to Oxidative Stress Generated by NMs

Nanomaterials might induce a strong increase in the amount of reactive oxygen species (ROS) as a byproduct of the plant detoxification mechanism. Production of ROS in response to nanomaterials and antioxidant activity is affected by the exposure time of NMs, their type and concentration, as well as the plant species (Dev et al. 2018). As is well known, defense mechanisms based on ROS scavenging in plant systems follow diverse enzymatic and non-enzymatic pathways and are necessary to protect plants from stress factors (Ma et al. 2015). There are three types of superoxide dismutase (SOD) in plants based on the type of catalytic ion, including Fe-SOD, Mn-SOD, and Cu-Zn-SOD. These enzymes participate in the detoxification process by converting $O_2^{\cdot-}$ to the less toxic H_2O_2 and further induction of antioxidant pathways, for instance, catalase (CAT) converts H_2O_2 to H_2O and O_2. In some cases, H_2O_2 can be involved in a Fenton reaction of metal ions (Fe^{2+}, Cu^{2+}) and result in the formation of the hydroxyl radical HO^-, highly reactive ROS (hROS), that can cause irreversible changes in biomolecules, such as lipids, proteins, and nucleic acids (Ma et al. 2015). Another issue is the ascorbate-glutathione (Asada-Halliwell) pathway, where a reaction is mediated by ascorbate peroxidase (APX), which converts H_2O_2 to H_2O by oxidizing ascorbate to two products, monodehydroascorbate (MDHA) and dehydroascorbate (DHA). Finally, glutathione peroxidase (GPX), which is responsible for the production of oxidized disulfide glutathione form (GSSG) with direct conversion of H_2O_2 to H_2O, and glutathione reductase, which reduces the GSSG to monomeric glutathione (GSH) (Dev et al. 2018; Hasanuzzaman et al. 2019). Moreover, non-enzymatic oxidative stress response includes biomolecules, such as anthocyanin, glutathione, ascorbic acid, carotenoids, phenolic compounds, or proline (Ahmad et al. 2010). The activity of antioxidant enzymes is a parameter that represents the oxidative stress defense mechanisms induced by the exposure of plants to nanomaterials. Significantly higher activity of antioxidants enzymes was observed in *Salvia verticillata* after foliar treatment with multiwalled-carbon nanotubes (MWCNTs) at a concentration > 1,000 mg mL^{-1}. Results displayed two times the increased activity of SOD, 2.1-fold higher activity of peroxidase, and 3.9-fold enhancement in catalase activity compared to control samples. In addition, increased content of malondialdehyde (MDA) and proline were also detected (Rahmani et al. 2020). The activity of ROS-scavenging enzymes indicates defense mechanisms that remove H_2O_2 to protect cells from nanomaterial-induced damage (Fan et al. 2018). In another study, the content of anthocyanin, a pigment that scavenges free radicals and chelates metal ions significantly increased in plants, such as *Arabidopsis, Solanum tuberosum*, and *Brassica rapa* ssp. *rapa* after treatment with metal nanoparticles (Qian et al. 2013; Bagherzadeh Homaee and Ehsanpour 2015; Thiruvengadam et al. 2015).

Adaptation to Nanomaterials and Detoxification

The plant responses at the phenotypic and physiological levels are altered by gene expression and protein synthesis. A number of studies were conducted on plants exposed to nanomaterials and the main results are summarized in Table 3. Proteomic analysis of *Triticum aestivum* after AgNPs treatment displayed important changes in the expression of proteins depending on targeted plant tissue. It implicated various responses to AgNPs-stress in leaves and root cells. Increased content of α-amylases, a fructose-bisphosphate aldolase, and aconitate hydratases accumulation in roots and leaves were observed, respectively. Authors suggest that by altering the levels of enzymes involved in energy metabolism, cells can produce more reducing substances to facilitate the AgNPs-stress response. Moreover, the results revealed higher production of enzymes, namely methionine synthase and S-adenosylmethionine synthetase, which are committed to sulfur amino acid biosynthesis (Vannini et al. 2014). While in another study, in both ZnO-NPs and Ag-NPs stressed soybean (*Glycine max* L.) roots more than two-fold increase in the content of methionine gamma-lyase was observed (Hossain et al. 2016). The product of the reaction catalyzed by these enzymes, S-adenosylmethionine, is employed in processes, including the synthesis of the plant hormone ethylene, cell wall, chlorophylls, secondary metabolites, DNA replication, and methylation

Table 3. Plant responses at different levels to various types of nanomaterials.

Tested Plant	Type of Nanomaterial	Concentration	Effects of NMs on Plants	Biochemical and Molecular Mechanisms for the Adverse Effects of NMs on Plants	References
Tomato (*Lycopersicon esculentum*)	AgNPs	10 mg L⁻¹, 20 mg L⁻¹, and 30 mg L⁻¹	Accumulation of Ag in plant tissues reduced plant biomass caused morphological changes in xylem cells and shifted the balance of water and nutrient dynamics	Upregulation by 23.50%, 52.09%, and 7.6% of genes related to membrane transporters H+-ATPase, potassium transporter, and sulfate transporter	Noori et al. 2020
Wheat (*Triticum aestivum*)	AgNPs-PVP coated	10 mg L⁻¹	Adversely affected seedling growth and induced morphological modifications in root tip cells	Altered expression of several proteins mainly involved in primary metabolism and cell defense	Vannini et al. 2014
Tobacco *Nicotiana tabacum*	AgNPs	100 µM	n.a.	Accumulation of pathogenesis-related (PR) proteins involved in modifications of cell wall	Peharec Štefanić 2019
Arabidopsis (*Arabidopsis thaliana*)	AgNPs	1.0 mg L⁻¹, 2.5 mg L⁻¹	Decreased plant fresh weight, inhibited photosynthesis, and delayed flowering	Stimulation of activity of molecules associated with the tricarboxylic acid cycle and in sugar metabolism, reduction of transcription by 25–40% of flowering key genes including AP1, LFY, FT, and SOC1,	Ke et al. 2018
Broad bean (*Vicia faba*)	ZnONPs	10 mg L⁻¹, 25 mg L⁻¹, 100 mg L⁻¹, and 200 mg L⁻¹	Seed germination and seedling growth were enhanced at lower concentrations (10 mg/mL and 25 mg/mL) and inhibited at higher concentrations (100 mg/mL and 200 mg/mL)	Altered expression of peroxidase isoenzymes and α- and β-esterase isoenzymes, induction of chromosomal aberration, and formation of micronuclei and vacuolated nuclei	Youssef and Elamawi 2020
Soyabean (*Glycine max* cultivar Enrei)	ZnONPs	10 ppm	Improved root and hypocotyl growth, higher biomass production	Increased biosynthesis of proteins (273) related to metabolism, stress, photosynthesis, transport, and amino acid metabolism, decreased level of proteins (198) associated with the cell wall, RNA metabolism, and lipid metabolism	Komatsu et al. 2022
Cucumber (*Cucumis sativus*)	CuNPs	50 mg L⁻¹, 100 mg L⁻¹, and 200 mg L⁻¹	Adverse phenotypic alterations, reduced biomass and decreased levels of the photosynthetic pigments in a concentration-dependent manner	Induction of copper-zinc superoxide dismutase (Cu-Zn SOD) gene expression, CuNP at all tested concentrations caused genomic alterations confirmed by RAPD analysis	Mosa et al. 2018
Mustard (*Brassica rapa*)	CuO NPs	50 mg L⁻¹, 250 mg L⁻¹, and 500 mg L⁻¹	Decrease of root and shoot length, significantly reduced after treatment of 250 and 500 mg L⁻¹	Decreased chlorophyll, carotenoid, and sugar content and increased proline, anthocyanins, and malondialdehyde. Upregulation of genes related to oxidative stress enzymes, glucosinolate, and phenolic compounds	Chung et al. 2019

Coriander (*Coriandrum sativum*)	CuNPs	200 mg L^{-1}, 400 mg L^{-1} and 800 mg L^{-1}	Decreased biomass and root length	Induced oxidative stress (higher H$_2$O$_2$ and MDA content) and genotoxicity in genomic DNA confirmed by RAPD technique analysis	Al Quraidi et al. 2019
Lilac sage (*Salvia erticillate*)	MWCNT	50 mg L^{-1}, 100 mg L^{-1}, 250 mg L^{-1}, 500 mg L^{-1}, and 1,000 mg L^{-1}	After foliar application of MWCNT at a concentration > 100 mg mL^{-1} decreased content of photosynthetic pigments	> 100 mg mL^{-1} MWCNT caused increased antioxidant enzymes (SOD, CAT, and POD) activities, overproduction of proline, two-fold higher protein content as well as higher expression of rosmarinic acid synthase and RA accumulation	Rahmani et al. 2020
Arabidopsis (*Arabidopsis thaliana*)	MWCNT	1,500 mg L^{-1}, 2,500 mg L^{-1}, or 3,500 mg L^{-1}	Dose-depended inhibition of root growth and leaf development	Increased H$_2$O$_2$ and malondialdehyde (MDA) content, alternated SOD, CAT, POD activity, hypermethylation of genes associated with stilbenoid, diarylheptanoid, and gingerol biosynthesis, tyrosine metabolism, phosphatidylinositol signaling system, and inositol phosphate metabolism and hypomethylation of genes related to ABC transporters, starch, and sucrose metabolism, and plant hormone signal transduction	Yang et al. 2021

(Hossain et al. 2016). Moreover, S-adenosylmethionine is essential for polyamine and glutathione (GSH) synthesis. GSH is one of the major molecules behind the sequestration of metals from inside the cell (Hossain et al. 2012). Interestingly, some studies indicate the upregulation of membrane transporters genes (Noori et al. 2020) and enhanced synthesis of defense compounds, protecting against pests and infections, such as rosmaric acid after exposure to AgNPs and MWCNT (Peharec Štefanić 2019; Rahmani et al. 2020).

Furthermore, the uptake of nanomaterials might be limited by exudates and mucilage (highly hydrated components from molecules including polysaccharides, proteins, and phenolic acids), promoting stabilization of nanomaterials in the rhizosphere solution (Lin and Xing 2008). Avellan et al. (2017) have observed that positively charged AuNPs stimulated the border cells on the root cap of *Arabidopsis thaliana* to release high-density mucus. The mucilage suppressed the translocation of AuNPs into the root by entrapment and immobilization of NPs. Interestingly, no similar effect was observed after treatment with negatively charged AuNPs, which were mostly observed in line with the cell wall, indicating apoplastic transport inside root tissue. In contrast, metabolomic analysis of root exudates of cucumber exposed to the CuNPs (at concentrations of 10 mg mL^{-1} and 20 mg mL^{-1}) revealed stimulation in the synthesis of amino acids, phenolic compounds, and ascorbic acid. Enhanced production and release of these low-molecular-weight substances have been implicated as a defense against metal stress by participating in the sequestration or exclusion of CuNPs or Cu ions, as well as improving antioxidant efficiency (Zhao et al. 2016). Additionally, plant growth-promoting bacteria may play a supporting role in preventing nanomaterials from entering plant tissues. As reported by Ahmed et al. (2021b) overproduction of EPS by *Azotobacter salinestris* significantly reduced the penetration of metal oxide (ZnO, CuO, Al$_2$O$_3$, and TiO$_2$) nanoparticles into tomato (*Solanum lycopersicum* L.).

The mechanisms of metal detoxification in plants include transport to storage parts, compartmentalization in subcellular parts, chelate formation, and elimination from the plant body (Rajput et al. 2019c). The transport and accumulation of ZnONPs in *Phaseolus vulgaris* were studied by Cruz et al. (2017, 2019). There was a decreasing Zn content from root to shoot after ZnONPs application, indicating that Zn in nano-form is stored, in lower tissues, contrary to the application of aqueous ZnSO$_4$ solution, where increased content was noted in higher tissues. In addition, the transcriptional analysis indicated increased expression levels of metal transporters located in the tonoplasts, implying that the mechanism for the tendency to accumulate Zn in lower tissues may be related to increased compartmentalization of Zn in vacuoles (Cruz et al. 2017, 2019). Furthermore, the high metal tolerance and absorption capacity of some plant species make them useful for environmental remediation. The main phytoremediation strategies include stabilization, degradation, extraction, and volatilization of contaminants and are based on processes naturally occurring in plants, such as uptake and sorption, translocation, metabolism, and evapotranspiration (da Conceição Gomes et al. 2016). Zhang and coworkers (2015) showed limited transport and potential to root accumulation of ZnONPs in *Schoenoplectus tabernaemontani*. The Zn nanoparticles and their aggregates were observed in cells of the root epidermis and intercellular space. A similar accumulation of nanoparticles was observed in the roots of wheat, rice, oat, and cucumber (Zhao et al. 2017a; Cai et al. 2017; Asgari et al. 2018; Alsuwayyid et al. 2022). Nowadays, with the expanding prevalence of NMs in soil or water, research efforts increasingly focus on the potential of plants for the phytoremediation of environments contaminated with nanomaterials (Ebrahimbabaie et al. 2020).

Techniques for Evaluation of Toxicity to Plants

As mentioned in the previous paragraph, nanomaterials influence plants in multiple ways, affecting individual cells and tissues, thereby altering their growth and development (Parsons et al. 2010; Ma et al. 2011; Dev et al. 2018; Wang et al. 2019). Mostly, the phytotoxicity of nanomaterials was tested using well-studied model plants, including onion (*Allium cepa*), thale cress (*Arabidopsis thaliana*), cucumber (*Cucumis sativus*), corn (*Zea mays*) rice (*Oryza sativa*), tobacco (*Nicotiana tabacum*), oilseed rape (*Brassica oleracea*), and lettuce (*Lactuca sativa*). The assays involve both evaluations of physiological and morphological parameters, as well as assessment of the implications for cell cycle or DNA damage after treatment with nanomaterials at specified periods and doses (Wang et al. 2012; Zhu et al. 2012; Ke et al. 2018; Peharec Štefanić 2019; Mosa et al. 2018; Zhang et al. 2021).

Techniques for Nanoparticle Detection

Various types of electron microscopy (such as transmission electron microscopy/scanning electron microscopy, and scanning transmission electron microscopy) have been applied to the investigation of nanomaterial uptake and localization in plants (Miralles et al. 2012; Ye et al. 2021). For instance, transmission electron microscopy (TEM) was used for the detection of TiO_2NPs in the cytoplasm of wheat roots cortex cells (Du et al. 2011), while scanning electron microscopy (SEM) coupled with energy dispersive X-ray (EDX) was employed for investigating nanoscale zero-valent iron (nZVI) uptake in *Leonurus cardiaca* leaves (Jafari and Hatami 2022). The inductively coupled plasma (ICP)-based techniques (such as ICP-optical emission spectrometry, ICP-mass spectrometry, and single-particle-ICP-mass spectrometry) are also used to determine the accumulation of nanomaterials in plant tissues, and the main advantage of this type of analysis is low detection threshold. Hence, sp-ICP-MS were used to obtain details on the concentration and particle size distribution of CuONPs accumulated in leaves of kale, lettuce, and collard green (Keller et al. 2018). Other techniques frequently used for the analysis of the morphology, chemical composition, and distribution of NMs in plants are the synchrotron-based X-ray approaches (Zhu et al. 2014; Castillo-Michel et al. 2017). Zhang et al. (2019) used TEM and synchrotron-based X-ray absorption near edge spectroscopy (XANES) for the detection of cerium nanoparticles in four plant species: monocots (corn and wheat) and dicots (soybean and cabbage). In another study reported by Larue et al. (2014), results from SEM and micro-X-ray fluorescence (μXRF) revealed that foliar-applied AgNPs penetrated leave cells of *Lactuca sativa*, attached to the cell wall and formed agglomerates with size 2 μm. In addition, ICP-MS analysis showed that the total content of AgNPs after 7 days of exposure was 240 μg per seedling. Recently, ICP-MS and X-ray fluorescence imaging (XFI) were used for the evaluation of biodistribution and real-time tracking of AuNPs in *Matricaria chamomilla*. The authors pointed out that XFI mapping by deep tissue imaging can be successfully exploited to assess the fate and interactions of NPs in plants, avoiding tissue destruction (Liu et al. 2022).

Approaches to Elucidate the Molecular Mechanism of Plant Response to Nanomaterials

The genotoxicity and cytotoxicity of nanomaterials are assessed by a wide range of approaches, including genomic, proteomic, and metabolomic methods. Currently, microscopic observation, electrophoresis-based methods, molecular markers, and polymerase chain reaction-based techniques are used to identify plant responses to nanomaterials at the molecular level (Barabadi et al. 2019; Rico et al. 2020). A number of studies on the genotoxic effect of nanomaterials focus on chromosomal aberrations, micronucleus formation, genomic (gDNA damage), mutational events, and copy number variation (Marmiroli et al. 2022). In a study reported by Banerjee et al. (2021) increase in micronuclei formation, a decrease in mitotic index, chromosomal abbreviation (chromosome breaks, anaphase-telophase bridges or multipolar anaphase-telophase cells), and DNA damage (four-fold increase in tail DNA detected in comet assay) were observed in *Allium cepa* root tip cells treated with CdSe quantum dots (QDs) at a concentration of 50 nM. Nevertheless, the authors suggest that the genotoxic effect is related to ROS generation, as evidenced by an increased antioxidant defense response. Moreover, in order to better understand the mechanism of toxicity of nanomaterials high-throughput methods, such as quantitative real-time PCR (qRT-PCR) or cDNA microarrays, are used to analyze changes in gene expression (Marmiroli et al. 2022). Results from global gene expression analysis of *Nicotiana tabacum* L. cv. Bright Yellow-2 (BY-2) cells exposed to 12 mg mL^{-1} of CuNPs demonstrated altered expression of 2692 genes. Performing gene ontology (GO) and Kyoto Encyclopedia of Genes and Genomes (KEGG) analysis specified that the genes were associated with oxidative stress (Dai et al. 2018). The phytotoxic effect of CuNPs on cucumber (*Cucumis sativus*) was evaluated by Mosa and coworkers (2018). The high accumulation level of CuNPs in plant roots was confirmed by X-ray fluorescence (XRF), atomic absorption spectroscopy (AAS), and SEM analysis. Results from the random amplified polymorphic DNA (RAPD) technique showed that treatment with CuNPs at a concentration of 200 mg L^{-1} induced genomic alterations in *C. sativus* and enhanced expression of copper-zinc superoxide dismutase (Cu-Zn SOD) gene (qRT-PCR analysis). In addition, changes in membrane permeability, an increase in H$_2$O$_2$ and MDA content, and a decrease in chlorophyll content (a and b) were recorded. All of these shifts resulted in the inhibition of seedling growth and a decrease in their biomass (Mosa et al. 2018). Considering that changes in gene expression and molecule biosynthesis can be induced by very negligible doses of contaminants, the above techniques are important in assessing toxicity associated with chronic exposure to nanomaterials. In addition, toxicological results enable the development of screening methods based on specific gene mutations and biomarkers for detecting alleged nanomaterial toxicity and associated risks (Jha and Pudake 2016).

Translocation of Nanomaterials in Plants

Uptake of Nanomaterials

In fact, there are two basic pathways of plant exposure to nanoparticles: foliar and root. The bioavailability of nanomaterials to plants and further translocation or accumulation is affected by factors, including the physicochemical properties of NMs, plant physiology, and environmental conditions (Lv et al. 2019). However, with regard to the various factors affecting the interaction of nanomaterials with plants, there is still a need for research to better understand such nanostructures. A number of studies showed that the uptake of NMs by plants and further transport to other plant tissues depends on their shape, size, and chemical composition as well as the anatomy of the exposed plants (Zhao et al. 2017b; Lyu et al. 2022; Zong et al. 2022).

Roots

Adsorption or uptake by the root surface is the initial stage of NMs translocation by plants from the soil. The efficiency of these processes highly depends on the size and surface charge of the nanomaterial, as indicated by several reports (Xia et al. 2013; Taylor et al. 2014; Hu et al. 2018; Lyu et al. 2022). Observations by TEM of wheat roots showed the penetration of smaller TiO$_2$ NPs with a size of 20 nm into the cortex cells, while larger ones with a diameter of 50 nm could not penetrate but adhered to the wall of the periderm cells (Du et al. 2011). Recently, Zong et al. (2022) reported higher translocation of nanosized (43 nm) CuONPs to cucumber roots, compared with copper oxide particles with a size of 510 nm. Nevertheless, NMs may undergo a wide range of biotransformation in the soil environment, affecting their bioavailability to plant roots. Several reports showed that after exposure to positively charged NPs root cap border cells are stimulated to increase mucilage production, which immobilizes positively charged NPs and acts as a barrier to root cells (Li et al. 2016; Avellan et al. 2017). According to Zhu et al. (2012), negatively charged (–24 mV) AuNPs were taken up with higher efficiency at 43–86% than similarly sized, positively charged (+ 24 mV) AuNPs at 13–25% uptake efficiency. The results from the above studies indicated also species-dependent uptake of NPs with the lowest uptake by rice, followed by pumpkin, radish, and ryegrass seedlings.

Leaves

It should be noted that the cuticle acts as a protective layer of plant tissues and plays a key role in the translocation of NMs from the leaf surface to the interior of the plant. According to the literature, small hydrophilic structures can be translocated through the aqueous pores present in the cuticle and stomatal apparatus. In the case of hydrophobic molecules, the translocation occurs by diffusion through the hydrophobic cuticle (Eichert et al. 2008). For instance, 60 nm AgNPs were immobilized on the surface of the cuticle and not observed on its inner side. Further analysis displayed that the adsorption of AgNPs was mainly facilitated by cutin (specifically with aliphatic hydroxy acids groups) and prevented by epicuticular waxes and pectin (Marciano et al. 2008). Furthermore, nanoparticles may enter plant organisms through stomata, hydathodes, and trichomes (Zhao et al. 2017a; Bombo et al. 2019; Spielman-Sun et al. 2020). Spielman-Sun et al. (2020) identified that the surface properties of AuNPs affected their distribution on the leaves. Citrate-reduced AuNPs were randomly distributed on leaves surface, while AuNPs coated with LM6M-antibody (specific for α-1,5-arabinan in stomata) and bovine serum albumin (BSA) were observed around stomata and trichomes, respectively. The results indicated that appropriate surface modification of the nanoparticles contributes to their targeted accumulation in areas particularly vulnerable to pathogen attacks and leads to an improvement in their anti-pathogenic efficacy (Spielman-Sun et al. 2020).

Translocation of Nanomaterials in Plants

After entering the plant, the transport of NMs through tissues is conducted through the apoplast and symplast pathways. Symplastic transport is based on transfer between the cytoplasm of cells across plasmodesmata or sieve plates. In contrast, apoplastic transport is defined as the movement beyond the plasma membrane in intercellular spaces through the cell walls of adjacent individual cells as well as xylem vessels (Pérez-de-Luque 2017). Indeed, some mechanisms have been identified to enable NPs to penetrate plant cells, including endocytosis-like or non-endocytic permeation, association with carrier proteins via aquaporins, or induction of new and large pore assembly (Palocci et al. 2017; Dai et al. 2018; Akdemir 2021; Dong et al. 2022). For instance, clathrin-independent endocytosis was identified as the main pathway for poly(lactic-co-glycolic) acid (PLGA) NPs internalization by grapevine cells. It has also been shown that the cell wall acts predominantly as a size-exclusion filter for the nanoparticle's uptake (Palocci et al. 2017). Gold nanoparticles with a size of 3.5 nm penetrated the tobacco root cells, while 18 nm AuNPs formed aggregates on the root surface. It is suggested that the transport of smaller particles occurred through pores (3.3–5.2 nm in size) in the cell walls (Carpita et al. 1979; Sabo-Attwood et al. 2012; Milewska-Hendel et al. 2017). It has been studied that small NMs after entering the cortex cells may form aggregates ~ 2 µm and their further transport is limited (Zhang et al. 2017; Dong et al. 2022). Size-dependent influx into plant cells was observed for SeNPs in wheat; exposure to 40 nm NPs resulted in 1.8-fold and 2.2-fold higher accumulation compared to SeNPs with diameter at 140 and 240 nm, respectively (Hu et al. 2018). However, certain nanomaterials can interact and modify the cell wall, thereby supporting their entry into the cell (Molnár et al. 2020). Hence, AuNPs affect the arrangement of specific pectin and arabinogalactan protein (AGP) epitopes in root cell walls, changing the chemical composition of the cell wall (Mielewska-Hendel et al. 2021). Whereas, Ag ions released from silver nanoparticles may bind to hydroxyl groups and change the cellulose structure in the cell wall, enabling their transport into the cytoplasm (Paiva Pinheiro et al. 2021).

Long-Distance Transport

The vascular tissues of plants play a significant role in the long-distance transport of NMs (Deepa et al. 2015; Hasaneen et al. 2016; Ma et al. 2017). In maize (*Zea mays* L.) CuNPs were translocated from the root to the shoot via the xylem and relocated back to the root via the phloem. In addition, microscopic observation indicated an apoplastic way through epidermal cortical cells by endocytosis up to the endodermis and the xylem (Wang et al. 2012). Nevertheless, the translocation of NPs from endodermis may be limited by the Casparian stripe. The crossing of this tissue has not been fully evaluated; however, it is suggested that in seedlings the incomplete development of this stripe at the top of the root facilitates penetration of NPs into the xylem (Wang et al. 2012). In turn, after foliar spraying, the chitosan nanoparticles were detected in sieve tubes of French bean phloem tissue (Hasaneen et al. 2016). In another study, Dong and coworkers (2022) observed graphene in the cytoplasm of the cortex, xylem, and mesophyll cells of *Triticum aestivum*. The results also displayed symplastic transport of graphene particles through plasmodesmata of adjacent, as well as further xylem and phloem-mediated transport. The long-distance transport of nanomaterials in plants is significantly influenced by the type of nanomaterials and application form but also by plant species and stage of development (Hasaneen et al. 2016; Ma et al. 2017).

Adverse Effect of NPs on Crop Yield and Quality

The increasing use of various nanomaterials (NMs) in medicine, agriculture, and industry, increases the threat to the environment, including agricultural areas. NMs can be absorbed from soils through plant roots or under foliar exposure and can reach all plant organs, including fruits and grains (El-Moneim et al. 2021). The interaction of nanomaterials with plants highly depends on the magnitude of their concentration, shape, and chemical nature of the NMs, as well as environmental conditions and plant species. Due to their unique properties and high reactivity nanomaterials might negatively affect basic cellular processes (including proliferation and metabolism) by interfering with transport across cell membranes, inducing oxidative stress, or altering gene regulation (de la Rosa et al. 2021). Furthermore, some studies indicated the negative effects of NMs, paying particular attention to the yield of economically important crops. In addition, the presence of nanomaterials in plant-origin foods poses a risk to human and animal health. The study of the interaction between NMs and plants, and their content in plant organs, as well as the associated risks to humans, is essential for the food production and medicine sectors (El-Moneim et al. 2021).

Effect of Concentrations

Seed germination is the first stage and most important phase for the growth of crop plants and yield quality (Rifna et al. 2019). Carbon nanotubes at low concentrations were recognized to stimulate seed germination and plant growth by improving water adsorption and transport such as by activation of water channels (Villagarcia et al. 2012; Hatami et al. 2017). Seedlings of wheat, maize, peanut, and garlic were stimulated by multiwalled-carbon nanotubes in a dose-dependent manner until the maximum tested concentration was at 50 µg mL^{-1} (Srivastava and Rao 2014). However, in another study, a tenfold increase in the concentration of single and multi-walled carbon nanotubes (from 0.1% to 1%) caused necrosis, disorders in root formation, and loss of turgor in plants (Basluk et al. 2019). Similar results were also presented by Gohari et al. (2020). They observed that MWCNT functionalized with carboxylic acid groups at a concentration of 50 µg mL^{-1}, under optimal conditions, promoted the growth of *Ocimum basilicum* seedlings. Additionally, under salinity stress, they act as protective factors and stimulated seedlings to synthesize chlorophyll, carotenoids, and also activity of the antioxidant system. However, treatments with higher concentrations (100 µg mL^{-1}) displayed toxic effects and lead to lowered photosynthetic efficiency or decreased membrane stability (Gohari et al. 2020). A toxic effect was observed in seeds of *Lupinus termis* treated with AgNPs at concentrations 400 ppm, 600 ppm, 800 ppm, and 900 ppm, where relative germination (RG) compared to control samples decreased by 22%, 33.4%, 44%, and 67%, respectively (Al-Huqail et al. 2018). Salehi et al. (2021) evaluated the effect of foliar-applied cerium oxide nanoparticles (CeO$_2$NPs) on bean plants (*Phaseolus vulgaris* L.) at a concentration range between 250–2000 mg L^{-1}. The findings pointed out that seeds from treated plants accumulated 45 µg Ce per 1 gram of dry mass while produced pollen grains sustained severe structural chromosome injuries under exposure to the maximum tested concentration and led to pollen abortion thereby consequently yield loss (Salehi et al. 2021).

Effect of Size and Agglomeration

There was a size-dependent toxic effect on the growth and development of *Salvinia minima* after treating AgNPs with an average size of 10 nm and 40 nm. Smaller AgNPs (10 nm) tended to form agglomerates more often than larger ones, but these structures were dynamic and not stable over long periods; this did not affect their dissolution ability. Higher relative growth inhibition was observed in seedlings treated with 10 nm AgNPs, than with 40 nm AgNPs. Chlorophyll content was differently modulated by these two sized NPs depending on the medium used. In moderately hard water (MHW) both AgNPs reduced the synthesis of chlorophyll while in natural organic matter (NOM) larger AgNPs inhibited and smaller ones stimulated the synthesis of chlorophyll compared to control samples (Thwala et al. 2021). Application of copper oxide nanoparticles (CuONPs) with an average size of 25 nm to soybeans significantly lowered the amount of seed yield compared to CuONPs with sizes of 50 nm and 250 nm and controls. With the use of the smallest CuONPs at a concentration of 500 mg mL^{-1}, increased hydrogen peroxide and malondialdehyde content, and improved activity of antioxidant system enzymes, namely SOD, catalase, peroxidase was observed (Yusefi-Tanha et al. 2020). It is suggested that the smaller size of CuONPs facilitates overcoming cellular barriers and enables them to move into the cell, thereby triggering higher oxidative stress. In contrast, larger-size CuONPs tend to be less surface reactive due to their smaller surface-to-volume ratio. As a result, they become potentially incapable of crossing cell barriers, displaying lower toxicity and not affecting crop plant productivity (Wang et al. 2012; Hong et al. 2015; Yusefi-Tanha et al. 2020).

Effect of Shapes

Another consideration in the uptake, translocation, and toxicity of nanomaterials is their shape, which determines chemical reactivity or possible biotransformation pathways in crop plants (Siddiqi et al. 2016). Zhang et al. (2017) reported shape-depended accumulation of CeO$_2$NPs in cucumber. Rod-like NPs were strongly accumulated in shoots and rapidly transformed Ce^{3+} ions when compared to octahedral, cubic, or irregularly shaped particles. Syu et al. (2014) demonstrated the shape-dependent effects of AgNPs on *Arabidopsis thaliana* seedlings. The results showed that AgNPs with triangular and decidual shapes mostly had a positive effect on seedling development, but treatment with spherical AgNPs led to reduced cotyledon growth. Moreover, increased anthocyanin content and SOD 2 activity suggested induction of oxidative stress by spherical AgNPs (Syu et al. 2014).

Effect of Surface Properties

Two-dimensional graphene oxide (GO) and reduced graphene oxide (rGO) nanosheets with negative and positive charges, respectively, demonstrated a contrasting impact on rice seedling growth. GO significantly inhibited shoot and root growth at concentrations of 100 mg L^{-1} and 250 mg L^{-1} while at the same dosage, rGO did not affect these parameters. The surface oxygen content was pointed as the main factor affecting their phytotoxicity (Zhang et al. 2020). Besides, noticeably more severe adverse effects on germination and growth of tobacco (*Nicotiana tabacum* L.) seedlings occurred after implementation of AgNPs coated with cetyltrimethylammonium bromide (CTAB) than polyvinylpyrrolidone (PVP). The research indicated that the toxic potency of AgNP-PVP was predominantly related to the release of Ag$^+$ ions, whereas the phytotoxicity of AgNP-CTAB was attributed to the surface coverage itself (Biba et al. 2020). Distinct levels of phytotoxicity on lettuce were also identified for graphene quantum dots (GQDs) subjected to various functionalizations, i.e., amination, carboxylation, and hydroxylation. The results confirmed the superior phytotoxicity of hydroxylated GQDs compared to the remaining two functionalizations, noting that aminated GQDs showed the lowest phytotoxicity. The surface property-dependent toxicity was observed by changes in growth parameters (reduced seedling length and amount of biomass) and by physiological responses, including disruption of photosynthesis and activation of antioxidant protection mechanism as well as regulation of phytohormone synthesis. In addition, GQDs affected mineral content in nutrient profiles at different levels, where hydroxylated GQDs showed the strongest effect (increase in Ca content and decrease in Mg, K, P, Mn, and Zn content) (Zhang et al. 2021).

Are Nanoparticles Transferred From One Generation to Another?

The comprehensive overview of NMs, i.e., transfer to subsequent generations of plants and NMs-induced transgenerational changes is lacking. Studies in this area are quite limited. Lin et al. (2009) showed that treatment of rice seedlings with carbon-based nanomaterials [fullerene C_{70} and multiwalled-carbon nanotubes (MWCNTs)] resulted in the accumulation of CNMs in seeds harvested from mature plants. Furthermore, C70 aggregates were observed in the leaf tissues of the next generation. In another study, Liu and coworkers (2018, 2019a, b) showed that CuONPs treatment in the previous generation of rice caused their accumulation in seeds but had no effect on Cu content in plants in the next generation. However, several studies suggested that nanomaterials translocated and accumulated in seeds may affect plant development in the next generation (Wang et al. 2013; Hernandez-Viezcas et al. 2013). For instance, irrigation of *Raphanus sativus* seedlings with ZnONPs and CuNPs suspension at a concentration of 1,000 mg mL^{-1} resulted in increased content of Zn (0.12 mg g^{-1}) and Cu (0.11 mg g^{-1}) obtained seeds comparing to untreated ones (Cu and Zn content, 0.03 and 0.04 mg g^{-1}, respectively). In a subsequent generation, reduced biomass, shoot, and root lengths were observed (Singh and Kumar 2018). Recently, Khan and coworkers (2022) investigated the impact of TiO$_2$NPs at a concentration range between 25–200 µg mL^{-1} on growth, yield quality, biochemical parameters, and heritable transgenerational alterations in the second generation of lentils (*Lens culinaris* Medik.) seedlings. Exposure of seeds in the first generation to NPs concentrations higher than 25 µg ml^{-1} led to reduced growth and development (number of branches and seeds), as well as lowered protein content in F2 generation seedlings in a dose-dependent manner. In addition, there was an increase in antioxidant enzyme activity (SOD and CAT), H$_2$O$_2$, and MDA content compared to control seedlings. However, subsequent generations displayed lower stress levels and relatively greater tolerance than in treated populations. In contrast, results presented by Medina-Velo (2018) showed that the parent plant's treatment with ZnONP not affected the yield, sugar, and protein content or nutrient profile of the second generation of bean (*Phaseolus vulgaris*) seeds. There was only a slightly lower Ni and increased Ca content in seeds from the second generation of plants. Indeed, finding this marginal effect highlights the feasibility of using ZnONPs in agricultural soils to improve crop quality. However, further studies at the molecular and genetic levels are required to provide the necessary insight into understanding the transgenerational effects of residual NPs (Medina-Velo et al. 2018).

Conclusions

Plants are exposed to nanomaterials that are introduced into the environment either intentionally (e.g., agricultural applications) or unintentionally (e.g., residues from industry). The growing interest in the use of nanomaterials in agriculture is due to their superior biological activities against plant pathogens but also as fertilizers. Despite the many positive aspects of the use of nanoformulations in agriculture, it has been observed, especially at higher doses, that NMs may bring negative consequences for plant organisms. Research indicates that the effects of NMs on plants depend on their properties but also the plant species and environmental conditions. Overall, in response to NMs toxicity, plants intensify the synthesis of biomolecules related to the stress and antioxidant systems. Uptake and biotransformation of NMs by plants are also limited by natural barriers, such as cuticle or mucilage. Nevertheless, the not fully understood mechanisms of NMS interactions with plants should be clarified by future research, especially on the effects on plant development and crop quality. Importantly for consumers, the risk of NMs accumulation in crop plants should be also assessed, as it poses a threat to animal and human health. In addition, non-standardized tests for properly assessing the safety of nanomaterials are another challenge.

Acknowledgement

J.T.W. acknowledges grant No. 2022/45/N/NZ9/01483 from National Science Centre, Poland.

References

Abdelkhalek, A. and Al-Askar, A.A. 2020. Green synthesized ZnO nanoparticles mediated by Mentha spicata extract induce plant systemic resistance against Tobacco mosaic virus. App. Sci. 10: 5054.

Ahmad, P., Jaleel, C.A., Salem, M.A., Nabi, G. and Sharma, S. 2010. Roles of enzymatic and nonenzymatic antioxidants in plants during abiotic stress. Crit. Rev. Biotechnol. 30: 161–175.

Ahmed, B., Syed, A., Rizvi, A., Shahid, M., Bahkali, A.H., Khan, M.S. and Musarrat, J. 2021b. Impact of metal-oxide nanoparticles on growth, physiology and yield of tomato (*Solanum lycopersicum* L.) modulated by *Azotobacter salinestris* strain ASM. Environ. Pollut. 269: 116218.

Ahmed, T., Noman, M., Luo, J., Muhammad, S., Shahid, M., Ali, M.A. and Li, B. 2021a. Bioengineered chitosan-magnesium nanocomposite: A novel agricultural antimicrobial agent against *Acidovorax oryzae* and *Rhizoctonia solani* for sustainable rice production. Int. J. Biol. Macromol. 168: 834–845.

Akdemir, H. 2021. Evaluation of transcription factor and aquaporin gene expressions in response to Al_2O_3 and ZnO nanoparticles during barley germination. Plant Physiol. Biochem. 166: 466–476.

Al-Huqail, A.A., Hatata, M.M., Al-Huqail, A.A. and Ibrahim, M.M. 2018. Preparation, characterization of silver phyto nanoparticles and their impact on growth potential of *Lupinus termis* L. seedlings. Saudi J. Biol. Sci. 25: 313–319.

AlQuraidi, A.O., Mosa, K.A. and Ramamoorthy, K. 2019. Phytotoxic and genotoxic effects of copper nanoparticles in coriander (*Coriandrum sativum*—Apiaceae). Plants 8: 19.

Alsuwayyid, A.A., Alslimah, A.S., Perveen, K., Bukhari, N.A. and Al-Humaid, L.A. 2022. Effect of zinc oxide nanoparticles on *Triticum aestivum* L. and bioaccumulation assessment using ICP-MS and SEM analysis. J. King Saud Univ. Sci. 34: 101944.

Anand, K. and Panigrahi, B. 2021. Green synthesized nanoparticles: a way to produce novel nano-biosensor for agricultural application. pp. 175–190. In Nanotechnology in Sustainable Agriculture, CRC Press.

Asgari, F., Majd, A., Jonoubi, P. and Najafi, F. 2018. Effects of silicon nanoparticles on molecular, chemical, structural and ultrastructural characteristics of oat (*Avena sativa* L.). Plant Physiol. Biochem. 127: 152–160.

Avellan, A., Schwab, F., Masion, A., Chaurand, P., Borschneck, D., Vidal, V. and Levard, C. 2017. Nanoparticle uptake in plants: gold nanomaterial localized in roots of *Arabidopsis thaliana* by X-ray computed nanotomography and hyperspectral imaging. Environ. Sci. Technol. 51: 8682–8691.

Babu, S., Singh, R., Yadav, D., Rathore, S.S., Raj, R., Avasthe, R. and Singh, V.K. 2022. Nanofertilizers for agricultural and environmental sustainability. Chemosphere 292: 133451.

Bagherzadeh Homaee, M. and Ehsanpour, A.A. 2015. Physiological and biochemical responses of potato (*Solanum tuberosum*) to silver nanoparticles and silver nitrate treatments under *in vitro* conditions. Indian J. Plant Physiol. 20: 353–359.

Banerjee, R., Goswami, P., Chakrabarti, M., Chakraborty, D., Mukherjee, A. and Mukherjee, A. 2021. Cadmium selenide (CdSe) quantum dots cause genotoxicity and oxidative stress in *Allium cepa* plants. Mutat. Res. Genet. Toxicol. Environ. Mutagen. 865: 503338.

Barabadi, H., Najafi, M., Samadian, H., Azarnezhad, A., Vahidi, H., Mahjoub, M.A. and Ahmadi, A. 2019 A systematic review of the genotoxicity and antigenotoxicity of biologically synthesized metallic nanomaterials: are green nanoparticles safe enough for clinical marketing? Medicina 55: 439.

Basiuk, V.A., Terrazas, T., Luna-Martínez, N. and Basiuk, E.V. 2019. Phytotoxicity of carbon nanotubes and nanodiamond in long-term assays with Cactaceae plant seedlings. Fullerenes, Nanotubes and Carbon Nanostructures 27: 141–149.

Biba, R., Matić, D., Lyons, D.M., Štefanić, P.P., Cvjetko, P., Tkalec, M. and Balen, B. 2020. Coating-dependent effects of silver nanoparticles on tobacco seed germination and early growth. Int. J. Mol. Sci. 21: 3441.

Bilesky-Jose, N., Maruyama, C., Germano-Costa, T., Campos, E., Carvalho, L., Grillo, R. and De Lima, R. 2021. Biogenic α-Fe_2O_3 nanoparticles enhance the biological activity of *Trichoderma* against the plant pathogen *Sclerotinia sclerotiorum*. ACS Sustain. Chem. Eng. 9: 1669–1683.

Bombo, A.B., Pereira, A.E.S., Lusa, M.G., de Medeiros Oliveira, E., de Oliveira, J.L., Campos, E.V.R. and Mayer, J.L.S. 2019. A mechanistic view of interactions of a nanoherbicide with target organism. J. Agric. Food Chem. 67: 4453–4462.

Cai, F., Wu, X., Zhang, H., Shen, X., Zhang, M., Chen, W. and Wang, X. 2017. Impact of TiO_2 nanoparticles on lead uptake and bioaccumulation in rice (*Oryza sativa* L.). NanoImpact 5: 101–108.

Cai, L., Chen, J., Liu, Z., Wang, H., Yang, H. and Ding, W. 2018. Magnesium oxide nanoparticles: effective agricultural antibacterial agent against *Ralstonia solanacearum*. Front. Microbiol. 9: 790.

Cao, X., Wang, C., Luo, X., Yue, L., White, J.C., Elmer, W. and Xing, B. 2021. Elemental sulfur nanoparticles enhance disease resistance in tomatoes. ACS Nano 15: 11817–11827.

Carpita, N., Sabularse, D., Montezinos, D. and Delmer, D.P. 1979. Determination of the pore size of cell walls of living plant cells. Science 205: 1144–1147.

Castillo-Michel, H.A., Larue, C., Del Real, A.E.P., Cotte, M. and Sarret, G. 2017. Practical review on the use of synchrotron based micro-and nano-X-ray fluorescence mapping and X-ray absorption spectroscopy to investigate the interactions between plants and engineered nanomaterials. Plant Physiol. Biochem. 110: 13–32.

Chung, I.M., Rekha, K., Venkidasamy, B. and Thiruvengadam, M. 2019. Effect of copper oxide nanoparticles on the physiology, bioactive molecules, and transcriptional changes in *Brassica rapa* ssp. *rapa* seedlings. Wat. Air Soil Poll. 230: 1–14.

da Conceição Gomes, M.A., Hauser-Davis, R.A., de Souza, A.N. and Vitória, A.P. 2016. Metal phytoremediation: General strategies, genetically modified plants and applications in metal nanoparticle contamination. Ecotoxicol. Environ. Saf. 134: 133–147.

da Cruz, T.N., Savassa, S.M., Gomes, M.H., Rodrigues, E.S., Duran, N.M., de Almeida, E. and de Carvalho, H.W. 2017. Shedding light on the mechanisms of absorption and transport of ZnO nanoparticles by plants via *in vivo* X-ray spectroscopy. Environ. Sci.: Nano 4: 2367–2376.

da Cruz, T.N., Savassa, S.M., Montanha, G.S., Ishida, J.K., de Almeida, E., Tsai, S.M. and Pereira de Carvalho, H.W. 2019. A new glance on root-to-shoot in vivo zinc transport and time-dependent physiological effects of ZnSO$_4$ and ZnO nanoparticles on plants. Sci. Rep. 9: 1–12.

Dai, Y., Wang, Z., Zhao, J., Xu, L., Xu, L., Yu, X. and Xing, B. 2018. Interaction of CuO nanoparticles with plant cells: internalization, oxidative stress, electron transport chain disruption, and toxicogenomic responses. Environ. Sci.: Nano 5: 2269–2281.

de la Rosa, G., Vázquez-Núñez, E., Molina-Guerrero, C., Serafín-Muñoz, A.H. and Vera-Reyes, I. 2021. Interactions of nanomaterials and plants at the cellular level: current knowledge and relevant gaps. Nanotechnol. Environ. Eng. 6: 1–19.

de Paiva Pinheiro, S.K., Miguel, T.B.A.R., de Medeiros Chaves, M., de Freitas Barros, F.C., Farias, C.P., de Moura, T.A. and Wu, H. 2021. Silver nanoparticles (AgNPs) internalization and passage through the *Lactuca sativa* (Asteraceae) outer cell wall. Funct. Plant Biol. 48: 1113–1123.

Deepa, M., Sudhakar, P., Nagamadhuri, K.V., Balakrishna Reddy, K., Giridhara Krishna, T. and Prasad, V. 2015. First evidence on phloem transport of nanoscale calcium oxide in groundnut using solution culture technique. Appl. Nanosci. 5: 545–551.

Dev, A., Srivastava, A.K. and Karmakar, S. 2018. Nanomaterial toxicity for plants. Environ. Chem. Lett. 16: 85–100.

Dhyani, K., Meenu, M., Bezbaruah, A.N., Kar, K.K. and Chamoli, P. 2022. Current prospective of nanomaterials in agriculture and farming. pp. 173–194. In Nanomaterials for Advanced Technologies, Springer.

Dong, S., Jing, X., Lin, S., Lu, K., Li, W., Lu, J. and Mao, L. 2022. Root hair apex is the key site for symplastic delivery of graphene into plants. Environ. Sci. Technol. 56: 12179–12189.

Du, W., Sun, Y., Ji, R., Zhu, J., Wu, J. and Guo, H. 2011. TiO$_2$ and ZnO nanoparticles negatively affect wheat growth and soil enzyme activities in agricultural soil. J. Environmen. Monitoring 13: 822–828.

Ebrahimbabaie, P., Meeinkuirt, W. and Pichtel, J. 2020. Phytoremediation of engineered nanoparticles using aquatic plants: Mechanisms and practical feasibility. J. Environ. Sci. 93: 151–163.

Eichert, T., Kurtz, A., Steiner, U. and Goldbach, H.E. 2008. Size exclusion limits and lateral heterogeneity of the stomatal foliar uptake pathway for aqueous solutes and water-suspended nanoparticles. Physiologia plantarum, 134: 151–160.

El-Abeid, S.E., Ahmed, Y., Daròs, J.A. and Mohamed, M.A. 2020. Reduced graphene oxide nanosheet-decorated copper oxide nanoparticles: A potent antifungal nanocomposite against fusarium root rot and wilt diseases of tomato and pepper plants. Nanomaterials 10: 1001.

El-Batal, A.I., El-Sayyad, G.S., Mosallam, F.M. and Fathy, R.M. 2020. *Penicillium chrysogenum*-mediated mycogenic synthesis of copper oxide nanoparticles using gamma rays for in vitro antimicrobial activity against some plant pathogens. J. Clust. Sci. 31: 79–90.

El-Moneim, D.A., Dawood, M.F., Moursi, Y.S., Farghaly, A.A., Afifi, M. and Sallam, A. 2021. Positive and negative effects of nanoparticles on agricultural crops. Nanotechnol. Environ. Eng. 6: 1–11.

El-Shetehy, M., Moradi, A., Maceroni, M., Reinhardt, D., Petri-Fink, A., Rothen-Rutishauser, B. and Schwab, F. 2021. Silica nanoparticles enhance disease resistance in *Arabidopsis plants*. Nature Nanotechnol. 16: 344–353.

Fan, X., Xu, J., Lavoie, M., Peijnenburg, W.J.G.M., Zhu, Y., Lu, T. and Qian, H. 2018. Multiwall carbon nanotubes modulate paraquat toxicity in *Arabidopsis thaliana*. Environ. Poll. 233: 633–641.

Fen, L.B., Rashid, A.H.A., Nordin, N.I., Hossain, M.M., Uddin, S.M.K., Johan, M.R. and Thangadurai, D. 2022. Applications of nanomaterials in agriculture and their safety aspect. pp. 243–299. In Biogenic Nanomaterials, Apple Academic Press.

Ghafari, J., Moghadasi, N. and Shekaftik, S.O. 2020. Oxidative stress induced by occupational exposure to nanomaterials: a systematic review. Industrial Health 58: 492–502.

Greenshields, M.W., Cunha, B.B., Coville, N.J., Pimentel, I.C., Zawadneak, M.A., Dobrovolski, S. and Hümmelgen, I.A. 2016. Fungi active microbial metabolism detection of *Rhizopus* sp. and *Aspergillus* sp. section nigri on strawberry using a set of chemical sensors based on carbon nanostructures. Chemosensors 4: 19.

Haggag, W.M. and Eid, M.M. 2022. Antifungal and antioxidant activities of Ag@ FeO-NPs@ Chitosan preparation by endophyte *Streptomyces aureofaciens* Int. J. Agric. Technol. 18: 535–548.

Hasaneen, M.N.A.G., Abdel-Aziz, H.M.M. and Omer, A.M. 2016. Effect of foliar application of engineered nanomaterials: carbon nanotubes NPK and chitosan nanoparticles NPK fertilizer on the growth of French bean plant. Biochem. Biotechnol. Res. 4: 68–76.

Hasanuzzaman, M., Bhuyan, M.B., Anee, T.I., Parvin, K., Nahar, K., Mahmud, J.A. and Fujita, M. 2019. Regulation of ascorbate-glutathione pathway in mitigating oxidative damage in plants under abiotic stress. Antioxidants 8: 384.

Hatami, M., Hadian, J. and Ghorbanpour, M. 2017. Mechanisms underlying toxicity and stimulatory role of single-walled carbon nanotubes in *Hyoscyamus niger* during drought stress simulated by polyethylene glycol. J. Hazard. Mater. 324: 306–320.

Hernandez-Viezcas, J.A., Castillo-Michel, H., Andrews, J.C., Cotte, M., Rico, C., Peralta-Videa, J.R. and Gardea-Torresdey, J.L. 2013. *In situ* synchrotron X-ray fluorescence mapping and speciation of CeO_2 and ZnO nanoparticles in soil cultivated soybean (*Glycine max*). ACS Nano 7: 1415–1423.

Hong, J., Rico, C.M., Zhao, L., Adeleye, A.S., Keller, A.A., Peralta-Videa, J.R. and Gardea-Torresdey, J.L. 2015. Toxic effects of copper-based nanoparticles or compounds to lettuce (*Lactuca sativa*) and alfalfa (*Medicago sativa*). Environ. Sci.: Process. Impacts 17: 177–185.

Hossain, M.A., Piyatida, P., da Silva, J.A.T. and Fujita, M. 2012. Molecular mechanism of heavy metal toxicity and tolerance in plants: central role of glutathione in detoxification of reactive oxygen species and methylglyoxal and in heavy metal chelation. J. Botany 2012: 872875.

Hossain, Z., Mustafa, G., Sakata, K. and Komatsu, S. 2016. Insights into the proteomic response of soybean towards Al_2O_3, ZnO, and Ag nanoparticles stress. J. Hazard. Mater. 304: 291–305.

Hu, T., Li, H., Li, J., Zhao, G., Wu, W., Liu, L. and Guo, Y. 2018. Absorption and bio-transformation of selenium nanoparticles by wheat seedlings (*Triticum aestivum* L.). Front. Plant Sci. 9: 597.

Irshad, M.A., Nawaz, R., Rehman, M.Z., Imran, M., Ahmad, J., Ahmad, S. and Ali, S. 2020. Synthesis and characterization of titanium dioxide nanoparticles by chemical and green methods and their antifungal activities against wheat rust. Chemosphere 258: 127352.

Jafari, A. and Hatami, M. 2022. Foliar-applied nanoscale zero-valent iron (nZVI) and iron oxide (Fe_3O_4) induce differential responses in growth, physiology, antioxidative defense and biochemical indices in *Leonurus cardiaca* L. Environ. Res. 215: 114254.

Jha, S. and Pudake, R.N. 2016. Molecular mechanism of plant–nanoparticle interactions. Plant Nanotechnol. 155–181.

Ke, M., Qu, Q., Peijnenburg, W.J.G.M., Li, X., Zhang, M., Zhang, Z. and Qian, H. 2018. Phytotoxic effects of silver nanoparticles and silver ions to *Arabidopsis thaliana* as revealed by analysis of molecular responses and of metabolic pathways. Sci. Total Environ. 644: 1070–1079.

Keller, A.A., Huang, Y. and Nelson, J. 2018. Detection of nanoparticles in edible plant tissues exposed to nano-copper using single-particle ICP-MS. J. Nanopart. Res. 20: 1–13.

Khan, A.U., Khan, M., Khan, A.A., Parveen, A., Ansari, S. and Alam, M. 2022. Effect of phyto-assisted synthesis of magnesium oxide nanoparticles (MgO-NPs) on bacteria and the root-knot nematode. Bioinorg. Chem. Appl. 2022: 3973841.

Khan, M., Khan, A.U., Bogdanchikova, N. and Garibo, D. 2021. Antibacterial and antifungal studies of biosynthesized silver nanoparticles against plant parasitic nematode *Meloidogyne incognita*, plant pathogens *Ralstonia solanacearum* and *Fusarium oxysporum*. Molecules 26: 2462.

Komatsu, S., Murata, K., Yakeishi, S., Shimada, K., Yamaguchi, H., Hitachi, K. and Fukuda, R. 2022. Morphological and proteomic analyses of soybean seedling interaction mechanism affected by fiber crosslinked with zinc-oxide nanoparticles. Int. J. Mol. Sci. 23: 7415.

Larue, C., Castillo-Michel, H., Sobanska, S., Cécillon, L., Bureau, S., Barthès, V. and Sarret, G. 2014. Foliar exposure of the crop *Lactuca sativa* to silver nanoparticles: evidence for internalization and changes in Ag speciation. J. Hazard. Mater. 264: 98–106.

Lei, R., Wu, P., Li, L., Huang, Q., Wang, J., Zhang, D. and Wang, X. 2021. Ultrasensitive isothermal detection of a plant pathogen by using a gold nanoparticle-enhanced microcantilever sensor. Sens. Actuators B: Chem. 338: 129874.

Li, H., Ye, X., Guo, X., Geng, Z. and Wang, G. 2016. Effects of surface ligands on the uptake and transport of gold nanoparticles in rice and tomato. J. Hazard. Mater. 314: 188–196.

Li, Z., Yu, T., Paul, R., Fan, J., Yang, Y. and Wei, Q. 2020. Agricultural nanodiagnostics for plant diseases: recent advances and challenges. Nanoscale Advances 2: 3083–3094.

Lin, D. and Xing, B. 2008. Root uptake and phytotoxicity of ZnO nanoparticles. Environ. Sci. Technol. 42: 5580–5585.

Lin, S., Reppert, J., Hu, Q., Hudson, J.S., Reid, M.L., Ratnikova, T.A. and Ke, P.C. 2009. Uptake, translocation, and transmission of carbon nanomaterials in rice plants. Small 5: 1128–1132.

Lipşa, F.D., Ursu, E.L., Ursu, C., Ulea, E. and Cazacu, A. 2020. Evaluation of the antifungal activity of gold–chitosan and carbon nanoparticles on *Fusarium oxysporum*. Agronomy 10: 1143.

Liu, J., Dhungana, B. and Cobb, G.P. 2018. Environmental behavior, potential phytotoxicity, and accumulation of copper oxide nanoparticles and arsenic in rice plants. Environ. Toxicol. Chem. 37: 11–20.

Liu, J., Wolfe, K. and Cobb, G.P. 2019a. Exposure to copper oxide nanoparticles and arsenic causes intergenerational effects on Rice (*Oryza sativa* japonica Koshihikari) seed germination and seedling growth. Environmental Toxicology and Chemistry 38: 1978–1987.

Liu, J., Wolfe, K., Potter, P.M. and Cobb, G.P. 2019b. Distribution and speciation of copper and arsenic in rice plants (*Oryza sativa* japonica 'Koshihikari') treated with copper oxide nanoparticles and arsenic during a life cycle. Environ. Sci. Technol. 53: 4988–4996.

Liu, Y., Kornig, C., Qi, B., Schmutzler, O., Staufer, T., Sanchez-Cano, C. and Parak, W.J. 2022. Size- and ligand-dependent transport of nanoparticles in *Matricaria chamomilla* as demonstrated by mass spectroscopy and X-ray fluorescence imaging. ACS Nano 16: 12941–12951.

Lv, J., Christie, P. and Zhang, S. 2019. Uptake, translocation, and transformation of metal-based nanoparticles in plants: recent advances and methodological challenges. Environmental Science: Nano 6: 41–59.

Lyu, L., Wang, H., Liu, R., Xing, W., Li, J., Man, Y.B. and Wu, F. 2022. Size-dependent transformation, uptake, and transportation of SeNPs in a wheat–soil system. J. Hazard. Mater. 424: 127323.

Ma, C., White, J.C., Dhankher, O.P. and Xing, B. 2015. Metal-based nanotoxicity and detoxification pathways in higher plants. Environ. Sci. Technol. 49: 7109–7122.

Ma, Y., He, X., Zhang, P., Zhang, Z., Guo, Z., Tai, R. and Chai, Z. 2011. Phytotoxicity and biotransformation of La_2O_3 nanoparticles in a terrestrial plant cucumber (*Cucumis sativus*). Nanotoxicology 5: 743–753.

Ma, Y., He, X., Zhang, P., Zhang, Z., Ding, Y., Zhang, J. and Yang, K. 2017. Xylem and phloem based transport of CeO_2 nanoparticles in hydroponic cucumber plants. Environ. Sci. Technol. 51: 5215–5221.

Marciano, A., Chefetz, B. and Gedanken, A. 2008. Differential adsorption of silver nanoparticles to the inner and outer surfaces of the agave americana cuticle. J. Phys. Chem. C 112: 18082–18086.

Marmiroli, M., Marmiroli, N. and Pagano, L. 2022. Nanomaterials Induced Genotoxicity in Plant: Methods and Strategies. Nanomaterials 12. 1650.

Medina-Velo, I.A., Zuverza-Mena, N., Tamez, C., Ye, Y., Hernandez-Viezcas, J.A., White, J.C. and Gardea-Torresdey, J.L. 2018. Minimal transgenerational effect of ZnO nanomaterials on the physiology and nutrient profile of *Phaseolus vulgaris*. ACS Sustain. Chem. Eng. 6: 7924–7930.

Milewska-Hendel, A., Sala, K., Gepfert, W. and Kurczyńska, E. 2021. Gold nanoparticles-induced modifications in cell wall composition in barley roots. Cells 10: 1965.

Miralles, P., Church, T.L. and Harris, A.T. 2012. Toxicity, uptake, and translocation of engineered nanomaterials in vascular plants. Environ. Sci. Technol. 46: 9224–9239.

Molnár, Á., Rónavári, A., Bélteky, P., Szőllősi, R., Valyon, E., Oláh, D. and Kolbert, Z. 2020. ZnO nanoparticles induce cell wall remodeling and modify ROS/RNS signalling in roots of *Brassica* seedlings. Ecotoxicol. Environ. Saf. 206: 111158.

Montes, A., Bisson, M.A., Gardella Jr, J.A. and Aga, D.S. 2017. Uptake and transformations of engineered nanomaterials: critical responses observed in terrestrial plants and the model plant *Arabidopsis thaliana*. Sci. Total Environ. 607: 1497–1516.

Mosa, K.A., El-Naggar, M., Ramamoorthy, K., Alawadhi, H., Elnaggar, A., Wartanian, S. and Hani, H. 2018. Copper nanoparticles induced genotoxicty, oxidative stress, and changes in superoxide dismutase (SOD) gene expression in cucumber (*Cucumis sativus*) plants. Front. Plant Sci. 9: 872.

Noori, A., Ngo, A., Gutierrez, P., Theberge, S. and White, J.C. 2020. Silver nanoparticle detection and accumulation in tomato (*Lycopersicon esculentum*). J. Nanopart. Res. 22: 1–16.

Oh, J.W., Chun, S.C. and Chandrasekaran, M. 2019. Preparation and *in vitro* characterization of chitosan nanoparticles and their broad-spectrum antifungal action compared to antibacterial activities against phytopathogens of tomato. Agronomy 9: 21.

Palocci, C., Valletta, A., Chronopoulou, L., Donati, L., Bramosanti, M., Brasili, E. and Pasqua, G. 2017. Endocytic pathways involved in PLGA nanoparticle uptake by grapevine cells and role of cell wall and membrane in size selection. Plant Cell Rep. 36: 1917–1928.

Parsons, J.G., Lopez, M.L., Gonzalez, C.M., Peralta-Videa, J.R. and Gardea-Torresdey, J.L. 2010. Toxicity and biotransformation of uncoated and coated nickel hydroxide nanoparticles on mesquite plants. Environ. Toxicol. Chem. 29: 1146–1154.

Peharec Štefanić, P., Jarnević, M., Cvjetko, P., Biba, R., Šikić, S., Tkalec, M. and Balen, B. 2019. Comparative proteomic study of phytotoxic effects of silver nanoparticles and silver ions on tobacco plants. Environ. Sci. Pollut. Res. 26: 22529–22550.

Pérez-de-Luque, A. 2017. Interaction of nanomaterials with plants: what do we need for real applications in agriculture? Front. Environ. Sci. 5: 12.

Pham, N.D., Duong, M.M., Le, M.V. and Hoang, H.A. 2019. Preparation and characterization of antifungal colloidal copper nanoparticles and their antifungal activity against *Fusarium oxysporum* and *Phytophthora capsici.* Comptes Rendus Chimie, 22: 786–793.

Qian, H., Peng, X., Han, X., Ren, J., Sun, L. and Fu, Z. 2013. Comparison of the toxicity of silver nanoparticles and silver ions on the growth of terrestrial plant model *Arabidopsis thaliana*. J. Environ. Sci. 25: 1947–1956.

Rahmani, N., Radjabian, T. and Soltani, B.M. 2020. Impacts of foliar exposure to multi-walled carbon nanotubes on physiological and molecular traits of *Salvia verticillata* L., as a medicinal plant. Plant Physiol. Biochem. 150: 27–38.

Rahmat, N., Yusof, N.A., Isha, A., Mui-Yun, W., Hushiarian, R. and Akanbi, F.S. 2020. Detection of stress induced by ganoderma boninense infection in oil palm leaves using reduced graphene oxide and zinc oxide nanoparticles screen-printed carbon electrode. IEEE Sensors Journal 20: 13253–13261.

Rajput, V.D., Minkina, T., Sushkova, S., Chokheli, V. and Soldatov, M. 2019. Toxicity assessment of metal oxide nanoparticles on terrestrial plants. Compr. Anal. Chem. 87: 189–207.

Rico, C.M., Wagner, D., Abolade, O., Lottes, B. and Coates, K. 2020. Metabolomics of wheat grains generationally-exposed to cerium oxide nanoparticles. Sci. Total Environ. 712: 136487.

Rifna, E.J., Ramanan, K.R. and Mahendran, R. 2019. Emerging technology applications for improving seed germination. Trends Food Sci. Technol. 86: 95–108.

Ruttkay-Nedecky, B., Krystofova, O., Nejdl, L. and Adam, V. 2017. Nanoparticles based on essential metals and their phytotoxicity. J. Nanobiotechnol. 15: 1–19.

Sabo-Attwood, T., Unrine, J.M., Stone, J.W., Murphy, C.J., Ghoshroy, S., Blom, D. and Newman, L.A. 2012. Uptake, distribution and toxicity of gold nanoparticles in tobacco (*Nicotiana xanthi*) seedlings. Nanotoxicology 6: 353–360.

Salehi, H., Chehregani Rad, A., Raza, A. and Chen, J.T. 2021. Foliar application of CeO$_2$ nanoparticles alters generative components fitness and seed productivity in Bean crop (*Phaseolus vulgaris* L.). Nanomaterials 11: 862.

Sathiyabama, M. and Muthukumar, S. 2020. Chitosan guar nanoparticle preparation and its *in vitro* antimicrobial activity towards phytopathogens of rice. Int. J. Biol. Macromol. 153: 297–304.

Sellappan, L., Manoharan, S., Sanmugam, A. and Anh, N.T. 2022. Role of nanobiosensors and biosensors for plant virus detection. pp. 493–506. In Nanosensors for Smart Agriculture Elsevier.

Shende, S., Bhagat, R., Raut, R., Rai, M. and Gade, A. 2021. Myco-fabrication of copper nanoparticles and its effect on crop pathogenic fungi. IEEE Trans. Nanobioscience 20: 146–153.

Shojaei, T.R., Salleh, M.A.M., Sijam, K., Rahim, R.A., Mohsenifar, A., Safarnejad, R. and Tabatabaei, M. 2016. Fluorometric immunoassay for detecting the plant virus *Citrus tristeza* using carbon nanoparticles acting as quenchers and antibodies labeled with CdTe quantum dots. Microchimica Acta 183: 2277–2287.

Siddiqi, K.S. and Husen, A. 2016. Engineered gold nanoparticles and plant adaptation potential. Nanoscale Res. Lett. 11: 1–10.

Singh, D. and Kumar, A. 2018. Investigating long-term effect of nanoparticles on growth of Raphanus sativus plants: a trans-generational study. Ecotoxicology 27: 23–31.

Spielman-Sun, E., Avellan, A., Bland, G.D., Clement, E.T., Tappero, R.V., Acerbo, A.S. and Lowry, G.V. 2020. Protein coating composition targets nanoparticles to leaf stomata and trichomes. Nanoscale 12: 3630–3636.

Srivastava, A. and Rao, D.P. 2014. Enhancement of seed germination and plant growth of wheat, maize, peanut and garlic using multiwalled carbon nanotubes. European Chemical Bulletin 3: 502–504.

Syu, Y.Y., Hung, J.H., Chen, J.C. and Chuang, H.W. 2014. Impacts of size and shape of silver nanoparticles on *Arabidopsis* plant growth and gene expression. Plant Physiol. Biochem. 83: 57–64.

Tahir, M.A., Bajwa, S.Z., Mansoor, S., Briddon, R.W., Khan, W.S., Scheffler, B.E. and Amin, I. 2018. Evaluation of carbon nanotube-based copper nanoparticle composite for the efficient detection of agroviruses. J. Hazard. Mater. 346: 27–35.

Taylor, A.F., Rylott, E.L., Anderson, C.W. and Bruce, N.C. 2014. Investigating the toxicity, uptake, nanoparticle formation and genetic response of plants to gold. PLOS One 9: e93793.

Thiruvengadam, M., Gurunathan, S. and Chung, I.M. 2015. Physiological, metabolic, and transcriptional effects of biologically-synthesized silver nanoparticles in turnip (*Brassica rapa* ssp. *rapa* L.). Protoplasma 252: 1031–1046.

Thwala, M., Klaine, S. and Musee, N. 2021. Exposure media and nanoparticle size influence on the fate, bioaccumulation, and toxicity of silver nanoparticles to higher plant salvinia minima. Molecules 26: 2305.

Tran, T.T., Clark, K., Ma, W. and Mulchandani, A. 2020. Detection of a secreted protein biomarker for citrus Huanglongbing using a single-walled carbon nanotubes-based chemiresistive biosensor. Biosens. Bioelectron. 147: 111766.

Vannini, C., Domingo, G., Onelli, E., De Mattia, F., Bruni, I., Marsoni, M. and Bracale, M. 2014. Phytotoxic and genotoxic effects of silver nanoparticles exposure on germinating wheat seedlings. J. Plant Physiol. 171: 1142–1148.

Vanti, G.L., Masaphy, S., Kurjogi, M., Chakrasali, S. and Nargund, V.B. 2020. Synthesis and application of chitosan-copper nanoparticles on damping off causing plant pathogenic fungi. Int. J. Biol. Macromol. 156: 1387–1395.

Villagarcia, H., Dervishi, E., de Silva, K., Biris, A.S. and Khodakovskaya, M.V. 2012. Surface chemistry of carbon nanotubes impacts the growth and expression of water channel protein in tomato plants. Small 8: 2328–2334.

Wang, H., Ramnani, P., Pham, T., Villarreal, C.C., Yu, X., Liu, G. and Mulchandani, A. 2019. Gas biosensor arrays based on single-stranded DNA-functionalized single-walled carbon nanotubes for the detection of volatile organic compound biomarkers released by huanglongbing disease-infected citrus trees. Sensors 19: 4795.

Wang, Q., Ebbs, S.D., Chen, Y. and Ma, X. 2013. Trans-generational impact of cerium oxide nanoparticles on tomato plants. Metallomics 5: 753–759.

Wang, Z., Xie, X., Zhao, J., Liu, X., Feng, W., White, J.C. and Xing, B. 2012. Xylem-and phloem-based transport of CuO nanoparticles in maize (*Zea mays* L.). Environ. Sci. Technol. 46: 4434–4441.

Win, T.T., Khan, S. and Fu, P. 2020. Fungus-(*Alternaria* sp.) mediated silver nanoparticles synthesis, characterization, and screening of antifungal activity against some phytopathogens. J. Nanotechnol. 2020: 8828878.

Xia, B., Dong, C., Zhang, W., Lu, Y., Chen, J. and Shi, J. 2013. Highly efficient uptake of ultrafine mesoporous silica nanoparticles with excellent biocompatibility by *Liriodendron* hybrid suspension cells. Sci. China Life Sci. 56: 82–89.

Yadav, N., Chhillar, A.K. and Rana, J.S. 2020. Detection of pathogenic bacteria with special emphasis to biosensors integrated with AuNPs. Sensors Int. 1: 100028.

Yang, Z., Deng, C., Wu, Y., Dai, Z., Tang, Q., Cheng, C. and Yan, A. 2021. Insights into the mechanism of multi-walled carbon nanotubes phytotoxicity in *Arabidopsis* through transcriptome and m6A methylome analysis. Sci. Total Environ. 787: 147510.

Ye, Y., Cota Ruiz, K., Cantu, J.M., Valdes, C. and Gardea-Torresdey, J.L. 2021. Engineered nanomaterials fate assessment in biological matrices: recent milestones in electron microscopy. ACS Sustain. Chem. Eng. 9: 4341–4356.

Youssef, M.S. and Elamawi, R.M. 2020. Evaluation of phytotoxicity, cytotoxicity, and genotoxicity of ZnO nanoparticles in *Vicia faba*. Environ. Sci. Pollut. Res. 27: 18972–18984.

Yusefi-Tanha, E., Fallah, S., Rostamnejadi, A. and Pokhrel, L.R. 2020. Particle size and concentration dependent toxicity of copper oxide nanoparticles (CuONPs) on seed yield and antioxidant defense system in soil grown soybean (*Glycine max* cv. Kowsar). Sci. Total Environ. 715: 136994.

Zhan, F., Wang, T., Iradukunda, L. and Zhan, J. 2018. A gold nanoparticle-based lateral flow biosensor for sensitive visual detection of the potato late blight pathogen, *Phytophthora infestans*. Anal. Chim. Acta 1036: 153–161.

Zhang, D., Hua, T., Xiao, F., Chen, C., Gersberg, R.M., Liu, Y. and Tan, S.K. 2015. Phytotoxicity and bioaccumulation of ZnO nanoparticles in *Schoenoplectus tabernaemontani*. Chemosphere 120: 211–219.

Zhang, P., Xie, C., Ma, Y., He, X., Zhang, Z., Ding, Y. and Zhang, J. 2017. Shape-dependent transformation and translocation of ceria nanoparticles in cucumber plants. Environ. Sci. Technol. Lett. 4: 380–385.

Zhang, P., Ma, Y., Xie, C., Guo, Z., He, X., Valsami-Jones, E. and Zhang, Z. 2019. Plant species-dependent transformation and translocation of ceria nanoparticles. Environ. Sci. Nano 6: 60–67.

Zhang, P., Guo, Z., Luo, W., Monikh, F.A., Xie, C., Valsami-Jones, E. and Zhang, Z. 2020. Graphene oxide-induced pH alteration, iron overload, and subsequent oxidative damage in rice (*Oryza sativa* L.): A new mechanism of nanomaterial phytotoxicity. Environ. Sci. Technol. 54: 3181–3190.

Zhang, P., Wu, X., Guo, Z., Yang, X., Hu, X. and Lynch, I. 2021. Stress response and nutrient homeostasis in lettuce (*Lactuca sativa*) exposed to graphene quantum dots are modulated by particle surface functionalization. Adv. Biol. 5: 2000778.

Zhao, J., Ren, W., Dai, Y., Liu, L., Wang, Z., Yu, X. and Xing, B. 2017a. Uptake, distribution, and transformation of CuO NPs in a floating plant *Eichhornia crassipes* and related stomatal responses. Environ. Sci. Technol. 51: 7686–7695.

Zhao, L., Huang, Y., Adeleye, A.S. and Keller, A.A. 2017b. Metabolomics reveals Cu (OH)$_2$ nanopesticide-activated anti-oxidative pathways and decreased beneficial antioxidants in spinach leaves. Environ. Sci. Technol. 51: 10184–10194.

Zhao, L., Huang, Y., Hu, J., Zhou, H., Adeleye, A.S. and Keller, A.A. 2016. 1H NMR and GC-MS based metabolomics reveal defense and detoxification mechanism of cucumber plant under nano-Cu stress. Environ. Sci. Technol. 50: 2000–2010.

Zhu, Y., Cai, X., Li, J., Zhong, Z., Huang, Q. and Fan, C. 2014. Synchrotron-based X-ray microscopic studies for bioeffects of nanomaterials. Nanomedicine: Nanotechnol. Biol. Med. 10: 515–524.

Zhu, Z.J., Wang, H., Yan, B., Zheng, H., Jiang, Y., Miranda, O.R. and Vachet, R.W. 2012. Effect of surface charge on the uptake and distribution of gold nanoparticles in four plant species. Environ. Sci. Technol. 46: 12391–12398.

Zong, X., Wu, D., Zhang, J., Tong, X., Yin, Y., Sun, Y. and Guo, H. 2022. Size-dependent biological effect of copper oxide nanoparticles exposure on cucumber (*Cucumis sativus*). Environ. Sci. Technol. Poll. Res. 1–10.

Index

Editors' Biography

Dr. Mahendra Rai is presently a visiting Professor at the Department of Microbiology, Nicolaus Copernicus University, Torun, Poland. Formerly, he was a Professor and Head of the Department of Biotechnology at SGB Amravati University in Maharashtra, India. He has also held positions as a visiting scientist at several prestigious institutions, including the University of Geneva, Debrecen University in Hungary, the University of Campinas in Brazil, VSB Technical University of Ostrava in the Czech Republic, the National University of Rosario, in Argentina, and the University of Sao Paulo. Dr. Rai has an impressive publication record, with over 425 research papers published in national and international journals with 82 h-index. In addition, he has edited/authored more than 70 books and holds 6 patents. Notably, he has been recently featured in Stanford's list of the top 2% of scientists in nanoscience and nanotechnology.

Dr. Graciela Avila-Quezada is a Professor at Faculty of Agrotechnology Science, Universidad Autónoma de Chihuahua, in Mexico. She has been conducting research related to plant pathogens for 30 years, beginning with her undergraduate thesis work focused on the control of *Puccinia graminis* f. sp. *avenae*. She has held positions as a visiting scientist in various eminent institutions, including the Volcani Center in Israel, Colegio de Postgraduados in Texcoco, Mexico, SENASICA Laboratories (Ministry of Agriculture) in Tecamac, State of Mexico, the California Polytechnic State Univ-San Luis Obispo California (Calpoly), Institut de Recherche pour le Développement IRD Montpellier, France, Universidad Autónoma Chapingo in the State of Mexico, and Universidad de Córdoba in Spain. Dr. Avila has also held several notable positions, including President of the Mexican Society of Phytopathology, Vice-president of the National Phytosanitary Advisory Council (CONACOFI), Executive Secretary of the National System for Research and Technology Transfer (SNITT) of the Minister of Agriculture (Sagarpa), and Director of Academic Cooperation, in the Coordination of Science and Technology at the Presidency of Mexico. Additionally, she has conducted more than 12 research projects and published over 70 research papers in national and international journals focused on plant pathogens.

For Product Safety Concerns and Information please contact our EU
representative GPSR@taylorandfrancis.com
Taylor & Francis Verlag GmbH, Kaufingerstraße 24, 80331 München, Germany

* 9 7 8 1 0 3 2 4 5 0 3 8 4 *